General, Organic, and Biochemistry for Nursing and Allied Health

General, Organic, and Biochemistry for Nursing and Allied Health

Second Edition

George Odian, Ph.D.

Professor of Chemistry
The College of Staten Island
City University of New York

Ira Blei, Ph.D.

Professor of Chemistry
The College of Staten Island
City University of New York

Schaum's Outline Series

New York Chicago San Francisco Lisbon London
Madrid Mexico City Milan New Delhi San Juan
Seoul Singapore Sydney Toronto

The McGraw·Hill Companies

GEORGE ODIAN holds M.A. (1956) and Ph.D. (1959) degrees in organic chemistry from Columbia University, where his teaching career began. Formerly Chairperson of the Division of Science and Engineering and Dean of Faculty at Richmond College of CUNY (now The College of Staten Island of CUNY), he has been Professor of Chemistry at that institution since 1976. Dr. Odian has 5 years of industrial experience as research chemist and research director, respectively, at Thiokol Chemical Co. and Radiation Applications, Inc. He is the author of over 70 research papers and the text "Principles of Polymerization," 4th edition, Wiley-Interscience (2004). Dr. Odian is also the coauthor with Ira Blei of "General, Organic, and Biochemistry," Media Update edition, Freeman (2009); "Organic and Biochemistry," 2nd edition, Freeman (2006); "An Introduction to General Chemistry," 2nd edition, Freeman (2006).

IRA BLEI received his M.A. degree in physical chemistry from Brooklyn College of CUNY (1954) and his Ph.D. in physical biochemistry from Rutgers University (1957). Formerly Deputy Chairperson for the Division of Science and Engineering and Dean of Administrative Planning at Richmond College of CUNY (now The College of Staten Island of CUNY), he has been Professor of Chemistry at The College of Staten Island since 1976. Dr. Blei has 10 years of industrial experience as research chemist and director of research and development, respectively, at Lever Bros. Co. and Melpar, Inc. His research interests are in the application of colloid and surface chemistry to biological problems. Dr. Blei is the coauthor with George Odian of "General, Organic, and Biochemistry," Media Update edition, Freeman (2009); "Organic and Biochemistry," 2nd edition, Freeman (2006); "An Introduction to General Chemistry," 2nd edition, Freeman (2006).

Schaum's Outline of GENERAL, ORGANIC, AND BIOCHEMISTRY FOR NURSING AND ALLIED HEALTH

2 3 4 5 6 7 8 9 CUS CUS 1 4 3 2 1

ISBN 978-0-07-161165-7
MHID 0-07-161165-7

McGraw-Hill Books are available at special quantity discounts to use as premiums and sales promotions, or for use in corporate training programs. To contact a representative please e-mail us at bulksales@mcgraw-hill.com.

Library of Congress Cataloging-in-Publication Data is on file with the Library of Congress.

Preface

This book is intended for students who are preparing for careers in health fields such as nursing, physical therapy, podiatry, medical technology, agricultural science, public health, and nutrition. The chemistry courses taken by these students typically include material covered in general chemistry, organic chemistry, and biochemistry, compressed into a 1-year period. Because of this broad requirement, many students feel overwhelmed. To help them understand and assimilate so much diverse material, we offer the outline of these topics presented in the text that follows. Throughout we have kept theoretical discussions to a minimum in favor of presenting key topics as questions to be answered and problems to be solved. The book can be used to accompany any standard text and to supplement lecture notes. Studying for exams should be much easier with this book at hand.

The solved problems serve two purposes. First, interspersed with the text, they illustrate, comment on, and support the fundamental principles and theoretical material introduced. Second, as additional solved problems and supplementary problems at the end of each chapter, they test a student's mastery of the material and, at the same time, provide step-by-step solutions to the kinds of problems likely to be encountered on examinations.

No assumptions have been made regarding student knowledge of the physical sciences and mathematics; such background material is provided where required. SI units are used as consistently as possible. However, non-SI units that remain in common use, such as liter, atmosphere, and calorie, will be found where appropriate.

The first chapter emphasizes the current method employed in mathematical calculations, viz., factor-label analysis. In the section on chemical bonding, although molecular orbitals are discussed, VSEPR theory (valence-shell electron-pair repulsion theory) is emphasized in characterizing three-dimensional molecular structure. The discussion of nuclear processes includes material on modern spectroscopic methods of noninvasive anatomical visualization.

The study of organic chemistry is organized along family lines. To simplify the learning process, the structural features, physical properties, and chemical behavior of each family are discussed from the viewpoint of distinguishing that family from other families with an emphasis on those characteristics that are important for the consideration of biologically important molecules.

The study of biochemistry includes chapters on the four important families of biochemicals—carbohydrates, lipids, proteins, and nucleic acids—with an emphasis on the relationship between chemical structure and biological function for each. These chapters are followed by others on intermediary metabolism and human nutrition. In the discussion of all these topics we have emphasized physiological questions and applications where possible.

We would like to thank Charles A. Wall (Senior Editor), Anya Kozorez (Sponsoring Editor), Kimberly-Ann Eaton (Associate Editor), Tama Harris McPhatter (Production Supervisor), and Frank Kotowski, Jr. (Editing Supervisor) at McGraw-Hill Professional, and Vasundhara Sawhney (Project Manager) at International Typesetting & Composition for their encouragement and conscientious and professional efforts in bringing this book to fruition.

The authors welcome comments from readers at irablei@bellsouth.net and odian@mail.csi.cuny.edu.

GEORGE ODIAN
IRA BLEI

Contents

Chemistry and Measurement

1.1 INTRODUCTION

Chemistry is the study of matter and energy and the interactions between them. This is an extremely broad and inclusive definition, but quite an accurate one. There is no aspect of the description of the material universe which does not depend on chemical concepts, both practical and theoretical.

Although chemistry is as old as the history of humankind, it remained a speculative and somewhat mysterious art until about 300 years ago. At that time it became clear that matter comes in many different forms and kinds; therefore some kind of classification was needed, if only to organize data. There was red matter and white matter, liquid matter and solid matter, but it did not take long to realize that such broad qualitative descriptions, although important, were not sufficient to differentiate one kind of matter from another. Additional criteria, now called *properties*, were required. It was found that these properties could be separated into two basic classes: *physical* and *chemical*. Changes in physical properties involve only changes in form or appearance of a substance; its fundamental nature remains the same. For example, the freezing of water involves only its conversion from liquid to solid. The fact that its fundamental nature remains the same is easily demonstrated by melting the ice. By passing an electric current through water, however, two new substances are created: hydrogen and oxygen. The fundamental nature of water is changed—it is no longer water, but has been transformed into new substances through chemical change.

Without knowing anything about the fundamental nature of matter, chemists were also able to establish that matter could be separated into simpler and simpler substances through physical separation methods (e.g., distillation, solubility) and through chemical reactivity. They developed methods for measuring physical properties such as density, hardness, color, physical state, and melting and boiling points to help them decide when these operations could no longer change the nature of the substance. From these considerations, another classification scheme emerged, based on *composition*. In this scheme, matter is divided into two general classes: pure substances and mixtures.

There are two kinds of *pure substances*: elements and compounds. An *element* is a substance that cannot be separated into simpler substances by ordinary chemical methods. Nor can it be created by combining simpler substances. All the matter in the universe is composed of one or more of these fundamental substances. When elements are combined, they form *compounds*—substances having definite, fixed proportions of the combined elements with none of the properties of the individual elements, but with their own unique set of new physical and chemical properties.

In contrast to the unique properties of compounds, the properties of *mixtures* are variable and depend on composition. An example is sugar in water. The most recognizable property of this mixture is its sweetness, which varies depending on its composition (the amount of sugar dissolved in the water). A mixture is then composed of at least two pure substances. In addition, there are two kinds of mixtures. *Homogeneous mixtures*, or *solutions*, are visually uniform (microscopically as well) throughout the sample. *Heterogeneous mixtures* reveal visual differences throughout the sample (pepper and salt, sand and water, whole blood).

1.2 MEASUREMENT AND THE METRIC SYSTEM

Most of the above considerations depended upon the establishment of the quantitative properties of matter. This required a system of units, and devices for measurement. The measuring device most familiar to you is probably the foot ruler or yardstick, now being replaced by the centimeter ruler and the meter stick, both of which measure length. Other devices measure mass, temperature, volume, etc. The units for these measures have been established by convention and promulgated by authority. This assures that a meter measured anywhere in the world is the same as any other meter. This standardization of units for measurement is fundamental to the existence of modern technological society. Imagine the consequences if a cubic centimeter of insulin solution in Albuquerque were not the same as a cubic centimeter of insulin solution in Wichita!

The standardized measurement units used in science and technology today are known as the *metric system*. It was originally established in 1790 by the French National Academy, and has undergone changes since then. The fundamental or base units of the modern metric system (SI for Système International d'Unités) are found in Table 1-1. In chemistry, for the most part, you will encounter the first five of these. All other units are derived from these fundamental units. For example:

$$\text{square meters } (\text{m}^2) = \text{area}$$

$$\text{cubic meters } (\text{m}^3) = \text{volume}$$

$$\text{density} = \text{kilograms/cubic meters } (\text{kg/m}^3)$$

$$\text{velocity} = \text{meters/second } (\text{m/s})$$

Older, non-SI units are in common use, and some of these are shown in Table 1-2.

Table 1-1. Fundamental Units of the Modern Metric System

Fundamental Quantity	Unit Name	Symbol
Length	meter	m
Mass	kilogram	kg
Temperature	kelvin	K
Time	second	s
Amount of substance	mole	mol
Electric current	ampere	A
Luminous intensity	candela	cd

Table 1-2. Non-SI Units in Common Use

Quantity	Unit	Symbol	SI Definition	SI Name
Length	Angstrom	Å	10^{-10} m	0.1 nanometers (nm)
Volume	Liter	L	10^{-3} m^3	1 decimeter^3 (dm^3)
Energy	Calorie	cal	$\text{kg} \cdot \text{m}^2/\text{s}^2$*	4.184 joules (J)

*A dot (center point) will be used in this book to denote multiplication in derived units.

All of these units can be expressed in parts of or multiples of 10. The names of these multiples are created by the use of prefixes of Greek and Latin origin. This is best illustrated by Table 1-3. These symbols can be used with any kind of unit to denote size, e.g., nanosecond (ns), millimol (mmol), kilometer (km). Some of the properties commonly measured in the laboratory will be discussed in detail in a following section.

Problem 1.1. Express (*a*) 0.001 second (s); (*b*) 0.99 meter (m); (*c*) 0.186 liter (L) in more convenient units.

Ans. (*a*) 1 millisecond (ms); (*b*) 99 centimeters (cm); (*c*) 186 milliliters (mL).

Table 1-3. Names Used to Express Metric Units in Multiples of 10

Multiple or Part of 10	Prefix	Symbol
1,000,000	mega	M
1000	kilo	k
100	hecto	h
10	deka	da
0.1	deci	d
0.01	centi	c
0.001	milli	m
0.000001	micro	μ
0.000000001	nano	n

1.3 SCIENTIFIC NOTATION

It is inconvenient to be limited to decimal representations of numbers. In chemistry, very large and very small numbers are commonly used. The number of atoms in about 12 grams (g) of carbon is represented by 6 followed by 23 zeros. Atoms typically have dimensions of parts of nanometers, i.e., 10 decimal places. A far more practical method of representation is called *scientific* or *exponential notation*. A number expressed in scientific notation is a number between 1 and 10 which is then multiplied by 10 raised to a whole number power. The number between 1 and 10 is called the *coefficient*, and the factor of 10 raised to a whole number is called the *exponential factor*.

Problem 1.2. Express the numbers 1, 10, 100, and 1000 in scientific notation.

Ans. We must first choose a number between 1 and 10 for each case. In this example, the number is the same for all, 1. This number must be multiplied by 10 raised to a power which is a whole number. There are two rules to remember; (*a*) any number raised to the zero power is equal to one (1), and (*b*) when numbers are multiplied, the exponents must be added. Examples are:

$$1 = 1 \times 10^0$$

$$10 = 1 \times 10^1$$

$$100 = 1 \times 10^1 \times 10^1 = 1 \times 10^2$$

$$1000 = 1 \times 10^1 \times 10^1 \times 10^1 = 1 \times 10^3$$

Note that in each case, the whole number to which 10 is raised is equal to the number of places the decimal point was moved to the left.

Problem 1.3. Express the number 4578 in scientific notation.

Ans. In this case, the number between 1 and 10 must be 4.578, which is also the result of moving the decimal point three places to the left. In scientific notation, 4578 is written 4.578×10^3.

Problem 1.4. Express the numbers 0.1, 0.01, 0.001, and 0.0001 in scientific notation.

Ans. The process leading to scientific notation for decimals involves expressing these numbers as fractions, then recalling the algebraic rule that the reciprocal of any quantity X (which includes units), that is $1/X$, may be expressed as X^{-1}. For example, $1/5 = 5^{-1}$, and $1/\text{cm} = \text{cm}^{-1}$.

Therefore, each of the above cases may be written

$$0.1 = 1 \times \frac{1}{10^1} = 1 \times 10^{-1}$$

$$0.01 = 1 \times \frac{1}{10^1} \times \frac{1}{10^1} = 1 \times \frac{1}{10^2} = 1 \times 10^{-2}$$

$$0.001 = 1 \times \frac{1}{10^1} \times \frac{1}{10^1} \times \frac{1}{10^1} = 1 \times \frac{1}{10^3} = 1 \times 10^{-3}$$

etc.

Note that in each case, the whole number to which 10 is raised is equal to the number of places the decimal point is moved to the right.

Therefore, moving the decimal point to the right requires a minus sign before the power of 10, and moving the decimal point to the left requires a plus sign before the power of 10.

Problem 1.5. Express the number 0.00352 in scientific notation.

Ans. To obtain a number between 1 and 10, we move the decimal point three places to the right. This yields 3.52, which is then multiplied by 10 raised to (−3) since we moved the decimal to the right: 3.52×10^{-3}. (It is useful to remember that any number smaller than 1.0 must be raised to a negative power of 10, and conversely any number greater than 1.0 must be raised to a positive power of 10.)

Problem 1.6. How many ways can the number 0.00352 be represented in scientific notation?

Ans. Since the value of the number must remain constant, the product of the coefficient and the exponential factor must remain constant, but the size of the coefficient can be varied as long as the value of the exponential factor is also properly modified. The modification is accomplished by either multiplying the coefficient by 10 and dividing the exponent by 10, or dividing the coefficient by 10 and multiplying the exponential factor by 10. Either process leaves the value of the number unchanged.

$$0.00352 = 0.00352 \times 10^0 = 0.0352 \times 10^{-1} = 0.352 \times 10^{-2} = 3.52 \times 10^{-3}$$

Problem 1.7. Add 2.0×10^3 and 3.4×10^4.

Ans. When adding (or subtracting) exponential numbers, first convert all exponents to the same value, then add (or subtract) the coefficients and multiply by the now common exponential factor. If we write the numbers in nonexponential form, it is easy to see why the rule works:

$$3.4 \times 10^4 = 34 \times 10^3 = 34{,}000$$

$$2.0 \times 10^3 = 2000$$

$$34{,}000 + 2000 = 36{,}000$$

$$(34 \times 10^3) + (2.0 \times 10^3) = (34 + 2.0) \times 10^3 = 36 \times 10^3 = 3.6 \times 10^4$$

Problem 1.8. Multiply the numbers 2.02×10^3 and 3.20×10^{-2}.

Ans. When exponential numbers are multiplied, the coefficients are multiplied and the exponents are added. A simplified calculation will show the derivation of this rule:

$$10^2 \times 10^3 = 10^1 \times 10^1 \times 10^1 \times 10^1 \times 10^1 = 10^5$$

Therefore

$$(2.02 \times 10^3) \times (3.20 \times 10^{-2}) = 6.46 \times 10^1 = 64.6$$

Problem 1.9. Divide 1.6×10^5 by 2.0×10^2.

Ans. When exponential numbers are divided, the coefficients are divided and the exponents are subtracted. Here again, we will illustrate this rule with a simplified calculation. First, remember that 0.01 in exponential form is 1×10^{-2}. Writing 0.01 in fractional form:

$$0.01 = \frac{1}{100}$$

Now we write numerator and denominator in exponential form:

$$\frac{1}{100} = \frac{10^0}{10^2} = 1 \times 10^{(0-2)} = 1 \times 10^{-2}$$

Therefore

$$(1.6 \times 10^5)/(2.0 \times 10^2) = 0.80 \times 10^3 = 8.0 \times 10^2$$

1.3.1 Logarithms

From the above discussion, we can say that any number may be expressed as

$$y = 10^x$$

The number x to which 10 must be raised is called the *logarithm* of y, and is written $x = \log y$. The logarithm (log) of a number is obtained by using either a calculator or a log table. To use a calculator, for example, to find the log of 472, enter 472, press the log key, and read the answer as 2.6739. If you use a logarithm table, the number y must be expressed in scientific notation—for example, the number 472 must be expressed as 4.72×10^2, and 0.00623 as 6.23×10^{-3}. The logarithm of a number in scientific notation is written as follows:

$$\log ab = \log a + \log b$$

so that the log of 4.72×10^2 is $\log 4.72 + \log 10^2$, or $\log 4.72 + 2.000$. Log tables are set up so that x, called the *mantissa*, is always a positive number between 1 and 10. From the logarithm table, $\log 4.72 = 0.6739$, and $\log 4.72 \times 10^2 = 2.6739$.

Problem 1.10. What is the logarithm of 6.23×10^{-3}?

Ans. To use a calculator, enter 6.23, press the EE key, press 3, press the change sign key, press the log key, and read the answer as –2.2055. To use a log table instead of a calculator:

$$\text{Log}\, 6.23 \times 10^{-3} = \log 6.23 + \log 10^{-3} = \log 6.23 + (-3)$$

$$= 0.7945 + (-3) = -2.2055$$

We often find it necessary in chemical calculations to find the number whose logarithm is given: That is, find $y = 10^x$, where x is given. That number is called the *antilogarithm*.

Problem 1.11. What is the number whose logarithm is 2.6532, that is, what is the antilogarithm of 2.6532?

Ans. To use a calculator, enter 2.6532, press the 10^x (or antilog) key, and read the answer as 450. To use a log table to obtain the antilog, we express y as $y = 10^a \times 10^n$, where a is always a positive number. Then the number is

$$y = 10^{2.6532} = 10^{0.6532} \times 10^2$$

We then examine the log table to find that the number whose logarithm is 0.6532 is 4.50. Then $y = 4.50 \times 10^2$ or 450.

Problem 1.12. What is the number whose logarithm is a negative number, -5.1367?

Ans. To use a calculator, enter 5.1367, press the change sign key, press the 10^x (or antilog) key, and read the answer as 7.30×10^{-6}. To use a log table, proceed as follows:
To write the number as $y = 10^a \times 10^n$, where a must be a positive number, we set $-5.1367 = 0.8633 - 6$, and $y = 10^{0.8633} \times 10^{-6}$. Next, we find the antilog of 0.8633, which is 7.30. The number is 7.30×10^{-6}.

1.4 SIGNIFICANT FIGURES

Most of us are familiar with sales taxes. You purchase an item for $12.99, for example, and you pay an additional 7.25% in tax. By calculator, the result is

$$\$12.99 \times 0.0725 = \$0.941775$$

The clerk then adds $0.94 to the bill.

Your accountant finds that you owe the Internal Revenue Service (IRS) $161.74 and instructs you to send a check for $162 to the IRS.

Your favorite shortstop came to the plate 486 times last season and hit successfully 137 times. His average is, by calculator, $137/486 = 0.28189$, and is reported as 0.282.

In each of these cases, a process called *rounding off* was employed. In this process two decisions must be made: (1) how many places make sense or are desired in the final answer after a calculation is made, and (2) what rule shall be followed in determining the value of the final digit.

The second decision is technical and straightforward. In common practice, if the digit following the one we want to retain is greater than 5, we increase the value of the digit we want to retain by 1 and drop the trailing digits. If its value is 4 or less, we retain the value of the digit and drop the trailing digits. In doing calculations, one usually keeps as many digits as possible until the calculation is completed. It is only the final result that is rounded off.

The first decision is not so simple. There is some kind of rule or policy involved, but it clearly depends on the situation and its unique requirements. In scientific work, the policy adopted goes under the name of *significant figures*.

When a measurement of any kind is made, it is made with a measuring device. The device is equipped with a scale of units which are usually divided into parts of units down to some practical limit. Typical examples are a centimeter ruler divided into millimeters, or a thermometer with a range from −10 °C to 150 °C divided into celsius degrees. It is quite rare that a measurement with such devices results in a value which falls precisely on a division of the scale. More likely, the measured value falls between scale divisions, and one must make an estimate of the value. The next person reading your value of the measurement must know what you had in mind when you made that estimate. Estimation means that there is always some degree of *uncertainty* in any measurement. The reported number of significant figures in any measured value reflects that uncertainty. We therefore need a set of rules which ensures proper interpretation of a measured value as written. Basically, these rules make clear our understanding of the reliability of a reported measurement.

Problem 1.13. What is the meaning of the value 17.45 cm?

Ans. We must assume that the last digit is the result of rounding off. That means that the measured value must have been between 17.446 and 17.454. If the number had been estimated as 17.455 then it would have been reported as 17.46. If it had been estimated as 17.444 it would have been reported as 17.44. The fundamental meaning of reporting the measurement as 17.45 cm is that it could only have been measured with a ruler graduated in 0.01-mm (millimeter) divisions. Alternatively, 17.45 must have been estimated as about half way between the smallest scale divisions. There are 4 significant figures in the numerical value of the measurement.

RULE 1. All digits of a number in which no zero appears are significant, and the last one is an estimate derived by rounding off.

Problem 1.14. What is the meaning of the value 17.40 cm?

Ans. Since the last digit is an estimate, the zero in the second place is significant. It means that the value was not 17.394 cm or less or 17.405 cm or more but somewhere between those two values.

Furthermore, as in Problem 1.13, the measurement must have been made with a ruler graduated in 0.01-mm divisions. There are 4 significant figures in the value 17.40 cm.

RULE 2. A zero at the end of a number with a decimal in it is significant.

Problem 1.15. What is the meaning of the value 17.05 cm?

Ans. Here again, since the last digit is assumed to be an estimate, the measured value must have been less than 17.054 and more than 17.045. The measurement must have been made with a ruler graduated in 0.01-mm divisions. There are therefore 4 significant figures in the value 17.05.

RULE 3. Zeros between nonzero digits are significant.

Problem 1.16. What is the meaning of the value 0.05 cm?

Ans. The last digit is the result of rounding off and therefore must have been less than 0.055 and more than 0.046. The zero in this case is used to locate the number on the ruler (scale), but it has no significance with respect to the estimation of the value of the number itself. There is therefore only 1 significant figure.

RULE 4. Zeros preceding the first nonzero digit are not significant. They only locate the position of the decimal point.

Problem 1.17. What is the meaning of the value 20 cm?

Ans. Since there is no estimate beyond the zero in 20 cm, we might infer that the number is approximately 20, say 20 ± 5. The number of significant numbers in such a case is uncertain. (Ask yourself whether the quantity 19 cm has more significant figures than the quantity 20 cm.) We might also mean 20 cm exactly. In that case there are unequivocally 2 significant figures. It is possible at times to deduce the number of significant figures from a statement of the problem. However, it is the responsibility of the writer to be clear about the number of significant figures. In this case with no other information available, one would write the number in exponential notation as 2×10^1.

RULE 5. Zeros at the end of a number having no decimal point may be significant; their significance depends upon the statement of the problem.

Problem 1.18. What is the meaning of the value 17.4 cm?

Ans. The last digit must be an estimate, and the implication is that its value is the result of rounding off. The measured value must have been between 17.44 cm or less, and 17.36 or more. The estimate was in the range of 0.01 mm, and therefore it is reasonable to assume that, in this case, the ruler must have been graduated in 0.1-mm divisions. There are 3 significant figures in the value 17.4 cm.

Rule 1 applies here. Every nonzero digit of a measured number is significant. We must assume that the last digit is an estimate and reflects the uncertainty of the measurement.

1.5 SIGNIFICANT FIGURES AND CALCULATIONS

In arithmetic operations, the final results must not have more significant figures than the least well-known measurement.

Problem 1.19. A room is measured for a rug and is found to be 89 inches by 163 inches (in). The area is calculated to be

$$89 \text{ in} \times 163 \text{ in} = 14507 \text{ inches}^2 \text{ (in}^2\text{)}$$

The calculated result cannot be any more precise than the least precisely known measurement, so the answer cannot contain more than 2 significant figures. If we want to retain 2 significant figures, we must go from 5 digits to 2 by rounding off.

Ans. The rug area should be reported in scientific notation as 1.5×10^4 in^2.

Problem 1.20. At 92 kilometers/hour (km/h) how long would it take to drive 587 km?

Ans. The calculation is simply

$$587 \text{ km} \left(\frac{1.00 \text{ h}}{92 \text{ km}}\right) = 6.38 \text{ h} = 6.4 \text{ h}$$

The result was rounded off to 2 significant figures since the least well-known quantity, 92 km, had 2 significant figures.

RULE 6. In multiplication or division, the calculated result cannot contain more significant figures than the least well-known measurement.

Problem 1.21. What is the sum of the following measured quantities? 14.83 g, 1.4 g, 282.425 g.

Ans. The least well-known of these quantities has only 1 figure after the decimal point, so the final sum cannot contain any more than that. We add all the values, and round off after the sum is made as follows:

$$\begin{array}{r} 14.83\ \text{g} \\ 1.4\ \ \ \text{g} \\ 282.425\ \text{g} \\ \hline 298.655\ \text{g} = 298.7\ \text{g} \end{array}$$

Problem 1.22. What is the result of the following subtraction? 5.753 g − 2.32 g

Ans. The least well-known quantity has 2 significant figures after the decimal point, so that the result cannot contain any more than that. As in addition, we round off the result after subtraction.

$$\begin{array}{r} 5.753\ \text{g} \\ -2.32\ \ \text{g} \\ \hline 3.433\ \text{g} = 3.43\ \text{g} \end{array}$$

RULE 7. In addition and subtraction, the calculated result cannot contain any more decimal places than the number with the fewest decimal places.

1.6 MEASUREMENT AND ERROR

Since the last place in a measured quantity is an estimate, it may have occurred to you that one person's estimate may not be the same as another's. Furthermore it is not likely that if you made the same measurement two or three times that you would record precisely the same value. This kind of variability is not a mistake or blunder. No matter how careful you might be, it is impossible to avoid it. This unavoidable variability is called the *indeterminate error*. In reporting our result in measurement with the centimeter ruler as 17.45 cm, the last place was the result of rounding off. It could have been 17.446 or 17.454. The difference between the two extremes was 0.008 cm. We therefore should have reported the value as 17.45 ± 0.004 cm. The variability of 0.004 cm occurs at any point along the ruler: it is a constant error. How serious an error is it when the size of the object measured is 17.45 cm compared with one of 1.75 cm or less? This is answered by calculating the *relative error*, i.e., how large

a part of the whole is the variability? This is calculated as the ratio of the constant variability to the actual size of the measurement.

$$\frac{0.004}{17.45} = 0.0002 \quad \text{or} \quad 0.02\%$$

$$\frac{0.004}{1.75} = 0.002 \quad \text{or} \quad 0.2\%$$

$$\frac{0.004}{0.18} = 0.02 \quad \text{or} \quad 2\%$$

It is easy to see that as the size of the item being measured decreases, the seriousness of the error, the relative error, increases. This means that the seriousness of an error for example, in weighing, can be minimized by increasing the size of the sample.

Problem 1.23. **If the indeterminate error in weighing on a laboratory balance is 0.003 g, what size sample should you take to keep the relative error to 1.0%?**

Ans.

$$1.0\,\%\ \text{Relative error} = \frac{1\ \text{part in weight}}{100\ \text{total parts in weight}} \times 100\%$$

$$\frac{1}{100} = \frac{0.0030\ \text{g}}{X\ \text{g of sample}}$$

$$X = 0.30\ \text{g of sample}$$

1.7 FACTOR-LABEL METHOD

For all problem solving in this book, we will employ a time-tested systematic approach called the *factor-label method*. It is also referred to as the *unit-factor* or *unit-conversion method* or *dimensional analysis*.

The underlying principle is the conversion of one type of unit to another by the use of a conversion factor:

$$\text{Unit}_1 \times \text{conversion factor} = \text{unit}_2$$

To take a simple example, suppose you want to know how many seconds there are in 1 minute (min). First write the unit you want to convert, and then multiply it by an appropriate conversion factor:

$$1\,\cancel{\text{min}} \times \frac{60\ \text{s}}{\cancel{\text{min}}} = 60\ \text{s}$$

The conversion factor had two effects: (1) it introduced a new unit, and (2) it allowed the cancellation of the old unit (units, as well as numbers, can be canceled).

Let us look at the reverse of the problem; how many minutes are there in 60 s? We write the unit we want to convert and multiply it by an appropriate conversion factor:

$$\cancel{60\ \text{s}} \times \frac{1\ \text{min}}{\cancel{60\ \text{s}}} = 1\ \text{min}$$

The reason we can use either form of the conversion factor can be seen by writing it as a solution to an algebraic expression:

$$1\ \text{min} = 60\ \text{s}$$

So $$\frac{1\ \text{min}}{60\ \text{s}} = 1$$

And $$\frac{60\ \text{s}}{1\ \text{min}} = 1$$

Therefore, $$\frac{1\ \text{min}}{60\ \text{s}} = \frac{60\ \text{s}}{1\ \text{min}}$$

A conversion factor does not change the size of a number or measurement, only its name, as in Fig. 1-1. This systematic approach does not eliminate the requirement for thoughtfulness on your part. You must deduce the form of the appropriate conversion factor. To do this, you must become familiar with the conversion factors listed in Appendix A.

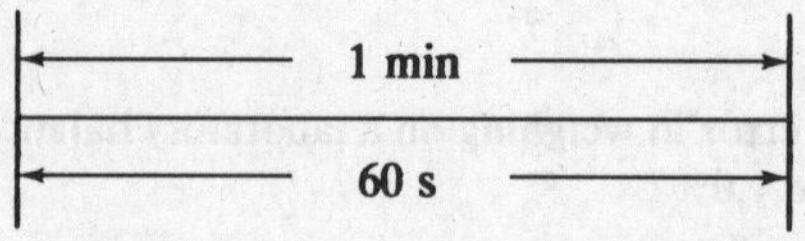

1 min

60 s

Fig. 1-1 Illustration of the equivalence in size of the same two quantities with different names.

Problem 1.24. How many centimeters are there in 14.0 in?

Ans. First decide on the unit to be converted, then deduce the proper form of the factor.

Step 1. Inches → centimeters

Step 2. Inches × conversion factor = centimeters

Step 3. Conversion factor must be $= \frac{\text{centimeters}}{\text{inches}}$

We then require the value for the conversion factor, which in this case is 2.54 cm/in. Then step 2 is implemented:

$$14.0\ \text{in} \times \frac{2.54\ \text{cm}}{\text{inch}} = 35.56\ \text{cm} = 35.6\ \text{cm}$$

The answer was rounded off because there were only 3 significant figures in the data.

Sometimes it becomes necessary to employ more than one conversion step, as in the following problem.

Problem 1.25. How many centimeters are there in 4.10 yards (yd)?

Ans. The problem here is to work out the conversion of yards to centimeters, which means a conversion factor with dimensions, centimeters/yards. This conversion factor does not exist, but must be constructed using a series of factors which when multiplied together produce the desired result. It is difficult to visualize the intervening steps from yards to centimeters, but they become clear when we go about it backward. What might be the best route to go from centimeters to yards?

centimeters ⟶ inches ⟶ feet ⟶ yards

Notice that this is the reverse of step 1 in Problem 1.24. From this analysis, it seems best to first multiply yards by a factor to get feet (ft); next, to multiply feet by a factor to get inches; and finally, to multiply inches by a factor to get centimeters:

$$4.10\ \text{yd} \times \frac{3.00\ \text{ft}}{\text{yd}} \times \frac{12.0\ \text{in}}{\text{ft}} \times \frac{2.54\ \text{cm}}{\text{in}} = 374.9\ \text{cm} = 375\ \text{cm}$$

There are 3 significant figures in the answer since the starting point, 4.10 yd, has only 3 significant figures.

Problem 1.26. How many milliliters are there in 1.3 gallons (gal)?

Ans. We will use the same backward approach here as in the previous example:

$$\text{milliliters} \longrightarrow \text{liters} \longrightarrow \text{quarts (qt)} \longrightarrow \text{gallons}$$

Therefore

$$1.3 \text{ gal} \times \frac{4.0 \text{ qt}}{\text{gal}} \times \frac{946 \text{ mL}}{\text{qt}} = 4919.2 \text{ mL} = 4.9 \times 10^3 \text{ mL}$$

Since there are only 2 significant figures in the data, there can be only 2 significant figures in the answer.

Problem 1.27. How many milligrams (mg) are there in 3.2 pounds (lb)?

Ans.

$$\text{milligrams} \longrightarrow \text{grams} \longrightarrow \text{kilograms} \longrightarrow \text{pounds}$$

$$3.2 \text{ lb} \times \frac{1 \text{ kg}}{2.2 \text{ lb}} \times \frac{1000 \text{ g}}{\text{kg}} \times \frac{1000 \text{ mg}}{\text{g}} = 1.5 \times 10^6 \text{ mg}$$

1.8 MASS, VOLUME, DENSITY, TEMPERATURE, HEAT, AND OTHER FORMS OF ENERGY

1.8.1 Mass

Mass is a measure of the quantity of matter. The device used for measuring mass is called a balance. A balance "balances" a known mass against an unknown mass, so that mass is not an absolute quantity, but determined by reference to a standard. Strictly speaking, one determines weight, not mass. Since all mass on earth is contained in the earth's gravitational field, what we measure is the force of gravity on the mass of interest. Remember that, out of the influence of a gravitational field (in space), objects are weightless, but they obviously still have mass. However, because mass is determined by "balance" gravity operates equally on both known and unknown mass, so we feel justified in using mass and weight interchangeably. The units of mass determined in the laboratory are grams.

1.8.2 Volume

Liquid volume is a commonly measured quantity in the chemical laboratory. It is measured using vessels calibrated with varying degrees of precision, depending upon the particular application. These vessels are calibrated in milliliters or liters, and are designed to contain (volumetric flasks, graduated cylinders) or to deliver (pipettes, burettes) desired volumes of liquids. According to our usage, 1 mL is 1/1000th of a liter; therefore there are 1000 mL in 1.0 L. Furthermore, 1.0 mL is equal to 1.0 cm^3 (cubic centimeter or cc).

Problem 1.28. Express the quantity 242 mL in liters.

Ans. Using the factor-label method, first write the quantity, then multiply by the appropriate conversion factor:

$$242 \text{ mL} \times \frac{1 \text{ L}}{1000 \text{ mL}} = 0.242 \text{ L}$$

1.8.3 Density

It is important to note that mass does not change with change in temperature, but volume does. If you look carefully at the volume designations on flasks or pipettes, you will see that the vessels are calibrated at one particular temperature. Is it possible to know the "true" volume at a temperature different from the temperature at calibration?

The answer to that question lies in the definition of a quantity derived from mass and volume. This quantity or property is called *density*, and is defined as the mass per unit volume:

$$\text{Density} = \text{grams/cubic centimeter (g/cm}^3\text{)}$$

This property was recognized in ancient times (remember the story of Archimedes who was almost arrested for public indecency) as an excellent indicator of the identity of a pure substance, regardless of color, texture, etc. You can see from the definition that the density must vary as temperature is changed. The volumes of liquids increase as the temperature increases, and therefore the density of liquids must decrease as the temperature increases. The density, as defined, provides us with a useful conversion factor for the conversion of grams to milliliters or milliliters to grams.

Problem 1.29. What is the mass of 124 mL of a solution which has a density of 1.13 g/mL?

Ans. Using the factor-label method, first write the quantity to be converted then multiply by the correct conversion factor:

$$124 \text{ mL} \times \frac{1.13 \text{ g}}{\text{mL}} = 140.12 \text{ g} = 1.40 \times 10^2 \text{g}$$

Problem 1.30. What is the volume in cubic centimeters of 10.34 g of a liquid whose density is 0.861 g/cm^3?

Ans. By the factor-label method:

$$10.34 \text{ g} \times \frac{1 \text{ cm}^3}{0.861 \text{ g}} = 12.0 \text{ cm}^3$$

Solids can also be characterized by their densities. Most solids have densities greater than liquids, the most notable exception being that of solid water (ice), which floats in liquid water. This effect provides us with a method of determining densities of unknown liquids. We use a device called a *hydrometer*. The hydrometer is a hollow, sealed glass vessel consisting of a bubble at the end of a narrow graduated tube. The bubble is filled with sufficient lead shot so that the vessel floats vertically. The position of the liquid surface with respect to the markings on the graduated portion of the vessel can be read as density. Hydrometers in common use in the clinical lab are graduated not in density units but in values of specific gravity. *Specific gravity* is defined as the ratio of the density of the test liquid to the density of a reference liquid:

$$\text{Specific gravity} = \frac{\text{density of test liquid}}{\text{density of reference liquid}}$$

Notice that specific gravity has no units since it is a ratio of densities. The reference liquid for aqueous solutions is water at 4 °C, the temperature at which its density is at its maximum, 1.000 g/cm^3. Specific gravities of blood or urine are reported, for example, as 1.048(d^{20}/d^4). This means that the sample's density was measured at 20 °C and was compared to the density of water at 4 °C and should be read as meaning that the sample's density was 1.048 times the density of water at 4 °C.

1.8.4 Temperature

We know that if we place a hot piece of metal on a cold piece of metal, the hot metal will cool and the cool metal will become warmer. We describe this by saying that "heat" flowed from the hot body to the cold body. We use temperature as a means of measuring how hot or cold a substance is. We regard heat as a form of energy which can flow, and which is ultimately a reflection of the degree of motion of a substance's constituent atoms. So it is clear that the flow of heat and the concept of temperature are inextricably related, but are not the same thing.

To measure temperature, a device and a scale are required. There are many ways of measuring temperature. The usual device is the common thermometer, a glass tube partially filled with a fluid which expands with increase in temperature. The tube is usually calibrated at two reproducible temperatures, the freezing and boiling points of pure water. There are three temperature scales in general use. On the *Celsius scale*, the freezing point of water in Celsius (°C) degrees is called 0 °C, and the boiling point, 100 °C. On the *Fahrenheit scale*, the freezing point in Fahrenheit (°F) degrees is called 32 °F, and the boiling point 212 °F. The SI temperature scale is in units called *kelvins*, symbol K (no degree symbol is used with kelvin). The uppercase T is reserved for use as a symbol in equations where kelvin temperature occurs as a variable. Where the temperature variable is in Celsius degrees, the lowercase t is used. Look for this lowercase symbol in Sec. 1.8.5 of this chapter. The size of the kelvin is identical to that of the Celsius degree, but the kelvin scale recognizes a lower temperature limit called *absolute zero*, the lowest theoretically attainable temperature, and gives it the value of 0 K. On the kelvin scale, the freezing point of water is 273.15 K, which we will round off to 273 K for convenience. To convert Celsius degrees to kelvins, simply add 273 to the Celsius value. Conversion between the Celsius and Fahrenheit scales is a bit more complicated. Figure 1-2 outlines the problem of conversion. The sizes of the degrees are different, and the reference points for freezing and boiling points do not coincide. There are (180/100) or (9/5) as many Fahrenheit divisions as Celsius divisions between freezing and boiling points. Another way to look at this is to consider that a Celsius degree is larger than a Fahrenheit degree; there are almost 2 (9/5) Fahrenheit degrees per 1 Celsius degree. We can write two equations to illustrate that idea:

$$°\text{F} = (9\ °\text{F}/5\ °\text{C}) \times °\text{C}$$

and

$$°\text{C} = (5\ °\text{C}/9\ °\text{F}) \times °\text{F}$$

These two reciprocal relationships allow us to convert from one size of degree to another. We must, however, recognize that the two scales do not numerically coincide at the reference points (freezing and boiling points of water).

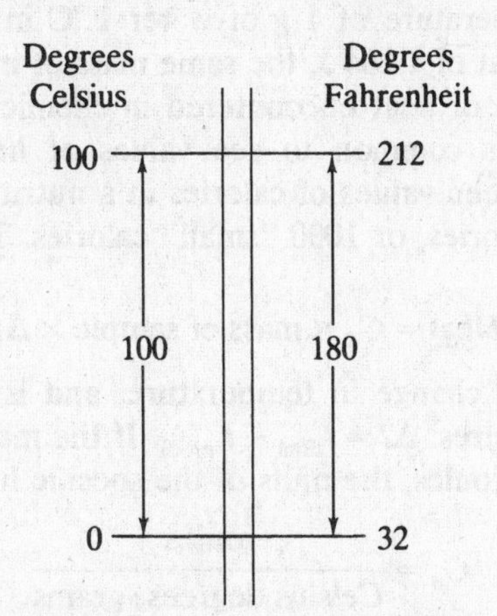

Fig. 1-2 Comparison of the Celsius and Fahrenheit temperature scales.

Problem 1.31. Convert 50 °C to Fahrenheit degrees.

Ans. Examining Fig. 1-2, the temperature scale diagram, we can see that 50 °C is halfway between freezing and boiling points, on the Celsius scale, and 90 °F higher than the Fahrenheit freezing point of 32 °F. The reading opposite the Celsius reading on the Fahrenheit scale is then 122 °F

(90 °F + 32 °F). So, in order to convert Celsius to Fahrenheit, we must multiply °C by 9/5, but we must also add 32 to the result. To convert Fahrenheit to Celsius, the first operation is to subtract 32, and only then to multiply by 5/9. These ideas can be summarized by three equations:

$$°C = (°F - 32)\ (5\ °C/9\ °F)$$

$$°F = (°C)(9\ °F/5\ °C) + 32$$

$$K = °C + 273$$

Problem 1.32. Convert 42 °F to Celsius degrees and to kelvins.

Ans. **To convert to Celsius degrees, we first subtract 32 and then multiply by 5/9:**

$$42\ °F - 32\ °F = 10\ °F$$

$$10\ °F \times \frac{5\ °C}{9\ °F} = 5.6\ °C$$

$$K = 5.6 + 273 = 278.6\ K$$

Problem 1.33. Convert 375 K to Celsius and Fahrenheit degrees.

Ans. **Since K = °C + 273, the temperature in Celsius must be**

$$375\ K - 273\ K = 102\ °C$$

$$°F = (102\ °C \times 9/5) + 32 = 215.6\ °F$$

1.8.5 Heat

Each substance has a different capacity to absorb heat. The quantitative characterization of the absorption of heat requires specification of the rise in temperature of a given mass of material for the input of a fixed amount of heat. The heat input has the SI units of *joules*, symbol J. The older non-SI unit is the *calorie*, abbreviation cal. The relationship between the two is that 4.184 J = 1 cal. It takes 4.184 J or 1 cal to raise the temperature of 1 g of water 1 °C in the temperature range 14.5 °C to 15.5 °C. For example, with an input of 4.184 J, the same mass of iron will experience a rise of almost 10 °C. It is usual for the amounts of heat encountered in chemical processes to be in the range of thousands of joules, so that it is common to see values of heat in kJ or kcal (kilojoules and kilocalories). You may also have seen values of calories in a nutritional context, which are written as Calories. These are in fact kilocalories, or 1000 "small" calories. The relationship between heat and temperature rise is

$$\text{Heat} = C_p \times \text{mass of sample} \times \Delta t$$

C_p is the *specific heat*. Δt is the change in temperature, and is the difference in Celsius degrees between final and initial temperatures: $\Delta t = t_{\text{final}} - t_{\text{initial}}$. If the mass is in grams, the temperature on the Celsius scale, and the heat in joules, the units of the specific heat are

$$C_p = \frac{\text{joules}}{\text{Celsius degrees} \cdot \text{grams}}$$

Problem 1.34. What is the specific heat of a substance if the temperature of 12 g of the substance rises from 20 °C to 35 °C when 6 J of heat are added to it?

Ans.

$$C_P = \frac{6.0\ J}{12.0\ g \times 15.0\ °C} = 0.033\ \frac{J}{g \cdot °C}$$

Specific heat describes the capacity of 1 g of a substance to absorb heat. A related term, *heat capacity*, depends on how much of the substance is under study. Experience tell us that if we add the same amount of heat (in joules) to two samples of the same substance, one being twice the mass of the other, it is clear that the smaller quantity will have a higher temperature.

Problem 1.35. How much heat is necessary to raise the temperature of 100.0 g of water from 25.00 °C to 100.00 °C?

Ans. We use the equation:

$$\text{Heat} = C_P \times \text{mass of sample} \times \Delta t$$

$$\Delta t = (t_{\text{FINAL}} - t_{\text{INITIAL}})$$

$$\text{Joules} = \left(4.184 \frac{\text{J}}{°\text{C} \times \text{g}}\right)(100.00\ \text{g})(100.00\ °\text{C} - 25.00\ °\text{C})$$

$$= 31380\ \text{J} = 31.38\ \text{kJ}$$

The heat capacity of 100.0 g of water heated from 25.00 °C to 100.0 °C is 31.38 kJ.

1.8.6 Other Forms of Energy

The concepts of energy and work are interchangeable. To measure energy, we measure the work done on a system. In the previous examples, the heat required to raise the temperature of water was expressed in joules. If we were to calculate the work required to lift a mass up to a certain height from a lower position, it would also be expressed in joules. When electricity travels in a wire, electrical work is done to move electrons through the wire, and is expressed in joules.

A mass which is stationary but has been lifted to a position so that it is capable of spontaneously falling if released is said to possess *potential energy*. When the mass is released and begins to fall, the potential energy becomes converted to a new form characteristic of moving masses, called *kinetic energy*. The stationary mass is capable of doing work only because work or energy has been "invested" in it. If a negative charge, say an electron, is moved away from a positive charge, the attraction of opposite charges must have been overcome by doing work to separate them. The negative charge now has a higher potential energy. When wood or coal is burned, heat and light (also a form of energy) are emitted. The source of this released energy is the chemical bonds of the substances which are broken and then rejoin to form other substances. The potential energy of the chemical components has been thereby reduced. We will use these ideas about work, potential energy, heat, and light in the description of atomic structure and chemical bonds.

ADDITIONAL SOLVED PROBLEMS

INTRODUCTION

Problem 1.36. Classify the following changes in properties as either chemical or physical:

(*a*) Water melts at 0 °C.
(*b*) Iron rusts.
(*c*) Maple syrup is made by boiling sugar maple tree sap.

(*d*) Paper burns.
(*e*) Magnesium dissolves in hydrochloric acid, with the production of bubbles of hydrogen gas.
(*f*) Bright, shiny copper wire, heated in a flame, turns black.
(*g*) Yellow sulfur powder, heated, melts and evolves a colorless gas which is suffocating.
(*h*) In a boiled egg, the liquid contents become solid and will not dissolve again.
(*i*) Glass heated in a hot flame glows and becomes fluid.

Ans.

(*a*) Physical change. Water is transformed to a new form but is still the same substance.

(*b*) Chemical change. The rust is a new substance, iron oxide, the result of the chemical combination of iron with oxygen.

(*c*) Physical change. The syrup is concentrated tree sap made by boiling off or distilling off the water in the sap.

(*d*) Chemical change. In the presence of oxygen, paper will burn and the process produces new substances, carbon dioxide and water.

(*e*) Chemical change. The production of a new substance, hydrogen, means that a chemical change took place.

(*f*) Chemical change. The copper reacts with oxygen and forms a new compound, copper oxide.

(*g*) Chemical change. Sulfur reacts with oxygen in the air to form sulfur dioxide, a corrosive gas.

(*h*) Chemical change. A chemical reaction occurs and as a result, the physical properties of the modified proteins in the egg are changed, and they cannot be returned to their former state.

(*i*) Physical change. The glass is merely melting. When the temperature is reduced it returns to its previous physical and chemical state.

Problem 1.37. Classify the following as pure substances or mixtures: (*a*) air; (*b*) mercury; (*c*) aluminum foil; (*d*) table salt; (*e*) sea water; (*f*) milk; (*g*) tobacco smoke; (*h*) blood.

Ans.

(*a*) Air is a mixture of gases, oxygen and nitrogen, which is homogeneous throughout. It is a solution.

(*b*) Mercury is homogeneous and pure. Since it cannot be physically or chemically changed into simpler substances, it is an element.

(*c*) Aluminum foil is homogeneous and pure. Since it cannot be physically or chemically changed into simpler substances, it is also an element.

(*d*) Table salt is homogeneous and pure. It can be separated into simpler substances which are elements, sodium and chlorine. Therefore table salt is a pure compound called sodium chloride.

(*e*) Sea water is homogeneous and contains a number of pure substances such as sodium chloride, magnesium chloride, and calcium carbonate. Sea water is therefore a mixture, and it is also a solution.

(*f*) Milk is heterogeneous, consisting of a continuous fluid phase and fat droplets which can be visualized microscopically. It is a heterogeneous mixture of liquids since its contents are both visualizable and variable in composition.

(*g*) Tobacco smoke is heterogeneous and contains particles and gases in variable amounts. It is therefore a heterogeneous mixture of solids and gases.

(*h*) Blood is similar to milk in character, a continuous fluid phase of dissolved substances in variable amounts, along with visually distinguishable particulate components. Milk, smoke, and blood are also called *suspensions*, because the sizes of the particulates are so small that they settle very slowly in earth's gravitational field.

MEASUREMENT AND THE METRIC SYSTEM

Problem 1.38. What are the fundamental quantities in the SI system which are encountered in chemistry?

Ans.

Fundamental Quantity	Unit Name	Symbol
Length	meter	m
Mass	kilogram	kg
Temperature	kelvin	K
Time	second	s
Amount of substance	mole	mol

Problem 1.39. What are some examples of units which can be derived from these five fundamental units?

Ans.

Velocity = meters/second = m/s

Molecular mass = kilogram/mole = kg/mol

Volume = meter × meter × meter = $\text{meter}^3 = \text{m}^3$

Energy = (kilogram · meter^2)/(seconds^2) = joules = J

Density = kilogram/meter^3 = kg/m^3

Problem 1.40. Express the following quantities in more convenient form: (*a*) 0.054 m; (*b*) 0.00326 kg; (*c*) 0.00000023 s; (*d*) 1253 g.

Ans.

(*a*) $$0.054\ \text{m}\left(\frac{100\ \text{cm}}{\text{m}}\right) = 5.4\ \text{cm}$$

(*b*) $$0.00326\ \text{kg}\left(\frac{1000\ \text{g}}{\text{kg}}\right) = 3.26\ \text{g}$$

(*c*) $$0.000000230\ \text{s}\left(\frac{1 \times 10^9\ \text{ns}}{\text{s}}\right) = 2.30 \times 10^2\ \text{ns (nanoseconds)}$$

(*d*) $$1253\ \text{g}\left(\frac{1\ \text{kg}}{1000\ \text{g}}\right) = 1.253\ \text{kg}$$

SCIENTIFIC NOTATION

Problem 1.41. Express the following numbers in scientific notation: (*a*) 156000; (*b*) 2301; (*c*) 0.000325; (*d*) 0.0000104.

Ans.

(*a*) $156000 = 1.56 \times 10 \times 10 \times 10 \times 10 \times 10 = 1.56 \times 10^5$

(*b*) $2301 = 2.301 \times 10 \times 10 \times 10 = 2.301 \times 10^3$

(*c*) $0.000325 = 3.25 \times 10^{-1} \times 10^{-1} \times 10^{-1} \times 10^{-1} = 3.25 \times 10^{-4}$

(*d*) $0.0000104 = 1.04 \times 10^{-1} \times 10^{-1} \times 10^{-1} \times 10^{-1} \times 10^{-1} = 1.04 \times 10^{-5}$

Problem 1.42. Express the following quantities in scientific notation, with appropriate metric abbreviations: (*a*) 12,500 years; (*b*) 2530 kg; (*c*) 0.000035 s; (*d*) 1,543 cm.

Ans.

(*a*) $12{,}500 \text{ years} = 1.25 \times 10^1 \times 10^1 \times 10^1 \times 10^1 = 1.25 \times 10^4 \text{ years}$

(*b*) $2530 \text{ kg} = 2.530 \times 10^1 \times 10^1 \times 10^1 = 2.53 \times 10^3 \text{ kg}$

(*c*) $0.000035 \text{ s} = 3.5 \times 10^{-1} \times 10^{-1} \times 10^{-1} \times 10^{-1} \times 10^{-1} = 3.5 \times 10^{-5} \text{ s}$

(*d*) $1543 \text{ cm} = 1.543 \times 10^1 \times 10^1 \times 10^1 = 1.543 \times 10^3 \text{ cm}$

Problem 1.43. Calculate the product $(1.3 \times 10^3) \times (2.0 \times 10^{-5})$.

Ans. When numbers in exponential or scientific notation are multiplied, the coefficients are multiplied and the exponents are added:

(1) $1.3 \times 2.0 = 2.6$

(2) $3 + (-5) = 2$

Therefore

(3) $(1.3 \times 10^3) \times (2.0 \times 10^{-5}) = 2.6 \times 10^{-2}$

Problem 1.44. What is the result of dividing 3.62×10^{-3} by 1.81×10^{-2}?

Ans. When numbers in exponential or scientific notation are divided, the coefficients are divided and the exponents are subtracted:

(1) $3.62/1.81 = 2.00$

(2) $-3 - (-2) = -1$

Therefore

(3) $(3.62 \times 10^{-3})/(1.81 \times 10^{-2}) = 2.00 \times 10^{-1} = 0.200$

Problem 1.45. What is the result of adding 5.6×10^{-3} and 2.4×10^{-2}?

Ans. First convert the exponential terms to the same value, then add the coefficients:

(1) $2.4 \times 10^{-2} = 24 \times 10^{-3}$

(2) $(24 \times 10^{-3}) + (5.6 \times 10^{-3}) = (24 + 5.6) \times 10^{-3} = 29.6 \times 10^{-3}$

(3) $29.6 \times 10^{-3} = 2.96 \times 10^{-2}$

Problem 1.46. What is the result of subtracting 1.4×10^{-4} from 3.60×10^{-3}?

Ans. First convert the exponential terms to the same value, then subtract the coefficients:

(1) $3.60 \times 10^{-3} = 36.0 \times 10^{-4}$

(2) $(36.0 \times 10^{-4}) - (1.4 \times 10^{-4}) = (36.0 - 1.4) \times 10^{-4} = 34.6 \times 10^{-4}$

(3) $34.6 \times 10^{-4} = 3.46 \times 10^{-3}$

SIGNIFICANT FIGURES

Problem 1.47. How many significant figures are contained in the following quantities? (The concept of significance applies only to measured quantities, the units are omitted here.) (*a*) 23.0; (*b*) 0.00132; (*c*) 20; (*d*) 17.23; (*e*) 1.009

Ans. (*a*) 3; (*b*) 3; (*c*) 1; (*d*) 4; (*e*) 4

Problem 1.48. How many significant figures are contained in the following quantities? (*a*) 156.00; (*b*) 1.450; (*c*) 2.0092; (*d*) 0.0104; (*e*) 12.8

Ans. (*a*) 5; (*b*) 4; (*c*) 5; (*d*) 3; (*e*) 3

SIGNIFICANT FIGURES AND CALCULATIONS

Problem 1.49. What is the result of multiplying 1.45 cm by 12.02 cm?

Ans. By calculator, $1.45\ \text{cm} \times 12.02\ \text{cm} = 17.429\ \text{cm}^2$. However, since the least well-known quantity has only 3 significant figures, the answer must also contain no more than 3 significant figures: $17.4\ \text{cm}^2$.

Problem 1.50. What is the result of multiplying $2.36 \times 0.0012 \times 4.2$?

Ans. By calculator, $2.36 \times 0.0012 \times 4.2 = 0.0118944$. The least well-known quantities contain only 2 significant figures. Therefore the answer can contain no more than that, and the product must be reported as: 0.012.

Problem 1.51. What is the result of dividing 13.87 by 1.23?

Ans. By calculator, $13.87/1.23 = 11.27642276$. The least well-known quantity contains 3 significant figures, and the answer can contain no more. The result should be reported as: 11.3.

Problem 1.52. What is the result of dividing 0.0023 by 2.645?

Ans. By calculator, $0.0023/2.645 = 0.000869565$. The least well-known number contains 2 significant figures, so the reported result can have no more: 0.00087.

Problem 1.53. What is the result of adding 12.786 to 1.23?

Ans. By calculator: $12.786 + 1.23 = 14.016$. The answer can have no more than 2 digits after the decimal point. Therefore the sum is: 14.02.

Problem 1.54. What is the result of subtracting 2.763 from 3.91?

Ans. By calculator: $3.91 - 2.763 = 1.147$. The answer can have no more than 2 digits after the decimal point. Therefore the difference is: 1.15.

MEASUREMENT AND ERROR

Problem 1.55. What is the relative error of a measurement reported as $12.3\ \text{mL} \pm 0.1\ \text{mL}$?

Ans. The relative error is defined as the ratio of the constant variability to the actual measurement:

$$0.1\ \text{mL}/12.3\ \text{mL} = 0.008$$

Problem 1.56. What is the percent error of a measurement reported as $12.3\ \text{mL} \pm 0.1\ \text{mL}$?

Ans. The percent error is defined as the relative error multiplied by 100:

$$(0.1\ \text{mL}/12.3\ \text{mL}) \times 100\% = 0.8\%$$

FACTOR-LABEL METHOD

Problem 1.57. At 45 miles per hour (mi/h) how long would it take to drive 250 miles (mi)?

Ans. The quantity given is 250 mi. This must be multiplied by the proper conversion factor (correct units) to give the answer in hours:

$$250\text{ mi}\left(\frac{1.0\text{ h}}{45\text{ mi}}\right) = 5.6\text{ h}$$

Why are there two significant figures in the answer?

Problem 1.58. How many milliliters are contained in 4.2 gal?

Ans. We require a factor for conversion from English to metric volume, i.e., 1 qt = 946 mL.

$$4.2\text{ gal}\left(\frac{4.0\text{ qt}}{\text{gal}}\right)\left(\frac{946\text{ mL}}{\text{qt}}\right) = 15892.8\text{ mL}$$

Problem 1.59. Is the answer in question 1.58 correct?

Ans. Since the least well-known quantity, 4.2 gal, has only 2 significant figures, the answer should have been 1.6×10^4 mL.

Problem 1.60. How many milligrams are there in 4.00 ounces (oz)?

Ans. We require a factor for conversion from English to metric weight, i.e., 1 lb = 454 g.

$$4.00\text{ oz}\left(\frac{1\text{ lb}}{16\text{ oz}}\right)\left(\frac{454\text{ g}}{\text{lb}}\right)\left(\frac{1000\text{ mg}}{1\text{ g}}\right) = 113500\text{ mg} = 1.14 \times 10^5\text{ mg}$$

Why are there only 3 significant figures in the answer?

Problem 1.61. The foot race known as the marathon is run over a distance of greater than 26 mi over an open course. Express 26 mi in kilometers.

Ans. This is a conversion from English to metric length. The conversion factor that most people seem to best remember is centimeters/inch; 2.54 cm = 1 in (exactly), so let us express the problem using that factor. We shall convert miles to yards, then yards to feet, then feet to inches, then inches to centimeters, and from there to kilometers:

$$26\text{ mi}\left(\frac{1760\text{ yd}}{\text{mile}}\right)\left(\frac{3\text{ ft}}{\text{yd}}\right)\left(\frac{12\text{ in}}{\text{ft}}\right) = 1{,}647{,}360\text{ in}$$

$$1{,}647{,}360\text{ in}\left(\frac{2.540\text{ cm}}{\text{in}}\right)\left(\frac{1\text{ m}}{100\text{ cm}}\right)\left(\frac{1\text{ km}}{1000\text{ m}}\right) = 41.84\text{ km} = 42\text{ km}$$

Problem 1.62. World class times in the marathon are about 2 h and 10 s. How many minutes is that?

Ans. First, reduce time to a common unit, in this case, seconds, then convert seconds to minutes:

$$2.00\text{ h}\left(\frac{60\text{ min}}{\text{h}}\right)\left(\frac{60\text{ s}}{\text{min}}\right) = 7200\text{ s}$$

$$7200\text{ s} + 10\text{ s} = 7210\text{ s}$$

$$7210\text{ s}\left(\frac{1\text{ min}}{60\text{ s}}\right) = 120.17 = 120.2\text{ min}$$

In this case we know that 2 h is exactly 120 min, so it would not make sense to round off at 3 figures, since the time to be converted is greater than 2 h by 10 s. The most practical alternative seems to be to express the time to at least tenths of a minute. If we had considered the hours to be 2 h exactly, the result could have been left as 120.17 min.

MASS, VOLUME, DENSITY, TEMPERATURE, HEAT, AND OTHER FORMS OF ENERGY

Problem 1.63. How many grams are there in 1.0 lb? There is no way that you can answer this question without a knowledge of the metric-to-English conversion factor for weight. Very few of us keep such information in our heads, so one generally must look up the values in a table somewhere in the book. Now is as good a time as any to become familiar with the use of the index at the end of your textbook.

Ans. The metric-to-English conversion factor for weight is 1.0 kg = 2.2 lb (avoirdupois). The avoirdupois pound contains 16 oz (as opposed to the troy pound used in weighing precious metals, e.g., gold, platinum).

$$1.0\ \text{lb}\left(\frac{1.0\ \text{kg}}{2.2\ \text{lb}}\right)\left(\frac{1000\ \text{g}}{\text{kg}}\right) = 454.5\ \text{g}$$

Problem 1.64. How many grams are there in 1 oz?

Ans. There are 16 oz/lb; therefore

$$1\ \text{oz}\left(\frac{1\ \text{lb}}{16\ \text{oz}}\right)\left(\frac{454.5\ \text{g}}{\text{lb}}\right) = 28.4\ \text{g}$$

Problem 1.65. How many milligrams are there in 1 oz?

Ans.

$$1\ \text{oz} = 28.4\ \text{g}$$

Therefore $$28.4\ \text{g}\left(\frac{1000\ \text{mg}}{\text{g}}\right) = 28400\ \text{mg} = 2.84 \times 10^4\ \text{mg}$$

Problem 1.66. What are the dimensions of a cube of volume 1.0 L?

Ans. Volume in the shape of a cube is the product of length, width, and height. In the case of a cube, those three dimensions are equal. One liter contains 1000 mL, which is equivalent to 1000 cm^3. Because 1000 is the result of the product $10 \times 10 \times 10$, 1.0 L must be contained in a cube which is 10 cm on a side.

Problem 1.67. What is the SI equivalent of 1.0 L?

Ans. Volume in SI units is in cubic meters. We must convert 1000 cm^3 to cubic meters.

$$1.0\ \text{L} = 1000\ \text{mL} = 1000\ \text{cm}^3$$

$$1\ \text{m} = 100\ \text{cm}$$

$$(1.0\ \text{m})^3 = 1.0\ \text{m}^3 = (100.0\ \text{cm})^3 = 1 \times 10^6\ \text{cm}^3$$

Therefore $$1000\ \text{cm}^3\left(\frac{1.0\ \text{m}^3}{1 \times 10^6\ \text{cm}^3}\right) = 1 \times 10^{-3}\ \text{m}^3$$

Note the difference between $(1.0\ \text{m})^3$, which means 1.0 m cubed, and 1.0 m^3, which means 1.0 cubic meter.

Problem 1.68. Is there another SI equivalent for 1.0 L?

Ans. Since 1.0 L is a cube of 10 cm on a side, and 10 cm is one-tenth of a meter, we may call 10 cm 1 decimeter, symbol dm.

Then $$1000 \text{ cm}^3 = 1 \text{ dm} \times 1 \text{ dm} \times 1 \text{ dm} = 1 \text{ dm}^3$$

So $$1.0 \text{ L} = 1000 \text{ mL} = 1000 \text{ cm}^3 = 1 \text{ dm}^3 = 1 \times 10^{-3} \text{ m}^3$$

Problem 1.69. What is the mass of a slab of iron, density 7.87 g/cm^3, whose dimensions are 2.20 cm × 3.50 cm × 7.00 mm?

Ans. Density = mass/volume. Therefore, mass = density × volume. Before we can do that calculation, we must convert all dimensions to the same units.

$$2.20 \text{ cm} \times 3.50 \text{ cm} \times 0.700 \text{ cm} = 5.39 \text{ cm}^3$$

$$5.39 \text{ cm}^3 \left(\frac{7.87 \text{ g}}{\text{cm}^3}\right) = 42.4 \text{ g}$$

Problem 1.70. What is the volume of 100 g of ethyl alcohol whose density is 0.789 g/cm^3?

Ans.

$$100 \text{ g} \left(\frac{1 \text{ cm}^3}{0.789 \text{ g}}\right) = 126.7 \text{ cm}^3 = 127 \text{ cm}^3$$

Problem 1.71. Convert 75 °F to Celsius degrees.

Ans.

$$75 \text{ °F} - 32 \text{ °F} = 43 \text{ °F}$$

$$43 \text{ °F} \left(\frac{5 \text{ °C}}{9 \text{ °F}}\right) = 23.9 \text{ °C} = 24 \text{ °C}$$

Problem 1.72. Convert 98.6 °F to Celsius degrees.

Ans.

$$(98.6 - 32)\left(\frac{5 \text{ °C}}{9 \text{ °F}}\right) = 37 \text{ °C}$$

Problem 1.73. Express 37 °C in kelvins.

Ans.

$$\text{K} = \text{°C} + 273$$

$$37 + 273 = 310 \text{ K}$$

Problem 1.74. Express 125 °C in Fahrenheit degrees.

Ans.

$$125 \text{ °C} \left(\frac{9 \text{ °F}}{5 \text{ °C}}\right) + 32 = 257 \text{ °F}$$

Problem 1.75. What is the specific heat of a substance if 13.2 g of it requires 6.0 J to raise its temperature 10.0 °C?

Ans.

$$C_p = \frac{\text{joules}}{\text{Celsius degrees} \cdot \text{grams}}$$

$$C_p = \frac{6.0\ \text{J}}{10\ °\text{C} \cdot 13.2\ \text{g}} = 0.045\ \frac{\text{J}}{°\text{C} \cdot \text{g}}$$

Problem 1.76. How many joules are necessary to raise the temperature of 1.500 kg of a substance whose specific heat is 0.850, from 25.0 °C to 75.0 °C?

Ans. Solving the heat capacity relationship for joules,

$$\text{Joules} = C_p \times \Delta t \times \text{grams}$$

$$= 0.850 \times 50.0\ °\text{C} \times 1500\ \text{g} = 63750\ \text{J} = 63.8\ \text{kJ}$$

SUPPLEMENTARY PROBLEMS

Problem 1.77. How many significant figures are there in the following numbers? (*a*) 0.0254; (*b*) 67.4; (*c*) 103; (*d*) 0.00203.

Ans. (*a*) 3; (*b*) 3; (*c*) 3; (*d*) 3

Problem 1.78. Perform the following calculations to the correct number of significant digits: (*a*) 14.45 cm − 0.245 cm; (*b*) 32.8 cm − 3.25 cm; (*c*) 672.21 cm − 83.2 cm; (*d*) 2.035 cm − 1.29 cm.

Ans. (*a*) 14.21 cm; (*b*) 29.6 cm; (*c*) 589.0 cm; (*d*) 0.75 cm

Problem 1.79. Perform the following calculations to the correct number of significant digits: (*a*) 6.83 g + .056 g; (*b*) 0.3526 g + 0.854 g; (*c*) 123 g + 0.738 g; (*d*) 32.97 g + 3.273 g.

Ans. (*a*) 6.89 g; (*b*) 1.207 g; (*c*) 124 g; (*d*) 36.24 g

Problem 1.80. Perform the following calculations to the correct number of significant digits: (*a*) 12.853 cm × 2.91 cm; (*b*) 1.8 cm × 2.9132 cm; (*c*) 43.92 g ÷ 27.2 mL; (*d*) 2.729 g ÷ 3.2 mL.

Ans. (*a*) 37.4 cm^2 (3 significant figures); (*b*) 5.2 cm^2 (2 significant figures); (*c*) 1.61 g/mL (3 significant figures); (*d*) 0.85 g/mL (2 significant figures)

Problem 1.81. How many (*a*) kilograms, (*b*) grams, (*c*) milligrams are there in 1.5 lb calculated to the correct number of significant digits? The conversion factor is 453.59 g = 1 lb.

Ans. (*a*) 0.68 kg; (*b*) 6.8×10^2 g; (*c*) 6.8×10^5 mg

Problem 1.82. How many (*a*) kilometers, (*b*) meters, (*c*) centimeters, (*d*) millimeters are there in 0.800 mi calculated to the correct number of significant digits? The conversion factor is 1 mi = 1.6093 km.

Ans. (*a*) 1.29 km; (*b*) 1.29×10^3 m; (*c*) 1.29×10^5 cm; (*d*) 1.29×10^6 mm

Problem 1.83. Express the following numbers in exponential notation: (*a*) 0.00527; (*b*) 23,900; (*c*) 152; (*d*) 0.000021.

Ans. (*a*) 5.27×10^{-3}; (*b*) 2.39×10^{4}; (*c*) 1.52×10^{2}; (*d*) 2.1×10^{-5}

Problem 1.84. Convert the following numbers to nonexponential form: (*a*) 1.2×10^{-3}; (*b*) 3.6×10^{2}; (*c*) 1.53×10^{-4}; (*d*) 4.32×10^{4}.

Ans. (*a*) 0.0012; (*b*) 360; (*c*) 0.000153; (*d*) 43200

Problem 1.85. Perform the following calculations. Answers must reflect the correct number of significant digits.

(*a*) $$(2.30 \times 10^{2}) + (1.62 \times 10^{4})$$

(*b*) $$(1.74 \times 10^{-2}) + (3.10 \times 10^{-3})$$

(*c*) $$(5.00 \times 10^{-4}) + (4.92 \times 10^{-3})$$

Ans. (*a*) $16430 = 1.64 \times 10^{4}$
(*b*) $0.0205 = 2.05 \times 10^{-2}$
(*c*) $0.00542 = 5.42 \times 10^{-3}$

Problem 1.86. Perform the following calculations. Answers must reflect the correct number of significant digits.

(*a*) $$(3.71 \times 10^{-2}) - (4.72 \times 10^{-3})$$

(*b*) $$(5.81 \times 10^{3}) - (1.67 \times 10^{2})$$

(*c*) $$(9.62 \times 10^{-4}) - (2.93 \times 10^{-3})$$

Ans. (*a*) $0.03238 = 3.24 \times 10^{-2}$
(*b*) $5643 = 5.64 \times 10^{3}$
(*c*) $-0.001968 = -1.97 \times 10^{-3}$

Problem 1.87. Perform the following calculations. Answers must reflect the correct number of significant digits.

(*a*) $$(3.49 \times 10^{-5}) \times (2.75 \times 10^{3})$$

(*b*) $$(2.73 \times 10^{12}) \times (4.91 \times 10^{-2})$$

(*c*) $$(4.82 \times 10^{42}) \times (6.19 \times 10^{-31})$$

Ans. (*a*) $0.095975 = 9.60 \times 10^{-2}$
(*b*) 1.34×10^{11}
(*c*) 2.98×10^{12}

Problem 1.88. Perform the following calculations. Answers must reflect the correct number of significant digits.

(*a*) $$(5.92 \times 10^{-8}) \div (3.75 \times 10^{-3})$$

(*b*) $$(3.52 \times 10^{2}) \div (5.27 \times 10^{-3})$$

(*c*) $$(6.29 \times 10^{-2}) \div (2.97 \times 10^{3})$$

Ans. (*a*) $0.000015786 = 1.58 \times 10^{-5}$
(*b*) $66793.17 = 6.68 \times 10^{4}$
(*c*) $0.000021178 = 2.12 \times 10^{-5}$

Problem 1.89. Perform the following calculations. Answers must reflect the correct number of significant digits.

(*a*) (4.5 cm + 2.06 cm) × 12.54

(*b*) (3.92 cm − 0.954 cm) ÷ 1.9

(*c*) (43.92 g + 3.624 g) × 0.0162

(*d*) (256.2 mL − 58.29 mL) ÷ 13.9

Ans. (*a*) 82 cm; (*b*) 1.6 cm; (*c*) 0.770 g; (*d*) 14.2 mL

Problem 1.90. A glass vessel calibrated to contain 9.76 mL of water at 4.00 °C was found to weigh 22.624 g when empty and dry. Filled with a sodium chloride solution at the same temperature, it was found to weigh 32.770 g. Calculate the solution's density.

Ans. 1.04 g/mL

Problem 1.91. A volume of glucose solution weighed 43.782 g at 4.00 °C. An equal volume of water at the same temperature weighed 42.953 g. At 4.00 °C, the density of water is exactly 1.0000. Calculate the density of the glucose solution.

Ans. 1.0193 g/mL

Problem 1.92. The density of gold is 19.32 g/cm^3. Calculate the volume of a sample of gold that weighs 1.449 kg.

Ans. 75.00 cm^3

Problem 1.93. A urine sample has a density of 1.04 g/mL. What volume will 10.0 g of the sample occupy?

Ans. 9.62 mL

Problem 1.94. The average man requires 8400 kJ of energy per day. How many pounds of carbohydrate at 16.74 kJ/g must be consumed to obtain that energy?

Ans.

$$\left(\frac{8400\ \text{kJ}}{\text{day}}\right)\left(\frac{1\ \text{g}}{16.74\ \text{kJ}}\right)\left(\frac{1\ \text{lb}}{454\ \text{g}}\right) = \frac{1.11\ \text{lb}}{\text{day}}$$

Problem 1.95. Calculate the specific heat of 10.3 g of an unknown metal whose temperature increased 25.0 °C when 49.5 J of heat energy were absorbed by the metal.

Ans. 0.192 J/g · °C

Problem 1.96. Thermometer glass has a heat capacity of 0.84 J/g · °C. How much heat in joules would be required to raise the temperature of a 100-g thermometer 100 °C?

Ans. 8.4 kJ

CHAPTER 2

Atomic Structure and the Periodic Table

2.1 THE ATOMIC THEORY

In Chap. 1, pure substances were described as existing in two forms: elements and compounds. Elements cannot be decomposed into simpler substances but compounds can be broken down into elements. Elements are therefore the building blocks of all substances. Currently 108 different elements are known, and each is represented by a universally accepted symbol. A list of these will be found in Appendix B. The method for arriving at these symbols was, where possible, to use the first letter of the elements' name, capitalized. However, since there are only 26 letters of the alphabet, where the first letters of two elements were the same, the second letter, lowercase, was added to the first. The symbol for carbon is C, while the symbol for calcium is Ca. Unfortunately, the symbol for iron is not I or Ir but Fe. In addition to iron, there are other elements whose symbols are not derived from their common names but from names given to them much earlier in human history when the language of scholarship was Latin or German. Some of the more common of these are listed in Table 2-1, with their common names, symbols, and names from which the symbols were derived.

Table 2-1. Common Names of Some Elements, Their Symbols, and Symbol Origin

Common Name	Symbol	Symbol Origin
Silver	Ag	Argentum
Gold	Au	Aurum
Iron	Fe	Ferrum
Mercury	Hg	Hydrargyrum
Potassium	K	Kalium
Sodium	Na	Natrium
Lead	Pb	Plumbum
Antimony	Sb	Stibium
Tin	Sn	Stannum
Tungsten	W	Wolfram

By the late 1700s, several fundamental truths or laws of chemistry had become established. These were the following:

The law of conservation of mass. The mass that enters into a chemical reaction remains unchanged as a result of the reaction. In precise form: mass is neither created nor destroyed in a chemical reaction.

The law of constant composition. The elements that a compound is composed of are present in fixed and precise proportion by mass.

Problem 2.1. What are the amounts of iron and oxygen in iron oxide?

Ans. In 100 g of a particular compound of iron and oxygen, 77.7 g of iron and 22.3 g of oxygen are present in any sample of this compound prepared or isolated from any source.

The law of multiple proportions. When the same elements can form two different compounds, the ratio of masses of one of the elements in the two compounds is a small whole number relative to a given mass of the other element.

Problem 2.2. Can iron and oxygen form another compound in addition to the one described in Problem 2.1?

Ans. Iron and oxygen form a compound in addition to the one referred to above. In 100 g of this second compound, there are 69.9 g of iron and 30.1 g of oxygen. The ratio of masses of iron in the two compounds relative to 1.00 g of oxygen are

$$1.00 \text{ g Oxygen} \left(\frac{77.7 \text{ g Fe}}{22.3 \text{ g O}}\right) = 3.48 \text{ g Fe/g O}$$

$$1.00 \text{ g Oxygen} \left(\frac{69.9 \text{ g Fe}}{30.1 \text{ g O}}\right) = 2.32 \text{ g Fe/g O}$$

$$\frac{3.48 \text{ g Fe/g O}}{2.32 \text{ g Fe/g O}} = \frac{1.50}{1.00} = \frac{3.00}{2.00}$$

These laws were explained by a simple model or mental picture proposed by John Dalton in 1808. His basic idea was that all matter was composed of infinitesimally small particles called *atoms*, which were indestructible. Further, that the atoms of any one element were identical, and that what chiefly distinguished the atoms of one element from another was that they had different masses. Compounds were seen as combinations of atoms. Last, there was the necessary assumption regarding how many atoms of one kind combined with atoms of another kind to form compounds. Dalton made a basic assumption that atoms combine in 1:1 ratios. In cases where more than one compound could be formed by two elements, he proposed that the compound with the 1:1 ratio of atoms is the most stable compound.

The law of conservation of mass could then be interpreted by viewing a chemical reaction as a process in which the atoms of one element were combined with those of another, or atoms combined in one compound were separated and rearranged with those in another. No mass could be lost, but new partners could be established.

The law of constant composition could be explained by the following proposition. If a compound were composed of atoms of one kind having a particular mass, and atoms of another kind having a different mass, then the measured mass ratio of the elements determined by quantitative analysis must be that of the mass ratio of the constituent atoms.

When elements can form more than one compound, since they must combine as atoms, and therefore in whole numbers, the ratios of the masses of one of the elements with respect to 1 g (a fixed mass) of the other must also result in whole numbers. Hence the law of multiple proportions.

2.2 ATOMIC MASSES

Dalton's ideas were based on the assumption that atoms of different elements possess different masses. However, there was no way at that time to measure those masses directly. The solution to that problem was to measure the relative masses. For example, the mass of a basketball relative to a baseball is simple to determine. One weighs both and expresses the relative mass as a ratio, as 16 (baseballs) = 1 (basketball), or 1/16 basketball = 1 baseball. Notice that the mass units are irrelevant as long as we stick to the same mass units. Suppose that the objects of interest were much too small to be handled, much less seen. Under such circumstances we would have to weigh large numbers of them. However, as long as we were sure that equal numbers had been weighed, then the ratio of the weighed masses would be in the ratio of the masses of individual objects. The relative masses of atoms could only be determined when they were both present in the same compound, and assumed to be present in 1:1 ratios, as for example, in carbon monoxide. Here, the ratio of masses of carbon to oxygen is 12/16. This means that if a large number of carbon atoms weigh 12.0 g, the same number of oxygen atoms weigh 16.0 g.

This approach was partially successful, but many difficulties and contradictions arose as more and more cases were examined by chemists. It took another 50 years to realize that the assumption of a universal 1:1 ratio of atoms in a stable compound was not correct. Once that assumption was discarded, new methods for determining the correct ratios and therefore the relative masses of the elements were established. These relative masses are called the *atomic masses* of the elements.

The modern atomic weight system is based on the mass of the most common form of the element carbon. The mass of this form of carbon is defined to be exactly 12 atomic mass units, abbreviated as amu. On this scale, for example, hydrogen has an atomic weight of 1.0078 amu. The atomic masses of all the elements appear in Appendix B as the average masses relative to "carbon twelve." We will examine the precise meaning of average mass in Sec. 2.4.

2.3 ATOMIC STRUCTURE

Dalton's atom had no features other than mass. Since the time of his proposal, many discoveries have demonstrated that the atom is not featureless or indestructible, but is composed of other parts, namely, subatomic particles, both electrically charged and uncharged types. There are principally two types of charged particles, a light and negatively charged *electron*, a much heavier positively charged *proton*, and an electrically uncharged (neutral) particle of about the same mass as the proton called a *neutron*. The protons and neutrons, and hence virtually all the mass of the atom, reside in an infinitesimally small volume of the atom called the *nucleus*. The electrons take up most of the volume of the atom. These particles and their properties are listed in Table 2-2.

Electrical charges interact in a simple way: Like charges repel each other, and unlike charges attract each other. Protons attract electrons, electrons repel electrons, and protons repel protons.

Table 2-2. Properties of Subatomic Particles

Subatomic Particles	Electrical Charge	Mass, amu	Location
Proton	+1	1.00728	Nucleus
Neutron	0	1.00894	Nucleus
Electron	−1	0.0005486	Outside nucleus

Since atoms are electrically neutral, the number of protons an atom contains is balanced by exactly the same number of electrons. Even though electrons and protons have very different masses, the magnitude of their charges is the same, although opposite. The electron charge is 1.6×10^{-19} coulombs (C).

Problem 2.3. Calculate the positive charge on a nucleus which has (*a*) 8 protons and 8 neutrons, and (*b*) 4 protons and 7 neutrons.

Ans. (*a*) $8(+1) + 8(0) = +8$; (*b*) $4(+1) + 7(0) = +4$

Problem 2.4. Calculate the mass to the nearest atomic-mass unit of an atom which contains (*a*) 9 protons and 10 neutrons, (*b*) 4 protons and 5 neutrons.

Ans. (*a*) 9(1 amu) + 10(1 amu) = 19 amu; (*b*) 4(1 amu) + 5(1 amu) = 9 amu

Problem 2.5. How would the mass of the electrons in an atom of 4 protons and 5 neutrons affect the actual total mass of the atom?

Ans.

$$\text{Total mass} = 4(1.00728) + 5(1.00894) + 4(0.0005486) = 9.076$$

$$\text{Mass} - \text{electron mass} = 9.074$$

$$\%\ \text{Difference} = [(9.076 - 9.074)/9.076] \times 100\% = 0.02200\,\%$$

For this reason, chemists feel justified in considering the mass of an atom to be the sum of protons and neutrons.

Problem 2.6. The element which has a mass of about 9 amu is beryllium (Be), atomic number 4. What is the charge on (*a*) the Be nucleus, and (*b*) the Be atom?

Ans. (*a*) The beryllium nucleus has a charge of +4. (*b*) The beryllium atom is neutral and has no electrical charge.

The chemical properties of an element are determined by the number of electrons surrounding the atomic nucleus. The numbers of these electrons are equal to the numbers of protons in the nucleus. The number of protons in the nucleus is called the *atomic number*, and is unique for each element. In Problem 2.4(b), we could have called beryllium element number 4 or beryllium. Either is correct, since no other element has 4 and only 4 protons in its nucleus.

2.4 ISOTOPES

In a random sample of any element found in nature, it is common to find atoms with slightly differing masses. Since they are the same element, they must have the same number of protons; therefore the mass differences must reside in differing numbers of neutrons. These slightly different family members are called *isotopes*. Isotopes have different nuclear properties and masses, but have the same chemical properties. The sum of the protons and neutrons is called the *mass number* of the isotope, and the mass number is used to distinguish isotopes from each other. By expressing the mass number and atomic or proton number of an element, we can distinguish in written material between isotopes. This is done by writing the mass number as a superscript, and the atomic or proton number as a subscript.

Problem 2.7. What is the symbolic notation for two of the isotopes of carbon?

Ans. Two isotopes of carbon have mass numbers of (*a*) 12 and (*b*) 14, and the same atomic or proton number, 6. The notation is (*a*) $^{12}_{6}C$ and (*b*) $^{14}_{6}C$.

Problem 2.8. What is the sum of the protons and neutrons, and what is the number of neutrons in (*a*) $^{12}_{6}C$ and (*b*) $^{14}_{6}C$?

Ans. The mass number consists of the *sum* of protons and neutrons, so for (*a*):

Protons + neutrons = 12

Protons = 6

Therefore Neutrons = 12 − 6 = 6

For (*b*): Protons + neutrons = 14

Protons = 6

Therefore Neutrons = 14 − 6 = 8

The atomic weights in Appendix B are the average atomic masses of naturally occurring elements.

Problem 2.9. What is (*a*) the total mass an elevator must support if it contains three men who weigh 150 lb, two men who weigh 200 lb, and four women who weigh 125 lb; and (*b*) what is the average mass per person in the elevator?

Ans. (*a*) Total mass = $(3 \times 150) + (2 \times 200) + (4 \times 125) = 1350$ lb

(*b*) Average mass = total mass/total number of persons

$= 1350/9 = 150$ lb/person

Another approach is to use the fraction of people in each category and thus do the calculation in one step:

$$\text{Average} = (3/9)(150) + (2/9)(200) + (4/9)(125) = 150 \text{ lb/person}$$

Problem 2.10. Neon has three naturally occurring isotopes of mass number (to the closest atomic mass unit) of 20.0, 21.0, and 22.0. Their abundances (percent of those found in nature) are 90.92 percent, 0.257 percent, and 8.82 percent. What is the atomic weight of neon?

Ans.

$$(20.0 \times 0.9092) + (21.0 \times 0.00257) + (22.0 \times 0.0882) = 20.2 \text{ amu}$$

The average mass is the sum of the masses of each isotope times its fractional abundance.

2.5 THE PERIODIC TABLE

Soon after the establishment of atomic weights, the *periodic law* was established. It states that the properties of the elements repeat periodically if the elements are arranged in order of increasing atomic number. The law is best illustrated by focusing on chemical reactivity.

There is a group of elements, helium, neon, argon, krypton, xenon, and radon, which are relatively inert. That is, in contrast to most of the other elements, they do not easily form compounds with other elements. These elements are commonly called *noble gases*. In Fig. 2-1 the first five of these noble gases are approximately located with respect to the other elements along a horizontal line representing the elements in order of increasing atomic number.

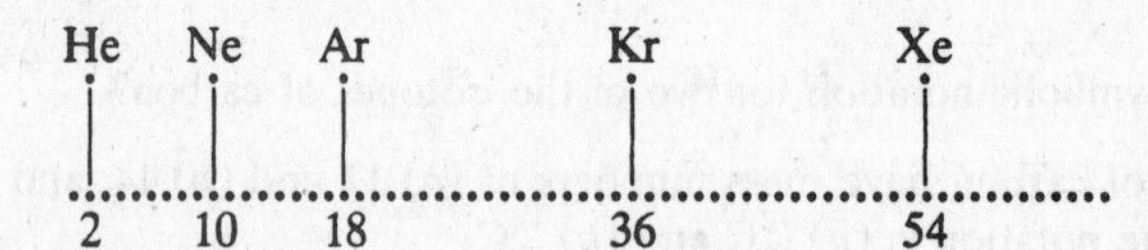

Fig. 2-1 Horizontal representation of the periodic reappearance of inertness of the noble gases.

The elements which directly follow the noble gases in position along the line of atomic number—lithium, sodium, potassium, rubidium, and cesium, atomic numbers 3, 11, 19, 37, and 55, respectively—possess very similar chemical properties, and are called *metals*. Metals are shiny, and are excellent conductors of electricity and heat. These metals react vigorously with water to produce gaseous hydrogen, and basic or caustic solutions, and are therefore called the *alkali metals*.

The elements which directly precede the noble gases—fluorine, chlorine, bromine, and iodine, atomic numbers 9, 17, 35, and 53, respectively—as a group have similar properties. The members of this group are called the *halogens*. They have none of the properties of metals, and are therefore known as *nonmetals*. Other groups of similar chemical properties can be identified in the same way, e.g., those following the noble gases by 2 atomic numbers are called *alkaline earths*—beryllium, magnesium, calcium, strontium, and barium.

In order to emphasize the similarities of chemical properties of the various groups, the *Periodic Table* is not arranged along a continuous horizontal line, but in repeating horizontal sequences of increasing atomic number, each beneath a preceding sequence. The result is that elements of similar chemical properties become aligned in vertical columns called *groups*. There are eight groups of similar chemical properties starting with the alkali metals and ending with the noble gases. This horizontal arrangement is called a *period*. When a period is completed, a new period, one down from the preceding one, is begun by placing the next member of the alkali metals directly under the first member of that group. The alkali metals comprise Group I, the noble gases Group VIII, with six intervening groups. The first period contains only two elements—hydrogen in I, and helium in VIII. The second and third periods each contain eight elements, but upon reaching the fourth period, a new feature emerges: An additional 10 elements are found whose properties lie between those in Group II and those in Group III. These are called the *transition elements*, and there are four series of transition elements of increasing complexity appearing in Periods 4, 5, 6, and 7. The elements appearing in the groups denoted by Roman numerals are called *main group* or *representative elements*. In the complete table in Appendix B, the lanthanide transition elements, numbers 57 to 70, and the actinide transition elements, numbers 89 to 102, have been placed below the table for a more compact arrangement. The chemical properties of each transition series are so similar that they are typically assigned to a single position in the Periodic Table. We will examine in some detail the first four periods in which the elements most closely associated with life processes are located. These periods are represented in Fig. 2-2, a shortened and simplified version of the Periodic Table in which the transition elements are identified.

Group

Period	I	II	Transition Elements										III	IV	V	VI	VII	VIII
1	1 H																	2 He
2	3 Li	4 Be											5 B	6 C	7 N	8 O	9 F	10 Ne
3	11 Na	12 Mg											13 Al	14 Si	15 P	16 S	17 Cl	18 Ar
4	19 K	20 Ca	21 Sc	22 Ti	23 V	24 Cr	25 Mn	26 Fe	27 Co	28 Ni	29 Cu	30 Zn	31 Ga	32 Ge	33 As	34 Se	35 Br	36 Kr
5	37 Rb																53 I	54 Xe
6	55 Cs																85 At	86 Rn

Metals
Metalloids or Semimetals
Nonmetals

Fig. 2-2 Abbreviated form of the Periodic Table.

In Fig. 2-2, only the atomic number is listed above the symbol for the element. The metals occupy most of the left-hand side of the table. Metals are elements which are lustrous, shiny, and good conductors of electricity. The nonmetals occupy the extreme right-hand side of the table, with a few elements having properties between metals and nonmetals, called *metalloids* or *semimetals*, separating the two large classes. The virtue of the Periodic Table is that if one is familiar with the chemistry of one member of a group, the chemistry of the other members of the group is known at least in broad outline.

Problem 2.11. Why are the alkali metals placed in the same group (Group I) of the Periodic Table?

Ans. They all react vigorously with water to produce hydrogen and form basic solutions. When these basic solutions are neutralized with hydrochloric acid, the product can be isolated by removal of water. For each of the alkali metals, the product is a white solid. The white solids are very soluble in water, and such solutions are good conductors of electricity.

Problem 2.12. Why are the halogens placed in the same group (Group VII) of the Periodic Table?

Ans. All the halogens react with hydrogen to form compounds which are very soluble in water, and result in acid solutions. These solutions can be neutralized with potassium hydroxide. The products can be isolated by removal of water. For each of the halogens, the product is a white potassium salt, quite soluble in water, and in solution an excellent conductor of electricity.

2.6 ATOMIC STRUCTURE AND PERIODICITY

The Periodic Table has been shown to be the result of a fundamental feature of the structure of the atom. The first model of atomic structure which successfully explained such physical observations as the emission spectrum of hydrogen was proposed by Niels Bohr, a Danish physicist. It was essentially a mathematical idea, but he attempted to explain his results in terms of a mental picture. The emission spectrum of an element is established by heating the element in a flame, thus "exciting" the atoms from the most stable state, the *ground state*, and observing the light emitted (the spectrum) as a result of "relaxing" back from the excited state to the ground state. The typical spectrum of an element consists of a series of lines of different wavelength. Each wavelength of light can be associated with a unique value of energy. Bohr's mathematical model of the atom explained the origin of that effect. He proposed that the atom resembled a miniature solar system, with electrons moving about the nucleus as planets revolve around the sun. The principal idea was that electrons are limited in the permissible values of their energies, and are therefore forced to remain in fixed "orbits" (as Bohr put it) about the nucleus, and can occupy no other portion of the volume of the atom. The distance from the nucleus reflects the energy state of the electrons. The farther away from the nucleus, the greater their energy. These orbits make the atom resemble a kind of "onion" made up of what Bohr called "shells." An electron can move from shell to shell only by absorbing or losing a unique amount of energy, no more and no less. This is what gives rise to the line spectra emitted by "excited" atoms. The energy states were characterized by an integer called the *principal quantum number*.

Bohr's ideas led to the development of modern quantum mechanics, the laws of motion which govern the behavior of atomic and subatomic particles. This is a mathematical theory even more complex than Bohr's mathematical model. In quantum mechanical language, the concept of an electron as an identifiable particle had to be given up, and in its place one can only estimate the probability of finding an electron in a given region of space. In quantum mechanics Bohr's principal quantum number which described the energy states of the atom was retained. But Bohr's shells now were shown to consist of *subshells*. Furthermore, electrons in these subshells were shown to reside in *orbitals* in which only specific regions of space are available to the electrons. Some of these are spherical, some of other shapes. The orbitals are therefore defined as regions of space in which there are high probabilities of finding electrons. As energy states grow larger, the number of subshells increases, and the orbitals assume more and more complex shapes.

The results of the application of quantum mechanics to the analysis of atomic structure can be summarized by a number of rules. We will use these rules to examine the underlying structure of the Periodic Table.

Electrons within atoms are characterized by four quantum numbers:

1. The principal quantum number, n, determines the energy state of an electron. It can have integer values of 1, 2, 3, up to n.

2. The subshell number, l, defines the orbital shape. Its values start at $(n - 1)$, and become smaller by integer values, ending at zero.
3. An orbital number, m, which specifies the spatial orientation of an orbital. It has integer values going from $+1$ through 0 to -1.
4. Finally a spin quantum number, s, which can have values of $+\frac{1}{2}$ or $-\frac{1}{2}$, and does not depend upon the values of n, l, or m. The electron within an atom behaves as though it spins on its own axis.

Furthermore, there is a strict limitation, called the *Pauli exclusion principle*, which states that no two electrons in an atom can have the same 4 quantum numbers. This means that each electron must have its own unique set of 4 quantum numbers. Two electrons in an atom may have the same values of n, l, and m, but the 4th quantum number, s, the spin number, must be different.

Problem 2.13. If an electron is in an energy state of principal quantum number 3, what possible values of the 4 quantum numbers can it have?

Ans.

$$n = 3$$
$$l = (n - 1) \text{ and down to 0, therefore } l = 2, 1, \text{ and } 0$$
$$m = +l, \text{ down to 0, then down to } -l \text{ or } 2, 1, 0, -1, -2$$
$$s = +\tfrac{1}{2} \text{ or } -\tfrac{1}{2}$$

An electron in principal quantum state 3 can have l values of 2, 1, and 0, m values of $+2$, $+1$, 0, -1, or -2, and s values of $+$ or $-\frac{1}{2}$.

Problem 2.14. How can one distinguish between n and l, since they both are expressed as integer values?

Ans. The n states are expressed as integers, but the l states are represented by lowercase letters:

$l = 0$ is designated as an s subshell
$l = 1$ is designated as a p subshell
$l = 2$ is designated as a d subshell
$l = 3$ is designated as an f subshell
$l = 4$ is designated as a g subshell

Problem 2.15. What kinds of orbitals are possible for electrons in principal quantum state 3?

Ans. For $n = 3$, $l = 2$, 1, and 0. Therefore, s, p, and d orbitals are possible. In applying quantum mechanical rules, we will see that different numbers of electrons are allowed in different kinds of orbitals.

Problem 2.16. Are there other significant differences in s, p, and d orbitals aside from differences in numbers of electrons allowed in these orbitals?

Ans. The s, p, and d orbitals have significant differences in spatial orientation and shapes. They can each be described by a three-dimensional envelope enclosing a region of space in which there is a high probability of finding an electron. The s orbital is spherical, indicating that in traveling outward from the nucleus there is an equal probability of finding an s electron no matter which direction one takes. The p and d orbitals are not spherical. They have a marked spatial orientation such that, in traveling from the nucleus outward in certain directions, no electrons can be found. In quantum mechanical language, we say that for electrons in these orbitals, there are regions of space which are forbidden. See Fig. 2-4.

In order to build atoms using quantum mechanical rules, by starting with one electron (atomic number 1), and adding one electron at a time, there is an additional requirement. We must be able to specify the order of filling of available orbitals. The order in which orbitals fill is based on the relative energy levels of orbitals. These data are shown in Fig. 2-3. Each circle in the diagram represents an orbital. The orbitals fill in the following order: $1s, 2s, 2p, 3s, 3p, 4s, 3d, 4p$ (we will not consider higher energy states). It is important to note that the $4s$ orbital fills before electrons enter the $3d$ orbital.

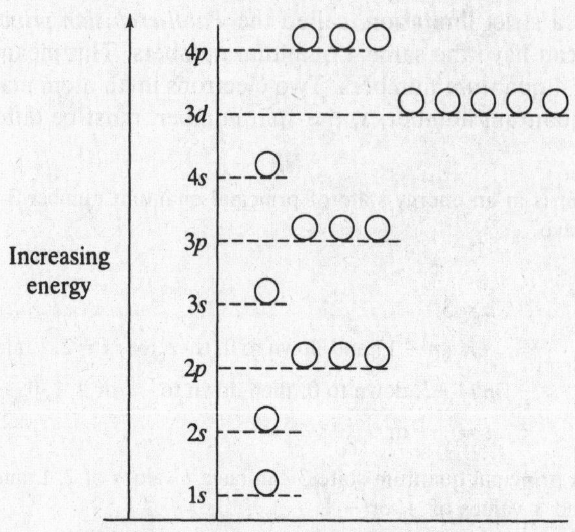

Fig. 2-3 Energy levels of atomic orbitals.

Using these quantum mechanical rules, we can now explore the underlying basis of the Periodic Table. To define the ground state configuration of an atom, we add electrons beginning with the lowest energy subshell until the correct number of electrons (equal to the elements' atomic number) have been added. This procedure is called the *aufbau* ("building-up") *principle*.

Problem 2.17. What are the 4 quantum numbers of the first two electrons added in building an atom?

Ans. We will construct a table to show these characteristic values.

Electron	n	l	m	s
1	1	0	0	$+\frac{1}{2}$
2	1	0	0	$-\frac{1}{2}$

No more electrons can go into quantum state $n = 1$, because a third electron would have the same 4 quantum numbers as one of the electrons already in the $n = 1$ state. Therefore electron 3 must go into the $n = 2$ state. The notation for the first electron, or element 1, hydrogen, is $1s^1$. When the second electron enters $n = 1$, the s subshell becomes complete, and represents element 2, helium, designated $1s^2$. These designations are called the *electron configurations* of the elements.

Problem 2.18. What are the electron configurations of the third through the tenth elements?

Ans. Since the $n = 1$ energy state is filled with the first two electrons, the next electron, the third, must enter the $n = 2$ state. In this state, more than one subshell can exist, since l can now have values of

1 and 0. The $n = 2$ state consists of two subshells, an s shell with $l = 0$, and a p subshell with $l = 1$. We will construct a table using all 4 quantum numbers.

n	l	m	s	Electrons per Orbital	Electrons per Subshell
2	0	0	$\pm\frac{1}{2}$	2	2
	1	+1	$\pm\frac{1}{2}$	2	
	1	0	$\pm\frac{1}{2}$	2	
	1	−1	$\pm\frac{1}{2}$	2	6

Total electrons for quantum state $n = 2$: eight.

Problem 2.19. How many electrons can fit into the $n = 3$ shell?

Ans. The $n = 3$ state consists of three subshells, an s subshell for $l = 0$, a p subshell for $l = 1$, and a d subshell for $l = 2$. We will construct a table as in Problem 2.18.

n	l	m	s	Electrons per Orbital	Electrons per Subshell
3	0	0	$\pm\frac{1}{2}$	2	2
	1	+1	$\pm\frac{1}{2}$	2	
	1	0	$\pm\frac{1}{2}$	2	
	1	−1	$\pm\frac{1}{2}$	2	6
	2	+2	$\pm\frac{1}{2}$	2	
	2	+1	$\pm\frac{1}{2}$	2	
	2	0	$\pm\frac{1}{2}$	2	
	2	−1	$\pm\frac{1}{2}$	2	
	2	−2	$\pm\frac{1}{2}$	2	10

Total electrons for quantum state $n = 3$: 18.

Problem 2.20. What is the maximum number of electrons per orbital?

Ans. Note that in the above examples, each orbital was defined by the quantum number m, and in each case the maximum number of electrons in each orbital was two. The important property of this pair of electrons was that their spins were opposite, $+\frac{1}{2}$ and $-\frac{1}{2}$. The two electrons in each m state are said to be "spin-paired." If we designate the spin by an arrow ↑ for a positive spin, and an arrow ↓ for a negative spin, then an electron pair in the same orbital is designated by a pair of arrows in opposite directions, as ↑ ↓. This will be of great significance when the nature of chemical bonds is discussed in Chap. 3.

Problem 2.21. What is the difference between the p orbitals of the $l = 1$ subshell?

Ans. There are six electrons in a filled $2p$ subshell. Only two electrons are permitted in one orbital; therefore there must be three different and distinguishable $2p$ orbitals. These three orbitals are fixed in space so as to be perpendicular to each other in three dimensions. Because they are arrayed perpendicularly to each other in three-dimensional cartesian coordinates, they are called the $2p_x$, $2p_y$, and $2p_z$ orbitals. The $2p_x$ orbitals appear as a teardrop pair along the x axis with the nucleus located at the origin. The $2p_y$ and $2p_z$ orbitals are arrayed similarly along the y and z axes. These are illustrated in Fig. 2-4, along with a sketch of the spherically symmetric $1s$ orbital.

$1s$ + $2p_x$ + $2p_y$ + $2p_z$

Fig. 2-4 Shapes of atomic orbitals.

Problem 2.22. The electron configuration of element 5, boron, is $1s^2 2s^2 2p^1$. To form the next element, carbon, an additional electron must be added to the $2p$ subshell to give the configuration $1s^2 2s^2 2p^2$. Does it go into the same or a different p orbital?

Ans. There are three p orbitals available to the second p electron, $2p_x$, $2p_y$, and $2p_z$. Because electrons have the same charge, they repel each other. Electrons can minimize the energy of repulsion, by entering separate orbitals. Only when each p orbital contains one electron will a fourth electron entering the p state pair with another electron in one of the half-filled orbitals. The electron configuration of carbon can be written as $1s^2 2s^2 2p_x^1 2p_y^1$, and that for the next element, nitrogen, as $1s^2 2s^2 2p_x^1 2p_y^1 2p_z^1$. This effect is described by another quantum mechanical principal called *Hund's rule*. In short it describes the fact that, within certain limitations, electrons within an atom have the tendency to spread out as much as possible so as to reduce the degree of repulsion of the like charges. The following is an alternative method of writing the electronic configuration of nitrogen which emphasizes the impact of Hund's rule:

$1s$	$2s$	$2p_x$	$2p_y$	$2p_z$
↑↓	↑↓	↑	↑	↑

This type of diagram shows the $1s$ and $2s$ electrons spin-paired, but the p electrons occupy different orbitals and are not paired. The diagram only illustrates the state of electron spins in each orbital, and does not indicate the relationship between energy levels. The next element, oxygen, will have one of the p orbitals spin-paired, with two p orbitals singly occupied:

$1s$	$2s$	$2p_x$	$2p_y$	$2p_z$
↑↓	↑↓	↑↓	↑	↑

Problem 2.23. Is there a more direct way of determining the total number of electrons in each shell or principal quantum state?

Ans. In the above examples we saw that the numbers of electrons per shell were as shown in the following table:

n	Total Numbers of Electrons	Rule
1	2	$2 = 2 \times (1)^2$
2	8	$8 = 2 \times (2)^2$
3	18	$18 = 2 \times (3)^2$

The general rule is for principal quantum state, n, the total number of electrons possible for that state is $2 \times n^2$. Note carefully, $2(n^2)$, *not* $(2n)^2$.

Problem 2.24. What is the electron configuration for the second, tenth, and eighteenth elements and where are they located in the Periodic Table?

Ans.

Atomic Number	Electron Configuration	Period	Group
2	$1s^2$	1	VIII
10	$1s^2 2s^2 2p^6$	2	VIII
18	$1s^2 2s^2 2p^6 3s^2 3p^6$	3	VIII

Problem 2.25. What is the electron configuration for the first, third, eleventh, and nineteenth elements, and where are they located in the Periodic Table?

Ans.

Atomic Number	Electron Configuration	Period	Group
1	$1s^1$	1	I
3	$1s^2 2s^1$	2	I
11	$1s^2 2s^2 2p^6 3s^1$	3	I
19	$1s^2 2s^2 2p^6 3s^2 3p^6 4s^1$	4	I

Problem 2.26. Is there an alternative method of writing the electronic configurations of the elements?

Ans. The electronic configuration of element 19, potassium, is $\underline{1s^2 2s^2 2p^6 3s^2 3p^6} 4s^1$. Notice that the underlined portion of the configuration is that of the previous noble gas, argon, which is abbreviated Ar. If we express the argon core (nucleus plus electrons) as [Ar], we can then write the potassium configuration as [Ar]$4s^1$. In this way, lithium is [He]$2s^1$, sodium is [Ne]$3s^1$, and calcium is [Ar]$4s^2$.

Although each of the elements described in Problems 2.24 and 2.25 has a unique electron configuration, each has the same outermost electron configuration. Since they are in the same group of the Periodic Table, and therefore have similar chemical properties, quantum mechanics has explained, most importantly, that the chemical properties of the elements depend not on the total number of electrons (or atomic number) but principally upon the configuration of the electrons in the outermost shell. This means the group number is in fact the number of electrons in the outer shell of the representative elements, and the principal quantum number is identified with the period.

In the case of the noble gases, in contrast to the other elements, the outermost shell is filled or complete. Thus the "inertness" or stability of the noble gases has led chemists to the proposition that a filled outermost electron shell is the most stable configuration of an atom. A logical corollary to that proposition is that an element's reactivity is related to its tendency to either gain or lose outermost electrons so as to achieve the same configuration as one of the noble gases. Since the number of electrons in this outermost shell is eight, this concept is called the *octet rule*. The outermost electron shell is called the *valence shell*. The electrons in the outer shell are called the *valence electrons*.

Problem 2.27. What are the consequences of completion of the valence shell of the elements?

Ans. When electrons are lost from the outer valence shell of the metals, a positively charged species is formed called a *cation*. Its charge reflects the number of electrons lost. Group I elements, which have one electron in the valence shell, form cations of charge +1 when that electron is lost. Group II elements, which have two electrons in the valence shell, lose two electrons to form cations or charge +2 when those two electrons are lost. In the case of the nonmetals, for example, Group VII elements, which have seven electrons in the valence shell, gain one electron to form negatively charged ions, called *anions*, of charge −1. The processes involved in the loss and gain of electrons leading to the completion of the other valence shell and consequent formation of chemical bonds will be more fully discussed in Chap. 3.

According to quantum mechanical rules, Period 3 should not be complete until all 10 *d* electrons are in place. However, if reference is made to Fig. 2-3, the order of orbital filling, you will note that the 4*s* subshell fills before the 3*d* subshell. This places element 19, potassium, in Period 4, and in Group I. The 4*s* orbital is complete in the next element, calcium, atomic number 20. The 3*d* subshell begins to fill with element 21, scandium, and as it fills, the 10 elements of the *transition series* of Period 4 are created. After the 3*d* subshell is filled, the 4*p* subshell begins to fill with element 31, gallium, which is placed in Group V, because it has the same valence shell configuration as aluminum in Period 3. The fact that *s* shells in the period one higher than the preceding subshell begin to fill before that preceding subshell leads to the result that all outermost shells have at most eight electrons.

Problem 2.28. Is there a more convenient way to represent the electron configuration central to the chemical reactivity of atoms—the valence shell?

Ans. Chemists use a symbolic representation called an *electron dot diagram* to emphasize the valence shell of an atom. The nucleus and inner shell electrons are represented by the elements' symbol, and dots around it represent the valence electrons. The electron dot diagrams for elements in Period 2 of the Periodic Table are

Li· ·Be· :B· ·C̈· ·N̈· ·Ö· :F̈· :N̈e:

For example, the formation of a lithium ion can be represented as

$$\text{Li}\cdot \longrightarrow \text{Li}^+ + e$$

ADDITIONAL SOLVED PROBLEMS

THE ATOMIC THEORY

Problem 2.29. What is the difference between vitamin B_1 obtained from yeast, and vitamin B_1 obtained by synthesis?

Ans. There is no difference. The law of constant composition tells us that a pure compound obtained from any source, either by extraction or synthesis, has the same composition.

Problem 2.30. What is the mass ratio of oxygen per gram of carbon found in two compounds of carbon: carbon monoxide, and carbon dioxide?

Ans.

	Grams Carbon	Grams Oxygen
Carbon monoxide	1.00	1.33
Carbon dioxide	1.00	2.66

Ratio of grams of oxygen, per gram of carbon in the two compounds carbon dioxide and carbon monoxide: $2.66/1.33 = 2/1$.

Problem 2.31. What is the mass of chlorine relative to the mass of hydrogen in hydrogen chloride?

Ans. Quantitative analysis of hydrogen chloride shows the mass percentages of hydrogen and chlorine to be 2.76 percent and 97.24 percent, respectively. Using 100 g of hydrogen chloride, and assuming a 1 : 1 ratio of the elements in this compound,

$$\frac{\text{Mass of a chlorine atom}}{\text{Mass of a hydrogen atom}} = \frac{97.24}{2.76} = \frac{35.2}{1}$$

or mass of a chlorine atom = 35.2 × mass of a hydrogen atom.

Problem 2.32. What is the mass of fluorine relative to the mass of hydrogen in hydrogen fluoride?

Ans. Quantitative analysis of hydrogen fluoride shows the mass percentages of hydrogen and fluorine to be 5.04 percent and 95.01 percent, respectively. Using 100 g of hydrogen fluoride, and assuming a 1 : 1 ratio of the elements in this compound,

$$\frac{\text{Mass of a fluorine atom}}{\text{Mass of a hydrogen atom}} = \frac{95.01}{5.04} = \frac{18.9}{1}$$

or, mass of a fluorine atom = 18.9 × mass of a hydrogen atom.

Problem 2.33. What is the mass of a chlorine atom relative to that of a fluorine atom?

Ans. From the previous problems,

$$\left(\frac{\text{Mass chlorine}}{\text{Mass hydrogen}}\right)\left(\frac{\text{mass hydrogen}}{\text{mass fluorine}}\right) = \left(\frac{35.2}{1}\right)\left(\frac{1}{18.9}\right) = \frac{1.86}{1}$$

So the mass of a chlorine atom = 1.86 × mass of a fluorine atom.

ATOMIC STRUCTURE

Problem 2.34. What is the atomic number of element 22?

Ans. The atomic number of element 22 is 22, and is equal to the number of protons in the atomic nucleus. Element 22 is called *titanium*.

ISOTOPES

Problem 2.35. The atomic mass of element 22 is 47.88 amu. The atomic mass of protons is about 1 amu (1.00728 amu). What makes up the difference in mass?

Ans. The difference in mass is caused by the presence of electrically neutral subatomic particles in the nucleus called neutrons whose atomic mass is also almost exactly 1 amu (1.00894 amu).

Problem 2.36. The mass difference between atomic number and atomic mass of element 22 is 25.88. How can particles of mass 1 amu add up to 25.88 amu?

Ans. The apparent fractional number of neutrons is caused by the presence of isotopes. Isotopes consist of atomic species having the same number of protons, but with differing numbers of neutrons in the nucleus.

Problem 2.37. How does the presence of isotopes account for titanium's mass of 47.88?

Ans. Titanium has five naturally occurring isotopes. The reported mass is the mean of the mass of all the isotopes:

% Natural Abundance	Atomic Mass
7.930	45.95
7.280	46.95
73.94	47.95
5.510	48.95
5.340	49.95

Thus

$$(0.07930 \times 45.95) + (0.07280 \times 46.95) + (0.7394 \times 47.95) + (0.05510 \times 48.95) + (0.05340 \times 49.95) = 47.88$$

Problem 2.38. Carbon, atomic number 6, has two principal isotopes, with atomic mass 12 and 13, respectively. What are the similarities and differences between these isotopes?

Ans. Both have the same atomic numbers, hence the same numbers of protons in the nucleus, and therefore the same chemical properties. The difference in mass is caused by differing numbers of neutrons in the nucleus. Carbon 12 has six neutrons and carbon 13 has seven neutrons in their respective nuclei.

THE PERIODIC TABLE

Problem 2.39. In what family or group is element 5?

Ans. We can map out the Periodic Table by considering (1) what atomic numbers begin and end each period, (2) there are eight groups of the representative elements in each period, and (3) there are 10 transition elements in Period 4.

Period	Group								
	I	II		III	IV	V	VI	VII	VIII
1	1								2
2	3								10
3	11								18
4	19	20	[Transition elements]	31					36

Element 5 is between elements 3 and 10, and therefore in Period 2 and in Group III.

Problem 2.40. In what family is element 14?

Ans. Element 14 is between elements 11 and 18, and therefore in Period 3 and in Group IV.

Problem 2.41. In what family is element 32?

Ans. Element 32 is between elements 19 and 36, and therefore in Period 4. The transition elements appear in Period 4—elements 21 through 30. Therefore, we can count forward from 31, or backward from the end of the period starting with element 36. Element 32 is in Group IV.

Problem 2.42. Identify the elements having the following symbols: (*a*) H; (*b*) K; (*c*) Be; (*d*) Zn; (*e*) Ca; (*f*) Cl; (*g*) Al; (*h*) P.

Ans. (*a*) H: hydrogen; (*b*) K: potassium; (*c*) Be: beryllium; (*d*) Zn: zinc; (*e*) Ca: calcium; (*f*) Cl: chlorine; (*g*) Al: aluminum; (*h*) P: phosphorus.

Problem 2.43. For which elements is it easiest to predict the number of electrons in the outermost or valence shell?

Ans. The representative or main group elements. The number of electrons in their outermost or valence shell is equal to their group number.

ATOMIC STRUCTURE AND PERIODICITY

Problem 2.44. Can one observe the emission spectrum of an atom in the ground state?

Ans. No. An atom's emission spectrum can only be seen if the atom has been promoted to a higher energy level by the absorption of energy. Once in the higher energy state, the atom will achieve the more stable ground state by the loss of energy in the form of emitted light.

Problem 2.45. Will the emission spectrum of an atom contain all wavelengths of light?

Ans. No. The emission spectrum of an atom consists only of specific wavelengths corresponding to unique values of energy.

Problem 2.46. How did Bohr interpret the nature of atomic emission spectra?

Ans. Bohr considered that the presence of only unique values of energy in atomic emission spectra meant that electrons could only occupy unique positions distant from the atomic nucleus. He called these distant positions shells, and postulated that emission spectra arose when electrons were promoted from the ground state to a shell higher in energy (more distant from the nucleus) and fell back to a shell of lower energy (closer to the nucleus).

Problem 2.47. The Bohr atomic model has been replaced by the modern quantum mechanical atomic model. Are there any features of the Bohr model which have been retained?

Ans. The Bohr concept of electron shells associated with the principal quantum number, n, has been incorporated into the modern atomic model.

Problem 2.48. In addition to the concept of electron shells whose energy increases with principal quantum number, n, what new features define the quantum mechanical atomic model?

Ans. A most important new feature is that electrons can no longer be thought of as discrete particles capable of precise location within the atom. Electrons within an atom are now thought of as occupying a region of space called an orbital. An orbital is a region of space in which there is a high probability of finding an electron.

Problem 2.49. What additional features characterize the modern atomic model?

Ans. Three new quantum numbers appear, which characterize atomic structure in finer detail. These are l, the subshell number, a quantum number which specifies orbital shape, m, which specifies the orbitals' orientation in space, and s, the spin quantum number, which describes the fact that electrons appear to rotate or spin on their axes.

Problem 2.50. How are these new quantum numbers related to the principal quantum number, n?

Ans. Their values depend on the value of n in the following way:

l can have values of $0, 1, 2, \ldots (n-1)$.

m can have values from $+1$, down to 0, up to -1.

s can only have one of two values, $+\frac{1}{2}$ or $-\frac{1}{2}$, and is independent of l and m.

Problem 2.51. What are the possible values of the subshell quantum number l if $n = 3$, and since both n and l are characterized by integer values, how can we differentiate the n from the l values?

Ans. The values of n are left in integer form, but the integer values of l are represented by lowercase letters.

n	l	Letter Designation of l
3	0	s
3	1	p
3	2	d

Problem 2.52. On what does the orbital quantum number m depend?

Ans. The values which the orbital quantum number can take depend upon the values of the subshell number l, as in the following chart. The orbital quantum number does not directly depend upon the principal quantum number n.

Subshell	m Values	Number of Orbitals	Number of Electrons
s	0	1	2
p	$-1, 0, +1$	3	6
d	$-2, -1, 0, +1, +2$	5	10

Problem 2.53. Write the detailed electronic configuration for the following elements: (*a*) C; (*b*) O; (*c*) Na; (*d*) P.

Ans. (*a*) $1s^22s^22p^2$; (*b*) $1s^22s^22p^4$; (*c*) $1s^22s^22p^63s^1$; (*d*) $1s^22s^22p^63s^23p^3$

Problem 2.54. Write the shorthand configuration for the following elements: (*a*) C; (*b*) O; (*c*) Na; (*d*) P.

Ans. (*a*) [He]$2s^22p^2$; (*b*) [He]$2s^22p^4$; (*c*) [Ne]$3s^1$; (*d*) [Ne]$3s^23p^3$

Problem 2.55. Write the detailed electronic configuration of (*a*) the sodium ion, Na^+, and (*b*) the calcium ion, Ca^{2+}.

Ans. (*a*) Na is $1s^22s^22p^63s^1$. One electron is lost to form the positive sodium ion; therefore, Na^+ is $1s^22s^22p^6$, or [Ne]. (*b*) Ca is $1s^22s^22p^63s^23p^64s^2$. Two electrons are lost to form the positive calcium ion; therefore Ca^{2+} is $1s^22s^22p^63s^23p^6$, or [Ar].

Problem 2.56. What elements or ions are represented by the following electronic configuration? $1s^22s^22p^6$

Ans. This is the electronic configuration of element 10, neon, the positive sodium ion, Na^+, and the negative fluoride ion, F^-.

Problem 2.57. What is the octet rule?

Ans. The octet rule is illustrated in Problem 2.55. When the outermost s and p subshells of elements in any period are filled, with a total of eight electrons, the elements are in the greatest state of stability. Elements other than the noble gases can achieve that state by either gaining electrons to form negative ions, or by losing electrons to form positive ions.

Problem 2.58. Is an *s* orbital aligned along any specific direction?

Ans. No. An *s* orbital is spherically symmetric. There is equal probability of finding an *s* electron in any direction outward from the nucleus.

Problem 2.59. Is a *p* orbital aligned along any specific direction?

Ans. Yes. There are three *p* orbitals aligned perpendicularly to each other along the *x*, *y*, and *z* axes with origin at the nucleus.

SUPPLEMENTARY PROBLEMS

Problem 2.60. What is the relative mass ratio of carbon to oxygen in the compound carbon monoxide, which has the composition CO?

Ans. 0.75

Problem 2.61. The mass difference between atomic number and atomic mass of element 37 is 48.47 amu. How can particles of mass 1 amu account for a fractional difference of 48.47 amu?

Ans. The apparent fractional difference arises from the presence of isotopes.

Problem 2.62. In what family or group is element 34?

Ans. Group VI.

Problem 2.63. For which elements do the number of electrons in their outer or valence shell correspond to their group number?

Ans. The representative or main group elements.

Problem 2.64. Write the complete electronic configuration for the following elements: (*a*) Mg; (*b*) S; (*c*) K.

Ans. (*a*) $1s^22s^22p^63s^2$; (*b*) $1s^22s^22p^63s^23p^4$; (*c*) $1s^22s^22p^63s^23p^64s^1$

Problem 2.65. Write the shorthand configuration for the following elements: (*a*) B; (*b*) F; (*c*) Ca; (*d*) Sr.

Ans. (*a*) $[He]2s^22p^1$; (*b*) $[He]2s^22p^5$; (*c*) $[Ar]4s^2$; (*d*) $[Kr]5s^2$.

Problem 2.66. Describe the spatial orientation of an *s* orbital.

Ans. An *s* orbital is spherically symmetrical.

Problem 2.67. Is there a directional character to the spatial orientation of *p* orbitals?

Ans. Yes. They indicate electron density only along the *x*, *y*, and *z* axes of a three dimensional Cartesian coordinate system.

Problem 2.68. What is the maximum number of electrons that can fit into a *p* orbital?

Ans. Six electrons.

Problem 2.69. Define the term *isoelectronic*, and give two examples of isoelectronic structures.

Ans. Isoelectronic refers to atoms or ions which have the same electronic configuration. Examples are Na^+, Ne, and F^-.

Problem 2.70. Which group of the Periodic Table is known as the alkali metals?

Ans. Group I.

Problem 2.71. Which group of the Periodic Table is known as the halogens?

Ans. Group VII.

Problem 2.72. Which group of the Periodic Table is known as the alkaline earths?

Ans. Group II.

Compounds and Chemical Bonding

3.1 INTRODUCTION

Any electrically neutral aggregate of atoms, held together strongly enough so as to behave as a unit, is called a *molecule*. The force holding any two atoms together in such an aggregate is called a *chemical bond*. Chemical bonds are of two general types: covalent bonds and ionic bonds. These represent extremes of the tendency of atoms to achieve filled valence shells. For example, Group I metals can achieve a noble gas configuration by losing one outermost electron, and Group VII halogens can achieve a noble gas configuration by gaining one electron. Bonds between Group I elements and Group VII elements are ionic bonds. In other cases, elements can achieve the noble gas configuration by sharing electrons. Such bonds are called *covalent bonds*.

Before we proceed with the details of the formation of ionic and covalent bonds, it will be useful to introduce the systematic methods employed in the naming of ionic and covalent compounds.

3.2 NOMENCLATURE

3.2.1 Binary Ionic Compounds

The underlying structure of the Periodic Table has led to our understanding of why sodium chloride is NaCl, and not $NaCl_2$. We are now able to systematize the many possibilities regarding binary compound formation between the metals and the nonmetals. In reactions between metals and nonmetals, ions are formed, whose charges depend on group identity and the octet rule. If we describe the results of reaction between metal X and nonmetal Y, as a compound X_aY_b, we can deduce all the possibilities of the combining ratios, $a:b$, by constructing a matrix as in Fig. 3-1. The critical constraint which must be adhered to is that the compounds must be electrically neutral, i.e., the sum of positive and negative charges must be equal to zero. Expressed quantitatively:

$$(\text{Cation charge} \times \text{number of cations}) = (\text{anion charge} \times \text{number of anions})$$

In Fig. 3-1, there are three columns representing the cations formed from elements in Groups I, II, and III. There are three rows, representing the anions formed from elements in Groups VII (first row), VI (second row), and V (third row). The compounds are listed at the juncture of row and column.

In these binary compounds the metals take the name of the element, and the anion's name begins with the element but takes the ending *−ide*. For example, potassium chloride (KCl), aluminum oxide (Al_2O_3), calcium fluoride (CaF_2).

Y \ X	+1	+2	+3
−1	XY	XY_2	XY_3
−2	X_2Y	XY	X_2Y_3
−3	X_3Y	X_3Y_2	XY

Fig. 3-1 Matrix illustrating ionic compound formation. Columns represent the cations from Groups I, II, and III. Rows represent anions from Groups VII (first row), VI (second row), and V (third row). Ionic compounds are listed at the juncture of row and column.

When we consider the elements in Period 4, in which the transition elements appear, the situation becomes more complex because these elements can form ions of more than one charge type. For example, iron can form ions of 2+ and 3+, and copper can form ions of 1+ and 2+, so that the chlorides of these metals can have the compositions $FeCl_2$, $FeCl_3$, CuCl, and $CuCl_2$. The older, and sometimes commonly used names of these compounds are ferrous chloride, ferric chloride, cuprous chloride, and cupric chloride, respectively. The modern systematized method is simply iron(II) chloride, iron(III) chloride, copper(I) chloride, and copper(II) chloride, respectively. Given the names of these compounds, and knowing the charge of the anion, one can deduce the combining ratios, and hence their formulas.

3.2.2 Polyatomic Ions

There are a number of ions consisting of more than one atom, and some of the more important of these are listed in Table 3-1.

Table 3-1. Some Important Polyatomic Ions

Ion	Name	Ion	Name
NH_4^+	Ammonium	CO_3^{2-}	Carbonate
NO_3^-	Nitrate	HCO_3^-	Hydrogen carbonate
SO_4^{2-}	Sulfate	PO_4^{3-}	Phosphate
OH^-	Hydroxide	HPO_4^{2-}	Hydrogen phosphate
$C_2H_3O_2^-$	Acetate	$H_2PO_4^-$	Dihydrogen phosphate

Problem 3.1. What are the formulas of the following compounds? (*a*) sodium acetate, (*b*) ammonium phosphate, (*c*) potassium hydrogen phosphate, (*d*) calcium dihydrogen phosphate.

Ans. (*a*) $NaC_2H_3O_2$, (*b*) $(NH_4)_3PO_4$, (*c*) K_2HPO_4, (*d*) $Ca(H_2PO_4)_2$. Note: (1) The number of positive and negative charges must be equal, and therefore, as in any ionic compound, the sum of positive and negative charges must be zero; (2) subscripts after parentheses refer to the number of ions within parentheses. If only one ion is present, no parentheses are necessary, as in answers (*a*) and (*c*). Finally, the cation is written first, followed by the anion.

3.2.3 Covalent Compounds

Some pairs of elements bound by covalent bonds can form more than one compound, i.e., CO and CO_2. One of the most interesting of these pairs are the compounds of nitrogen and oxygen. The system adopted for naming these compounds uses Greek prefixes. Table 3-2 illustrates the systematic method. Note that NO is called *nitrogen monoxide*, not *mononitrogen monoxide*, while CO is called

Table 3-2. Systematic Names of Covalent Compounds

Nitrogen / Oxygen Compound	Systematic Name
N_2O	Dinitrogen monoxide
NO	Nitrogen monoxide
NO_2	Nitrogen dioxide
N_2O_3	Dinitrogen trioxide
N_2O_4	Dinitrogen tetroxide
N_2O_5	Dinitrogen pentoxide

carbon monoxide, not *monocarbon monoxide*. The prefix *mono-* is never used for naming the first element.

Problem 3.2. Name the following compounds: (*a*) CCl_4, (*b*) PCl_3, (*c*) PCl_5, (*d*) SO_2, (*e*) SO_3.

Ans. (*a*) carbon tetrachloride, (*b*) phosphorus trichloride, (*c*) phosphorus pentachloride, (*d*) sulfur dioxide, (*e*) sulfur trioxide

3.3 IONIC BONDS

Ionic bonds are chemical bonds which are the result of the electrical attraction between positively charged and negatively charged ions. Removing an electron of negative charge from a neutral atom leaves behind a positively charged nucleus. Therefore, to remove an electron from the outer shell of an element, a significant amount of energy is required to overcome the attraction of positive charge for negative charge. The energy for this process is called the *ionization energy*, and is measured by the *ionization potential* of the element. In general the ionization potential decreases as one moves down in any group of the Periodic Table. This is because the valence shell electrons are farther and farther away from the influence of the positively charged nucleus as the periods (atomic number) increase. This process produces a positively charged species called a *cation*, whose chemical properties bear no relationship to those of the neutral atom.

Problem 3.3. How does the tendency to form cations vary within Group I?

Ans. Within Group I, the ease of forming cations is in the following order:

Cesium > rubidium > potassium > sodium > lithium

Because atomic number (nuclear charge) increases from Group I to Group VII, the ionization potential also increases as one proceeds across the Periodic Table from left to right. It is easier to remove one electron from a lithium atom (atomic number 3) than from a beryllium atom (atomic number 4). The ease of forming cations within Period 2 is Li > Be > B.

The addition of an electron to a neutral atom is described quantitatively by its *electron affinity*. This process produces a negatively charged particle called an *anion*, whose chemical properties bear no relationship to those of the neutral atom. In adding an electron to a neutral atom the principal force which operates is the attractive force of the positive nucleus. The attractive force of the positive nucleus decreases as the periods increase, because as additional electron shells are added to the atom, the outer electrons become more insulated from the nucleus by the intervening electron shells. It is easier to add an electron to the valence shell of fluorine (Period 2) than to that of chlorine (Period 3), and easier yet to add an electron to the valence shell of chlorine (Period 3) than to that of bromine (Period 4). The ease of forming anions within Group VII of the Periodic Table is F > Cl > Br > I.

Problem 3.4. What specific changes in atomic structure occur when ions are formed?

Ans. Let us focus our attention on the second period, and particularly on the first two and last two elements in the period, Li, Be, O, and F. The octet rule requires that, in order to achieve chemical stability, elements gain or lose electrons to gain noble gas configuration. Therefore, Li and Be lose one and two electrons, respectively, to achieve the helium configuration, and O and F gain two and one electrons, respectively, to achieve the neon configuration. Table 3-3 illustrates these processes. Within a period, the elements exhibit two tendencies: Those closest to Group I tend to lose electrons to form positive ions—cations—and those closest to Group VII tend to gain electrons to form negative ions—anions.

Table 3-3. Formation of Ions to Achieve Noble Gas Configuration

			Noble Gas Configuration	Ion Formed
$Li(1s^2 2s^1)$	$- 1e$	=	$1s^2$ (He)	Li^+
$Be(1s^2 2s^2)$	$- 2e$	=	$1s^2$ (He)	Be^{2+}
$O(1s^2 2s^2 2p^4)$	$+ 2e$	=	$1s^2 2s^2 2p^6$ (Ne)	O^{2-}
$F(1s^2 2s^2 2p^5)$	$+ 1e$	=	$1s^2 2s^2 2p^6$ (Ne)	F^-

Problem 3.5. Do all the elements in Period 2 form ions?

Ans. Not generally. For example, for carbon to form an ion to achieve noble gas configuration, it must lose or gain four electrons. The electrical work to accomplish that is very large, and almost never available in ordinary chemical reactions. In those cases where the necessary electrical work to remove electrons from the valence shell is not available, electrons will be shared in a covalent bond. In Sec. 3.4 we will discuss a concept related to electron affinity, called *electronegativity*, which offers a quantitative method for deciding whether or not an ionic bond will form in the reaction between two selected elements.

Problem 3.4 illustrated the atomic processes involved in the formation of either cations or anions. Now let us examine the reaction between the elements lithium and fluorine, in which ions are formed as the result of a chemical reaction.

Li can achieve the He configuration and F can achieve the Ne configuration by the loss of an electron from the valence shell of Li with the simultaneous addition of that electron to the valence shell of F. The removal of the electron from Li cannot occur unless energy for the work of removal is available for that process. That energy is supplied by the affinity of the electron for the F atom. This is the principal driving force for the reaction which results in the formation of lithium ions, designated Li^+, with the concomitant formation of fluoride ions designated F^-. The reaction can be written as the sum of two processes as

$$Li - e^- \longrightarrow Li^+$$

$$F + e^- \longrightarrow F^-$$

The compound which forms has the formula LiF, and the chemical bond which unites the ions is an *ionic bond*. An ionic bond is identified as a bond which unites ions of opposite charge. Ions are formed by an electron(s) leaving the valence shell of one atom and entering the valence shell of its partner. This is most likely to happen when the difference in ability to attract electrons of the two partners is large, as in the case of lithium and fluorine.

Problem 3.6. How do magnesium and fluorine react, and what kind of compound do they form?

Ans. Magnesium, atomic number 12, can achieve the Ne configuration, atomic number 10, by the loss of two electrons. Fluorine requires only one electron to achieve the Ne configuration. This transfer can be effected if two fluorine atoms react with one magnesium atom:

$$2F + 2e \longrightarrow 2F^-$$

$$Mg - 2e \longrightarrow Mg^{2+}$$

The combination of one Mg^{2+} ion with two F^- ions results in an electrically neutral aggregate with the formula MgF_2. The bonds holding this aggregate together are ionic. We denote the charge of an ion with a superscript, and the number of atoms of one kind in a compound with a subscript. We say that the charge of the magnesium ion is +2, and that of the fluoride ion is −1. In general the charge of a cation is the same as its group number in the Periodic Table. The charge of an anion is its group number minus 8. For example the valence for fluorine is −1 = VII − 8, and for oxygen, −2 = VI − 8.

Problem 3.7. Can the reactions described in Problems 3.5 and 3.6 be characterized in another way?

Ans. Reactions in which electrons lost by one of the reaction components are gained by another component are known as *oxidation-reduction* or *redox reactions*. The component which loses electrons is *oxidized*, and the component which gains electrons is *reduced*. Further, the component supplying the electrons is called the *reducing agent* or the *reductant*, and the component receiving the electrons is called the *oxidizing agent* or the *oxidant*.

Problem 3.8. Do the formulas LiF or MgF_2 describe molecules having those constitutions?

Ans. No. The chemical bonds in LiF and MgF_2 are ionic. Ionic bonds are the result of the strong attraction of opposite electrical charges. A positive charge attracts as many negative charges as can fit around it, and the same is true of negative charges attracting positive charges. In an ionic crystal, each positive ion (cation) is surrounded by negatively charged ions (anions). In the same way, each anion is surrounded by cations. The fundamental pattern repeats throughout the three-dimensional crystal so there is no way to identify a particular LiF unit. We cannot speak of an LiF molecule, or use the word *molecule* for any ionic compound. The formulas of ionic compounds only describe the combining ratios of the elements composing those compounds.

3.4 COVALENT BONDS

In many chemical reactions, particularly those between the nonmetals, ionization potentials and electron affinities are not large enough to produce ions, and therefore valence shells are filled by the sharing of electrons. In this sharing process, electrons are neither gained nor lost, but become part of the octets of both of the bonded partners.

The element hydrogen is a gas consisting of aggregates of two atoms with the formula H_2. Since the atoms are identical, their affinity for electrons and ease of ionization are identical. Therefore, the forces leading to the loss and concomitant gain of electrons, which we have seen in the above examples lead to ionic or electrical attractions, are not present. The alternative to the complete transfer of electrons from the valence shell of one atom of a pair to the valence shell of the other atom is to share the electrons. Both hydrogen atoms can achieve the He configuration, $1s^2$, by the compromise of sharing the electron pair. The electrons belong to neither of the atoms, but to both. The force holding this aggregate together is called a *covalent bond*. The electrons in a covalent bond must fit into available orbitals of the atom pair. Remember that no more than two electrons can exist in an orbital, and these must be of opposite spin. Therefore, a single covalent bond consists of a shared pair of electrons of opposite spin.

Problem 3.9. Are the electron orbitals of the hydrogen molecule the same as the atomic orbitals of the individual hydrogen atoms?

Ans. No. There are now two nuclei in the hydrogen molecule with two electrons forming a chemical bond. The rules governing shapes and orientations of atomic orbitals must now be modified to take into account the new and very different electrical environment existing in a molecule consisting of a number of individual positive nuclei held in some permanent array by a number of electrons. The electrons in a hydrogen molecule spend most of their time between the two nuclei, thus allowing the nuclei to approach closely. Remember also that (Hund's rule) the potential energy of electrons is reduced (enter a more stable state) when they can occupy a larger space. This occurs when electrons in an atomic orbital combine to form a covalent bond. The electrons in the hydrogen atom now occupy a *molecular orbital* in the shape of an ellipsoid (a short, fat cigar shape) called a *sigma orbital*, enclosing both nuclei. The symbol for this sigma bond formed by the combination of two atomic *s* orbitals is σ_s.

Problem 3.10. Can fluorine combine with itself?

Ans. The element fluorine is a gas consisting of aggregates of two atoms with the formula F_2. Since the atoms are identical, their affinity for electrons and ease of ionization are identical. As in the case of hydrogen, the identical nature of the atoms permits only a covalent bond to be formed. The electron configuration of fluorine is $1s^22s^22p^5$. Remember that there are three distinct *p* orbitals, so that we may rewrite the electron configuration as $1s^22s^22p_x^22p_y^22p_z^1$. Only one electron from each fluorine need be shared to complete the outer shell. In this case the single covalent bond formed consists of a shared pair of electrons formed from colinear *p* orbitals.

Problem 3.11. What are colinear *p* orbitals?

Ans. Recall that in an atom there are three atomic *p* orbitals arranged perpendicularly to each other. When the *p* orbitals of two reacting atoms approach each other along the same axis (both are p_x or p_y or p_z), they are said to be *colinear*. A covalent bond formed by the combination of colinear *p* orbitals is also called a *sigma bond*, as was the case of the combination of two atomic *s* orbitals. The symbol for a sigma bond formed by the combination of colinear *p* orbitals is σ_p.

Problem 3.12. Can fluorine combine with hydrogen?

Ans. Fluorine can complete its $2p^5$ outershell by sharing hydrogen's $1s^1$ electron. The HF bond is a single covalent bond, formed between a 1*s* and 2*p* orbital. It is significantly different from covalent bonds formed between identical atoms.

Just as the tendency of atoms to attract electrons into their valence shells to form anions is described by the concept of electron affinity, the ability of an atom within a molecule to attract electrons is characterized by the concept of *electronegativity*, to which we will give the symbol EN. Electronegativity describes the tendency of an atom to attract electrons shared in a covalent bond. All elements are assigned a numerical value of electronegativity. Table 3-4 is an abbreviated table of values of electronegativities of the elements. In binary compounds, when the difference in electronegativities is greater than 2.0, the bond will be ionic, and when the difference is 1.5 or less, the bond will be covalent. Furthermore, when electronegativities of covalent compounds differ, the electrons

Table 3-4. Abbreviated Table of Electronegativities

H 2.2							He
Li 0.98	Be 1.6	B 2.0	C 2.6	N 3.0	O 3.4	F 4.0	Ne
Na 0.93	Mg 1.2	Al 1.6	Si 1.9	P 2.2	S 2.6	Cl 3.2	Ar
K 0.88	Ca 1.0	Ga 1.8	Ge 2.0	As 2.2	Se 2.6	Br 2.8	Kr
Rb 0.82	Sr 0.95	In 1.8	Sn 2.0	Sb 1.9	Te 2.1	I 2.7	Xe
Cs 0.79	Ba 0.89	Tl 1.8	Pb 1.9	Bi 1.9	Po 2.0	At 2.2	Rn

bonding the two nuclei together do not spend equal time on each atom, but spend more time near the atom with the stronger attraction for electrons. This kind of covalent bond is said to be *polar*, in contrast to a covalent bond between atoms of identical electronegativity, which is called *nonpolar*. The bond between hydrogen atoms in H_2, and fluorine atoms in F_2, is a single nonpolar covalent bond, but the bond between hydrogen and fluorine in HF is a single polar covalent bond.

Problem 3.13. Are the properties of binary compounds characterized by polar covalent bonds different from substances in which nonpolar bonds are formed?

Ans. In the H_2 and F_2 molecules, the electrons spend most of their time midway between the positive nuclei, because their electronegativities are identical. The H_2 and F_2 molecules are electrically neutral from two points of view: (*a*) Their overall charge is zero, i.e., equal numbers of protons and electrons; (*b*) the "center of gravity" of negative and positive charges coincide with each other.

The center of gravity is a single point in any object which behaves as though all the mass of the body is located at that point. A seal balancing a ball on its nose, or a juggler balancing a chair on her chin makes sure that the center of gravity of the object is precisely above the point of support. Thus the "center of gravity" of electrical charge is a single point in any system which behaves as though either all the negative or all of the positive electrical charge is located at that point.

In the HF molecule, in contrast to the H_2 and F_2 molecules, because the covalent bond is polar, the electrons spend more of their time in the vicinity of the fluorine nucleus than near the hydrogen nucleus. Therefore, the center of gravity of negative charge does not coincide with the center of gravity of positive charge. The consequence of this is that the HF molecule is *dipolar*, a molecule which has an overall neutral charge, but which is not internally neutral—it has a relatively positive end and a relatively negative end. The electrical charges are not full but partial charges which are indicated by lowercase Greek symbols $\delta+$ or $\delta-$. The existence of partial charges on the atoms in the HF molecule is expressed by writing the formula $\overset{\delta+}{H} - \overset{\delta-}{F}$. The magnitude of a dipole is described quantitatively by its *dipole moment*, which depends principally on the difference in electronegativities of the combining atoms. It is customarily represented by an arrow pointing along the bond from the positive end toward the negative end as follows: $\leftarrow\!\!\!+\!\!\!\longrightarrow$

Problem 3.14. Arrange the following molecules in order of their dipole moments, largest first: (*a*) HI, (*b*) HCl, (*c*) HBr, (*d*) HF.

Ans. The magnitude of the dipole moment depends chiefly upon the difference in electronegativities. We must therefore calculate these differences using the values in Table 3-4.

(*a*) $\text{EN(I)} - \text{EN(H)} = 2.7 - 2.2 = 0.5$

(*b*) $\text{EN(Cl)} - \text{EN(H)} = 3.2 - 2.2 = 1.0$

(*c*) $\text{EN(Br)} - \text{EN(H)} = 2.8 - 2.2 = 0.6$

(*d*) $\text{EN(F)} - \text{EN(H)} = 4.0 - 2.2 = 1.8$

The order of the dipole moments in these binary compounds largest first is HF > HCl > HBr > HI.

The consequences of molecular polarity are expressed in both physical and chemical properties. Molecules with large dipole moments tend to interact strongly with each other, and with other molecules possessing dipole moments. For example, substances composed of polar molecules have higher melting and boiling points than those composed of nonpolar molecules. At room temperature, methane, CH_4, which has no dipole moment, is a gas, but water, H_2O, which has a large dipole moment, is a liquid at the same temperature. Water is also an excellent solvent for polar substances. An understanding of many aspects of physical and chemical properties of organic compounds depends upon the concept of polarity of chemical bonds.

Problem 3.15. Is there a general method for determining the atomic ratios of the different elements in a covalent compound?

Ans. In the compound formed from carbon (Group IV) and chlorine (Group VII), C and Cl require four and one electrons, respectively, to complete their valence shells. One of the four electrons in carbon's valence shell will fit into the valence shell of a Cl atom. An electron in a chlorine atom's valence shell will fit into the valence shell of the C atom. Since C needs a total of four electrons to complete its valence shell, four Cl atoms combine with one C atom to form CCl_4. The formula CCl_4 is a *molecular formula*, which describes only the numbers of atoms of each element in the molecule. The number of electrons needed for the atom of an element to complete its valence shell is called the *combining power*. The combining power of C, N, O, H, and Cl is 4, 3, 2, 1, and 1, respectively. The combining power is also the number of bonds an atom forms when it combines with other atoms to form a covalent compound. Sec. 3.5 presents a method for predicting the way atoms in covalent compounds are joined together, i.e., the *structural formula*.

3.5 LEWIS STRUCTURES

Chemists use a number of different methods to represent chemical bonds. In Chap. 2 we introduced the concept of electron dot notation to represent the structure of the valence shells of the elements of Period 2 of the Periodic Table. These were

Li·   ·Be·   :B·   ·C·   ·N·   ·O·   :F: :Ne:

The symbol of the element represents the nucleus and all of the electrons except the outer valence shell. The valence electrons are represented by dots, and sometimes by crosses or circles. A key concept in building Lewis structural formulas is to satisfy the *octet rule* which states that many elements achieve stability by forming covalent bonds in order to fill their outer shell with eight electrons.

This method of representing molecular structure will be illustrated by examining how fluorine atoms react with each other to form a diatomic molecule. In this reaction, each fluorine atom can achieve the neon configuration by sharing a pair of electrons between them. Intersecting circles have been drawn around the fluorine atoms joined by the covalent bond so that you can see that the shared pair of electrons are part of the octet around each of the fluorine atoms:

:F· + ·F: ⟶   (:F(:)F:)

The reacting atoms are represented with numbers of valence electrons equal to their group numbers. In the fluorine molecule, each atom is surrounded by a completed octet. The electron dot picture of the molecule, or Lewis formula, can be simplified by representing the bonding pair of electrons by a line between atoms, and the other pairs as dots surrounding the atoms: :F—F:. The pairs of electrons not shared in the covalent bond are called *nonbonded electrons* or *lone pairs*.

Problem 3.16. Are there exceptions to the octet rule?

Ans. Yes. An exception we will be concerned with involves the hydrogen atom. Hydrogen requires only two electrons, or a *duet*, to achieve the nearest noble gas configuration, helium. Because of this, hydrogen will always form a covalent bond to only one other atom, and therefore will appear as a *terminal* atom in Lewis formulas. To illustrate this, consider the formation of the hydrogen molecule:

·H + ·H ⟶ H:H   or   H—H

Each hydrogen atom has achieved the helium configuration by sharing each other's single electron.

The other exceptions to this rule involve compounds of boron and beryllium, which form compounds like BeH_2 and BF_3, in which there are four and six electrons, respectively, in their completed valence shells. However, the octet rule applies to all of the other elements in the first and second periods of the Periodic Table, on whose compounds we will now focus our attention.

There is a systematic way of deriving a Lewis structural formula given the molecular formula, which will be illustrated by building the Lewis formula for ammonia, NH_3.

Step 1. *Arrange the bonded atoms next to each other*. The arrangement is guesswork to a certain extent, but in this case, as well as in many others, there is only one type of atom among the rest, and it is a good guess that it is the central atom to which the rest are bonded.

```
    H

H   N   H
```

Step 2. *Determine the total number of valence electrons in the molecule*. This is done by adding up the valence electrons of all the atoms in the molecule.

Element	Valence Electrons	Atoms / Molecule	Number of Electrons
N	5	1	5
H	1	3	3
		Total valence electrons available =	8

Step 3. Represent electron pair bonds by drawing a line between *bonded atoms*. For ammonia:

```
    H
    |
H—N—H
```

Step 4. The remaining valence electrons must now be arranged as lone pairs around each atom so as to *satisfy the octet rule*.

```
    H
    |
H—N—H
   ··
```

Problem 3.17. How can one write a Lewis formula for a compound in which there is no unique atom, as in ethane, C_2H_6?

Ans. Step 1. We recognize again that hydrogen must be terminal, which implies that a hydrogen atom will not be inserted between the two carbon atoms. The best first guess is that the carbon atoms are joined together with hydrogens terminal:

```
  H  H

H  C  C  H

  H  H
```

Step 2. Calculate the total number of valence electrons.

Element	Valence Electrons	Atoms / Molecule	Number of Electrons
Carbon	4	2	8
Hydrogen	1	6	6
		Total valence electrons available =	14

Step 3. These electrons are all accommodated by drawing seven single bonds between all atoms of the molecule.

Step 4. Because there are no lone pairs, the final structure is

```
    H  H
    |  |
H—C—C—H
    |  |
    H  H
```

In many cases, there will not be a sufficient number of valence electrons to complete octets around each atom by using only single bonds. When we try to write the Lewis formula for the compound ethylene, C_2H_4, we will find that to be the case.

Step 1. We assume the carbon atoms to be joined, with hydrogen atoms terminal:

```
    H H

H   C   C   H
```

Step 2. Calculate the total number of valence electrons:

Element	Valence Electrons	Atoms / Molecule	Number of Electrons
Carbon	4	2	8
Hydrogen	1	4	4
		Total valence electrons available =	12

Step 3. We can accommodate 10 of the 12 electrons by drawing single bonds between all atoms, with the result that two electrons are left over:

```
   H  H
   |  |
H—C—C—H
```

Step 4. The hydrogen atom duets are satisfied, but we would need two additional electrons to satisfy the octet rule for the carbon atoms. Since these are not available, we can satisfy the octet rule by adding the two remaining electrons as another pair between the carbon atoms to form what is called a *double bond*:

```
   H  H
   |  |
H—C=C—H
```

An even more interesting case arises when we try to write the Lewis formula for the nitrogen molecule, N_2.

Step 1. Nitrogen is a diatomic molecule whose structure must consist of two nitrogen atoms joined by a covalent bond:

N—N

Step 2. Calculate the total number of valence electrons.

Element	Valence Electrons	Atoms / Molecule	Number of Electrons
Nitrogen	5	2	10
		Total valence electrons available =	10

Step 3. Adding one bond between the atoms, and placing the remaining electrons as lone pairs on each nitrogen atom, :N̈—N̈:, we find we are short four electrons to complete the octets around each nitrogen atom.

Step 4. However, we can place two of the lone pairs so as to form two more bonds between the nitrogen atoms, and add the remaining four electrons as lone pairs on each of the atoms to obtain :N≡N:. The bond between the nitrogen atoms is called a *triple bond*. Other molecules which contain triple bonds are HCN, hydrogen cyanide, H—C≡N, and acetylene, C_2H_2, H—C≡C—H. When you study organic chemistry, you will find that many important molecules contain single, double, or triple bonds.

We can also write Lewis formulas to account for chemical reactions in which ions are formed when electrons are transferred (oxidation), for example, from lithium to fluorine (reduction):

$$\text{Li}\cdot + \cdot\ddot{\underset{..}{\text{F}}}\text{:} \longrightarrow \text{Li}^+ + [\text{:}\ddot{\underset{..}{\text{F}}}\text{:}]^-$$

Since lithium need lose only one electron and fluorine gain one electron to achieve noble gas configuration, the atoms react in a 1:1 ratio. After the reaction, both ions have the stable noble gas configuration, the fluoride ion with a completed octet, and lithium with the helium duet, the stable $1s^2$ filled shell of the first period.

We can also use electron dot notation to show how we can determine the composition (formula) of the ionic compound formed, for example, between calcium and chlorine. The valence electrons of the atoms can be determined by noting that calcium is in Group II and chlorine is in Group VII of the Periodic Table. Calcium therefore has two, and chlorine seven, valence electrons. It requires two chlorine atoms to accept the two valence electrons from calcium:

$$\text{Ca:} + 2\,\text{:}\ddot{\underset{..}{\text{Cl}}}\cdot \longrightarrow \text{Ca}^{2+} + 2\,[\text{:}\ddot{\underset{..}{\text{Cl}}}\text{:}]^-$$

The formula of calcium chloride is $CaCl_2$.

The Lewis structures for polyatomic ions are developed in the same way as in the illustrated problems. However, in the case of a positive ion like NH_4^+, there is one less than the total number of valence electrons, and in the case of a negative ion like CO_3^{2-}, there are two more than the total number of valence electrons. Lewis structures of typical polyatomic ions are illustrated by the ammonium ion, $\left[\begin{array}{c} \text{H} \\ | \\ \text{H—N—H} \\ | \\ \text{H} \end{array}\right]^+$, and the carbonate ion, $\left[\begin{array}{c} \text{O} \diagdown \diagup \text{O} \\ \text{C} \\ \| \\ \text{O} \end{array}\right]^{2-}$

Polyatomic ions do not have their origin through direct electron transfer from element to element which we called oxidation-reduction reactions. It is important to recognize that ions can be formed in other types of chemical reactions. Most of the ions of interest to us arise from reactions in solutions, in which substances are dissolved in water. All the polyatomic ions described in Sec. 3.2.2 arose by reactions in water, either by reaction with the water itself or some other substance like the hydroxide ion, OH^-. For example, two substances HCl, hydrogen chloride, and NH_3, ammonia, a base, react with water in the following way:

$$\text{H—Cl} + \text{H—O—H} \rightarrow \left[\begin{array}{c} \text{H} \\ | \\ \text{H—O—H} \end{array}\right]^+ + \text{Cl}^-$$

$$\begin{array}{c} \text{H} \\ | \\ \text{H—N—H} \end{array} + \text{H—O—H} \rightarrow \left[\begin{array}{c} \text{H} \\ | \\ \text{H—N—H} \\ | \\ \text{H} \end{array}\right]^+ + \text{O—H}^-$$

The H_3O^+ ion, the hydrated proton or *hydronium ion*, is often abbreviated as H^+. The source of the H^+ ion is in general from one of a group of compounds called acids. As indicated in the above reactions, the bonds uniting all atoms in these polyatomic ions are covalent. The newly formed covalent bond created from a lone pair contributed by the central atom is known as a *coordinate covalent bond*. However, once formed, it cannot be differentiated from any of the other covalent bonds. We will explore the details of the chemistry of acids and bases in Chap. 9.

Can Lewis structures provide us with information regarding the actual spatial arrangements of atoms within molecules? To explore this question, let us invent a triatomic molecule and call it XY_2, in which X is the central atom with bonds to each Y atom. Each bond is polar since the joined atoms have different electronegativities. Therefore each bond contributes to the overall dipole moment of the molecule. Consider two possibilities for the molecular shape, linear and bent:

Y—X—Y

X
Y Y

In the case of the linear shape, the two internal dipoles act equally and opposite to each other and therefore cancel each other's effect. The linear molecule has no overall dipole moment. However, in the case of the bent molecule, the internal dipoles act in concert and do not cancel each other, with the result that the bent molecule has an overall dipole moment. The important point to remember is that although Lewis formulas allow us to decide which atoms are connected, and the nature of the bonds holding them together, they cannot be used to suggest the three-dimensional arrangements of the atoms.

3.6 THREE-DIMENSIONAL MOLECULAR STRUCTURES

To predict the shapes of a variety of molecules we will use a theory which postulates that the shapes of molecules depend on the total number of bonded, and nonbonded or lone electron pairs surrounding a central atom. The chief concept is that electrons tend to repel each other, and that the mutual repulsion of all the electron pairs results in the molecules' shape. The name of the method is therefore the *valence-shell electron-pair repulsion theory*, or the *VSEPR theory*.

Problem 3.18. What does VSEPR theory predict for the structure of methane, CH_4?

Ans. The Lewis formula for methane shows a flat or planar array of atoms. When we visualize the disposition of the H atoms in a plane, the H atoms are only 90° away from each other as:

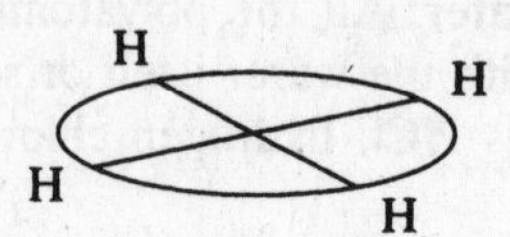

However, there is another arrangement in which the H atoms (and their bonding pairs) can separate even further. That structure is a tetrahedron:

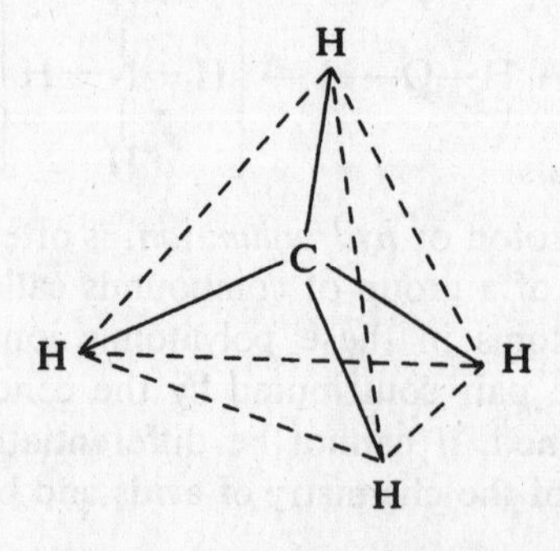

A tetrahedron is a three-dimensional figure called a *regular polyhedron*. It has four faces consisting of equilateral triangles. The carbon atom is placed centrally, with single bonds, separated by angles of 109.5 °, to each of four hydrogen atoms. The dotted lines indicate the outline of the hypothetical solid figure enclosing the carbon atom, and the solid lines represent the carbon-hydrogen single bonds. Although each C—H bond is somewhat polar ($\Delta EN = 0.4$), the four identical bonds are arranged symmetrically, and this means that the dipoles add to zero. Experiments have shown that the structure of methane is tetrahedral, and that it has no dipole moment.

Problem 3.19. What does VSEPR theory predict for the structure of ammonia?

Ans. The Lewis structure for ammonia, NH_3 is H—N̤(—H)—H (H above N), again a flat or planar array. We know that there are three bonding pairs of electrons and one lone or nonbonding pair. VSEPR theory requires us to find a three-dimensional figure which allows for maximal separation of all electron pairs. Since there are four electron pairs, we predict the shape to be tetrahedral. In this case however, one of the pairs is nonbonded, which is different from the methane case. We know that electron pairs in bonds are localized because they are fixed between two positive nuclei. Nonbonded pairs are not subject to the localizing action of two positively charged nuclei, and therefore can repel each other to a greater extent than a bonded pair. The result is that they occupy a greater volume of space than a bonded pair. In doing so, they force the bonded pairs somewhat closer to each other than the tetrahedral geometry would predict. The bond angles between the central nitrogen atom and the three hydrogens have been determined by experiment to be 107 °, as opposed to the 109.5 ° of a perfect tetrahedron. Although the lone pair of electrons contribute to considerations of the overall shape of the molecule, they cannot be seen in the final structure. Therefore, the ammonia molecule forms a shallow pyramid with three hydrogen atoms forming the base, and the lone pair extending above them. This shape is called *trigonal pyramidal* because the pyramid has a triangle for its base. Because of the asymmetric arrangement of bonds and lone pair electrons, this molecule has a strong dipole moment.

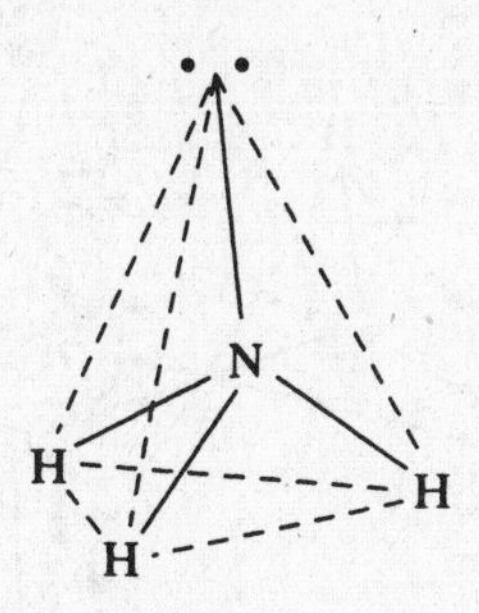

Problem 3.20. What does VSEPR theory predict for the structure of water, H_2O?

Ans. The Lewis structure for water is H—Ö—H, with two lone pairs on the oxygen atom. From VSEPR theory we know that the two bonded and two nonbonded or lone pairs must be accommodated in a geometric figure which allows maximal separation of the four electron pairs, which is the tetrahedron. However, in this case we have two lone pairs, which will repel each other even more than in the case of one lone pair. The two hydrogen atoms are squeezed even closer together than in the ammonia molecule, with the two lone electron pairs directed at the other two corners of a distorted tetrahedron. The bond angle between the oxygen atom and the two hydrogen atoms has been determined experimentally to be 104.5 °. Here again, the lone pairs cannot be seen;

therefore, the water molecule appears to be bent. This molecule has an even greater overall polarity than ammonia, and has one of the largest dipole moments ever measured.

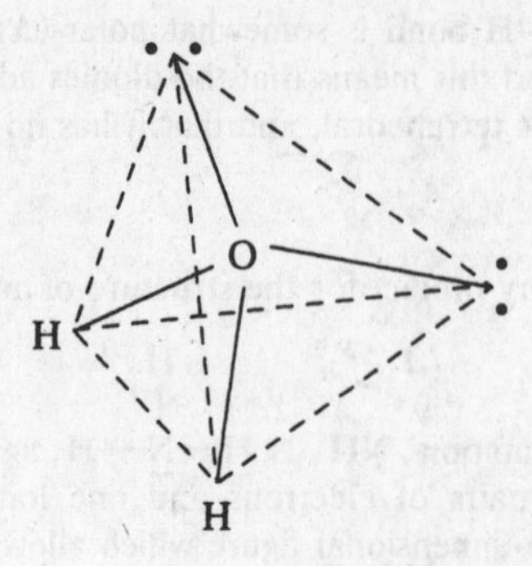

Problem 3.21. Can VSEPR theory account for the shape of ethylene, C_2H_4?

Ans. **The Lewis structure for ethylene is H—C(H)=C(H)—H. The VSEPR theory treats double and triple bonds as though they were single bonds. Therefore we can conclude that there are three bonding pairs around each carbon atom to be accommodated in some symmetrical structure. That structure is the equilateral triangle. VSEPR theory predicts that ethylene consists of two equilateral triangles connected at one corner. The carbon atoms are connected by a single line which we know to be the double bond. The angles between all atoms are equal, and are 120 °. Because of the symmetry of the molecule, it has no dipole moment.**

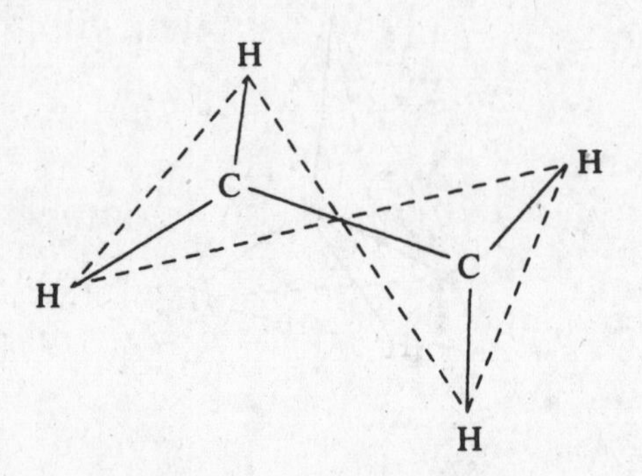

Problem 3.22. Can VSEPR theory account for the structure of acetylene, C_2H_2?

Ans. **The Lewis structure for acetylene is H—C≡C—H. The VSEPR theory treats double and triple bonds as though they were single bonds. Therefore we can conclude that there are two bonding pairs to be accommodated around each carbon atom. The geometric arrangement which allows maximal separation of two bonding pairs is a straight line separating the bonding pairs by 180 °. The VSEPR structure is H—C—C—H, a linear molecule with all atoms in a straight line, verified** by experiment (it has no dipole moment). The line connecting the carbon atoms represents the triple bond.

The steps taken to determine the shapes of molecules are the following:

1. Determine the Lewis structure.
2. Use VSEPR theory to decide what symmetrical three-dimensional figure will accommodate all bonded and nonbonded electron pairs.
3. We consider the shape of the molecule to be determined by the arrangement of only the joined atoms. That is why, although the CH_4, NH_3 and H_2O molecules each have four pairs of electrons around the central atom, methane has the tetrahedral shape, ammonia is trigonal pyramidal, and water is bent.

Problem 3.23. In Problem 3.18, methane, with the composition CH_4, was shown to have a three-dimensional structure known as a *tetrahedron*. We know that the valence shell of carbon consists of four electrons in two different kinds of orbitals, 2*s* and 2*p* orbitals. Knowing this, how can we explain the fact that carbon forms four equivalent covalent bonds, bonds of identical character, with the 1*s* electrons of hydrogen?

Ans. Recall that in the formation of a molecule, the atomic orbitals undergo changes to allow the bonding electrons to achieve the most stable configuration, that is, to get as far away from each other as possible. This is accomplished by the electrons in the carbon atom, as well as in other atoms, by a process called *hybridization*. This means literally a mixing of the *s* and *p* orbitals to form a new type of orbital. The four electrons from carbon and the four electrons from the four hydrogen atoms combine to form four equivalent pairs or bonds. These new orbitals are called sp^3 *hybrid orbitals*, because they are the result of combining one of the 2*s* and all three 2*p* orbitals of carbon. The energy level of the sp^3 orbitals is higher than the atomic *s* orbitals, but lower than the atomic *p* orbitals. This compromise allows the formation of the hybrid orbitals, and therefore the symmetrical tetrahedral structure of the methane molecule. Other types of hybrid orbitals are sp^2, in which three equivalent hybrid orbitals are formed directed to the corners of an equilateral triangle in a flat plane, which explains the VSEPR structure of ethylene, and *sp*, in which two equivalent hybrid orbitals are formed directed 180 ° from each other in a line, which explains the VSEPR structure of acetylene. This is a very important feature of bond formation which accounts for the many kinds of structural symmetry encountered in organic and inorganic compounds.

ADDITIONAL SOLVED PROBLEMS

INTRODUCTION

Problem 3.24. What is a chemical bond?

Ans. A chemical bond is a force which holds a group of atoms (at least two) together so as to form an electrically neutral aggregate.

Problem 3.25. What is the reason for the formation of chemical bonds?

Ans. The most stable elements are the noble gases. Their stability is based on the fact that they have filled outer (valence) shells. All the other elements have an unfilled outer valence shell and are therefore unstable. They can achieve the same stability by filling their valence shells with the appropriate number of electrons. This is achieved through chemical reaction and consequent chemical bond formation. Therefore, the principal driving force for the formation of chemical bonds is that most elements are inherently unstable.

Problem 3.26. What is the octet rule?

Ans. The filled outer (valence) shell of the noble gases consists of eight electrons. The octet rule states that elements achieve maximum stability when their outer (valence) shell contains eight (an octet) of electrons.

Problem 3.27. What are the ways elements can satisfy the octet rule?

Ans. There are two ways to achieve valence shell completion: (1) electrons can be completely transferred from the valence shell of one element to the valence shell of another element, and (2) electrons can be shared between two elements, and therefore become part of the valence shells of the two elements simultaneously.

Problem 3.28. What are the results of these two ways of satisfying the octet rule?

Ans. The transfer of electrons from one element to another results in the formation of charged atoms called ions. Positively charged ions are called cations, and negatively charged ions are called anions. The electrical attraction of oppositely charged ions results in the formation of a chemical bond called an ionic bond. The sharing of electrons between elements results in the formation of a covalent bond.

NOMENCLATURE

Problem 3.29. Write the systematic name of each of the following compounds: (*a*) K_2SO_4, (*b*) $Mn(OH)_2$, (*c*) $Fe(NO_3)_2$, (*d*) KH_2PO_4, (*e*) $Ca(C_2H_3O_2)_2$, (*f*) Na_2CO_3.

Ans. (*a*) Potassium sulfate, (*b*) manganese hydroxide, (*c*) iron(II) nitrate, (*d*) potassium dihydrogen phosphate, (*e*) calcium acetate, (*f*) sodium carbonate

Problem 3.30. Write the formulas of the compounds formed by the cations and the anions in the following table:

	Chloride	Hydroxide	Sulfate
Sodium	*a*	*b*	*c*
Calcium	*d*	*e*	*f*
Iron(II)	*g*	*h*	*i*

Ans. (*a*) NaCl; (*b*) NaOH; (*c*) Na_2SO_4; (*d*) $CaCl_2$; (*e*) $Ca(OH)_2$; (*f*) $CaSO_4$; (*g*) $FeCl_2$; (*h*) $Fe(OH)_2$; (*i*) $FeSO_4$

Problem 3.31. In the blank space to the right of the ion given, write the name of the ion:

	Ion	Name
(*a*)	NH_4^+	
(*b*)	NO_3^-	
(*c*)	SO_4^{2-}	
(*d*)	PO_4^{3-}	
(*e*)	$C_2H_3O_2^-$	

Ans. (*a*) ammonium, (*b*) nitrate, (*c*) sulfate, (*d*) phosphate, (*e*) acetate

Problem 3.32. In the left-hand column, write the formula of the ion named to the right:

Ion	Name
(*a*)	Carbonate
(*b*)	Hydrogen carbonate
(*c*)	Hydroxide
(*d*)	Hydrogen phosphate
(*e*)	Dihydrogen phosphate

Ans. (*a*) CO_3^{2-}, (*b*) HCO_3^-, (*c*) OH^-, (*d*) HPO_4^{2-}, (*e*) $H_2PO_4^-$

Problem 3.33. Write the formulas of the compounds formed in the reactions between aluminum and (*a*) chlorine, (*b*) sulfur, and (*c*) nitrogen.

Ans. (*a*) $AlCl_3$, (*b*) Al_2S_3, (*c*) AlN

Problem 3.34. What are the formulas and names of the compounds formed from the following elements: (*a*) lithium and oxygen; (*b*) calcium and bromine; (*c*) aluminum and oxygen; (*d*) sodium and sulfur.

Ans. (*a*) Li_2O, lithium oxide; (*b*) $CaBr_2$, calcium bromide; (*c*) Al_2O_3, aluminum oxide; (*d*) Na_2S, sodium sulfide

IONIC BONDS

Problem 3.35. What are the combining ratios of elements in Group I with elements in Group VII of the Periodic Table?

Ans. All the elements in Group I have one electron in their valence shells. All the elements in Group VII have seven electrons in their valence shells. Transfer of one electron from the valence shell of elements in Group I to the valence shell of elements in Group VII results in (1) octet formation for both, and (2) formation of ions having a charge of $+1$ and -1, respectively. The combining ratio is 1 : 1. This ratio of ions results in a neutral aggregate with the formula XY, where X stands for the cations of Group I elements, and Y stands for the anions of Group VII.

Problem 3.36. What are the combining ratios of elements in Group I with elements of Group VI of the Periodic Table?

Ans. Group I cations are monovalent (a single positive charge), X^+. Elements in Group VI require two electrons to fill their valence shell to form divalent anions, Y^{2-}. Therefore the combining ratios are 2 : 1, and the formula is X_2Y.

Problem 3.37. What are the combining ratios of elements in Group II with elements in Group VII?

Ans. Group II cations are divalent, and Group VII anions are monovalent. The combining ratios are 1 : 2, and the formula is XY_2.

Problem 3.38. What are the combining ratios of elements in Group III with elements in Group VI?

Ans. The elements in Group III form trivalent cations, X^{+3}, and the elements of Group VI form divalent anions, Y^{-2}. Therefore, the combining ratios are 2 : 3, and the formula is X_2Y_3.

Problem 3.39. Write and name the compounds which form by reaction of calcium with the first three elements of Group VI.

Ans. The elements in Group VI require two electrons to complete their octets, but calcium, Group II, must lose two electrons to complete its octet. The first three elements in Group VI are oxygen, O; sulfur, S; and selenium, Se. The compounds are CaO, calcium oxide; CaS, calcium sulfide; and CaSe, calcium selenide.

Problem 3.40. Write and name the compounds which form by reaction of aluminum with the first three elements of Group VI.

Ans. The elements in Group VI require two electrons to complete their octets, but aluminum, Group III, must lose three electrons to complete its octet. The first three elements in Group VI are oxygen, O; sulfur, S; and selenium, Se. The compounds are Al_2O_3, aluminum oxide; Al_2S_3, aluminum sulfide; and Al_2Se_3, aluminum selenide.

Problem 3.41. Write and name the compounds which form by reaction of aluminum with the first four elements of Group VII.

Ans. The elements in Group VII require one electron to complete their octets, but aluminum, Group III, must lose three electrons to complete its octet. The first four elements in Group VII are fluorine, F; chlorine, Cl; bromine, Br; and iodine, I. The compounds are AlF_3, aluminum fluoride; $AlCl_3$, aluminum chloride; $AlBr_3$, aluminum bromide; and AlI_3, aluminum iodide.

Problem 3.42. Indicate what kinds of bonds will be formed by the reaction of the following pairs of elements: (*a*) Si and O; (*b*) Be and C; (*c*) C and N; (*d*) Ca and Br; (*e*) B and N.

Ans. If the difference in electronegativities of the element pairs is greater than 2.0, ionic bonds will form. If the difference in electronegativities is 1.5 or less, covalent bonds will form.

	Element Pairs	EN Difference	Bond Type
(*a*)	Si and O	1.5	Covalent
(*b*)	Be and C	1.0	Covalent
(*c*)	C and N	0.4	Covalent
(*d*)	Ca and Br	1.8	Ionic
(*e*)	B and N	1.0	Covalent

COVALENT BONDS AND LEWIS STRUCTURES

Problem 3.43. How many electrons are shared in a covalent bond?

Ans. The maximum number of electrons occupying an orbital is two. Their quantum numbers are identical except for the spin quantum number. The two electrons must have opposite spins; therefore, a single covalent bond consists of a shared pair of electrons of opposing spins.

Problem 3.44. What is the Lewis structure of carbon tetrachloride, CCl_4?

Ans. Step 1. We begin an analysis of the possible Lewis structure by arranging the atoms next to each other, and assume that since there is only one carbon atom in this molecule, we will

```
      Cl
Cl    C    Cl
      Cl
```

consider it to be the central atom, as: Cl C Cl. When there is only one kind of atom among the rest of the atoms in a molecule, we define it as the *unique atom*.

Step 2. We determine the number of valence electrons in CCl_4:

Element	Valence Electrons	Atoms / Molecule	Number of Electrons
Carbon	4	1	4
Chlorine	7	4	28
		Total valence electrons available =	32

Step 3. We represent the molecular structure with a single bond between the joined atoms, as:

```
      Cl
      |
  Cl—C—Cl.
      |
      Cl
```

Step 4. The other 24 electrons, (32 − 8 = 24) must be distributed about the Cl atoms as lone pairs as follows:

```
       ..
      :Cl:
   ..  |  ..
  :Cl—C—Cl:
   ..  |  ..
      :Cl:
       ..
```

In this way each atom is surrounded by an octet of electrons, and the octet rule is satisfied.

Problem 3.45. Write the Lewis formula for chloroform, $CHCl_3$.

Ans. Step 1. Consider the probable arrangement of atoms. In this situation, there are three different types of atoms. However, we note that hydrogen, an exception to the octet rule, is terminal in a Lewis formula, and that carbon is unique among the remaining atoms. A good first guess is that carbon is the central atom to which all others are bonded:

```
      Cl
  H   C   Cl
      Cl
```

Step 2. Calculate the number of available valence electrons:

Element	Valence Electrons	Atoms / Molecule	Number of Electrons
Carbon	4	1	4
Chlorine	7	3	21
Hydrogen	1	1	1
		Total valence electrons available =	26

Step 3. We will use four pairs (eight electrons) to form four bonds, as:

```
      Cl
      |
  H—C—Cl
      |
      Cl
```

Step 4. The remaining electrons (26 − 8 = 18) can be distributed as lone pairs around the 3 chlorine atoms, as:

```
      ..
     :Cl:
      |    ..
   H—C—Cl:
      |    ..
     :Cl:
      ..
```

This structure satisfies the octet rule for all the atoms.

Problem 3.46. What is the Lewis structure of OF_2?

Ans. Steps 1 and 2. There is one unique atom, O; assume it is the central atom, and draw single bonds between the atoms:

$$F - O - F$$

Step 3. Count the number of available valence electrons:

$$[2(F) \times 7] + [1(O) \times 6] = 20$$

Step 4. Four electrons have been placed in the two single bonds, so 16 must be added as lone pairs to complete the octets on all atoms:

```
 ..  ..  ..
:F—O—F:
 ..  ..  ..
```

Problem 3.47. What is the Lewis structure of NCl_3?

Ans. Steps 1 and 2. The first guess at the atomic arrangement will have nitrogen as the central atom because it is unique, and at the same time we will draw single bonds from the nitrogen atom to each of the chlorine atoms:

```
       Cl
       |
   Cl—N—Cl
```

Step 3. Count the number of available valence electrons:

$$[1(N) \times 5] + [3(Cl) \times 7] = 26$$

Step 4. Six electrons have been placed in single bonds; therefore, 20 electrons must be placed as lone pairs to complete the octets on all atoms:

```
        ..
       :Cl:
  ..    |   ..
 :Cl—N—Cl:
  ..    ..  ..
```

Problem 3.48. Write the Lewis structure of carbon dioxide, CO_2.

Ans. Steps 1 and 2. Assume carbon is central, and connect the atoms with single bonds:

$$O - C - O$$

Step 3. Count the number of available valence electrons:

$$[1(C) \times 4] + [2(O) \times 6] = 16$$

Step 4. Four electrons have been placed in single bonds; therefore 12 electrons must be used to complete all octets. If we fill the oxygen octets by placing electrons as lone pairs around the O atoms, carbon will be deficient:

$$:\ddot{\underset{\cdot\cdot}{O}}-C-\ddot{\underset{\cdot\cdot}{O}}:$$

We can solve the problem by moving two of the lone pairs on oxygen so as to form double bonds linking the oxygen atoms to the carbon atom:

$$O{=}C{=}O$$

Problem 3.49. Hydrocarbons are compounds which contain only carbon and hydrogen. There is a type of hydrocarbon in which the carbon atoms are connected to each other to form a chain. Write the Lewis structure for C_3H_8.

Ans. Steps 1 and 2. We consider the hydrogen atoms to be terminal, and the carbon atoms to be connected in a chain:

```
    H   H   H
    |   |   |
H — C — C — C — H
    |   |   |
    H   H   H
```

Step 3. Count the number of available valence electrons:

$$[3(C) \times 4] + [8(H) \times 1] = 20$$

Step 4. Ten single bonds have been drawn, utilizing all 20 electrons. Since the octet rule is satisfied for all atoms, the structure drawn is the Lewis structure for this hydrocarbon.

Problem 3.50. An organic compound containing an OH group is called an *alcohol*. What is the Lewis structure for CH_3OH?

Ans. Step 1. The first step will be to make carbon the central atom. The hydrogen atoms must be terminal not only on the carbon atom, but also on the oxygen atom.

Step 2. The first trial structure is

```
    H
    |
H — C — O — H
    |
    H
```

Step 3. Count the number of available valence electrons:

$$[1(C) \times 4] + [4(H) \times 1] + [1(O) \times 6] = 14$$

Step 4. Five single bonds were drawn using 10 of the available valence electrons. The last four electrons must be arranged as lone pairs on the oxygen atom in order to satisfy the octet rule:

```
    H
    |   ••
H — C — O — H
    |   ••
    H
```

THREE-DIMENSIONAL MOLECULAR STRUCTURES

Problem 3.51. Use VSEPR theory to predict the angular orientation of fluorine atoms around the central boron atom in the polyvalent ion, BF_4^-.

Ans. To predict the spatial organization of this ion, we must first construct the Lewis structure in order to identify bound pairs and lone pairs of electrons.

Steps 1 and 2. We make the unique atom central, attach the four fluorine atoms to it, and indicate the negative ionic charge:

$$\left[\begin{array}{c} F \\ | \\ F-B-F \\ | \\ F \end{array}\right]^-$$

Step 3. Count the number of available valence electrons:

$$[1(B) \times 3] + [4(F) \times 7] + [1(e)] = 32$$

Step 4. Eight electrons are arranged as single bonds; therefore we place $32 - 8 = 24$ electrons arranged as lone pairs on the fluorine atoms. This satisfies the octet rule for all atoms. We note that there are four equivalents bonds from the central boron atom to each of the fluorine atoms. The geometric structure which permits four electron pairs to be equidistant from each other in three dimensions is the tetrahedron:

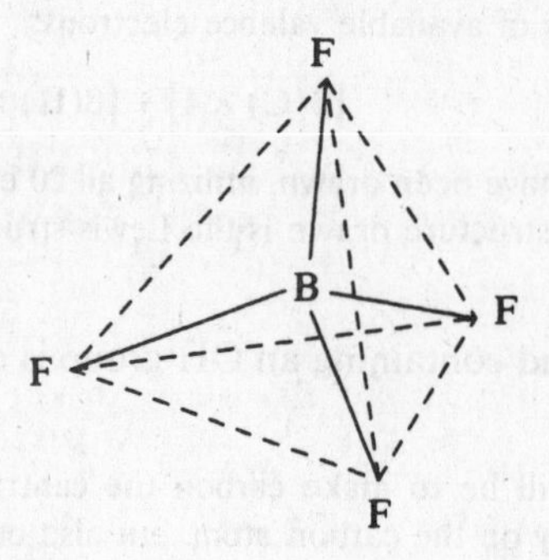

Problem 3.52. Use VSEPR theory to predict the three-dimensional structure of OF_2.

Ans. The Lewis structure for OF_2 was constructed in Problem 3.46. It was: $:\ddot{\underset{..}{F}}-\ddot{\underset{..}{O}}-\ddot{\underset{..}{F}}:$. In this molecule, there are four pairs of electrons around the central oxygen atom, two in single bonds and two in lone pairs. Four pairs of electrons must be accommodated by a tetrahedral structure. However, note carefully that two of the electron pairs are *lone pairs*. This means that they take up more space than the bound pairs with the result that the tetrahedron must be distorted very much like the VSEPR structure of water:

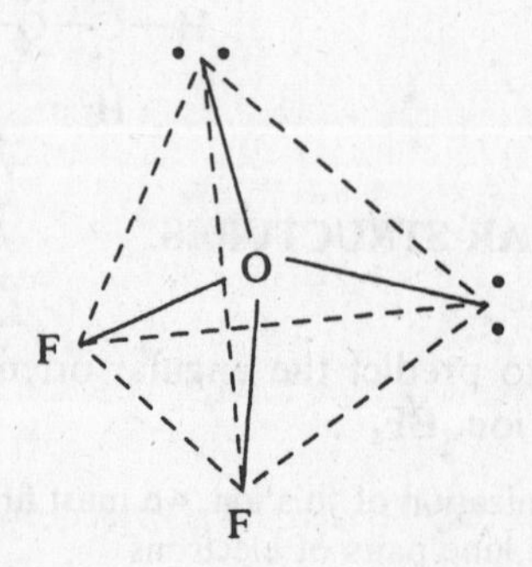

Problem 3.53. For each of the following molecules (*a*) describe the molecular shape, (*b*) describe the type of molecular orbitals present, and (*c*) predict whether or not the molecule has a dipole moment: (*a*) CCl_4, (*b*) BF_4^-, (*c*) CO_2, (*d*) $CHCl_3$, (*e*) NCl_3, (*f*) H_2O.

Ans.

Molecule	Molecular Shape	Orbitals	Dipole Moment
CCl_4	Tetrahedron	sp^3 Hybrids	None
BF_4^-	Tetrahedron	sp^3 Hybrids	None
CO_2	Linear	sp Hybrids	None
$CHCl_3$	Distorted tetrahedron	sp^3 Hybrids	Yes
NCl_3	Trigonal pyramidal	sp^3 Hybrids	Yes
H_2O	Bent or angular	sp^3 Hybrids	Yes

SUPPLEMENTARY PROBLEMS

Problem 3.54. What is a covalent chemical bond?

Ans. It is a chemical bond in which electrons are shared between two atoms.

Problem 3.55. How many electrons are there in a covalent bond?

Ans. There are only two electrons in a covalent bond whose spins are opposite.

Problem 3.56. What is an ionic bond?

Ans. It is a bond formed by the electrostatic attraction of ions of opposite charge formed by electron transfer.

Problem 3.57. Write the systematic name of each of the following compounds: (*a*) $AgNO_3$; (*b*) $HgCl_2$; (*c*) Na_2CO_3; (*d*) $Ca_3(PO_4)_2$.

Ans. (*a*) Silver nitrate; (*b*) mercury(II) chloride; (*c*) sodium carbonate; (*d*) calcium phosphate.

Problem 3.58. Write the formulas of the compounds formed in the reactions between magnesium, and (*a*) chlorine, (*b*) sulfur, and (*c*) nitrogen.

Ans. (*a*) $MgCl_2$; (*b*) MgS; (*c*) Mg_3N_2

Problem 3.59. Write and name the compounds which form by reaction of barium with the first three elements of Group VI.

Ans. (*a*) BaO, barium oxide; (*b*) BaS, barium sulfide; (*c*) BaSe, barium selenide.

Problem 3.60. Write and name the compounds which form by reaction of potassium with the first 3 elements of Group V.

Ans. (*a*) K_3N, potassium nitride; (*b*) K_3P, potassium phosphide; (*c*) K_3As, potassium arsenide.

Problem 3.61. What is the Lewis structure of carbon monoxide?

Ans. :C≡O:

Problem 3.62. Use VSEPR theory to predict the angular orientation of the atoms in SBr_2.

Ans. Directed to the corners of a distorted tetrahedron, angles somewhat less than 109.5 °.

Problem 3.63. What kind of chemical bond will form in binary compounds where the electronegativity difference between atoms is greater than 2.0?

Ans. Ionic bonds.

Problem 3.64. What kind of chemical bond will form in binary compounds where the electronegativity difference between atoms is less than 1.5?

Ans. Covalent bonds.

Chemical Calculations

An understanding of chemical processes invariably means the establishment of an explanation based on atomic or molecular considerations. As your study of chemistry proceeds, you must consistently attempt to explain chemical issues in terms of atoms and molecules. In this chapter we will explore the quantitative basis of the law of conservation of mass and law of constant composition described in Chap. 2. You will learn how to put the atomic theory and the concept of molecules to work.

4.1 CHEMICAL FORMULAS AND FORMULA MASSES

In Chap. 3 we pointed out that the formulas for both ionic and covalent compounds ignored the nature of the chemical bonds of compounds. A formula is a statement of the combining ratios or the relative numbers of atoms of each kind in the simplest unit representing the composition of the compound. We also learned in Chap. 2 that the atoms of the elements have different masses, which were expressed as relative masses, and that the quantity of any element which is numerically equal to its atomic mass must contain the same number of atoms as the corresponding quantity of any other element. The table of atomic masses assures us that 55.85 g of iron contains the same number of atoms as 12.00 g of the ^{12}C isotope of carbon. Can we expand the concept of relative mass to compounds as well as elements? For example, how many grams of water will contain the same number of fundamental units as 12 g of ^{12}C carbon? We know that the fundamental unit of water is a molecule with the composition H_2O. Its relative mass must be equal to the sum of the masses of the atoms in the molecule:

$$2 \times 1.008 \text{ amu (H)} + 1 \times 16.00 \text{ amu(O)} = 18.02 \text{ amu}(H_2O)$$

Therefore, 18.02 g of water will have the same number of fundamental units as 12.00 g of ^{12}C carbon. How many grams of sodium chloride will have the same number of fundamental units as 12.00 of ^{12}C carbon? The fundamental unit of sodium chloride is not a molecule, but is described by the smallest unit representing its composition, NaCl:

$$1 \times 22.99 \text{ amu (Na)} + 1 \times 35.45 \text{ amu(Cl)} = 58.44 \text{ amu(NaCl)}$$

Therefore, 58.44 grams of sodium chloride will have the same number of fundamental units as 32.00 grams of O_2, and 18.02 grams of H_2O whose formulas describe molecules. So, in the case of substances which exist as molecules, we can call the sum of atomic masses the *molecular mass.* But how shall we name the mass of sodium chloride equivalent to those masses. We know that the designation, molecule, for NaCl is not appropriate. However, if we describe the mass of the fundamental unit representing the composition of any compound by using the expression *formula mass,* such difficulties are avoided. Table 4-1 lists the formulas and formula masses of several representative compounds.

Note that the atomic masses will continue to be rounded off to whole numbers, so as to simplify calculations. However, when we use an atomic mass like that of chlorine, 35.45, or mercury, 200.59, we

Table 4-1. Formulas and Formula Masses of Several Compounds

Substance	Formula Mass
Methane, CH_4	$(1 \times 12) + (4 \times 1) = 16$
Carbon dioxide, CO_2	$(1 \times 12) + (2 \times 16) = 44$
Ammonia, NH_3	$(1 \times 14) + (3 \times 1) = 17$
Calcium chloride, $CaCl_2$	$(1 \times 40) + (2 \times 35.5) = 111$
Iron(III) nitrate, $Fe(NO_3)_3$	$(1 \times 56) + 3[(1 \times 14) + (3 \times 16)] = 242$

generally round off such numbers to 35.5 and 200.6, and use those in calculating formula mass. This is done to avoid significant errors in quantative calculations. For example, the formula mass of HCl is 36.5 using a rounded off value for chlorine. If we rounded the atomic mass of chlorine to 36, the formula mass of HCl would be 37. The percent error would be

$$\left(\frac{37 - 36.5}{36.5}\right) \times 100 = 1.4\%$$

Errors introduced by weighing on an analytical balance are rarely greater than 0.1 percent, so that a 1.4 percent error caused by rounding off is unacceptable. Table 4-2 lists elements whose atomic masses may be rounded off to 1 decimal place. This list is not exhaustive, but it should provide a basis for judgement in those cases not listed.

Table 4-2. Acceptable Rounded-Off Values of Atomic Masses for Several Elements

Element	Precise Atomic Mass	Acceptable Rounded-Off Value
Barium	137.33	137.3
Boron	10.811	10.8
Chlorine	35.453	35.5
Copper	63.54	63.5
Magnesium	24.312	24.3
Mercury	200.59	200.6
Strontium	87.62	87.6
Zinc	65.41	65.4

4.2 THE MOLE

The formula mass in grams of any element or compound must contain the same number of fundamental units as the formula mass of any other element or compund. Chemists call that formula mass, when expressed in grams, the *mole*. The unit for the mole is mol, introduced in Chap. 1, Table 1-1. The mole has two meanings: (1) one formula mass of an element or compound, and (2) a unique number of fundamental units of that element or compound. That number has been determined by experiment to be 6.02205×10^{23} mol^{-1}, and has been given the name *Avogadro's number*, in honor of Amadeo Avogadro, whose ideas were used to solve many of the problems involved in mass relationships in chemical reactions.

The mass in grams of 1 mol of a compound is called the *molar mass*. The term *molar mass* applies to both covalent and ionic compounds. Keep in mind that we use the terms formula mass and molecular mass to refer to ionic and covalent compounds, respectively (Section 4.1). Molar mass is given in units of grams while formula mass and molecular mass are given in units of amu. (Note that the terms molar mass and formula mass apply to elements as well as compounds.)

Problem 4.1. How can we determine the number of moles in 52.0 g of oxygen?

Ans. The formula mass of oxygen is 32.0; therefore there are 32.0 g per mole of oxygen. To determine the number of moles of oxygen in 52.0 g of oxygen, we must convert grams of oxygen to moles of oxygen. To do this we require a unit conversion factor:

$$1 = \frac{1.00 \text{ mol } O_2}{32.0 \text{ g } O_2}$$

Therefore

$$\text{mol } O_2 = 52.0 \text{ g } O_2\left(\frac{1.00 \text{ mol } O_2}{32.0 \text{ g } O_2}\right) = 1.63 \text{ mol } O_2$$

The answer is expressed in moles and 3 significant figures.

Problem 4.2. How many grams are there in 3.20 mol of calcium chloride, $CaCl_2$?

Ans. The formula mass of calcium chloride (using rounded-off values of atomic masses) is 111.0. The mass of $CaCl_2$ in 3.20 mol is

$$\text{g CaCl}_2 = 3.20 \text{ mol CaCl}_2\left(\frac{111.0 \text{ g CaCl}_2}{1.00 \text{ mol CaCl}_2}\right) = 355 \text{ g CaCl}_2$$

The units yield grams of $CaCl_2$, and the result is expressed in 3 significant figures.

Problem 4.3. Calculate the number of moles in 0.85 g of sucrose, $C_{12}H_{22}O_{11}$ (about one teaspoon of table sugar).

Ans. The formula mass of sucrose (using rounded-off atomic masses) is

$$(12 \times 12) + (22 \times 1) + (11 \times 16) = 342$$

The number of moles in 0.85 grams of sucrose is

$$\text{mol C}_{12}\text{H}_{22}\text{O}_{11} = 0.85 \text{ g C}_{12}\text{H}_{22}\text{O}_{11}\left(\frac{1 \text{ mol C}_{12}\text{H}_{22}\text{O}_{11}}{342 \text{ g C}_{12}\text{H}_{22}\text{O}_{11}}\right) = 2.5 \times 10^{-3} \text{ mol}$$

Note significant figures and units. Note also that in order to calculate the number of moles in the mass of a compound one must know the formula of the compound. *A mole is defined only in terms of the formula of a compound.*

4.3 AVOGADRO'S NUMBER

Avogadro's number can be used to express the idea that a mole of any kind of matter consists of 6.02205×10^{23} fundamental units or entities. We will simplify this number to 6.02×10^{23}, and use that value as exact. Let us represent Avogadro's number by the symbol N_A. For example, there are N_A electrons in a mole of electrons, as well as N_A oranges in a mole of oranges or N_A molecules of sucrose in a mole of sucrose. To get an idea of the difficulty in imagining the size of that number, consider the following question. How high would a stack of an Avogadro's number of coins, each 1 mm thick, be? Further, if you could travel at the speed of light, 3×10^5 km/s, how long would it take to travel from one end of the stack to the other?

First we will calculate the height of the stack in kilometers, and then the time to traverse it in years:

$$1 \text{ mm } (6.02 \times 10^{23})\left(\frac{1 \text{ m}}{1000 \text{ mm}}\right)\left(\frac{1 \text{ km}}{1000 \text{ m}}\right) = 6.02 \times 10^{17} \text{ km}$$

$$6.02 \times 10^{17} \text{ km}\left(\frac{1 \text{ s}}{3 \times 10^5 \text{ km}}\right)\left(\frac{1 \text{ min}}{60 \text{ s}}\right)\left(\frac{1 \text{ h}}{60 \text{ min}}\right)\left(\frac{\text{day}}{24 \text{ h}}\right)\left(\frac{1 \text{ year}}{365 \text{ days}}\right) = 63{,}631 \text{ years}$$

Each of those years is called a light-year, the distance light will travel in 1 year. Astronomical distances are so great that they are reckoned in light-years. To gain an appreciation of the distance represented by 63,631 light-years, the star nearest to our solar system is Proxima Centauri, which is 4.3 light-years away.

One of the most useful earthly applications of Avogadro's number is that it allows us to calculate the mass of one atom or one molecule.

Problem 4.4. Calculate the mass of one oxygen molecule.

Ans. Oxygen exists as a diatomic molecule, O_2. Its formula mass is 32.0 amu, and so there are N_A oxygen molecules in 32.0 g of oxygen. 32.0 g O_2 = mass of one oxygen molecule × Avogadro's number.

Therefore,

$$\text{Mass of one } O_2 \text{ molecule} = \frac{32.00 \text{ g of } O_2}{6.02 \times 10^{23} \text{ molecules of } O_2}$$

and the mass of one oxygen molecule = 5.316×10^{-23} g/molecule of O_2.

Problem 4.5. How many molecules are there in 9.00 g of ammonia?

Ans. The formula mass of ammonia is $(1 \times 14.0) + (3 \times 1.00) = 17.0$. Therefore, 9.00 g of ammonia is

$$\text{mols } NH_3 = 9.00 \text{ g } NH_3 \left(\frac{1 \text{ mol } NH_3}{17.0 \text{ g } NH_3} \right) = 0.529 \text{ mol } NH_3$$

Because 1 mol of ammonia contains N_A molecules of ammonia, 0.529 mol must contain

$$0.529 \text{ mol } NH_3 \left(\frac{6.02 \times 10^{23} \text{ molecules of } NH_3}{1 \text{ mol } NH_3} \right) = 3.19 \times 10^{23} \text{ } NH_3 \text{ molecules}$$

Furthermore, since there are three hydrogen atoms and one nitrogen atom in each molecule of nitrogen, the number of N atoms is

$$3.19 \times 10^{23} \text{ molecules of } NH_3 \left(\frac{1 \text{ N atom}}{1 \text{ } NH_3 \text{ molecule}} \right) = 3.19 \times 10^{23} \text{ N atoms}$$

The number of H atoms is

$$3.19 \times 10^{23} \text{ molecules of } NH_3 \left(\frac{3 \text{ H atoms}}{1 \text{ } NH_3 \text{ molecule}} \right) = 9.57 \times 10^{23} \text{ H atoms}$$

4.4 EMPIRICAL FORMULAS AND PERCENT COMPOSITION

We can use the concept of the mole to find the simplest chemical formula of a compound from its composition. Chemical analysis of magnesium oxide reveals that it is 60.3% Mg and 39.7% O by mass. It is useful to convert those percentages to grams, and this is done by assuming we are dealing with 100 g of the compound, so that 100 g of magnesium oxide consists of 60.3 g of magnesium and 39.7 g of oxygen. Furthermore, we can say that 60.3 g of magnesium combines with 39.7 g of oxygen, or, that 60.3 g of magnesium is equivalent to 39.7 g of oxygen. The next step is to calculate the number of moles of magnesium in 60.3 g, and the number of moles of oxygen in 39.7 g.

$$\text{mol Mg} = 60.3 \text{ g Mg} \left(\frac{1 \text{ mol Mg}}{24.3 \text{ g Mg}} \right) = 2.48 \text{ mol Mg}$$

$$\text{mol O} = 39.7 \text{ g O} \left(\frac{1 \text{ mol O}}{16.0 \text{ g O}} \right) = 2.48 \text{ mol O}$$

Therefore, 2.48 mol Mg combines with 2.48 mol O in magnesium oxide, or 1 mol Mg combines with 1 mol O in magnesium oxide. If we divide both quantities by Avogadro's number, we can say that one atom of magnesium combines with one atom of oxygen in magnesium oxide. Therefore, the simplest chemical formula of magnesium oxide is MgO. This formula is called the simplest or ***empirical formula*** because chemical mass analysis yields only the ratios of atoms in a compound. It cannot tell us if the formula is MgO, Mg_2O_2, or $Mg_{21}O_{21}$.

Problem 4.6. Chemical analysis of phenol (a potent germicide), shows that its composition is 76.6% C, 6.38% H, and 17.0% O. Calculate its empirical formula.

Ans. We first determine the number of mols of each constituent in 100 g of the compound:

$$\text{mol C} = 76.6 \text{ g C}\left(\frac{1 \text{ mol C}}{12.0 \text{ g C}}\right) = 6.38 \text{ mol C}$$

$$\text{mol O} = 17.0 \text{ g O}\left(\frac{1 \text{ mol O}}{16.0 \text{ g O}}\right) = 1.06 \text{ mol O}$$

$$\text{mol H} = 6.38 \text{ g H}\left(\frac{1 \text{ mol H}}{1.01 \text{ g H}}\right) = 6.32 \text{ mol H}$$

We must find a simple whole number relationship among these values, so we divide each by the smallest value, 1.06. The result is that the ratio of C:H:O is 6.02:5.96:1.00. We recognize that 6.02 and 5.96 can be rounded off to 6.0, and that is probably within experimental error of the chemical analysis. The results of our calculation indicate that phenol consists of 6 mol of carbon atoms, 6 mol of hydrogen atoms, and 1 mol of oxygen atoms. If we divide those moles by Avogadro's number, the phenol molecule consists of six atoms of carbon, six atoms of hydrogen, and one atom of oxygen. Therefore its empirical formula is C_6H_6O.

Problem 4.7. Calculate the mass percentage of nitrogen in the rocket fuel hydrazine, N_2H_4.

Ans. The mass percent of nitrogen in this compound requires calculating the mass of nitrogen per formula mass of the compound. Knowing the formula of hydrazine, we calculate the formula mass as

$$(2 \times 14.01) + (4 \times 1.008) = 32.05$$

Nitrogen's contribution to this mass is $(2 \times 14.01) = 28.02$. Therefore the mass percent of nitrogen is

$$\left(\frac{\text{Mass of nitrogen}}{\text{Mass of hydrazine}}\right) \times 100\% = \left(\frac{28.02}{32.05}\right) \times 100\% = 87.43\%$$

4.5 MOLECULAR FORMULA FROM EMPIRICAL FORMULA AND MOLECULAR MASS

An empirical formula is the chemical formula which represents the basic combining ratios of the compound's elements. It cannot tell us the actual numbers of each kind of atom in the molecule. That information is contained in the *molecular formula* of a compound. For example, the percent composition of glucose is the following: carbon, 40.00 percent; hydrogen, 6.714 percent; oxygen, 53.29 percent. From these data we can calculate the empirical formula:

Carbon: $$40.00 \text{ g C}\left(\frac{1 \text{ mol C}}{12.01 \text{ g C}}\right) = 3.331 \text{ mol C}$$

Hydrogen: $$6.714 \text{ g H}\left(\frac{1 \text{ mol H}}{1.008 \text{ g H}}\right) = 6.661 \text{ mol H}$$

Oxygen: $$53.29 \text{ g O}\left(\frac{1 \text{ mol O}}{16.00 \text{ g O}}\right) = 3.331 \text{ mol O}$$

Dividing those numbers by the smallest to obtain the smallest ratio of whole numbers, the ratio of C:H:O is 1:2:1. From this we conclude that the empirical formula is CH_2O. However, it is known that the molecular mass of glucose, determined by an independent method, is 180. The formula mass of CH_2O is 30. We know that the ratio of C:H:O is correct, so that the molecular mass of 180 must be some whole number multiple of the empirical formula mass of 30. Table 4-3 contains multiples of CH_2O along with the corresponding molecular mass. This table will permit us to select the correct molecular formula corresponding to the appropriate molecular mass. The correct molecular formula corresponding to the molecular mass of 180 is $C_6H_{12}O_6$.

Table 4-3. Molecular Masses Corresponding to Various Molecular Formulas

Multiple	Molecular Formula	Corresponding Molecular Mass
$(CH_2O) \times 1$	CH_2O	30
$(CH_2O) \times 2$	$C_2H_4O_2$	60
$(CH_2O) \times 3$	$C_3H_6O_3$	90
$(CH_2O) \times 4$	$C_4H_8O_4$	120
$(CH_2O) \times 5$	$C_5H_{10}O_5$	150
$(CH_2O) \times 6$	$C_6H_{12}O_6$	180
$(CH_2O) \times 7$	$C_7H_{14}O_7$	210

4.6 BALANCING CHEMICAL EQUATIONS

The law of conservation of mass states that no mass can be lost as the result of a chemical reaction. Chemists express these ideas in chemical shorthand by writing the reactions as equations, in which conservation of mass is implicitly noted by an arrow separating components which react, called *reactants*, from components resulting from the reaction, called *products*. There are two fundamental types of chemical reactions: type I, or reactions in which the valence shells of the reactants remain unchanged after conversion to products, and type II, or reactions in which the valence shells of the reactants are modified by the gain or loss of electrons. Type II reactions were described in Chap. 3 as oxidation-reduction or redox reactions. The method used in taking into account the law of conservation of mass for type I reactions requires equating the numbers of atoms of each type on the reactant side with the number of atoms of each type on the product side. Type II reactions are subject not only to conservation of mass, but also to conservation of electrons. That is, in any redox reaction, the number of electrons lost by the reductant must be equal to the number of electrons gained by the oxidant. In order to decide whether a reaction is type I or type II, chemists have devised a method which assigns a number to an atom defining its oxidation state called an *oxidation number*. If that number undergoes a change as a result of a chemical reaction, the reaction is a redox or type II reaction. If there is no change, the reaction must be type I. Many type II as well as all type I reactions can be handled by the methods presented in this chapter for balancing reactions. However, there are redox reactions which require the use of oxidation numbers in order to account for conservation of mass. We will postpone considerations of oxidation numbers until Chap. 9, where redox reactions in solution will be introduced and discussed in more detail.

Let us examine the calcium chloride/sodium carbonate reaction, and write it in the form of a chemical equation:

$$\underset{\text{Reactants}}{CaCl_2 + Na_2CO_3} \longrightarrow \underset{\text{Products}}{CaCO_3 + NaCl}$$

The formulas of reactants and products are correct, but the number of atoms on either side of the arrow are not. The number of calcium atoms and carbonate ions are the same on both sides, but only half the number of sodium and chlorine atoms appear as products. To correct this we place a 2 before the formula for NaCl as follows:

$$CaCl_2 + Na_2CO_3 \longrightarrow CaCO_3 + 2\,NaCl$$

The factor of 2, which is called a *coefficient*, may be considered a multiplier which modifies the number of moles of the substance it is placed before. The equation now reads: 1 mol of $CaCl_2$ reacts with 1 mol of Na_2CO_3 to form 1 mol of $CaCO_3$ and 2 mol of NaCl. This procedure is called ***balancing the equation***, and requires that one know the formulas of both reactants and products.

This reaction is one of a type known as a *double-replacement* or an *exchanging-partner reaction*. Other types of chemical reactions include the following:

Combination reactions: for example, $2\,Mg + O_2 \rightarrow 2\,MgO$, where two components react to form one new component.

Decomposition reactions: for example, $2KClO_3 \rightarrow 2KCl + 3O_2$, where a reactant decomposes to its component parts.

Displacement reactions: for example, $2NaBr + Cl_2 \rightarrow 2NaCl + Br_2$, in which one of the reactants displaces one of the components of another reactant.

Problem 4.8. Balance the equation for the production of ammonia from the reaction of hydrogen with nitrogen, a combination reaction.

Ans. The formula for the product, ammonia, is NH_3. Both the reactants exist as diatomic molecules, H_2, and N_2. The reaction may be written as

$$N_2 + H_2 \longrightarrow NH_3$$

In the following table, the reactants are arranged on the left and products on the right, along with their formulas. Next to the formulas we count the atoms of each element according to the formula and list them as the Numbers. Next to those numbers of atoms we leave room for a multiplier, explained below.

Reactants				Products			
Compound	Atoms	Number per Compound	Multiplier	Compound	Atoms	Number per Compound	Multiplier
N_2	N	2		NH_3	N	1	
H_2	H	2			H	3	

The multipliers have been left out so that we can examine the number of atoms on each side of the equation prior to balancing. There are two N atoms on the left-hand side, and one N atom on the right-hand side. We can make the numbers of atoms equal by multiplying the number of N atoms in the product by 2. There are two H atoms on the left, and three H atoms on the right. The numbers on the left and right can be made equal by multiplying the left-hand-side H atoms by 3, and by multiplying the right-hand-side H atoms by 2.

Reactants				Products			
Compound	Atoms	Number per Compound	Multiplier	Compound	Atoms	Number per Compound	Multiplier
N_2	N	2	1	NH_3	N	1	2
H_2	H	2	3		H	3	2

The multipliers or *coefficients* are now placed before the reaction components to achieve a balanced equation:

$$N_2 + 3H_2 \longrightarrow 2NH_3$$

Note that the 2 before the NH_3 multiplies both the N and H atoms in the ammonia molecule. The equation now reads: 1 mol of N_2 reacts with 3 mol of H_2 to produce 2 mol of NH_3. (The hydrogen molecule does not consist of six H atoms, nor does the ammonia molecule consist of two N atoms and six H atoms.)

Problem 4.9. Balance the equation for the reaction between aluminum hydroxide, $Al(OH)_3$, and sulfuric acid, H_2SO_4, a double-replacement reaction.

Ans. The reaction equation is:

$$Al(OH)_3 + H_2SO_4 \longrightarrow Al_2(SO_4)_3 + H_2O$$

In cases of this kind, where polyatomic ions are involved, we can treat the SO_4^{2-} ion as a unit. Let us organize our data in the form of a table:

Reactants				Products			
Compound	Atoms	Number per Compound	Multiplier	Compound	Atoms	Number per Compound	Multiplier
$Al(OH)_3$	Al	1	?	$Al_2(SO_4)_3$	Al	2	1
H_2SO_4	SO_4	1	?		SO_4	3	1
$Al(OH)_3$	OH	3	?	H_2O	OH	1	?
H_2SO_4	H	2	?		H	1	?

Notice that the reactants have been entered twice. This was done to make clear the connection between the OH and H in both reactants, and the H_2O in the products. Balancing here requires a careful examination of the products. There are 2 mol of Al and 3 mol of SO_4^{2-} in $Al_2(SO_4)_3$, which forces us to place a multiplier or coefficient of 2 before the $Al(OH)_3$, and a coefficient of 3 before the H_2SO_4. However, the coefficients before the $Al(OH)_3$ and the H_2SO_4 then multiply the numbers of H and OH on the reactant side. Since $Al(OH)_3$ has been multiplied by 2, there are six O atoms and six H atoms from that compound on the left (6 OH). There are an additional six H atoms on the left because of the multiplication of H_2SO_4 by 3. Since all of the H (6 H) and OH (6 OH) must be contained in the product, H_2O, the coefficient before the H_2O is 6.

Reactants				Products			
Compound	Atoms	Number per Compound	Multiplier	Compound	Atoms	Number per Compound	Multiplier
$Al(OH)_3$	Al	1	2	$Al_2(SO_4)_3$	Al	2	1
H_2SO_4	SO_4	1	3		SO_4	3	1
$Al(OH)_3$	OH	3	2	H_2O	OH	1	6
H_2SO_4	H	2	3		H	1	6

The balanced equation is

$$2\,Al(OH)_3 + 3\,H_2SO_4 \longrightarrow Al_2(SO_4)_3 + 6\,H_2O$$

4.7 STOICHIOMETRY

We are now in a position to consider the mass relationships in chemical reactions, or what is known as *stoichiometry*. Let us examine the quantitative aspects of the reaction in which ammonia is formed from nitrogen and hydrogen:

$$N_2 + 3\,H_2 \longrightarrow 2\,NH_3$$

Problem 4.10. How many moles of hydrogen will react with 0.50 mol of ammonia?

Ans. The balanced equation tells us that regardless of the actual amounts of reactants used, the ratios of all balancing coefficients must remain constant. We can use the ratio of moles of hydrogen to moles of nitrogen as one of several possible unit conversion factors to calculate the number of moles of hydrogen corresponding to 0.50 mol of nitrogen in this reaction:

$$0.50 \text{ mol } N_2\left(\frac{3 \text{ mol } H_2}{1 \text{ mol } N_2}\right) = 1.5 \text{ mol } H_2$$

Problem 4.11. How many moles of ammonia will be formed from 0.50 mol of N_2 and 1.5 mol of H_2?

Ans. **In this case, we will create two conversion factors from the balancing coefficients of both reactants with that of the product and do two equivalent calculations to determine the amount of product, one based on the initial amount of N_2, and the second based on the initial concentration of H_2:**

$$0.50 \text{ mol } N_2\left(\frac{2 \text{ mol } NH_3}{1 \text{ mol } N_2}\right) = 1 \text{ mol } NH_3$$

$$1.5 \text{ mol } H_2\left(\frac{2 \text{ mol } NH_3}{3 \text{ mol } H_2}\right) = 1 \text{ mol } NH_3$$

Problem 4.12. How many grams of ammonia can be formed from 42 g of nitrogen and how many grams of hydrogen are required to completely convert that amount of nitrogen?

Ans. **We examine the balanced equation: $N_2 + 3H_2 \rightarrow 2NH_3$. To solve this type of problem, first convert the mass of nitrogen to moles, then solve for the corresponding numbers of moles of hydrogen or ammonia using the balancing coefficients as unit conversion factors. Then finally convert back to mass, as follows:**

$$42 \text{ g } N_2\left(\frac{1 \text{ mol } N_2}{28 \text{ g } N_2}\right) = 1.5 \text{ mol } N_2$$

$$1.5 \text{ mol } N_2\left(\frac{2 \text{ mol } NH_3}{1 \text{ mol } N_2}\right) = 3 \text{ mol } NH_3$$

$$3 \text{ mol } NH_3\left(\frac{17 \text{ g } NH_3}{1 \text{ mol } NH_3}\right) = 51 \text{ g } NH_3$$

$$1.5 \text{ mol } N_2\left(\frac{3 \text{ mol } H_2}{1 \text{ mol } N_2}\right) = 4.5 \text{ mol } H_2$$

$$4.5 \text{ mol } H_2\left(\frac{2.0 \text{ g } H_2}{1 \text{ mol } H_2}\right) = 9.0 \text{ g } H_2$$

Check for conservation of mass:

$$42 \text{ g } N_2 + 9.0 \text{ g } H_2 = 51 \text{ g } NH_3$$

An alternative method for solving this problem is to note that, since we know the formulas and molar ratios of reactants and products, we can express the balancing coefficients not only as whole numbers (moles) but also as formula masses multiplied by the balancing coefficients (the total numbers of moles of reactants and products):

$$(2 \times 14) + 3(2 \times 1) = 2[(1 \times 14) + (3 \times 1)]$$

$$28 \text{ g } N_2 + 6 \text{ g } H_2 = 34 \text{ g } NH_3$$

Using these as conversion factors, we can calculate the required masses:

$$42 \text{ g } N_2\left(\frac{34 \text{ g } NH_3}{28 \text{ g } N_2}\right) = 51 \text{ g } NH_3$$

$$51 \text{ g } NH_3\left(\frac{6.0 \text{ g } H_2}{34 \text{ g } NH_3}\right) = 9.0 \text{ g } H_2$$

ADDITIONAL SOLVED PROBLEMS

CHEMICAL FORMULAS AND FORMULA MASSES

Problem 4.13. Calculate the formula masses of the following compounds using rounded off values of atomic masses: (*a*) Ag_2CrO_4, (*b*) $Ba(OH)_2$, (*c*) $Ce(HSO_4)_4$, (*d*) $K_2Cr_2O_7$, (*e*) C_2H_5OH (ethyl alcohol)

Ans.

(*a*) $[2(Ag) \times 108] + [1(Cr) \times 52] + [4(O) \times 16] = 332$

(*b*) $[1(Ba) \times 137] + [2(O) \times 16] + [2(H) \times 1] = 171$

(*c*) $[1(Ce) \times 140] + [4(H) \times 1] + [4(S) \times 32] + [16(O) \times 16] = 528$

(*d*) $[2(K) \times 39] + [2(Cr) \times 52] + [7(O) \times 16] = 294$

(*e*) $[2(C) \times 12] + [6(H) \times 1] + [1(O) \times 16] = 46$

Problem 4.14. Calculate the formula masses of the following compounds using rounded off values of atomic masses: (*a*) NaSCN, (*b*) $NaC_2H_3O_2$, (*c*) Hg_2Cl_2, (*d*) $Fe(NH_4)_2(SO_4)_2$, (*e*) C_3H_8O.

Ans.

(*a*) $[1(Na) \times 23] + [1(S) \times 32] + [1(C) \times 12] + [1(N) \times 14] = 81$

(*b*) $[1(Na) \times 23] + [2(C) \times 12] + [3(H) \times 1] + [2(O) \times 16] = 82$

(*c*) $[2(Hg) \times 200.6] + [2(Cl) \times 35.5] = 472.2$ (rounding off here would result in significant error)

(*d*) $[1(Fe) \times 56] + [2(N) \times 14] + [8(H) \times 1] + [2(S) \times 32] + [8(O) \times 16] = 284$

(*e*) $[3(C) \times 12] + [8(H) \times 1] + [1(O) \times 16] = 60$

THE MOLE

Problem 4.15. How many moles are there in 42.5 g of sodium chloride?

Ans. First determine the formula mass:

$$[1(Na) \times 23] + [1(Cl) \times 35.5] = 58.5$$

Therefore 1 mol of sodium chloride contains 58.5 g. Using that value for a conversion factor:

$$42.5 \text{ g NaCl}\left(\frac{1 \text{ mol NaCl}}{58.5 \text{ g NaCl}}\right) = 0.726 \text{ mol NaCl}$$

Problem 4.16. How many moles are there in 125.0 g of sodium thiocyanate, NaSCN?

Ans. The formula mass of NaSCN is 81.0. Using that value for a conversion factor:

$$125.0 \text{ g NaSCN}\left(\frac{1 \text{ mol NaSCN}}{81.0 \text{ g NaSCN}}\right) = 1.54 \text{ mol NaSCN}$$

Problem 4.17. How many grams of sodium acetate, $NaC_2H_3O_2$, are there in 2.650 mol of that compound?

Ans. The formula mass of $NaC_2H_3O_2$ is 82.00. Using that value for a conversion factor:

$$2.650 \text{ mol } NaC_2H_3O_2\left(\frac{82.00 \text{ g } NaC_2H_3O_2}{1 \text{ mol } NaC_2H_3O_2}\right) = 217.3 \text{ g } NaC_2H_3O_2$$

Problem 4.18. How many grams are there in 0.557 mol of silver chromate, Ag_2CrO_4?

Ans. The formula mass of silver chromate is 332. Using that value for a conversion value:

$$0.557 \text{ mol } Ag_2CrO_4\left(\frac{332 \text{ g } Ag_2CrO_4}{1 \text{ mol } Ag_2CrO_4}\right) = 185 \text{ g } Ag_2CrO_4$$

AVOGADRO'S NUMBER

Problem 4.19. How many silver ions are present in 185 g of Ag_2CrO_4?

Ans. From problem 4.17, we know that 185 g of Ag_2CrO_4 is equal to 0.557 mol of Ag_2CrO_4. Therefore there are $2 \times 0.557 = 1.11$ mol of Ag^+ ion present. Since there is one Avogadro's number worth of particles per mole of any particle, there are

$$1.11 \text{ mol } Ag^+\left(\frac{6.02 \times 10^{23} \text{ } Ag^+}{1 \text{ mol } Ag^+}\right) = 6.68 \times 10^{23} \text{ } Ag^+ \text{ ions}$$

Problem 4.20. How many H atoms are there in 30 g of water?

Ans. We must first determine the number of moles of water in 30 g of water. The formula for water is H_2O; therefore its formula mass is 18. The number of moles in that mass is

$$30.0 \text{ g}\left(\frac{1 \text{ mol } H_2O}{18.0 \text{ g } H_2O}\right) = 1.67 \text{ mol } H_2O$$

There are 2 mol of H atoms per mol of H_2O; therefore there are $2 \times 1.67 = 3.34$ mol H atoms in 30 g of water, and

$$3.34 \text{ mol H atoms}\left(\frac{6.02 \times 10^{23} \text{ H atoms}}{1 \text{ mol H atoms}}\right) = 2.01 \times 10^{24} \text{ H atoms}$$

Problem 4.21. What is the mass in grams of 9.23×10^{24} molecules of CO_2?

Ans. First calculate the number of moles of CO_2 in 9.23×10^{24} molecules of CO_2, and then convert to grams of CO_2:

$$9.23 \times 10^{24} \text{ molecules of } CO_2\left(\frac{1 \text{ mol } CO_2}{6.02 \times 10^{23} \text{ molecules } CO_2}\right) = 15.3 \text{ mol } CO_2$$

$$15.3 \text{ mol } CO_2\left(\frac{44.0 \text{ g } CO_2}{1 \text{ mol } CO_2}\right) = 673 \text{ g } CO_2$$

EMPIRICAL FORMULAS AND PERCENT COMPOSITION

Problem 4.22. By chemical analysis, the composition of bismuth oxide is 89.70% Bi and 10.30% O. Calculate its empirical formula.

Ans. First, we assume that we have 100.0 g of the compound. Therefore the mass of bismuth is 89.70 g, and the mass of oxygen is 10.30 g. The next step is to determine the number of moles of each element contained in those masses. The atomic mass of bismuth is 209.0 and that of oxygen is 16.00. Using these as unit conversion factors:

$$89.70 \text{ g Bi}\left(\frac{1 \text{ mol Bi}}{209.0 \text{ g Bi}}\right) = 0.4292 \text{ mol Bi}$$

$$10.30 \text{ g O}\left(\frac{1 \text{ mol O}}{16.00 \text{ g O}}\right) = 0.6438 \text{ mol O}$$

To obtain the ratio of moles in small whole numbers, divide both answers by the smaller of the two:

$$\frac{0.6438 \text{ mol O}}{0.4292 \text{ mol Bi}} = \frac{1.5}{1.0} = \frac{3}{2}$$

It is possible to have 1.5 mol of any element, but never 1.5 atoms. The ratio of 1.5/1 must be multiplied by 2 so as to reflect the fact that the ratio is one of numbers of atoms, not moles. Therefore the empirical formula of bismuth oxide is Bi_2O_3.

Problem 4.23. By chemical analysis, the composition of potassium phosphate is 55.19% K, 14.62% P, and 30.19% O. Determine its empirical formula.

Ans. Converting percent to grams by assuming 100 g of the compound, we determine the number of moles in each mass using atomic masses of each as unit conversion factors:

$$55.19 \text{ g K}\left(\frac{1 \text{ mol K}}{39.10 \text{ g K}}\right) = 1.412 \text{ mol K}$$

$$14.62 \text{ g P}\left(\frac{1 \text{ mol P}}{30.97 \text{ g P}}\right) = 0.4721 \text{ mol P}$$

$$30.19 \text{ g O}\left(\frac{1 \text{ mol O}}{16.00 \text{ g O}}\right) = 1.887 \text{ mol O}$$

Dividing each value by the smallest one, the ratio of all three becomes: 2.991 (K): 1.000 (P): 3.997 (O). Rounding off, the ratio of K : P : O becomes 3 : 1 : 4. The empirical formula of potassium phosphate is K_3PO_4.

MOLECULAR FORMULA FROM EMPIRICAL FORMULA AND MOLECULAR MASS

Problem 4.24. By chemical analysis, an oxide of nitrogen is found to be composed of 30.43% N and 69.57% O. Its molecular mass is found to be 92. What is its molecular formula?

Ans. First determine its empirical formula. Assuming we have 100 g of the oxide, the number of moles of each component is

$$30.43 \text{ g N}\left(\frac{1 \text{ mol N}}{14.01 \text{ g N}}\right) = 2.172 \text{ mol N}$$

$$69.97 \text{ g O}\left(\frac{1 \text{ mol O}}{16.00 \text{ g O}}\right) = 4.373 \text{ mol O}$$

Next obtain the smallest whole number ratio of moles:

$$\frac{4.373 \text{ mol O}}{2.172 \text{ mol N}} = \frac{2.0 \text{ mol O}}{1.0 \text{ mol N}}$$

From this we deduce the empirical formula, NO_2, and a formula mass of 46. The molecular mass of the oxide was found to be 92. We construct a table of multiples of the empirical formula and corresponding molecular masses:

Multiples of Empirical Formula	Corresponding Mass
NO_2	46
N_2O_4	92
N_3O_6	138

From this we may conclude that the molecular formula is N_2O_4.

Problem 4.25. By chemical analysis, a hydrocarbon (a compound consisting of only carbon and hydrogen) was found to be composed of 85.71% C and 14.29% H. Its molecular mass was found to be 84. Determine its molecular formula.

Ans. First determine its empirical formula. Assuming we have 100 g of the oxide, the number of moles of each component is

$$85.71 \text{ g C}\left(\frac{1 \text{ mol C}}{12.01 \text{ g C}}\right) = 7.137 \text{ mol C}$$

$$14.29 \text{ g H}\left(\frac{1 \text{ mol H}}{1.000 \text{ g H}}\right) = 14.29 \text{ mol H}$$

Next obtain the smallest whole number ratio of moles:

$$\frac{14.29 \text{ mol H}}{7.137 \text{ mol C}} = \frac{2.0 \text{ mol H}}{1.0 \text{ mol C}}$$

From this we deduce the empirical formula, CH_2, and a formula mass of 14. The molecular mass of the hydrocarbon was found to be 84. We construct a table of multiples of the empirical formula and corresponding molecular masses:

Multiples of Empirical Formula	Corresponding Mass
CH_2	14
C_2H_4	28
C_4H_8	56
C_6H_{12}	84
C_8H_{16}	112

From this we may conclude that the molecular formula is C_6H_{12}.

BALANCING CHEMICAL REACTIONS

Problem 4.26. Balance the following equations:
(*a*) $N_2 + I_2 \rightarrow NI_3$; (*b*) $Al + O_2 \rightarrow Al_2O_3$; (*c*) $HCl + Ca(OH)_2 \rightarrow CaCl_2 + H_2O$; (*d*) $ZnCl_2 + H_2S \rightarrow ZnS + HCl$; (*e*) $H_2O_2 \rightarrow H_2O + O_2$.

Ans. (*a*)

Reactants				Products			
Compound	Atoms	Number per Compound	Multiplier	Compound	Atoms	Number per Compound	Multiplier
N_2	N	2	1	NI_3	N	1	2
I_2	I	2	3		I	3	2

The balanced equation is $N_2 + 3I_2 \rightarrow 2NI_3$.

(*b*)

Reactants				Products			
Compound	Atoms	Number per Compound	Multiplier	Compound	Atoms	Number per Compound	Multiplier
O_2	O	2	3	Al_2O_3	O	3	2
Al	Al	1	4		Al	2	2

To get the numbers of O atoms correct, it is necessary to multiply the left-hand-side O atoms by 3, and the right-hand-side O atoms by 2. By multiplying the product side by 2, we are forced to place four Al atoms on the reactant side. The balanced equation is $4\,Al + 3\,O_2 \rightarrow 2\,Al_2O_3$.

(*c*)

Reactants				Products			
Compound	Atoms	Number per Compound	Multiplier	Compound	Atoms	Number per Compound	Multiplier
$Ca(OH)_2$	Ca OH	1 2	1 1	$CaCl_2$	Ca Cl	1 2	1 1
HCl	Cl H	1 1	2 2	H_2O	O H	1 2	2 2

The balanced equation is $Ca(OH)_2 + 2\,HCl \rightarrow CaCl_2 + 2\,H_2O$. (*Note:* H_2O can sometimes be more easily recognized as HOH, and there are 2 H and 2 OH on the reactant side.)

(*d*)

Reactants				Products			
Compound	Atoms	Number per Compound	Multiplier	Compound	Atoms	Number per Compound	Multiplier
$ZnCl_2$	Zn Cl	1 2	1 1	ZnS	Zn S	1 1	1 1
H_2S	S H	1 2	1 1	HCl	Cl H	1 1	2 2

The balanced equation is $ZnCl_2 + H_2S \rightarrow ZnS + 2\,HCl$.

(*e*)

Reactants				Products			
Compound	Atoms	Number per Compound	Multiplier	Compound	Atoms	Number per Compound	Multiplier
H_2O_2	O H	2 2	2 2	H_2O	O H	1 2	2 2
				O_2	O	2	1

The balanced equation is $2\,H_2O_2 \rightarrow 2\,H_2O + O_2$.

STOICHIOMETRY

Problem 4.27. What are the unit conversion factors which can be derived from the following balanced equation?

$$3\,NaOH + H_3PO_4 \longrightarrow Na_3PO_4 + 3\,H_2O$$

Ans. There are 12 different unit conversion factors; each of the following 6 can be used inversely:

$$\frac{3 \text{ mol NaOH}}{1 \text{ mol } H_3PO_4} \qquad \frac{3 \text{ mol NaOH}}{1 \text{ mol } Na_3PO_4} \qquad \frac{3 \text{ mol NaOH}}{3 \text{ mol } H_2O}$$

$$\frac{1 \text{ mol } H_3PO_4}{1 \text{ mol } Na_3PO_4} \qquad \frac{1 \text{ mol } H_3PO_4}{3 \text{ mol } H_2O} \qquad \frac{1 \text{ mol } Na_3PO_4}{3 \text{ mol } H_2O}$$

Problem 4.28. How many moles of Na_3PO_4 can be prepared from 2 mol of H_3PO_4 using sufficient NaOH?

Ans. Using an appropriate unit conversion factor from the above list:

$$2 \text{ mol } H_3PO_4\left(\frac{1 \text{ mol } Na_3PO_4}{1 \text{ mol } H_3PO_4}\right) = 2 \text{ mol } Na_3PO_4$$

Problem 4.29. How many moles of NaOH are required to completely react with 1.57 mol of H_3PO_4?

Ans.

$$1.57 \text{ mol } H_3PO_4\left(\frac{3 \text{ mol NaOH}}{1 \text{ mol } H_3PO_4}\right) = 4.71 \text{ mol NaOH}$$

Problem 4.30. How many moles of H_3PO_4 must be used to produce 6.3 mol of water?

Ans.

$$6.3 \text{ mol } H_2O\left(\frac{1 \text{ mol } H_3PO_4}{3 \text{ mol } H_2O}\right) = 2.1 \text{ mol } H_3PO_4$$

Problem 4.31. How many grams are 2 mol of Na_3PO_4?

Ans. The unit conversion factor here must be created using the formula mass of Na_3PO_4:

$$2 \text{ mol } Na_3PO_4\left(\frac{164 \text{ g } Na_3PO_4}{1 \text{ mol } Na_3PO_4}\right) = 328 \text{ g } Na_3PO_4$$

Problem 4.32. How many moles of Na_3PO_4 can be prepared from 147 g of H_3PO_4, using sufficient NaOH?

Ans. Two conversion factors are required, grams to moles, and the ratio of moles from the balanced equation:

$$147 \text{ g } H_2PO_4\left(\frac{1 \text{ mol } H_3PO_4}{98.0 \text{ g } H_3PO_4}\right) = 1.50 \text{ mol } H_3PO_4$$

$$1.50 \text{ mol } H_2\ PO_4\left(\frac{1 \text{ mol } Na_3PO_4}{1 \text{ mol } H_3PO_4}\right) = 1.50 \text{ mol } Na_3PO_4$$

Problem 4.33. How many grams are there in 1.50 mol of Na_3PO_4?

Ans.

$$1.50 \text{ mol } Na_2\ PO_4\left(\frac{164 \text{ g } Na_3PO_4}{1 \text{ mol } Na_3PO_4}\right) = 246 \text{ g } Na_3PO_4$$

Problem 4.34. How many grams of Na_3PO_4 can be prepared from 147 g of H_3PO_4, using sufficient NaOH?

Ans. The calculation here requires the use of three conversion factors, (1) to convert H_3PO_4 to moles, (2) to convert moles of H_3PO_4 to moles of Na_3PO_4, and (3) to convert moles of Na_3PO_4 to grams:

$$147\text{ g } H_3PO_4\left(\frac{1\text{ mol } H_3PO_4}{98.0\text{ g } H_3PO_4}\right)\left(\frac{1\text{ mol } Na_3PO_4}{1\text{ mol } H_3PO_4}\right)\left(\frac{164\text{ g } Na_3PO_4}{1\text{ mol } Na_3PO_4}\right) = 246\text{ g } Na_3PO_4$$

Problem 4.35. Is there an alternative method for calculating the number of grams of Na_3PO_4 which can be prepared from 147 g of H_3PO_4 using sufficient NaOH?

Ans. Yes, we can use a unit conversion factor consisting of the ratio of molar masses equal to the formula masses multiplied by the accompanying balancing coefficients:

$$147\text{ g } H_3PO_4\left(\frac{164\text{ g } Na_3PO_4}{98.0\text{ g } H_3PO_4}\right) = 246\text{ g } Na_3PO_4$$

Notice that this conversion factor is the product of the three factors in Problem 4.34. It is another interpretation of the balanced equation which states that 1 formula weight of H_3PO_4 will produce 1 formula weight of Na_3PO_4.

Problem 4.36. How many grams of NaOH are required to completely convert 147 g of H_3PO_4 to Na_3PO_4?

Ans.

$$147\text{ g } H_3PO_4\left(\frac{1\text{ mol } H_3PO_4}{98.0\text{ g } H_3PO_4}\right) = 1.50\text{ mol } H_3PO_4$$

$$1.50\text{ mol } H_3PO_4\left(\frac{3\text{ mol NaOH}}{1\text{ mol } H_3PO_4}\right) = 4.50\text{ mol NaOH}$$

$$4.50\text{ mol NaOH}\left(\frac{40\text{ g NaOH}}{1\text{ mol NaOH}}\right) = 180\text{ g NaOH}$$

SUPPLEMENTARY PROBLEMS

Problem 4.37. Calculate the formula masses of the following compounds, using precise atomic masses: (*a*) $BaCO_3$; (*b*) $K_2Cr_2O_7$; (*c*) $C_{10}H_{12}$; (*d*) $(NH_4)_2C_2O_4$; (*e*) $Zn_2P_2O_7$.

Ans. (*a*) 197.35; (*b*) 294.19; (*c*) 132.2; (*d*) 124.1; (*e*) 304.68

Problem 4.38. Calculate the formula masses of the following compounds, using rounded-off atomic masses: (*a*) $AgNO_3$; (*b*) Cr_2O_3; (*c*) HBr; (*d*) $ZnCl_2$; (*e*) BF_3.

Ans. (*a*) 170; (*b*) 152; (*c*) 81; (*d*) 136.4; (*e*) 67.8

Problem 4.39. Calculate the formula masses of the following compounds, using rounded-off atomic masses: (*a*) $HgSO_4$; (*b*) $SrCl_2$; (*c*) $Zn(C_2H_3O_2)_2$; (*d*) $MgC_8H_4O_4$; (*e*) CuCl.

Ans. (*a*) 296.6; (*b*) 158.6; (*c*) 183.4; (*d*) 188.3; (*e*) 99

Problem 4.40. Using rounded-off formula masses, calculate the number of moles contained in the given amounts of the following compounds: (*a*) 102.6 g of $BaCO_3$; (*b*) 60.75 g of HBr; (*c*) 148.5 g of CuCl; (*d*) 50.4 g of HNO_3; (*e*) 65 g of $C_6H_{12}O_6$.

Ans. (*a*) 0.5199 mol; (*b*) 0.7508 mol; (*c*) 1.500 mol; (*d*) 0.800 mol; (*e*) 0.36 mol

Problem 4.41. Using rounded-off formula masses, calculate the number of moles contained in the given amounts of the following compounds: (*a*) 45 g of H_2SO_4; (*b*) 133 g of $Ca(OH)_2$; (*c*) 35 g of C_2H_6O; (*d*) 110 g of C_4H_{10}; (*e*) 210 g of $FeCl_2$.

Ans. (*a*) 0.46 mol; (*b*) 1.80 mol; (*c*)) 0.76 mol; (*d*) 1.89 mol; (*e*) 1.66 mol

Problem 4.42. How many formula units are there in 133 g of $Ca(OH)_2$?

Ans.

$$(1.8 \text{ mol})(6.02 \times 10^{23} \text{ formula units/mol}) = 1.1 \times 10^{24} \text{ formula units}$$

Problem 4.43. How many molecules are there in 35 g of C_2H_6O (ethyl alcohol)?

Ans.

$$(0.76 \text{ mol})(6.02 \times 10^{23} \text{ molecules/mol}) = 4.6 \times 10^{23} \text{ molecules}$$

Problem 4.44. How many electrons are lost by 42 g of sodium when its atoms are oxidized to ions?

Ans.

$$(1.83 \text{ mol})(6.02 \times 10^{23} \text{ electrons/mol}) = 1.10 \times 10^{24} \text{ electrons}$$

Problem 4.45. How many molecules are there in 1.00 kg of water?

Ans.

$$(55.5 \text{ mol})(6.02 \times 10^{23} \text{ molecules/mol}) = 3.34 \times 10^{25} \text{ molecules}$$

Problem 4.46. By chemical analysis, silver chloride was found to be 75.26% Ag and 24.74% Cl. Determine its empirical formula.

Ans. AgCl

Problem 4.47. By chemical analysis, an oxide of copper was found to be 88.81% Cu and 11.19% O. Determine its empirical formula.

Ans. Cu_2O

Problem 4.48. Chemical analysis of methyl ether showed it to be composed of 52.17% C, 13.05% H, and 34.78% O. Determine its empirical formula.

Ans. C_2H_6O

Problem 4.49. The empirical formula of glycerin is $C_3H_8O_3$. Determine its percent composition.

Ans. 39.13% C, 8.7% H, and 52.17% O

Problem 4.50. The percent composition of a hydrocarbon was 89.36% C and 10.64% H. Its molecular mass determined by an independent method was 188. Determine its molecular formula.

Ans. $C_{14}H_{20}$

Problem 4.51. Two organic acids have the same percent compositions: 50.00% C, 5.56% H, and 44.44% O. The molecular mass of a sample having that composition was found to be 144. What is the molecular formula of the compound?

Ans. $C_6H_8O_4$.

Problem 4.52. Balance the following equations:

(*a*) $Ca(OH)_2 + FeCl_3 \longrightarrow Fe(OH)_3 + CaCl_2$

(*b*) $KI + Br_2 \longrightarrow KBr + I_2$

(*c*) $K + H_2O \longrightarrow KOH + H_2$

(*d*) $S + O_2 \longrightarrow SO_3$

(*e*) $IBr + NH_3 \longrightarrow NI_3 + NH_4Br$

Ans.

(*a*) $3\,Ca(OH)_2 + 2\,FeCl_3 \longrightarrow 2\,Fe(OH)_3 + 3\,CaCl_2$

(*b*) $2\,KI + Br_2 \longrightarrow 2\,KBr + I_2$

(*c*) $2\,K + 2\,H_2O \longrightarrow 2\,KOH + H_2$

(*d*) $2\,S + 3\,O_2 \longrightarrow 2\,SO_3$

(*e*) $3\,IBr + 4\,NH_3 \longrightarrow NI_3 + 3\,NH_4Br$

Problem 4.53. Balance the following equations:

(*a*) $C_4H_{10} + O_2 \longrightarrow CO_2 + H_2O$

(*b*) $C_6H_{12}O_6 + O_2 \longrightarrow CO_2 + H_2O$

(*c*) $PbCl_2 + Na_2SO_4 \longrightarrow PbSO_4 + NaCl$

(*d*) $AgCl + KOH + NH_4Cl \longrightarrow Ag(NH_3)_2Cl + KCl + H_2O$

(*e*) $Cu(NO_3)_2 + KBr \longrightarrow Br_2 + CuBr + KNO_3$

Ans.

(*a*) $2\,C_4H_{10} + 13\,O_2 \longrightarrow 8\,CO_2 + 10\,H_2O$

(*b*) $C_6H_{12}O_6 + 6\,O_2 \longrightarrow 6\,CO_2 + 6\,H_2O$

(*c*) $PbCl_2 + Na_2SO_4 \longrightarrow PbSO_4 + 2\,NaCl$

(*d*) $AgCl + 2\,KOH + 2\,NH_4Cl \longrightarrow Ag(NH_3)_2Cl + 2\,KCl + 2\,H_2O$

(*e*) $2\,Cu(NO_3)_2 + 4\,KBr \longrightarrow Br_2 + 2\,CuBr + 4\,KNO_3$

Problem 4.54. Balance the following equations:

(*a*) $Ca_3(PO_4)_2 + H_3PO_4 \longrightarrow Ca(H_2PO_4)_2$

(*b*) $FeCl_2 + (NH_4)_2S \longrightarrow FeS + NH_4Cl$

(*c*) $KClO_3 \longrightarrow KCl + O_2$

(*d*) $O_2 \longrightarrow O_3$

(*e*) $C_6H_6 + O_2 \longrightarrow CO_2 + H_2O$

Ans.

(*a*) $Ca_3(PO_4)_2 + 4H_3PO_4 \longrightarrow 3Ca(H_2PO_4)_2$

(*b*) $FeCl_2 + (NH_4)_2S \longrightarrow FeS + 2NH_4Cl$

(*c*) $2KClO_3 \longrightarrow 2KCl + 3O_2$

(*d*) $3O_2 \longrightarrow 2O_3$

(*e*) $2C_6H_6 + 15O_2 \longrightarrow 12CO_2 + 6H_2O$

Problem 4.55. Given the following equation:

$$2Cu(NO_3)_2 + 4KBr \longrightarrow Br_2 + 2CuBr + 4KNO_3$$

(*a*) How many moles of $Cu(NO_3)_2$ will react with 0.56 mol of KBr?
(*b*) How many moles of KBr must react to produce 3.2 mol of CuBr?
(*c*) How many grams of $Cu(NO_3)_2$ will react with 10.0 g of KBr?
(*d*) How many grams of KNO_3 will be produced from 10.0 g of KBr?

Ans. (*a*) 0.28 mol; (*b*) 6.4 mol; (*c*) 7.88 g; (*d*) 8.49 g

Problem 4.56. Given the following equation:

$$3CaCl_2 + 2K_3PO_4 \longrightarrow Ca_3(PO_4)_2 + 6KCl$$

(*a*) How many moles of K_3PO_4 will react with 0.69 mol of $CaCl_2$?
(*b*) How many moles of $Ca_3(PO_4)_2$ can be produced from 47.0 g of $CaCl_2$?
(*c*) How many grams of KCl can be produced from 54.0 g of K_3PO_4?
(*d*) How many grams of $Ca_3(PO_4)_2$ can be produced from 0.280 mol of K_3PO_4?

Ans. (*a*) 0.46 mol; (*b*) 0.141 mol; (*c*) 56.9 g; (*d*) 43.4 g

Physical Properties of Matter

5.1 INTRODUCTION

When a laboratory procedure calls for 3.2000 g of a solid like glucose or sodium chloride, it is a simple matter of weighing out the solid on an analytical balance. No attention need be paid to temperature or atmospheric pressure. If 10 mL of water is required, one can add the liquid to an appropriate vessel calibrated in units of volume. The liquid will then fill the vessel up to the correct calibration mark. In working with gases, however, the temperature and pressure are the principal variables in defining the volume (therefore the mass) of the gas. A solid retains its shape and undergoes virtually no change in volume when pressure is applied. The transition to the liquid state is called *melting*, and when this occurs, there is usually only a slight change in volume. The liquid's volume, as is the case with the solid, undergoes no perceptible changes with increase in pressure. However, when it is converted to the gaseous state (or is vaporized) there is a large change in volume. A mole of liquid water at 100 °C occupies slightly less than 19 mL, but a mole of water vapor at the same temperature has a volume of about 30000 mL.

The properties of matter in their different states can be rationalized by the following mental picture or model. Because they are virtually incompressible, it is assumed that the molecules of solids and liquids are held together by cohesive forces, and are in contact so that their volumes cannot be reduced by increased pressure. The difference between them is that the molecules of a solid are fixed in space and may vibrate about those positions, but cannot move from them. In the liquid state, the molecules are just about as close to each other as in the solid state, but the cohesive forces are not quite as strong as in the solid state, so that they are not fixed in space and are free to move about. The transition from liquid to gaseous state entirely changes this picture. The molecules now occupy such a large volume and are therefore so far apart that, for all intents and purposes, cohesive forces no longer operate among them. They are now free to move about with only the walls of their container to restrain them.

5.2 KINETIC-MOLECULAR THEORY

The ideas presented above are expressed in a general theory of the physical properties of matter called the *kinetic-molecular theory*. The theory is based on the following propositions:

1. All matter is composed of particles, i.e., atoms or molecules, whose characteristic sizes and shapes remain unchanged whether in the solid, liquid, or gaseous state.
2. These particles are in ceaseless, random motion (vibration and translation). This motion is a measure of their kinetic energy; hence the name *kinetic-molecular theory*. Motion is limited in the solid and liquid states by cohesive forces. The energy of cohesion is potential energy. This is the energy which must be overcome when solids are melted and liquids are vaporized.
3. Kinetic energy is a function of temperature. As the temperature increases, so does kinetic energy, and as a result, the particles move faster. In hot systems, the particles move about much faster than in cold systems.
4. Collisions are elastic. This means that, in a collision, kinetic energy is transferred from one particle to another with no overall loss in energy. One particle loses and the other gains an equal amount of kinetic energy.

5.3 COHESIVE FORCES

Cohesive forces among molecules are called *intermolecular forces*, and are electrical in nature. The ones we will be most interested in arise from molecular polarity. Permanent molecular polarity, as a consequence of electronegativity differences and molecular shape, and quantified by the dipole moment, was described in Chap. 3. However, even in molecules like methane which have no permanent polarity, transitory dipoles exist for brief moments, i.e., 1×10^{-10} to 1×10^{-13} s. This is because erratic motions of electrons lead to uneven distributions of electrical charge within the molecule over these brief time intervals. These give rise to very weak cohesive forces called *dispersion*, or *London forces*. These are the forces at work when gases like methane or helium are condensed to liquids. All molecules, those with and those without permanent dipoles, interact by dispersion forces. Because dispersion forces arise from uneven intramolecular distributions of electrons, molecules with large numbers of electrons develop larger dispersion forces. Numbers of electrons are associated with molecular mass, so nonpolar molecules of large mass exert larger cohesive force than nonpolar molecules of smaller mass. Compare the boiling points of methane and hydrogen given in Table 5-2. Molecules with permanent dipole moments tend to interact strongly with each other, and with other molecules possessing dipole moments. The attractive force between two dipolar molecules, for example, CH_3Cl, methyl chloride, and $CHCl_3$, chloroform, are called *dipole-dipole bonds*. Substances composed of polar molecules have higher melting and boiling points than those composed of nonpolar molecules. At room temperature, methane, CH_4, which has no permanent dipole moment, is a gas, but water, H_2O, a molecule of about the same mass, which has a large dipole moment, is a liquid at the same temperature.

A covalent bond between hydrogen and any one of the elements oxygen, nitrogen, or fluorine possesses a unique character. The differences in electronegativity, and therefore polarity, between any of those pairs is very large. This leads to very strong cohesive forces among molecules like H_2O, NH_3, and HF, called *hydrogen bonding*. A hydrogen bond, sometimes abbreviated H-bond, is a special case of a dipole-dipole bond because it is significantly stronger than any other dipolar intermolecular bond. It not only forms between molecules of the same kind like water or ammonia, but also between molecules of different kinds, e.g., water and ammonia. The following is a sketch of hydrogen bonding between two water molecules. The intermolecular hydrogen bond appears as a dashed line to differentiate it from the intramolecular covalent bonds of the water molecules.

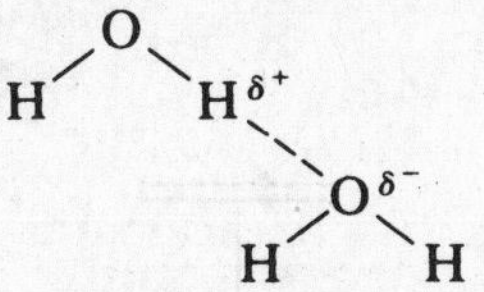

O
H H$^{\delta+}$
O$^{\delta-}$
H H

The hydrogen atom of one of the water molecules with its intense partial positive charge interacts strongly with a lone electron pair of the oxygen atom in the other water molecule. The partial charges on all other atoms are ignored so as to focus on the intermolecular hydrogen-oxygen interaction. The hydrogen bond does not bond hydrogen atoms to each other, but acts as the "glue" between the oxygen, nitrogen, or fluorine atoms in two different molecules. The bond, though strong, is much weaker than a covalent bond and is therefore longer than the covalent bonds. The importance of water in biological systems depends in large part upon its hydrogen-bonding properties. The characteristics of many important biological substances—e.g., proteins, nucleic acids, carbohydrates—are strongly dependent upon their ability to form hydrogen bonds within their own molecules, and external hydrogen bonds with other molecules, particularly water.

5.4 THE GASEOUS STATE

5.4.1 Gas Pressure

The volume of a gas depends upon the amount of the gas (moles), its pressure, and its temperature. Pressure is defined as force/unit area, and is measured by reference to atmospheric pressure. Atmospheric pressure is the result of the weight of the gases surrounding the earth against the surface of the earth. It is measured by means of a barometer which consists of a mercury-filled closed-end tube inverted in a bath of mercury which is exposed to the atmosphere. The pressure of the atmosphere supports a column of mercury about 760 mm high. Atmospheric pressure varies from place to place and in time. However, the standard unit of pressure, called the ***atmosphere***, abbreviated atm, is defined as the pressure that will support a vertical column of mercury to a height of exactly 760 mm at 0 °C. Pressure can be expressed either in atmospheres, mm of mercury (mmHg), or torr. The unit called *torr*, which is equal to 1 mmHg, is in honor of the inventor of the barometer, Torricelli.

Problem 5.1. (*a*) What is the pressure in atmospheres that a gas exerts if it supports a 380 mm column of mercury? (*b*) What is the equivalent of 0.75 atm in millimeters of mercury?

Ans.

(*a*) $$380\ \text{mmHg}\left(\frac{1\ \text{atm}}{760\ \text{mmHg}}\right) = 0.500\ \text{atm}$$

(*b*) $$0.750\ \text{atm}\left(\frac{760\ \text{mmHg}}{1\ \text{atm}}\right) = 570\ \text{mmHg}$$

5.4.2 The Gas Laws

Gases are compressible, expand when heated, and take up room proportional to the amount (moles) of gas present. In order to visualize the mathematical consequences of the physical manipulation of gases, we will use as a model system, a cylinder with a moveable piston sketched in Fig. 5-1, which will illustrate changes in the gas properties as temperature, volume, and mass are varied. Exerting force on the piston handle will compress the gas. If the force on the piston remains constant, and the gas is heated, the piston will rise as the gas expands. If the gas is cooled at constant pressure, it will contract and the piston will fall. If we can add gas to the gas chamber, and both the temperature and pressure are maintained constant, the piston will rise to accommodate the increased amount of gas.

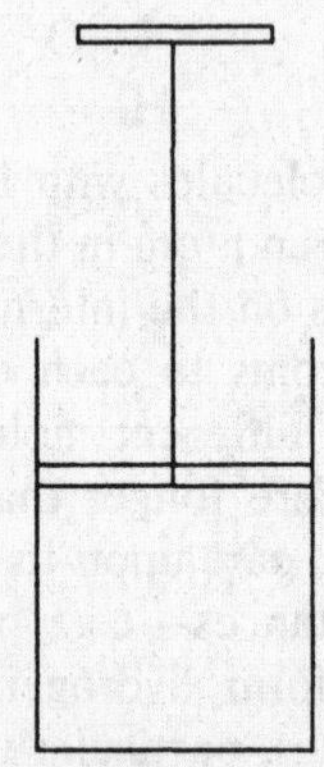

Fig. 5-1 Gas-containing cylinder fitted with moveable piston.

5.4.3 Boyle's Law

The first study of the physical properties of gases was done by Robert Boyle (1627–1691) some 300 years ago. He studied the variation in volume of a gas with changes in pressure. He found that the product of the pressure and volume was a constant at constant temperature. In Fig. 5-2, the pressure of a gas confined in a cylinder with a moveable piston is plotted versus its volume at three different

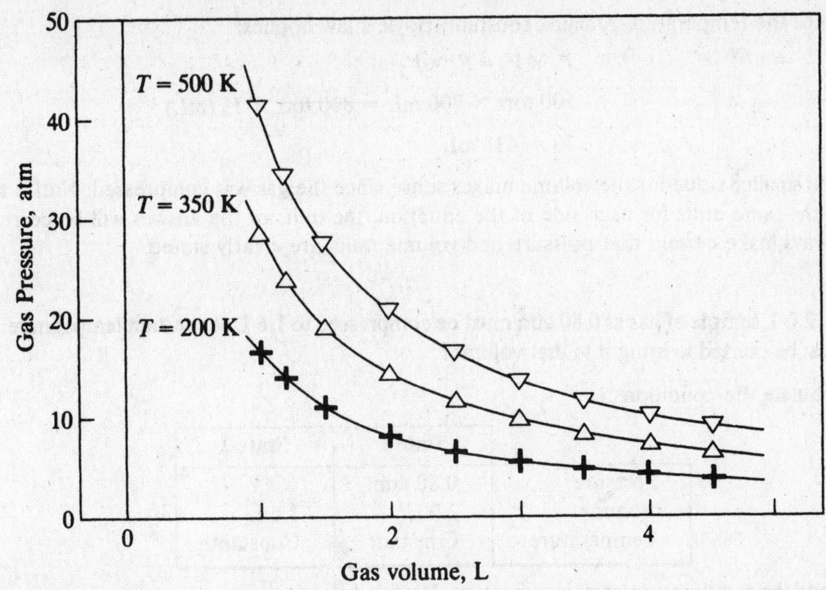

Fig. 5-2 Illustration of Boyle's law. Variation of gas pressure with change in volume, at three temperatures.

temperatures. As the pressure increases, the volume decreases. This inverse relationship is expressed mathematically as

$$P \times V = \text{constant}$$

The typical results of variation of volume of a fixed amount of gas with variation of pressure at 0 °C are found in Table 5-1. There are two things to note: (1) The PV product is a constant and (2) the constant has units which reflect the choice of pressure and volume units. If we compare the condition of the gas at state 1 to that of the gas at state 2, then

$$P_1 \times V_1 = 2.0\,\text{L} \cdot \text{atm} = P_2 \times V_2$$

or, since both are equal to the same constant, Boyle's law at constant temperature becomes

$$P_1 \times V_1 = P_2 \times V_2$$

Table 5-1. Variation of Pressure and Volume at Constant Temperature

State	Pressure, atm	Volume, L	$P \times V$, L · atm
1.	5.3	0.38	2.0
2.	2.7	0.75	2.0
3.	1.3	1.6	2.0
4.	0.67	3.0	2.0

Problem 5.2. A 700.0-mL sample of gas at 500.0 torr pressure is compressed at constant temperature until its final pressure is 800.0 torr. What is its final volume?

Ans. Tabulate the conditions:

	State 1	State 2
Pressure	500.0 torr	800.0 torr
Volume	700.0 mL	?
Temperature	Constant	Constant

Since the temperature remains constant, Boyle's law applies:

$$P_1 \times V_1 = P_2 \times V_2$$

$$500 \text{ torr} \times 700 \text{ mL} = 800 \text{ torr} \times V_2 \text{ (mL)}$$

$$V_2 = 438 \text{ mL}$$

The smaller value for the volume makes sense since the gas was compressed. Notice that if we hold to the same units for each side of the equation, the units of the answer will be correct. Therefore, always make certain that pressure and volume units are clearly stated.

Problem 5.3. **A 2.0-L sample of gas at 0.80 atm must be compressed to 1.6 L at constant temperature. What pressure in millimeters must be exerted to bring it to that volume?**

Ans. Tabulate the conditions:

	State 1	State 2
Pressure	0.80 atm	?
Volume	2.0 L	1.6 L
Temperature	Constant	Constant

Since the temperature remains constant, Boyle's law applies:

$$P_1 \times V_1 = P_2 \times V_2$$

$$0.80 \text{ atm} \times 2.0 \text{ L} = P_2\text{(atm)} \times 1.6 \text{ L}$$

$$P_2 = 1.0 \text{ atm}$$

Here again, the pressure must be increased to make the volume smaller, and although the units are different from those in Problem 5.2, they are the same on both sides of the equation, and the units of the answer are correct.

5.4.4 Charles' Law

Boyle's law applies to changes in volume with changes in pressure only when the temperature remains constant. Gas volume increases when the temperature is raised. The quantitative relationship between gas volume and temperature with the pressure held constant was first studied by J. Charles. He showed that there is a linear relationship between gas volume and Celsius temperature. The variation of volume of the same amount of a gas with changes in temperature in Celsius degrees at three different pressures is plotted in Fig. 5-3. The significant discovery made obvious with this treatment of the data was that extrapolation of the data for all pressures intersected the temperature axis at −273.15 °C, which corresponds to a volume of zero. To simplify our work, we will use the value

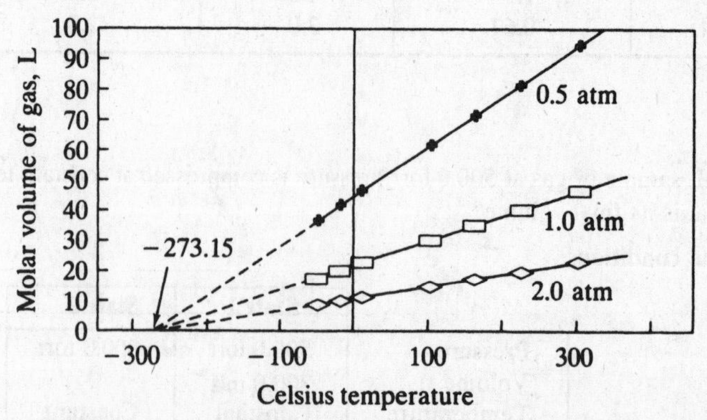

Fig. 5-3 Illustration of Charles' law. Variation of gas volume with change in temperature, at three pressures.

of −273 °C. By adding 273° to the Celsius scale, a new absolute temperature scale was created—the now familiar Kelvin scale. Charles' law now becomes

$$V = k(t + 273)$$

where k is a proportionality constant which depends upon the pressure and the amount of gas. Since $(t + 273) = T$, the final form of Charles' law is $V = kT$. This form of the law clearly tells us for example, that if the Kelvin temperature is doubled, the volume is doubled, and when it is halved, the volume is also halved.

As was the case with Boyle's law, the constant k takes the same value at all temperatures if the quantity of gas and its pressure remain constant. Therefore, we can compare the properties of a sample of gas at two different temperatures by noting that

$$\frac{V_1}{T_1} = k = \frac{V_2}{T_2}$$

and that since the ratios of volume to Kelvin temperature are both equal to the same constant, they are then equal to each other.

Problem 5.4. A 500-mL sample of gas in a cylinder with a moveable piston at 0.00 °C is heated at a constant pressure to 41.0 °C. What is its final volume?

Ans. Tabulate the conditions:

	State 1	State 2
Pressure	Constant	Constant
Volume	500 mL	?
Celsius temperature	0.00 °C	41.0 °C
Kelvin temperature	273 K	314 K

The pressure is constant, so Charles' law applies:

$$\frac{V_1}{T_1} = \frac{V_2}{T_2}$$

It is necessary to convert Celsius temperature to kelvins:

$$\frac{500 \text{ mL}}{273 \text{ K}} = \frac{V_2 \text{ mL}}{314 \text{ K}}$$

and

$$V_2 = \left(\frac{314 \text{ K}}{273 \text{ K}}\right) 500 \text{ mL} = 575 \text{ mL}$$

Common sense tells us that, if the temperature increased, the volume had to increase. Therefore, you can see that we had to multiply the initial volume by a fraction formed by the initial and final temperatures which is greater than 1.0, i.e., 314/273.

Problem 5.5. A 1.0-L sample of gas at 27 °C is heated so that it expands at constant temperature to a final volume of 1.5 L. What is its final temperature?

Ans. Tabulate the conditions.

	State 1	State 2
Pressure	Constant	Constant
Volume	1.0 L	1.5 L
Celsius temperature	27 °C	? °C
Kelvin temperature	300 K	? K

The pressure is constant, so Charles' law applies:

$$\frac{V_1}{T_1} = \frac{V_2}{T_2}$$

It is necessary to convert Celsius temperature to kelvins:

$$\frac{1.0\ \text{L}}{300\ \text{K}} = \frac{1.5\ \text{L}}{T_2\ \text{K}}$$

$$T_2 = \left(\frac{1.5\ \text{L}}{1.0\ \text{L}}\right) 300\ \text{K} = 450\ \text{K}$$

Common sense tells us that if the volume increased, the temperature had to increase. Therefore we multiplied the initial temperature by a fraction formed by the initial and final volumes which is greater than 1.0, i.e., 1.5/1.0.

5.4.5 Combined Gas Laws

Boyle's and Charles' laws can be written as proportionalities:

$$V \propto \frac{1}{P} \qquad \text{and} \qquad V \propto T$$

These read, respectively: "Gas volume is inversely proportional to pressure," and "Gas volume is directly proportional to the absolute temperature." The two relationships can be combined into one as

$$V \propto \frac{T}{P}$$

Then,

$$\frac{PV}{T} = \text{constant} = k \qquad \text{or} \qquad PV = kT$$

Regardless of the state of the gas, k remains constant if the amount of gas remains constant. So as we did with both Boyle's and Charles' laws, we can write the equation to reflect two different states of the gas:

$$\frac{P_1V_1}{T_1} = k = \frac{P_2V_2}{T_2}$$

or

$$\frac{P_1V_1}{T_1} = \frac{P_2V_2}{T_2}$$

This allows us to calculate the new value of one of the variables, for example V, when both T and P are changed.

Problem 5.6. A 1.2-L sample of gas at 27 °C and 1.0 atm pressure is heated to 177 °C and 1.5 atm final pressure. What is its final volume?

Ans. Tabulate the conditions:

	State 1	State 2
Pressure	1.0 atm	1.5 atm
Volume	1.2 L	? L
Temperature	27 °C	177 °C
Temperature	300 K	450 K

We can use the combined gas law

$$\frac{P_1V_1}{T_1} = \frac{P_2V_2}{T_2}$$

and solve it for V_2:

$$V_2 = V_1\left(\frac{T_2}{T_1}\right)\left(\frac{P_1}{P_2}\right)$$

substituting appropriate values:

$$V_2 = 1.2\text{ L}\left(\frac{450\text{ K}}{300\text{ K}}\right)\left(\frac{1.0\text{ atm}}{1.5\text{ atm}}\right) = 1.2\text{ L}$$

Notice that increasing the temperature should increase the volume, but the increase in pressure was just enough to keep the volume constant.

5.4.6 The Ideal Gas Law

In Chap. 4 we introduced the concept of the mole, the mass of a substance which contains the same number of fundamental units as a mole of any other substance. The original statement of that idea is called Avogadro's law, and applied to gases in which the fundamental units are molecules, it states: Equal volumes of gases at the same pressure and temperature have the same number of molecules or moles, designated n. This adds a third gas law to the previous two:

$$V = k_A n$$

We can combine all three laws as we previously did with Boyle's and Charles' laws to give

$$\frac{PV}{T} = nR$$

or as it is more commonly written:

$$PV = nRT.$$

This is known as the *ideal gas law*, called ideal because, under conditions of low pressure and high temperature, all gases obey it regardless of types or mass of molecule. It is universally applicable. You will find it necessary to use this law in gas problems which involve moles or masses of gas. To use this equation, it is necessary to evaluate the ideal gas constant R. Experimentally it has been found that at 1.0 atm, and 273 K, (standard temperature and pressure or STP), 1.0 mol of a gas occupies 22.4 L. Solving the equation for R, and inserting those values:

$$R = \left(\frac{1.00\text{ atm}}{1.00\text{ mol}}\right)\left(\frac{22.4\text{ L}}{273\text{ K}}\right) = 0.0821\ \frac{\text{L}\cdot\text{atm}}{\text{K}\cdot\text{mol}}$$

To avoid using the units of R as a fraction, we will use it as follows: $R = 0.0821\text{ L}\cdot\text{atm}\cdot\text{K}^{-1}\cdot\text{mol}^{-1}$. Note carefully the units of R: pressure is in atmospheres, temperature in kelvins, volume in liters, and n in moles. To use this equation for solving gas law problems you must pay strict attention to all units.

Problem 5.7. Inhaled air contains O_2 at a pressure of 160 torr. The capacity of a human lung is about 3.0 L. Body temperature is 37 °C. Calculate the mass of O_2 filling a lung upon inspiration.

Ans. We are given P, T, and V, and are required to find mass, so the ideal gas law must be used. However, we must convert pressure from torr to atmospheres, and Celsius degrees to kelvins before it can be used:

$$160\text{ torr}\left(\frac{1.0\text{ atm}}{760\text{ torr}}\right) = 0.21\text{ atm}$$

$$37 + 273 = 310\text{ K}$$

Solving the ideal gas law for n:

$$n = \frac{PV}{RT} = \frac{0.21\text{ atm} \times 3.0\text{ L}}{0.0821\text{ L}\cdot\text{atm}\cdot\text{mol}^{-1}\cdot\text{K}^{-1} \times 310\text{ K}} = 0.025\text{ mol}$$

Finally,

$$\text{grams } O_2 = 0.025\text{ mol}\left(\frac{32.0\text{ g } O_2}{1.0\text{ mol } O_2}\right) = 0.80\text{ g}$$

5.4.7 The Ideal Gas Law and Molar Mass

The ideal gas law provided the first method for determining molar masses. The number of moles in the ideal gas law, n, is equal to the mass in grams divided by the molecular mass, M:

$$n = \frac{PV}{RT} = \frac{g}{M}$$

Solving for M:

$$M = \frac{gRT}{PV}$$

Problem 5.8. The density of CO_2 at 25 °C and 1.0 atm is 1.80 g/L. Calculate its molar mass.

Ans. The density of the gas is the factor g/V in the above equation. Thus

$$M = \frac{1.80\text{ g} \times 0.0821\text{ L}\cdot\text{atm}\cdot\text{K}^{-1}\cdot\text{mol}^{-1} \times 298\text{ K}}{1.0\text{ atm} \times 1.0\text{ L}}$$

$$= 44\text{ g}\cdot\text{mol}^{-1}$$

5.4.8 Dalton's Law of Partial Pressures

The gas laws tell us that when one doubles the amount of a gas in a fixed volume, the pressure doubles, and at a fixed pressure, the volume doubles. Since all gases obey the ideal gas law, there is no difference between doubling the amount of a gas, or adding an equal amount of another gas. Dalton's law of partial pressures states that the total pressure of a mixture of gases is equal to the sum of the individual pressures of all gases in the mixture. These individual pressures are called *partial pressures*. It is the pressure each gas would exert if it were present alone instead of in a mixture. In the previous problem the pressure of O_2 in air was specified as 160 torr. Since air is principally a mixture of O_2 and N_2, 160 torr is the partial pressure of O_2.

Problem 5.9. What is the partial pressure of N_2 in air?

Ans. Dalton's law of partial pressures can be written

$$P_{AIR} = P_{O_2} + P_{N_2}$$

Since atmospheric pressure (pressure of air) is about 760 torr, we rearrange the equation and solve for P_{N_2}:

$$P_{N_2} = 760\text{ torr} - 160\text{ torr} = 600\text{ torr}$$

Problem 5.10. Gases are collected in the laboratory by bubbling them into an inverted bottle filled with water, which is displaced by the entering gas. The experimental arrangement assures that the trapped gas is present at atmospheric pressure. However, since liquid water is present, some water in the form of gas is also present along with the collection of gas. If the atmospheric pressure is 745 torr and the partial pressure of water is 23.0 torr, what is the pressure of a gas collected under those conditions?

Ans.

$$P_{\text{ATMOSPHERE}} = P_{\text{GAS}} + P_{\text{WATER}}$$

$$745 \text{ torr} = P_{\text{GAS}} + 23.0 \text{ torr}$$

$$P_{\text{GAS}} = 745 \text{ torr} - 23.0 \text{ torr} = 722 \text{ torr}$$

Problem 5.11. The partial pressure of carbon dioxide in exhaled air is 28 torr. If the total pressure is 760 torr, the temperature 37 °C, and the volume of a lung is 3.0 L, what is the mass of CO_2 exhaled by both lungs?

Ans. Remember, the partial pressure is the pressure the CO_2 would exert if it were present alone. Therefore we will use the ideal gas law with the given variables to determine the mass of CO_2. However, we must convert pressure from torr to atmospheres, and from Celsius degrees to kelvins, before it can be used:

$$28 \text{ torr}\left(\frac{1.0 \text{ atm}}{760 \text{ torr}}\right) = 0.037 \text{ atm}$$

$$37 + 273 = 310 \text{ K}$$

There is also the capacity of 2 lungs to consider; therefore:

$$n = \frac{PV}{RT} = \frac{0.037 \text{ atm} \times 3.0 \text{ L} \times 2}{0.0821 \text{ L} \cdot \text{atm} \cdot \text{mol}^{-1} \cdot \text{K}^{-1} \times 310 \text{ K}} = 0.0087 \text{ mol}$$

Finally

$$0.0087 \text{ mol } CO_2\left(\frac{44 \text{ g } CO_2}{1 \text{ mol } CO_2}\right) = 0.38 \text{ g } CO_2$$

5.5 LIQUIDS

The cohesive forces which hold molecules together in condensed states are electrical in nature, and are effective only over very small distances, perhaps 1 or 2 molecular diameters. When molecules are in the gaseous state, they are so far apart that cohesive forces have no effect on the properties of the gas. Even if two molecules were to collide, or approach very closely, their kinetic energy would be large enough to overcome any attractions, so that they would rebound and go on their separate ways. However, if the temperature, which is a measure of the gas' kinetic energy, were continuously lowered, a point would be reached where the cohesive forces could overcome the kinetic forces, and the gas would condense to a liquid.

5.5.1 Liquids and Vapor Pressure

Referring to Fig. 5-1, the picture we would see when condensation occurs is that some liquid would be present along with some of the gas phase. Whenever a substance is present in the liquid state, and some of it is also present in the gaseous state, that gaseous state is known as its *vapor*. Since vapors are gases, their properties and behavior are properly described by the gas laws.

One of the principal methods for the characterization of liquids is through the study of vapor pressure. A liquid's vapor pressure increases as its temperature increases. When the liquid is in a vessel open to the atmosphere and its vapor pressure is equal to atmospheric pressure, it boils. The temperature at which that occurs is called the *normal boiling point*. Differences in boiling points among liquids reflect differences in the cohesive forces operating among the molecules of the liquids. Some molecules such as water have a very strong cohesive character. For example, the normal boiling point of H_2O is 100 °C compared to −61 °C for its sister molecule H_2S. The difference is due to the presence of hydrogen bonding in water which is absent in H_2S. Other gases like H_2S, e.g., methane,

Table 5-2. Heats of Vaporization and Boiling Points of Several Compounds

Compound		Heat of vaporization, kcal / mol	Normal boiling point, °C
Water	H_2O	9.72	100.0
Ethyl alcohol	C_2H_5OH	9.22	78.7
Sulfur dioxide	SO_2	5.96	−9.87
Methane	CH_4	1.96	−161.3
Hydrogen	H_2	0.22	−252.6
Helium	He	0.02	−268.8

have a very weak cohesive character. Nevertheless, all gases will condense to the liquid state under the appropriate conditions of temperature and pressure. A quantitative measure of a liquid's cohesive character is the amount of heat required to vaporize a mole of the liquid at its normal boiling point. This quantity is called the *molar heat of vaporization*. The heat evolved when a mole of a gas condenses to a liquid is identical to the molar heat of vaporization. The normal boiling point is a good qualitative indication of the cohesive forces within the liquid. Table 5-2 lists some representative heats of vaporization, along with their normal boiling points.

Problem 5.12. What is the normal boiling point of a liquid?

Ans. The normal boiling point of a liquid is the temperature at which its vapor pressure is equal to atmospheric pressure. The vessel containing the liquid may be open to the atmosphere, in which case the atmosphere acts as the piston in a cylinder. The vessel containing the liquid may be closed, in which case the internal gas pressure must be equal to the atmospheric pressure if it is to boil at its normal boiling point.

Problem 5.13. Can a liquid boil at any pressure?

Ans. Yes. The boiling point of a liquid is defined as the temperature at which its vapor pressure is equal to the pressure external to the liquid. The vapor pressure of water at 25 °C is 23.8 torr. If water is contained in a closed vessel, and the gas pressure in the vessel is reduced to 23.8 torr, the water will boil at 25 °C.

Problem 5.14. A sample of gaseous water is contained in a closed vessel at 125° and 1.0 atm. What will happen if the temperature is reduced to 100 °C?

Ans. At 1.0 atm, the boiling point of water is 100 °C. Therefore at its normal boiling point, gaseous water will condense to form a two-phase system of liquid and vapor. This temperature is called the *dew point*, and is identical to the normal boiling point. Evaporation and condensation occur simultaneously at the same temperature.

A very important aspect of the vapor pressure of a liquid is that it generally decreases when mixed with other substances to form solutions. This effect will be discussed in detail in Chap. 7, "Solutions."

5.5.2 Viscosity of Liquids

Viscosity is a property of liquids which describes their resistance to flow. Glycerol is a more viscous fluid than water, which is more viscous than acetone (nail polish remover). Viscosity is quantified by a viscosity coefficient, η. Table 5-3 lists some representative values measured at 20 °C. Glycerol, whose viscosity is about the same as maple syrup, is over 1000 times more viscous than

Table 5-3. Viscosity Coefficients of Several Compounds

Compound	η, Viscosity Coefficient, $g \cdot cm^{-1} \cdot s^{-1}$
Glycerol	10.690
Water	0.0101
Acetone	0.0033

water. It contains 3 OH groups/molecule, compared with water with 1 OH group/molecule, and acetone with none. The trend in viscosities reflects the H-bonding capacities of these liquids.

5.5.3 Surface Tension

Just as the properties of molecules in a liquid differ from the properties of molecules in a gas, so do the properties of molecules at a surface between a liquid and a gas. That boundary is called the *interphase*. To increase the surface area of a liquid sample, or in other words, to bring molecules from the interior of a liquid to the surface, requires work to overcome the cohesive force operating between the molecules in the surface. This force is called the *surface tension*. Surface tension opposes the expansion of the surface of the liquid. The shape of any object which minimizes the amount of surface for a given volume is a sphere. Surface tension forces a droplet of any liquid small enough to escape the influence of gravity to assume the shape of a sphere. Table 5-4 lists the surface tension, given the symbol γ, of several liquids at 20 °C. The units of γ are force per unit length compared with the units of pressure, force per unit area.

Table 5-4. Surface Tensions of Some Representative Compounds

Compound	γ, Surface Tension, $dyne \cdot cm^{-1}$
Water, H_2O	72.75
Chloroform, $CHCl_3$	27.14
Acetone, CH_3COCH_3	23.7

*A dyne is a unit of force, so that $dyne \cdot cm^{-1}$ should be read as force per unit length.

Water possesses the largest cohesive forces and consequently, the largest surface tension of the commonly encountered liquids. The molecules which form biological membranes have unique properties which allow them to reduce the surface tension at a cell's aqueous interface, and thereby increase the surface area of the cell and its internal organs, e.g., the Golgi apparatus.

5.6 SOLIDS

Molecules (or ions) in a solid occupy fixed positions in space. In a crystalline solid, the distance between each molecule is the same. If you could stand on a molecule in a crystalline solid, and jumped some distance which was an integral multiple of the distance between two molecules, you would land on an identical distant molecule. This uniform spatial array is called a *crystalline lattice*. When a solid melts, the molecules remain almost as close to each other as in the solid, but the uniformity of the crystal lattice is destroyed. The molecules are no longer restricted to fixed positions in space, but can now move about freely.

One of the most useful methods for characterizing the cohesive forces in solids is the determination of the amount of energy in the form of heat necessary to melt them, and the temperature at which melting occurs. These are precise analogues of the heats of vaporization and the normal boiling

Table 5-5. Melting Points and Heats of Fusion of Several Compounds

Compound	Melting point, °C	Heat of fusion, kJ / mol
Potassium chloride	772.3	23.3
Glycerol	18	18.3
Water	0	6.0
Acetone	−95	5.2

points of liquids, and are called the *heat of fusion*, and the *melting point*, respectively. The melting point and the freezing point are identical, and the amount of heat necessary to melt a solid is equal to the amount of heat evolved when a solid freezes. Table 5-5 lists several compounds along with their melting (freezing) points and heats of fusion. The crystalline forces within solid KCl are ionic chemical bonds, so the melting point is correspondingly quite high. The trends in melting points and heats of fusion reflect the strength of the cohesive forces within the solids.

Many solids can evaporate directly to the gaseous state without melting to an intermediate liquid state. Examples are solid CO_2 (dry ice), and solid I_2, which is always present in its container with some purple vapor. The direct transition from solid to gas is called *sublimation*. The energy required for this transition is the sum of the energies for melting and evaporation:

$$\Delta H_{\text{SUBLIMATION}} = \Delta H_{\text{FUSION}} + \Delta H_{\text{VAPORIZATION}}$$

ADDITIONAL SOLVED PROBLEMS

KINETIC-MOLECULAR THEORY

Problem 5.15. How does the kinetic-molecular theory account for gas pressure?

Ans. According to the theory, gas molecules are in constant random motion, and change direction only upon collision with each other or the walls of their container. In their collisions with container walls, they exert a force on the wall. The total force of these wall collisions per unit time is the origin of gas pressure.

Problem 5.16. How does the kinetic-molecular theory account for the evaporation of a liquid?

Ans. Evaporation occurs when sufficient energy is accumulated by a molecule near the surface of the liquid through collisions with other nearby molecules so that the cohesive forces within the liquid can be overcome, and the molecule can then escape into the gaseous phase.

Problem 5.17. How does the kinetic-molecular theory account for the melting of a solid?

Ans. The molecules of a solid are held in place in an orderly lattice by strong cohesive forces. By increasing the kinetic energy of the molecules, the cohesive forces can be overcome, and this allows the molecules to escape from their lattice positions to move about more freely in the liquid state.

COHESIVE FORCES

Problem 5.18. Describe the condition of matter if there were no cohesive forces operating between molecules.

Ans. All matter would exist only in the gaseous state. No matter would be present in either liquid or solid form.

Problem 5.19. How can molecules in the gaseous state condense to a liquid, if their molecules are not polar?

Ans. Even though molecules are not polar, transient and temporary dipoles form in all molecules. These are called *dispersion forces*, and though weak, lead to condensation of nonpolar molecules.

Problem 5.20. Compare the solubilities of ethyl alcohol and methane in water. Their structures are

```
            H  H
            |  |
Ethanol:  H—C—C—O—H
            |  |
            H  H

            H
            |
Methane:  H—C—H
            |
            H
```

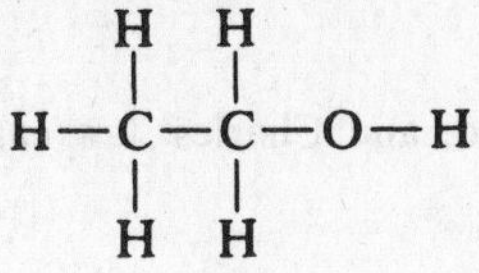

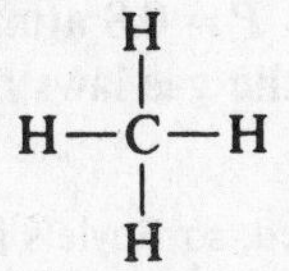

Ans. The OH group in the alcohol allows it to form a hydrogen bond with water, and this weak bond formation leads to solubility. Methane's structure does not permit H-bond formation. Ethyl alcohol is soluble in water, but methane is not.

THE GASEOUS STATE

Problem 5.21. Convert 735 torr to (*a*) mmHg, and (*b*) atm.

Ans.

(*a*) $$735 \text{ torr} = 735 \text{ mmHg}$$

(*b*) $$735 \text{ torr}\left(\frac{1 \text{ atm}}{760 \text{ torr}}\right) = 0.967 \text{ atm}$$

Problem 5.22. Convert 160 mmHg to (*a*) torr, and (*b*) atm.

Ans.

(*a*) $$160 \text{ mmHg} = 160 \text{ torr}$$

(*b*) $$160 \text{ torr}\left(\frac{1 \text{ atm}}{760 \text{ torr}}\right) = 0.211 \text{ atm}$$

Problem 5.23. Convert (*a*) 1.20 atm to torr, (*b*) 0.450 atm to mmHg, (*c*) 850 mmHg to atm.

Ans.

(*a*) $$1.20 \text{ atm}\left(\frac{760 \text{ mmHg}}{1 \text{ atm}}\right) = 912 \text{ mmHg}$$

$$912 \text{ mmHg} = 912 \text{ torr}$$

(*b*) $$0.450 \text{ atm}\left(\frac{760 \text{ mmHg}}{1 \text{ atm}}\right) = 342 \text{ mmHg}$$

(*c*) $$850 \text{ mmHg}\left(\frac{1 \text{ atm}}{760 \text{ mmHg}}\right) = 1.12 \text{ atm}$$

Problem 5.24. The initial state of a gas was $P = 0.8$ atm, $t = 25$ °C, $V = 1.2$ L. The conditions were changed to $P = 1.0$ atm, $t = 50$ °C, and $V = ?$ L. Can Boyle's law be used to find the new volume?

Ans. No. The temperature was changed, and Boyle's law can only be used when the temperature remains constant.

Problem 5.25. The initial state of a gas was $P = 0.8$ atm, $t = 25$ °C, $V = 1.2$ L. The conditions were changed to $P = 1.0$ atm, $t = ?$ °C, and $V = 2.0$ L. Can Charles' law be used to find the new temperature?

Ans. No. The pressure was changed, and Charles' law can only be used when the pressure remains constant.

Problem 5.26. The initial state of a gas was $P = 0.8$ atm, $t = 25$ °C, $V = 1.2$ L. If the conditions were changed to $P = 1.6$ atm, $t = 25$ °C, which of the gas laws must be used to solve this problem, and what is the final volume?

Ans. The temperature remains unchanged, so Boyle's law may be used to calculate the final volume:

(1) $$P_1 \times V_1 = P_2 \times V_2$$

(2) $$0.8 \text{ atm} \times 1.2 \text{ L} = 1.6 \text{ atm} \times V_2(\text{L})$$

(3) $$V_2 = \frac{0.8 \text{ atm} \times 1.2 \text{ L}}{1.6 \text{ atm}} = 0.6 \text{ L}$$

Problem 5.27. The initial state of a gas was $P = 1.5$ atm, $t = 25$ °C, $V = 1.2$ L. The conditions were changed to $t = 50$ °C and $P = 1.5$ atm. Which of the gas laws must be used to solve this problem, and what is the final volume?

Ans. The pressure remains unchanged, so Charles' law may be used to calculate the final volume:

$$\frac{V_1}{V_2} = \frac{T_1}{T_2}$$

$$T_1 = 273 + 25 = 298 \text{ K}$$

$$T_2 = 273 + 50 = 323 \text{ K}$$

$$\frac{1.2 \text{ L}}{V_2} = \frac{298 \text{ K}}{323 \text{ K}}$$

$$V_2 = 1.2 \text{ L}\left(\frac{323 \text{ K}}{298 \text{ K}}\right) = 1.3 \text{ L}$$

Problem 5.28. The volume of a 3.6-L sample of gas at 27 °C is reduced at constant pressure to 3.0 L. What is its final temperature?

Ans. Solve this problem using Charles' law since the pressure remains constant:

$$\frac{T_2}{T_1} = \frac{V_2}{V_1}$$

$$T_2 = T_1\left(\frac{V_2}{V_1}\right) = 300 \text{ K}\left(\frac{3.0 \text{ L}}{3.6 \text{ L}}\right) = 250 \text{ K} = -23° \text{ C}$$

Problem 5.29. The initial state of a gas was $P = 1.25$ atm, $t = 27$ °C, $V = 1.5$ L. The conditions were changed to $V = 1.0$ L, $t = 77$ °C. Which of the gas laws must be used to solve this problem, and what is the final pressure?

Ans. Two of the three variables have been changed, so we must use the combined gas law to calculate the final pressure:

$$\frac{P_1V_1}{T_1} = \frac{P_2V_2}{T_2}$$

$$T_1 = 273 + 27 = 300 \text{ K}$$

$$T_2 = 273 + 77 = 350 \text{ K}$$

$$P_2 = P_1\left(\frac{T_2}{T_1}\right)\left(\frac{V_1}{V_2}\right)$$

$$P_2 = 1.25 \text{ atm}\left(\frac{350 \text{ K}}{300 \text{ K}}\right)\left(\frac{1.5 \text{ L}}{1.0 \text{ L}}\right) = 2.2 \text{ atm}$$

Problem 5.30. The pressure on a 3.00 L sample of gas is doubled, and its absolute temperature (K) is increased by 50 percent. What is its final volume?

Ans. P_2 was changed to $(2 \times P_1)$, and T_1 was changed to $(1.5 \times T_1)$, so we can use those values as factors in the combined gas law to calculate the final volume:

$$V_2 = V_1\left(\frac{P_1}{P_2}\right)\left(\frac{T_2}{T_1}\right)$$

$$= 3.00 \text{ L}\left(\frac{P_1}{2 \times P_1}\right)\left(\frac{1.5 \times T_1}{T_1}\right) = 2.3 \text{ L}$$

Problem 5.31. A sample of gas at 0.83 atm and 25 °C has a volume of 2.0 L. What will be its volume at STP?

Ans. STP means standard temperature and pressure, i.e., 0 °C and 1.0 atm pressure. We can use the combined gas law to calculate the new volume:

$$V_2 = 2.0 \text{ L}\left(\frac{0.83 \text{ atm}}{1.0 \text{ atm}}\right)\left(\frac{273 \text{ K}}{298 \text{ K}}\right) = 1.5 \text{ L}$$

Problem 5.32. Ammonia gas occupies a volume of 5.0 L at 4 °C and 760 torr. Find its volume at 77 °C and 800 torr.

Ans.

$$V_2 = 5.0 \text{ L}\left(\frac{760 \text{ torr}}{800 \text{ torr}}\right)\left(\frac{350 \text{ K}}{277 \text{ K}}\right) = 6.0 \text{ L}$$

Why was it not necessary to convert torr to atmospheres?

Problem 5.33. A helium weather balloon filled at ground level and atmospheric pressure of 1.0 atm contains 50 L of gas at 20°C. When it rises to the stratosphere, the external pressure will be 0.40 atm, and the temperature will be −50 °C. What should the capacity of the balloon be at ground level?

Ans.

$$V_2 = 50.0 \text{ L}\left(\frac{1.0 \text{ atm}}{0.40 \text{ atm}}\right)\left(\frac{223 \text{ K}}{293 \text{ K}}\right) = 95 \text{ L}$$

The helium will occupy over 95 L in the stratosphere, so the capacity of the balloon must be about 100 L to be on the safe side. You may have noticed how limp stratospheric balloons appear as they are released from ground level.

Problem 5.34. Calculate the value of R at STP when volume is in milliliters and pressure is in millimeters of Hg.

Ans.

$$R = \frac{PV}{nT} = \frac{760 \text{ mmHg} \times 22400 \text{ mL}}{1.00 \text{ mol} \times 273 \text{ K}} = 6.24 \times 10^4 \text{mL} \cdot \text{mmHg} \cdot \text{K}^{-1} \cdot \text{mol}^{-1}$$

Problem 5.35. Calculate the value of R at STP when the volume is in milliliters and the pressure is in atmospheres.

Ans.

$$R = \frac{PV}{nT} = \frac{1.00 \text{ atm} \times 22400 \text{ mL}}{1.00 \text{ mol} \times 273 \text{ K}} = 82.1 \text{ mL} \cdot \text{atm} \cdot \text{K}^{-1} \cdot \text{mol}^{-1}$$

Problem 5.36. Fill in the correct values in the empty places in the following table:

	P, atm	*V*, L	*T*, K	*n*, mol
(*a*)	0.800	___	300	0.900
(*b*)	___	1.6	225	2.0
(*c*)	0.900	15.0	___	0.500
(*d*)	4.0	10.0	490	___

Ans. These must all be solved using the ideal gas law. Simply rearrange the equation for each problem, e.g., $P = nRT/V$, $V = nRT/P$, etc.:

(*a*) $$V = \frac{nRT}{P} = \frac{0.900 \text{ mol} \times 0.0821 \text{ L} \cdot \text{atm} \cdot \text{mol}^{-1} \cdot \text{K}^{-1} \times 300 \text{ K}}{0.800 \text{ atm}} = 27.7 \text{ L}$$

(*b*) $$P = \frac{nRT}{V} = \frac{2.0 \text{ mol} \times 0.0821 \text{ L} \cdot \text{atm} \cdot \text{mol}^{-1} \cdot \text{K}^{-1} \times 225 \text{ K}}{1.6 \text{ L}} = 23 \text{ atm}$$

(*c*) $$T = \frac{PV}{nR} = \frac{0.900 \text{ atm} \times 15.0 \text{ L}}{0.500 \text{ mol} \times 0.0821 \text{ L} \cdot \text{atm} \cdot \text{mol}^{-1} \cdot \text{K}^{-1}} = 329 \text{ K}$$

(*d*) $$n = \frac{PV}{RT} = \frac{4.0 \text{ atm} \times 10.0 \text{ L}}{490 \text{ K} \times 0.0821 \text{ L} \cdot \text{atm} \cdot \text{mol}^{-1} \cdot \text{K}^{-1}} = 1.0 \text{ mol}$$

Problem 5.37. Calculate the volume occupied by 8.00 g of oxygen at STP.

Ans. The ideal gas law is not required since the volume occupied by a mole of any gas at STP is 22.4 L. However, the mass of oxygen must be first converted to moles. Oxygen is a diatomic molecule whose molar mass is 32.0 g/mol; therefore, 8.00 g represents 0.25 mol. The volume occupied by 8.00 g of oxygen at STP is

$$0.25 \text{ mol}\left(\frac{22.4 \text{ L}}{\text{mol}}\right) = 5.6 \text{ L}$$

Problem 5.38. Compare the volumes occupied by 4.0 g of helium and 44.0 g of carbon dioxide at STP.

Ans. Since the volume occupied by 1 mol of any two gases at STP is 22.4 L/mol, the ratio of moles will be equal to the ratio of volumes:

$$4.0 \text{ g He}\left(\frac{1 \text{ mol He}}{4.0 \text{ g He}}\right) = 1.0 \text{ mol He}$$

$$44.0 \text{ g } CO_2\left(\frac{1 \text{ mol } CO_2}{44.0 \text{ g } CO_2}\right) = 1.0 \text{ mol } CO_2$$

Therefore, the volumes are the same. If the amount of CO_2 had been 22.0 g, the ratio of volumes of He/CO_2 would have been 2/1.

Problem 5.39. The volumes of hydrogen and ammonia at STP are 11.2 L and 5.6 L, respectively. What masses are contained in those volumes?

Ans. We need to calculate the number of moles of those gases, and then the masses:

$$11.2 \text{ L } H_2\left(\frac{1 \text{ mol}}{22.4 \text{ L}}\right) = 0.500 \text{ mol } H_2$$

$$5.6 \text{ L } NH_3\left(\frac{1 \text{ mol}}{22.4 \text{ L}}\right) = 0.25 \text{ mol } NH_3$$

$$0.500 \text{ mol } H_2\left(\frac{2.0 \text{ g}}{1 \text{ mol } H_2}\right) = 1.0 \text{ g } H_2$$

$$0.25 \text{ mol } NH_3\left(\frac{17.0 \text{ g } NH_3}{1 \text{ mol } NH_3}\right) = 4.3 \text{ g } NH_3$$

Problem 5.40. A 6.13 g sample of a hydrocarbon occupied 5.0 L at 25 °C and 1.0 atm. Calculate its molar mass.

Ans. The molar mass of a gas is calculated from the rearranged ideal gas law:

$$M = \frac{gRT}{PV}$$

$$= \frac{6.13 \text{ g} \times 0.0821 \text{ L} \cdot \text{atm} \cdot \text{K}^{-1} \cdot \text{mol}^{-1} \times 298 \text{ K}}{1.0 \text{ atm} \times 5.0 \text{ L}} = 30 \text{ g/mol}$$

Problem 5.41. The total pressure in a vessel containing oxygen collected over water at 25 °C was 723 torr. The vapor pressure of water at that temperature is 23.8 torr. What was the pressure of the gas?

Ans. The total pressure of gas in the vessel was the sum of the partial pressures of the collected gas and water vapor:

$$P_{\text{TOTAL}} = P_{\text{OXYGEN}} + P_{\text{WATER}}$$

$$723 \text{ torr} = P_{\text{OXYGEN}} + 23.8 \text{ torr}$$

$$P_{\text{OXYGEN}} = 723 \text{ torr} - 23.8 \text{ torr} = 699 \text{ torr}$$

Problem 5.42. The volume of a sample of hydrogen collected over water at 25 °C was 6.0 L. The total pressure was 752 torr. How many moles of hydrogen were collected?

Ans. First, calculate the partial pressure of hydrogen:

$$P_{\text{HYDROGEN}} = 752 \text{ torr} - 23.8 \text{ torr} = 728 \text{ torr}$$

Now use the ideal gas law to calculate moles, making sure to convert torr to atmospheres and Celsius degrees to kelvins:

$$T = 273 + 25 = 298 \text{ K}$$

$$728.2 \text{ torr}\left(\frac{1 \text{ atm}}{760 \text{ torr}}\right) = 0.96 \text{ atm}$$

$$n = \frac{PV}{RT} = \frac{0.96 \text{ atm} \times 6.0 \text{ L}}{0.0821 \text{ L} \cdot \text{atm} \cdot \text{K}^{-1} \cdot \text{mol}^{-1} \times 298 \text{ K}} = 0.24 \text{ mol}$$

LIQUIDS

Problem 5.43. The normal boiling point of liquid A is 60 °C, and its melting point is −10 °C. The normal boiling point of liquid B is −10 °C and its melting point is −60 °C. Compare the vapor pressures of A and B at 60 °C and −10 °C, respectively.

Ans. **Since those temperatures are the normal boiling points of both liquids, their vapor pressures must be the same, and are equal to atmospheric pressure.**

Problem 5.44. Compare the vapor pressures of liquid A and liquid B in Problem 5.43 if both liquids are at −5 °C.

Ans. **At −5 °C, liquid B is close to its normal boiling point, and liquid A is just above its melting point and very far from its normal boiling point. Therefore, the vapor pressure of liquid B must be greater than the vapor pressure of liquid A at that temperature.**

Problem 5.45. Compare the normal boiling points of water (hydrogen oxide) and hydrogen sulfide. The structures of H_2O and H_2S are

$$\mathrm{H{-}O{-}H} \quad \text{and} \quad \mathrm{H{-}S{-}H}$$

Ans. **Hydrogen bonds form when hydrogen is covalently bound to oxygen, nitrogen, or fluorine. Of these two compounds, only water can form H-bonds. Therefore water must have the higher normal boiling point.**

SOLIDS

Problem 5.46. Compare the melting points and heats of fusion of calcium chloride, and water.

Ans. **Calcium chloride is an ionic compound whose solid form is maintained by ionic chemical bonds. The solid form of water is maintained by hydrogen bonds which are significantly weaker than chemical bonds. Therefore the melting point of calcium chloride and its heat of fusion must be much greater than those of water.**

SUPPLEMENTARY PROBLEMS

Problem 5.47. What is the meaning of STP, and why is it used?

Ans. STP stands for ***standard temperature and pressure*** of 0 °C and 1 atm pressure. It is under those fixed conditions that the volume of 1 mol of any gas is 22.4 L.

Problem 5.48. Calculate the mass in grams of hydrogen that will be produced by 36.5 g of magnesium in the following reaction: $Mg + 2\,HCl = MgCl_2 + H_2$.

Ans. 3.03 g of hydrogen

Problem 5.49. A gas sample occupies 2.3 L at 500 mm pressure. Calculate its volume if the pressure is changed at constant temperature to (*a*) 720 mm; (*b*) 2.1 atm; (*c*) 82 torr; (*d*) 256 mm.

Ans. (*a*) 1.6 L; (*b*) 0.72 L; (*c*) 14 L; (*d*) 4.5 L

Problem 5.50. A 1.6-L sample of gas at 25 °C was heated and expanded at constant pressure to 1.9 L. Calculate its final temperature.

Ans. 81 °C

Problem 5.51. A 7.0-L sample of gas at 0 °C was cooled to −133 °C at constant pressure. Calculate its final volume.

Ans. 3.6 L

Problem 5.52. The volume of a 23.0-L sample of gas at 250 °C was reduced at constant pressure to 1.00 L. What was its final temperature?

Ans. −250 °C

Problem 5.53. Fill the blanks in the following table with the correct values:

P_1	P_2	V_1	V_2	t_1	t_2
(1) 500 torr	390 torr	0.50 L	___ L	25 °C	100 °C
(2) 1.3 atm	_____ atm	2.3 L	2.3 L	0.0 °C	250 °C
(3) 1.2 atm	1596 torr	8.0 L	2.9 L	55 °C	___ °C

Ans. (1) 1.6 L; (2) 2.5 atm; (3) −65 °C

Problem 5.54. Fill the blanks in the following table with the correct values:

P	V	t	n
(1) 1.6 atm	___ L	37 °C	1.2 mol
(2) 558 torr	146 L	___	4.5 mol
(3) ___ torr	47.0 L	125 °C	1.60 mol
(4) 160 torr	6.0 L	37 °C	___ mol

Ans. (1) 19 L; (2) 17 °C; (3) 845 torr; (4) 0.050 mol

Problem 5.55. A 2.5-L sample of oxygen was collected over water at 25 °C, and a total pressure of 745 torr. How many grams of oxygen were present?

Ans. 3.2 g

Problem 5.56. A 2.00-L sample of N_2 at 550 torr and 4.00 L of O_2 at 436 torr were put into a 12.0-L container. Calculate the total final pressure of the mixture.

Ans. 237 torr

Problem 5.57. What is the density of a sample of CO_2 at 380 torr and 25 °C?

Ans. 0.90 g/L

Problem 5.58. Between which of the following pairs of molecules can hydrogen bonds be formed?

```
(a)        H
           |
        H—C—H     and       O
           |              /   \
           H             H     H

(b)                          H
            O                |
          /   \     and      N
         H     H           /   \
                          H     H

                             H
                             |
(c)         S                |
          /   \     and  Cl—C—Cl
         H     H             |
                             Cl

                             H
                             |
(d)       H—F       and      N
                           /   \
                          H     H

           H              H   H
           |              |   |
(e)     H—C—H     and  H—C—C—H
           |              |   |
           H              H   H
```

Ans. (*b*) and (*d*)

Problem 5.59. Compare the viscosities water and chloroform. The structures of water and chloroform are

```
                             Cl
            O                |
          /   \     and   H—C—Cl
         H     H             |
                             Cl
```

Ans. Water can form hydrogen bonds, but chloroform cannot. Therefore, the cohesive forces in water are greater than those in chloroform. On that basis, one would predict that the viscosity of water would be greater than that of chloroform.

Problem 5.60. Compare the surface tensions of water and ethane. The structure of water and ethane are

```
                          H   H
            O             |   |
          /   \     and  H—C—C—H
         H     H          |   |
                          H   H
```

Ans. Surface tension depends upon the strength of cohesive forces between molecules in the interphase (the surface boundary between gas and liquid). Water can form hydrogen bonds but ethane cannot (why?). Therefore one would predict that water has the higher surface tension.

Concentration and Its Units

6.1 INTRODUCTION

Before we can discuss the properties of solutions, it is necessary to develop the quantitative concept of concentration.

The simplest kind of solution consists of a mixture of two substances, the *solute*, the one in smaller amount, dissolved in the *solvent*, the component in larger amount. *Concentration* is a measure of the amount of solute dissolved in a given amount of solvent. It is a kind of density. No matter how small or large an amount of a solution is observed, the amount of dissolved solute found in any fixed volume will remain the same. The amount of solute can be expressed in physical mass units, i.e., grams, or chemical mass units, i.e., moles. The amount of solvent may be expressed in units of volume, e.g., liters, or physical mass units, e.g., kilograms.

6.2 PERCENT CONCENTRATION

Definition: Parts of solute per 100 parts of solution. The parts may be expressed either in mass or volume; specifically, the following are most useful: w/w is (weight solute)/(weight solution), w/v is (weight solute)/(volume solution), v/v is (volume solute)/(volume solution). The solution consists of both solute and solvent.

Problem 6.1. How would you prepare a 10% w/w aqueous solution of NH_4Cl?

Ans. The *w/w* designation means that the amounts of both solvent and solute are expressed in physical mass units, i.e., grams.

$$\% \text{ w/w} = \frac{\text{grams solute}}{\text{grams solute} + \text{grams solvent}} \times 100\%$$

Where grams solute + grams solvent = 100 g

Since the units of % w/w designation cancel, it is a true fraction.

To prepare a 10% w/w aqueous NH_4Cl solution, dissolve 10 g of NH_4Cl in 90 g of solvent. The solution then consists of 10 g of salt and 100 g of solution.

Why is it necessary to multiply this ratio by 100? The ratio stands for grams of solute per gram of solution. It must be multiplied by 100 to express the concentration as parts per hundred parts.

Problem 6.2. How many grams of NaCl are present in 1 g of a 2.5% w/w NaCl solution?

Ans.

$$2.5\% \text{ w/w NaCl solution} = \frac{2.5 \text{ g NaCl}}{100 \text{ g solution}} = \frac{0.025 \text{ g NaCl}}{\text{gram of solution}}$$

Problem 6.3. How many grams of NaCl are contained in 22 g of a 2.5% w/w NaCl solution?

Ans. There are 2.5 g of NaCl in 100 g of solution. Using the factor-label method, first write the quantity, then multiply by the conversion factor:

$$22 \text{ g solution}\left(\frac{2.5 \text{ g NaCl}}{100 \text{ g solution}}\right) = 0.55 \text{ g NaCl}$$

Since the ratio of grams of solute/grams of solution is a constant, one can also solve the problem by setting up two equivalent ratios:

$$\frac{2.5 \text{ g NaCl}}{100 \text{ g solution}} = \frac{X \text{ g NaCl}}{22 \text{ g solution}}$$

$$X = 0.55 \text{ g NaCl}$$

Problem 6.4. How would you prepare an aqueous 10% v/v ethanol solution?

Ans. The *v/v* designation stands for volume/volume, and means that the amounts of both solvent and solute are expressed in volume units, i.e., milliliters. It is the most practical way to express the method of preparation of solutions of liquids in liquids.

$$\% \text{ v/v} = \frac{\text{milliliters solute}}{100 \text{ mL solution}} \times 100\%$$

Therefore, an aqueous 10% v/v solution of ethanol is prepared by dissolving 10.00 mL of ethanol in 100.00 mL of solution. Note that this ratio is also multiplied by 100 in order to express the fraction as parts per hundred parts.

Problem 6.5. How would you prepare a 10% w/v aqueous NH_4Cl solution?

Ans. The *w/v* designation means that the amount of solute is expressed in physical mass units, i.e., grams, and the amount of solution (not solvent) is expressed in volume units, i.e., milliliters.

$$\% \text{ w/v} = \frac{\text{grams solute}}{100 \text{ mL solution}} \times 100\%$$

This solution is prepared by dissolving 10 g of salt in 100 mL of solution. The mass units in a % w/v solution are grams, but the volume units are milliliters of solution, not grams of solution. Therefore the units do not cancel, and the % w/v is not a true fraction or percent. Rather than writing, say, 5% NaCl w/v, it would be more meaningful to express the concentration as 5 g NaCl/100 mL (of solution). Mathematically, it is handled in the same way as % w/w or % v/v problems, either by the factor-label or ratio methods.

Problem 6.6. Calculate the number of grams contained in 55 mL of a 12% w/v aqueous glucose solution.

Ans. Using the factor-label method, first write the quantity, and then multiply the conversion factor:

$$55 \text{ mL solution} \left(\frac{12 \text{ g glucose}}{100 \text{ mL solution}}\right) = 6.6 \text{ g glucose}$$

Another expression of concentration commonly used in the clinical laboratory is mg %. This is defined as

$$\text{mg } \% = \frac{\text{milligrams solute}}{100 \text{ mL solution}}$$

Problem 6.7. Cholesterol concentrations in blood are reported, for example, as 175 mg %. What does this mean in terms of mass relationship?

Ans.

$$175 \text{ mg } \% = \frac{\text{milligrams cholesterol}}{100 \text{ mL blood}}$$

Cholesterol concentration = 1.75 mg/mL of blood.

It is sometimes convenient when describing very dilute solutions to employ the designation ppm (sometimes spelled p.p.m.). This is an abbreviation for parts per million, or more precisely, parts of solute per million parts of solution, a weight/volume designation. Let us examine the case in which we have prepared a 0.0001% w/v solution of some salt. This concentration may be written as 0.0001 g salt/100 mL. Let us convert this ratio to one of whole numbers:

$$\frac{0.0001 \text{ g}}{100 \text{ mL}}\left(\frac{10{,}000}{10{,}000}\right) = \frac{1 \text{ g}}{1{,}000{,}000 \text{ mL}}$$

The solution can then be seen to contain 1 g of solute per million mL of solution or to have a concentration of 1.0 ppm. It is more common to express parts per million as the amount of solute contained in a liter of solution:

$$1\ \text{ppm} = \frac{\left(\dfrac{1\ \text{g}}{1000}\right)}{\left(\dfrac{1{,}000{,}000\ \text{mL}}{1000}\right)} = \frac{0.001\ \text{g}}{1000\ \text{mL}}$$

$$\left(\frac{0.001\ \text{g}}{1000\ \text{mL}}\right)\left(\frac{1000\ \text{mg}}{1\ \text{g}}\right) = \frac{1\ \text{mg}}{1000\ \text{mL}}$$

So that

$$\left(\frac{1\ \text{mg}}{1000\ \text{mL}}\right)\left(\frac{1000\ \text{mL}}{\text{L}}\right) = \frac{1\ \text{mg}}{\text{L}}$$

$$1\ \text{ppm} = \frac{1\ \text{mg}}{\text{L}}$$

Problem 6.8. What is the concentration in parts per million of a 0.00175% w/v solution of mercuric nitrate?

Ans.

$$0.00175\%\ \text{w/v} = \frac{0.00175\ \text{g}}{100\ \text{mL}}\left(\frac{10}{10}\right) = \frac{0.0175\ \text{g}}{\text{L}}$$

$$\frac{0.0175\ \text{g}}{\text{L}}\left(\frac{1000\ \text{mg}}{1.0\ \text{g}}\right) = \frac{17.5\ \text{mg}}{\text{L}} = 17.5\ \text{ppm}$$

In certain applications, e.g., environmental toxicology, trace quantities are sometimes designated as ppb, or parts per billion. That would be equivalent to 0.001 mg/L of solution.

6.3 MOLARITY

Molarity is a concentration term based on chemical mass units. It is defined as moles of solute/liter of solution. The unit of molarity is *molar*, and given the symbol *M*. For example, a 2.0 *M* solution is a 2.0 molar solution and its molarity is 2.0 mol/L.

Problem 6.9. How would you prepare a 0.5 *M* aqueous solution of NH_4Cl?

Ans. Weigh out 26.75 g (0.5 mol) of the salt, and dissolve it in sufficient water so that the final volume of solution is one liter. The dilution is done in a volumetric flask calibrated to contain 1000 mL.

Problem 6.10. How many moles of solute are contained in 3.0 L of a 1.8 *M* solution?

Ans. Using the factor-label method, first write the quantity, and then multiply it by the conversion factor:

$$3.0\ \text{L}\left(\frac{1.8\ \text{mol}}{\text{L}}\right) = 5.4\ \text{mol}$$

Problem 6.11. What volume of a 2.0 *M* solution will contain 0.50 mol of solute?

Ans.

$$0.50\ \text{mol}\left(\frac{1.0\ \text{L}}{2.0\ \text{mol}}\right) = 0.25\ \text{L}$$

Problem 6.12. What will be the molarity of a NaCl solution in which 175 g of NaCl is dissolved in 750 mL of solution?

Ans. Since molarity is defined as moles per liter, we must first calculate the number of moles in 175 g of NaCl, then convert 750 mL to liters, and finally calculate M:

$$175 \text{ g NaCl}\left(\frac{1.0 \text{ mol}}{58.5 \text{ g NaCl}}\right) = 3.0 \text{ mol NaCl}$$

$$750 \text{ mL}\left(\frac{1 \text{ L}}{1000 \text{ mL}}\right) = 0.75 \text{ L}$$

$$\frac{3.0 \text{ mol}}{0.75 \text{ L}} = \frac{4.0 \text{ mol}}{\text{L}} = 4.0\,M$$

It is sometimes convenient to express dilute concentrations in terms of millimoles, given the symbol mmol. The units are related as follows:

$$1000 \text{ millimol} = 1 \text{ mol}$$

Therefore:

$$\frac{1 \text{ mmol}}{\text{mL}} = \frac{1 \text{ mmol}\left(\dfrac{1 \text{ mol}}{1000 \text{ mmol}}\right)}{1 \text{ mL}\left(\dfrac{1 \text{ L}}{1000 \text{ mL}}\right)} = \frac{1 \text{ mol}}{\text{L}}$$

Problem 6.13. What is the concentration of a solution of KCl containing 7.4 mmol in 2.0 mL of solution?

Ans. Since mmol/mL = mol/L, the concentration must be

$$\frac{7.4 \text{ mmol}}{2.0 \text{ mL}} = \frac{3.7 \text{ mmol}}{\text{mL}} = \frac{3.7 \text{ mol}}{\text{L}} = 3.7\,M$$

It is also convenient to express the formula weight of a compound in milligrams and millimoles. From Problem 6.12, you can see that mg/mmol = g/mol.

Problem 6.14. How many millimoles are contained in 0.0585 g of NaCl?

Ans.

$$0.0585 \text{ g NaCl}\left(\frac{1000 \text{ mg}}{\text{g}}\right) = 58.5 \text{ mg NaCl}$$

$$58.5 \text{ mg NaCl}\left(\frac{1 \text{ mmol NaCl}}{58.5 \text{ mg NaCl}}\right) = 1 \text{ mmol NaCl}$$

Problem 6.15. What is the concentration of a 4.0 M solution of NaCl in millimoles per milliliter?

Ans.

$$4.0\,M = \frac{4.0 \text{ mol}}{\text{L}} = \frac{4.0 \text{ mmol}}{\text{mL}}$$

6.4 MOLALITY

Molality is also a concentration unit based on chemical mass units, i.e., moles, but differs in the specification of the amount of solvent in which the solute is dissolved. Rather than moles of solute/liter of solution, the definition of molality is moles of solute/kilogram of solvent. Note carefully: The denominator is the mass of solvent not the volume of solution. Molality is given the symbol *m*. This concentration designation is used in calculating freezing point depression and boiling point elevation of solutions, which we will take up in Chap. 7.

Problem 6.16. How would you prepare a 0.5 *m* aqueous solution of NH_4Cl?

Ans. Weigh out 26.75 g of salt (0.5 mol), and dissolve it in 1 kg (1000 g) of solvent (water).

ADDITIONAL SOLVED PROBLEMS

PERCENT COMPOSITION

Problem 6.17. Calculate the number of grams contained in 55 g of a 12% w/w aqueous glucose solution.

Ans. The factor-label method requires use of the % w/v as a conversion factor:

$$55 \text{ g solution}\left(\frac{12 \text{ g glucose}}{100 \text{ g solution}}\right) = 6.6 \text{ g glucose}$$

Problem 6.18. How would you prepare an isotonic saline solution?

Ans. A medical dictionary states that an isotonic saline solution is a 0.92% NaCl w/v solution. We need to weigh out 0.92 g of NaCl and dissolve it in enough distilled water so that the final solution volume is 100 mL (not grams).

Problem 6.19. How much 5.0% w/v NaCl solution can be prepared from 3.0 g of NaCl?

Ans.

$$3.0 \text{ g NaCl}\left(\frac{100 \text{ mL solution}}{5.0 \text{ g NaCl}}\right) = 60 \text{ mL solution}$$

Problem 6.20. What mass of KCl is required to prepare 120 g of a 6.40% w/w solution of KCl?

Ans.

$$120 \text{ g solution}\left(\frac{6.40 \text{ g KCl}}{100 \text{ g solution}}\right) = 7.68 \text{ g KCl}$$

Problem 6.21. What volume of a 4.60% w/v $CaCl_2$ solution contains 2.50 g of salt?

Ans.

$$2.50 \text{ g CaCl}_2\left(\frac{100 \text{ mL solution}}{4.60 \text{ g CaCl}_2}\right) = 54.4 \text{ mL solution}$$

Problem 6.22. How much 7.50% v/v solution of ethanol in water can be prepared with 17.6 mL of ethanol?

Ans.

$$17.6 \text{ mL ethanol}\left(\frac{100 \text{ mL solution}}{7.50 \text{ mL ethanol}}\right) = 235 \text{ mL solution}$$

Problem 6.23. What is the concentration in mg % of a KCl solution which contains 0.0245 g KCl in 35 mL of solution?

Ans. mg % = mg KCl/100 mL solution. Therefore, determine the number of mg of KCl in 35 mL of solution, and then the equivalent number in 100 mL:

$$\frac{0.0245 \text{ g KCl}}{35 \text{ mL solution}}\left(\frac{1000 \text{ mg}}{\text{g}}\right) = \frac{0.70 \text{ mg KCl}}{\text{mL solution}}$$

$$\left(\frac{100}{100}\right)\left(\frac{0.70 \text{ mg KCl}}{\text{mL solution}}\right) = \frac{70 \text{ mg KCl}}{100 \text{ mL solution}} = 70 \text{ mg \%}$$

Problem 6.24. What is the concentration in parts per million of a solution which contains 0.0245 g of KCl in 35 mL?

Ans. ppm = mg KCl/L of solution. Therefore, determine the number of mg in 35 mL and then the equivalent number in 1000 mL or 1.0 L. From Problem 6.7:

$$\frac{70 \text{ mg}}{100 \text{ mL}}\left(\frac{1000 \text{ mL}}{\text{L}}\right) = \frac{700 \text{ mg}}{\text{L}} = 700 \text{ ppm}$$

Problem 6.25. What is the concentration in % w/v of a $MgCl_2$ solution whose concentration is 25 ppm?

Ans. % w/v = g ($MgCl_2$)/100 mL solution. We begin by noting that 25 ppm = 25 mg/L. Then

$$\frac{25 \text{ mg}}{\text{L}}\left(\frac{\text{g}}{1000 \text{ mg}}\right) = \frac{0.025 \text{ g}}{\text{L}}$$

$$\frac{0.025 \text{ g}}{\text{L}}\left(\frac{0.1 \text{ L}}{100 \text{ mL}}\right) = \frac{0.0025 \text{ g}}{100 \text{ mL}} = 0.0025\% \text{ w/v}$$

MOLARITY

Problem 6.26. What is the volume of a 2.0 *M* solution that contains 0.50 mol of solute?

Ans. A 2.0 *M* solution contains 2.0 mol of solute/liter of solution. Using the factor-label method with the molarity as factor:

$$0.50 \text{ mol}\left(\frac{1.0 \text{ L}}{2.0 \text{ mol}}\right) = 0.25 \text{ L} = 250 \text{ mL}$$

Problem 6.27. How many moles of solute are there in 3.0 L of a 0.50 *M* solution?

Ans.

$$3.0 \text{ L}\left(\frac{0.50 \text{ mol}}{\text{L}}\right) = 1.5 \text{ mol}$$

Problem 6.28. Which contains more solute, 3.0 L of a 0.50 *M* KCl solution or 2.0 L of a 0.75 *M* KCl solution?

Ans.

$$3.0 \text{ L}\left(\frac{0.50 \text{ mol}}{\text{L}}\right) = 1.5 \text{ mol}$$

$$2.0 \text{ L}\left(\frac{0.75 \text{ mol}}{\text{L}}\right) = 1.5 \text{ mol}$$

Both solutions contain the same amount of solute.

Problem 6.29. What are the molarities of solutions containing (*a*) 0.33 mol in 1.75 L, (*b*) 0.50 mol in 2.5 L, (*c*) 1.3 mol in 2.6 L?

Ans.

(*a*) $$\frac{0.33 \text{ mol}}{1.75 \text{ L}} = \frac{0.19 \text{ mol}}{\text{L}} = 0.19\ M$$

(*b*) $$\frac{0.5 \text{ mol}}{2.5 \text{ L}} = \frac{0.2 \text{ mol}}{\text{L}} = 0.2\ M$$

(*c*) $$\frac{1.3 \text{ mol}}{2.6 \text{ L}} = \frac{0.50 \text{ mol}}{\text{L}} = 0.50\ M$$

Problem 6.30. Calculate the number of moles contained in (*a*) 2.0 L of a 0.65 *M* solution, (*b*) 1.2 L of a 0.60 *M* solution, (*c*) 0.750 L of a 1.75 *M* solution.

Ans.

(*a*) $$2.0 \text{ L}\left(\frac{0.65 \text{ mol}}{\text{L}}\right) = 1.3 \text{ mol}$$

(*b*) $$1.2 \text{ L}\left(\frac{0.60 \text{ mol}}{\text{L}}\right) = 0.72 \text{ mol}$$

(*c*) $$0.750 \text{ L}\left(\frac{1.75 \text{ mol}}{\text{L}}\right) = 1.31 \text{ mol}$$

Problem 6.31. Calculate the volumes containing the specified number of moles for the following solutions: (*a*) 0.6 mol of a 1.5 *M* solution, (*b*) 1.5 mol of a 2.0 *M* solution, (*c*) 2.7 moles of a 0.675 *M* solution.

Ans.

(*a*) $$0.6 \text{ mol}\left(\frac{1.0 \text{ L}}{1.5 \text{ mol}}\right) = 0.4 \text{ L}$$

(*b*) $$1.5 \text{ mol}\left(\frac{1.0 \text{ L}}{2.0 \text{ mol}}\right) = 0.75 \text{ L}$$

(*c*) $$2.7 \text{ mol}\left(\frac{1.0 \text{ L}}{0.675 \text{ mol}}\right) = 4.0 \text{ L}$$

Problem 6.32. What is the molarity of a solution containing 14.63 g of NaCl in 658 mL?

Ans. Since molarity is defined as moles per liter, the mass of NaCl must be expressed in moles, and the volume in liters:

$$14.63 \text{ g NaCl}\left(\frac{1 \text{ mol NaCl}}{58.5 \text{ g NaCl}}\right) = 0.250 \text{ mol}$$

$$658 \text{ mL}\left(\frac{1.0 \text{ L}}{1000 \text{ mL}}\right) = 0.658 \text{ L}$$

$$\frac{0.250 \text{ mol}}{0.658 \text{ L}} = \frac{0.380 \text{ mol}}{\text{L}} = 0.380\ M$$

Problem 6.33. How many grams of KCl are contained in 575 mL of a 0.40 *M* solution?

Ans. The concentration is given in molarity, and hence in moles. Since we are required to determine the mass of KCl in grams, the first step is to convert the moles to grams. Furthermore, because molarity is in terms of liters, the volume given in milliliters should be converted to liters. We will use two factors to calculate the number of moles in that volume, and in the second step convert moles to grams, using the formula weight of KCl as a factor:

$$575\ \text{mL}\left(\frac{1\ \text{L}}{1000\ \text{mL}}\right)\left(\frac{0.40\ \text{mol}}{\text{L}}\right) = 0.23\ \text{mol}$$

$$0.23\ \text{mol}\left(\frac{74.5\ \text{g KCl}}{\text{mol KCl}}\right) = 17\ \text{g KCl}$$

Problem 6.34. What is the volume of a 0.50 *M* KCl solution containing 25 g of KCl?

Ans. We must use the molarity as a factor to determine the required volume, and therefore we must first convert grams of KCl to moles:

$$25\ \text{g KCl}\left(\frac{1\ \text{mol KCl}}{74.5\ \text{g KCl}}\right) = 0.34\ \text{mol}$$

$$0.34\ \text{mol}\left(\frac{1.0\ \text{L}}{0.50\ \text{mol}}\right) = 0.68\ \text{L} = 680\ \text{mL}$$

Problem 6.35. How many milligrams of NaCl are contained in 50.0 mL of a 1.20 *M* solution?

Ans. Since we want to calculate milligrams, and we are given milliliters, we might as well work with millimoles. First determine the number of millimoles, then convert to milligrams:

$$50.0\ \text{mL}\left(\frac{1.20\ \text{mmol}}{\text{mL}}\right) = 60.0\ \text{mmol}$$

$$60.0\ \text{mmol}\left(\frac{58.5\ \text{mg NaCl}}{1\ \text{mmol NaCl}}\right) = 3.51 \times 10^3\ \text{mg NaCl}$$

Problem 6.36. What is the molarity of 125 mL of a solution containing 37.25 mg of KCl?

Ans.

$$37.25\ \text{mg KCl}\left(\frac{1\ \text{mmol KCl}}{74.5\ \text{mg KCl}}\right) = 0.500\ \text{mmol KCl}$$

$$\frac{0.500\ \text{mmol}}{125\ \text{mL}} = \frac{0.00400\ \text{mol}}{\text{mL}} = \frac{0.00400\ \text{mmol}}{\text{L}} = 0.00400\ M$$

MOLALITY

Problem 6.37. What is the molality of a 10% w/w NaCl solution?

Ans. Molality is a ratio of masses. It is therefore necessary to convert the % w/w to moles and kilograms. Since 10% w/w = 10 g NaCl/100 g solution, the solution composition is 10 g NaCl and 90 g water. First convert 10 g NaCl to the number of moles in 90 g of water, then calculate the number of moles of NaCl in 1 kg of water:

$$10\ \text{g NaCl}\left(\frac{1\ \text{mol NaCl}}{58.5\ \text{g NaCl}}\right) = 0.17\ \text{mol NaCl}$$

$$90\ \text{g H}_2\text{O}\left(\frac{1\ \text{kg}}{1000\ \text{g}}\right) = 0.090\ \text{kg H}_2\text{O}$$

$$\frac{0.17\ \text{mol NaCl}}{0.090\ \text{kg H}_2\text{O}} = \frac{1.89\ \text{mol NaCl}}{\text{kg H}_2\text{O}} = 1.9\ m$$

Problem 6.38. What is the molality of a 10% w/v NaCl solution?

Ans. % w/v is not a ratio of masses. It does not allow calculation of the mass of water in the solution. Therefore the question cannot be answered.

SUPPLEMENTARY PROBLEMS

Problem 6.39. Calculate the % w/w of the following solutions: (*a*) 12.5 g $CaCl_2$ in 150 g H_2O, (*b*) 13 g of ethyl alcohol in 175 g H_2O, (*c*) 1.5 g of procaine hydrochloride in 12.5 g of ethylene glycol.

Ans. (*a*) 7.69% w/w, (*b*) 6.9% w/w, (*c*) 11% w/w

Problem 6.40. Given an aqueous stock solution which is 15 % w/v in mercury(II) chloride, how many milliliters must you take to obtain 5.25 g of $HgCl_2$?

Ans. 35 mL

Problem 6.41. How much of each compound is present in the following mixtures? (*a*) 25.0 g of a 4.50% w/w solution of glucose, (*b*) 125 g of a 5.00% w/w solution of ammonium sulfate, (*c*) 63.0 g of a 1.20% w/w solution of insulin.

Ans. (*a*) 1.13 g, (*b*) 6.25 g, (*c*) 0.756 g

Problem 6.42. Calculate the volume percent of the following mixtures: (*a*) 12 mL of ethyl alcohol made up to 125 mL of solution with water, (*b*) 4.20 mL of chloroform made up to 25.0 mL with ethyl alcohol, (*c*) 11.6 mL of propylene glycol made up to 32.0 mL with water.

Ans. (*a*) 9.6% v/v, (*b*) 16.8% v/v, (*c*) 36.3% v/v

Problem 6.43. Calculate the weight/volume percent of the following mixtures made up to final volume with water: (*a*) 3.75 g of KCl in 45.0 mL, (*b*) 12.4 g of $NaHCO_3$ in 250 mL, (*c*) 2.5 g of $CaCl_2$ in 85 mL.

Ans. (*a*) 8.33 % w/v, (*b*) 4.96% w/v, (*c*) 2.9% w/v

Problem 6.44. Calculate the molarity of the following solutions: (*a*) 0.25 mol in 1.2 L, (*b*) 0.120 mol NaCl in 0.650 L, (*c*) 2.36 mol $NaHCO_3$ in 1.87 L, (*d*) 0.175 mol HCl in 0.084 L.

Ans. (*a*) 0.21 *M*, (*b*) 0.185 *M*, (*c*) 1.26 *M*, (*d*) 2.1 *M*

Problem 6.45. Calculate the number of moles in each of the following solutions: (*a*) 35 mL of 1.5 *M* $CaCl_2$ solution, (*b*) 1.2 L of 0.56 *M* KNO_3 solution, (*c*) 0.435 L of 0.450 *M* LiCl solution, (*d*) 74.2 mL of 0.252 *M* $NaHCO_3$ solution.

Ans. (*a*) 0.053 mol, (*b*) 0.67 mol, (*c*) 0.196 mol, (*d*) 0.0187 mol

Problem 6.46. Calculate the number of grams of solute in each of the following solutions: (*a*) 125 mL of 1.20 *M* KCl, (*b*) 0.0450 L of 0.650 *M* LiCl, (*c*) 1.50 L of 0.320 *M* NaSCN, (*d*) 50.0 mL of 15.0 *M* H_2SO_4.

Ans. (*a*) 11.2 g, (*b*) 1.24 g, (*c*) 38.9 g, (*d*) 73.5 g

Problem 6.47. How many milliliters of a 0.375 *M* solution of glucose will contain the following number of moles? (*a*) 0.560 mol, (*b*) 2.52 mol, (*c*) 0.960 mol, (*d*) 0.680 mol.

Ans. (*a*) 1.49×10^3 mL, (*b*) 6.72×10^3 mL, (*c*) 2.56×10^3 mL, (*d*) 1.81×10^3 mL

Problem 6.48. How many grams of $KHCO_3$ must you add to 0.300 L of water to prepare a 0.450 *M* solution?

Ans. 13.5 g

Problem 6.49. How would you prepare a 2.60 *m* aqueous solution of KBr?

Ans. Add 309 g KBr to 1.00 kg water.

Problem 6.50. What is the % w/w of a 2.60 *m* solution of KBr?

Ans. 23.6 percent

Problem 6.51. How many grams of a 2.60 *m* solution of KBr would contain 5.00 g KBr?

Ans. 21.2 g

Problem 6.52. What mass of a 2.10% w/w solution of glucose must you take to obtain 0.500 g of glucose?

Ans. 23.8 g

Problem 6.53. Calculate the % w/w of the following solutions: (*a*) 25 g $CaCl_2$ in 150 g H_2O, (*b*) 13 g NaCl in 215 g H_2O, (*c*) 1.6 g C_2H_5OH in 55.4 g H_2O.

Ans. (*a*) 14 percent, (*b*) 5.7 percent, (*c*) 2.8 percent

Problem 6.54. What volume of physiological saline, 0.900% w/v NaCl, can be prepared from 150 mL of 1.50% w/v NaCl solution?

Ans. 250 mL

Problem 6.55. Calculate the grams of solute present in each of the following solutions: (*a*) 2.50 L of 0.500 *M* NaCl, (*b*) 425 mL of 0.65 *M* $Mg(NO_3)_2$, (*c*) 600 mL of 1.60 *M* K_2CO_3.

Ans. (*a*) 73.1 g, (*b*) 41 g, (*c*) 133 g

Problem 6.56. Calculate the molarity of the following solutions: (*a*) 35.0 g $K_2C_2O_4$ in 1.20 L of solution, (*b*) 152 g $C_6H_{12}O_6$ in 750 mL of solution, (*c*) 48 g $BaCl_2$ in 3.0 L of solution.

Ans. (*a*) 0.176 *M*, (*b*) 1.12 *M*, (*c*) 0.077 *M*.

Problem 6.57. Given 25.0 mL of a solution of 2.50 *M* HCl, to what volume must it be diluted to obtain a concentration of 0.500 *M*?

Ans. 125 mL

Problem 6.58. Calculate the volumes of the following concentrated solutions required to prepare solutions of the final concentrations indicated: (*a*) 15 *M* H_2SO_4 to prepare 2.0 L of 3.0 *M* H_2SO_4, (*b*) 2.5 *M* KCl to prepare 150 mL of 1.5 *M* KCl, (*c*) 0.800 *M* glucose to prepare 2.00 L of 0.500 *M* glucose.

Ans. (*a*) 0.40 L, (*b*) 0.090 L, (*c*) 1.25 L

Problem 6.59. Calculate the molality of the following solutions: (*a*) 9.4 g C_6H_5OH in 200 g H_2O, (*b*) 11.5 g C_2H_5OH in 125 g H_2O, (*c*) 25.6 g $C_2H_4(OH)_2$ in 2.0 kg H_2O.

Ans. (*a*) 0.50 *m*, (*b*) 2.00 *m*, (*c*) 0.21 *m*

CHAPTER 7

Solutions

7.1 SOLUTIONS AS MIXTURES

The term "mixture" describes a system whose composition is variable, and whose components retain their chemical identity. Remember that the composition of a pure substance is fixed and invariant. A solution is a special kind of mixture in which substances are intermixed so intimately (down to the molecular level) that they cannot be observed as separate components. There is no rigid requirement with respect to the dimensions of the dispersed components of solutions. Particles larger than ordinary molecular dimensions can form solutions, and these solutions comprise a special category called *colloids*.

Problem 7.1. How many kinds of solutions are possible?

Ans. Since there are three states of matter—gas, liquid and solid—nine kinds of solutions are theoretically possible. However, neither liquids nor solids can be dissolved in gases. We can illustrate that with a matrix in which solvent is represented by vertical columns and solute by horizontal rows. The most commonly encountered solutions are found in the central column of Fig. 7-1—gases, liquids, and solids dissolved in liquids.

	Solvent		
Solute	**Gas**	**Liquid**	**Solid**
Gas	G/G	G/L	G/S
Liquid	Mist or fog	L/L	L/S
Solid	Suspension	S/L	S/S

Fig. 7-1 Matrix describing the kinds of solutions which can be formed by mixtures of gases (G), liquids (L), and solids (S).

7.2 SOLUBILITY

The principal component of liquid solutions is called the *solvent*. It is the component in which a smaller quantity of another substance, called the *solute*, is dissolved. In the case of gas mixtures and certain liquid mixtures, e.g., ethanol and water, solvent and solute can be mixed in any proportions. Such liquid/liquid systems are called *miscible*.

There is generally a limit to how much of a particular solute can be dissolved in a specific solvent. When less than that limit or amount is dissolved, the solution is said to be *unsaturated*. When that limit is reached, the solution is said to be *saturated*. The limiting amount of solute which will dissolve in a given amount of solvent at a given temperature is called the *solubility*. Solubility is a quantitative term, not to be confused with the qualitative term, *soluble*. All substances are soluble to some extent, but the designation, soluble, is reserved for those instances in which significant amounts of a substance can be dissolved in a solvent.

The solubility of most substances is dependent upon temperature. Hot liquids generally can dissolve more solids than cold liquids, but cold liquids dissolve more gases than hot liquids.

7.2.1 Solubility of Gases

The amount of a gas which will dissolve in a liquid chiefly depends upon the gas pressure—the greater the gas pressure, the greater its solubility. Gas solubility will be even greater if the gas reacts with the solvent, e.g., ammonia and carbon dioxide in water. Gas solubility is quantitatively described by *Henry's law*, which states that the amount of a gas dissolved in a liquid at a fixed temperature depends only upon the gas pressure. This is true for both pure gases and mixtures of gases. The solubilities of gases in a mixture depend upon the pressure each gas would exert if it were present alone, i.e., its *partial pressure*. In quantitative terms, Henry's law states

$$\left(\frac{\text{Amount of gas dissolved}}{\text{Within a unit volume of solvent}}\right) = K \text{ (gas partial pressure)}$$

K is the Henry's law constant, characteristic of the gas at fixed temperature. Let us assume that the units of gas dissolved and the volume of solvent are given in milliliters, and that the gas pressure is given in atmospheres. Gas pressure may also be given in millimeters of mercury. Remember the conversion factor: 1 atm = 760 mmHg. We can now solve the equation for K and define its units:

$$K = \left(\frac{\text{mL gas}}{\text{mL solvent} \cdot \text{atm}}\right)$$

One must know or be able to calculate this constant from provided data in order to calculate the solubility of a gas.

Problem 7.2. What is the oxygen content of pure water in the presence of air at 20 °C?

Ans. The Henry's law constant for pure oxygen at 20 °C is (0.031 mL O_2/mL H_2O) at 1.0 atm total pressure. Air consists of about 20% O_2, which means that, at 1 atm, the O_2 content of air is 0.20 atm. We now can write Henry's law:

Solubility of O_2 in water in the presence of air

$$= 0.20 \text{ atm}\left(\frac{0.031 \text{ mL } O_2}{\text{mL } H_2O \cdot 1 \text{ atm}}\right) = 0.0062 \text{ mL } O_2/\text{mL } H_2O$$

7.2.2 Solubility of Solids

Problem 7.3. At 20 °C, no more than 37.2 g of ammonium chloride can be dissolved in 100 g of water. At 40 °C the same amount of water can dissolve no more than 45.8 g. If all you had in your laboratory were beakers and a thermometer, could you prepare a saturated solution of ammonium chloride at room temperature of 25 °C?

Ans. The data indicate that the solubility (saturated solutions) of the salt increases with increase in temperature, and conversely decreases with decrease in temperature. To prepare the required solution, add salt to water at 40 °C until some salt remains suspended, or undissolved. Then lower the temperature to 25 °C. As the solubility decreases salt will precipitate and the resultant solution will beyond a doubt be saturated.

7.3 WATER

Liquid solutions form because of interaction between the molecules of solvent and solute. These can be very weak forces, as between xylene and benzene, or very strong, as between NaCl and water. In any case, for solution to occur, the interaction between solvent and solute must be at least as strong as or stronger than those among solvent molecules and those among solute molecules. When solute-solvent interactions prevail, solution will occur. These ideas are often stated in the simpler form that "like dissolves like." Water is the best example of the kind of solvent in which solution occurs because of the strong interaction between solvent and solute. Its role as a solvent is based on its polarity (a bent molecule with strongly polar bonds), and on its ability to form hydrogen bonds. Therefore water will interact with, *solvate*, and dissolve, substances which are electrically charged, polar, or can form hydrogen bonds. Hydrogen bonds are unique, strong electrical interactions between molecules in which hydrogen is covalently bound to one of the three most electronegative elements, fluorine, oxygen, and nitrogen, and interacts with partner molecules containing one of those same three elements. They give rise to the unique properties of water, e.g., its high boiling point, large heat capacity, high surface tension.

Problem 7.4. Will carbon tetrachloride dissolve in water?

Ans. *Analysis:* Water is strongly polar, and forms hydrogen bonds.

Is carbon tetrachloride electrically charged?

Is it a polar molecule?

Can it form hydrogen bonds?

Carbon tetrachloride is not ionic, but has four polar bonds. However, they are arranged symmetrically around the central carbon atom. Therefore, they effectively cancel the overall polarity of the molecule. Carbon tetrachloride is insoluble in water.

Problem 7.5. Will ethyl alcohol dissolve in water?

Ans. *Analysis:* Water is strongly polar, and forms hydrogen bonds.

Is ethyl alcohol electrically charged?

Is it a polar molecule?

Can it form hydrogen bonds?

Ethyl alcohol is not ionic, but is polar due to an hydroxyl group at one end of the molecule. Furthermore, the hydroxyl group is a hydrogen bond former (hydrogen covalently bonded to electronegative oxygen). Ethyl alcohol is soluble in water.

Problem 7.6. Will sodium chloride dissolve in water?

Ans. *Analysis:* Water is strongly polar, and forms hydrogen bonds.

Is sodium chloride electrically charged?

Is it a polar molecule?

Can it form hydrogen bonds?

Sodium chloride is an ionic solid. Therefore, its components are electrically charged, and consequently it is soluble in water.

7.4 DILUTION

To prepare for instance, a liter of a 1×10^{-6} *M* solution, one is faced with the need to weigh out 1×10^{-6} mol of the solute, a delicate task at best. However, there is a more practical way to accomplish the same thing, and that is to prepare a more concentrated solution and dilute it to the desired concentration. To carry out this process of dilution, we first note the required amount of solute in the final dilute solution, and then calculate the volume of the concentrated solution which will contain that amount of solute.

Problem 7.7. What volume of a 0.01 *M* solution is required to prepare 1 L of a 1×10^{-6} *M* solution?

Ans. In a 1×10^{-6} *M* solution there are 1×10^{-6} mol of solute per liter. So, to prepare 1 L of that solution we require 1×10^{-6} mol of solute from a concentrated solution, say, a 0.01 *M* solution. We must therefore calculate the volume of the concentrated solution that contains 1×10^{-6} mol of solute:

$$1 \times 10^{-6}\ \text{mol}\left(\frac{1\ \text{L}}{0.01\ \text{mol}}\right) = 1 \times 10^{-4}\ \text{L} = 0.1\ \text{mL}$$

To prepare the desired solution, add sufficient water to 0.1 mL of the concentrated solution to a final solution volume of 1 L.

It is difficult to measure out 0.1 mL with accuracy, so in this case it is best to carry out the dilution process in two steps. First prepare a 1×10^{-4} *M* solution by diluting the concentrated solution, then dilute that solution to 1×10^{-6} *M*. The mathematical procedure describing the dilution procedure can be greatly simplified by noting that, whatever the concentration of the final solution, the amount of solute in the final solution must be the same as the amount of solute taken from the more concentrated solution. So that

$$\text{mol}_{\text{initial}} = \text{mol}_{\text{final}}$$

and $$\text{mol} = \text{L}\left(\frac{\text{mol}}{\text{L}}\right)$$

Then $$M_A \times V_A = M_B \times V_B$$

or $$\text{Concentration}_A \times \text{Volume}_A = \text{Concentration}_B \times \text{Volume}_B$$

It is very important to note that any concentration units and volume units may be used as long as the right-hand-side units are identical to the left-hand-side units. Let us repeat the problem by using the last equation: $M_A \times L_A = M_B \times L_B$. Then

First dilution: $$(1 \times 10^{-2}\ M)_A(\text{L})_A = (1 \times 10^{-4}\ M)_B(1\ \text{L})_B$$
$$\text{L}_A = 0.01\ \text{L} = 10\ \text{mL}$$

Second dilution: $$(1 \times 10^{-4}\ M)_A(\text{L})_A = (1 \times 10^{-6}\ M)_B(1\ \text{L})_B$$
$$\text{L}_A = 0.01\ \text{L} = 10\ \text{mL}$$

To prepare the 1×10^{-6} *M* solution, take 10 mL of the 0.01 *M* solution and dilute to 1 L to prepare a 1×10^{-4} *M* solution. Then take 10 mL of the 1×10^{-4} *M* solution and dilute it to 1 L to prepare the desired 1×10^{-6} *M* solution.

Problem 7.8. What are the molar concentrations of the components of Ringer's solution? The recipe for Ringer's solution is 0.7% NaCl, 0.03% KCl, 0.025% $CaCl_2$.

Ans. The concentrations are given in w/v %. Therefore, recalling the discussion in Chap. 6, for NaCl:

$$\left(\frac{0.7\ \text{g}}{100\ \text{mL}}\right)\left(\frac{10}{10}\right) = \frac{7\ \text{g}}{1000\ \text{mL}}$$

$$\frac{7\ \text{g}}{1000\ \text{mL}}\left(\frac{1000\ \text{mL}}{1\ \text{L}}\right) = \frac{7\ \text{g}}{\text{L}}$$

$$\frac{7\ \text{g}}{\text{L}}\left(\frac{1\ \text{mol}}{58.5\ \text{g}}\right) = \frac{0.12\ \text{mol}}{\text{L}} = 0.12\ M$$

We will combine all steps for the KCl and $CaCl_2$ calculations:

KCl: $$\frac{0.03\ \text{g}}{100\ \text{mL}}\left(\frac{10}{10}\right)\left(\frac{1000\ \text{mL}}{\text{L}}\right)\left(\frac{1\ \text{mol}}{74.5\ \text{g}}\right) = 0.004\ M$$

$CaCl_2$: $$\frac{0.025\ \text{g}}{100\ \text{mL}}\left(\frac{10}{10}\right)\left(\frac{1000\ \text{mL}}{\text{L}}\right)\left(\frac{1\ \text{mol}}{111\ \text{g}}\right) = 0.0023\ M$$

Since ions in solution are independent of each other, their concentrations may be individually specified.

	Cations	Anions
Na^+	0.12 *M*	0.12 *M*
K^+	0.004 *M*	0.004 *M*
Ca^{2+}	0.0023 *M*	0.005 *M*
Sum of Charges	0.129 *M*	0.129 *M*

7.5 NEUTRALIZATION AND TITRATION

In Problem 7.8, the ionic components of Ringer's solution were described as independent entities, whose concentrations could be individually specified. However, there are other possible outcomes of mixing ionic substances. One of the most important of these is the formation of water from the reaction of an acid with a base. This reaction is called *neutralization*. In this reaction, a hydrogen ion (H^+) reacts with a hydroxyl ion (OH^-) to form H_2O, while the accompanying anions and cations remain unaffected by the reaction. The anions and cations which are unaffected by a reaction in solution are called *spectator ions*. One cannot add H^+ ions or OH^- ions to a solution. They must be added as some acid, say HCl, or some base, say NaOH. Since the Na^+ and the Cl^- ions are unaffected by the reaction, and appear on both sides of the reaction equation, chemists express that fact by leaving them out of the equation entirely. The resulting equation is called the ***net ionic equation***, consisting of only the ions which undergo reaction: $H^+ + OH^- = H_2O$.

Problem 7.9. Calculate the concentration of all the ions in solution after mixing 0.50 L of 1.0 *M* HCl with 1.0 L of 1.0 *M* KOH.

Ans. We begin by writing the neutralization reaction:

$$HCl + KOH = KCl + H_2O$$

The next step recognizes the stoichiometric requirement that 1 mol of HCl reacts with one and only 1 mol of KOH. We must therefore calculate the numbers of moles of reaction components in the solutions prior to reaction so that we can calculate concentrations after the reaction. This is sometimes called the *mole method*.

HCL: $$0.50\ \text{L}\left(\frac{1.0\ \text{mol}}{\text{L}}\right) = 0.50\ \text{mol of H}^+ \text{ and Cl}^-$$

KOH: $$1.0\ \text{L}\left(\frac{1.0\ \text{mol}}{\text{L}}\right) = 1.0\ \text{mol of OH}^- \text{ and K}^+$$

We can infer from the 1 : 1 molar ratio of H^+ to OH^- in neutralization that OH^- is in excess, and that, after the reaction of 0.5 mol of H^+ with 0.5 mol of OH^- to form water, there will be 0.5 mol of OH^- left over. In addition we must recognize that the final volume of the reaction mixture will be the sum of added volumes, 1.5 L. Therefore the final concentrations of all ions will be

OH^-: $$\frac{0.50\ \text{mol}}{1.5\ \text{L}} = 0.33\ M$$

K^+: $$\frac{1.0\ \text{mol}}{1.5\ \text{L}} = 0.67\ M$$

Cl^-: $$\frac{0.50\ \text{mol}}{1.5\ \text{L}} = 0.33\ M$$

There are two things to note: (1) The sum of positive and negative ions before and after reaction is zero, and (2) since the number of moles of water is 55.6 per liter, we make the assumption that the total volume of solution is not materially affected by the 0.5 mol of water produced by the reaction.

The neutralization reaction is used as an analytical technique to determine the concentrations of acids or bases in solutions of unknown concentration. This method is called *titration*.

Problem 7.10. What is the concentration of OH^- in 100 mL of a KOH solution, which is completely neutralized by 50 mL of 0.1 M HCl?

Ans. Since there is a 1:1 ratio of H^+ to OH^- in the neutralization reaction, we can determine the number of moles of OH^- ion neutralized by calculating the number of moles of H^+ ion required (added) for neutralization of all the OH^- ion:

$$\text{mol } H^+{}_{\text{ADDED}} = 0.05 \text{ L}\left(\frac{0.1 \text{ mol}}{1.0 \text{ L}}\right) = 0.005 \text{ mol} = \text{mol } OH^-{}_{\text{NEUTRALIZED}}$$

Therefore, the concentration of the KOH solution must have been

$$\frac{0.005 \text{ mol}}{0.1 \text{ L}} = 0.05\ M$$

Since mol $H^+{}_{\text{ADDED}}$ = mol $OH^-{}_{\text{NEUTRALIZED}}$, we could have done this calculation in one step using the equation $M_A \times V_A = M_B \times V_B$:

$$(0.1\ M_{\text{HCl}})(0.05 \text{ L}_{\text{HCl}}) = (M_{\text{KOH}})(0.1 \text{ L}_{\text{KOH}})$$

$$M_{\text{KOH}} = 0.05\ M$$

7.6 COLLIGATIVE PROPERTIES, DIFFUSION, AND MEMBRANES

As in the case of gases, there are properties of solutions which do not depend on the nature of the solute, but only on the number (not the kind) of solute molecules per unit volume—the concentration. These are called *colligative properties*, and include *osmotic pressure*, *freezing point depression*, and *boiling point elevation*.

7.6.1 Osmotic Pressure

Diffusion describes the process by which perfume from a bottle opened in one corner of a large lecture hall will be smelled almost immediately throughout the hall. In more precise language, molecules which are free to move in the space in which they are located will tend, in time, to distribute themselves uniformly throughout that space: that is, to move from regions of high concentration to regions of lower concentration, whether they are in the gaseous, liquid, or solid state. Our interests will focus on diffusion and its consequences in solutions.

Membranes are structural barriers which place certain limitations on diffusion. Their properties are such that only certain sized molecules or molecules of a certain electrical charge may be able to pass through. Such a membrane is called *semipermeable*. Those molecules larger than that limiting size or of incorrect electrical charge are then restrained from diffusing through the membrane even though they may be at a higher concentration on their side of the membrane. However, in the case of water and biological membranes, there is no restriction on the flow of solvent through the membrane. Therefore water (the solvent) will diffuse from the low-solute-concentration side to the high-solute-concentration side of the membrane. The reason it takes place is that on the low-solute side of the membrane the water is at a higher concentration than on the high-solute side of the membrane. This diffusion process is called *osmosis*. The driving force for this flow is called the *osmotic pressure*. Therefore, flow of water into a membrane-bounded system will occur when the *external* solution has a lower concentration of solute particles than the *internal* solution. This state of concentration is called *hypotonic*. The inflow causes an increase in the hydrostatic pressure within the system, and consequent swelling. The flow may continue until the membrane-bounded system bursts. For example this occurs in hypotonic solutions of red blood cells. When a solution external to the membrane-bounded system has a higher solute concentration than the solution inside the membrane, it is called *hypertonic*. This situation causes the membrane-bounded system to lose solvent and consequently to shrink. Red blood cells in hypertonic solution shrink and appear crumpled or *crenate*. When the solute concentrations external and internal to a membrane-bounded system are equal, no flow will be observed, and the external solution is said to be *isotonic*. The osmotic pressure depends on the sum of molar concentrations of all independent particles in solution.

Problem 7.11. What is the osmotic pressure of a 0.0100 M solution of glucose at 25 °C?

Ans. The relationship which describes osmotic pressure is analogous to the ideal gas law, $P = (n/V)RT$, with the exception of an additional factor, i:

$$\Pi = i\left(\frac{n}{V}\right)RT$$

where Π = the osmotic pressure, atm
R = the gas constant, 0.0821 ($L \cdot atm \cdot K^{-1} \cdot mol^{-1}$)
T = the temperature, K
n = normally the number of moles present
n/V = the molar concentration, mol/L

However, we must take into account the fact that certain substances when dissolved in solution yield more than one particle per mole, e.g., ionic solids such as NaCl. The factor i accounts for this. In the case of glucose, since it does not ionize in solution, $i = 1$, for NaCl, $i = 2$, for $CaCl_2$, $i = 3$, etc. So, for glucose, the total molar concentration is 0.01 M. Then

$$\Pi = \left(\frac{0.0100 \text{ mol}}{\text{L}}\right)\left(\frac{0.0821 \text{ L} \cdot \text{atm}}{\text{K} \cdot \text{mol}}\right)298 \text{ K}$$

$$= 0.244 \text{ atm}$$

Since 1 atm = 760 mmHg:

$$\Pi = 0.244 \text{ atm}\left(\frac{760 \text{ mmHg}}{\text{atm}}\right)$$

$$= 186 \text{ mmHg}$$

Problem 7.12. What is the osmotic pressure of a 0.900% w/v solution of NaCl at 25.0 °C?

Ans. In order to calculate the osmotic pressure, the solute concentration must be expressed in moles per liter. Thus

$$0.900\% \text{ w/v NaCl} = \left(\frac{0.900 \text{ g}}{100 \text{ mL}}\right)$$

$$\left(\frac{0.900 \text{ g}}{100 \text{ mL}}\right)\left(\frac{10^3 \text{ mL}}{1.00 \text{ L}}\right)\left(\frac{1 \text{ mol}}{58.5 \text{ g}}\right) = 0.154\ M$$

Since each mole of NaCl provides 2 mol of ions in solution, the osmotic pressure is

$$\Pi = 2\left(\frac{0.154 \text{ mol}}{\text{L}}\right)\left(\frac{0.0821 \text{ L} \cdot \text{atm}}{\text{K} \cdot \text{mol}}\right)(298 \text{ K}) = 7.54 \text{ atm} = 5.73 \times 10^3 \text{ mmHg}$$

Problem 7.13. **What is the osmotic pressure of a 0.900% w/v solution of a protein consisting of a single polypeptide chain ($i = 1$) with molar mass 58500 at 25 °C? °C?**

Ans. The molar concentration of the protein is

$$0.900\% \text{ w/v protein} = \left(\frac{0.900 \text{ g}}{100 \text{ mL}}\right)$$

$$\left(\frac{0.900 \text{ g}}{100 \text{ mL}}\right)\left(\frac{10^3 \text{ mL}}{\text{L}}\right)\left(\frac{1 \text{ mol}}{58500 \text{ g}}\right) = 0.000154\ M$$

$$\Pi = \left(\frac{0.000154 \text{ mol}}{\text{L}}\right)\left(\frac{0.0821 \text{ L}\cdot\text{atm}}{\text{K}\cdot\text{mol}}\right)(298 \text{ K}) = 0.00377 \text{ atm}$$

$$0.00377 \text{ atm} = 2.86 \text{ mmHg}$$

The Ringer's solution described in Problem 7.8 was designed to keep the volume of living cells constant. Its ionic molarity is virtually identical to the ionic molarity of living cells, even though the kinds of molecules inside the cell are not the same as in the Ringer's solution. Because the ionic molarities are the same, the rates of flow of water into and out of the cell are the same, and the cell volume remains constant. The solution is called isotonic. Because the identity of the ions is not critical, only their osmotic effects, their total concentration is called the ***osmolarity***, given the symbol Osm. Note that $i(n/V) = i(M) =$ osmolarity.

Problem 7.14. What is the osmolarity of a 0.01 M glucose solution?

Ans. Glucose is a water soluble compound which does not form ions. Therefore the number of moles per liter is identical with its osmolarity, $i(M) = 1(0.01\ M) = 0.01$ Osm.

Problem 7.15. What are the osmolarities of 0.01 M NaCl and 0.01 M $CaCl_2$ solutions?

Ans. NaCl is an ionic substance which yields 2 mol of ions per mole of salt when dissolved in water. NaCl osmolarity $= i(M) = 2(0.01\ M) = 0.02$ Osm.

$CaCl_2$ is an ionic solid which yields 3 mol of ions per mole of salt when dissolved in water. Therefore the osmolarity of a 0.01 M $CaCl_2$ solution is $i(M) = 3(0.01\ M) = 0.03$ Osm. Remember, that in calculating osmotic pressures, the correct concentration term used is the osmolarity, not the molarity.

7.6.2 Freezing Point Depression and Boiling Point Elevation

The vapor pressure of liquids is lowered when mixed with solute to form solutions. The result of this effect is that solutions have lower freezing points and higher boiling points than their pure solvents. The extent to which these effects occur depends, just as in the case of osmotic pressure, on the number of dissolved particles, not on their masses or types. The freezing and boiling point temperature changes of solvents per mole of dissolved solute are unique for each solvent, and are given the symbol $-K_f$ for freezing point depression (note the minus sign), and $+K_b$ for boiling point elevation (note the plus sign). A mole of solute per kilogram of water lowers the freezing point by 1.86 °C and raises the boiling point by 0.512 °C. We must also recognize that since these effects depend upon all the particles in solution, it is necessary to include the same factor, i, the number of particles derived from each mole of dissolved solute, which we introduced in the discussion of osmotic pressure. We express these ideas in the form of the following equations.

For freezing point depression:

$$\Delta t = -iK_f m$$

For boiling point elevation:

$$\Delta t = +iK_b m$$

In both cases, m = molality.

Problem 7.16. What is the freezing point of a 0.100 m (molal) solution of glucose?

Ans. Using the equation for freezing point depression, and K_f for water, we can calculate the number of degrees the 0.100 m solution will lower the freezing point:

$$\Delta t = (1)(0.100\ m)\left(\frac{-1.86\ ^\circ\text{C}}{1\ m}\right)$$

$$= -0.186\ ^\circ\text{C}$$

Since water freezes at 0 °C, the freezing point will be

$$0.000\ ^\circ\text{C} - 0.186\ ^\circ\text{C} = -0.186\ ^\circ\text{C}$$

Problem 7.17. What is the freezing point of a 0.100 m solution of NaCl?

Ans. The value of i for a NaCl solution is 2 (2 mol of ions per mole of salt). Therefore, the freezing point lowering is

$$\Delta t = (2)(0.100\ m)\left(\frac{1.86\ ^\circ\text{C}}{1\ m}\right) = -0.372\ ^\circ\text{C}$$

$$= -0.372\ ^\circ\text{C}$$

Freezing point:

$$0\ ^\circ\text{C} - 0.372\ ^\circ\text{C} = -0.372\ ^\circ\text{C}$$

Problem 7.18. What is the boiling point of an aqueous 0.100 m NaCl solution?

Ans. Using the equation for boiling point elevation, and $+K_b$, whose value is 0.512 °C/molal, and noting that 1 mol of NaCl yields 2 mol of ions, the boiling point elevation is

$$\Delta t\ 2(0.100\ m)\left(\frac{0.512\ ^\circ\text{C}}{1\ m}\right) = 0.102\ ^\circ\text{C}$$

$$= 0.1\ ^\circ\text{C}$$

Boiling point:

$$100\ ^\circ\text{C} + 0.1\ ^\circ\text{C} = 100.1\ ^\circ\text{C}$$

ADDITIONAL SOLVED PROBLEMS

SOLUTIONS AS MIXTURES

Problem 7.19. Sixty-five milliliters of ethanol is dissolved in 45 mL of water. Which component is the solute and which is the solvent?

Ans. The component in excess is considered the solvent. Hence, the ethanol should be labeled as solvent and the water as solute.

Problem 7.20. Would you describe air as a mixture or a solution?

Ans. Air is of variable composition and therefore a mixture of gases, approximately 20% O_2 and 79% N_2, with small amounts of other gases. It is also a solution, because its components cannot be discerned as separate entities and are dispersed molecularly.

Problem 7.21. Is blood a mixture or a solution?

Ans. Blood is not a solution but a suspension because it contains components, e.g., red blood cells, platelets, which can be discerned as separate from the liquid medium. However, those visible components are immersed in a complex solution, called *plasma*, consisting of a mixture of molecules of very different sizes and masses. Along with substances of molecular dimensions comparable to the solvent (e.g., NaCl, glucose) are substances of much greater sizes and masses, that is, proteins of molecular mass up to 250,000. Therefore, plasma is classified as a colloidal solution.

SOLUBILITY

Problem 7.22. What is the solubility of nitrogen in water at 20 °C which is in equilibrium with air?

Ans. The Henry's law constant for nitrogen at 20 °C and 1 atm total pressure is 0.0152 mL N_2/mL H_2O. Air consists of about 80% N_2. The solubility of N_2 depends upon its partial pressure, 0.80 atm, (80 percent of 1 atm).

$$\left(\frac{\text{mL N}_2}{\text{mL H}_2\text{O}}\right)_{\text{AIR}} = 0.80\ \text{atm}\left(\frac{0.0152\ \text{mL N}_2}{\text{mL H}_2\text{O}\cdot\text{atm}}\right) = \frac{0.012\ \text{mL N}_2}{\text{mL H}_2\text{O}}$$

Note that in this calculation, the units of atmosphere cancel. If the partial pressure of a gas is given in millimeters of mercury, the pressure in atmospheres can be calculated from the relationship 1 atm = 760 mmHg = 760 torr.

Problem 7.23. What is the solubility of carbon dioxide in arterial blood?

Ans. The partial pressure of CO_2 in contact with human arterial blood is 40 torr. The Henry's law constant for CO_2 in blood plasma at 37 °C is 0.51 mL CO_2/mL plasma at 1 atm or 760 torr total pressure. Therefore the solubility of $CO_{2\,(\text{ARTERIAL BLOOD})}$ is

$$40\ \text{torr}\left(\frac{0.51\ \text{mL CO}_2}{\text{mL plasma}\cdot 760\ \text{torr}}\right) = \frac{0.027\ \text{mL CO}_2}{\text{mL plasma}}$$

Problem 7.24. What is the molar concentration of CO_2 in human arterial blood?

Ans. We must convert mL CO_2/mL blood to mol CO_2/L of blood.

(1)
$$\left(\frac{0.027\ \text{mL CO}_2}{\text{mL blood}}\right)\left(\frac{1000\ \text{mL}}{\text{L}}\right) = \frac{27\ \text{mL CO}_2}{\text{L blood}}$$

(2) There are 22400 mL/mol of gas at 273 K but we must correct the volume since blood's temperature is 310 K.

$$\left(\frac{22400\ \text{mL CO}_2}{\text{mol CO}_2}\right)\left(\frac{310\ \text{K}}{273\ \text{K}}\right) = \frac{25436\ \text{mL CO}_2}{\text{mol CO}_2}$$

(3)
$$\left(\frac{27\ \text{mL CO}_2}{\text{mol CO}_2}\right)\left(\frac{1\ \text{mol CO}_2}{25436\ \text{mL CO}_2}\right) = 0.0011\ M$$

Problem 7.25. An unknown amount of NH_4Cl was added to 500 mL of water at 40 °C to form a clear solution. The temperature was lowered to 30 °C, and the solution remained clear. Was the solution at 40 °C saturated?

Ans. No. The solubility of NH_4Cl decreases as the temperature is lowered. If the 40 °C solution had been saturated, lowering the temperature would have caused precipitation of the salt. The solution remained clear; therefore, it could not have been saturated.

Problem 7.26. When urea dissolves in water, the solution becomes cold. What would you expect to happen to an aqueous solution of urea with additional solid urea in the beaker if it were heated?

Ans. Urea solutions become cold because dissolution is accompanied by the absorption of heat. Therefore, by providing heat to a saturated solution, more urea could be dissolved.

Problem 7.27. When lithium carbonate is dissolved in water, the solution becomes hot. What would you expect to happen to a saturated solution of this salt if it were heated?

Ans. Lithium carbonate solutions become hot because dissolution is accompanied by the evolution of heat. Therefore, by heating a saturated solution of lithium carbonate, salt would precipitate.

NEUTRALIZATION AND TITRATION

Problem 7.28. A chemist has 50 mL of 0.20 *M* HCL available, and requires 200 mL of 0.075 *M* HCl for titration. Will she be able to produce the required quantity?

Ans. The problem can be solved in one step using the following relationship:

$$M_A \times V_A = M_B \times V_B$$

$$L_{HCl}\left(\frac{0.20\text{ mol}}{L}\right) = 0.20\text{ L}\left(\frac{0.075\text{ mol}}{L}\right)$$

$$= 0.075\text{ L} = 75\text{ mL}$$

To produce the required volume she needs 75 mL of 0.20 *M* HCl, and has only 50 mL on hand. The solution cannot be prepared. A two-step approach, sometimes called the *mole method,* involves calculating the moles availabe, and the moles required:

$$\text{Moles available:}\quad 0.050\text{ L}\left(\frac{0.20\text{ mol}}{L}\right) = 0.0050\text{ mol}$$

$$\text{Moles required:}\quad 0.20\text{ L}\left(\frac{0.075\text{ mol}}{L}\right) = 0.015\text{ mol}$$

The required amount of solution cannot be prepared because the number of moles required is greater than that available.

Problem 7.29. The concentration of sulfuric acid, H_2SO_4, is 36.0 *M*. What volume is required to prepare 5.00 L of 1.00 *M* acid?

Ans.

$$L_{H_2SO_4}\ \frac{36.0\text{ mol}}{L} = 5.00\text{ L}\ \frac{1.00\text{ mol}}{L}$$

$$L_{H_2SO_4} = 0.139\text{ L} = 139\text{ mL}$$

Problem 7.30. 250 mL of a 5.00% w/v saline solution is available in stock. How much 0.900% w/v saline solution can be prepared from it?

Ans.

$$250\text{ mL}_{5\%\text{ SALINE}}\left(\frac{5.00\text{ g}}{100\text{ mL}}\right) = \text{mL}_{0.9\%\text{ SALINE}}\left(\frac{0.900\text{ g}}{100\text{ mL}}\right)$$

$$\text{mL}_{0.9\%\text{ SALINE}} = 1.39 \times 10^3\text{ mL}$$

Problem 7.31. Two liters of a 3.0% w/v solution of glucose was prepared from 0.75 L of a glucose stock solution. What was the concentration of the stock solution?

Ans.

$$(2.0 \text{ L})(3.0\%) = (0.75 \text{ L})(X\%)$$

$$\text{Concentration of stock solution} = 8.0\%$$

Problem 7.32. What is the molarity of a solution prepared by diluting 200 mL of 0.5 *M* KOH to a final volume of 500 mL?

Ans.

$$(M)(500 \text{ mL}) = (0.5\ M)(200 \text{ mL})$$

$$M = 0.2$$

Problem 7.33. The volume of 0.2 *M* HCl required to neutralize 100 mL of a KOH solution was 20 mL. What was the molarity of the KOH solution?

Ans.

$$(20 \text{ mL})(0.2\ M) = (100 \text{ mL})(M)$$

$$M = 0.04$$

COLLIGATIVE PROPERTIES, DIFFUSION, AND MEMBRANES

Problem 7.34. Calculate the osmolarity of the following aqueous solutions: (*a*) 0.30 *M* glucose, (*b*) 0.15 *M* NaCl, (*c*) 0.10 *M* $CaCl_2$.

Ans. We use the relationship that osmolarity = $i(M)$, and determine the factor i in each case.

(*a*) Glucose is a water-soluble compound which does not form ions in solution; therefore, $i = 1$, and its osmolarity is (1)(0.30 *M*) = 0.30 Osm, identical to its molarity.

(*b*) In water, NaCl yields 2 mol of ions per mole of salt. Therefore, $i = 2$, and its osmolarity is (2)(0.15 *M*) = 0.30 Osm.

(*c*) In water, $CaCl_2$ yields 3 mol of ions per mole of salt. Therefore, $i = 3$, and its osmolarity is (3)(0.10 *M*) = 0.30 Osm.

Problem 7.35. Calculate the osmotic pressure of the three solutions in the previous problem.

Ans. Since the osmolarities are identical for the three solutions, their osmotic pressures must be the same, so one calculation will do for all three:

$$\Pi = \text{Osm}(R)(T)$$

$$= 0.30\left(\frac{0.082 \text{ L} \cdot \text{atm}}{\text{K} \cdot \text{mol}}\right)298 \text{ K}$$

$$= 7.3 \text{ atm} = 5.6 \times 10^3 \text{ mmHg}$$

Problem 7.36. A nonionizing solid dissolved in water changed the freezing point to −2.79 °C. Calculate its concentration.

Ans. The freezing point depression constant for water is 1.86 °C/molal. Because the solute was a nonionizing solid, the factor $i = 1$. Using those factors in the equation for freezing point depression, $\Delta t = -iK_f m$:

$$-2.79\ ^\circ\text{C} = (1)(-1.86\ ^\circ\text{C})(m)$$

Solving for m:

$$m = 1.50$$

Using the factor-label method, first write the quantity, and multiply it by the appropriate conversion factor with $i = 1$:

$$-2.79\ ^\circ\text{C}\left(\frac{i \cdot 1\ m}{-1.86\ ^\circ\text{C}}\right) = 1.50\ m$$

Problem 7.37. A sample of $CaCl_2$ dissolved in water changed the freezing point to $-2.79\ ^\circ$C. Calculate its concentration.

Ans. The molality of this solution appears to be identical to that in Problem 7.36. The molality of 1.5 measured by a freezing point depression, a colligative property, must be that of the sum of molalities of all components, and we know that 1 mol of $CaCl_2$ yields 3 mol of ions in aqueous solution. Therefore, $i = 3$, and the molal concentration of $CaCl_2$ is one-third of its colligative molality:

$$i(m) = 1.5$$

$$3(m) = 1.5$$

$$m = \frac{1.5}{3} = 0.50$$

Problem 7.38. Calculate the boiling point of a 0.44 m aqueous solution of $AlCl_3$.

Ans. We note that 1 mol of $AlCl_3$ yields 4 mol of ions in aqueous solution, and, that the boiling point of water is elevated to 0.512 °C/m. Using that constant as a factor in the equation $\Delta t = +iK_b m$. Then

$$\Delta t = (4)\left(+\frac{0.512\ ^\circ\text{C}}{m}\right)(0.44\ m) = 0.90\ ^\circ\text{C}$$

$$\text{Boiling point} = 100.9\ ^\circ\text{C}$$

SUPPLEMENTARY PROBLEMS

Problem 7.39. What volume of 0.100 HCl can be prepared from 1.00 L of 0.875 M HCl?

Ans.

$$(\text{L})(0.100\ M) = (1.00\ \text{L})(0.875\ M)$$

$$\text{L} = 8.75$$

Problem 7.40. What volume of 0.200 M KOH is required to prepare 500 mL of 0.0800 M KOH?

Ans.

$$(\text{mL})(0.200\ M) = (500\ \text{mL})(0.0800\ M)$$

$$\text{mL} = 200$$

Problem 7.41. What was the concentration of stock saline solution if 100 mL of it was used to prepare 500 mL of 0.90% w/v saline?

Ans.

$$(100\ \text{mL})(\%\ \text{w/v}) = (500\ \text{mL})(0.90\%\ \text{w/v})$$

$$\%\ \text{w/v} = 4.5$$

Problem 7.42. Calculate the molarity of a solution prepared by diluting 200 mL of 1.0 M KOH to a final volume of 1250 mL.

Ans.

$$(M)(1250\text{ mL}) = (1.0)(200\text{ mL})$$

$$M = 0.16$$

Problem 7.43. Twenty milliliters of 0.15 M KOH was required to neutralize 47.5 mL of HCl. Calculate the concentration of the HCl solution.

Ans.

$$(M)(47.5\text{ mL}) = (0.15\ M)(20\text{ mL})$$

$$M = 0.063$$

Problem 7.44. The osmotic pressure of a glucose solution was found to be 371.5 torr at 25 °C. Calculate its concentration.

Ans.

$$371.5\text{ torr}\left(\frac{1\text{ atm}}{760\text{ torr}}\right) = (1)(M)\left(\frac{0.082\text{ L}\cdot\text{atm}}{\text{K}\cdot\text{mol}}\right)298\text{ K}$$

$$M = 0.020$$

Problem 7.45. An NaCl solution was found to have an osmotic pressure of 77.3 torr at 37 °C. Calculate its molarity.

Ans.

$$77.3\text{ torr}\left(\frac{1\text{ atm}}{760\text{ torr}}\right) = (2)(M)\left(\frac{0.082\text{ L}\cdot\text{atm}}{\text{K}\cdot\text{mol}}\right)310\text{ K}$$

$$M = 0.0020$$

Problem 7.46. Calculate the freezing point of a 0.11 m aqueous solution of glucose.

Ans.

$$\Delta t = (1)(0.11\ m)\left(\frac{-1.86\ ^\circ\text{C}}{1.0\ m}\right) = -0.20\ ^\circ\text{C}$$

Freezing point: 0.00 °C − 0.20 °C = −0.20 °C

Problem 7.47. Calculate the boiling point of a 1.25 m $CaCl_2$ aqueous solution.

Ans.

$$i = 3$$

$$\Delta t = 3(1.25\ m)\left(\frac{0.512\ ^\circ\text{C}}{1.00\ m}\right) = 1.92\ ^\circ\text{C}$$

Boiling point: 100.00 °C + 1.92 °C = 101.92 °C

Problem 7.48. Calculate the concentration of a glucose solution whose freezing point was found to be $-0.52\ °C$.

Ans.

(1) $$\Delta t = i(-K_f)(m)$$

(2) $$i = 1$$

(3) $$m = \frac{\Delta t}{(-K_f)}$$

Therefore:

$$-0.52\ °C\left(-\frac{1\ m}{1.86\ °C}\right) = 0.28\ m$$

Chemical Reactions

8.1 INTRODUCTION

In this chapter we will consider some additional fundamental aspects of chemical reactions, i.e., how they occur, the driving forces behind them, and their dependence upon specifics of molecular structure and conditions of reaction, such as temperature and concentration. There are two aspects of chemical reactions which, though interrelated, are dealt with as separate topics. The first of these is the study of the reaction from its initiation to the point where the system seems to undergo no further change, called *chemical kinetics*. The second deals with the system after all apparent change has stopped, and is called *chemical equilibrium*. Following those topics we will examine some specific aspects of chemical equilibria involving oxidation-reduction reactions in aqueous solutions and combustion reactions.

8.2 CHEMICAL KINETICS

Chemical kinetics deals with the speed with which chemical reactions occur, and also attempts to account for the energetic and molecular structural characteristics of the chemical systems undergoing change.

8.2.1 Collision Theory

It is generally considered that, in order for a chemical reaction to occur, a molecular collision must take place between the reactants. Many such collisions take place in chemical systems; however, not all of them will lead to a chemical reaction. If a chemical reaction, in other words a successful or effective collision, is to occur, the chemical bonds of both partners must break and then reform into products. The atoms in molecules are bound together by chemical bonds, but they are in constant motion. Since they cannot escape from each other, that motion is restricted to vibration. In a successful collision, the kinetic energy of the colliding molecules is converted into internal energy of vibration. If the transferred kinetic energy and consequent vibrations are great enough, the bonds will break and new bonds, hence new molecules, may form. The kinetic energy which is converted to internal energy of vibration great enough to cause bond breakage is called the *activation energy*. In general, the greater the activation energy, the slower the reaction, and the smaller the activation energy, the faster the reaction.

A successful collision also requires that molecules be properly oriented with respect to each other. A glancing blow may set a molecule spinning, but not vibrating. However, a direct blow along the line of centers connecting two atoms will be effective in causing atoms to vibrate.

In a successful collision, the colliding molecules do not simply bounce off each other. The electrons and nuclei of both molecules merge to form a short-lived, unstable structure. It consists of the atoms of all the reactants connected to each other, and vibrating over distances larger than the stable bonds of either the reactants or products. This unstable, distorted state is called the *transition complex* or *activated state*. It is the molecular stage between reactant and product. It is neither reactant nor product, but may be considered both. It can dissociate into either product or reactant. The activated state may be reached from either direction, starting either with reactants or products. The activation energy required when starting with reactants is called E_a (forward), the activation energy for the forward reaction. The activation energy for the reverse or backward reaction is called E_a (reverse). Let us examine the progress of a collision between two molecules, A and B to form a product C, illustrated in Figs. 8-1 and 8-2. The reaction is written as occurring in two steps:

$$A + B \longrightarrow AB^*$$
$$AB^* \longrightarrow C$$

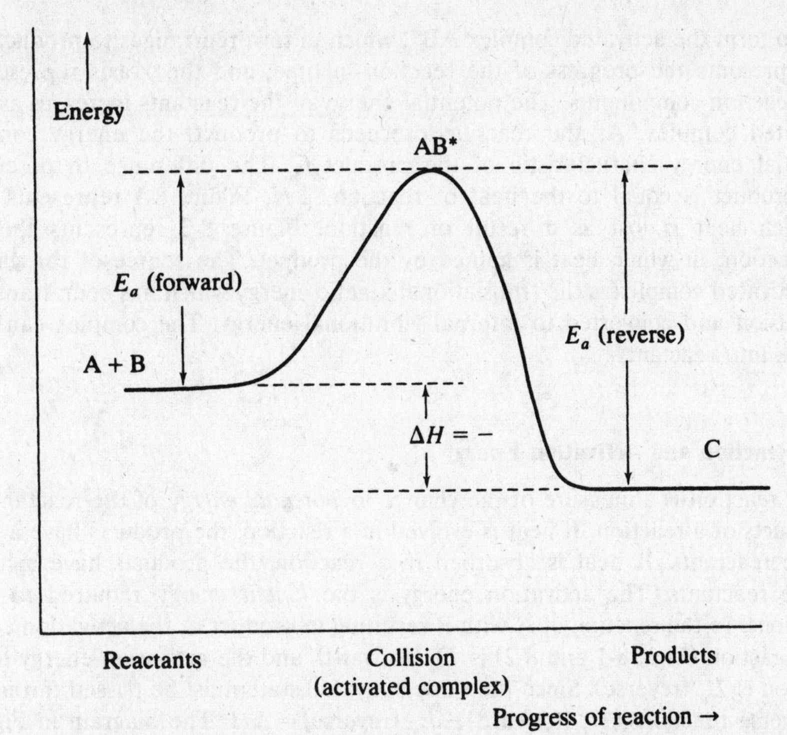

Fig. 8-1 Energy relationships in an exothermic reaction. (*After J. L. Rosenberg and L. M. Epstein, Schaum's Outline of Theory and Problems of College Chemistry, 7th ed., McGraw-Hill, New York, 1990.*)

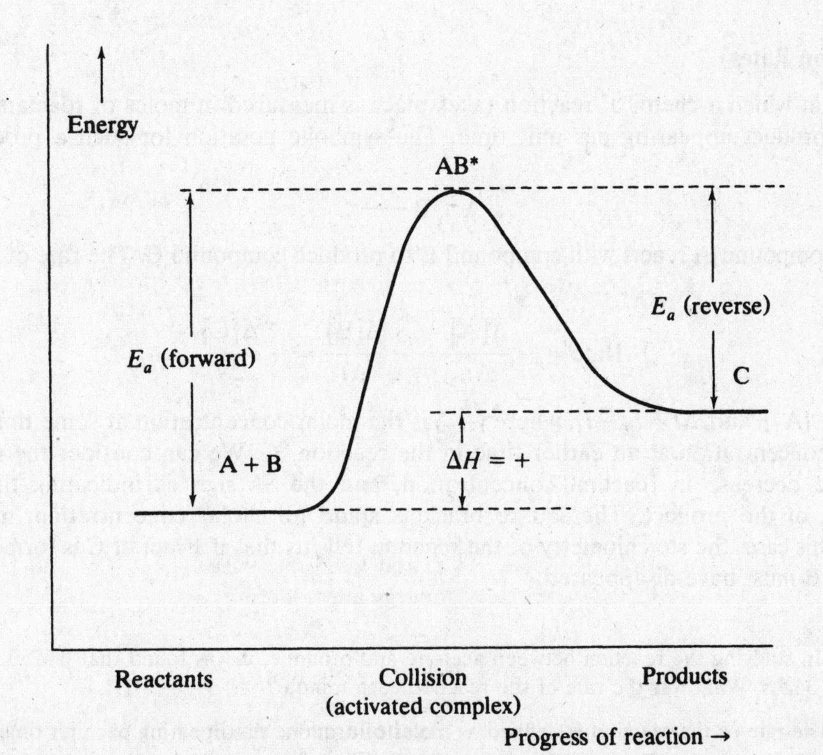

Fig. 8-2 Energy relationships in an endothermic reaction. (*After J. L. Rosenberg and L. M. Epstein, Schaum's Outline of Theory and Problems of College Chemistry, 7th ed., McGraw-Hill, New York, 1990.*)

A and B react to form the activated complex AB*, which in turn rearranges to product C. The X-axis in the figure represents the progress of the reaction in time, and the Y-axis represents the energy content of the reaction components. The potential energy of the reactants increases as they collide to form the activated complex. As the reaction proceeds to product, the energy content falls to a different potential energy characteristic of the product C. The difference in potential energy of reactants and product is equal to the heat of reaction, ΔH. Figure 8-1 represents an exothermic reaction in which heat is lost as a result of reaction. Figure 8-2 represents the course of an endothermic reaction, in which heat is gained by the product. The source of the gain in potential energy of the activated complex is the translational kinetic energy which has been transferred into the complex by collision and converted to internal vibrational energy. The complex can dissociate into products or back into reactants.

8.2.2 Heat of Reaction and Activation Energy

The heat of reaction is a measure of the change in *potential energy* of the reactants compared to that of the products of a reaction. If heat is evolved in a reaction, the products have a lower potential energy than the reactants. If heat is absorbed in a reaction, the products have a higher potential energy than the reactants. The activation energy is the *kinetic energy* required to cause effective molecular collisions. In the reaction of A with B resulting in product C, the activation energy for 1 mol of the forward reaction (Figs. 8-1 and 8-2) is E_a (forward), and the activation energy for 1 mol of the backward reaction is E_a (reverse). Since the same activated state must be passed through for both the forward and reverse reactions, E_a (forward) $-$ E_a (reverse) $= \Delta H$. The diagram in Fig. 8-1 indicates that ΔH is negative, i.e., the product has a lower potential energy than the reactants. This means that heat was evolved (lost), and the reaction was exothermic. Figure 8-2 describes an endothermic (heat gained by the product) reaction. In this reaction, E_a (forward) is greater than E_a (reverse).

8.2.3 Reaction Rates

The rate at which a chemical reaction takes place is measured in moles of reactant disappearing or moles of product appearing per unit time. The symbolic notation for such a process is usually written as

$$A + B \longrightarrow C$$

which reads compound A reacts with compound B to produce compound C. The rate of the reaction is written as

$$\text{Rate} = -\frac{\Delta[A]}{\Delta t} = -\frac{\Delta[B]}{\Delta t} = +\frac{\Delta[C]}{\Delta t}$$

$\Delta[A] = [A_2] - [A_1]$, and $\Delta t = t_2 - t_1$, where $[A_2]$ is the molar concentration at some time t_2, and $[A_1]$ is the molar concentration at an earlier time in the reaction, t_1. We can consider the minus signs as indicating the decrease in reactant concentration, and the + sign as indicating the increase in concentration of the product. The square brackets stand for molar concentration, moles per liter (mol/L). In this case, the stoichiometry of the reaction tells us that if 1 mol of C is formed, 1 mol of A and 1 mol of B must have disappeared.

Problem 8.1. In studying the reaction between acetone and bromine, it was found that 0.0250 mol of bromine disappeared in 34.5 s. What was the rate of the reaction each minute?

Ans. The rate of the reaction is defined as moles of bromine disappearing per unit time. Therefore

$$\text{Rate} = \left(\frac{-0.0250\ \text{mol}}{34.5\ \text{s}}\right)\left(\frac{60.0\ \text{s}}{\text{min}}\right) = -0.0435\frac{\text{mol}}{\text{min}}$$

8.2.4 Effect of Concentration on Reaction Rate

The likelihood of collision increases as the numbers of molecules increase, provided that the volume containing them remains the same. The probability of finding a molecule A in some volume of its container is the same as its concentration, i.e., double its concentration and you double the probability of finding it. The same goes for a molecule B, with which it may react. The probability of collision increases as the concentration of reactants increases. In general, increasing the concentrations of reactants will increase the rate of reaction.

8.2.5 Effect of Temperature on Reaction Rate

The kinetic theory informs us that the average kinetic energy of an assembly of molecules is directly proportional to temperature. That is why raising the temperature of reacting systems increases the rate of reaction. It increases the number of collisions per unit time. In order to double the average kinetic energy of a reacting system, one must double the absolute temperature, say, from 200 K to 400 K. However, experience has shown that a 10-K rise in temperature will about double the rate of a reaction. This is because only a small percentage of molecules, perhaps 1 to 2 percent, in any large assembly has sufficient kinetic energy to collide successfully. It is this small proportion of molecules which undergoes reaction per unit time. Raising the temperature a few degrees significantly increases the proportion of those molecules above the minimum energy, the activation energy, that must be available if a reaction is to occur.

8.2.6 Effect of Catalysts on Reaction Rate

The rates of reactions can also be effected by chemical or physical agents called catalysts which enter into the reaction, but are not consumed. These agents speed up the rate at which the reaction proceeds by reducing the activation energy for the reaction. This is accomplished by modifying the chemical route to the final product, i.e., by changing the nature of the transition complex as enzymes do, or introducing new steps into the reaction, which are faster overall than the uncatalyzed reaction. However, they do not change the final results, i.e., the amounts of products and reactants present when the reaction appears to proceed no further.

Problem 8.2. Sulfur dioxide, SO_2, along with traces of nitrogen monoxide, is produced by the burning of coal. Its oxidation to SO_3 by O_2 is ordinarily a very slow reaction. Unfortunately, its conversion to SO_3 which dissolves in water to form sulfuric acid, H_2SO_4, occurs very rapidly in the atmosphere. What is the explanation for its rapid oxidation?

Ans. The direct and slow oxidation of sulfur dioxide by oxygen is represented by the following equation:

$$2\,SO_2 + O_2 \longrightarrow 2\,SO_3$$

In the presence of NO, the reaction takes place in two steps:

(1) $$2\,NO + O_2 \longrightarrow 2\,NO_2$$

(2) $$2\,SO_2 + 2\,NO_2 \longrightarrow 2\,SO_3 + 2\,NO$$

(3) $$2\,SO_2 + O_2 \longrightarrow 2\,SO_3$$

Equation (3) is the sum of Eq. (1) and (2). The NO_2 produced by the first very rapid reaction is used up in the second rapid reaction leading to the regeneration of NO, the reactant in reaction (1). When we add reactions (1) and (2), since NO and NO_2 appear on both sides of the equations, they will algebraically cancel, and the result is identical to what occurs in the absence of NO. NO is a catalyst. It enters into the reaction, speeds it up by offering an alternate pathway of lower activation energy, but is not consumed by the reaction. Reactions (1) and (2) represent a ***reaction mechanism***, an explanation of the stoichiometry of a chemical reaction in terms of a series of elementary chemical reactions. The NO_2, produced in reaction 1, and consumed in reaction (2) is called a *reaction intermediate*. The same scheme is involved in enzymatic catalysis, the generation of a reaction intermediate followed by the evolution of product.

8.3 CHEMICAL EQUILIBRIUM

Up to now, chemical reactions have been written as though reactants are completely converted to products, or as is usually said, go to completion. However, there are many cases in which, as products are formed, they react with each other to reform the reactants. These situations are characterized by the simultaneous occurrence of two opposed chemical reactions. In these cases, some product is formed, but some reactants are also present at the same time. The reactions proceed to a point where no further change in concentrations can be detected, and at that time, a state called *chemical equilibrium* has been reached. Even those cases where reactions go to completion are equilibrium systems in which the amounts or reactant left over are too small to be of practical importance. Nitrogen reacts with oxygen to produce nitrogen monoxide, and the reaction can be written

$$N_2 + O_2 \longrightarrow 2\,NO$$

We will call this the *forward* reaction. When sufficient NO has been produced, it reacts with itself to produce N_2 and O_2 as

$$2\,NO \longrightarrow N_2 + O_2$$

and this is the *reverse* reaction. The reaction is said to be *reversible*. When the rate of the forward reaction is equal to the rate of the backward reaction, no further change in concentrations can occur, and the system is said to be in chemical equilibrium. We can indicate an equilibrium system by writing the equation with two arrows pointing in opposite directions as

$$N_2 + O_2 \rightleftharpoons 2\,NO$$

8.3.1 Equilibrium Constants

There is a mathematical relationship derived from experiment and verified by theory which describes the equilibrium state and is called the *equilibrium constant expression*. For the reaction of N_2 with O_2, it takes the form

$$K_{eq} = \frac{[NO]^2}{[N_2][O_2]}$$

There are several conventions which apply very strictly to this expression.

1. The products (in terms of the reaction equation as written) always appear in the numerator, with reactants in the denominator. The above reaction is reversible, and therefore could have been written $2\,NO \rightleftharpoons N_2 + O_2$, in which case the equilibrium concentrations of all components would have been the same, but the equilibrium constant expression written for that process would have had a different numerical value, and would have been written

$$K'_{eq} = \frac{[N_2][O_2]}{[NO]^2}$$

where $K'_{eq} = 1/K_{eq}$.
2. The brackets indicate that only molar concentrations may be used.
3. The molar concentrations are raised to a power equal to the stoichiometric coefficients of the stoichiometrically balanced equation.
4. We must begin to use some new notation which has been left out up to now for simplicity, but is essential to complete the characterization of a chemical process. From now on, we will indicate the physical state of each component in the reaction equation. Pure liquids and solids do not appear in the equilibrium constant expression for a reaction in which they take part. The reason for this is that the concentration, in moles per liter, of a pure solid or liquid cannot be varied, and therefore cannot affect the equilibrium state. For example, in the above reaction, all components are in the gaseous state, so the reaction should have been written

$$N_2(g) + O_2(g) \rightleftharpoons 2\,NO(g)$$

There is good reason for this notation. In this reaction all components are gases. Variation in pressure or volume will change the concentration of all the components which will affect the equilibrium. The molar concentration of each component must therefore appear in the equilibrium expression.

$$K_{eq} = \frac{[NO]^2}{[O_2][N_2]}$$

Carbon in the solid state reacts with oxygen to form carbon dioxide, as

$$C(s) + O_2(g) \rightleftharpoons CO_2(g)$$

In this reaction the concentration of carbon, a pure solid, cannot be changed. It therefore can have no effect on the equilibrium, and the equilibrium constant for that reaction is

$$K_{eq} = \frac{[CO_2]}{[O_2]}$$

C(s) does not appear in the equilibrium constant expression.

The reaction of barium ions and sulfate ions in water (aqueous media) to form an insoluble precipitate of barium sulfate is written

$$Ba^{2+}(aq) + SO_4^{\ 2-}(aq) \rightleftharpoons BaSO_4(s)$$

If we reverse this reaction, the equation describes the dissolution of the salt in water:

$$BaSO_4(s) \rightleftharpoons Ba^{2+}(aq) + SO_4^{\ 2-}(aq)$$

Since the concentration of the barium sulfate, a pure solid, cannot be changed, the equilibrium constant for this process is

$$K_{eq} = [Ba^{2+}][SO_4^{\ 2-}]$$

This equilibrium constant has a special name, the solubility product constant, K_{sp}.

The reaction of hydrogen ion with hydroxyl ion in water, called *neutralization*, is written

$$H^+(aq) + OH^-(aq) \rightleftharpoons H_2O(l)$$

Since this reaction takes place in water, the pure liquid water formed by the neutralization reaction adds to the large volume of pure liquid water already present. Its concentration cannot be varied, and therefore it does not appear in the equilibrium constant expression. The equation written in reverse describes the dissociation or ionization of water:

$$H_2O(l) \rightleftharpoons H^+(aq) + OH^-(aq)$$

The equilibrium constant expression for this process is

$$K_{eq} = [H^+][OH^-]$$

Water does not appear in the expression. This equilibrium constant also has a special name, the ion product of water, K_w.

Problem 8.3. Write the equilibrium constant expression for the reaction

$$N_2(g) + 3\ H_2(g) \rightleftharpoons 2\ NH_3(g)$$

Ans. All components are present in the gaseous state, and must appear in the equilibrium constant expression:

$$K_{eq} = \frac{[NH_3]^2}{[N_2][H_2]^3}$$

Problem 8.4. Write an equilibrium constant expression for the following reaction:

$$PCl_3(l) + Cl_2(g) \rightleftharpoons PCl_5(s)$$

Ans. Because the concentrations of pure liquids and solids cannot be varied, those components cannot appear in the equilibrium constant expression:

$$K_{eq} = \frac{1}{[Cl_2]}$$

Problem 8.5. Write an equilibrium constant expression for the following reaction:

$$C(s) + H_2O(g) \rightleftharpoons CO(g) + H_2(g)$$

Ans. Carbon is present in the solid state and its concentration cannot be varied, so it does not appear in the equilibrium constant expression:

$$K_{eq} = \frac{[CO][H_2]}{[H_2O]}$$

The equilibrium constant K_{eq} always has a constant numerical value except when the temperature is changed. The temperature has a significant effect on the value of the equilibrium constant. The changes depend on whether the process is exothermic or endothermic. These effects will be discussed in the section on the Le Chatelier principle. Changes in concentration may occur, e.g., one could add some NO to a system of N_2 and O_2 which has already reached equilibrium, in which case, a new state of equilibrium will be produced. All concentrations will readjust to new values so as to maintain constant the numerical value of the equilibrium constant. The magnitude of the equilibrium constant tells us a good deal about the reaction. If the value is large, say 100, we know that at equilibrium much more product is present than reactant. If it is small, say, 1×10^{-3}, very little reaction has taken place and much more reactant is present than product. If the constant is about 1.0, roughly equal amounts of product and reactant are present. We can use the equilibrium constant to determine the extent of conversion of reactants into products.

Problem 8.6. Consider the reaction: $2\,HI(g) \rightleftharpoons H_2(g) + I_2(g)$. The equilibrium constant for this reaction is 10. If the reaction is started with 2 mol of HI, how many moles of HI will be left at equilibrium?

Ans. We first write the equilibrium constant expression:

$$K_{eq} = \frac{[H_2][I_2]}{[HI]^2}$$

We know that at equilibrium, some of the HI will have been converted to product, and therefore some I_2 and H_2 will have appeared. The stoichiometry tells us that for every two molecules of HI disappearing, one molecule each of I_2 and H_2 will appear. We set up a system to account for the changes in concentration leading to equilibrium concentrations for all components. We let $2x$ mol of HI disappear, and x mol each of I_2 and H_2 appear as in the following diagram:

	HI	I_2	H_2
Start	2.0	0.0	0.0
Change	$-2x$	$+x$	$+x$
Final	$2.0 - 2x$	x	x

We now substitute the final concentrations into the equilibrium constant expression:

$$K_{eq} = \frac{[x]^2}{[2.0 - 2x]^2} = 10$$

Since both sides are perfect squares, we take the square root of both sides, and the solve for x:

$$\frac{[x]}{[2.0-2x]} = 3.16$$

$$[x] = 0.86$$

Therefore, at equilibrium:

$$[\text{HI}] = 2.0 - 2(0.86) = 0.28M$$

$$[\text{I}_2] = [\text{H}_2] = x = 0.86M$$

8.4 LE CHATELIER PRINCIPLE

A system in chemical equilibrium can be subjected to changes in conditions that displace it from equilibrium. The shift will be from left to right or right to left depending on the nature of the stress. We do not need to know the equilibrium constant expression or its numerical value to predict the direction of the shift. It can be predicted by applying the *Le Chatelier principle*: If a chemical reaction at equilibrium is subjected to a change in conditions which displaces it from equilibrium, the chemical system will readjust to a new equilibrium state which tends to reduce the stress.

For example, let us consider the reaction

$$N_2(g) + O_2(g) \rightleftharpoons 2\ NO(g)$$

to be at equilibrium. If we were to add some N_2 to this system, its concentration would be higher than its previous equilibrium concentration. To restore equilibrium, some of the N_2 must be used up to form NO, and to do this, some O_2 must also be used up. The final concentration of N_2 would be lower than the added N_2 plus the N_2 previously present, but higher than the original N_2 concentration. The final O_2 concentration would be lower than the original O_2 concentration, and the final NO concentration would be somewhat higher than the original concentration. The reaction would have shifted to the right. If we had added some NO to the equilibrium mixture, using the same reasoning, the reaction would have shifted to the left.

Problem 8.7. Consider the following reaction:

$$N_2O_4(g) \rightleftharpoons 2\ NO_2(g)$$

What would be the result of adding some NO_2 to the equilibrium mixture?

Ans. The stress would be relieved by reducing the NO_2 concentration. The result would be to produce more N_2O_4. The reaction would shift to the left.

In chemical reactions, heat, called the *heat of reaction*, may be either produced or absorbed as products are formed. A reaction in which heat is produced is called an *exothermic reaction*. One in which heat is absorbed is called an *endothermic reaction*. The reaction of N_2 with H_2 to produce NH_3 is accompanied by the evolution of heat, i.e., it is an exothermic reaction. The complete equation for this process is

$$N_2 + 3\ H_2 \rightleftharpoons 2\ NH_3 + 98\ \text{kJ}$$

The heat of reaction, $\Delta H = -98$ kJ. The minus sign indicates that, as a result of reaction, the system has lost heat. Adding heat to this equilibrium system places a stress on it which can be relieved by the absorption of the heat. In that case, the equilibrium will shift to the left. Removing heat will cause the equilibrium to shift to the right.

Problem 8.8. Consider the following equilibrium system:

$$CaCO_3 + 158\ kJ \rightleftharpoons CaO + CO_2$$

In which direction will the equilibrium shift if heat is (*a*) added, and (*b*) removed from the system?

Ans. The heat of reaction, $\Delta H = +158$ kJ. The plus sign indicates that, as a result of reaction, the products have gained (absorbed) heat. (*a*) Adding heat to the system will shift the equilibrium in the direction of products, to the right. (*b*) Removing heat will shift the equilibrium in the direction of reactants, to the left. These effects explain the changes in value of equilibrium constants upon change in temperature.

8.5 OXIDATION-REDUCTION REACTIONS

In Chap. 3, chemical reactions were described as being either one of two general types: type I reactions, in which the valence shells of the reactants remain unchanged after conversion to products, and type II reactions, in which the valence shells of the reactants are modified by the gain or loss of electrons. Type II reactions are oxidation-reduction or redox reactions, and are the subject of this discussion. Balancing type I reactions was outlined in Chap. 5, and involves application of the law of conservation of mass, i.e., equating the numbers of atoms of each type on the reactant side with the number of atoms of each type on the product side. Redox reactions are subject not only to conservation of mass, but also to conservation of electrons. In a redox reaction, the number of electrons lost by the reductant must be equal to the number of electrons gained by the oxidant. In order to decide whether a reaction is type I or type II, we will use a method which assigns a number to an atom defining its oxidation state called an *oxidation number*. If that number undergoes a change as a result of a chemical reaction, the reaction is a redox reaction. Furthermore the changes in oxidation number of reductant and oxidant will be used to balance redox reaction equations.

8.5.1 Oxidation States

The following are a set of rules which will allow you to assign an oxidation number to atoms whether in ionic or covalent compounds. They are listed in priority of application, i.e., Rule 1 takes priority over Rule 2, which takes priority over Rule 3, etc.

RULE 1. Free elements are assigned an oxidation number of zero. The modern usage is to call an oxidation number an *oxidation state*, so that free elements have an oxidation state of zero.

RULE 2. The sum of oxidation states of all the atoms in a species (compound or polyatomic ion) must be equal to the net charge of the species.

RULE 3. In compounds, the Group I metals, Na, K, etc., are assigned an oxidation state of +1. Note that the charge of an ion is written as 1+ or 2+. The oxidation states are written +1, +2, etc., so as to differentiate them from ionic charge.

RULE 4. In its compounds, fluorine is always assigned an oxidation state of −1, since it is the most electronegative element.

RULE 5. The Group II metals, Ca, Mg, etc., are always assigned an oxidation state of +2, and Group III ions, Al, Ga, are assigned oxidation state of +3. These states are identical to the ionic charges of these elements in compounds.

RULE 6. Hydrogen in compounds is assigned an oxidation state of +1.

RULE 7. Oxygen in compounds is assigned an oxidation state of −2.

These rules were developed by considering numbers of electrons and relative electronegativities, so that with ions like Na^+ or Ca^{2+} the ionic charges and the oxidation states are identical. However, in covalent compounds like NO_2 or CH_4, this is not the case. Even though the combined atoms have no real charge, they are assigned oxidation states which imply an electrical charge.

Problem 8.9. Determine the oxidation states of each atom in the following compounds: (*a*) $CaCl_2$, (*b*) NaH, (*c*) SO_2, (*d*) H_2O_2.

Ans. (*a*) Ca is assigned an oxidation state of +2 (Rule 5). Because $CaCl_2$ is a neutral species, the oxidation state of Cl must be −1. Using Rule 2, we set up an algebraic equation with one unknown quantity to solve for the oxidation state of Cl in $CaCl_2$:

$$(+2) + (2\ \text{Cl}) = 0$$

$$\text{Cl} = -1$$

The oxidation state of Cl in $CaCl_2$ is −1.

(*b*) Na is assigned an oxidation state of +1 (Rule 3), and Rule 2 informs us that in NaH, H must have an oxidation state of −1. Rule 6 is not violated since it is of lower priority than Rule 3. Hydrogen forms compounds with metals of Groups I and II called *hydrides.*

(*c*) Oxygen is assigned an oxidation state of −2 (Rule 7). The oxidation state of sulfur (Rule 2) is therefore

$$(\text{S}) + (2 \times -2) = 0$$

$$\text{S} = +4$$

The oxidation state of sulfur in SO_2 is +4.

(*d*) Rule 6 requires H to have an oxidation state of +1. Therefore by Rule 2:

$$(2 \times +1) + 2\ \text{O} = 0$$

$$\text{O} = -1$$

The oxidation state of oxygen in H_2O_2 is −1, typical of peroxides. There is no conflict because Rule 6 has priority over Rule 7.

Oxidation states are used for two purposes: to determine if a redox reaction has occurred, and to balance the redox equations. In Chap. 3, oxidation was defined as a loss of electrons from the valence shell of the metals, and reduction was defined as the gain of electrons by the valence shell of a nonmetal. The reaction of sodium with chlorine was written as two separate processes, to emphasize the loss of electrons by sodium with the simultaneous gain of electrons by chlorine, as

(1) $$2\ \text{Na} \longrightarrow 2\ \text{Na}^+ + 2\,e^-$$

(2) $$\text{Cl}_2 + 2\,e^- \longrightarrow 2\ \text{Cl}^-$$

Reaction (1) is an oxidation, and reaction (2) is a reduction. The rules tell us that the oxidation state of sodium changed from 0 to +1, and that of the chlorine changed from 0 to −1. We can generalize these observations to say that any change in oxidation state to a more positive state is an oxidation, and any change in oxidation state to a less positive (or more negative) state is a reduction.

Problem 8.10. What is the oxidation state of carbon in each of the following compounds, and how are these compounds related to each other in terms of their oxidation states? (*a*) CH_4, (*b*) CH_3OH, (*c*) CH_2O, (*d*) HCOOH, (*e*) CO_2.

Ans. (*a*) By Rule 6, H is assigned an oxidation state of +1. Using Rule 2:

$$(\text{C}) + (4 \times +1) = 0$$

$$\text{C} = -4$$

In methane, carbon has an oxidation state of −4.

(*b*) Rule 6 assigns an oxidation state of +1 to H, and Rule 7 assigns an oxidation state of −2 to 0. Using Rule 2:

$$(\mathrm{C}) + (4 \times +1) + (-2) = 0$$
$$\mathrm{C} = -2$$

In methyl alcohol, carbon has an oxidation state of −2.

(*c*) By Rule 2, using the previous assignments of oxidation states to hydrogen and oxygen,

$$(\mathrm{C}) + (2 \times +1) + (-2) = 0$$
$$\mathrm{C} = 0$$

In formaldehyde, carbon has an oxidation state of 0.

(*d*) By Rule 2, using the previous assignments of oxidation states to hydrogen and oxygen,

$$(\mathrm{C}) + (2 \times +1) + (2 \times -2) = 0$$
$$\mathrm{C} = +2$$

In formic acid, carbon has an oxidation state of +2.

(*e*) By Rule 2, using the previous assignments of oxidation state to oxygen,

$$(\mathrm{C}) + (2 \times -2) = 0$$
$$\mathrm{C} = +4$$

In carbon dioxide, carbon has an oxidation state of +4.

This series of compounds represents the range of oxidation states of carbon in compounds with hydrogen and oxygen. It is in its maximally reduced state in methane and maximally oxidized state in carbon dioxide. This analysis will be of more significance when you begin your study of organic chemistry.

8.5.2 Balancing Redox Reactions

Let us reexamine the reaction between sodium and chlorine, but rewrite it as one reaction:

$$\mathrm{Na} + \mathrm{Cl_2} \longrightarrow \mathrm{Na^+} + 2\,\mathrm{Cl^-}$$

This equation is not balanced, but by inspection it is easily balanced:

$$2\,\mathrm{Na} + \mathrm{Cl_2} \longrightarrow 2\,\mathrm{Na^+} + 2\,\mathrm{Cl^-}$$

When we write the reaction as two *half-reactions*, we can see that not only are the atoms balanced, but the electrons transferred are also balanced.

(1) $$2\,\mathrm{Na} \longrightarrow 2\,\mathrm{Na^+} + 2\,e^-$$

(2) $$\mathrm{Cl_2} + 2\,e^- \longrightarrow 2\,\mathrm{Cl^-}$$

Balancing this redox reaction is deceptively simple, since the atom balance and electron balance are the same. Guessing the balancing coefficients for the typical and more complex redox reactions can be a frustrating experience, so we will use a systematic procedure which will permit a straightforward approach to solving these problems. It is called the *ion-electron* or *half-reaction method*. The keys to balancing redox reactions of any level of complexity are first to recognize the changes in oxidation states of oxidant and reductant, and second to make certain that the numbers of electrons lost by the reductant are equal to the numbers of electrons gained by the oxidant. In the following material, the equations will be written as the net ionic equations. Spectator ions will not appear, so you may assume that when you see Fe^{3+}(aq), some soluble iron salt like $FeCl_3$ had been added.

Problem 8.11. Balance the following equation:

$$Fe^{2+}(aq) + MnO_4^-(aq) \xrightarrow{H^+(aq)} Fe^{3+}(aq) + Mn^{2+}(aq)$$

Ans. It is important to note that this reaction takes place in an acidic solution. The first step is to separate the oxidation half-reaction from the reduction half-reaction:

Step 1.

$$Fe^{2+} \longrightarrow Fe^{3+} \qquad \text{oxidation}$$
$$MnO_4^- \longrightarrow Mn^{2+} \qquad \text{reduction}$$

Step 2. Make certain that both equations are balanced with respect to all elements other than oxygen and hydrogen. In this case, they are balanced.

Step 3. Balance the oxygen atoms. This is done by noting that the oxygen of MnO_4^- must appear as H_2O, one of the products of the reduction step. In this case, only the manganese half-reaction involves oxygen. There are four oxygen atoms on the left, and none on the right, so we balance the oxygens by adding 4 H_2O to the right-hand side of the manganese half-reaction:

$$Fe^{2+} \longrightarrow Fe^{3+}$$
$$MnO_4^- \longrightarrow Mn^{2+} + 4\,H_2O$$

Step 4. Balance with respect to hydrogen. There are eight hydrogen atoms on the right-hand side of the MnO_4^- half-reaction, so we add 8 H^+ ions to the left-hand side of that half-reaction:

$$Fe^{2+} \longrightarrow Fe^{3+}$$
$$8\,H^+ + MnO_4^- \longrightarrow Mn^{2+} + 4\,H_2O$$

The two half-reactions are now balanced with respect to atoms, but not with respect to charge.

Step 5. Now balance each half-reaction with respect to charge by adding electrons to the side with *excess* positive charge. The iron half-reaction has a charge of +2 on the left, and +3 on the right, so we add one electron to the right-hand side:

$$Fe^{2+} \longrightarrow Fe^{3+} + e^-$$

The manganese half-reaction has a net charge of +7 on the left-hand side, and +2 on the right-hand side, so we add five electrons to the left-hand side:

$$5\,e^- + 8\,H^+ + MnO_4^- \longrightarrow Mn^{2+} + 4\,H_2O$$

The situation at this stage is that both equations are completely balanced with respect to atoms and charge. It is now necessary to recognize that there must be equal exchanges of electrons in a redox reaction; numbers of electrons consumed must be equal to the numbers of electrons produced. The oxidation of 1 mol of Fe^{2+} produces 1 mol of electrons. The reduction of 1 mol of MnO_4^- requires 5 mol of electrons. This is accounted for by multiplying the iron half-reaction by 5:

$$5\,Fe^{2+} \longrightarrow 5\,Fe^{3+} + 5\,e^-$$
$$5e^- + 8\,H^+ + MnO_4^- \longrightarrow Mn^{2+} + 4\,H_2O$$

Step 6. Obtain the complete balanced equation by adding the two half-reactions and canceling any terms that appear on both sides:

$$5\,Fe^{2+} \longrightarrow 5\,Fe^{3+} + 5\,e^-$$
$$5\,e^- + 8\,H^+ + MnO_4^- \longrightarrow Mn^{2+} + 4\,H_2O$$
$$\overline{5\,Fe^{2+} + MnO_4^- + 8\,H^+ \longrightarrow 5\,Fe^{3+} + Mn^{2+} + 4\,H_2O}$$

Note that in the complete equation, the electrons do not appear. They were equal on both sides of the equation, and therefore canceled. Properly done, electrons will never appear in the complete balanced equation. The last step is to rewrite the complete equation indicating appropriate physical states:

$$5\ Fe^{2+}(aq) + MnO_4^-(aq) + 8\ H^+(aq) \longrightarrow 5\ Fe^{3+}(aq) + Mn^{2+}(aq) + 4\ H_2O(l)$$

Problem 8.12. The breatholyzer test for blood alcohol requires breathing into a tube containing a gel impregnated with an acidic solution of potassium dichromate. Alcohol in the breath will be oxidized to CO_2, and the yellow dichromate will be reduced to the green chromium(III) ion. Balance the following equation which represents the redox process:

$$C_2H_5OH(aq) + Cr_2O_7^{2-} \xrightarrow{H^+(aq)} CO_2(g) + Cr^{3+} + H_2O(l)$$

Ans. Identify the oxidation and reduction half-reactions, and write them separately.

The oxidation state of chromium changes from +6 in $Cr_2O_7^{2-}$ to +3 in Cr^{3+}. The oxidation state of carbon changes from −2 in ethyl alcohol to +4 in CO_2. Therefore dichromate is reduced, and carbon is oxidized. The atomically balanced half-reactions are

$$Cr_2O_7^{2-} \longrightarrow 2\ Cr^{+3} \qquad \text{reduction}$$

$$C_2H_5OH \longrightarrow 2\ CO_2 \qquad \text{oxidation}$$

The reduction half-reaction balanced with respect to oxygen is

$$Cr_2O_7^{2-} \longrightarrow 2\ Cr^{+3} + 7\ H_2O$$

In order to balance the oxidation half-reaction with respect to oxygen we must add oxygen in the form of H_2O to the left-hand side:

$$C_2H_5OH + 3\ H_2O \longrightarrow 2\ CO_2$$

Both equations must now be balanced with respect to hydrogen. In the case of the reduction, $7\ H_2O$ appears on the right-hand side, so $14\ H^+$ must appear on the left-hand side. In the oxidation half-reaction, the 12 hydrogen atoms on the left-hand side must appear as $12\ H^+$ on the right-hand side:

$$Cr_2O_7^{2-} + 14\ H^+ \longrightarrow 2\ Cr^{+3} + 7\ H_2O$$

$$C_2H_5OH + 3\ H_2O \longrightarrow 2\ CO_2 + 12\ H^+$$

We now examine the charge relationships. There is an excess positive charge of +6 on the right-hand side of the oxidation half-reaction, so we add six electrons to that side. There is an excess positive charge of +12 on the right-hand side of the oxidation half-reaction, so we add 12 electrons to that side:

$$Cr_2O_7^{2-} + 14\ H^+ + 6\ e^- \longrightarrow 2\ Cr^{+3} + 7\ H_2O$$

$$C_2H_5OH + 3\ H_2O \longrightarrow 2\ CO_2 + 12\ H^+ + 12\ e^-$$

In order for the electron exchange to be equal, we must multiply the oxidation half-reaction by 2. The equations are now balanced with respect to atoms and charge, and may now be added together, canceling all terms which appear on both sides:

$$2\ Cr_2O_7^{2-} + 28\ H^+ + 12\ e^- \longrightarrow 4\ Cr^{+3} + 14\ H_2O$$

$$C_2H_5OH + 3\ H_2O \longrightarrow 2\ CO_2 + 12\ H^+ + 12\ e^-$$

$$C_2H_5OH(aq) + 2\ Cr_2O_7^{2-}(aq) + 16\ H^+ \longrightarrow 2\ CO_2(g) + 4\ Cr^{3+}(aq) + 11\ H_2O(l)$$

The electrons cancel as expected. The H^+ appearing on the left-hand side of the final balanced equation is the difference between $28\ H^+$ on the left-hand side of the reduction half-reaction, and the $12\ H^+$ on the right-hand side of the oxidation half-reaction. The water appearing on the right-hand side is the difference between $14\ H_2O$ on the right-hand side, and $3\ H_2O$ on the left-hand side. The net charge on the left-hand side of the final equation is equal to the net charge on the right-hand side. The atoms on the left are equal to the atoms on the right.

8.5.3 Combustion Reactions

The direct reaction of many elements and compounds with oxygen accompanied by the familiar sight of burning are called *combustion reactions*. These elements include carbon, hydrogen, sulfur, and nitrogen. Compounds containing carbon react with oxygen to produce CO_2 if sufficient O_2 is available. If less than sufficient O_2 is available, the combustion product will be CO, carbon monoxide. For example if a hydrocarbon, C_3H_8, is burned in excess O_2, the products will be CO_2 and H_2O. These equations can be balanced as though they were type I reactions, i.e., it is necessary to account only for the numbers of atoms.

ADDITIONAL SOLVED PROBLEMS

CHEMICAL KINETICS

Problem 8.13. In the reaction $3\ O_2 \longrightarrow 2\ O_3$ (ozone), the rate of appearance of ozone was 1.50 mol/min. What was the rate of disappearance of oxygen?

Ans. The stoichiometry tells us that for every 2 mol of O_3 that appears, 3 mol of O_2 must disappear. Using that information, create a conversion factor of the proper units:

$$1.50\left(\frac{\text{mol } O_3}{\text{min}}\right)\left(\frac{-3 \text{ mol } O_2}{2 \text{ mol } O_3}\right) = -2.25\left(\frac{\text{mol } O_2}{\text{min}}\right)$$

Problem 8.14. The rate of reaction between $H_2(g)$ and $I_2(g)$ doubled when the concentration of $I_2(g)$ was doubled. Explain.

Ans. The rate of reaction depends on the number of successful collisions per unit time between reactant molecules. Doubling the number of I_2 molecules effectively doubled the number of collisions per unit time, and consequently doubled the rate of reaction.

Problem 8.15. When the temperature of the reaction mixture of $N_2(g) + O_2(g)$ was raised 10 K, the reaction rate doubled. Explain.

Ans. Raising the temperature 10 K virtually doubled the population of molecules possessing sufficient kinetic energy (activation energy) to form an activated complex. It did not double the average kinetic energy of all the molecules in the reaction mixture.

Problem 8.16. The rate of a reaction of 20 °C is 0.25 mol/min. If the rate of the reaction doubles for a 10 °C rise in temperature, what is the rate of the reaction at 60 °C?

Ans. There is a 40 °C rise in temperature, so there must have been four doublings of the reaction rate (one doubling for each 10 °C rise). The new rate of reaction is

$$0.25\frac{\text{mol}}{\text{min}}(2^4) = 4.0\frac{\text{mol}}{\text{min}}$$

Problem 8.17. In the presence of the enzyme catalase, the rate of decomposition of hydrogen peroxide, H_2O_2, was increased by a factor of over 1×10^6, with no change in the concentration or properties of the enzyme. Explain.

Ans. The enzyme, catalase, is a catalyst. As such it speeds up the rate of the reaction, but is not consumed in the reaction. It forms an activated complex with H_2O_2, thereby providing a new and different reaction pathway possessing a significantly lower activation energy compared with the decomposition of H_2O_2 alone.

Problem 8.18. ΔH of a reaction is +15 kJ, and the activation energy of the forward reaction is E_a (forward) = 72 kJ. What is the activation energy of the reverse reaction, E_a (reverse)?

Ans. $\Delta H = E_a$ (forward) - E_a (reverse). Therefore

$$E_a \text{ (reverse)} = E_a \text{ (forward)} - \Delta H$$
$$= 72 \text{ kJ} - 15 \text{ kJ} = 57 \text{ kJ}$$

CHEMICAL EQUILIBRIUM

Problem 8.19. Calcium carbonate is heated in a crucible open to the atmosphere. The reaction is a decomposition and the equation for the reaction is $CaCO_3(s) \rightleftharpoons CaO(s) + CO_2(g)$. Is the reaction reversible as performed?

Ans. Since the reaction is performed in a crucible open to the atmosphere, the CO_2 evolved can escape into the atmosphere and will not be available to react with the CaO. Under these conditions, the reaction will not be reversible. Only when carried out in a closed container will this reaction be reversible.

Problem 8.20. Which is the forward and which is the reverse reaction for the decomposition of $CaCO_3$?

Ans. The forward reaction is

$$CaCO_3(s) \longrightarrow CaO(s) + CO_2(g)$$

The reverse reaction is

$$CaO(s) + CO_2(g) \longrightarrow CaCO_3(s)$$

Problem 8.21. For the reaction $CaO(s) + CO_2(g) \rightleftharpoons CaCO_3(s)$, which is the forward reaction, and which is the reverse reaction?

Ans. The forward reaction is

$$CaO(s) + CO_2(g) \longrightarrow CaCO_3(s)$$

The reverse reaction is

$$CaCO_3(s) \longrightarrow CaO(s) + CO_2(g)$$

Problem 8.22. Write the equilibrium constant expression for the reaction

$$CaCO_3(s) \rightleftharpoons CaO(s) + CO_2(g)$$

Ans. Pure solids and pure liquids do not appear in the equilibrium constant expression; therefore, $K_{eq} = [CO_2]$.

Problem 8.23. What is the meaning of the square brackets around CO_2 in the equilibrium constant expression above?

Ans. The square brackets around CO_2 mean that the concentration of CO_2 must be expressed in moles per liter.

Problem 8.24. In the reaction $H_2(g) + I_2(g) \rightleftharpoons 2\ HI(g)$ the equilibrium concentrations were found to be $[H_2] = [I_2] = 0.86\ M$, and $[HI] = 0.27\ M$. Calculate the value of the equilibrium constant.

Ans. We first require the form of the equilibrium constant expression:

$$K_{eq} = \frac{[HI]^2}{[H_2][I_2]}$$

Substituting the molar concentrations as given:

$$K_{eq} = \frac{[0.27]^2}{[0.86][0.86]} = 0.099$$

Problem 8.25. Pure liquid water ionizes to a small extent as shown in the following reaction: $H_2O(l) \rightleftharpoons H^+(aq) + OH^-(aq)$. The equilibrium concentration of $H^+(aq) = OH^-(aq) = 1 \times 10^{-7}\,M$ at 25 °C. Calculate the value of the equilibrium constant.

Ans. The equilibrium constant expression is $K_{eq} = K_w = [H^+][OH^-]$ (pure liquids do not appear in the equilibrium constant expression). Substituting the molar concentrations as given:

$$K_{eq} = K_w = (1 \times 10^{-7})^2 = 1 \times 10^{-14}$$

Problem 8.26. The equilibrium constant for the reaction

$$NH_4Cl(s) \rightleftharpoons NH_3(g) + HCl(g)$$

at 25 °C is $K_{eq} = 1.38 \times 10^{-5}$. Calculate the concentrations of the components of the equilibrium mixture.

Ans. Pure solids do not appear in the equilibrium constant expression. Therefore, $K_{eq} = [NH_3][HCl]$. The stoichiometry of the reaction requires that, whatever the final concentrations of the equilibrium components, they must be equal to each other. We will let each concentration equal x, substitute that unknown into the equilibrium constant expression, and solve for the unknown:

$$K_{eq} = 1.38 \times 10^{-5} = [x][x] = x^2$$

$$[NH_3] = [HCl] = x = \sqrt{1.38 \times 10^{-5}} = 3.71 \times 10^{-3}\,M$$

LE CHATELIER PRINCIPLE

Problem 8.27. For the reaction $N_2(g) + 3\,H_2(g) \rightleftharpoons 2\,NH_3(g)$, in which direction will the equilibrium shift if some NH_3 is removed from the equilibrium mixture?

Ans. The reduced ammonia concentration will require some replacement by the reaction of the hydrogen and nitrogen. The equilibrium will shift to the right.

Problem 8.28. ΔH for the reaction $N_2(g) + 3\,H_2(g) \rightleftharpoons 2\,NH_3(g)$ is -92 kJ. What will be the effect of increasing the temperature on the extent of the reaction?

Ans. $\Delta H = -92$ kJ means the reaction is exothermic, i.e., evolves heat. Raising the temperature favors the absorption of heat. Therefore the equilibrium will shift toward reactants and reduce the extent of the reaction.

Problem 8.29. Magnesium hydroxide, $Mg(OH)_2$ is slightly soluble in water. The solubility can be described by the following equation:

$$Mg(OH)_2(s) \rightleftharpoons Mg^{2+}(aq) + 2\,OH^-(aq)$$

What will happen to the equilibrium if hydrogen ion, H^+, is added to the solution?

Ans. The H^+ will react with the OH^- to produce H_2O, thereby reducing the OH^- concentration. This will cause the equilibrium to shift toward product, and more $Mg(OH)_2$ will dissolve.

OXIDATION-REDUCTION REACTIONS

Problem 8.30. Using the rules described in Sec. 8.7.1, assign oxidation states to the elements in the following compounds: (*a*) KCl, (*b*) $AlCl_3$, (*c*) $AgNO_3$, (*d*) H_3PO_4, (*e*) C_6H_6.

Ans. (*a*) K = +1, Cl = −1, (*b*) Cl = −1, Al = +3, (*c*) O = −2, Ag = +1, N = +5, (*d*) H = +1, O = −2, P = +5, (*e*) H = +1, C = −1

Problem 8.31. Assign oxidation states to the elements in the following compounds: (*a*) $CaCO_3$, (*b*) $Ce(SO_4)_2$, (*c*) $KMnO_4$, (*d*) $NaC_2H_3O_2$, (*e*) $C_6H_{12}O_6$.

Ans. (*a*) Ca = +2, O = −2, C = +4; (*b*) O = −2, Ce = +4, S = +6; (*c*) O = −2, K = +1, Mn = +7; (*d*) Na = +1, H = +1, O = −2, C = 0; (*e*) H = +1, O = −2, C = 0

Problem 8.32. What is the oxidation state of sulfur in the following compounds? (*a*) H_2S, (*b*) S_8, (*c*) SO_2, (*d*) $Na_2S_2O_3$, (*e*) H_2SO_4.

Ans. (*a*) −2, (*b*) 0, (*c*) +4, (*d*) +2, (*e*) +6

Problem 8.33. In each of the following reactions identify the half-reactions representing oxidation and reduction:

(*a*) $3\,Mg(s) + N_2(g) \longrightarrow Mg_3N_2(s)$

(*b*) $2\,Fe(s) + 3\,Cl_2(aq) \longrightarrow 2\,FeCl_3(aq)$

(*c*) $2\,AgCl(aq) + Cu(s) \longrightarrow CuCl_2(aq) + 2\,Ag(s)$

Ans.

	Oxidized	Reduced
(*a*)	$Mg(s) \longrightarrow Mg(+2)$	$N_2(g) \longrightarrow N(-3)$
(*b*)	$Fe(s) \longrightarrow Fe^{3+}(aq)$	$Cl_2(g) \longrightarrow Cl^-(aq)$
(*c*)	$Cu(s) \longrightarrow Cu^{2+}(aq)$	$Ag^+(aq) \longrightarrow Ag(s)$

Because magnesium nitride is an insoluble solid, we identify the states of its components by using oxidation numbers, in contrast to identifying the states of the ions in (*b*) and (*c*) by their charge.

Problem 8.34. In each of the following reactions identify the atoms undergoing oxidation and reduction:

(*a*) $Br_2(aq) + OH^-(aq) \longrightarrow BrO_3^-(aq) + Br^-(aq)$

(*b*) $Zn(s) + OH^-(aq) + H_2O(l) \longrightarrow Zn(OH)_4^{2-}(aq) + H_2(g)$

(*c*) $Fe^{2+}(aq) + Cr_2O_7^{2-}(aq) + H^+(aq) \longrightarrow Fe^{3+}(aq) + Cr^{3+}(aq) + H_2O(l)$

Ans.

	Oxidized	Reduced
(*a*)	$Br_2(aq)$	$Br_2(aq)$
(*b*)	$Zn(s)$	H_2O
(*c*)	Fe^{2+}	$Cr_2O_7^{2-}$

Problem 8.35. Use the method of half-reactions to complete and balance the following equation:

$$C_2O_4^{2-}(aq) + MnO_2(s) \longrightarrow Mn^{2+}(aq) + CO_2(g) \qquad \text{(acidic)}$$

Ans. The half-reactions are

$$C_2O_4^{2-}(aq) \longrightarrow 2\ CO_2(g) + 2e^-$$
$$MnO_2(s) + 4\ H^+(aq) + 2e^- \longrightarrow Mn^{2+}(aq) + 2\ H_2O(l)$$

Summing:

$$MnO_2(s) + 4\ H^+(aq) + C_2O_4^{2-} \longrightarrow Mn^{2+}(aq) + 2\ CO_2(g) + 2\ H_2O(l)$$

Problem 8.36. Use the method of half-reactions to complete and balance the following equation:

$$BrO_3^-(aq) + F_2(g) \longrightarrow BrO_4^-(aq) + F^-(aq)$$

Ans. The half-reactions are

$$BrO_3^-(aq) + H_2O \longrightarrow BrO_4^-(aq) + 2\ H^+(aq) + 2e^-$$
$$F_2(g) + 2e^- \longrightarrow 2\ F^-(aq)$$

Summing:

$$BrO_3^-(aq) + F_2(g) + H_2O(l) \longrightarrow BrO_4^-(aq) + 2\ F^-(aq) + 2\ H^+(aq)$$

Problem 8.37. Complete and balance the following combustion reaction:

$$C_6H_6 + O_2 \longrightarrow$$

Ans. The products must be CO_2 and H_2O. All the carbon from benzene must be in the CO_2, and all the hydrogen from benzene must be in the H_2O. First attempt:

$$C_6H_6(l) + 6\ O_2(g) \longrightarrow 6\ CO_2(g)$$
$$C_6H_6(l) + 1.5\ O_2(g) \longrightarrow 3\ H_2O(g)$$

To eliminate the fractional coefficient of the O_2 in the second equation, multiply both equations by a factor of 2:

$$2\ C_6H_6(l) + 12\ O_2(g) \longrightarrow 12\ CO_2(g)$$
$$2\ C_6H_6(l) + 3\ O_2(g) \longrightarrow 6\ H_2O(g)$$

Two mol of benzene requires 12 mol of O_2 to produce 12 mol of CO_2, and 3 additional mol of O_2 to produce 6 mol of H_2O:

$$2\ C_6H_6(1) + 15\ O_2(g) \longrightarrow 12\ CO_2(g) + 6\ H_2O(g)$$

Problem 8.38. Complete and balance the following combustion reaction:

$$C_6H_{12}O_6(s) + O_2(g) \longrightarrow$$

Ans.

$$C_6H_{12}O_6(s) + 3\ O_2(g) \longrightarrow 6\ CO_2(g)$$
$$C_6H_{12}O_6(s) + 3\ O_2(g) \longrightarrow 6\ H_2O(g)$$

One mol of glucose requires 3 mol of O_2 to produce 6 mol of CO_2 and 3 more mol of O_2 to produce 6 mol of H_2O. The balanced equation is

$$C_6H_{12}O_6(s) + 6\ O_2(g) \longrightarrow 6\ CO_2(g) + 6\ H_2O(g)$$

SUPPLEMENTARY PROBLEMS

Problem 8.39. Write the equilibrium constant expressions for each of the following reactions:

(*a*) $$C(s) + O_2(g) \rightleftharpoons CO_2(g)$$

(*b*) $$COCl_2(g) \rightleftharpoons CO(g) + Cl_2(g)$$

(*c*) $$CH_4(g) + H_2O(g) \rightleftharpoons CO(g) + 3\ H_2(g)$$

(*d*) $$C_6H_{12}O_6(s) + 6\ O_2(g) \rightleftharpoons 6\ CO_2(g) + 6\ H_2O(g)$$

Ans.

(*a*) $$K_{eq} = \frac{[CO_2]}{[O_2]}$$

(*b*) $$K_{eq} = \frac{[CO][Cl_2]}{[COCl_2]}$$

(*c*) $$K_{eq} = \frac{[CO][H_2]^3}{[CH_4][H_2O]}$$

(*d*) $$K_{eq} = \frac{[CO_2]^6[H_2O]^6}{[O_2]^6}$$

Problem 8.40. Write equilibrium constant expressions for each of the following reactions:

(*a*) $$S_2(g) + C(s) \rightleftharpoons CS_2(g)$$

(*b*) $$ZnO(s) + CO(g) \rightleftharpoons Zn(s) + CO_2(g)$$

(*c*) $$NH_2COONH_4(s) \rightleftharpoons 2\ NH_3(g) + CO_2(g)$$

(*d*) $$2\ Cu(s) + 0.5\ O_2(g) \rightleftharpoons Cu_2O(s)$$

Ans.

(*a*) $$K_{eq} = \frac{[CS_2]}{[S_2]}$$

(*b*) $$K_{eq} = \frac{[CO_2]}{[CO]}$$

(*c*) $$K_{eq} = [NH_3]^2[CO_2]$$

(*d*) $$K_{eq} = \frac{1}{[O_2]^{0.5}}$$

Problem 8.41. Given the reaction

$$C(s) + H_2O(g) \rightleftharpoons CO(g) + H_2(g), \quad \Delta H_{reaction} = +131\ kJ$$

complete the following table:

	Change	Effect on equilibrium
(*a*)	Increase in temperature	
(*b*)	Increase in concentration of $H_2O(g)$	
(*c*)	Increase in concentration of products	
(*d*)	Decrease in temperature	
(*e*)	Adding a catalyst	

Ans.

	Change	Effect on equilibrium
(*a*)	Increase in temperature	Shift to products
(*b*)	Increase in concentration of $H_2O(g)$	Shift to products
(*c*)	Increase in concentration of $CO(g)$	Shift to reactants
(*d*)	Decrease in temperature	Shift to reactants
(*e*)	Adding a catalyst	No effect

Problem 8.42. The activation energies characteristic of the reversible chemical reaction $A + B \rightleftarrows C$, are E_a (forward) = 98 kJ, and E_a (reverse) 123 kJ. Calculate ΔH, the heat of the reaction. Is the reaction exothermic or endothermic?

Ans. $\Delta H = E_a$ (forward) − E_a (reverse) = 98 kJ − 123 kJ = −25 kJ. The reaction is exothermic.

Problem 8.43. The activation energies characteristic of the reversible chemical reaction $A + B \rightleftarrows C$, are E_a (forward) = 74 kJ, and E_a (reverse) = 68 kJ. Calculate ΔH, the heat of the reaction. Is the reaction exothermic or endothermic?

Ans. $\Delta H = E_a$ (forward) − E_a (reverse) = 74 kJ − 68 kJ = +6 kJ. The reaction is endothermic.

Problem 8.44. Assign an oxidation state to nitrogen in each of the following compounds: (*a*) NO_2, (*b*) N_2O, (*c*) N_2O_5, (*d*) N_2O_3.

Ans. (*a*) +4, (*b*) +1, (*c*) +5, (*d*) +3

Problem 8.45. Assign an oxidation state to chlorine in each of the following compounds: (*a*) Cl_2O, (*b*) Cl_2O_3, (*c*) ClO_2, (*d*) Cl_2O_6.

Ans. (*a*) +1, (*b*) +3, (*c*) +4, (*d*) +6

Problem 8.46. For the reaction

$$2\,Ag^+(aq) + 2\,S_2O_3^{2-}(aq) \longrightarrow S_4O_6^{2-}(aq) + 2\,Ag(s)$$

identify (*a*) the electron donor, (*b*) the electron acceptor, (*c*) the oxidizing agent, (*d*) the reducing agent, (*e*) the species oxidized, (*f*) the species reduced.

Ans. (*a*) $S_2O_3^{2-}$, (*b*) Ag^+, (*c*) Ag^+, (*d*) $S_2O_3^{2-}$, (*e*) $S_2O_3^{2-}$, (*f*) Ag^+

Problem 8.47. Complete and balance the following equations. The reactions all take place in acidic solutions.

(*a*) $$N_2H_4(aq) + I^-(aq) \longrightarrow NH_4^+(aq) + I_2(s)$$

(*b*) $$Fe^{2+}(aq) + Ce^{4+}(aq) \longrightarrow Fe^{3+}(aq) + Ce^{3+}(aq)$$

(*c*) $$Ce^{4+}(aq) + SO_3^{2-}(aq) \longrightarrow Ce^{3+}(aq) + SO_4^{2-}(aq)$$

Ans.

(*a*) $$N_2H_4(aq) + 2\,I^-(aq) + 4\,H^+(aq) \longrightarrow 2\,NH_4^+(aq) + I_2(s)$$

(*b*) $$Fe^{2+}(aq) + Ce^{4+}(aq) \longrightarrow Fe^{3+}(aq) + Ce^{3+}(aq)$$

(*c*) $$2\,Ce^{4+}(aq) + SO_3^{2-}(aq) + H_2O \longrightarrow 2\,Ce^{3+}(aq) + SO_4^{2-}(aq) + 2H^+(aq)$$

Problem 8.48. Complete and balance the following equations. The reactions all take place in acidic solutions.

(*a*) $CrO_4^{2-}(aq) + S^{2-}(aq) \longrightarrow CrO_2^-(aq) + S(s)$

(*b*) $Zn(s) + NO_3^-(aq) \longrightarrow Zn^{2+}(aq) + NH_4^+(aq)$

(*c*) $I^-(aq) + HNO_2(aq) \longrightarrow I_2(s) + NO(g)$

Ans.

(*a*) $2\ CrO_4^{2-}(aq) + 3\ S^{2-}(aq) + 8\ H^+(aq) \longrightarrow 2\ CrO_2^-(aq) + 3\ S(s) + 4\ H_2O(l)$

(*b*) $4\ Zn(s) + NO_3^-(aq) + 10\ H^+(aq) \longrightarrow 4\ Zn^{2+}(aq) + NH_4^+(aq) + 3\ H_2O(l)$

(*c*) $2\ I^-(aq) + 2\ HNO_2(aq) + 2\ H^+(aq) \longrightarrow I_2(s) + 2\ NO(g) + 2\ H_2O(l)$

Problem 8.49. Complete and balance the following equation:

$$C_4H_8O_2(s) + O_2(g) \longrightarrow CO_2(g) + H_2O(g)$$

Ans. $C_4H_8O_2(s) + 5\ O_2(g) \longrightarrow 4\ CO_2(g) + 4\ H_2O(g)$

Problem 8.50. Complete and balance the following equation:

$$C_3H_7(OH)_3(s) + O_2(g) \longrightarrow CO_2(g) + H_2O(g)$$

Ans. $C_3H_7(OH)_3(s) + 4\ O_2(g) \longrightarrow 3\ CO_2(g) + 5\ H_2O(g)$

CHAPTER 9

Aqueous Solutions of Acids, Bases, and Salts

9.1 ACID-BASE THEORIES

9.1.1 Acids and Bases according to Arrhenius

The first useful theory of acids and bases was proposed by Svante Arrhenius. His idea was that acids were substances which produced hydrogen ions, H^+, in aqueous solutions. Bases were substances which produced hydroxide ions, OH^-, when dissolved in water. Some acids and bases were found to be 100 percent ionized in solution, and were called *strong acids* or *bases*. Examples of strong acids are HCl and HNO_3, which are completely ionized in aqueous solution to form H^+ and Cl^- and H^+ and NO_3^-, respectively. Others were found to be much less ionized in solution, and were called *weak acids* or *bases*. The process of ionization was called *dissociation*, and for the weak acids and bases this process is reversible and can be characterized by an equilibrium constant. In the case of weak acids, for example, acetic acid, $HC_2H_3O_2$, the equilibrium constant is called K_a, and for the ionization reaction

$$HC_2H_3O_2 \rightleftharpoons H^+ + C_2H_3O_2^-$$

the equilibrium constant expression is

$$K_a = \frac{[H^+][C_2H_3O_2^-]}{[HC_2H_3O_2]}$$

Examples of strong bases are NaOH and $Ca(OH)_2$, which are completely ionized in aqueous solution to form Na^+ and OH^-, and Ca^{2+} and OH^- ions, respectively. Ammonia, NH_3, is an example of a weak base. When dissolved in water it produces OH^- ions, but it is considered a weak base because the OH^- concentration is considerably less than the ammonia concentration.

9.1.2 Acids and Bases according to Brønsted and Lowry

In the Brønsted-Lowry scheme, protons have the central role. Acids are defined as substances which donate protons, and bases are substances which accept protons. In the Brønsted-Lowry model the solvent may act as a proton acceptor. An acid therefore becomes a base after donation of its proton, since the process may be reversed, and a base becomes an acid for the same reason. This process may be represented symbolically as

$$HA + B \rightleftharpoons A^- + HB^+$$

The two pairs of protonated and deprotonated substances, HA and A^-, and HB^+ and B, are called *conjugate acid-base pairs*. The charges that appear are the consequence of loss and gain of the positive charge of the proton. If we substitute acetic acid for HA, in contrast to the Arrhenius theory, water becomes an integral component of the ionization process as

$$HC_2H_3O_2 + H_2O \rightleftharpoons H_3O^+ + C_2H_3O_2^-$$

Water as a proton acceptor acts as a base, and in turn becomes an acid, H_3O^+, now a proton donor. H_3O^+ is also called the *hydronium ion*. H_2O–H_3O^+ is a conjugate acid-base pair, and $HC_2H_3O_2$–$C_2H_3O_2^-$ is the other conjugate acid-base pair. The acid ionization constant expression for this process is

$$K_a = \frac{[H_3O^+][C_2H_3O_2^-]}{[HC_2H_3O_2]}$$

Note that water, a pure liquid, does not appear in the equilibrium constant expression. The Brønsted-Lowry ionization process for the weak base ammonia is written

$$NH_3 + H_2O \rightleftharpoons NH_4^+ + OH^-$$

and the ionization constant expression is

$$K_b = \frac{[NH_4^+][OH^-]}{[NH_3]}$$

Note that for a reaction which produces protonated water, H_3O^+, the constant is denoted as K_a, whereas for a process which produces deprotonated water, OH^-, the constant is denoted as K_b. Furthermore, the reader should understand that $[H^+]$ and $[H_3O^+]$ are equivalent to each other. The simpler $[H^+]$ will be used henceforth for two reasons: (1) It will simplify mathematical expressions, and (2) one ought not emphasize the difference between the hydration of all ionic substances, e.g., Na^+, Mg^{2+}, and the hydration of the proton. The values of K_a's and K_b's indicate the extents of ionization of weak acids and bases. The larger the constants, the stronger the acids or bases.

9.1.3 Lewis Acids and Bases

The most general view of acids and bases was advanced by G. N. Lewis. In this model, acids are substances which have an affinity for lone electron pairs, and bases are substances which possess lone electron pairs. Water and ammonia are the most common substances which possess lone electron pairs, and therefore behave as bases in the Lewis scheme. The reaction of silver ion, Ag^+ with cyanide ion, CN^-, and boron trifluoride, BF_3 (an electron-deficient compound), with ammonia, NH_3, are two examples of Lewis acid-base reactions. The Lewis acid-base concept is most useful in chemical reactions in nonaqueous solvents. We will not find it useful in our study of ionic equilibria in water.

9.2 WATER REACTS WITH WATER

Water behaves as either an acid or a base, and therefore in aqueous solution, a process known as *autoionization* takes place. In this process, an H_2O molecule transfers a proton to another H_2O molecule, even if other acids or bases are present, as

$$H_2O + H_2O \rightleftharpoons H_3O^+ + OH^-$$

The ionization constant expression for this process is

$$K_w = [H^+][OH^-]$$

The value of this constant in pure water at 25 °C was given in Chap. 8 as 1×10^{-14}. In pure water, which is denoted as a neutral solution, $[H^+] = [OH^-] = \sqrt{1.0 \times 10^{-14}} = 1 \times 10^{-7}$. Therefore, an acidic solution will have a $[H^+]$ greater than 1×10^{-7}, and a $[OH^-]$ less than 1×10^{-7}. The opposite is true in a basic solution.

9.3 ACIDS AND BASES: STRONG VERSUS WEAK

Experiments have shown that there are essentially no undissociated HCl molecules in a dilute aqueous solution of HCl. The only components of HCl which can be detected are $H_3O^+(aq)$, and $Cl^-(aq)$. Acids that are completely dissociated in water are called *strong acids*. The term *strong* reflects the proton transferability of the acid. Strong acids transfer *all* of their dissociable protons to water.

Problem 9.1. Calculate the H_3O^+, OH^-, and Cl^- concentrations in a 0.20 *M* aqueous solution of HCl.

Ans. HCl is a strong acid and therefore is completely dissociated in water. [HCl] = 0.20, but if it is completely dissociated, then, $[H_3O^+] = 0.20$, and also, $[Cl^-] = 0.20$. The concentration of OH^- can be computed by use of the dissociation constant, K_w, of water:

$$K_w = [H_3O^+][OH^-] = 1 \times 10^{-14}$$

$$[OH^-] = \frac{1 \times 10^{-14}}{0.20} = 5 \times 10^{-14}$$

The very small value of $[OH^-]$ means that the solution is very acidic. It also means that the $[H_3O^+]$ contributed by the ionization of water is also 5×10^{-14}, and is negligible compared with the $[H_3O^+]$ supplied by the dissociation of HCl.

Experiments have also shown that KOH in aqueous solution is completely dissociated. In water, there are only $K^+(aq)$ and $OH^-(aq)$ ions, and no undissociated KOH molecules. This substance is a base, because OH^- is a proton acceptor:

$$H_3O^+(aq) + OH^-(aq) \rightleftharpoons 2\,H_2O(l)$$

Bases like KOH which are completely dissociated in water are called *strong bases.*

Problem 9.2. Calculate the OH^-, K^+, and H_3O^+ concentrations in a 0.20 *M* aqueous solution of KOH.

Ans. KOH is a strong base and therefore is completely dissociated in water. [KOH] = 0.20, but if it is completely dissociated, then $[OH^-] = 0.20$, and also, $[K^+] = 0.20$. The concentration of H_3O^+ can be computed by use of the dissociation constant, K_w, of water:

$$K_w = [H_3O^+][OH^-] = 1 \times 10^{-14}$$

$$[H_3O^+] = \frac{1 \times 10^{-14}}{0.20} = 5 \times 10^{-14}$$

The very small value of $[H_3O^+]$ means that the solution is very basic. It also means that the $[OH^-]$ contributed by the ionization of water is also 5×10^{-14}, and is negligible compared with the $[OH^-]$ supplied by the KOH.

Most acids and bases in aqueous solution are only partially dissociated. There are only a few strong acids and bases. These are listed in Table 9-1. It is important to memorize these few strong acids and bases. It is then only necessary to consult this short list to determine whether a problem in

Table 9-1. Names and Formulas of All the Strong Acids and Bases

Strong Acids		Strong Bases	
Formula	**Name**	**Formula**	**Name**
$HClO_4$	Perchloric acid	LiOH	Lithium hydroxide
HNO_3	Nitric acid	NaOH	Sodium hydroxide
H_2SO_4	Sulfuric acid*	KOH	Potassium hydroxide
HCl	Hydrochloric acid	RbOH	Rubidium hydroxide
HBr	Hydrobromic acid	CsOH	Cesium hydroxide
HI	Hydroiodic acid	TlOH	Thallium(I) hydroxide
		$Ca(OH)_2$	Calcium hydroxide
		$Sr(OH)_2$	Strontium hydroxide
		$Ba(OH)_2$	Barium hydroxide

**Note:* Only the first proton in sulfuric acid is completely dissociable. The first product of dissociation, HSO_4^-, is a weak acid.

acid-base chemistry involves a strong or weak acid or base.

All inorganic acids not listed above are weak acids, e.g., HCN, hydrocyanic acid, HNO_2, nitrous acid, HF, hydrofluoric acid. All organic compounds containing the carboxyl group, general formula RCOOH, are weak acids. In the general formula RCOOH, R can be an H or CH_3, or any other carbon-containing group. In most of our acid-base problems we will use as an example of a typical weak acid, acetic acid, whose structural formula is

$$CH_3-C\begin{matrix}\diagup\!\!\diagup O\\ \diagdown OH\end{matrix}$$

The hydrogen of the OH of the carboxyl group is the dissociable hydrogen. The reaction of acetic acid with water is

$$CH_3-C\begin{matrix}\diagup\!\!\diagup O\\ \diagdown OH\end{matrix}(aq) + H_2O(l) \rightleftharpoons CH_3-C\begin{matrix}\diagup\!\!\diagup O\\ \diagdown O^-\end{matrix}(aq) + H_3O^+(aq)$$

The reaction of acetic acid with OH^- will yield the same acetate anion, so that the products of the reaction of acetic acid with KOH will be the ionic salt, potassium acetate, and water. The formula of acetic acid used previously. $HC_2H_3O_2$, is a simplified version of the complete structural formula, which emphasizes its acidic properties, and that there is only one dissociable proton in the molecule.

Ammonia, NH_3, is a weak base. It is a base because it reacts with water to produce OH^- ions, and it is weak because the proportion of OH^- is far less than unreacted NH_3 molecules in water. Its properties as a base depend on its lone pair of electrons which is able to bond with the proton donated by water. This is best illustrated by Lewis formulas:

$$H-\underset{H}{\overset{H}{|}}{N}: + \underset{H\quad H}{\ddot{O}} \rightleftharpoons \left[H-\underset{H}{\overset{H}{N}}-H\right]^{\oplus} + \left[:\ddot{\underset{..}{O}}-H\right]^{\ominus}$$

The same type of process occurs when NH_3 reacts with HCl to form NH_4Cl. Many organic compounds containing nitrogen are weak bases, and can be considered as derivatives of ammonia, with other groups substituting for the three hydrogens of ammonia. Examples are CH_3NH_2, methyl amine, pyridine, C_5H_5N, morphine, $C_{17}H_{19}O_3N$, and cocaine, $C_{17}H_{21}O_4N$. Many physiologically active nitrogen bases such as morphine are not soluble in water, and are typically made water-soluble by reaction with HCl to form the halogen salt, e.g., morphine hydrochloride.

9.4 pH, A MEASURE OF ACIDITY

Because the concentration of H^+ can vary over wide ranges, it is convenient to express its concentration in logarithmic form called *pH*. It is defined as

$$pH = -\log[H^+] \quad \text{or} \quad [H^+] = 10^{-pH}$$

The smaller the pH, the greater the acidity.

Problem 9.3. What is the pH of a solution whose $[H^+] = 0.001$?

Ans. The quantity 0.001 can be expressed in exponential notation as 1×10^{-3}. Thus

$$pH = -\log[H^+] = -\log[1 \times 10^{-3}] = 3$$

or

$$[H^+] = 10^{-pH}$$

$$0.001 = 10^{-3}$$

Therefore pH = 3.

Problem 9.4. What is the pH of a solution whose $[H^+] = 0.00263$?

Ans. In order to solve this problem one must use either log tables or a hand calculator which has a logarithm function key. With a calculator, enter the value of 0.00263, and press the log key. The result will read -2.58. Now press the change sign key, $\boxed{+/-}$, and the answer will read 2.58, which is the pH.

A logarithmic function is also defined for $[OH^-]$ as

$$pOH = -\log[OH^-] \quad \text{or} \quad [OH^-] = 10^{-pOH}$$

The smaller the pOH, the greater the alkalinity. It is calculated precisely as is the pH.

Just as $-\log[H^+] = pH$, equilibrium constants can be used in a more convenient form as $-\log K_{eq} = pK_{eq}$. Using these definitions, we can write the equilibrium constant for water in a new way:

$$K_w = [H^+][OH^-] = 1.0 \times 10^{-14}$$

$$-\log K_w = -\log[H^+] - \log[OH^-] = -\log(1 \times 10^{-14})$$

$$pK_w = pH + pOH = 14$$

Problem 9.5. What is the pH and the pOH of a solution whose hydroxide ion concentration is 0.001 M?

Ans.

$$[OH^-] = 0.001$$

$$0.001 = 10^{-3}$$

$$pOH = 3$$

$$pH + pOH = 14$$

$$pH = 14 - 3 = 11$$

Problem 9.6. The pH of an HCl solution is 2.7. Calculate the concentration of the HCl solution.

Ans. HCl is a strong acid and consequently is 100 percent ionized, which means that $[HCl] = [H^+]$. Therefore, if we calculate $[H^+]$, we will have calculated the concentration of HCl. The pH is related to H^+ concentration by

$$pH = -\log[H^+] \quad \text{or} \quad [H^+] = 10^{-pH}$$

The problem resolves itself into the question: What is the antilog of -2.7, or what is the numerical value of 10 raised to the power of -2.7? Your calculator should have a 10^x, or antilog key. It is used as follows: first enter 2.7, then press the change sign key +/–. Next press the 10^x key and read the answer, 0.002 M. If your calculator does not have an antilog function, do the calculation as described in Chap. 1 on the use of a logarithm table.

9.5 pH AND WEAK ACIDS AND BASES

The calculation of pH of aqueous weak acid or basic solutions must taken into account that the dissociation is incomplete and leads to the formation of equilibrium mixtures.

Problem 9.7. Calculate the pH of a 0.00500 M solution of acetic acid.

Ans. We can calculate the H^+ concentration of a weak acid solution in the same way we calculated the equilibrium concentrations of the components of a reaction in Chap. 8.

We first write the reaction equation:

$$HC_2H_3O_2 + H_2O \rightleftharpoons H_3O^+ + C_2H_3O_2^-$$

We know that we have added 0.00500 mol of acetic acid to 1.00 L of solution, some of it has ionized to form the yet to be determined or x mol of H^+ and an equal amount of $C_2H_3O_2^-$. We will use the accounting method developed in Chap. 8 to describe the various equilibrium concentrations of the reaction mixture:

	$HC_2H_3O_2$	H^+	$C_2H_3O_2^-$
Start	0.00500	0	0
Change	$-x$	$+x$	$+x$
Final	$0.00500 - x$	x	x

The equilibrium constant is:

$$K_a \frac{[H^+][C_2H_3O_2^-]}{[HC_2H_3O_2]} = 1.74 \times 10^{-5}$$

Substituting appropriate values from the table:

$$1.74 \times 10^{-5} = \frac{[x][x]}{[0.00500 - x]} = \frac{[x^2]}{[0.00500 - x]}$$

This is a quadratic equation in one unknown, and can be solved using the standard solution for the general equation $ax^2 + bx + c = 0$:

$$x = \frac{-b \pm \sqrt{b^2 - 4ac}}{2a}$$

noting that a negative solution is not possible. However, we can use our chemical knowledge to simplify the solution. From the magnitude of the equilibrium constant, we can see that the size of x must be quite small. That means that the value of $0.00500 - x$ must be approximately the size of 0.00500. By using 0.00500 in place of $0.00500 - x$, the arithmetic is greatly simplified to

$$1.74 \times 10^{-5} = \frac{[x^2]}{[0.00500]}$$

$$x^2 = 8.70 \times 10^{-8}$$

$$x = [H^+] = 2.95 \times 10^{-4}\ M$$

$$pH = 3.53$$

The quadratic solution is $[H^+] = 2.86 \times 10^{-4}$, pH = 3.54. The approximate solution is about 3 percent greater than the precise value for $[H^+]$, which is acceptable.

9.6 POLYPROTIC ACIDS

Some acids have more than one dissociable proton, and are called *polyprotic acids*. Sulfuric acid, H_2SO_4, and carbonic acid, H_2CO_3, are diprotic acids, phosphoric acid, H_3PO_4, is a triprotic acid. Other diprotic acids such as succinic, fumaric, and malic acids and the triprotic acid, citric acid, are important components of intermediary metabolism. In the case of carbonic acid, 1 mol of the acid can neutralize 2 mol of KOH:

$$H_2CO_3(aq) + 2\ KOH(aq) \rightleftharpoons K_2CO_3(aq) + 2\ H_2O(l)$$

There are, however, two distinct stages in the dissociation of the acid:

$$H_2CO_3(aq) + H_2O(l) \rightleftharpoons H_3O^+(aq) + HCO_3^-(aq)$$

$$K_{a_1} = \frac{[H_3O^+][HCO_3^-]}{[H_2CO_3]} = 4.45 \times 10^{-7}$$

$$HCO_3^-(aq) + H_2O(l) \rightleftharpoons H_3O^+(aq) + CO_3^{2-}(aq)$$

$$K_{a_2} = \frac{[H_3O^+][CO_3^{2-}]}{[HCO_3^-]} = 4.7 \times 10^{-11}$$

As a general rule, successive dissociations of polyprotic acids are between 1×10^{-4} and 1×10^{-5} smaller than the preceding one. This is because succeeding positively charged protons must leave an already negatively charged ion. The significance of this is that the pH of a polyprotic acid solution can be approximated very well by ignoring the contributions to the overall $[H^+]$ of acid dissociations after the first one.

Problem 9.8. Calculate the pH of a 0.0100 M solution of H_2CO_3, carbonic acid, a diprotic acid.

Ans. We will handle this problem as we did just above, by first writing the first dissociation reaction:

$$H_2CO_3(aq) + H_2O(l) \rightleftharpoons H_3O^+(aq) + HCO_3^-(aq)$$

$$K_{a_1} = \frac{[H_3O^+][HCO_3^-]}{[H_2CO_3]} = 4.45 \times 10^{-7}$$

Now we will develop our accounting scheme:

	H_2CO_3	H^+	HCO_3^-
Start	0.0100	0	0
Change	$-x$	$+x$	$+x$
Final	$0.0100 - x$	x	x

Because K_{a_1} is so small, we may safety assume that $[0.01 - x] \approx [0.01]$, and substitute the accounting entries into the dissociation constant expression:

$$K_{a_1} = \frac{[x][x]}{[0.0100]} = \frac{[x^2]}{[0.0100]} = 4.45 \times 10^{-7}$$

$$x = [HCO_3^-] = [H^+] = 6.67 \times 10^{-5}\ M$$

$$pH = 4.18$$

Because the second dissociation constant is 1×10^{-4} smaller than the first, it is a safe assumption that the contribution of the second dissociation to the overall H^+ concentration can be ignored. To verify that assumption, let us calculate the CO_3^{2-} concentration which is equal to the H^+ concentration from the second dissociation:

$$[CO_3^{2-}] = \frac{[HCO_3^-]K_{a_2}}{[H^+]} = \frac{[6.67 \times 10^{-5}][4.7 \times 10^{-11}]}{[6.67 \times 10^{-5}]} = 4.7 \times 10^{-11}$$

The small value of $[CO_3^{2-}]$, which is equal to the $[H^+]$ ion from the second dissociation compared with the value of $[H^+]$ from the first ionization, verifies the assumption that dissociation of HCO_3^- contributes a negligible amount of H^+ ion to the solution.

9.7 SALTS AND HYDROLYSIS

When NaCl is added to pure water, the solution remains neutral. When sodium acetate, $NaC_2H_3O_2$, is added to pure water, the solution becomes slightly basic. When ammonium nitrate, NH_4NO_3, is added to pure water, the solution becomes slightly acidic. To understand these effects, let us first write the ionization reaction for acetic acid in reverse:

$$H_3O^+ + C_2H_3O_2^- \rightleftharpoons HC_2H_3O_2 + H_2O$$

The equation for the ionization of water is

$$H_2O + H_2O \rightleftharpoons H_3O^+ + OH^-$$

Since these two equilibria occur simultaneously in water, we can add these two equations to get the following:

$$C_2H_3O_2^- + H_2O \rightleftharpoons HC_2H_3O_2 + OH^-$$

This tells us that the anion of a weak acid reacts with water to produce the protonated form of the acid and an hydroxide ion. Consequently, such a solution becomes slightly basic. This process is called *hydrolysis*.

Because the conjugate base of any weak acid is itself a weak base, we can generalize the above example by denoting the weak acid as HB, and the conjugate weak base as B^-, and write the two equilibria as

(1)
$$HB + H_2O \rightleftharpoons B^- + H_3O^+$$

where
$$K_a = \frac{[H_3O^+][B^-]}{[HB]}$$

(2)
$$B^- + H_2O \rightleftharpoons HB + OH^-$$

where
$$K_b = \frac{[HB][OH^-]}{[B^-]}$$

Let us multiply K_a by K_b:

$$K_a \times K_b = \frac{[H_3O^+][B^-]}{[HB]} \times \frac{[HB][OH^-]}{[B^-]} = [H_3O^+][OH^-] = K_w$$

This defines a new relationship between K_a and K_b:

$$K_w = K_a \cdot K_b$$

If we take negative logarithms of both sides of this equation (consistent with the definition of pH),

$$-\log K_w = -\log K_a - \log K_b$$

this becomes

$$pK_w = pK_a + pK_b = 14$$

Problem 9.9. What is the equilibrium constant for the hydrolysis of acetate ion at 25 °C?

Ans. K_a for acetic acid is 1.74×10^{-5} at 25 °C. K_w for water at 25 °C is 1.00×10^{-14}. Therefore,

$$K_b = \frac{1.00 \times 10^{-14}}{1.74 \times 10^{-5}} = 5.75 \times 10^{-10}$$

The small value of K_b tells us that acetate ion is not a very strong base. In general, the conjugate base of a weak acid is itself a weak base in aqueous solution.

Problem 9.10. Compute the pK_a for the ionization of acetic acid and the pK_b for the hydrolysis of acetate ion in water at 25 °C.

Ans.

$$pK_a = -\log K_a = -\log(1.74 \times 10^{-5}) = 4.76$$
$$pK_b = -\log K_b = -\log(5.75 \times 10^{-10}) = 9.24$$

Problem 9.11. What is the sum of the pK_a for ionization of acetic acid and the pK_b for hydrolysis of acetate ion at 25 °C in water?

Ans. Remember, $K_w = K_a \cdot K_b$. Therefore,

$$-\log K_w = -\log K_a - \log K_b$$
$$pK_w = pK_a + pK_b$$
$$4.76 + 9.24 = 14.00$$

From this you can see that the relationship

$$pH + pOH = 14$$

is a special case of the more general relationship

$$pK_a + pK_b = 14$$

The sum of the pK_a and pK_b values for a conjugate acid-base pair must equal 14 at 25 °C.

Table 9-2 lists values of K_a, pK_a, K_b, and pK_b for some conjugate acid base pairs at 25 °C.

Table 9-2. Values of K_a, pK_a, K_b, and pK_b for Selected Conjugate Acid-Base Pairs

Name	Acid	K_a	pK_a	Base	K_b	pK_b
Nitrous acid	HNO_2	4.47×10^{-4}	3.35	NO_2^-	2.24×10^{-11}	10.65
Cyanic acid	HCNO	2.19×10^{-4}	3.66	CNO^-	4.57×10^{-11}	10.34
Formic acid	$HCHO_2$	1.78×10^{-4}	3.75	CHO_2^-	5.62×10^{-11}	10.25
Acetic acid	$HC_2H_3O_2$	1.74×10^{-5}	4.76	$C_2H_3O_2^-$	5.75×10^{-10}	9.24
Carbonic acid	H_2CO_3	4.45×10^{-7}	6.35	HCO_3^-	2.24×10^{-8}	7.65
Dihydrogen phosphate ion	$H_2PO_4^-$	6.32×10^{-8}	7.20	HPO_4^{2-}	1.58×10^{-7}	6.80
Ammonium ion	NH_4^+	5.71×10^{-10}	9.24	NH_3	1.75×10^{-5}	4.76
Hydrogen carbonate ion	HCO_3^-	4.72×10^{-11}	10.33	CO_3^{2-}	2.19×10^{-4}	3.67
Hydrogen phosphate ion	HPO_4^{2-}	4.84×10^{-13}	12.32	PO_4^{3-}	2.07×10^{-2}	1.68

The cations and anions of salts derive from the reactions of acids with bases. The conjugate bases of weak acids are the anions of many salts, e.g., acetate, carbonate, phosphate. All these anions are themselves weak bases, so that in water they hydrolyze and produce basic solutions. Certain cations are weak acids,e.g., NH_4^+, and in water will hydrolyze to produce acidic solutions. There are no basic cations. All Group I and Group II cations, e.g., Na^+, Ca^{2+}, are neutral in aqueous solution. However, salts of the cations Al^{3+}, Pb^{2+}, Sn^{2+} and of the transition metals Fe^{3+}, etc., form acidic solutions. In these metal salts, water is covalently bound to the central cation. A hydrogen ion from the bound water dissociates and causes aqueous solutions of these salts to be acidic. For example, $Fe(H_2O)_6^{3+}$ is an acidic cation. The acid dissociation reaction in water is

$$Fe(H_2O)_6{}^{3+}(aq) + H_2O(l) \rightleftharpoons Fe(OH)(H_2O)_5{}^{2+}(aq) + H_3O^+(l)$$

$$K_a = 1 \times 10^{-3}$$

It is important to note that the proton's origin is the bound water.

The pH of a 1 *M* aqueous solution of $FeCl_3$ is about 3.5.

Problem 9.12. Predict whether the following salts will produce an acidic, basic, or neutral aqueous solution: $MgCl_2$, K_2CO_3, $AlCl_3$, NH_4Cl.

Ans.

Salt	Cation	Anion	Solution
$MgCl_2$	Neutral	Neutral	Neutral
K_2CO_3	Neutral	Basic	Basic
$AlCl_3$	Acidic	Neutral	Acidic
NH_4Cl	Acidic	Neutral	Acidic

Problem 9.13. Predict whether $(NH_4)_2HPO_4$, ammonium monohydrogen phosphate, will form an acidic or a basic solution.

Ans. Since both anion and cation will hydrolyze, it is necessary to compare basic and acidic ionization constants in order to determine the extent of hydrolysis. The ammonium ion will produce H_3O^+, and monohydrogen phosphate will produce OH^- in water.

For $NH_4{}^+$: $K_a = 5.7 \times 10^{-10}$

For $HPO_4{}^{2-}$: $K_b = 1.6 \times 10^{-7}$

Since K_b for the anion is so much greater than K_a for the cation, the products of its hydrolysis will prevail, and the solution will be basic.

9.8 BUFFERS AND BUFFER SOLUTIONS

A buffered solution is one in which changes in pH by the addition of H^+ or OH^- are resisted. A buffer system consists of a conjugate acid-base pair whose selection is based on the requirement of what value of pH is desired to be maintained constant. Our analysis begins with a restatement of the ionization constant of a weak acid which we will denote as HA. The ionization process is $HA \rightleftharpoons H^+ + A^-$, where A^- is the conjugate base of the acid HA, and the equilibrium constant expression is

$$K_a = \frac{[H^+][A^-]}{[HA]} = [H^+]\left(\frac{[A^-]}{[HA]}\right)$$

We next take the negative logarithm of both sides to obtain

$$-\log K_a = -\log[H^+] - \log\frac{[A^-]}{[HA]}$$

$$pK_a = pH - \log\frac{[A^-]}{[HA]}$$

We now rearrange as

$$pH = pK_a + \log\frac{[A^-]}{[HA]}$$

This equation is known as the Henderson-Hasselbalch (H-H) equation, and it describes the dependence of the pH as a function of the ratio of molar concentrations of *proton acceptor* to *proton donor* of a conjugate acid-base pair in aqueous solution. The size of the logarithmic ratio must be kept as close to 1.0 as possible. The logarithm changes very rapidly when the ratio is much larger or much smaller than 1.0, causing the pH to also change rapidly. In order to maintain the pH to ± 0.2 units, the ratio must be maintained from about 1.5 to about 0.6.

Problem 9.14. What is the pH of an aqueous solution consisting of 0.006 M acetic acid, and 0.008 M sodium acetate?

Ans. We require the pK_a (not the pK_b) of acetic acid, 4.76, and substitute all values in the H-H equation:

$$pH = 4.76 + \log \frac{[0.0080]}{[0.0060]}$$

$$= 4.76 + 0.13$$

$$= 4.89$$

Problem 9.15. Why does a solution of a conjugate acid-base pair behave as a buffer solution?

Ans. To answer this, let us examine the change in pH if the ratio of proton acceptor to proton donor is varied from 10/1 to 1/10. The ratio can be varied by adding H^+ or OH^- to the solution. The logarithm of 10/1 is +1, and the logarithm of 1/10 is −1. According to the H-H equation, the pH must then vary as

$$pH = pK_a \pm 1$$

Therefore, even when the concentrations of base to acid are varied over 100-fold, the pH varies by only ± 1 pH unit.

Problem 9.16. Consult Table 9-1, and decide how you would prepare a buffer to maintain the pH of a solution as close to pH 7 as possible.

Ans. First, let us consider what the H-H equation tells us if the concentrations of conjugate base to acid are equal:

$$pH = pK_a + \log \frac{[A^-]}{[HA]}$$

If they are equal, the ratio is 1, and the log of 1 = 0. Therefore, under those circumstances, $pH = pK_a$. Table 9-1 informs us that the conjugate acid-base pair, $H_2PO_4^- - HPO_4^{2-}$, is characterized by a pK_a of 6.80. The desired buffer will be prepared by dissolving equal molar quantities of KH_2PO_4 and K_2HPO_4 in water.

9.9 TITRATION

The concentration of acids and bases can be experimentally determined by neutralization reactions. The steps in the process involve careful measurement of a volume of, for example, an acid solution of unknown concentration. Then a basic solution of well-known concentration is slowly added until the acid is totally neutralized. This process is called *titration*. This analytical technique depends on the following conditions: (a) The reaction must go to completion, (b) the reaction must be fast, and very importantly, (c) there must be a way of detecting when the reaction is complete. How would one know when to stop adding titrating reagent? The reaction is stoichiometrically complete at what is called the *equivalence point*. The experimentally determined point of completion is called the *end point*. The goal is to get the theoretical equivalence point and the experimentally determined end point to be as close as possible. End points are determined by the use of dyes which change color at the pH of the equivalence point, or with an electronic device called a *pH meter*. The calculations involving neutralization reactions were described in Chap. 7, and we will review them here.

Problem 9.17. The neutralization of a 25.0-mL solution of HCl of unknown concentration required 16.7 mL of a 0.0101 *M* KOH standard solution. (A *standard solution* is one of precisely known concentration.) Calculate the concentration of the acid.

Ans. The number of moles of added base must be equal to the number of moles of acid.

$$M_{\text{base}} \times V_{\text{base}} = M_{\text{acid}} \times V_{\text{acid}}$$

$$0.0101 \times 16.7\ \text{mL} = M_{\text{acid}} \times 25.0\ \text{mL}$$

$$M_{\text{acid}} = 0.00674\ M$$

9.10 NORMALITY

A previously undefined unit of concentration, *normality*, symbol *N*, is often encountered in acid-base and redox titrations. Its meaning rests upon a quantity called *equivalent weight*, hereafter called *equivalent mass*. Normality is defined as N = equivalents/L.

In acid-base chemistry, one is concerned with, for example, the molar mass of a base like KOH, which is equivalent in moles to the molar mass of some acid. From the stoichiometry of the reaction between HCl and KOH, it is clear that 1 mol of KOH is equivalent to 1 mol of HCl. So that in that reaction, 1 mol/L of HCl is said to have a concentration of 1 *N*, or 1 equivalent mass/L of H^+ ion. If we look at the reaction between H_2SO_4 and KOH, we can see that 1 mol of H_2SO_4 contains 2 mol of H^+/mol, or that $\frac{1}{2}$ mol of H_2SO_4 is equivalent to 1 mol of KOH. To obtain a mass of H_2SO_4 equivalent to 1 mol of KOH in a neutralization reaction, divide the molar mass of H_2SO_4 by 2, i.e., 98/2 = 49 g/mol. The equivalent mass of H_2SO_4 is 49 g/mol. A 1.0 *N* solution of H_2SO_4 contains 49 g/L.

Problem 9.18. How would you prepare a 1.0 *N* solution of phosphoric acid, H_3PO_4?

Ans. First write the balanced equation for the neutralization of H_3PO_4 by a base like KOH:

$$H_3PO_4(aq) + 3\ KOH(aq) \rightleftharpoons K_3PO_4(aq) + 3\ H_2O(l)$$

Now rewrite the equation so that the acid reacts with only 1 mol of base:

$$\tfrac{1}{3}\ H_3PO_4(aq) + KOH(aq) \rightleftharpoons \tfrac{1}{3}\ K_3PO_4(aq) + H_2O(l)$$

You can see that $\frac{1}{3}$ mol of H_3PO_4 is equivalent to 1 mol of KOH in a neutralization reaction. To prepare a 1 *N* solution of the acid, weigh out $\frac{1}{3}$ mol of H_3PO_4, 98.0/3 = 32.7 g/equivalent, and dissolve it in 1.0 L of aqueous solution. The equivalent mass is the formula weight of the acid divided by the number of dissociable H^+ ions/mol of acid. The normality of a solution used in neutralization is

$$N = \frac{\text{equivalents(acid)}}{\text{L}} = \left(\frac{\text{mol(acid)}}{\text{L}}\right)\left(\frac{\text{equivalents(H}^+\text{)}}{\text{mol(acid)}}\right)$$

The meaning of equivalent mass or equivalents may be expanded to include the equivalent mass of charged species like Na^+ or Mg^{2+}. An equivalent mass of charged species is the mass of that species which contains an Avogadro number of charges. For example, the equivalent mass of Na^+ ion is its atomic mass. That for Ca^{2+} is one-half its atomic mass, and for Al^{3+} is one-third its atomic mass. Ion concentrations are reported in milliequivalents, or meq, where 1000 meq = 1 equivalent mass = 1 eq.

Problem 9.19. Typical concentrations of several ions in blood plasma are given in units of normality: Na^+, 142 meq/L; K^+, 5 meq/L; Ca^{2+}, 5 meq/L; Mg^{2+}, 3 meq/L. Calculate the molar concentrations of these ions.

Ans. Conversion of moles to equivalents (abbreviated eq.), or equivalents to moles, requires a factor-label conversion factor. For conversion of moles to equivalents, the factor is

$$\frac{\text{Equivalents}}{\text{L}} = \left(\frac{\text{mol}}{\text{L}}\right)\left(\frac{\text{equivalents}}{\text{mol}}\right)$$

For conversion of equivalents to moles, the factor is

$$\frac{\text{Moles}}{\text{L}} = \left(\frac{\text{equivalents}}{\text{L}}\right)\left(\frac{\text{mol}}{\text{equivalents}}\right)$$

For Na^+: $$\left(\frac{142\text{ meq}}{\text{L}}\right)\left(\frac{1\text{ eq}}{1000\text{ meq}}\right)\left(\frac{1\text{ mol}}{1\text{ eq of charge}}\right) = 0.142\ M$$

For K^+: $$\left(\frac{5\text{ meq}}{\text{L}}\right)\left(\frac{1\text{ eq}}{1000\text{ meq}}\right)\left(\frac{1\text{ mol}}{1\text{ eq of charge}}\right) = 0.005\ M$$

For Ca^{2+}: $$\left(\frac{5.0\text{ meq}}{\text{L}}\right)\left(\frac{1\text{ eq}}{1000\text{ meq}}\right)\left(\frac{1\text{ mol}}{2\text{ eq of charge}}\right) = 0.0025\ M$$

For Mg^{2+}: $$\left(\frac{3.0\text{ meq}}{\text{L}}\right)\left(\frac{1\text{ eq}}{1000\text{ meq}}\right)\left(\frac{1\text{ mol}}{2\text{ eq of charge}}\right) = 0.0015\ M$$

ADDITIONAL SOLVED PROBLEMS

ACID-BASE THEORIES

Problem 9.20. What is the difference between the Arrhenius and Brønsted-Lowry theories of acids and bases?

Ans. The Arrhenius model focuses on the acid or base molecule only, in that an acid is defined as a molecule which dissociates in water to produce protons, and a base is defined as a molecule which dissociates in water to produce hydroxyl ions. In the Brønsted-Lowry model, the solvent assumes a central role. This theory proposes that an acid is a compound which can donate protons to a base, and consequently, a base is a compound which can accept the acid's donated proton. Therefore, an acid cannot behave as an acid in the absence of a base. Water assumes a central role in these processes, since in the presence of an acid, it acts as a base, accepting the acid's proton. Furthermore, in the presence of a base, it can donate a proton, and act as an acid.

Problem 9.21. What is a conjugate acid-base pair?

Ans. A conjugate acid-base pair is two molecules or ions which are related to each other by the fact that the base is derived from the acid by the acid's loss of a proton. The acid loses its proton by donating it to some other base, which is, in turn, converted to its conjugate acid. Symbolically:

$$HB + A \rightleftharpoons AH^+ + B^-$$

The conjugate acid-base pairs are HB and B^-, and AH^+ and A.

Problem 9.22. What is a Lewis acid and what is a Lewis base?

Ans. A Lewis acid is a molecule or ion which has room for a lone pair of electrons in its valence shell. A Lewis base is a molecule or ion which possesses a lone pair of electrons in its valence shell. An example of a Lewis acid-base reaction is the reaction between ammonia and boron trifluoride:

$$\begin{array}{c}H\\|\\H-\ddot{N}:\\|\\H\end{array} + \begin{array}{c}F\\|\\B-F\\|\\F\end{array} \rightleftharpoons \begin{array}{c}H\ \ F\\|\ \ \ |\\H-N-B-F\\|\ \ \ |\\H\ \ F\end{array}$$

ACIDS AND BASES: STRONG VERSUS WEAK

Problem 9.23. What is the definition of a strong acid? Give examples.

Ans. A strong acid is a proton-donating compound which is completely dissociated in water to produce only $H_3O^+(aq)$ ions and its accompanying anions. Some examples are HCl, HNO_3, H_2SO_4. For others see Table 9-1

Problem 9.24. Calculate the $[H^+]$, and $[NO_3^-]$, of a 0.175 *M* solution of nitric acid, HNO_3.

Ans. Nitric acid is a strong acid and is therefore 100 percent ionized in aqueous solution. Therefore, $[HNO_3] = [H^+] = [NO_3^-] = 0.175\ M$.

Problem 9.25. What is the definition of a strong base? Give examples.

Ans. A strong base is a compound which is completely dissociated in water to produce only $OH^-(aq)$ ions along with its accompanying cations. Examples are NaOH, $Ca(OH)_2$, KOH. For others see Table 9-1.

Problem 9.26. Calculate the $[OH^-]$, and $[Ca^{2+}]$, of a 0.01 *M* solution of $Ca(OH)_2$.

Ans. $Ca(OH)_2$ is a strong base and is therefore 100 percent ionized in aqueous solution. Since there are 2 mol of hydroxide ion per mole of $Ca(OH)_2$, $[OH^-] = 2 \times [Ca(OH)_2] = 0.02\ M$, and $[Ca(OH)_2] = [Ca^{2+}] = 0.01\ M$.

Problem 9.27. What is the definition of a weak acid? Give examples.

Ans. A weak acid is one which does not undergo complete ionization in aqueous solution. The hydrogen ion concentration in such a solution is a fraction of the total acid concentration. Examples are nitrous acid, HNO_2, hydrosulfuric acid, H_2S, acetic acid, $HC_2H_3O_2$.

Problem 9.28. What is the definition of a weak base? Give examples.

Ans. An aqueous solution of a weak base is one in which the hydroxide ion concentration is only a fraction of the concentration of the added base. Examples are ammonia, NH_3, methyl amine, CH_3NH_2, and the rocket fuel hydrazine, NH_2NH_2.

Problem 9.29. Why is water called an amphoteric substance?

Ans. An amphoteric substance is one which can behave both as an acid and as a base. Water reacts with acids, thus acting as a base, and will also donate a proton to a base, thus acting as an acid.

Problem 9.30. The concentration of hydrogen ion and hydroxide ion in pure water at 25 °C is 1×10^{-7} *M*. What is the value of the equilibrium constant for the ionization of water at that temperature?

Ans. The ionization reaction for water is

$$H_2O + H_2O \rightleftharpoons H_3O^+ + OH^-$$

The equilibrium constant is

$$K_{eq} = K_w = [H^+][OH^-] = [1 \times 10^{-7}]^2 = 1 \times 10^{-14}$$

Problem 9.31. Calculate the hydrogen ion concentration of a 0.01 *M* solution of $Ca(OH)_2$.

Ans. Hydrogen ion concentration may be calculated from the relationship $K_w = [H^+][OH^-]$. Calcium hydroxide is a strong base, and since there are 2 mol of OH^- per mole of $Ca(OH)_2$, $[OH^-] = 0.02\ M$. Therefore,

$$[H^+] = \frac{1 \times 10^{-14}}{0.02} = 5 \times 10^{-13}\ M$$

Problem 9.32. Calculate the hydroxide ion concentration in a 0.175 *M* solution of HCl.

Ans. Hydroxide ion concentration may be calculated from the relationship $K_w = [H^+][OH^-]$. HCl is a strong acid. Therefore, $[H^+] = 0.175\ M$. Thus

$$[OH^-] = \frac{1.00 \times 10^{-14}}{0.175} = 5.71 \times 10^{-14}\ M$$

Problem 9.33. The ionization constant for water increases with increases in temperature. Is the ionization of water an exothermic or an endothermic reaction?

Ans. An increase in the equilibrium constant for any reaction means a shift in equilibrium to products. A shift to products because of an application of heat means the reaction absorbs heat and is therefore endothermic.

Problem 9.34. The ionization constant for water is $K_w = 1.00 \times 10^{-13.60}$ at body temperature, 37 °C. What are the H_3O^+ and OH^- concentrations at that temperature?

Ans.

$$K_w = [H^+][OH^-] = 1.00 \times 10^{-13.60}$$

$$[H^+] = [OH^-] = \sqrt{1.00 \times 10^{-13.60}} = 1.58 \times 10^{-7}$$

pH, A MEASURE OF ACIDITY

Problem 9.35. What is the meaning of pH?

Ans. pH is a convenient way of expressing the hydrogen ion concentration of an aqueous solution. It is defined as

$$pH = -\log[H^+] \quad \text{or} \quad [H^+] = 10^{-pH}$$

Problem 9.36. What is the pH of pure water at 25 °C?

Ans. The hydrogen ion concentration of pure water at 25 °C is 1×10^{-7}. The simplest way to calculate pH in this case, since the preexponential factor is 1, is to use the alternative expression

$$[H^+] = 10^{-pH} = 10^{-7}$$

Therefore, pH = 7.

Problem 9.37. What is the pH of pure water at 37 °C?

Ans. The hydrogen ion concentration of pure water at 37 °C is 1.58×10^{-7}. Because the preexponential factor here is not 1.0, use your calculator, and calculate the pH using the expression

$$pH = -\log[H^+] = -\log[1.58 \times 10^{-7}] = 6.80$$

Problem 9.38. Calculate the pH of the following solutions at 25 °C: (*a*) 0.0026 *M* HCl, (*b*) 0.014 *M* HNO_3, (*c*) 0.00052 *M* HBr, (*d*) 0.092 *M* HI.

Ans. These are all strong acids. Therefore, $[\text{Acid}] = [H^+] = 10^{-pH}$. Because the preexponential factors are not simple, we must use the logarithmic relationship $pH = -\log[H^+]$. Compute the answers using the log key of your calculator: (*a*)2.59, (*b*) 1.85, (*c*) 3.28, (*d*) 1.04.

Problem 9.39. Calculate the pOH and the hydroxide ion concentrations, $[OH^-]$, of the solutions listed in in Problem 9.38.

Ans. Use the relationship pH + pOH = 14 to determine pOH. The second calculation is made using the relationship $[OH^-] = 10^{-pOH}$. This requires use of the 10^x or Y^x key of your calculator.

	pOH	$[OH^-]$
(*a*)	11.42	3.8×10^{-12} *M*
(*b*)	12.15	7.1×10^{-13} *M*
(*c*)	10.72	1.9×10^{-11} *M*
(*d*)	12.96	1.1×10^{-13} *M*

Problem 9.40. Calculate the pH of the following solutions at 25 °C: (*a*) 0.0028 *M* KOH, (*b*) 0.00036 *M* $Ca(OH)_2$, (*c*) 0.016 *M* Ba $(OH)_2$, (*d*) 0.0073 *M* TIOH.

Ans. All these compounds are strong bases, and are therefore 100 percent ionized in aqueous solution. Furthermore, two of them contain 2 mol of hydroxide ion per mol of base. First calculate the pOH as $-\log[OH^-]$, then simply add 14 to the resulting negative number. The results are (*a*) 11.45, (*b*) 10.86, (*c*) 12.51, (*d*) 11.86.

pH AND WEAK ACIDS AND BASES

Problem 9.41. Calculate the pH of a 0.015 *M* acetic acid, HC_2H_3O, at 25 °C.

Ans. Acetic acid is a weak acid and therefore reacts with water to form an equilibrium mixture as

$$HC_2H_3O_2(aq) + H_2O(l) \rightleftharpoons H_3O^+(aq) + C_2H_3O_2^-(aq)$$

$$K_a = 1.74 \times 10^{-5}$$

Use our accounting scheme to determine the equilibrium concentrations:

	$HC_2H_3O_2$	H^+	$C_2H_3O_2^-$
Start	0.015	0	0
Change	$-x$	$+x$	$+x$
Final	$0.015 - x$	x	x

Substitute the final entries in the equilibrium constant expression, and make the approximation $[0.015 - x] = [0.015]$:

$$1.74 \times 10^{-5} = \frac{[x]^2}{0.015}$$

$$x = [H^+] = 5.1 \times 10^{-4}\ M$$
$$pH = -\log[H^+] = 3.29$$

Problem 9.42. Calculate the pH of a 0.025 M solution of ammonia, NH_3.

Ans. Ammonia is a weak base, and therefore reacts with water to form an equilibrium mixture of ionized and nonionized species. Use the same approach as in Problem 9.41, but use the K_b of NH_3. Write the reaction equation:

$$NH_3(aq) + H_2O(l) \rightleftharpoons NH_4^+(aq) + OH^-(aq)$$
$$K_b = 1.75 \times 10^{-5}$$

Substitute the final entries in the accounting scheme in the equilibrium constant expression and make the approximation $[0.025 - x] = [0.025]$:

	NH_3	NH_4^+	OH^-
Start	0.025	0	0
Change	$-x$	$+x$	$+x$
Final	$0.025 - x$	x	x

$$1.75 \times 10^{-5} = \frac{[x]^2}{0.025}$$

$$x = [OH^-] = 6.6 \times 10^{-4}\ M$$

$$pOH = 3.18 \quad \text{and} \quad pH = 10.82$$

POLYPROTIC ACIDS

Problem 9.43. What is the definition of a polyprotic acid?

Ans. A polyprotic acid is an acid which can donate more than one proton to a base.

Problem 9.44. Write equations for the dissociation of carbonic acid, H_2CO_3, in water.

Ans.

$$H_2CO_3(aq) + H_2O(l) \rightleftharpoons HCO_3^-(aq) + H_3O^+(aq)$$
$$HCO_3^-(aq) + H_2O(l) \rightleftharpoons CO_3^{2-}(aq) + H_3O^+(aq)$$

Problem 9.45. Write equilibrium constant expressions for the dissociation of carbonic acid in water.

Ans.

$$K_{a_1} = \frac{[H_3O^+][HCO_3^-]}{[H_2CO_3]} = 4.45 \times 10^{-7}$$

$$K_{a_2} = \frac{[H_3O^+][CO_3^{2-}]}{[HCO_3^-]} = 4.72 \times 10^{-11}$$

Problem 9.46. In an aqueous solution of H_2CO_3, what proportion of the H_3O^+ present is contributed by HCO_3^-?

Ans. Because K_{a_2} is so much smaller than K_{a_1}, the proportion of H_3O^+ ion present in the solution of H_2CO_3 from the ionization of HCO_3^- must be negligible.

Problem 9.47. Calculate the pH and the concentration of HPO_4^{2-} present in a 0.015 *M* solution of KH_2PO_4.

Ans. The ion, $H_2PO_4^-$, may be the product of the first ionization of the weak acid, H_3PO_4, as well as the anion of the salt, KH_2PO_4. It is itself a weak acid which reacts with water as follows:

$$H_2PO_4^-(aq) + H_2O(l) \rightleftharpoons HPO_4^{2-}(aq) + H_3O^+(aq)$$

To calculate the pH of its aqueous solution, use the same approach as with acetic acid, making the same approximation, $[H_2PO_4 - x] = [H_2PO_4^-]$, and use as equilibrium constant, $K_{a_2} = 6.32 \times 10^{-8}$:

	$H_2PO_4^-$	H^+	HPO_4^{2-}
Start	0.015	0	0
Change	$-x$	$+x$	$+x$
Final	$0.015 - x$	x	x

$$6.32 \times 10^{-8} = \frac{[x]^2}{0.015}$$

$$x = [H^+] = [HPO_4^{2-}] = 3.1 \times 10^{-5}\,M$$

$$pH = -\log[H^+] = 4.51$$

The K_{a_3} for the ionization of HPO_4^{2-} is 4.84×10^{-13}, much smaller than K_{a_2}. Because K_{a_3} is about 1×10^{-5} smaller than K_{a_2}, it is reasonable to assume that the contribution of the second dissociation to the overall H^+ concentration can be ignored. To verify that assumption, let us calculate the PO_4^{3-} concentration which is equal to the H^+ concentration from the second dissociation:

$$[PO_4^{3-}] = \frac{[HPO_4^{2-}]K_{a3}}{[H^+]} = \frac{[3.1 \times 10^{-5}][4.84 \times 10^{-13}]}{[3.1 \times 10^{-5}]} = 4.8 \times 10^{-13}$$

The small value of $[PO_4^{3-}]$ which is equal to $[H^+]$ from the second stage of ionization, compared with the value of $[H^+]$ from the first stage of ionization, verifies the assumption that dissociation of HPO_4^{2-} contributes a negligible amount of H^+.

SALTS AND HYDROLYSIS

Problem 9.48. Predict whether aqueous solutions of the following salts are acidic, neutral, or basic: (*a*) NH_4Cl, (*b*) $Al_2(SO_4)_3$, (*c*) KI, (*d*) $NaHCO_3$, (*e*) K_2HPO_4, (*f*) $Ca(NO_3)_2$, (*g*) $NaNO_2$, (*h*) KCN, (*i*) $NaC_2H_3O_2$, (*j*) $MgSO_3$, (*k*) $MgSO_4$.

Ans. Cations can be only neutral or acidic, and anions can be only neutral or basic.

	Cation		Anion		
	Acidic	Neutral	Basic	Neutral	Solution
(*a*)	NH_4^+			Cl^-	Acidic
(*b*)	Al^{3+}			SO_4^{2-}	Acidic
(*c*)		K^+		I^-	Neutral
(*d*)		Na^+	HCO_3^-		Basic
(*e*)		K^+	HPO_4^{2-}		Basic
(*f*)		Ca^{2+}		NO_3^-	Neutral
(*g*)		Na^+	NO_2^-		Basic
(*h*)		K^+	CN^-		Basic
(*i*)		Na^+	$C_2H_3O_2^-$		Basic
(*j*)		Mg^{2+}	SO_3^{2-}		Basic
(*k*)		Mg^{2+}		SO_4^{2-}	Neutral

BUFFERS AND BUFFER SOLUTIONS

Problem 9.49. What is a buffered solution?

Ans. A buffered solution is one which resists changes in pH when either acid or base is added to it.

Problem 9.50. What does a buffered solution consist of?

Ans. A buffered solution contains a mixture of a conjugate acid-base pair of comparable concentrations. The conjugate acid-base pair may be a weak acid plus a salt of the weak acid, or a weak base and a salt of the weak base.

Problem 9.51. Give some examples of buffer mixtures.

Ans. $HC_2H_3O_2$–$NaC_2H_3O_2$, KH_2PO_4–K_2HPO_4, H_2CO_3–$NaHCO_3$, NH_3–NH_4Cl

Problem 9.52. Why must the concentrations of the conjugate acid base pair be comparable?

Ans. The pH of a buffered solution is described by the Henderson-Hasselbalch (H-H) equation

$$pH = pK_a + \log\frac{\text{proton acceptor}}{\text{proton donor}}$$

The logarithmic term will vary very rapidly if the ratio of proton acceptor to donor is not close to 1.0. This will cause the pH to also vary rapidly when acid or base is added to the solution.

Problem 9.53. What is the pH of a solution consisting of 0.050 M acetic acid, $HC_2CH_3O_2$, and 0.050 *M* sodium acetate, $NaC_2H_3O_2$?

Ans. This solution contains a conjugate acid-base pair of comparable concentrations, and is therefore a buffer solution. The proton acceptor is acetate, and the donor is acetic acid. Use the H-H equation with $K_a = 1.74 \times 10^{-5}$, $pK_a = 4.76$:

$$\log[(0.050)/(0.050)] = \log 1.00 = 0.00$$
$$pH = 4.76 + 0.00$$
$$= 4.76$$

Problem 9.54. Calculate the pH of a solution consisting of 0.080 M K_2HPO_4 and 0.050 M KH_2PO_4.

Ans. **This solution contains the conjugate acid-base pair $H_2PO_4^- - HPO_4^{2-}$, and since they are of comparable concentrations, the solution is buffered. The dissociation reaction is equivalent to the second dissociation of phosphoric acid, H_3PO_4, characterized by K_{a_2}:**

$$H_2PO_4^-(aq) + H_2O(l) \rightleftharpoons HPO_4^{2-}(aq) + H_3O^+(aq)$$

Therefore, use of the H-H equation, with $pK_{a_2} = 7.20$:

$$\log[(0.080)/(0.050)] = \log 1.60 = 0.20$$
$$pH = 7.2 + 0.20 = 7.4$$

Problem 9.55. Calculate the pH of a solution consisting of 0.050 M K_2HPO_4 and 0.080 M KH_2PO_4.

Ans. **This is the same as Problem 9.54, but with the concentrations of proton acceptor, HPO_4^{2-}, and donor, $H_2PO_4^-$, reversed. Therefore,**

$$\log[(0.050)/(0.080)] = \log[0.63] = -0.20$$
$$pH = 7.2 + (-0.20) = 7.0$$

TITRATION

Problem 9.56. What was the molar concentration of 51.4 mL of a solution of HNO_3 that required 29.20 mL of a 0.0160 M standard solution of KOH for complete neutralization?

Ans. **It is first necessary to determine the stoichiometry of the reaction**

$$HNO_3(aq) + KOH(aq) \rightleftharpoons H_2O(l) + K^+(aq) + NO_3^-(aq)$$

The net ionic equation is

$$H^+(aq) + OH^-(aq) \rightleftharpoons H_2O(l)$$

The stoichiometry is 1 HNO_3/1 KOH, and therefore

$$M_{acid} \times V_{acid} = M_{base} \times V_{base}$$
$$M_{acid} \times 51.4 \text{ mL} = 29.20 \text{ mL} \times 0.0160\ M$$
$$M_{acid} = 0.0091\ M$$

Problem 9.57. What was the molar concentration of 32.80 mL of an H_2SO_4 solution which required 31.72 mL of a 0.0240 M standard solution of KOH for complete neutralization?

Ans. **First determine the stoichiometry of the reaction**

$$H_2SO_4(aq) + 2\ KOH(aq) \rightleftharpoons 2\ H_2O(l) + 2\ K^+(aq) + SO_4^{2-}(aq)$$

The net ionic equation is

$$2\ H^+(aq) + 2\ OH^-(aq) \rightleftharpoons 2\ H_2O(l)$$

The stoichiometry is 1 H_2SO_4/2 KOH. We must therefore take this into account in our calculations. Our formula basically tells us that in a neutralization reaction

$$\text{Moles acid} = \text{moles base}$$

Let us rewrite the basic equation to include the stoichiometry of the neutralization reaction

$$\left(\frac{\text{mol(acid)}}{\text{L(acid)}}\right)[\text{L(acid)}] = \left(\frac{\text{mol(base)}}{\text{L(base)}}\right)[\text{L(base)}]\left(\frac{1\ \text{mol(acid)}}{2\ \text{mol(base)}}\right)$$

Thus $\text{m}M\text{(acid)/mL} \times 32.80\ \text{mL(acid)} = 0.0240\ \text{m}M\text{(base)/mL} \times 31.72\ \text{mL(base)} \times (1/2)$

$$\text{m}M\text{(acid)/mL} = M\text{(acid)} = 0.0116\ M$$

Problem 9.58. What was the molar concentration of 28.92 mL of an H_3PO_4 solution which required 91.21 mL of a 0.0390 *M* standard solution of KOH for complete neutralization?

Ans. Here again, the reaction stoichiometry is central:

$$H_3PO_4(aq) + 3\ KOH(aq) \rightleftharpoons 3\ H_2O(l) + 3\ K^+(aq) + PO_4^{3-}(aq)$$

The net ionic equation is

$$3\ H^+(aq) + 3\ OH^-(aq) \rightleftharpoons 3\ H_2O(l)$$

Taking the stoichiometry, 1 mol acid/3 mol base, into account:

$$(\text{m}M_{(\text{acid})}/\text{mL}) \times 28.92\ \text{mL}_{(\text{acid})} = (0.0390\ \text{m}M_{(\text{base})}/\text{mL}) \times 91.21\ \text{mL}_{(\text{base})} \times (1/3)$$

$$(\text{m}M_{(\text{acid})}/\text{mL}) = M_{(\text{acid})} = 0.0410\ M$$

NORMALITY

Problem 9.59. What is the normality of a 0.0116 *M* solution of H_2SO_4?

Ans. The definition of normality is equivalents/liter. In this case, we must decide what kind of reaction the acid is involved with. Since no reaction context was provided, the question cannot be answered.

Problem 9.60. What is the normality of a 0.0116 *M* solution of H_2SO_4 used for neutralization of a base?

Ans. The question of equivalents/liter can now be addressed, since we know that the H_2SO_4 reacts with a base to produce water. The stoichiometry of this reaction was determined in Problem 9.57. Its meaning in terms of equivalents is that 1 mol of H_2SO_4 contributes 2 mol of H^+ per mol of H_2SO_4, and is equivalent to 2 mol of hydroxide ion. We have

$$N = \frac{\text{equivalents(acid)}}{\text{L}} = \left(\frac{\text{mol(acid)}}{\text{L}}\right)\left(\frac{\text{equivalents}(H^+)}{\text{mol(acid)}}\right)$$

Therefore, $N = 0.0116\ M \times 2\ \text{eq}\ H^+/\text{mol acid} = 0.0232\ N$.

Problem 9.61. A solution contains (*a*) 27.34 mmol of K^+, (*b*) 5.27 mmol of Ca^{2+}, and (*c*) 1.63 mmol of Al^{3+} ions, all in 47.2 mL. Compute the normality of this solution with respect to each ionic species.

Ans. Normality in this case is defined as equivalents of charge/liter. To convert molarity to normality, use

$$N = \frac{\text{meq}}{\text{mL}} = \left(\frac{\text{mmol}}{\text{mL}}\right)\left(\frac{\text{meq}}{\text{mmol}}\right)$$

(*a*) $$N = \left(\frac{27.34\ \text{mmol}\ K^+}{47.2\ \text{mL}}\right)\left(\frac{1\ \text{meq of charge}}{\text{mmol}\ K^+}\right) = 0.579\ N$$

(*b*) $$N = \left(\frac{5.27\ \text{mmol}\ Ca^{2+}}{47.2\ \text{mL}}\right)\left(\frac{2\ \text{meq of charge}}{\text{mmol}\ Ca^{2+}}\right) = 0.223\ N$$

(*c*) $$N = \left(\frac{1.63\ \text{mmol}\ Al^{3+}}{47.2\ \text{mL}}\right)\left(\frac{3\ \text{meq of charge}}{\text{mmol}\ Al^{3+}}\right) = 0.104\ N$$

SUPPLEMENTARY PROBLEMS

Problem 9.62. Calculate the pH of the following solutions: (*a*) 0.0024 *M* HCl, (*b*) 1.0 *M* HNO_3, (*c*) 0.0056 *M* HI, (*d*) 0.027 *M* HBr, (*e*) 0.12 *M* HCl.

Ans. (*a*) 2.25, (*b*) 0.00, (*c*) 2.25, (*d*) 1.57, (*e*) 0.92

Problem 9.63. Calculate the pOH of the solutions in Problem 9.62.

Ans. (*a*) 11.38, (*b*) 14, (*c*) 11.75, (*d*) 12.43, (*e*) 13.08

Problem 9.64. Calculate the pOH of the following solutions: (*a*) 0.0047 *M* KOH, (*b*) 0.00063 *M* $Ca(OH)_2$, (*c*) 0.012 *M* $Ba(OH)_2$, (*d*) 1.0 *M* NaOH, (*e*) 0.010 *M* KOH

Ans. (*a*) 2.33, (*b*) 2.90, (*c*) 1.62, (*d*) 0.00, (*e*) 2.00

Problem 9.65. Calculate the pH of the solutions in Problem 9.64.

Ans. (*a*) 11.67, (*b*) 11.1, (*c*) 12.38, (*d*) 14.00, (*e*) 12.00

Problem 9.66. Are the following solutions of salts acidic, neutral, or basic? (*a*) $Fe_2(SO_4)_3$, (*b*) $CaCl_2$, (*c*) NaBr, (*d*) $NaNO_2$, (*e*) NH_4NO_3, (*f*) $Mg(CN)_2$.

Ans. (*a*) acidic, (*b*) neutral, (*c*) neutral, (*d*) basic, (*e*) acidic, (*f*) basic

Problem 9.67. Calculate the pH of a solution which is 0.060 *M* in H_2CO_3 and 0.025 *M* in $KHCO_3$.

Ans. pH = 5.97

Problem 9.68. What is the molar ratio of $[HPO_4^{2-}/H_2PO_4^-]$ in a buffered solution whose pH is 6.8?

Ans.

$$\left[\frac{HPO_4^{2-}}{H_2PO_4^-}\right] = \frac{1.0}{2.5}$$

Problem 9.69. What conjugate acid-base pair would you employ to buffer a solution at pH 9.0?

Ans. NH_4Cl–NH_3

Problem 9.70. Compute the molarity of 34.7 mL of nitric acid solution which required 43.6 mL of 0.051 *M* KOH for complete neutralization.

Ans. 0.064 *M*

Problem 9.71. Calculate the normality of 26.34 mL of a $Ca(OH)_2$ solution which required 52.68 mL of a 0.015 *M* HCl solution for complete neutralization.

Ans. 0.030 *N*

Nuclear Chemistry and Radioactivity

In the chemical reactions we have studied to this point, the focus has been on the changes in the arrangements of valence electrons. The identities of the elements involved in the chemical changes have remained unchanged. However, there are many elements whose atomic nuclei are unstable. Their instability is evidenced by the emission of radiation and subatomic particles of high energy. Often this emission of particles results in the transformation into new elements. The process of spontaneous nuclear decomposition is called *radioactivity*. Radioactive elements find application to problems in biology, chemistry, physics, medicine, agriculture, geology, forensic science, and archeology. Nuclear processes are also the basis for the generation of significant amounts of electricity in many parts of the world, and the construction of nuclear weapons.

10.1 RADIOACTIVITY

The concept of isotopes was introduced in Chap. 2, and it was used to describe a family of nuclei of the same element which have the same atomic number (number of protons), but different numbers of neutrons. The atomic masses used in chemical calculations are averages of the atomic masses of an element consisting of a number of isotopes with varying abundances in nature. In discussing nuclear processes, we refer to a particular isotope because we are interested in the precise number of *nucleons* (the name for either the protons or neutrons in the nucleus) and any changes in that number as a result of particulate emissions. We use the name *nuclide* for referring to particular isotopes of different elements.

10.1.1 Radioactive Emissions

All elements with more than 83 protons are radioactive. The symbol for isotopes (Chap. 2) consists of the element's symbol with a subscript denoting the atomic number, and a superscript denoting the atomic mass which is the sum of protons and neutrons. Another method for denoting a specific isotope is to write the name of the element with a hyphen followed by the nuclear mass. The atomic number is implied by the element's name.

Problem 10.1. What are the alternative forms for the names of some isotopes of uranium?

Ans.

Subscript-Superscript Form	Alternative Form
$^{234}_{92}U$	Uranium-234
$^{235}_{92}U$	Uranium-235
$^{238}_{92}U$	Uranium-238

The radioactive element uranium-238 emits helium nuclei, $^{4}_{2}He$. When a $^{238}_{92}U$ nucleus emits a helium nucleus, the mass number decreases by four and the atomic number decreases by two. The change in atomic number reveals that a new element has been produced. The process is described by a *nuclear equation*.

Problem 10.2. What is the form of the nuclear equation which describes the ejection of a helium nucleus from uranium-238?

Ans.

$$^{238}_{92}\text{U} \longrightarrow {}^{234}_{90}\text{X} + {}^{4}_{2}\text{He}$$

Note carefully that the sum of the mass numbers of the products, i.e., the sum of protons plus neutrons, is equal to the mass number of the reactant. In addition, the sum of atomic numbers, i.e., numbers of protons only, of the products is equal to the atomic number, the protons, of the reactant.

Problem 10.3. How is the new element, $^{234}_{90}\text{X}$, identified?

Ans. Examination of the table of atomic numbers tells us that the new element has an atomic number of 90, and is therefore thorium.

Since the uranium-238 spontaneously disintegrates, it is a *radioisotope* or *radionuclide*. Henri Becquerel discovered that uranium-238 was a source of radiation which he called *α-rays*. These rays were shown later to be helium-4 nuclei, or *α-particles*. Uranium-238 is called an *α-emitter*.

Other emissions of interest are *β-rays*, *positrons*, and *γ-rays*. β-particles have been shown to be electrons, and in nuclear reactions are given the symbol $_{-1}^{0}e$. The superscript denotes the infinitesimal mass of an electron relative to a nucleon, and the subscript refers to the electron's charge.

Problem 10.4. What is an example of β-particle emission, and what is the nature of the product?

Ans. Thorium-234 is radioactive and emits a β-particle as

$$^{234}_{90}\text{Th} \longrightarrow {}^{234}_{91}\text{Pa} + {}_{-1}^{0}e$$

The mass number of the product is the same as thorium, but the atomic number has increased by 1. The increase in atomic number is the result of the nucleus losing a negative charge, and thereby increasing its net positive charge. Examination of the table of atomic numbers indicates that the product of atomic number 91 is protactinium.

Problem 10.5. How can the nucleus which contains only protons and neutrons lose an electron?

Ans. There are no electrons in atomic nuclei. The β-particle is the result of the conversion of a neutron into a proton within the nucleus as

$$^{1}_{0}n \longrightarrow {}^{1}_{1}\text{H} + {}_{-1}^{0}e$$

This is followed by the ejection of the energetic β-particle.

The emission of an α- or β-particle often leaves the nucleus of the product in an excited state. Recall that when an atom is in an excited state, it can return to its ground state by losing the energy of excitation as emitted light, or a photon. Atomic nuclei in an excited state undergo a similar process by emitting a photon of much greater energy than that of visible light. This radiation is called *γ-radiation*, or *γ-rays*. Because γ-ray emission does not alter the mass of a radioactive nuclide, it is usually not included in nuclear equations.

Some nuclei emit a positron. This is a particle which has the same mass as an electron, but with a positive rather than a negative charge. The symbol for a positron in nuclear reactions is $_{+1}^{0}e$.

Problem 10.6. What is an example of positron emission?

Ans. A radioactive isotope of potassium, $^{38}_{19}\text{K}$, is a positron emitter, and the nuclear reaction for the process is

$$^{38}_{19}\text{K} \longrightarrow {}^{38}_{18}\text{Ar} + {}_{+1}^{0}e$$

The product has the same mass number as potassium, but its atomic number has decreased by 1. The table of atomic numbers reveals the product to be a new element, argon.

Problem 10.7. If the nucleus contains only protons and neutrons, how can a "positive" electron be emitted?

Ans. As in the case of the β-particle, the emission of a positron is the result of the conversion of a proton into a neutron as

$$_{1}^{1}\text{H} \longrightarrow {}_{0}^{1}n + {}_{+1}^{0}e$$

A positron is very rapidly annihilated by reacting with any available electron to produce two γ-rays as

$$_{-1}^{0}e + {}_{+1}^{0}e \longrightarrow 2\,{}_{0}^{0}\gamma$$

Table 10-1 summarizes the properties and results of the various nuclear emissions.

Table 10-1. A Summary of the Properties and Results of Nuclear Emissions

Emission	Symbol	Change in Nucleus: Mass Number	Change in Nucleus: Atomic Number
α	$_{2}^{4}\text{He}$	Decreases by 4	Decreases by 2
β	$_{-1}^{0}e$	No change	Increases by 1
γ	$_{0}^{0}\gamma$	No change	No Change
Positron	$_{+1}^{0}e$	No change	Decreases by 1

10.1.2 Radioactive Decay

A radioactive nucleus which emits a particle to become transformed to another nucleus is described as *decaying* to that nucleus. Such a radioactive event is called *radioactive decay*. Radionuclides decay at different rates. Some can decay in millionths of a second, others take millions of years. Decay is independent of all the variables which affect chemical reactions such as temperature, pressure, and concentration. This poses particular difficulty with regard to the disposal of nuclear wastes. The rate of radioactive decay is characterized by the loss of a constant percent per unit time, not a constant number of moles per unit time. We therefore characterize the decay rate by specifying the time required for 50 percent of the original material to decay. This period of time is called the *half-life*, given the symbol, $t_{1/2}$. The constant percent change means that 50 percent will be lost during the first half-life, 50 percent of what is left after the first half-life will decay over the second half-life, etc.

Problem 10.8. How much of a radioisotope will be left after 4 half-lives?

Ans. Let us put this into tabular form:

End of Half-Life	Amount Left, %
0	100.00
1	50.00
2	25.00
3	12.50
4	6.25

Problem 10.9. Is there a method for calculating the amount of radioisotope remaining after any number of half-lives?

Ans. We can calculate the amount of radioisotope left after some number of half-lives using the following relationship:

$$\frac{N}{N_0} = \left(\frac{1}{2}\right)^n$$

where N = amount left after a number, n, of half-lives, and N_0 is the amount of isotope at the start of the process. Let us recalculate the amount of a radioisotope left after 4 half-lives:

$$\frac{N}{N_0} = \left(\frac{1}{2}\right)^4 = \frac{1}{16}$$

$$N = N_0\left(\frac{1}{16}\right)$$

$$= \frac{100\%}{16} = 6.25\%$$

Problem 10.10. Calculate the amount of radioisotope remaining after 3.2 half-lives.

Ans. Using the same relationship as above:

$$\frac{N}{N_0} = \left(\frac{1}{2}\right)^{3.2}$$

$$\text{Log}\,\frac{N}{N_0} = 3.2 \log 0.5$$

$$= -0.9633$$

$$\frac{N}{N_0} = 10^{-0.9633} = 0.1088$$

$$N = N_0 \times 0.11$$

$$= 100\% \times 0.11 = 11\%$$

Different radioisotopes have different half-lives. It is as characteristic of a particular nuclide as the melting point is of a pure compound. Table 10-2 lists the half-lives of some biomedically useful radioisotopes.

Table 10-2. Half-Lives and Biomedical Applications of Some Radioisotopes

Nuclide	Half-Life	Application
Barium-131	11.6 days	Detection of bone tumors
Iodine-131	8.05 days	Measurement of thyroid uptake of iodine
Phosphorus-32	14.3 days	Detection of breast carcinoma
Technetium-99	6.0 h	Detection of blood clots

10.1.3 Radioactive Series

The decay of radioisotopes found in nature often results in the formation of products called *daughter nuclei*, which may or may not be radioactive. If the resulting products are nonradioactive (stable), the decay stops. However, the products may be radioactive, and decay will continue to produce new elements until a stable state is achieved. A series of nuclear reactions which begins with an unstable nucleus and ends with the formation of a stable one is called a *nuclear decay* or *nuclear disintegration series*. There are three such naturally occurring series. One begins with uranium-238, and terminates with lead-206, taking 14 steps; another begins with uranium-235, and terminates with lead-207; and the last begins with thorium-232, and terminates with lead-208.

10.1.4 Transmutation

Transmutation refers to a nuclear process in which a nuclide is transformed to a new element by interaction with a high-energy nuclear particle.

Problem 10.11. How was transmutation discovered?

Ans. The first artificial conversion of one nucleus to another was performed by E. Rutherford in 1919. He bombarded nitrogen-14 with α-particles emitted by radium and produced oxygen-17. The reaction is

$$^{14}_{7}N + ^{4}_{2}He \longrightarrow ^{17}_{8}O + ^{1}_{1}H$$

Since that time, hundreds of radioisotopes have been produced in the laboratory. In order for charged particles like α-particles to penetrate a positively charged nucleus, or the electron cloud surrounding the nucleus, they must be moving at extremely high velocities. Such velocities are achieved in particle accelerators such as the cyclotron. All the elements above atomic number 92 have been created in this manner. Most radioisotopes used in medicine have been produced using neutrons as subatomic projectiles.

Problem 10.12. What is an example of the production of a radioactive isotope used in medicine?

Ans. Cobalt-60, a γ-emitter, used in radiation therapy for cancer is prepared by bombarding iron-58 with neutrons. The reactions are

$$^{58}_{26}Fe + ^{1}_{0}n \longrightarrow ^{59}_{26}Fe$$

$$^{59}_{26}Fe \longrightarrow ^{59}_{27}Co + ^{0}_{-1}e$$

$$^{59}_{27}Co + ^{1}_{0}n \longrightarrow ^{60}_{27}Co$$

10.1.5 Nuclear Fission

In the previous examples of nuclear processes, mass and atomic numbers changed by only a few units. In 1939, it was found that bombardment of uranium by neutrons caused its fragmentation into large nuclei, e.g., barium, krypton, cerium, and lanthanum. This was accompanied by the release of very large quantities of energy. This new process was given the name of *nuclear fission*. High-energy neutrons are absorbed by uranium, thus producing an unstable nucleus which responds by spontaneously decomposing into more stable products.

Problem 10.13. Are the same products always produced by nuclear fission?

Ans. No. There are a large number of possible products. Over 200 isotopes of 35 different elements have been found as the result of the fission of uranium-235, two of which are illustrated by the following reactions:

$$^{235}_{92}U + ^{1}_{0}n \longrightarrow ^{139}_{56}Ba + ^{94}_{36}Kr + 3\,^{1}_{0}n$$

$$^{235}_{92}U + ^{1}_{0}n \longrightarrow ^{144}_{54}Xe + ^{90}_{38}Sr + 2\,^{1}_{0}n$$

Note carefully that the impact of one neutron results in the yield of at least two neutrons from the fission. This can result in subsequent fission reactions when those fission-produced neutrons are absorbed by other uranium nuclei. The multiple yield can result in an explosive chain reaction which yields extremely large amounts of energy, which is the operating principle for the atomic bomb. By reducing the rate of neutron production from spontaneous decomposition of uranium, the fission reaction can be controlled. This is done in nuclear reactors by the presence of neutron capturing material in the form of cadmium control rods. Nuclear reactors produce electrical energy by capturing in a circulating fluid the heat released in fission reactions.

10.1.6 Nuclear Fusion

Large amounts of energy are emitted as the result of another type of nuclear process called *nuclear fusion*. In nuclear fusion, nuclei of two light elements collide at velocities high enough to overcome the mutual repulsion of the nuclear protons, and fuse to form a nucleus of larger atomic number.

Problem 10.14. Does nuclear fusion occur naturally?

Ans. Yes. The reactions in stars like our sun consist of collisions among protons, deuterons, ${}^{2}_{1}H$, and helium nuclei as follows:

$$ {}^{1}_{1}H + {}^{1}_{1}H \longrightarrow {}^{2}_{1}H + {}^{0}_{+1}e $$

$$ {}^{1}_{1}H + {}^{2}_{1}H \longrightarrow {}^{3}_{2}He $$

$$ {}^{3}_{2}He + {}^{3}_{2}He \longrightarrow {}^{4}_{2}He + 2\,{}^{1}_{1}H $$

$$ {}^{3}_{2}He + {}^{1}_{1}H \longrightarrow {}^{4}_{2}He + {}^{0}_{+1}e $$

Problem 10.15. What are the conditions for initiating and sustaining nuclear fusion reactions?

Ans. Temperatures of about 4×10^7 K are required to overcome the mutual repulsions of the various nuclei, but when fusion occurs, it releases vast quantities of energy, more than enough to sustain stellar fusion processes. Research on the use of nuclear fusion has been in progress in the hope that it will be a major source of energy in the foreseeable future. The chief problem is to design a system which will produce and maintain in a controlled manner the extraordinarily high temperatures required.

10.1.7 Nuclear Energy

The energy changes in nuclear reactions are much greater than those associated with chemical reactions. This is because there are mass changes in nuclear processes. Mass and energy are related by the Einstein equation

$$ E = \Delta mc^2 $$

where Δm is the change in mass and c is the velocity of light, 3×10^8 m/sec. It is the large value of the speed of light which produces large energies in the event of a small change in mass. The masses of nuclei are always less than the sum of the masses of the constituent nucleons.

Problem 10.16. What is the mass of a helium-4 nucleus compared to the sum of its nucleons?

Ans. The mass of a helium-4 nucleus is 4.00150 amu. The proton mass is 1.00728 amu, and that of the neutron is 1.00867 amu. Calculating the mass of 2 protons and 2 neutrons:

Mass of 2 protons $= 2 \times 1.00728$ amu =	2.01456 amu
Mass of 2 neutrons $= 2 \times 1.00867$ amu =	2.01734 amu
Total mass =	4.03190 amu
Mass of a helium-4 nucleus =	4.00150 amu
Difference in mass =	0.03040 amu

This difference in mass is called the *mass defect*.

Problem 10.17. What is the energy change associated with the fusion of protons and neutrons into a helium-4 nucleus?

Ans. The energy for this process is calculated using the Einstein relationship in the form

$$E = \Delta m \times c^2$$

To obtain the energy in joules, we must use conversion factors which convert atomic mass units to kilograms:

$$0.0304 \text{ amu}\left(\frac{3 \times 10^8 \text{ m}}{\text{s}}\right)^2\left(\frac{1.0 \text{ g}}{6.02 \times 10^{23} \text{ amu}}\right)\left(\frac{1 \text{ kg}}{1000 \text{ g}}\right) = 4.54 \times 10^{-12} \frac{\text{kg} \cdot \text{m}^2}{\text{s}^2}$$

$$= 4.54 \times 10^{-12} \text{ J}$$

Disintegration of a mole of helium-4 requires the input of

$$\left(6.02 \times 10^{23} \frac{\text{nuclei}}{\text{mole}}\right)\left(4.54 \times 10^{-12} \frac{\text{joule}}{\text{nucleus}}\right) = 2.73 \times 10^{12} \frac{\text{joule}}{\text{mole}}$$

The fusion of protons and neutrons to form a helium nucleus will produce the same amount of energy. This is one million times the energy released in chemical reactions which may reach upper values of 1 to 2×10^6 J/mol.

The energy change calculated from the mass defect of a nucleus is called the *binding energy*. It is the energy which must be supplied in order to separate the nucleus into its constituent nucleons. It is also the energy released in nuclear fission, when a heavy nucleus splits into lighter ones, and in nuclear fusion when light nuclei fuse to form a heavier nucleus.

10.2 EFFECTS OF RADIATION

Nuclear emissions and x-rays are called *ionizing radiation*. They may have energies in the millions of electron volts (MeV) compared to the 2 to 3 electron volts (eV) of visible light. Table 10-3 is a list of some radioactive isotopes used as radiation sources along with the types of radiation emitted and their energy.

Table 10-3. Radioactive Isotopes Used as Radiation Sources

Isotope	Half-Life	Type of Emission	Energy, MeV
Polonium-210	138 days	α-Particle	5.3
Radium-226	1620 years	α-Particle	4.8
Cesium-137	30 years	β-Particle	1.2
Cobalt-60	5.27 years	γ-Ray	1.33

Problem 10.18. What is the meaning of eV?

Ans. eV stands for electron-volt. An electron-volt is the energy imparted to an electron accelerated in an electrical field of 1 volt (V). The charge of the electron is 1.602×10^{-19} C, so that 1 eV = 1.602×10^{-19} J.

Problem 10.19. Why are nuclear emissions called ionizing radiation?

Ans. Nuclear emissions and x-rays are called ionizing radiation because, as this extremely energetic radiation passes through matter, it creates ions by collision with electrons.

Problem 10.20. Are there differences in the ionization patterns produced by different particles?

Ans. Particles of the same energy produce different patterns of ionization. α-particles are massive and highly charged. They interact strongly with nuclei in their paths, are stopped quickly, and produce extensive centers of ionization. β-particles are much lighter; therefore they penetrate more deeply, and create ions along a thin track, much less dense than in the case of α-particles.

The study of the effects of radiation on matter is called *radiochemistry*, or *radiation chemistry*. The two major considerations of radiochemistry are (1) how effective is radiation in producing a chemical event, and (2) what are the specific chemical changes induced by the radiation.

With respect to the effectiveness in producing chemical effects, a principal concern is the penetrating power of the radiation. The mass and charge of the α-particle causes many atomic collisions, and intense interactions with bound electrons along its path. Therefore, the massive and highly charged α-particle can only penetrate matter to a depth of a few millimeters before it discharges its energy and is stopped. β-particles are charged, but have the minute mass of the electron, and therefore have greater penetrating power than α-particles. γ-rays have no mass or charge, and therefore have very large penetrating power.

Problem 10.21. What are the comparative distances of penetration of nuclear particles in human tissue?

Ans. α-particles may penetrate to a depth of 0.05 mm. β-particles may penetrate from 5 to 10 mm. γ-rays cannot be stopped by tissue, but their penetration is attenuated by about 10 to 20 percent after passing through tissue. Similar attenuation is experienced by x-rays.

High-energy radiation produces ions and free radicals along its penetration tracks. Ionization is the result of electrons torn from the valence shells of atoms in the paths of the particles. A free radical is the result of the rupture of a covalent bond. The bond breaks in such a way that each of the two new molecular species bears an unpaired electron. Free radicals are very reactive and will rapidly find new bonding partners.

Problem 10.22. What is an example of the formation of free radicals caused by incident radiation?

Ans. We will write the equation for the results of radiation on methyl alcohol:

$$CH_3OH + h\nu \longrightarrow \cdot CH_3 + \cdot OH$$

$h\nu$ represents the energy of the incident radiation, $\cdot CH_3$ and $\cdot OH$ represent the methyl and the hydroxyl free radicals, respectively, showing the single unpaired electron characteristic of a free radical as a dot on the appropriate atom.

Problem 10.23. What new species form when water is irradiated with high-energy radiation?

Ans. The irradiation of water with high-energy radiation can be represented by the following reaction:

$$H_2O + h\nu \longrightarrow H_2O^+ + e^-_{aq}$$

where e^-_{aq} is a free but hydrated electron. This reaction is called the *primary radiation event*, and it is rapidly followed by a number of *secondary chemical processes* resulting in free radical formation:

$$H_2O^+ + H_2O \longrightarrow H_3O^+ + \cdot OH$$
$$e^-_{aq} + H_2O \longrightarrow H\cdot + OH^-$$

Problem 10.24. What is an example of a reaction which will occur between hydrogen or hydroxyl free radicals and biomolecules?

Ans. Either of these free radicals may abstract an electron from a donor biomolecule which produces a biomolecule free radical as

$$\cdot OH + R\text{-}NH_2 \longrightarrow HOH + R\text{-}\dot{N}H$$

The new free radical may now combine with itself to produce a dimer:

$$R\text{-}\dot{N}H + R\text{-}\dot{N}H \longrightarrow RNH\text{-}NHR$$

When DNA is irradiated, one of the results is the formation of a dimer of thymine.

Direct absorption of high-energy radiation of tissue by a polypeptide or nucleic acid can result in a variety of other chemical events which lead to the alteration of the chemical identity of the biomolecule.

In light of these considerations, high-energy radiation may have a host of effects on living tissue, ranging from repairable damage to lethality. The primary damage sites are within cells, so that organ or tissue damage is the result of significant amounts of cell death.

Problem 10.25. Does all tissue respond to ionizing radiation in the same way.

Ans. The immediate chemical results of cellular irradiation are likely to be the same. However, the consequences of radiation exposure are different in different tissue. Radiation effects become apparent in tissue which undergoes rapid cell division, e.g., bone marrow, intestinal epithelium. This is because of alterations in DNA structure with consequent derangements in cell reproduction.

Nonlethal exposure (exposure which does not result in death) therefore results in a number of disorders, e.g., nausea, drop in white blood cell count. These disorders are called *radiation sickness*. Since children's tissues are in a more rapid mitotic state, the susceptibility of children to radiation sickness is greater than that of adults.

Problem 10.26. Can radiation effects be overcome if the organism survives?

Ans. Even if cells survive, there may be chemical alterations of the nucleic acids, particularly DNA, which can result in chromosome damage or mutations. In somatic tissue, this can induce cancer. For example, it can lead to cancer in bone tissue directly, or to leukemia from effects on bone marrow. If these changes occur in germ tissue, the radiation effects may not show up until in later generations.

Humans can protect themselves from the effects of high-energy radiation in two ways: (1) by the use of an absorbing barrier of some kind, and (2) by remaining as far away from the radiation source as is practicable. The absorbing barriers commonly used by x-ray technicians are aprons composed of layers of lead. The reason for putting distance between you and a radiation source resides in the fact that radiation of all kinds follows an inverse square law with respect to distance.

Problem 10.27. If the distance between a source and target is 4 m, how far should the source be from the target if it is desired to decrease the radiation intensity by a factor of 4?

Ans. An inverse square law with respect to distance means that the radiation intensity is proportional to the reciprocal of the square of the distance. That is,

$$I \propto \frac{1}{d^2}$$

We can compare the ratio of two intensities with the inverse ratio of their corresponding distances as

$$\frac{I_{new}}{I_{old}} = \frac{d^2_{old}}{d^2_{new}}$$

Substituting appropriate values into these ratios:

$$\frac{1}{4} = \frac{16 \text{ m}^2}{d^2_{new}}$$

$$d^2_{new} = 64 \text{ m}^2$$

$$d_{new} = 8 \text{ m}$$

By doubling the distance the radiation intensity decreases by a factor of 4.

10.3 DETECTION

Nuclear emissions are quite energetic, and as each penetrates matter it leaves behind a trail of ions. The detection of nuclear radiation is therefore the detection of the results of the interaction of emissions with the matter through which they travel. The most common and inexpensive method of detection is by the use of photographically sensitive film. One can tell the degree of exposure to the radiation by the extent of darkening of the film. γ-rays penetrate much more matter than β-particles, which in turn penetrate more matter than α-particles before they are stopped. This means that by covering photographic film with different thicknesses of material, one can differentiate among types of radiation. Other devices such as the Geiger counter use electronics to detect ions which are a consequence of radiation passing through matter. Salts called *phosphors* emit visible light when penetrated by high-energy radiation. These are incorporated into devices called *scintillation counters* which measure the dosage of radiation by the number of flashes of light per unit time.

10.4 UNITS

The detection of radiation can be quantitated by the use of a variety of units. The units used in quantitation depend on the context of measurement. X-rays and γ-rays are particularly effective in penetrating human tissue. However, α-rays are stopped by skin, and β-rays may penetrate to a depth of perhaps 1 cm and are therefore not ordinarily considered dangerous.

Problem 10.28. Are there situations in which α- or β-rays are dangerous?

Ans. Neither α- nor β-rays are particularly dangerous unless solids which emit these radiations happen to enter the body and are not excreted quickly. In that case, α-rays are very damaging because of the density of damage around the emitting particle.

Knowing how many disintegrations per unit time occur in a radioactive sample, or the number of ionizations in a volume of air, are not very helpful in characterizing the effects of radiation on human tissue and therefore the *rad* and the *rem* are commonly used to measure radiation dosage. Table 10-4 lists the types of radiation units, and the situations in which they are appropriately applied. A rad of α-rays will produce more damage than a rad of β-rays.

Table 10-4. Units of Radiation, Their Symbols, and Their Definitions

Unit	Symbol	Definition
Curie	Ci	The number of disintegrations per second (dps) in 1.0 g of pure radium, 3.7×10^{10} dps; 1 μCi = 3.7×10^4 dps.
Roentgen	R	The unit of exposure to γ or x-radiation based on the amount of ionization produced in air. 1 R will produce 1.61×10^{12} ion pairs in 1.0 g of air.
Rad	rad	A measure of the energy absorbed in matter as a result of exposure to any form of radiation. 1 rad = 0.01 J absorbed per kilogram of matter.
Gray	Gy	This is an SI unit which describes the energy absorbed by tissue. 1 Gy = 1 J absorbed per kilogram of tissue. 1 Gy = 100 rad.
Rem	rem	A measure of the energy absorbed in matter as a result of exposure to specific forms of radiation. This unit takes into account the fact that different forms of radiation produce different biological effects. 1 rem is equal to 1 rad times a factor dependent on the type of radiation. rem = roentgen equivalent for man.

Problem 10.29. How does one differentiate the biological damage caused by different radiation?

Ans. To measure the relative biological damage induced by radiation, the rad is multiplied by a factor known as the *RBE*, or *relative biological effectiveness*. The exact value of the RBE depends upon the type of tissue irradiated, the dose rate, and total dose, but it is approximately 1 for x-rays, γ-rays, and β-rays, and 10 for α-rays, protons, and neutrons. The product, rads $\times$ RBE = rems (*roentgen equivalent for man*).

The clinical effects of short-term exposure to radiation are listed in Table 10-5. All life forms on earth are continuously exposed to *background radiation*. This comes from natural sources such as cosmic rays and radioactivity from elements in the earth. Background radiation has probably played a large role in the evolution of life forms. Typical exposures to this radiation is about 125 millirem

Table 10-5. Effects of Short-Term Exposure to Radiation

Dose, rem	Clinical Effect
0–25	Nondetectible.
25–50	Temporary decrease in white blood cell count.
100–200	Large decrease in white blood cell count, nausea, fatigue.
200–300	Immediate nausea, vomiting; delayed appearance of appetite loss, diarrhea; probable recovery in 3 months.
500	Within 30 days of exposure, 50% of the exposed population will die.

(mrem) per year (1 mrem = 1×10^{-3} rem). Background radiation from x-rays, radiotherapy, and other humanly created sources is about 65 mrem/year.

10.5 APPLICATIONS

Radiation has seen wide application in medical and biological science. It is used in the form of x-rays and electron beams, or from radioactive emitters. Radioisotopes are employed extensively for the diagnosis of pathological conditions as well as for therapeutic use.

Problem 10.30. What are the important characteristics of a radionuclide useful in medicine?

Ans. The benefits of the radiation must outweigh the risks. This requires that (*a*) the radionuclide's half-life must be short because persistence in the body will cause damage; (*b*) the decay products must have either none or very little radioactivity; (*c*) for diagnostic use the radiation must have significant penetrating power in order to be accurately detected, i.e., it should be primarily a γ-emitter; (*d*) for therapeutic use, which involves causing damage to tissue, the isotope should be an α- or β-emitter, and if administered as a solution, it should be able to concentrate in a particular tissue. If the radionuclide cannot be selectively concentrated, it can be inserted as a package, i.e., in a metal or plastic tube, into the subject tissue.

Problem 10.31. What is an example of the use of a radioisotope in diagnosis?

Ans. Malfunction of the thyroid gland results in serious metabolic disorders, with cardiovascular consequences. The thyroid gland secretes thyroxin, an iodine derivative of tyrosine. The thyroid gland therefore concentrates iodine, and administration of iodine-123, which is a γ-emitter, will allow direct observation of the radionuclides' distribution in the gland, using radiation detection apparatus. Uneven distribution will reveal the presence and possibly the type of disorder.

Problem 10.32. What is an example of the therapeutic use of radioisotopes?

Ans. Iodine-131 is a β-emitter, and since the thyroid concentrates iodine, this radionuclide is used in the treatment of thyroid cancer.

Problem 10.33. How is ionizing radiation used in medicine?

Ans. Since γ-radiation completely penetrates and can be lethal to rapidly growing tissue, radiation has become a powerful weapon in the treatment of many forms of inoperable cancer. This is because cancer cells grow much more rapidly than normal tissue.

ADDITIONAL SOLVED PROBLEMS

RADIOACTIVITY

Problem 10.34. What is radioactivity?

Ans. Certain elements have nuclei which are unstable. In order to reach a stable state, they spontaneously emit radiation of high energy. Such elements are called *radioactive*.

Problem 10.35. What are some of the characteristics of radioactive emissions?

Ans. There are four principal types of identifiable radioactive emissions. These are the α-particle, which is a helium nucleus; the β-particle, an electron; the positron with the mass of an electron but with a unit positive charge; and the γ-ray, exceedingly energetic but with neither mass nor charge.

Problem 10.36. What is a daughter nucleus?

Ans. Some radioactive substances emit α- or β-particles, and thereby undergo a change in atomic number or mass or both. The resulting new atomic species is called a *daughter nucleus.*

Problem 10.37. Specify the missing component of the following nuclear equation:

$$^{240}_{95}\text{Am} + \text{X} \longrightarrow {}^{243}_{97}\text{Bk} + {}^{1}_{0}n$$

Ans.

$$X = {}^{4}_{2}\text{He}$$

Problem 10.38. Specify the missing component in the following nuclear equation:

$$^{238}_{92}\text{U} + {}^{12}_{6}\text{C} \longrightarrow {}^{244}_{98}\text{Cf} + \text{X}$$

Ans. $\text{X} = 6\,{}^{1}_{0}n$ (six neutrons)

Problem 10.39. Complete the following nuclear equation:

$$^{40}_{19}\text{K} \longrightarrow {}^{40}_{20}\text{Ca} + \text{X}$$

Ans.

$$\text{X} = {}^{0}_{-1}e \quad \text{(an electron)}$$

Problem 10.40. Complete the following nuclear reaction:

$$_{-1}^{0}e + \text{X} \longrightarrow 2\,\gamma$$

Ans.

$$\text{X} = {}^{0}_{+1}e \quad \text{(a positron)}$$

Problem 10.41. Does radioactive decay always result in transmutation?

Ans. No. Since γ-rays have no mass or charge, if a radionuclide is a γ-emitter, there will be no change in either atomic number or mass.

EFFECTS OF RADIATION

Problem 10.42. What are the primary effects of high-energy radiation?

Ans. The passage of high-energy radiation through matter produces ionization and fragmentation of molecules into free radicals.

Problem 10.43. Are there any other effects of high-energy radiation?

Ans. Yes. After the initial production of ions and/or free radicals, secondary chemical processes occur in which ions, free electrons, or free radicals react with nearby molecules to yield a variety of products.

DETECTION

Problem 10.44. How does a Geiger counter work?

Ans. The counter consists of a tube fitted with a thin window, and a positively charged electrode running down the center of the tube which is also filled with argon gas. When a photon or particle passes through the window it creates ions in the gas, which are collected by the positive electrode. Each burst of current signals the presence of an ionization event. This current can be measured on a meter, or heard audibly as a click.

Problem 10.45. How does a scintillation counter work?

Ans. A scintillation counter contains a phosphor, a substance that emits a flash of light when struck by a photon or particle. These phosphors are similar to the ones that cover the inner surfaces of television tubes. Typical phosphors are sodium and thallium iodides, and zinc sulfide. The scintillation counter not only counts particles, but since the intensity of the flash is proportional to the energy of radiation, it can be used to identify the source of radiation. The flash of light is converted to an electrical signal by a phototube.

UNITS

Problem 10.46. What is a rad?

Ans. A rad is a measure of the amount of high energy radiation absorbed by matter. 1 rad equals 0.01 J of energy absorbed by 1.0 kg of matter.

Problem 10.47. Does 1 rad of α-rays have the same effect on tissue as 1 rad of γ-rays?

Ans. No. α-rays are much more destructive than γ-rays. The results of different radiation are clarified by specifying the RBE of the radiation absorbed, and using that value, quantitating the effects in rems. The rem = rads $\times$ RBE. The RBE for α-rays is 10, and for γ-rays is 1.0.

APPLICATIONS

Problem 10.48. Would iodine-131, a β-emitter, be used for scanning the thyroid for disorders?

Ans. No. Iodine-131 is a β-emitter, and because β-rays are absorbed by tissue, their presence would be difficult to observe externally. In this application, iodine-126 is preferred since it is a γ-emitter, and such radiation easily penetrates tissue and can be easily detected externally.

SUPPLEMENTARY PROBLEMS

Problem 10.49. Technetium-99 is a radioisotope used to assess heart damage. Its half-life is 6.0 h. How much of a 1.0-g sample of technetium-99 will be left after 30 h?

Ans. Thirty hours represents 5 half-lives, and therefore we can use the relationship

$$\frac{N}{N_0} = \left(\frac{1}{2}\right)^n$$

$$= \left(\frac{1}{2}\right)^5 = \frac{1}{32} = 0.03125$$

$$N = N_0 \times 0.03125 = 1\text{ g} \times 0.03125 = 0.03\text{ g}$$

Problem 10.50. Iodine-126 has a half-life of 13.3 h. If a diagnostic dose is 10 nanograms (ng), how much is left after 2.5 days?

Ans. We can calculate the number of half-lives in 2.5 days and then use our half-life equation to calculate the remaining dose:

$$2.5\text{ days}\left(\frac{24\text{ h}}{\text{day}}\right)\left(\frac{1\text{ half-life}}{13.3\text{ h}}\right)=4.5\text{ half-lives}$$

$$\text{Log}\frac{N}{N_0}=4.5\times\log 0.5=-1.35$$

$$\frac{N}{N_0}=0.045$$

$$N=10\text{ ng}\times 0.045=0.45\text{ ng}$$

Problem 10.51. A sample of $Na_3{}^{32}PO_4$ had an activity of 6.7 millicuries (mCi). What does that mean?

Ans. 1 Ci is equal to 3.7×10^{10} dps (disintegrations per second). Therefore, since 6.7 mCi = 6.7×10^{-3} Ci, and 1 Ci = 3.7×10^{10} dps, the activity of the sample is

$$6.7\times10^{-3}\text{ Ci}\left(\frac{3.7\times10^{10}\text{ dps}}{1\text{ Ci}}\right)=2.5\times10^{8}\text{ dps}$$

Problem 10.52. The radiation in rads absorbed by tissue from an α-emitter was the same as that absorbed from a β-emitter. The radiation absorbed from the β-emitter was determined to be 15.2 mrem. What was the absorbed dose in rem from the α-emitter?

Ans. We must use the relationship: Rems = Rads × RBE, and solve for Rads. The RBE of α-absorption is 10, that of β-absorption is 1, and the mrads of radiation absorbed is the same for both radiations.

$$15.2\text{ mrem }(\beta\text{-radiation})=\text{mrad}\times 1$$
$$\text{mrad }(\beta\text{-radiation})=15.2$$
$$\text{mrad }(\alpha\text{-radiation})=\text{mrad }(\beta\text{-radiation})=15.2$$
$$\text{mrem }(\alpha\text{-radiation})=15.2\times 10=152$$

A dose of α-radiation equal to a dose of β-radiation causes 10 times more radiation damage to tissue.

Problem 10.53. The radioisotope fluorine-18 is used for medical imaging. It has a half-life of 110 min. What percentage of the original sample activity is left after 4 h?

Ans. 22 percent.

Problem 10.54. The intensity of x-rays at 2 m from a source was 0.5 R. How far from the source should the target be to reduce the intensity to 0.2 R?

Ans. Using the inverse square law for intensities:

$$\frac{I_{new}}{I_{old}}=\frac{d^2_{old}}{d^2_{new}}$$

$$\frac{0.2}{0.5}=\frac{4\text{ m}^2}{d^2_{new}}$$

$$d^2_{new}=10\text{ m}^2$$

$$d_{new}=3.2\text{ m}$$

Problem 10.55. What is the fundamental reason that nuclear fission reactions are characterized as chain reactions?

Ans. Neutron-induced fission reactions of the heavy elements have as their products two or more neutrons of high energy. This multiplies the effect of the initiating neutron, and can lead to an explosive chain reaction.

Problem 10.56. One of the products of radiation fallout from atmospheric bomb testing or nuclear accidents as at Chernobyl is iodine-131. People living in a region where iodine-131 was known to have been deposited were encouraged to use salt enriched with nonradioactive iodine-127. What was the basis of this treatment?

Ans. It was expected that the thyroid gland could not discriminate between iodine-127 and iodine-131. The rate of iodine uptake by the thyroid is dependent upon the iodine concentration. By flooding the body with the nonradioactive iodine, the body's concentration of iodine-131 would be significantly diluted, and its rate of uptake sharply reduced compared to the uptake of iodine-127.

Problem 10.57. Other byproducts of nuclear fission released through bomb testing are strontium-90 and cesium-137. Why are these considered particularly dangerous?

Ans. The properties of strontium are very similar to those of calcium, and therefore if radioactive strontium enters the body it will be incorporated into bone, particularly growing bone, as in children. Because of its insertion into bone, this radioisotope becomes a lifetime burden, and can produce cancer of the bone and/or leukemia. Cesium is an alkali metal so its properties are similar to those of potassium, whose cation concentrates inside cells. The presence of radioactive cesium in cells can inflict damage in a wide variety of tissues. However, since it is quite soluble, it is rapidly eliminated from the body.

Problem 10.58. What is the meaning of the symbol CT?

Ans. CT is an abbreviation for *computerized tomography*. This is a diagnostic technique in which radiation from inside the body is detected and the information treated by computer to generate an image of the body's inner structures.

Problem 10.59. What are some examples of the CT technique?

Ans. Short bursts of x-rays are directed toward some portion of the body, and a circular array of detectors is rotated around the body. The x-rays are absorbed differentially, and the emerging detected radiation is transformed by a computer program into an internal image. In another technique, glucose is labeled with carbon-11, a positron emitter. Glucose is one of the few compounds which can enter the brain. There the emitted positrons quickly combine with electrons to create γ-rays. These internally generated γ-rays are detected by arrays of detectors similar to those used in x-ray tomography. This technique is known as a PET scan, for *positron emission tomography*, and is used primarily to produce brain scans to detect disorders in metabolism indicative of a variety of pathologies. MRI scans are made with the same physical array of detectors, but use radiation of very low energy, radio waves, in the presence of a strong magnetic field. This radiation arrangement excites protons. All protons absorb the same amount of radio wave energy, but the total amount of energy absorbed depends upon the concentration of protons. There are differing concentrations of protons in different tissue and within regions of the same tissue, therefore the absorption of energy is not uniform. These non-uniformities in absorption can be detected externally and transformed by computer programs into an image of soft tissue. MRI stands for *Magnetic Resonance Imaging*.

Organic Compounds; Saturated Hydrocarbons

11.1 ORGANIC CHEMISTRY

Organic chemistry is the chemistry of compounds of carbon. There are several million organic compounds, 10-fold more than inorganic compounds. This occurs because carbon atoms easily bond to each other as well as to other elements, principally hydrogen, oxygen, halogen (fluorine, chlorine, bromine, iodine), nitrogen, sulfur, and phosphorus, in a variety of structural patterns.

Organic compounds typically contain covalent bonds while inorganics contain ionic or highly polar bonds. The result is that organic compounds possess lower melting and boiling points and lower solubility in water. Reactions are generally slower in organic compounds than in inorganic compounds since reactions between ions are fast while reactions of organic compounds require the slower breakage of covalent bonds.

11.2 MOLECULAR AND STRUCTURAL FORMULAS

The *molecular formula* of a compound shows the number of each type of atom that is present in a molecule of the compound. The *structural formula* shows the details of the bonding present in the molecule, i.e., which atoms are connected to which other atoms. The molecular formula for the compound methane is CH_4, while the structural formula is

```
    H
    |
H—C—H
    |
    H
```

The student who has difficulty with organic chemistry usually has difficulty in correctly reading and writing molecular and structural formulas. Much of the difficulty come from not understanding the combining power of different atoms (Problem 3.8). The combining power of an atom in a covalent compound is the number of bonds that it forms to complete its octet of valence electrons. Carbon has a combining power of 4 (is *tetravalent*), which means that all carbon atoms in all compounds have four bonds—not less than four and not more than four. Hydrogen and halogen are *monovalent* (one bond), oxygen and sulfur are *divalent* (two bonds), and nitrogen and phosphorus are *trivalent* (three bonds). No molecular or structural formula can represent a real compound unless the combining power of each atom in the structure is correct. Any structure which has an incorrect combining power for any atom is incorrect.

Problem 11.1. Which of the following are correct molecular formulas? (*a*) CH_4Cl, (*b*) CH_4O, (*c*) CH_4N.

Ans. CH_4O is correct but CH_4Cl and CH_4N are not correct molecular formulas. A molecular formula is correct only if we can draw a structural formula which contains all atoms of the molecular formula and the combining powers of all atoms are correct.

We can draw structural formulas such as

```
 H   H               H
  \ /                |
   C—Cl          H—C—Cl—H
  / \                |
 H   H               H

     I                II
```

for CH_4Cl but they violate the combining power requirement of one or another atom. Formula I shows carbon with five bonds but carbon is tetravalent. Formula II shows Cl with two bonds but Cl is monovalent. Formulas I and II and other variations of CH_4Cl do not represent real compounds.

We can draw structural formulas such as

```
H  H                 H
   |                 |
H—C—N—H          H—C—N—H
                     |
                     H

   III               IV
```

for CH_4N but they violate the combining power requirement of one or another atom. Formula III shows three, instead of four, bonds to C. Formula IV shows two, instead of three, bonds to N. CH_4N does not represent a compound.

The structural formula for CH_4O is

```
   H
   |
H—C—O—H
   |
   H
```

Each of the hydrogens has one bond, the oxygen has two bonds, and the carbon has four bonds. All combining powers are satisfied and we have accounted for all atoms (one carbon, one oxygen, and four hydrogens) in the molecular formula.

11.3 FAMILIES OF ORGANIC COMPOUNDS; FUNCTIONAL GROUPS

The study of organic compounds is facilitated by their organization into *families* (also referred to as *classes* or *homologous series*) of compounds. There are 20 or so different families that are of importance. Each family contains a very large number of different compounds which have a common chemical behavior. The common behavior within one family differs from that within another family. It is the behavior of each of the families that we study. The common behavior of all compounds within the same family is due to the presence of a common *functional group*. A functional group is a particular atom or type of bond or group of atoms in a particular bonding arrangement.

Table 11-1 shows many of the important classes of organic compounds. *Alkanes* contain only carbon-carbon and carbon-hydrogen single bonds. *Alkenes* contain a carbon-carbon double bond, i.e., there are two bonds between a pair of adjacent carbon atoms. *Alkynes* contain a carbon-carbon triple bond, i.e., there are three bonds between the same pair of carbons. *Aromatics* contain six carbons in a cyclic arrangement with alternating single and double bonds. *Alcohols* contain a *hydroxyl group*, an OH group attached to a carbon atom. *Ethers* contain an oxygen attached directly to two different carbons. *Amines* contain an *amine group*, an NH_2 group attached to a carbon atom. *Aldehydes*, *ketones*, *carboxylic acids*, *esters*, and *amides* all possess the *carbonyl group*, a carbon-oxygen double bond, but differ from each other in what other atom or group of atoms is connected to the carbon of the carbonyl group.

11.4 ALKANES

The alkane, alkene, alkyne, and aromatic families are members of a larger grouping referred to as *hydrocarbons*. Hydrocarbons are organic compounds that contain only carbon and hydrogen. The alkenes, alkynes, and aromatics are *unsaturated hydrocarbons* since they contain carbon-carbon multiple bonds. The alkanes are referred to as *saturated hydrocarbons* since they do not contain carbon-carbon multiple bonds, only carbon-carbon single bonds.

Alkanes contain only carbon-carbon and carbon-hydrogen single bonds and have the general formula C_nH_{2n+2} where n is an integer other than zero. The smallest member of the alkane family is

Table 11-1. Families of Organic Compounds

Family	Functional Group	Example
Alkane	C—C and C—H single bonds	$CH_3—CH_3$
Alkene	$>C=C<$	$CH_2=CH_2$
Alkyne	$—C\equiv C—$	$CH\equiv CH$
Aromatic	six-membered ring of C atoms with alternating double bonds (C=C), each C with one bond outward	six-membered ring of C atoms with alternating double bonds (C=C), each C bonded to H
Alcohol	$—\overset{\vert}{\underset{\vert}{C}}—O—H$	$CH_3CH_2—O—H$
Ether	$—\overset{\vert}{\underset{\vert}{C}}—O—\overset{\vert}{\underset{\vert}{C}}—$	$CH_3—O—CH_3$
Aldehyde	$—\overset{O}{\overset{\Vert}{C}}—H$	$CH_3—\overset{O}{\overset{\Vert}{C}}—H$
Ketone	$—\overset{\vert}{\underset{\vert}{C}}—\overset{O}{\overset{\Vert}{C}}—\overset{\vert}{\underset{\vert}{C}}—$	$CH_3—\overset{O}{\overset{\Vert}{C}}—CH_3$
Carboxylic acid	$—\overset{O}{\overset{\Vert}{C}}—OH$	$CH_3—\overset{O}{\overset{\Vert}{C}}—OH$
Ester	$—\overset{O}{\overset{\Vert}{C}}—O—\overset{\vert}{\underset{\vert}{C}}—$	$CH_3—\overset{O}{\overset{\Vert}{C}}—O—CH_3$
Amine	$—\overset{H}{\overset{\vert}{N}}—H$	$CH_3—\overset{H}{\overset{\vert}{N}}—H$
Amide	$—\overset{O}{\overset{\Vert}{C}}—\overset{H}{\overset{\vert}{N}}—H$	$CH_3—\overset{O}{\overset{\Vert}{C}}—\overset{H}{\overset{\vert}{N}}—H$

CH_4 ($n = 1$), followed by

$n = 2$	C_2H_6	$CH_3—CH_3$	Ethane
$n = 3$	C_3H_8	$CH_3—CH_2—CH_3$	Propane
$n = 4$	C_4H_{10}	$CH_3—CH_2—CH_2—CH_3$	Butane
$n = 5$	C_5H_{12}	$CH_3—CH_2—CH_2—CH_2—CH_3$	Pentane
$n = 6$	C_6H_{14}	$CH_3—CH_2—CH_2—CH_2—CH_2—CH_3$	Hexane

This continues all the way up to the plastic known as polyethylene, where n is a large number such as 1000 to 5000 or more, the exact number depending on the conditions of manufacture.

The main sources of hydrocarbons are *natural gas* and *petroleum*. Natural gas consists of alkanes with less than five carbons. Petroleum contains hundreds of different hydrocarbons with five carbons and more, mostly alkanes but also sizable amounts of aromatics. Natural gas and petroleum are critically important to our civilization. Burning of gas and petroleum heats buildings, fuels automobiles, trains, and aircraft, and generates electricity. Various components of petroleum are used as lubricating oils and greases, and as asphalt. Natural gas and petroleum are the source for producing more than 90 percent of the chemicals used to synthesize plastics, textiles, rubbers, drugs, detergents, dyes, and a host of other industrial products.

Carbon atoms in alkanes form bonds to each other and to hydrogen atoms through orbitals referred to as *sp^3 orbitals*. This results in a tetrahedral geometry for the four bonds of each carbon atom, i.e., the carbon atom is pictured in the center of a tetrahedron with its four bonds directed to the four corners of the tetrahedron (Fig. 11-1).

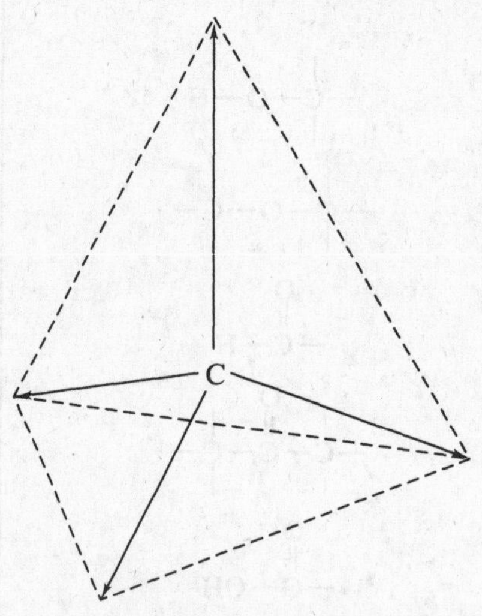

Fig. 11-1 Tetrahedral bond angles of carbon with sp^3 orbitals.

Problem 11.2. Why is it necessary to postulate that carbon forms bonds by using sp^3 orbitals?

Ans. The atomic orbitals in the valence shell of an isolated (*ground-state*) carbon atom contains two paired electrons in the $2s$ orbital and two unpaired electrons in each of two $2p$ orbitals. The $2s$ orbital is filled and there are two half-filled $2p$ orbitals together with an empty $2p$ orbital. This predicts incorrectly that the valence of carbon is 2. Carbon is tetravalent and, furthermore, the four bonds of carbon are equivalent. This is the situation for carbon atoms in alkanes (and also in diamond). The reality of carbon's four bonds requires that we postulate excitation of carbon to a different electronic structure prior to bond formation. The excitation involves two processes—*promotion* and *hybridization* (Fig. 11-2). *Promotion* is the unpairing of the two electrons in the $2s$ orbital by promotion of one of the electrons to a higher energy $2p$ orbital. The resulting electronic

$2p$ $\uparrow$ $\uparrow$ __ $\xrightarrow{\text{Promotion}}$ $2p$ $\uparrow$ $\uparrow$ $\uparrow$ $\xrightarrow{\text{Hybridization}}$ sp^3 $\uparrow$ $\uparrow$ $\uparrow$ $\uparrow$
$2s$ $\uparrow\downarrow$ $2s$ $\uparrow$

Ground state Excited state

Fig. 11-2 Formation of sp^3 orbitals for carbon.

structure of carbon correctly describes it as tetravalent, but the four bonds would not all be equivalent. Mixing (*hybridization*) of the $2s$ and three $2p$ orbitals with the total of four electrons produces four equivalent orbitals, referred to as sp^3 *hybrid orbitals*, each of which contains an electron.

Figure 11-3 shows an alternate representation of the hybridization process by which the sp^3 orbitals of carbon are formed. It emphasizes the shapes of the orbitals. The $2s$ orbital is spherical while each $2p$ orbitals consists of a pair of teardrop-shaped lobes. Mixing of the $2s$ and three $2p$ orbitals produces four sp^3 orbitals, each of which is teardrop-shaped. Carbon uses its four sp^3 orbitals when forming bonds to other atoms. Each sp^3 orbital forms a covalent bond with another atom by overlapping with an orbital of that atom. Bonding of carbon to another carbon involves overlapping of sp^3 orbitals from each carbon atom. Bonding of carbon to nitrogen and oxygen also involves overlapping of sp^3 orbitals since N and O, like C, bond through sp^3 orbitals. Bonding to hydrogen and halogen involves overlapping with s and p orbitals, respectively. The bond angle involving any tetrahedral (sp^3) carbon in any compound is 109.5°, referred to as the *tetrahedral bond angle*.

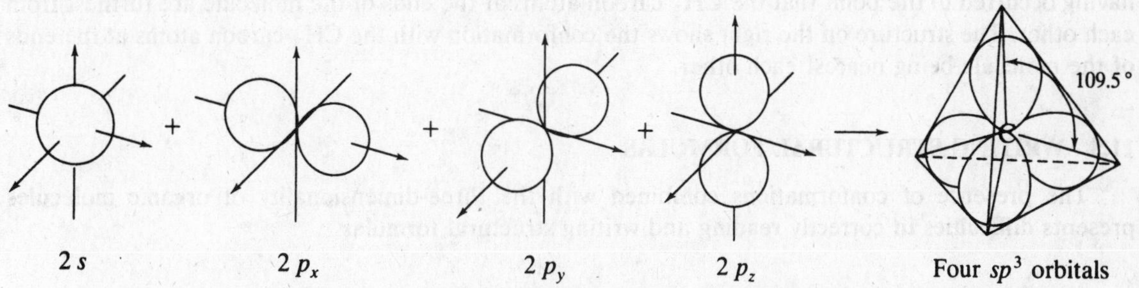

$2s$ + $2p_x$ + $2p_y$ + $2p_z$ → Four sp^3 orbitals (109.5°)

Fig. 11-3 Hybridization of one $2s$ and three $2p$ orbitals of carbon to form four sp^3 orbitals.

Problem 11.3. Give a pictorial representation of the orbitals involved when one carbon and four hydrogens form CH_4.

Ans. This is shown in Fig. 11-4. Each sp^3 orbital of carbon (containing one electron) overlaps with the $1s$ orbital of a hydrogen (also containing one electron).

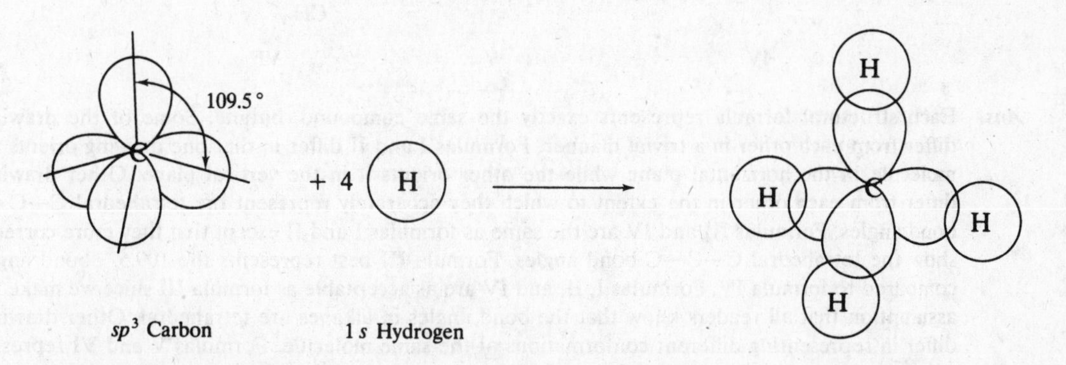

Fig. 11-4 Formation of CH_4 by overlap between $1s$ orbitals of hydrogen and sp^3 orbitals of carbon.

Single bonds between C, H, O, N, S, P, and halogen are free to rotate. (Exceptions to this generalization will be discussed as the need arises.) This results in different *conformations* or *conformational isomers* for the same molecule. The following are two of the different conformations of the compound butane (C_4H_{10}):

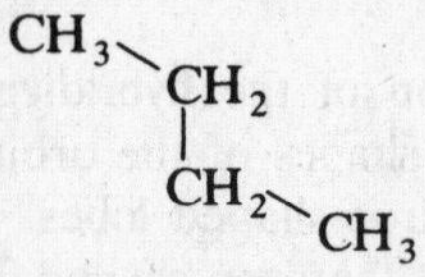

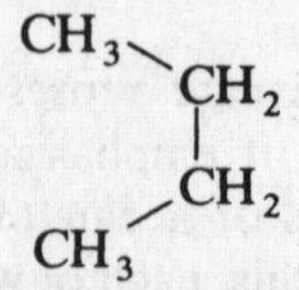

CH_3—CH_2—CH_2—CH_3 (left: end CH_3 groups on opposite sides) $\qquad$ CH_3—CH_2—CH_2—CH_3 (right: end CH_3 groups on the same side)

Conformations differ from each other in the position to which some single bond or set of bonds has rotated. For molecules such as butane, the various C—C and C—H single bonds are continuously undergoing rotation since there is no (or very little) energy barrier to that rotation. One conformation cannot be isolated from another, and at any one moment some molecules of butane are in one conformation and other molecules are in another conformation. The greater the total number of atoms in a molecule, the greater the number of different conformations for the molecule.

The two structures shown above represent conformations differing in the position of rotation about the middle C—C bond. The left-hand structure shows the conformation with the bond rotation having occurred to the point that the CH_3 carbon atoms at the ends of the molecule are furthest from each other. The structure on the right shows the conformation with the CH_3 carbon atoms at the ends of the molecule being nearest each other.

11.5 WRITING STRUCTURAL FORMULAS

The presence of conformations combined with the three-dimensionality of organic molecules presents difficulties in correctly reading and writing structural formulas.

Problem 11.4. What do each of the following structural formulas represent?

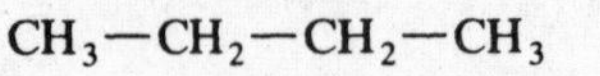

I: $CH_3—CH_2—CH_2—CH_3$

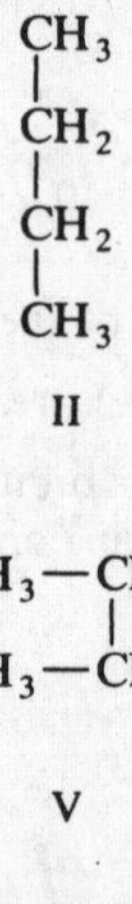

II: CH_3 | CH_2 | CH_2 | CH_3 (vertical)

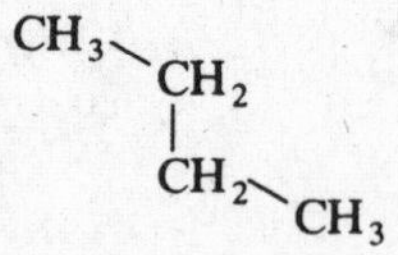

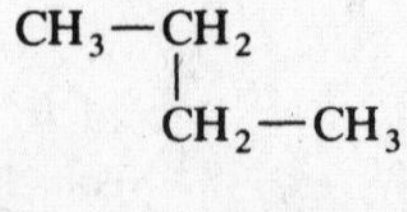

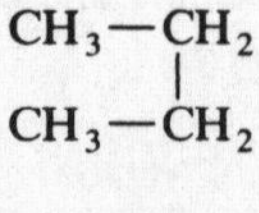

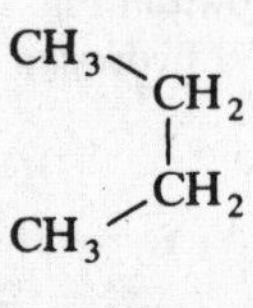

III: $CH_3\diagdown CH_2$ | $CH_2\diagdown CH_3$

IV: $CH_3—CH_2$ | $CH_2—CH_3$

V: $CH_3—CH_2$ | $CH_3—CH_2$

VI: $CH_3\diagdown CH_2$ | $CH_2\diagup CH_3$

Ans. Each structural formula represents exactly the same compound, butane. Some of the drawings differ from each other in a trivial manner. Formulas I and II differ in that one drawing orients the molecule in the horizontal plane while the other orients it in the vertical plane. Other drawings differ from each other in the extent to which they accurately represent the tetrahedral C—C—C bond angles. Formulas III and IV are the same as formulas I and II except that they more correctly show the tetrahedral C—C—C bond angles. Formula III best represents the 109.5° bond angles compared to formula IV. Formulas I, II, and IV are as acceptable as formula III since we make the assumption that all readers know that the bond angles in alkanes are tetrahedral. Other drawings differ in representing different conformations of the same molecule. Formulas V and VI represent the same conformation but a different conformation from that represented by formulas I, II, III, and IV.

Problem 11.5. What do each of the following structural formulas represent?

$$CH_3CH_2CH_2CH_3 \qquad CH_3-CH_2-CH_2-CH_3 \qquad \begin{array}{c} \text{H} \; \text{H} \; \text{H} \; \text{H} \\ | \;\;\; | \;\;\; | \;\;\; | \\ \text{H}-\text{C}-\text{C}-\text{C}-\text{C}-\text{H} \\ | \;\;\; | \;\;\; | \;\;\; | \\ \text{H} \; \text{H} \; \text{H} \; \text{H} \end{array}$$

I II III

Ans. Each of these structural formulas represents the same compound—the same compound as in the Problem 11.4. The three formulas differ in the extent to which various bonds in the molecule are drawn. Formula III is referred to as a *full* or *expanded structural formula* in contrast to more *condensed structural formulas* such as formulas I and II. Formulas I and II are condensed to different degrees. Formula I shows no bonds while formula II shows all carbon-carbon bonds but no carbon-hydrogen bonds.

All of this is confusing to the beginning student who often asks, "Why don't chemists always use expanded structural formulas?" The answer is that your textbooks would contain many more pages and be much more expensive. Formulas I and II in Problem 11.5 are shorthand versions of formula III. Once you understand the valence requirements of various atoms, you will read all three formulas as representing the same molecule. Depending on the text you use, you may find other shorthand versions of the same molecule:

$$C-C-C-C$$

Skeleton Line

The *skeleton structural formula* shows only the carbon atoms in the molecule. In the *line structural formula*, carbon atoms are present at the intersection of two or more lines and wherever a line begins or ends.

11.6 CONSTITUTIONAL ISOMERS

Up to now we have considered only *unbranched alkanes*, but there are also *branched alkanes* once we have four or more carbons in a molecule. There are two different C_4H_{10} compounds:

$$CH_3-CH_2-CH_2-CH_3 \qquad \begin{array}{c} CH_3-CH-CH_3 \\ | \\ CH_3 \end{array}$$

Butane Isobutane

Different compounds that have the same molecular formula are referred to as *isomers*. There are several different types of isomers. Here we have what is referred to as *constitutional* (or *structural*) *isomers*—compounds that differ from each other in *connectivity*. Connectivity is the order of attachment of atoms to each other.

The two constitutional isomers of C_4H_{10} differ in the number of carbon atoms successively connected to each other in their longest *continuous chain of carbon atoms*. To determine the length of the longest continuous chain, visualize the exercise of *walking the carbon atoms*. If one steps onto some carbon atom in the molecule and then proceeds to walk onto successive carbon atoms, what is the maximum number of carbon atoms that can be walked without retracing your steps? Note that you may walk from one carbon atom to another only if there is a bond between the two atoms. This is a trial and error exercise. You can walk four carbon atoms in butane if you start walking from an end carbon instead of an inner carbon. The maximum number of carbons that can be walked in isobutane is three, not four. You cannot walk onto the fourth carbon atom in isobutane without retracing a part of your path. Butane has a longest continuous chain of four carbons while isobutane has a longest continuous chain of three carbons. The fourth carbon of isobutane is a branch attached to the middle carbon of the longest continuous chain. The terms *branched* and *unbranched alkanes* distinguish these two situations. Unbranched alkanes are also referred to as *linear* or *straight-chain alkanes.*

As the total number of carbons increases, there is a corresponding increase in the number of constitutional isomers that exist.

Problem 11.6. How many isomeric hexanes exist, i.e., how many constitutional isomers exist for the molecular formula C_6H_{14}? Show one structural formula for each isomer. Do not show more than one structural formula for each isomer.

Ans. There are five different compounds (constitutional isomers) of C_6H_{14}:

$$\underset{\text{I}}{CH_3{-}CH_2{-}CH_2{-}CH_2{-}CH_2{-}CH_3} \qquad \underset{\text{II}}{CH_3{-}\overset{\overset{\displaystyle CH_3}{|}}{C}H{-}CH_2{-}CH_2{-}CH_3}$$

$$\underset{\text{III}}{CH_3{-}CH_2{-}\overset{\overset{\displaystyle CH_3}{|}}{C}H{-}CH_2{-}CH_3} \qquad \underset{\text{IV}}{CH_3{-}\overset{\overset{\displaystyle CH_3}{|}}{C}H{-}\overset{\overset{\displaystyle CH_3}{|}}{C}H{-}CH_3} \qquad \underset{\text{V}}{CH_3{-}\underset{\underset{\displaystyle CH_3}{|}}{\overset{\overset{\displaystyle CH_3}{|}}{C}}{-}CH_2{-}CH_3}$$

We start out with all six carbons connected in a continuous chain to give compound I. We then go to a continuous chain of five carbons and place a single carbon, referred to as a *methyl group*, as a branch on the five-carbon chain. There are two possibilities. Placing the sixth carbon as a branch on the second carbon from *either* end of the five carbon chain yields compound II. Placing the sixth carbon as a branch on the middle carbon of the chain yields compound III.

Next we go to a longest continuous chain of four carbons with two carbons to be branches on that chain. There are two possibilities—compounds IV and V.

What about including structure VI in the answer to this question? Structure VI is obtained by placing the two carbons together, to form what is referred to as an *ethyl group*, and then placing

$$\underset{\text{VI}}{CH_3{-}\underset{\underset{\underset{\underset{\displaystyle CH_3}{|}}{\displaystyle CH_2}}{|}}{C}H{-}CH_2{-}CH_3} \qquad \underset{\text{VII}}{CH_3{-}\underset{\underset{\underset{\underset{\underset{\underset{\displaystyle CH_3}{|}}{\displaystyle CH_2}}{|}}{\displaystyle CH_2}}{|}}{C}H{-}CH_3}$$

the ethyl group as a branch on an inner carbon of the four carbon chain. This is a mistake as it incorrectly states that there are six different compounds when there are only five in reality. Structure VI is a duplicate drawing of structure III. Structure VI, like structure III, has a longest continuous chain of five carbons with the sixth carbon as a branch at the middle carbon of the chain. One must see the longest continuous chain whether or not it is written horizontally. This is evident if one uses the *walking the carbon atoms* exercise. Either structure III or structure VI is a legitimate drawing of the real compound, although most chemists and publishers would tend to write structure III instead of structure VI in order to conserve space. To include both structures III and VI is a mistake which must be avoided. There are many other duplicate drawings that one will generate in the process of answering this question; for example, structure VII is a duplicate of structure II. One must be able to recognize a duplicate structure no matter the way it is drawn.

11.7 NOMENCLATURE

The *common nomenclature* system names unbranched alkanes by using a prefix denoting the total number of carbons followed by the ending *-ane*. Methane, ethane, propane, butane, pentane, hexane, heptane, octane, nonane, and decane are the names of the alkanes from C_1 to C_{10}. Note that the ending of each of the names is the ending of the family name. Starting with C_4, there is more than one compound due to constitutional isomerism and we use an additional prefix to distinguish the constitutional isomers—butane and isobutane, for example. But this approach becomes unmanageable very quickly as the number of constitutional isomers increases rapidly with increasing numbers of carbons. The number of different prefixes needed and the amount of memorizing of individual structures becomes impossible. An aternate nomenclature system, the *IUPAC system*, minimizes the problem. IUPAC stands for *International Union of Pure and Applied Chemistry*.

11.7.1 Alkyl Groups

The use of the IUPAC system requires that we first name certain commonly encountered *groups* (also referred to as *substituents*). Two different organic groups, the one carbon and two carbon groups referred to as *methyl* and *ethyl*, respectively, were briefly mentioned in Problem 11.6. The structures are

$$CH_3- \qquad CH_3-CH_2-$$

Methyl Ethyl

We derive (on paper, not in the laboratory) these groups from methane and ethane, respectively, by loss of a hydrogen atom. Organic groups such as methyl and ethyl and those described below in Problem 11.7 are referred to as *alkyl groups*.

Problem 11.7. What groups are derived from propane, butane, and isobutane?

Ans. Two groups are derived from propane. Removal of a hydrogen from one of the end carbons yields the *propyl group* while removal of a hydrogen from the central carbon yields the *isopropyl group*.

$$CH_3-CH_2-CH_3 \rightarrow CH_3-CH_2-CH_2- \text{ (Propyl)}$$

$$CH_3-CH_2-CH_3 \rightarrow CH_3-\overset{|}{C}H-CH_3 \text{ (Isopropyl)}$$

Butane yields the *butyl* and *s-butyl groups*, the former by removal of a hydrogen from one of the end carbons and the latter by removal of a hydrogen from one of the inner carbons.

$$CH_3-CH_2-CH_2-CH_3 \rightarrow CH_3-CH_2-CH_2-CH_2- \text{ (Butyl)}$$

$$CH_3-CH_2-CH_2-CH_3 \rightarrow CH_3-\overset{|}{C}H-CH_2-CH_3 \text{ (s-Butyl)}$$

Isobutane yields the *isobutyl group* by removal of a hydrogen from one of the methyl carbons. The *t-butyl group* is derived by removal of a hydrogen from the CH carbon.

$$CH_3-\underset{\displaystyle CH_3}{\underset{|}{CH}}-CH_3 \rightarrow CH_3-\underset{\displaystyle CH_3}{\underset{|}{CH}}-CH_2- \quad \text{Isobutyl}$$

$$CH_3-\underset{\displaystyle CH_3}{\underset{|}{CH}}-CH_3 \rightarrow CH_3-\underset{\displaystyle CH_3}{\underset{|}{\overset{|}{C}}}-CH_3 \quad t\text{-Butyl}$$

The prefixes *s*- and *t*- are abbreviations for *sec*- (or *secondary*-) and *tert*- (or *tertiary*-), respectively. These prefixes are italicized when used. The prefix iso- is neither italicized nor abbreviated by *i* (which appears to be an inconsistency but one that is universally adopted.) No prefix is used for the unbranched groups.

11.7.2 IUPAC Nomenclature

The rules for naming alkanes by the IUPAC system are:

1. Locate the longest continuous chain of carbon atoms. This chain gives the *base* or *parent name* of the compound.
2. Note the groups attached to the longest chain.
   (a) Place the names of the groups in front of the base name.
   (b) When there are two or more different types of groups, place them in alphabetical order (ignoring *s*- and *t*- prefixes but not ignoring iso- prefix).
   (c) If there are more than two of the same group, use prefixes such as di-, tri-, and tetra- to avoid naming the same group more than once. These prefixes are also ignored in alphabetizing the names of groups.
3. Number the longest continuous chain from the end nearest a group.
4. Place the appropriate number in front of the name of each group to indicate the group's location on the longest continuous chain.
5. A number is separated from a word by a dash while any two numbers are separated from each other by a comma.

The IUPAC nomenclature system is used not only for naming organic compounds but is also useful for recognizing that two structural drawings represent the same compound. Correct application of the IUPAC rules yields the same name for duplicate drawings of the same compound.

Problem 11.8. Name the following compound using the IUPAC system:

$$CH_3-CH_2-CH_2-\overset{\displaystyle CH_3}{\overset{|}{\overset{\displaystyle CH_2}{\overset{|}{CH}}}}-CH_2-\underset{\displaystyle CH_3}{\underset{|}{\underset{\displaystyle CH_2}{\underset{|}{CH}}}}-CH_3$$

Ans. The walking the carbon atoms exercise shows that the longest chain has eight carbons; the base name is octane. The eight carbons of that chain are numbered below:

$$\begin{array}{c} \text{CH}_3 \\ | \\ \text{CH}_2 \\ | \\ \underset{8}{\text{CH}_3}-\underset{7}{\text{CH}_2}-\underset{6}{\text{CH}_2}-\underset{5}{\text{CH}}-\underset{4}{\text{CH}_2}-\overset{3}{\text{CH}}-\text{CH}_3 \\ | \\ 2\,\text{CH}_2 \\ | \\ 1\,\text{CH}_3 \end{array}$$

Keep in mind that the longest chain in this and many other structures is often not drawn in a straight line. The number of carbons in the horizontally written chain contains only six carbons. The word *octane* accounts for the eight carbons of the longest continuous chain, but does not account for the ethyl and methyl groups (which are encircled). The name becomes ethylmethyloctane but we need numbers to show where the ethyl and methyl groups are located on the octane chain. The direction shown for numbering of the octane chain gives carbon-3 for the location of the nearest group (methyl). Numbering in the opposite direction is incorrect since it would give carbon-4 for location of the nearest group (ethyl):

$$\begin{array}{c} \text{CH}_3 \\ | \\ \text{CH}_2 \\ | \\ \underset{1}{\text{CH}_3}-\underset{2}{\text{CH}_2}-\underset{3}{\text{CH}_2}-\underset{4}{\text{CH}}-\underset{5}{\text{CH}_2}-\overset{6}{\text{CH}}-\text{CH}_3 \\ | \\ 7\,\text{CH}_2 \\ | \\ 8\,\text{CH}_3 \end{array}$$

The correct IUPAC name is 5-ethyl-3-methyloctane. Note the alphabetical order of substituents.

11.7.3 Other Names

CH_2 and CH groups are referred to as *methylene* and *methine* to complement methyl for CH_3.

Carbon atoms are referred to as *primary* (1°), *secondary* (2°), *tertiary* (3°), or *quaternary* (4°), depending on whether they are *directly* bonded to a total of one, two, three, or four other carbons. This means that the carbons of methyl, methylene, and methine groups are 1°, 2°, and 3° carbons, respectively. Hydrogens or other groups such as OH or Cl attached to 1°, 2°, and 3° carbons are in turn referred to as 1°, 2°, and 3°. A 4° hydrogen or other group is not possible since C is tetravalent.

Problem 11.9. Indicate the 1°, 2°, 3°, and 4° carbons in (*a*) the compound in Problem 11.8 and (*b*) compound V of Problem 11.6.

Ans.

$$(a)\quad \begin{array}{c} 1^\circ\,\text{CH}_3 \\ | \\ 2^\circ\,\text{CH}_2 \\ | \\ \overset{1^\circ}{\text{CH}_3}-\overset{2^\circ}{\text{CH}_2}-\overset{2^\circ}{\text{CH}_2}-\underset{3^\circ}{\text{CH}}-\overset{2^\circ}{\text{CH}_2}-\overset{3^\circ}{\text{CH}}-\overset{1^\circ}{\text{CH}_3} \\ | \\ 2^\circ\,\text{CH}_2 \\ | \\ 1^\circ\,\text{CH}_3 \end{array} \qquad (b)\quad \begin{array}{c} 1^\circ\,\text{CH}_3 \\ | \\ 1^\circ\,\text{CH}_3-\overset{4^\circ}{\text{C}}-\overset{2^\circ}{\text{CH}_2}-\overset{1^\circ}{\text{CH}_3} \\ | \\ 1^\circ\,\text{CH}_3 \end{array}$$

11.8 CYCLOALKANES

The molecules considered up to this point are *acyclic* or *open-chain*. If you perform the walking the carbon atoms exercise, you cannot come back to the original starting carbon atom because there is no bond connecting the first and last carbons of the longest chain. For any acyclic molecule, there is a corresponding *cyclic* or *ring molecule*—a molecule where you do come back to the original carbon atom because there is a bond between the first and last carbons. An example is cyclopentane:

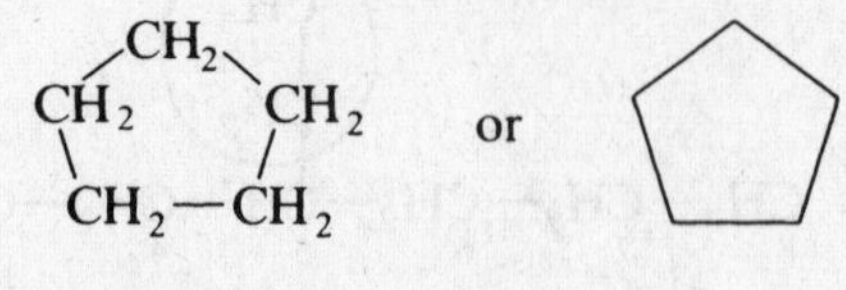

CH_2
CH_2 CH_2 or
CH_2-CH_2

Geometrical shapes such as the pentagon above are often used to show the structure of a cycloalkane. Recall that the intersection of two lines defines a carbon atom.

The general formula for cycloalkanes is C_nH_{2n} compared to C_nH_{2n+2} for the alkanes. A cycloalkane contains two hydrogen atoms less than the corresponding alkane. We can visualize the formation of a cycloalkane from an alkane by loss of a hydrogen from each of two carbons followed by bond formation between the two carbons:

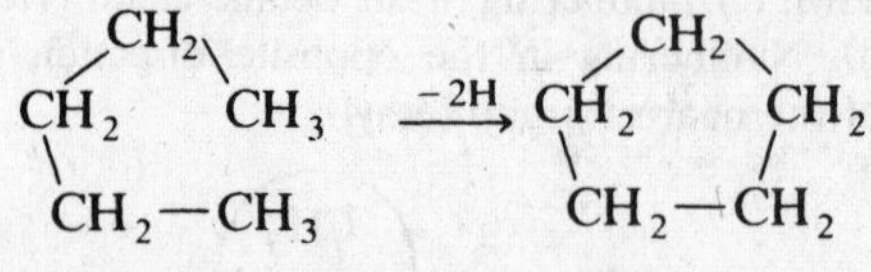

CH_2 CH_2
CH_2 CH_3 $\xrightarrow{-2H}$ CH_2 CH_2
CH_2-CH_3 CH_2-CH_2

This is not necessarily the way cycloalkanes are actually synthesized but it is a useful exercise for understanding the relationship between alkanes and cycloalkanes.

Problem 11.10. Give the IUPAC name for the following compound:

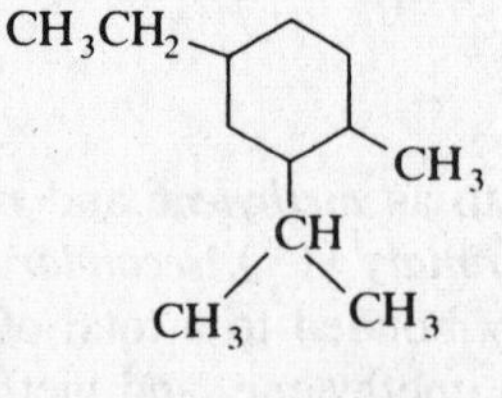

CH_3CH_2
CH_3
CH
CH_3 CH_3

Ans. Naming cycloalkanes follows from the nomenclature for alkanes. The base name is *cyclo-* followed by the name indicating the number of carbons in the ring structure, *-hexane* in this case. The carbons of the ring are numbered by starting at a carbon that contains a substituent and proceeding to the other carbons of the ring by the direction that yields the lowest number(s) for the other substituent(s). The method involves a trial and error approach of starting at any carbon with a substituent and proceeding first clockwise and then counterclockwise. For example, the shortest route starting from the ethyl group is to move counterclockwise to give 1, 3, 4 instead of clockwise to give 1, 4, 5. However, the set of absolute lowest numbers is 1, 2, 4, obtained by starting at the methyl group and moving clockwise to the ethyl group. The IUPAC name is 4-ethyl-2-isopropyl-1-methylcyclohexane.

Problem 11.11. How many constitutional isomers exist for the molecular formula C_5H_{10}? Show one structural formula for each isomer. Do not show more than one structural formula for each isomer.

Ans. C_5H_{10} fits the general formula for a cycloalkane ($n = 5$). There are five constitutional isomers of C_5H_{10}:

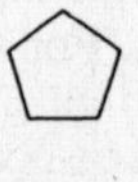

CH_3 CH_3
CH_3

Cyclopentane 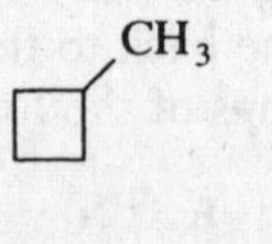 Methylcyclobutane 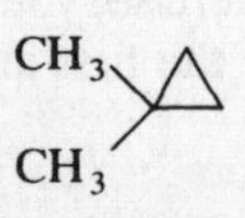 1,1-Dimethylcyclopropane

CH_3 CH_3 CH_2CH_3

1,2-Dimethylcyclopropane Ethylcyclopropane

Constitutional isomerism in cycloalkanes involves different ring sizes and/or different placements of the carbons that are outside the ring. Cyclopentane has all its carbons in the ring. There is one isomer, methylcyclobutane, when one of the carbon atoms is outside the ring. There are three isomers when two of the carbons are outside the ring. When the two carbons outside the ring are two methyl groups, there are two possible placements of the methyls relative to each other—1,1-dimethylcyclopropane and 1,2-dimethylcyclopropane. We have ethylcyclopropane when the two outside carbons are together as an ethyl group.

Note that there is no need to use "1-" in the names ethylcyclopropane and methylcyclobutane. With only one group present on the ring, "1-" is understood.

Problem 11.12. When geometrical shapes are used to show a cyclic structure, we usually do not draw in all of the hydrogens attached to the carbons of the ring. Redraw methylcyclobutane and 1,1-dimethylcyclopropane from Problem 11.11 to show all hydrogens attached to carbon atoms of the ring.

Ans. The number of hydrogens at each carbon of the ring is the number which is required to satisfy the tetravalency of carbon:

H H H H
H CH_3 CH_3 H
H H
H H CH_3 H

Two hydrogens are present at carbons holding no other groups. Carbons holding one group require one hydrogen. Carbons holding two groups require no hydrogens.

We draw ring structures as if the rings are flat, i.e., as if all the carbon atoms of the ring lie in the same plane. This is approximately correct for the 3-, 4-, and 5-membered rings. It is incorrect for the 6-membered and larger-sized rings which exist in a puckered, nonplanar shape. The 6-membered cyclohexane ring exists in the shape referred to as the *chair conformation*:

H H
H H
H H
H H
H H
H H

Problem 11.13. What is the significance of 3-, 4-, and 5-membered rings being approximately flat while 6-membered rings are puckered?

Ans. There is a large difference in stability among some of the ring sizes. Rings of 3 and 4 members have low stability because the bond angles resulting from the flat geometry are greatly distorted relative to the stable tetrahedral bond angles. The 5-membered ring has bond angles close to the tetrahedral bond angle and is much more stable than 3- and 4-membered rings. The 6-membered ring would be relatively unstable if it were flat due to distortions from the tetrahedral bond angle. The chair conformation allows the six atoms of the ring to be arranged so that the bond angles are exactly the tetrahedral bond angle. The overall result is that 6-membered rings are as stable as acyclic compounds, 5-membered rings are slightly less stable, while 3- and 4-membered rings are much less stable. Compounds with 6-membered rings and, to a lesser extent, those with 5-membered rings are often found in nature while 3- and 4-membered rings are almost never found.

We will use a flat hexagon for the 6-membered ring throughout most of this book, as is done in most of your textbooks. Failure to use the chair conformation does not present a problem at the level of our discussions. The important thing to remember is that 6-membered rings are highly stable.

11.9 PHYSICAL PROPERTIES

11.9.1 Boiling and Melting Points

The magnitude of the intermolecular attractive forces present in a compound determines whether its boiling and melting points are high or low and whether it dissolves in a polar solvent or nonpolar solvent. Recall from Section 5.3 that there are three types of intermolecular attractive forces possible in molecules—hydrogen bonding, dipole-dipole, and dispersion (London). Hydrogen bonding is the strongest intermolecular force and dispersion the weakest. All molecules possess dispersion forces. This is the only intermolecular force present in nonpolar molecules. Polar molecules have higher intermolecular attractive forces relative to nonpolar molecules since they possess dipole-dipole attractive forces in addition to dispersion forces. Hydrogen-bonding forces are present only in molecules possessing O—H, N—H, and F—H bonds.

Alkanes and cycloalkanes are nonpolar since the C—C and C—H bonds are nonpolar bonds. (The electronegativity difference between C and H is so small that the extent of polarity in the C—H bond is zero from a practical viewpoint.) Only London dispersion forces are present in alkanes and cycloalkanes. They have the lowest intermolecular forces of all organic families. Molecules of less than five carbons are gases at ambient temperatures. Among the unbranched alkanes, it is not until C_{15} and higher that the compounds are solids.

The physical change of a compound from solid to liquid (melting) or from liquid to gas (boiling) requires the input of thermal energy to allow molecules to overcome their intermolecular attractions for each other. Molecules with larger intermolecular forces are attracted to each other more strongly and have higher melting and boiling points.

Molecular mass and family determine the magnitude of the intermolecular forces. The family of the compound determines the types of chemical bonds present and this determines whether there are hydrogen-bonding, dipole-dipole, or dispersion forces. Intermolecular forces increase with increasing molecular mass since larger molecules contact and attract each other over a larger surface area. The ability to predict the order of increasing intermolecular forces for a set of compounds is good when only molecular mass or family differs among the compounds. Predictions are less reliable when both structural features differ.

Differences in boiling points of different compounds are used in the separation of the different components of a mixture. The most important example is the refining of petroleum by distillation. Petroleum is of little value without its separation into fractions useful as heating and cooking gas, gasoline, jet fuel and kerosene, lubricating oils, greases, waxes, and asphalt.

Problem 11.14. Place the following compounds in order of increasing boiling points and explain the order. Answer the same question with regard to melting points.

$$CH_4 \qquad CH_3CH_3 \qquad CH_3CH_2CH_3$$

Ans. The order of boiling points is $CH_3CH_2CH_3 > CH_3CH_3 > CH_4$ since the intermolecular attractive forces increase with increasing molecular mass in the same family. The order of melting points is the same since melting also requires overcoming intermolecular forces. The only difference between melting and boiling is the extent to which the intermolecular forces must be overcome. For melting, the energies of the molecules must be increased sufficiently that any molecule can move from one neighbor to another neighbor. For boiling, the molecular energies must be increased even further to the point that molecules move far away from each other.

Problem 11.15. The difference in the boiling points of H_2O and CH_4 is large, 100 °C versus −162 °C. Why?

Ans. Water and methane are very nearly the same molecular mass (16 versus 18). The difference in boiling points is due to the large difference in the intermolecular attractions present in the two compounds. Methane has only the weak attractions of dispersion forces. Water has the strongly polar 0—H bond and this results in the strongest type of intermolecular attractions—hydrogen bonding. The magnitude of hydrogen-bonding attractions compared to dispersion forces is evident by noting that it is not until heptane (C_7H_{16}) that the boiling point of an alkane is comparable (98.4 °C) to that of water.

Problem 11.16. Explain the differences in the boiling points of the following compounds:

$CH_3-CH_2-CH_2-CH_2-CH_3$ — Pentane, 36.1 °C

$CH_3-CH(CH_3)-CH_2-CH_3$ — 2-Methylbutane, 27.8 °C

$CH_3-C(CH_3)_2-CH_3$ — 2,2-Dimethylpropane, 9.5 °C

(pentagon) — Cyclopentane, 49 °C

Ans. The family is the same for all compounds. The acyclic compounds are constitutional isomers and have the same molecular mass. The very slightly lower molecular mass of cyclopentane compared to the acyclics (70 versus 72) is too small to be significant. The variable among the compounds is molecular shape (branching, cyclization) and its effect on intermolecular attraction. The molecular shape becomes more spherical and less cylindrical with increased branching, as we proceed from pentane to 2-methylbutane to 2,2-dimethylpropane. The intermolecular attractions decrease in that order since spherical molecules contact each other over less surface area. Think of two spheres, which contact each other only at a point, while two cylinders contact each other along a line. The effect of a cyclic structure is in the opposite direction from branching since cyclopentane has the highest boiling point. The cyclic compound is more rigid, which allows the molecules to approach each other more closely, resulting in more effective (stronger) intermolecular attractions. The linear alkane is far from rigid since there is free rotation around all of the C—C bonds. The molecule is continuously passing through various conformations (Sec. 11.3) which do not allow the molecules to approach each other as closely as do rigid cyclic molecules. (Our previous analogy of a linear molecule being in the shape of a cylinder works in comparing it to branched molecules but fails in comparing it to cyclic molecules.)

11.9.2 Solubility

Solubility of a solute in a solvent is successful only when the intermolecular attraction between solute and solvent molecules is at least as strong or stronger than the attraction of solute for solute and solvent for solvent. This occurs only when the types of attractive forces present in solute and solvent are similar, i.e., *like dissolves like*. Nonpolar compounds dissolve in other nonpolar compounds. Polar or hydrogen-bonding compounds dissolve in other polar or hydrogen-bonding compounds. Nonpolar compounds do not dissolve in polar or hydrogen-bonding compounds.

Problem 11.17. Will hexane dissolve in water? Why?

Ans. Hexane or any other alkane or cycloalkane will not dissolve in water since the two compounds are very different in the level of their intermolecular forces. Hexane is nonpolar while water undergoes hydrogen bonding. There is no attraction between the hydrogen-bonding water and the nonpolar alkane and mixing does not take place.

11.10 CHEMICAL REACTIONS

The terminology *saturated hydrocarbons* used to describe alkanes and cycloalkanes conveys that these compounds are generally unreactive. They are the least reactive of all organic compounds. Unlike other families, alkanes and cycloalkanes do not contain a bond or bonds which are relatively weak and give rise to special reactivity. However, alkanes and cycloalkanes do undergo two reactions —halogenation and combustion.

11.10.1 Halogenation

Halogenation is the reaction of an alkane or cycloalkane with molecular halogen in the presence of light or heat. A hydrogen atom in the alkane or cycloalkane is substituted by a halogen atom, e.g., for chlorination of methane:

$$CH_4 + Cl_2 \xrightarrow[\text{or heat}]{\text{light}} CH_3Cl + HCl$$

The reaction is classified as a *substitution reaction*. The equation

$$\text{H}-\underset{\text{H}}{\overset{\text{H}}{\text{C}}}-\text{H} + \text{Cl}-\text{Cl} \xrightarrow[\text{or heat}]{\text{light}} \text{H}-\underset{\text{H}}{\overset{\text{H}}{\text{C}}}-\text{Cl} + \text{H}-\text{Cl}$$

shows the details of the reaction in terms of bond breakages and bond formations. Two bonds are broken, the C—H bond of methane and the Cl—Cl bond of chlorine. Two new bonds are made, the C—Cl and H—Cl bonds of the two products. The organic product is referred to as an *alkyl halide*.

There are other ways of writing the equation to describe this reaction. Sometimes we generalize the halogen by using X_2 instead of Cl_2 or Br_2:

$$CH_4 + X_2 \xrightarrow[\text{or heat}]{\text{light}} CH_3X + HX$$

We can also generalize the alkane by using R—H where R represents an alkyl group such as methyl, ethyl, and so on:

$$R{-}H + X_2 \xrightarrow[\text{or heat}]{\text{light}} R{-}X + HX$$

This shows that any C—H bond in any alkane (or cycloalkane) can be converted to a C-halogen bond by reaction with molecular halogen. We learn the generalization to avoid having to learn separate equations for each of the thousands and thousands of alkanes and for the different halogens.

Halogenation by Cl_2 and Br_2 proceeds easily. I_2 reacts too slowly to be useful. The reaction with F_2 is limited for a different reason. It proceeds so fast and is so highly exothermic that the reaction is difficult to control (it is difficult to prevent explosion).

Organic chemists often write reactions in the more abbreviated form

$$R{-}H \xrightarrow[\text{light or heat}]{X_2} R{-}X$$

to place emphasis on the organic compounds. The reagent (X_2) and reaction conditions (light or heat) are placed above and below the arrow and the nonorganic product (HCl) is often omitted.

Problem 11.18. What organic products other than CH_3Cl are formed when CH_4 undergoes chlorination?

Ans. During the initial reaction period, some CH_3Cl is formed and the reaction mixture contains both CH_3Cl and CH_4. The halogenation reaction occurs with all C—H bonds, not just those in CH_4. The result is that some of the CH_3Cl undergoes chlorination to yield some CH_2Cl_2. Some of the CH_2Cl_2 reacts to form $CHCl_3$ and the $CHCl_3$ reacts further to form CCl_4:

$$CH_4 \longrightarrow CH_3Cl \longrightarrow CH_2Cl_2 \longrightarrow CHCl_3 \longrightarrow CCl_4$$

The relative amounts of CH_3Cl, CH_2Cl_2, $CHCl_3$, and CCl_4 present in the product depend on the amount of Cl_2 used relative to CH_4. One obtains more of the more highly chlorinated products by using a larger excess of Cl_2.

Problem 11.19. Considering only monohalogenation, what product(s) is (are) formed when propane undergoes chlorination?

Ans. A mixture of two different products is formed since halogenation proceeds in a relatively random fashion, i.e., any C — H bond in the molecule can react. Substitution of any one of the six hydrogens of the two CH_3 groups of the molecule gives the same product, 1-chloropropane. A different product, 2-chloropropane, is produced when either of the two hydrogens of the CH_2 group undergoes substitution.

$$CH_3-CH_2-CH_3 \xrightarrow{Cl_2} \underset{\text{1-Chloropropane}}{CH_3-CH_2-CH_2-Cl} + \underset{\text{2-Chloropropane}}{CH_3-\overset{\displaystyle Cl}{\overset{|}{C}H}-CH_3}$$

Note that the above equation looks unbalanced—there are twice as many carbons, hydrogens, and chlorines on the right-hand side as on the left-hand side. Equations for organic reactions are often written without concern for balancing. The emphasis is on a description of the organic reactant(s) used and the organic product(s) formed without indicating balance. Balance is of minimal concern because most organic reactions give less, often much less, than a 100 percent yield. For example, a careful analysis of the product mixture from the chlorination of propane might show that the yields of 1-chloropropane and 2-chloropropane are 38 percent and 41 percent, respectively, with smaller amounts of various dichloro-, trichloro-, and more highly chlorinated products. We show only the major products of the reaction in the typical equation. Such an equation is not to be read as a balanced equation.

Note the naming of the products, 1-chloropropane and 2-chloropropane. The IUPAC nomenclature rules are exactly the same as described in Sec. 11.7.2. Halogen substituents, named as *fluoro-*, *chloro-*, *bromo-*, and *iodo-*, are handled in the same manner as alkyl groups. The common nomenclature system names alkyl halides by using the name of the alkyl group connected to the halogen followed by the word *halide*. 1-Chloropropane and 2-chloropropane are named propyl chloride and isopropyl chloride. This is applicable only to alkyl halides with simple alkyl groups.

A variety of halogen-containing compounds find uses as refrigerants (Freon), dry-cleaning fluids, and anesthetics, and as solvents in which to carry out organic reactions.

Problem 11.20. Explain why the boiling point of an alkyl halide is much higher than that of the alkane from which it is synthesized. For example, the boiling points of propyl and isopropyl chlorides are 46.6 °C and 35.7 °C, respectively, compared to −42.1 °C for propane.

Ans. The higher boiling points (and also melting points) of alkyl halides arise from their higher molecular mass coupled with the presence of dipole-dipole attractive forces. The dipole-dipole attractions is a result of the polar C—Cl bond. Cl is more electronegative than C. The Cl end of the C—Cl bond is partially negative ($\delta -$) and the C end is partially positive ($\delta +$). Oppositely charged ends of adjacent dipoles attract each other as indicated by the dashed line:

$$(\delta + R-Cl\,\delta -) - - - - (\delta + R-Cl\,\delta -)$$

The dipole-dipole attractions in the two alkyl chlorides are stronger than the dispersion forces present in propane. Higher temperatures are required to overcome the intermolecular attractions in the alkyl chlorides.

The higher boiling point of propyl chloride compared to isopropyl chloride is the result of the branching effect described in Problem 11.16.

11.10.2 Combustion

Alkanes, cycloalkanes, and other hydrocarbons undergo *combustion* (burning). The reaction involves reaction with O_2 whereby all C—H and C—C bonds in the hydrocarbon undergo cleavage. In the presence of sufficient oxygen, the reaction involves a quantitative conversion of the carbon and hydrogen atoms of the hydrocarbon into carbon dioxide and water:

$$CH_3{-}CH_2{-}CH_2{-}CH_2{-}CH_3 + 8\,O_2 \longrightarrow 5\,CO_2 + 6\,H_2O + \text{heat energy}$$

Combustion equations are not always written to show heat as one of the products but it is the important product. The heat energy is used to heat our homes and workplaces, for cooking purposes, and to generate electricity. The other use for the combustion reaction is the generation of power in engines for automobiles, aircraft, and trains. Power is generated in engines because there is a large pressure buildup during combustion (a consequence of the temperature increase coupled with the increase in the number of moles of gases present after reaction).

The combustion reaction described above is *complete combustion*. When there is insufficient oxygen or combustion conditions (temperature, good mixing of alkane and oxygen) are not optimized, *incomplete combustion* occurs with the formation of varying amounts of carbon monoxide (CO) instead of carbon dioxide. This is detrimental in two ways—the amount of heat generated per mole of alkane is less and carbon monoxide is highly toxic.

Combustion is an *oxidation reaction*. Oxidation of organic compounds is defined as an increase in the oxygen content and/or a decrease in the hydrogen content of the carbon atoms in the compound. For complete combustion of pentane, we see that the starting reactant has 12 hydrogens bonded with its 5 carbons. After reaction, the five carbons no longer have any hydrogens attached and have become attached to a total of 10 oxygens. Each of the carbon atoms of the alkane has been oxidized to its highest oxidation state, i.e., each carbon ends up with the maximum number of bonds to oxygen—four—in forming CO_2.

Problem 11.21. Give the balanced equation for the complete combustion of 2,3,3-trimethylpentane:

$$\begin{array}{c} \quad CH_3\;\; CH_3 \\ \quad | \qquad | \\ CH_3{-}CH{-}C{-}CH_2{-}CH_3 \\ \qquad\quad | \\ \qquad\quad CH_3 \end{array}$$

Ans. The structural formula need not be used for this equation since the reaction involves complete dissection of the alkane. The molecular formula is sufficient and the same equation can be used for the combustion of any of the structural isomers of C_8H_{18}. The unbalanced equation is

$$C_8H_{18} + O_2 \longrightarrow CO_2 + H_2O$$

The procedure for balancing consists of balancing for C, then for H, and last for O. The eight C on the left are balanced by 8 CO_2 on the right. The 18 H on the left are balanced by 9 H_2O on the right. The 25 O on the right (16 from 8 CO_2 plus 9 from 9 H_2O) are balanced by $\frac{25}{2}$ O_2. The balanced equation is

$$C_8H_{18} + \tfrac{25}{2}\,O_2 \longrightarrow 8\,CO_2 + 9\,H_2O$$

Most texts prefer to multiply both sides of the equation by 2 to yield

$$2\,C_8H_{18} + 25\,O_2 \longrightarrow 16\,CO_2 + 18\,H_2O$$

as the final result since this avoids the presence of a fractional coefficient.

ADDITIONAL SOLVED PROBLEMS

MOLECULAR AND STRUCTURAL FORMULAS

Problem 11.22. Which of the following are correct molecular formulas? (*a*) C_4H_{10}, (*b*) $C_4H_{10}Br$, (*c*) C_4H_9Br.

Ans. The molecular formula is correct only if you can draw a structural formula in which the valence of each atom is correct. C_4H_{10} is correct and represents either of two constitutional isomers:

$$\underset{\text{Butane}}{CH_3{-}CH_2{-}CH_2{-}CH_3} \qquad \underset{\text{2-Methylpropane}}{CH_3{-}\underset{\displaystyle CH_3}{\underset{|}{CH}}{-}CH_3}$$

Each of the C and H atoms has the correct combining power. We also arrive at the answer by noting that C_4H_{10} fits the general formula for an alkane, $C_n H_{2n+2}$, with $n = 4$.

$C_4H_{10}Br$ cannot be correct since C_4H_{10} is correct. Br cannot be "added" to C_4H_{10} to yield $C_4H_{10}Br$ since there is no available empty bond vacancy for bond formation. The presence of Br or any other halogen in a compound comes about by Br replacing H since both Br and H are monovalent. No matter how hard you try, you cannot put together 4 carbons, 10 hydrogens, and 1 bromine to form a molecule in which each atom has its correct combining power.

C_4H_9Br is correct as it fits the general formula for an alkane where the total of H and Br is given by $(2n + 2)$. It involves the replacement of H by Br in C_4H_{10}. One constitutional isomer is 1-bromobutane, $CH_3CH_2CH_2CH_2Br$. The other constitutional isomers of C_4H_9Br are 2-bromobutane, 1-bromo-2-methylpropane, and 2-bromo-2-methylpropane.

Problem 11.23. Which of the following are correct molecular formulas? (*a*) C_4H_9, (*b*) C_4H_8.

Ans. C_4H_9 is incorrect as it fits neither the general formula of an alkane nor cycloalkane. C_4H_8 fits the general formula for a cycloalkane and is either cyclobutane or methylcyclopropane:

[cyclobutane square] [cyclopropane triangle]$-CH_3$

FAMILIES OF ORGANIC COMPOUNDS; FUNCTIONAL GROUPS

Problem 11.24. Identify the family of each of the following compounds by reference to Table 11-1. Clearly indicate how you arrive at the identification.

(*a*) $CH_3CH_2{-}OH$ (*b*) $CH_3CH_2{-}O{-}CH_2CH(CH_3)_2$

(*c*) $CH_3CH_2CH_2{-}\overset{\overset{\displaystyle O}{\|}}{C}{-}H$ (*d*) $CH_3CH_2CH_2{-}\overset{\overset{\displaystyle O}{\|}}{C}{-}CH_3$

(*e*) $CH_3CH_2{-}\overset{\overset{\displaystyle O}{\|}}{C}{-}NH_2$ (*f*) $CH_3\underset{\underset{\displaystyle CH_3}{|}}{CH}{-}\overset{\overset{\displaystyle O}{\|}}{C}{-}OH$

(*g*) $CH_3CH_2{-}\overset{\overset{\displaystyle O}{\|}}{C}{-}OCH_2CH_2CH_3$ (*h*) $CH_3CH_2{-}NH_2$

Ans. (*a*) alcohol, (*b*) ether. Both alcohols and ethers contain an oxygen atom in which each of the oxygen's two single bonds is attached to different atoms. Alcohols differ from ethers in that one of the attachments is to a hydrogen. The ether oxygen's two attachments are both to carbon (but different carbon atoms).

(*c*) aldehyde, (*d*) ketone, (*e*) amide, (*f*) carboxylic acid, (*g*) ester. Each of these families has the carbonyl group, C=O, but they differ in what else is attached to the carbon of the C=O group. The aldehyde carbonyl C has H and C directly attached while the ketone has two different carbons. The amide carbonyl carbon has a nitrogen attached. The ester and carboxylic acid have an oxygen attached to the carbonyl carbon, but differ in what is attached to that oxygen. The carboxylic acid's oxygen has a hydrogen while the ester has a carbon.

(*h*) amine. An amine has a nitrogen to which are directly attached any combination of three hydrogen or carbon atoms. Any carbon atom attached to N cannot be part of a carbonyl group.

Redrawing the compounds in a generalized manner with R or R′ representing any alkyl groups allows one to focus on the functional group characteristics of each compound:

(*a*) R—OH (*b*) R—O—R′

(*c*) R—C(=O)—H (*d*) R—C(=O)—R′

(*e*) $R—C(=O)—NH_2$ (*f*) R—C(=O)—OH

(*g*) R—C(=O)—OR′ (*h*) $R—NH_2$

WRITING STRUCTURAL FORMULAS; CONSTITUTIONAL ISOMERS

Problem 11.25. For each of the following structural drawings, indicate whether the pair represents (1) the same compound, (2) different compounds which are constitutional isomers, or (3) different compounds which are not isomers. (*Note*: A pair of drawings represents the same compound when the drawings represent different conformations, or they differ in the extent to which they are condensed or expanded, or they differ in the extent to which they show the exact bond angles which are present.)

(*a*) $(CH_3)_2CHCH_2Cl$ and $ClCH_2—CH(CH_3)—CH_3$

(*b*) $ClCH_2—CH(CH_3)—CH_3$ and $CH_3—CH_2—CH_2—CH_2Cl$

(*c*) $(CH_3)_3C—Cl$ and $CH_3—CH_2—CH(Cl)—CH_2—CH_3$

(*d*) cyclopentane bearing $—CH_3$ and $—CH(CH_3)—CH_3$ and cyclohexane bearing three CH_3 groups

(*e*) $CH_3—CH_2—CH_2—CH(CH_3)—CH_3$ and $CH(CH_3)(CH_3)—CH_2—CH_2—CH_3$

(*f*) cyclohexane ring and open six-carbon chain drawn in hexagon shape

Ans. (*a*) Same compound. Both drawings show a three-carbon chain with a methyl attached to the middle carbon and a Cl on an end carbon. The drawing on the right-hand side is an expanded version of the left-hand-side drawing and is rotated 180° in the plane of the paper.

(*b*) Constitutional isomers of C_4H_9Cl. One is linear and the other branched.

(*c*) Different compounds that are not isomers. One is C_4H_9Cl and the other is $C_5H_{11}Cl$.

(*d*) Constitutional isomers of C_9H_{18}. One has a 5-membered ring with a total of four other carbons attached. The other has a 6-membered ring with a total of three other carbons attached.

(*e*) Same compound. Both represent the isomer of C_6H_{14} in which there is a continuous chain of five carbons with a methyl group attached as a branch on a next-to-last carbon of the five-carbon chain.

(*f*) Different compounds that are not isomers, C_6H_{12} and C_6H_{14}.

Problem 11.26. How many constitutional isomers exist for the molecular formula C_7H_{16}? Show one structural formula for each isomer. Do not show more than one structural formula for each isomer.

Ans. The answer is given using the skeleton formulas. There are eight different constitutional isomers. There is the linear compound with all seven carbons being part of one continuous chain:

```
C—C—C—C—C—C—C
```

There are two compounds having a six-carbon chain with a methyl branch:

```
  C                   C
  |                   |
C—C—C—C—C—C       C—C—C—C—C—C
```

There are four compounds having a five-carbon chain with two methyl branches:

```
  C C             C   C            C                C
  | |             |   |            |                |
C—C—C—C—C       C—C—C—C—C        C—C—C—C—C      C—C—C—C—C
                                   |                |
                                   C                C
```

There is only one compound having a four-carbon chain:

```
  C C
  | |
C—C—C—C
  |
  C
```

NOMENCLATURE

Problem 11.27. Give the IUPAC name for the following compound:

```
                                   CH3
                                   |
      Cl        CH2—CH2—CH2—CH—CH3
      |         |
CH3—C—CH2—C—CH2—CH3
      |         |
      CH3       CH—CH3
                |
                CH3
```

Ans. The longest chain has nine carbons and is numbered as shown:

```
                 5     6     7    CH3
      Cl        CH2—CH2—CH2—CH—CH3
 1   2|   3   4 |                8   9
CH3—C—CH2—C—CH2—CH3
      |         |
      CH3       CH—CH3
                |
                CH3
```

The IUPAC name is 2-chloro-4-ethyl-2,8-dimethyl-4-isopropylnonane. Numbering in the opposite direction is incorrect since it gives the name 2-chloro-6-ethyl-2,8-dimethyl-6-isopropylnonane which has a set of larger numbers for the substituents. Note that the prefixes di- and iso- are ignored in alphabetizing the names of the substituents.

Problem 11.28. Draw the structural formula for 2-chloro-2-methylpentane and identify 1°, 2°, 3°, and 4° carbons.

Ans.

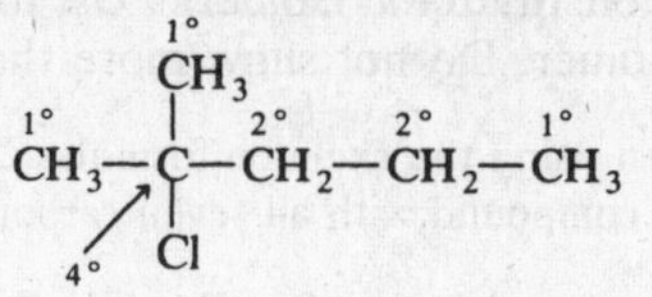

1°
CH_3
1° 2° 2° 1°
$CH_3-C-CH_2-CH_2-CH_3$
4° Cl

CYCLOALKANES

Problem 11.29. Give the IUPAC name for each of the following compounds:

(*a*) CH_3 and $CH_2CH_2CH_2CH_3$ on a cyclohexane ring

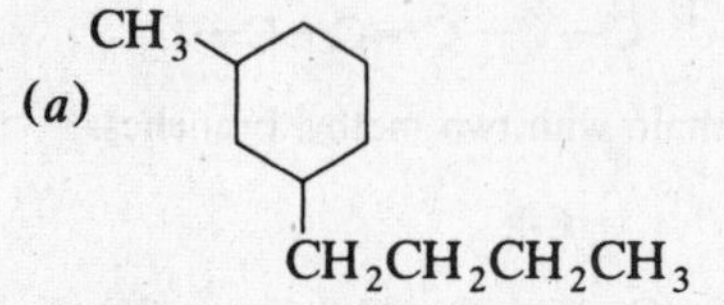

(*b*) CH_3, CH_3CH_2 and $CH_2CH_2CH_2CH_3$ on a cyclohexane ring

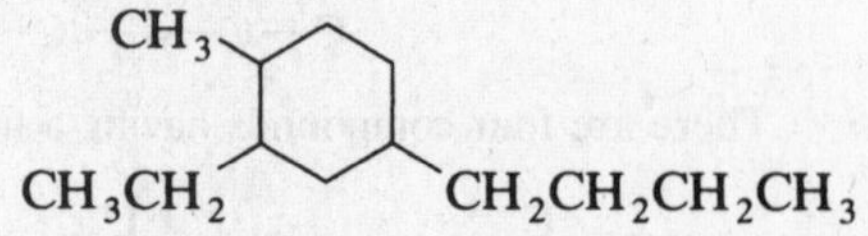

Ans. (*a*) The set of smallest numbers for moving between the methyl and butyl groups is 1,3. The name is either 1-butyl-3-methylcyclohexane or 3-butyl-1-methylcyclohexane, depending on whether we start numbering at the methyl or the butyl. Where do we start numbering? The answer is alphabetical order. We start at the butyl since butyl comes before methyl. The IUPAC name is 1-butyl-3-methylcyclohexane.
(*b*) The set of smallest numbers is 1,2,4 for moving counterclockwise from methyl to butyl. Starting from any substituent but methyl results in a set of larger numbers. The IUPAC name is 4-butyl-2-ethyl-1-methylcyclohexane.

Problem 11.30. What is the relationship between structures I and II below?

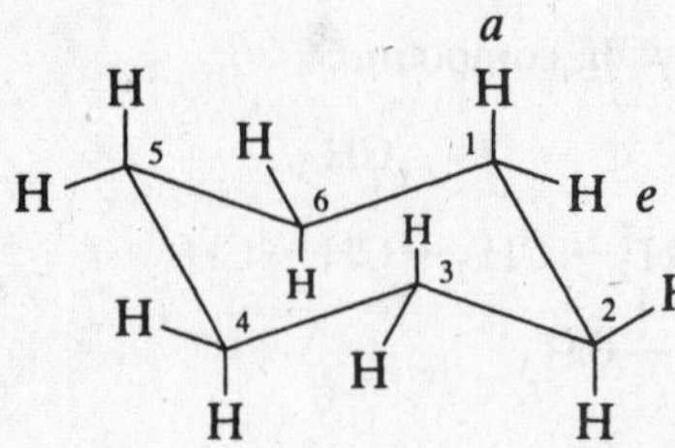

I 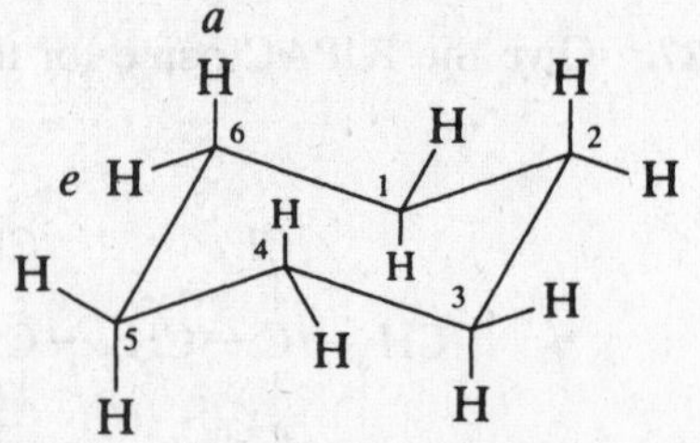 II

Ans. Structures I and II represent different *conformations* or *conformational isomers* (Sec. 11.4) of the same compound. One conformation converts into the other by the simultaneous rotation of all C — C bonds of the ring. (This is difficult to see without the use of molecular models.) Structures I and II are of equal stability.

Problem 11.31. What is the relationship between structures I and II below?

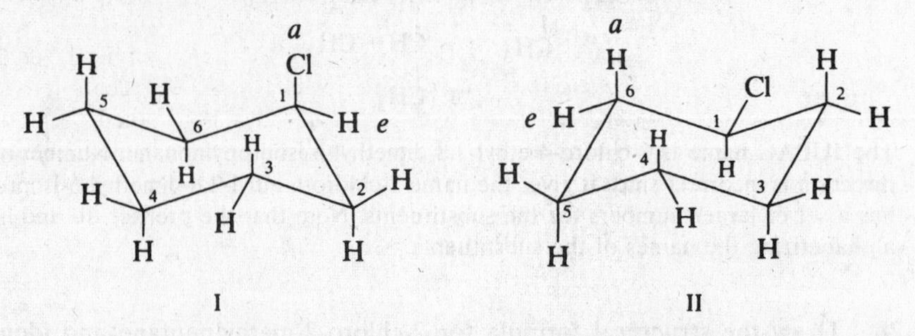

Ans. Structures I and II are conformational isomers of chlorocyclohexane but the two are not of equal stability. There are two types of bonds at each carbon of the ring—*axial* (*a*) and *equitorial* (*e*).

Axial bonds are those in the vertical plane. Equitorial bonds are those at an angle to the horizontal plane. Substituents larger than hydrogen are more stable in equitorial positions. This makes conformation II more stable than I since II has the large Cl in the less stable axial position while I has the smaller H in the axial position. Chlorocyclohexane consists primarily of conformation II.

PHYSICAL PROPERTIES

Problem 11.32. For each of the following pairs of compounds, indicate which compound in the pair has the higher boiling point (BP) and explain why:

(*a*) $CH_3-CH_2-CH_2-CH_2-CH_3$ (Pentane) or $CH_3-CH_2-CH_2-CH_3$ (Butane)

(*b*) $CH_3-CH_2-CH_2-CH_2-CH_3$ (Pentane) or $CH_3-CH(CH_3)-CH_2-CH_3$ (2-Methylbutane)

(*c*) $CH_3-CH_2-CH_2-CH_3$ (Butane) or $CH_3-CH(CH_3)-CH_2-CH_3$ (2-Methylbutane)

(*d*) $CH_3-CH_2-CH_2-CH_2-CH_3$ (Pentane) or $CH_3-CH_2-CH_2-CH_2-CH_2Cl$ (1-Chloropentane)

Ans. (*a*) Pentane has a higher BP than butane since it has the higher molecular mass. Intermolecular forces and BP increase with increasing molecular mass.

(*b*) Pentane has a higher BP than 2-methylbutane since branching lowers intermolecular attractions and BP at the same molecular mass.

(*c*) 2-Methylbutane has a higher BP than butane because, although similarly branched, it has a higher molecular mass. A useful analogy for comparing molecules with the same size of longest chain is that of cylindrical-shaped molecules. Butane and 2-methylbutane have the same cylindrical length but 2-methylbutane has the larger diameter. The larger diameter means a larger total surface area, larger intermolecular forces, and higher BP.

(*d*) 1-Chloropentane has a higher BP than pentane since it has a higher molecular mass and is more polar.

Problem 11.33. Explain why alkanes and other hydrocarbons are insoluble in water.

Ans. There is negligible intermolecular attraction between water and a hydrocarbon since hydrocarbons are nonpolar while water is hydrogen-bonded.

Problem 11.34. Why does an alkane such as hexane float on top of water instead of the reverse?

Ans. Low-density liquids (alkanes and other hydrocarbons) float on top of higher-density liquids such as water. Density, like BP, melting point (MP), and solubility, is dependent on intermolecular forces. Density increases with an increase in the number of molecules packed into a unit volume. Water molecules pack closer to each other because of their stronger intermolecular forces and this results in higher density for water.

CHEMICAL REACTIONS

Problem 11.35. Complete the following reaction:

$$CH_3-CH_2-CH_2-CH_2-CH_3 \xrightarrow[\text{light or heat}]{Cl_2} ?\ \text{(monochlorination)}$$

Ans. There are three products. Substitution of Cl for any one of the six methyl hydrogens yields 1-chloropentane, substitution for any one of the four CH_2 hydrogens of the next-to-last carbon from each end yields 2-chloropentane, and substitution at either of the two CH_2 hydrogens of the middle carbon yields 3-chloropentane:

$$CH_3{-}CH_2{-}CH_2{-}CH_2{-}CH_3 \xrightarrow[\text{light or heat}]{Cl_2} CH_3{-}CH_2{-}CH_2{-}CH_2{-}CH_2Cl$$

$$+ CH_3{-}\underset{\displaystyle Cl}{\underset{|}{CH}}{-}CH_2{-}CH_2{-}CH_3 + CH_3{-}CH_2{-}\underset{\displaystyle Cl}{\underset{|}{CH}}{-}CH_2{-}CH_3$$

Problem 11.36. Complete the following reaction:

$$CH_3{-}\overset{\displaystyle CH_3}{\overset{|}{\underset{\displaystyle CH_3}{\underset{|}{C}}}}{-}CH_3 \xrightarrow[\text{light or heat}]{Cl_2} \text{? (monochlorination)}$$

Ans. There is only one product. Substitution of Cl for any one of the 12 hydrogens yields the same product, 1-chloro-2,2-dimethylpropane:

$$CH_3{-}\overset{\displaystyle CH_3}{\overset{|}{\underset{\displaystyle CH_3}{\underset{|}{C}}}}{-}CH_3 \xrightarrow[\text{light or heat}]{Cl_2} CH_3{-}\overset{\displaystyle CH_3}{\overset{|}{\underset{\displaystyle CH_3}{\underset{|}{C}}}}{-}CH_2Cl$$

Problem 11.37. Complete each of the following reactions by writing the structure of the organic product. If no reaction occurs, write "No Reaction."

(*a*) $CH_3{-}CH_2{-}CH_2{-}CH_3 \xrightarrow{H_2SO_4} ?$ (*b*) $CH_3{-}CH_2{-}CH_2{-}CH_3 \xrightarrow{NaOH} ?$

(*c*) $CH_3{-}CH_2{-}CH_2{-}CH_3 \xrightarrow{H_2} ?$ (*d*) $CH_3{-}CH_2{-}CH_2{-}CH_3 \xrightarrow{HCl} ?$

Ans. The answer to each is No Reaction. Alkanes and cycloalkanes do not react with any of the reagents. This is not a trick question. It is just as important to know which reactions do not occur as to know which reactions do occur.

SUPPLEMENTARY PROBLEMS

Problem 11.38. Identify the family for each items *a* through *g* in the following compound:

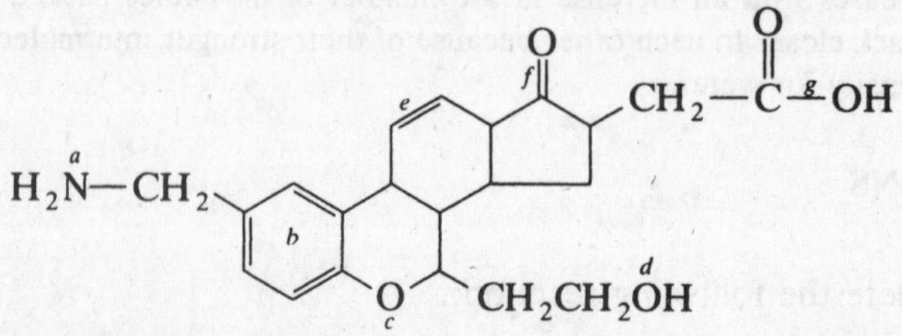

$H_2N{-}CH_2$ (a) ... b ... O (c) ... CH_2CH_2OH (d) ... e ... f (C=O) ... $CH_2{-}\overset{O}{\overset{\|}{C}}{-}OH$ (g)

Ans. (*a*) amine, (*b*) aromatic, (*c*) ether, (*d*) alcohol, (*e*) alkene, (*f*) ketone, (*g*) carboxylic acid.

Problem 11.39. How many constitutional isomers exist for the molecular formula $C_5H_{11}Cl$? Show one skeleton structural formula for each isomer. Do not show more than one structural formula for each isomer.

Ans.

```
Cl                    Cl                      Cl
|                     |                       |
C—C—C—C—C           C—C—C—C—C             C—C—C—C—C

  C—Cl            C                 C                 C
  |               |                 |                 |
C—C—C—C         C—C—Cl—C          C—C—C—C           C—C—C—C—Cl
                  |                 |
                  Cl                Cl

        C                    C
        |                    |
    C—C—C—C—Cl           C—C—C—Cl
                             |
                             C
```

Problem 11.40. Name the compounds in Problem 11.39.

Ans. **In the order given, left to right, the compounds are 1-chloropentane, 2-chloropentane, 3-chloropentane, 2-chloro-methylbutane, 2-chloro-2-methylbutane, 2-chloro-3-methylbutane, 1-chloro-3-methylbutane, 1-chloro-2-methylbutane, 1-chloro-2,2-dimethylpropane.**

Problem 11.41. How many constitutional isomers exist for the molecular formula C_6H_{12}? Show one structural formula for each isomer. Do not show more than one structural formula for each isomer.

Ans.

(cyclohexane); cyclopentane–CH_3; cyclobutane–CH_2CH_3; cyclobutane with CH_3 and CH_3 (1,2); cyclobutane with CH_3 and CH_3 (1,3); cyclobutane with CH_3, CH_3 on the same carbon;

cyclopropane–$CH_2CH_2CH_3$; cyclopropane–$CH(CH_3)_2$;

cyclopropane with CH_3CH_2 and CH_3 on the same carbon; cyclopropane with CH_3CH_2 and CH_3; cyclopropane with CH_3, CH_3, CH_3

Problem 11.42. Name the compounds in Problem 11.41.

Ans. **In the order given, left to right, the compounds are cyclohexane, methylcyclopentane, ethylcyclobutane, 1,2-dimethylcyclobutane, 1,3-dimethylcyclobutane, 1,1-dimethylcyclobutane, *n*-propylcyclopropane, isopropylcyclopropane, 1-ethyl-1-methylcyclopropane, 1-ethyl-2-methylcyclopropane, 1,2,3,-trimethylcyclopropane.**

Problem 11.43. Draw the structural formula for each of the following compounds: (*a*) 3-*t*-butylpentane, (*b*) 2-cyclohexyl-3,5-dimethylheptane.

Ans.

```
(a) CH3—CH2—CH—CH2—CH3          (b)        CH3       CH3
            |                                 |         |
            C(CH3)3                CH3—CH—CH—CH2—CH—CH2—CH3
                                        |
                                   (cyclohexyl)
```

Problem 11.44. Why is the compound in Problem 11.43(*b*) not named as a derivative of cyclohexane?

Ans. The nine-carbon group attached to the cyclohexane ring is not one of the simple substituents and has no simple name. Instead, we name the cyclic portion as a substituent attached to the acyclic portion of the molecule.

Problem 11.45. Identify the 1°, 2°, 3°, and 4° carbons in the compounds in Problem 11.43.

Ans.

(*a*) $\overset{1°}{CH_3}-\overset{2°}{CH_2}-\overset{3°}{CH}-\overset{2°}{CH_2}-\overset{1°}{CH_3}$, with $\overset{4°}{C}(\overset{1°}{CH_3})_3$ attached to the 3° CH

(*b*) $\overset{1°}{CH_3}-\overset{3°}{CH}-\overset{3°}{CH}-\overset{2°}{CH_2}-\overset{3°}{CH}-\overset{2°}{CH_2}-\overset{1°}{CH_3}$, with $\overset{1°}{CH_3}$ on the second CH and on the third CH, and a cyclohexyl ring (ring carbons: 3°, 2°, 2°, 2°, 2°, 2°) attached to the first CH

Problem 11.46. Which compound has the higher boiling point in each of the following pairs? (*a*) decane and 3-methylnonane, (*b*) 3-methylnonane and 2,2,3,3-tetramethylpentane, (*c*) 2-chlorobutane and 2-chloropentane, (*d*) 2-chlorobutane and 2-methylbutane.

Ans. (*a*) decane, (*b*) 3-methylnonane, (*c*) 2-chloropentane, (*d*) 2-chlorobutane

Problem 11.47. Why is sodium chloride soluble in water but insoluble in hexane?

Ans. NaCl dissolves in water because the Na^+ and Cl^- ions of the NaCl crystal lattice are strongly attracted to the polar water molecules. The $\delta+$ ends of water molecules are attracted to Cl^- while the $\delta-$ ends of water molecules are attracted to Na^+. Na^+ and Cl^- ions cannot form strong intermolecular attractions to hexane since hexane is nonpolar. Hexane has the ability to form only the weakest of intermolecular attractions, via dispersion forces. Water molecules will not separate from each other in order to form weak attractions to hexane.

Problem 11.48. Give the equation showing the monochlorination product(s) of each of the following compounds: (*a*) cyclohexane, (*b*) methylcyclohexane.

Ans.

(*a*) cyclohexane $\xrightarrow[\text{light or heat}]{Cl_2}$ chlorocyclohexane (Cl)

(*b*) methylcyclohexane (CH_3) $\xrightarrow[\text{light or heat}]{Cl_2}$ (CH_2Cl)-cyclohexane + 1-chloro-1-methylcyclohexane (CH_3, Cl) + 1-chloro-2-methylcyclohexane (CH_3, Cl) + 1-chloro-4-methylcyclohexane (CH_3, Cl) + 1-chloro-3-methylcyclohexane (Cl, CH_3)

Problem 11.49. Write the balanced equation for the complete combustion of (*a*) cyclohexane, (*b*) 2-methylpentane, and (*c*) 2,3-dimethylbutane.

Ans. (*a*)

$$C_6H_{12} + 9\ O_2 \longrightarrow 6\ CO_2 + 6\ H_2O$$

(*b*) and (*c*) The equation is the same since both compounds have the same molecular formula:

$$2\ C_6H_{14} + 19\ O_2 \longrightarrow 12\ CO_2 + 14\ H_2O$$

Problem 11.50. Which of the following is the most stable conformation of methylcyclohexane?

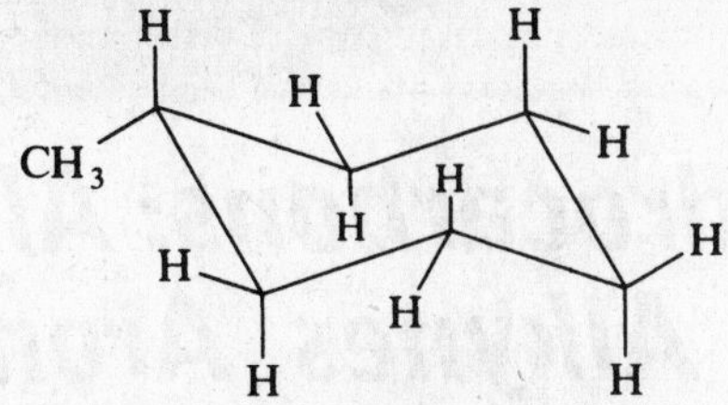

I 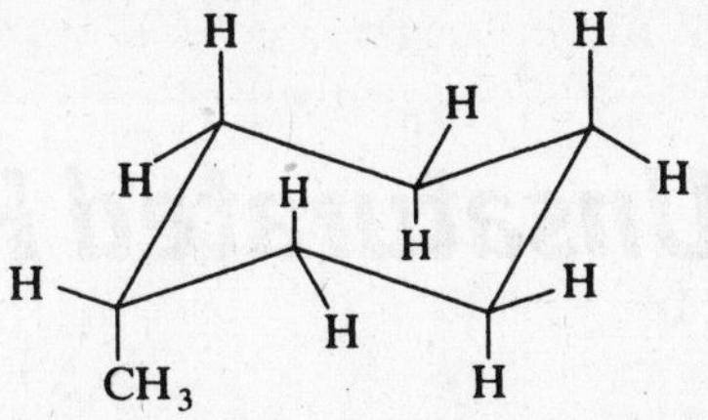 II

Ans. Conformation I since the methyl group is in the more stable equitorial position.

Problem 11.51. Place the following compounds in order of increasing boiling points and explain the order:

$$CH_3-CH_2-CH_2-CH_2-CH_3$$

Pentane

$$CH_3-\underset{}{\overset{CH_3}{\overset{|}{C}}}H-CH_2-CH_3$$

2-Methylbutane

$$CH_3-\overset{CH_3}{\overset{|}{C}}H-CH_3$$

2-Methylpropane

$$CH_3-CH_2-CH_2-CH_3$$

Butane

$$CH_3-CH_2-CH_2-CH_2-CH_2Cl$$

1-Chloropentane

Ans. This problem is difficult because the compounds differ in all three parameters (family, molecular mass, and branching) that determine intermolecular attractions. The approach to use is to compare one compound with another; do not compare more than two at a time. There is no "magic" place to start. Choose any compound and see if you quickly note a comparison with one of the other compounds where the two compounds differ in only one of the three parameters. For example, pentane has a higher BP than butane since it has a higher molecular mass. Intermolecular forces and BP increase with increasing molecular mass. Pentane also has a higher molecular mass and BP compared to 2-methylpropane:

Pentane > butane and 2-methylpropane

Butane has a higher BP than 2-methylpropane since branching (at the same molecular mass) lowers intermolecular attractions:

Pentane > butane > 2-methylpropane

2-Methylbutane has a higher BP than 2-methylpropane because it is similarly branched but has a higher molecular mass. 2-Methylbutane also has a higher BP than butane. A useful analogy for comparing molecules with the same size of longest chain is that of cylindrical-shaped molecules. Butane and 2-methylbutane have the same length of continuous chain but 2-methylbutane has the larger diameter because of branching. The larger diameter means a larger surface area, larger intermolecular forces, and higher BP:

Pentane > 2-methylbutane > butane > 2-methylpropane

1-Chloropentane has a higher BP than pentane since it has a higher molecular mass and is more polar. The overall order of BP is

1-Chloropentane > pentane > 2-methylbutane > butane > 2-methylpropane

Unsaturated Hydrocarbons: Alkenes, Alkynes, Aromatics

12.1 ALKENES

Alkenes are unsaturated hydrocarbons that contain a *carbon-carbon double bond*, i.e., two adjacent carbon atoms are joined by two bonds. Alkenes have the general formula C_nH_{2n}, where n is an integer greater than 1. The simplest member of the alkene family is C_2H_4 (IUPAC name: ethene; common name: ethylene)

$$\begin{array}{ccc} H & & H \\ & \diagdown\!\!\!\!\!\quad C=C \quad\!\!\!\!\!\diagup & \\ H & & H \end{array}$$

followed by

$n = 3$	C_3H_6	$CH_2{=}CH-CH_3$	Propene
$n = 4$	C_4H_8	$CH_2{=}CH-CH_2-CH_3$	1-Butene
$n = 5$	C_5H_{10}	$CH_2{=}CH-CH_2-CH_2-CH_3$	1-Pentene
$n = 6$	C_6H_{12}	$CH_2{=}CH-CH_2-CH_2-CH_2-CH_3$	1-Hexene

Problem 12.1. What is the structural relationship between an alkene and the corresponding alkane (i.e., the alkane with the same number of carbons)?

Ans. The general formula for an alkene (C_nH_{2n}) shows two fewer hydrogen atoms than that for an alkane (C_nH_{2n+2}). We can visualize the formation of an alkene by the loss of two hydrogen atoms from a pair of adjacent carbon atoms:

$$CH_3-CH_2-CH_2-CH_3 \xrightarrow{-2H} CH_2{=}CH-CH_2-CH_3$$

The loss of a hydrogen would leave each carbon atom with only three bonds if nothing else occurred. Recall that the combining power of carbon is 4, not 3. The need for four bonds to each carbon is satisfied by the formation of a second bond between the two carbon atoms. The conversion of an alkane to alkene is an important industrial reaction. It is carried out by heating the alkane in the presence of a metal catalyst. Alkenes are important as intermediate compounds used in the manufacture of a host of plastics, textiles, rubbers, drugs, detergents, and other commercial products.

Problem 12.2. What is the relationship between an alkene and a cycloalkane?

Ans. Alkenes and cycloalkanes have the same general formula and are constitutional isomers of each other. Knowing that the molecular formula of some unknown such as C_5H_{10} fits the general formula C_nH_{2n} tells us that it is not an alkane and that it is either an alkene or a cycloalkane. The molecular formula alone does not allow us to distinguish an alkene from a cycloalkane. However, alkenes and cycloalkanes have very different chemical properties. Cycloalkanes have similar chemical behavior as alkanes since both have the same kinds of bonds (C — C and C — H single bonds). The chemistry of alkenes is very different due to the presence of the carbon-carbon double bond (Sec. 12.6).

12.2 THE CARBON-CARBON DOUBLE BOND

The two bonds of the double bond are not equivalent. One of the bonds is considerably weaker than the other and this is the reason that alkenes have very different chemical behavior than alkanes. The weaker bond breaks in the presence of certain chemical reagents and various chemical reactions take place. The weaker and stronger bonds are referred to as π- (*pi*-) and σ- (*sigma*-) bonds, with bond energies of 250 kJ/mol (60 kcal/mol) and 355 kJ/mol (85 kcal/mol) respectively.

Similarly to our discussion of sp^3 orbitals, the properties of the double bond are incompatible with the ground-state electronic structure of carbon. Three of the four bonds formed by carbon are equivalent to each other. These are σ-bonds and are comparable in strength to the bonds in alkanes, which are also referred to as σ-bonds. One of the three sigma bonds is used to join one carbon to another in the formation of the double bond. The geometry of these three bonds is the *trigonal* geometry, i.e., the three bonds lie in a flat plane with bond angles of 120° (Fig. 12-1). The fourth bond of each carbon of the double bond, used to form the second or π-bond of the double bond, is weaker than a σ-bond.

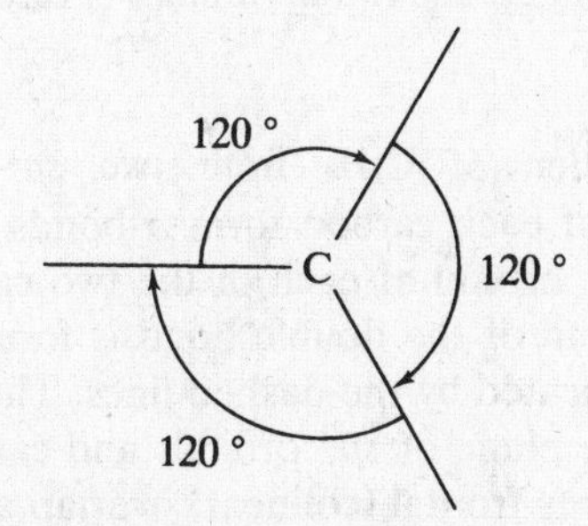

Fig. 12-1 Trigonal bond angles for carbon atoms of a double bond.

Problem 12.3. Describe the orbitals used by carbon in forming the carbon-carbon double bond in ethene, C_2H_4.

Ans. A theoretical description of the double bond that fits the experimental facts requires that we postulate the formation of sp^2 orbitals for carbon through an excitation process analogous to the formation of sp^3 orbitals. The promotion process is exactly the same as for sp^3 orbitals—unpairing of the two electrons in the $2s$ orbital by promotion of one of the electrons to the higher-energy $2p$ orbital (Fig. 12-2). The hybridization (mixing) process is different. It involves mixing of the $2s$ orbital and two (not three) of the $2p$ orbitals to yield three equivalent sp^2 orbitals, each containing one electron, arranged in the trigonal geometry. One of the $2p$ orbitals with one electron is left unhybridized.

Ground state				Excited state
$2p$ ↑ ↑ —	Promotion →	$2p$ ↑ ↑ ↑	Hybridization →	$2p$ ↑
$2s$ ↑↓		$2s$ ↑		sp^2 ↑ ↑ ↑

Fig. 12-2 Formation of sp^2 orbitals for carbon.

Figure 12-3 shows an alternative representation of the hybridization process by which the sp^2 orbitals of carbon are formed. It emphasizes the shapes of the orbitals and their spatial arrangements. The three sp^2 orbitals lie in a flat plane with 120° bond angles. This trigonal geometry is in line with valence shell electron-pair repulsion (VSEPR) theory (Sec. 3.6) since that geometry minimizes repulsions between electrons in the three sp^2 orbitals. Further, VSEPR requires that the unhybridized $2p$ orbital be perpendicular to the plane of the sp^2 orbitals to minimize repulsions between $2p$ and sp^2 orbitals.

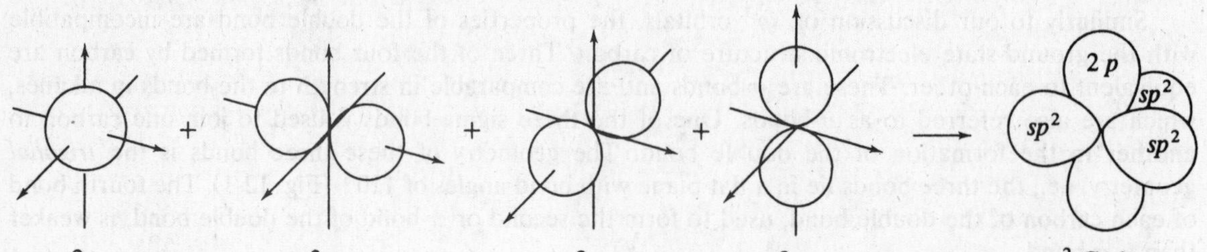

+ + + → 2 p, sp^2, sp^2, sp^2

$2s$ $2p_x$ $2p_y$ $2p_z$ sp^2 Carbon

Fig. 12-3 Hybridization of one $2s$ and two $2p$ orbitals of carbon to form three sp^2 orbitals.

Figure 12-4 shows the formation of C_2H_4 from two sp^2-hybridized carbon atoms and four hydrogen atoms. Two sp^2 orbitals of each carbon form σ-bonds with $1s$ orbitals of hydrogen atoms. Overlap between the remaining sp^2 orbital of each of the two carbons forms the stronger σ-bond of the double bond. The weaker π-bond of the double bond is formed by overlap of the $2p$ orbitals of the two carbons. This overlap is indicated by the dashed lines. The π-bond is relatively weak since the $2p$ orbitals are perpendicular to the plane of sp^2 orbitals and can only overlap in a sidewise manner. [Strong overlap of $2p$ orbitals requires frontal (colinear) overlap as in the formation of the bond in F_2 —see Problem 3.11.]

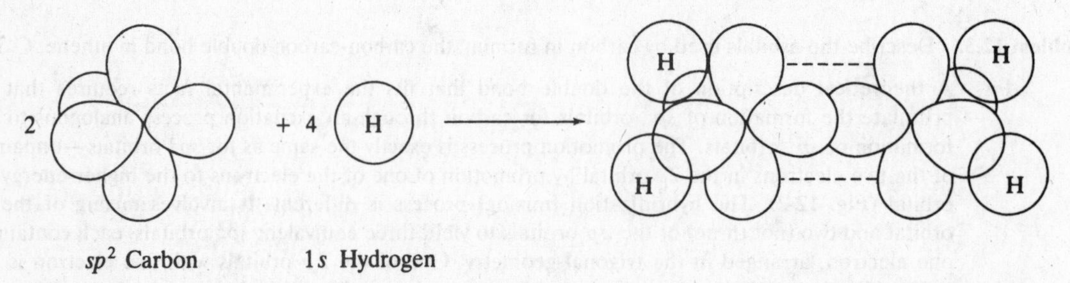

Fig. 12-4 Formation of ethene from two sp^2-hybridized carbons and four hydrogens with $1s$ orbitals.

Problem 12.4. Give the bond angles for A, B, C, and D:

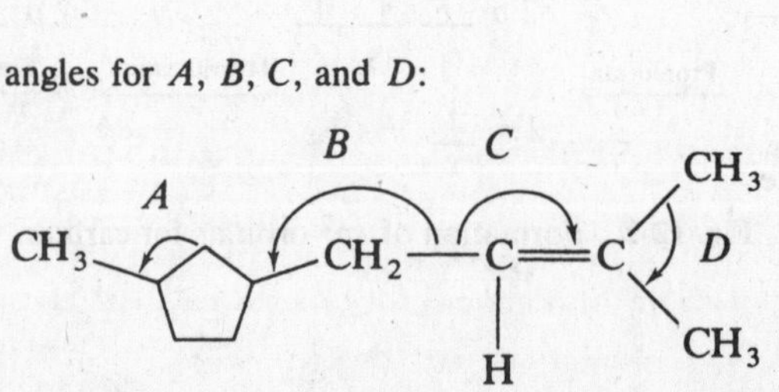

B C
A CH₃
CH_3—(cyclopentane)—CH_2—C═C D
H CH_3

Ans. A bond angle is defined by the angle between two bonds sharing a common or central atom. The bond angle is determined by the hybridization of the central atom. A and B are 109.5° since the central atom is an sp^3 carbon. C and D are 120° since the central atom is an sp^2 carbon.

12.3 CONSTITUTIONAL ISOMERISM IN ALKENES

Constitutional isomerism in alkenes is more complicated than in alkanes. There are two ways in which alkenes can differ in connectivity—not only different carbon skeletons as in alkanes but also different locations of the double bond within any carbon skeleton.

Problem 12.5. How many alkenes exist for the molecular formula C_4H_8? Show only one structural formula for each constitutional isomer.

Ans. Proceed in two steps. First, draw the carbon skeletons possible for molecules containing four carbons. Second, for each carbon skeleton, examine how many different compounds result from placement of the double bond in different locations. There are two different carbon skeletons possible for a C_4 compound:

$$\underset{\text{I}}{\text{C—C—C—C}} \qquad \underset{\text{II}}{\text{C—}\overset{\overset{\text{C}}{|}}{\text{C}}\text{—C}}$$

There are two possible locations for the double bond within skeleton I—between the first two carbons or between the second and third carbon:

$$\underset{\text{I}(a)}{CH_2{=}CH{-}CH_2{-}CH_3} \qquad \underset{\text{I}(b)}{CH_3{-}CH{=}CH{-}CH_3}$$

As always one must guard against drawing duplicate structural formulas of the same isomer. Location of the double bond between the far right pair of carbons gives a duplicate of I(a) while location of the double bond between the second and third carbons from the right-hand side gives a duplicate of I(b). There is only one constitutional isomer with skeleton II:

$$\underset{\text{II}(a)}{CH_2{=}\overset{\overset{CH_3}{|}}{C}{-}CH_3}$$

There are a total of three C_4 alkenes [I(a), I(b), II(a)]. This compares to only two C_4 alkanes (corresponding to skeletons I and II).

Problem 12.6. How many compounds exist for the molecular formula C_4H_8? Show only one structural formula for each isomer.

Ans. This question is broader than Problem 12.5 since it is not limited to alkenes. Alkenes and cycloalkanes are constitutional isomers since both follow the same general molecular formula C_nH_{2n}. The answer to this question consists of the alkenes in the answer to Problem 12.5 plus the cycloalkanes:

□ ▷—CH_3

12.4 NOMENCLATURE OF ALKENES

The IUPAC rules follow those for the alkanes with a few modifications:

1. The longest continuous chain must contain both carbons of the double bond.
2. The ending of the name is changed from *-ane* to *-ene*.
3. The longest continuous chain is numbered from the end which gives the smallest numbers for the carbons of the double bond. The smaller of the two numbers for the double bond is placed in front of the base name of the compound.

Aside from IUPAC names, the simple alkenes (2 to 4 carbons) are also known by *common names*. These names employ the ending *-ylene* instead of *-ene*, e.g., ethylene and propylene for ethene and propene, respectively.

Problem 12.7. Name the following compound by the IUPAC system:

$$\begin{array}{l} CH_2{=}C{-}CH_2{-}CH_3 \\ \quad\ \ | \\ \quad\ \ CH_2{-}CH_2{-}CH_3 \end{array}$$

Ans. The absolute longest continuous chain contains six carbons:

$$\begin{array}{l} \qquad\ \ {\scriptstyle 3}\qquad {\scriptstyle 2}\qquad\ {\scriptstyle 1} \\ CH_2{=}C{-}CH_2{-}CH_3 \\ \quad\ \ | \\ \quad\ \ CH_2{-}CH_2{-}CH_3 \\ \quad\ \ \ {\scriptstyle 4}\qquad\ {\scriptstyle 5}\qquad\ {\scriptstyle 6} \end{array}$$

But it does not contain both carbons of the double bond. The longest chain containing both carbons of the double bond contains five carbons. Numbering from left to right assigns numbers 1 and 2 to the carbons of the double bond. The smaller of the two numbers is used to give 1-pentene as the base name. Note that numbering from right to left is incorrect for this compound since it assigns larger numbers (4 and 5) to the double bond. The complete name is 2-ethyl-1-pentene.

$$\begin{array}{l} \ \ {\scriptstyle 1}\qquad\ {\scriptstyle 2} \\ CH_2{=}C{-}CH_2{-}CH_3 \\ \quad\ \ | \\ \quad\ \ CH_2{-}CH_2{-}CH_3 \\ \quad\ \ \ {\scriptstyle 3}\qquad\ {\scriptstyle 4}\qquad\ {\scriptstyle 5} \end{array}$$

12.5 *CIS-TRANS* ISOMERS

12.5.1 *Alkenes*

Certain alkenes possess an additional type of isomerism, referred to as cis-trans or *geometrical* isomerism. For example, there are two compounds, not one, with structure Kb) (2-butene) of Problem 12.5:

$$\begin{array}{ccc} H & & H \\ & \diagdown\ \ \diagup & \\ & C{=}C & \\ & \diagup\ \ \diagdown & \\ CH_3 & & CH_3 \end{array} \qquad\qquad \begin{array}{ccc} H & & CH_3 \\ & \diagdown\ \ \diagup & \\ & C{=}C & \\ & \diagup\ \ \diagdown & \\ CH_3 & & H \end{array}$$

cis-2-Butene　　　　*trans*-2-Butene

The two compounds, referred to as *cis-trans* (or *geometrical*) *stereoisomers*, are distinguished by using the prefixes *cis-* and *trans-* before the IUPAC name. Cis refers to the isomer where the two similar groups (hydrogens or methyls in this case) are on the same side of the double bond. The *trans* isomer has the two similar groups on opposite sides of the double bond.

Geometrical isomers are one type of stereoisomers; other types are discussed in Chap. 17.

Cis and *trans* isomers are not constitutional isomers since they have exactly the same connectivity (the order of attachment of atoms to each other). *Cis* and *trans* isomers differ only in their *configurations*, the arrangements of atoms or groups of atoms in space, at the carbons of the double bond. The carbons of the double bond are *stereocenters*. A stereocenter is defined as an atom bearing other atoms or groups of atoms whose identities are such that an interchange of two of the groups produces a stereoisomer. Thus, an interchange of the H and CH_3 groups on either carbon of the double bond converts *cis*-2-butene to *trans*-2-b\itene and vice versa. These stereocenters are more precisely called *trigonal stereocenters*.

Problem 12.8. Is the difference between *cis*-2-butene and *trans*-2-butene the same as that between the two conformations of butane shown below?

$$\begin{array}{ccc} & CH_2-CH_2 & \\ CH_3 \diagup & & \diagdown CH_3 \end{array} \qquad \begin{array}{ccc} & & CH_3 \\ & CH_2-CH_2 \diagup & \\ CH_3 \diagup & & \end{array}$$

Ans. The similarity is only superficial. The difference between a pair of different conformations and a pair of geometrical isomers is best seen by asking the question: "What is needed to convert one conformation to the other compared to converting the *cis* isomer to the *trans* isomer?" One conformation of butane is converted to the other by rotation about the middle carbon-carbon single bond. There is very little energy barrier to this rotation. The result is that there is only one compound butane. Any sample of butane contains molecules in which the carbon-carbon bond is undergoing constant rotation and each molecule of butane is passing in and out of the different conformations. The situation is very different for geometrical isomers. *Cis*-2-butene and *trans*-2-butene are different compounds. There is no rotation around the central carbon-carbon bond because that bond is a double bond. The *cis*- and *trans*-2-butenes do not convert one into the other by bond rotation. The only way to convert one into the other is to break bonds from the doubly bonded carbon to the H and CH_3 groups, switch the spatial locations of those groups and then make new bonds to the doubly bonded carbon.

Problem 12.9. What do the following structural formulas represent?

$$\begin{array}{ccc} CH_3 & & H \\ & C{=}C & \\ H & & CH_3 \end{array} \qquad \begin{array}{ccc} H & & CH_3 \\ & C{=}C & \\ CH_3 & & H \end{array}$$

Ans. Both represent the same compound, *trans*-2-butene. Only one of the structural drawings should be shown in answering questions such as "Draw *trans*-2-butene" or "Draw all isomers of C_4H_8." One structural formula is converted into the other by rotating 180° out of the plane of the paper because of the trigonal geometry. The carbons of the double bond and all atoms connected to those carbons lie in the same (flat) plane.

Problem 12.10. Are *cis* and *trans* isomers possible for all alkene structures?

Ans. No. *Cis* and *trans* isomers are possible for an alkene structure only when each carbon of the double bond is bonded to two different substituents. *Cis* and *trans* isomers are not possible when either doubly bonded carbon is bonded to two of the same substituents. Alkenes of the following types cannot exist as *cis* and *trans* isomers

$$\begin{array}{ccc} H & & H \\ & C{=}C & \\ H & & R^1 \end{array} \qquad \begin{array}{ccc} H & & R^2 \\ & C{=}C & \\ H & & R^1 \end{array} \qquad \begin{array}{ccc} H & & R^1 \\ & C{=}C & \\ R^2 & & R^1 \end{array} \qquad \begin{array}{ccc} R^3 & & R^1 \\ & C{=}C & \\ R^2 & & R^1 \end{array}$$

Problem 12.11. Which of the following exist as a pair of *cis* and *trans* isomers? (*a*) 1,1-dichloro-1-butene, (*b*) 1,2-dichloroethene, (*c*) 3-hexene. Show structures of the *cis* and *trans* isomers.

Ans. (*a*) 1,1-Dichloro-1-butene does not exist as *cis* and *trans* isomers since one of the doubly bonded carbons has two of a kind (two chlorines):

$$\begin{array}{ccc} Cl & & H \\ & C{=}C & \\ Cl & & CH_2CH_3 \end{array}$$

(*b*) cis-ClHC=CHCl and trans-ClHC=CHCl

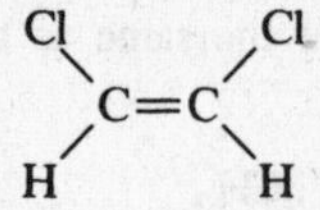

cis-1,2-Dichloroethene 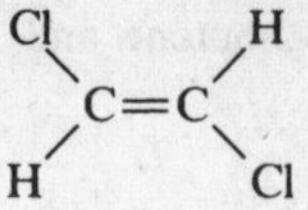 *trans*-1,2-Dichloroethene

(*c*) CH_3CH_2(H)C=C(H)CH_2CH_3 and CH_3CH_2(H)C=C(H)CH_2CH_3

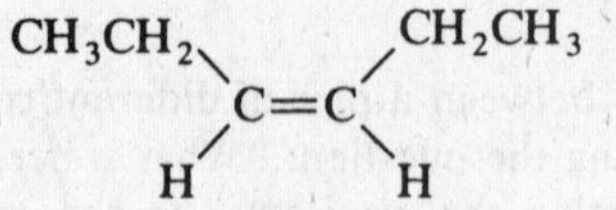

cis-Hexene 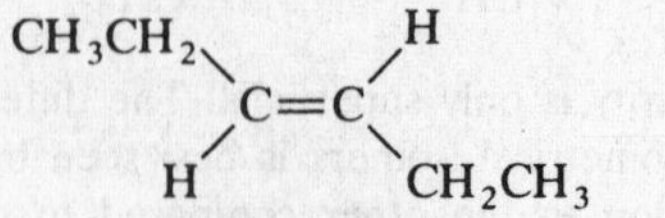 *trans*-Hexene

12.5.2 Cycloalkanes

Geometrical isomers are also possible for cycloalkanes when there are two stereocenters, i.e., when two of the ring carbons are each bonded to two different substituents. Each ring carbon has bonds to two substituents (in addition to the two bonds that are part of the ring). The ring is more or less planar and the substituents at any ring carbon are on opposite sides of the plane of the ring. Substituents on different ring carbons can be on the same (*cis*) or opposite (*trans*) sides of the ring. These stereocenters are more precisely called *tetrahedral stereocenters*.

Problem 12.12. Which of the following exist as *cis* and *trans* isomers? (*a*) 1-methylcyclobutane, (*b*) 1,1,2-trimethylcyclopentane, (*c*) 1,2-dimethylcyclopentane.

Ans. 1-Methylcyclobutane and 1,1,2-trimethylcyclopentane do not exist as *cis* and *trans* isomers since each has only one carbon with two different substituents. Two stereocenters are required for the existence of *cis-trans* isomerism. Three of the four ring carbons of 1-methylcyclobutane have two of the same substituents (two hydrogens). Four of the five ring carbons of 1,1,2-trimethylcyclopentane have two of the same substituents—C-1 has two methyl groups while C-3, C-4, and C-5 each have two hydrogens.

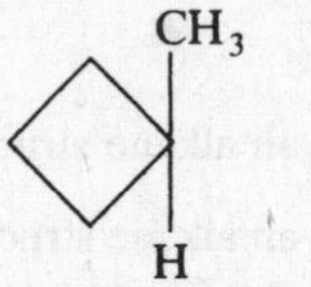

CH_3, H; CH_3, H, CH_3, CH_3

1-Methylcyclobutane 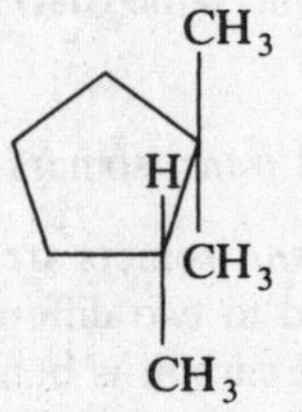 1,1,2-Trimethylcyclopentane

1,2-Dimethylcyclopentane exists as two different compounds since there are two stereocenters:

H, H, CH_3, CH_3; CH_3, H, H, CH_3

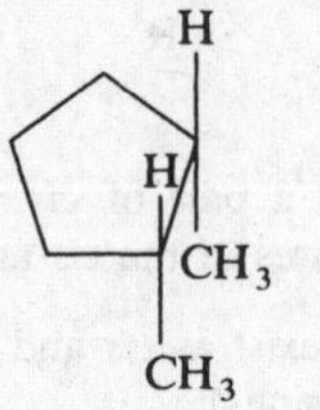

cis 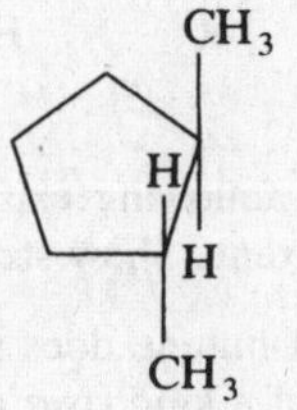 *trans*

Ring carbons C-1 and C-2 each have two different groups (one hydrogen and one methyl). Note that different books use other structural drawings to indicate more clearly that the H and CH_3

groups lie in the opposite plane from the ring. For example, *cis*-1,2-dimethylcyclopentane can be drawn as

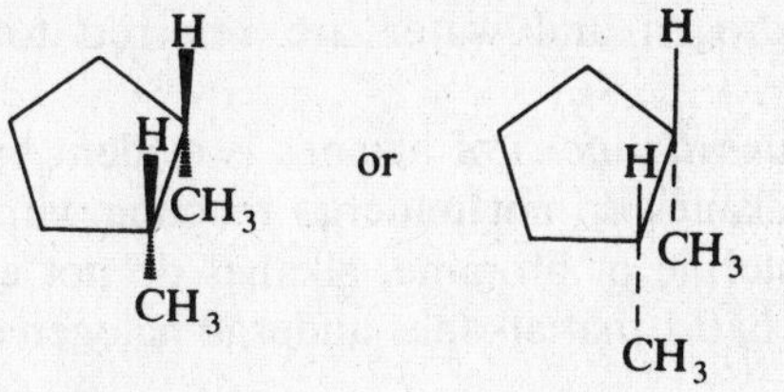

H H CH$_3$ CH$_3$ or H H CH$_3$ CH$_3$

12.6 CHEMICAL REACTIONS OF ALKENES

The physical properties (boiling and melting points and solubility) of an alkene are very similar to those of the corresponding alkane of the same number of carbons. Some differences are described in Section 19.3.1. The chemical properties are very different. Alkenes are highly reactive compared to alkanes due to the presence of the π-bond, which is significantly weaker than a σ-bond.

12.6.1 Addition

A variety of chemical reagents are sufficiently reactive to break the π-bond of an alkene and bring about an *addition reaction*. The reaction proceeds as follows:

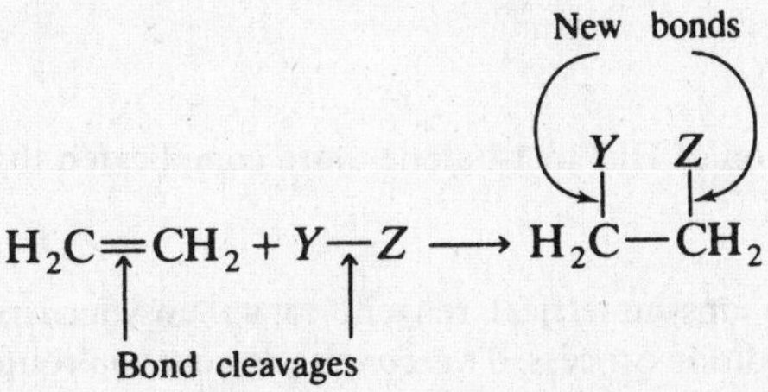

New bonds

$$H_2C{=}CH_2 + Y{-}Z \longrightarrow \underset{}{H_2\overset{Y}{C}{-}\overset{Z}{C}H_2}$$

Bond cleavages

where $Y{-}Z$ symbolizes the reagent. $Y{-}Z$ undergoes cleavage at some bond to produce two reactive fragments Y and Z, which force cleavage of the π-bond of the alkene. This results in the momentary loss of completed valence for Y and Z as well as for the two carbons of the former double bond. Completed valence for all atoms is reestablished by bonding of the Y and Z fragments to the two carbons.

Hydrogen (H_2), halogen (Cl_2 or Br_2), hydrogen halide (HCl, HBr, HI), water, and sulfuric acid (H_2SO_4) act as $Y{-}Z$ reagents:

$$H_2C{=}CH_2 + H{-}H \xrightarrow{Pt} H_2\overset{H}{C}{-}\overset{H}{C}H_2$$

$$H_2C{=}CH_2 + Br{-}Br \longrightarrow H_2\overset{Br}{C}{-}\overset{Br}{C}H_2$$

$$H_2C{=}CH_2 + H{-}Cl \longrightarrow H_2\overset{H}{C}{-}\overset{Cl}{C}H_2$$

$$H_2C{=}CH_2 + H{-}OSO_3H \longrightarrow H_2\overset{H}{C}{-}\overset{OSO_3H}{C}H_2$$

$$H_2C{=}CH_2 + H{-}OH \xrightarrow{H^+} H_2\overset{H}{C}{-}\overset{OH}{C}H_2$$

Pt and H^+ over the reaction arrow in the first and last reactions indicate that platinum is required as catalyst in the first reaction and acid is required as catalyst in the last reaction.

The addition of hydrogen and water are referred to as *hydrogenation* (or *reduction*) and *hydration*, respectively.

The high reactivity (*unsaturation*) of alkenes is evident by the fact that the reactions proceed at ambient temperatures. Alkanes do not undergo reaction with any of these reagents under the same conditions. Except for chlorine or bromine, alkanes do not even undergo reaction at high temperatures. (Recall from Sec. 11.10.1 that alkanes undergo halogen substitution only at high temperature or in the presence of light.)

The addition of bromine to the double bond is useful as a *simple chemical test* for alkenes. Br_2 is red while the product of its addition to an alkene is colorless. This allows one to easily detect by a color change whether some unknown sample is an alkene. If the red color of bromine persists when bromine is added to an unknown, the unknown does not contain a double bond. If the red color disappears, the unknown contains a double bond.

Note that two of the reagents, hydrogen and halogen, are *symmetrical molecules*, i.e., the two fragments Y and Z that add to the double bond are identical. The other reagents are *unsymmetrical*: Y and Z are not identical. It helps in remembering the unsymmetrical reagents to realize that one of the fragments is H for all three reagents. Only Z is different, being halogen, OSO_3H, or OH.

Alkenes are also classified as symmetrical or unsymmetrical depending on whether the substituents on one doubly bonded carbon are identical with those on the other doubly bonded carbon. $CH_2{=}CH_2$, $RCH{=}CHR$, and $R_2C{=}CR_2$ are symmetrical alkenes. $CH_2{=}CHR$, $CH_2{=}CR_2$, and $RCH{=}CR_2$ are unsymmetrical alkenes.

Problem 12.13. Why is the addition of HCl to 1-butene more complicated than the additions of Cl_2 to 1-butene and Cl_2 or HCl to 2-butene?

Ans. The addition of an unsymmetrical reagent to an unsymmetrical alkene is complicated by the *orientation* of the addition process. Two *competing* reaction routes exist. Two different products are possible depending on whether the H or the Cl adds to the end carbon. This type of competitive

$$CH_2{=}CH{-}CH_2{-}CH_3 \xrightarrow{HCl} \underset{\text{2-Chlorobutane}}{\overset{H\quad\ \ Cl}{CH_2{-}CH{-}CH_2{-}CH_3}} + \underset{\text{1-Chlorobutane}}{\overset{Cl\quad\ \ H}{CH_2{-}CH{-}CH_2{-}CH_3}}$$

situation occurs in a number of organic reactions. When two products are possible, some reactions proceed to give more or less equal amounts of the possible products. Other reactions proceed in a *selective* manner—one product is formed in a larger amount and is referred to as the *major product*. The product formed in smaller amount is the *minor product*. For additions of any of the unsymmetrical reagents to an unsymmetrical alkene, one product is formed almost exclusively. 2-Chlorobutane is the major product in the addition of HCl to 1-butene.

When either the reagent or alkene is symmetrical, orientation is not a complication since both orientations of the addition process yield the same product. Only one product is possible for the addition of a symmetrical reagent such as Cl_2 to a symmetrical alkene such as 1-butene or unsymmetrical alkene such as 2-butene:

$$CH_2{=}CH{-}CH_2{-}CH_3 \xrightarrow{Cl_2} \overset{Cl\quad\ \ Cl}{CH_2{-}CH{-}CH_2{-}CH_3}$$

$$CH_3{-}CH{=}CH{-}CH_3 \xrightarrow{Cl_2} CH_3{-}\overset{Cl}{CH}{-}\overset{Cl}{CH}{-}CH_3$$

Only one product is possible in the addition of an unsymmetrical reagent such as HCl to a symmetrical alkene such as 2-butene:

$$CH_3-CH{=}CH-CH_3 \xrightarrow{HCl} CH_3-\underset{}{\overset{H}{\overset{|}{C}}}H-\overset{Cl}{\overset{|}{C}}H-CH_3$$

Problem 12.14. How do we predict which product will be the major product in additions of unsymmetrical reagents such as H — Cl, H — OH, and H — OSO_3H to unsymmetrical alkenes?

Ans. *Markovnikoff's rule* is used: The H fragment from the reagent adds to that carbon of the double bond which already has more hydrogens. The rule is often stated as "The rich (in hydrogens) get richer." For addition of HCl to 1-butene, C-1 receives H from HCl since it has more hydrogens than C-2.

Problem 12.15. Show the product(s) formed when 2-methyl-1-propene reacts with H_2, Cl_2, HCl, H_2O, and H_2SO_4.

Ans. The alkene is unsymmetrical. Only one product is formed in the additions of the symmetrical reagents Cl_2 and H_2:

$$CH_2{=}C(CH_3)_2 \xrightarrow{H_2} \overset{H}{\overset{|}{C}}H_2-\overset{H}{\overset{|}{C}}(CH_3)_2$$

$$CH_2{=}C(CH_3)_2 \xrightarrow{Cl_2} \overset{Cl}{\overset{|}{C}}H_2-\overset{Cl}{\overset{|}{C}}(CH_3)_2$$

Two orientations of the addition process are possible for the reaction with the unsymmetrical reagents HCl, H_2O, and H_2SO_4. Markovnikoff's rule tells us that the major product is the one where the H of the reagent attaches to the carbon of the double bond that has two hydrogens attached. The other doubly bonded carbon has fewer (zero) hydrogens attached:

$$CH_2{=}C(CH_3)_2 \xrightarrow[H^+]{H-OH} \underset{\text{Major}}{\overset{H}{\overset{|}{C}}H_2-\overset{OH}{\overset{|}{C}}(CH_3)_2} + \underset{\text{Minor}}{\overset{OH}{\overset{|}{C}}H_2-\overset{H}{\overset{|}{C}}(CH_3)_2}$$

$$CH_2{=}C(CH_3)_2 \xrightarrow{H-Cl} \underset{\text{Major}}{\overset{H}{\overset{|}{C}}H_2-\overset{Cl}{\overset{|}{C}}(CH_3)_2} + \underset{\text{Minor}}{\overset{Cl}{\overset{|}{C}}H_2-\overset{H}{\overset{|}{C}}(CH_3)_2}$$

$$CH_2{=}C(CH_3)_2 \xrightarrow{H-OSO_3H} \underset{\text{Major}}{\overset{H}{\overset{|}{C}}H_2-\overset{OSO_3H}{\overset{|}{C}}(CH_3)_2} + \underset{\text{Minor}}{\overset{HO_3SO}{\overset{|}{C}}H_2-\overset{H}{\overset{|}{C}}(CH_3)_2}$$

12.6.2 Mechanism of Addition Reactions

The equations shown up to now are *stoichiometric equations* which show what happens in a reaction—what reactants are used up and what products are formed. They are not *mechanistic equations* which show how a reaction occurred. An understanding of a reaction mechanism usually allows one to better predict the outcome of the reaction.

The mechanisms of the addition of halogen, hydrogen halide, water, and sulfuric acid, but not H_2, proceed by a common mechanism. The mechanism involves three steps and is illustrated for the addition of HCl to ethene. First, HCl undergoes bond breakage to form H^+ and Cl^-:

$$H-Cl \longrightarrow H^+ + Cl^-$$

The two electrons in the bond between H and Cl go with Cl because Cl is the more electronegative element. This results in H and Cl becoming H^+ and Cl^-. Second, close approach of H^+ to ethene causes the electrons in the weak π-bond to move away from one of the doubly bonded carbons. The π-electrons bond the H^+ to the carbon of the former double bond:

$$CH_2{=}CH_2 + H^+ \longrightarrow \overset{+}{C}H_2{-}CH_3$$

Carbocation

The curved arrow is used to describe the flow of electrons in the process of bond formation. Since electrons have flowed away from one of the former doubly bonded carbons, that carbon is left deficient in electrons and is now shown as carrying a + charge. The positively charged species is referred to as a *carbocation*. Third, there is an attraction followed by bond formation between the positive carbocation and negative chloride ion:

$$\overset{+}{C}H_2{-}CH_3 + Cl^- \longrightarrow Cl{-}CH_2{-}CH_3$$

Carbocation

Carbocations are reactive intermediates formed in the course of many organic reactions. Competitive reactions proceeding via carbocations are understood when one recognizes that carbocations differ considerably in stability. The stability of carbocations is 3° > 2° > 1°:

$$\underset{3°}{R{-}\overset{+}{\underset{R}{C}}{-}R} > \underset{2°}{R{-}\overset{+}{\underset{H}{C}}{-}R} > \underset{1°}{R{-}\overset{+}{\underset{H}{C}}{-}H}$$

Tertiary, secondary, and primary carbocations are defined by classifying the carbon carrying the + charge according to whether there are three, two, or one atoms other than H directly attached to that carbon (Sec. 11.7.3). The more stable a carbocation, the more easily and the faster it is formed. Reactions that proceed via more stable carbocations occur at faster rates.

Problem 12.16. Show how carbocation stability explains Markovnikoff's rule in the addition of HCl to propene.

Ans. We start by writing the mechanism for each of the two orientations in the addition of HCl. One of the pathways proceeds via a secondary carbocation while the other proceeds via a primary carbocation. Two competing reactions occur. The faster reaction is the one that proceeds via the more stable carbocation. The overall result is that much more 2-chloropropane is formed than 1-chloropropane. This is the result predicted by Markovnikoff's rule.

$$CH_2{=}CH{-}CH_3 \xrightarrow{H^+} \underset{2°\ \text{Carbocation}}{CH_3{-}\overset{+}{C}H{-}CH_3} \xrightarrow{Cl^-} CH_3{-}\overset{Cl}{\overset{|}{C}H}{-}CH_3$$

$$CH_2{=}CH{-}CH_3 \xrightarrow{H^+} \underset{1°\ \text{Carbocation}}{\overset{+}{C}H_2{-}CH_2{-}CH_3} \xrightarrow{Cl^-} \overset{Cl}{\overset{|}{C}H_2}{-}CH_2{-}CH_3$$

12.6.3 Polymerization

Polymerization is the reaction whereby many hundreds or thousands of alkene molecules, referred to as *monomer*, add to each other to form high molecular weight molecules referred to as *polymer*. The weak π-bonds of the alkene molecules break and this is followed by bond formation between carbons of adjacent alkene molecules. A small portion (four monomer molecules long) of a polymerization is shown below for ethene:

$$\underset{\text{Monomer}}{CH_2{=}CH_2 + CH_2{=}CH_2 + CH_2{=}CH_2 + CH_2{=}CH_2}, \text{ etc.} \longrightarrow$$

$$\underset{\text{Polymer}}{\sim\sim CH_2{-}CH_2{-}CH_2{-}CH_2{-}CH_2{-}CH_2{-}CH_2{-}CH_2 \sim\sim}$$

The more usual way of showing a polymerization reaction is

$$\underset{\text{Ethylene}}{nCH_2{=}CH_2} \xrightarrow[\text{catalyst}]{\text{polymerization}} \underset{\text{Polyethylene}}{\left(CH_2{-}CH_2\right)_n}$$

where n is a large number.

Alkenes with various substituents also undergo polymerization. Examples include the following:

$$\underset{\text{Propylene}}{nCH_2{=}CH(CH_3)} \longrightarrow \underset{\text{Polypropylene}}{\left(CH_2{-}CH(CH_3)\right)_n}$$

$$\underset{\text{Vinyl chloride}}{nCH_2{=}CH(Cl)} \longrightarrow \underset{\text{Poly(vinyl chloride)}}{\left(CH_2{-}CH(Cl)\right)_n}$$

$$\underset{\text{Styrene}}{nCH_2{=}CH(C_6H_5)} \longrightarrow \underset{\text{Polystyrene}}{\left(CH_2{-}CH(C_6H_5)\right)_n}$$

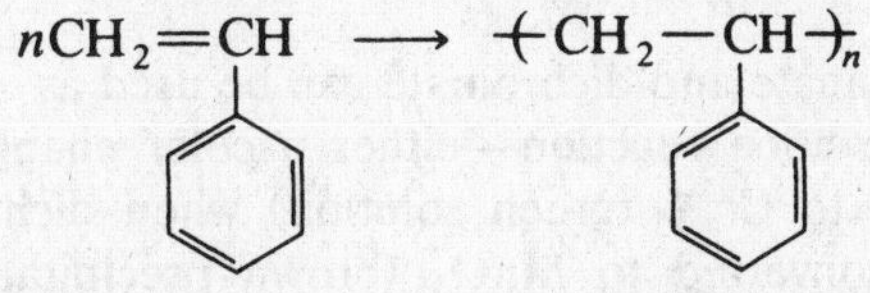

$$\underset{\text{Tetrafluoroethylene}}{nCF_2{=}CF_2} \longrightarrow \underset{\text{Polytetrafluoroethylene}}{\left(CF_2{-}CF_2\right)_n}$$

Polyethylene is used for milk, food, and detergent bottles. Polypropylene is used for outdoor carpeting for home and sports stadia. Poly(vinyl chloride) is used for home siding and plastic pipe. Polystyrene is used for hot drinking cups and fast-food containers. Polytetrafluoroethylene is used for nonstick cookware. Polymerization of these and other alkenes on an industrial scale is responsible for producing over 50 billion lb per year of plastic and elastomer products in the United States.

Polymerization requires the presence of a catalyst, the particular catalyst depending on the monomer being polymerized. The value of n, which determines the molecular mass of the polymer, depends on the reaction conditions. Polymerization occurs only when reagents such as hydrogen, halogen, hydrogen halide, water, or sulfuric acid are not present in large amounts. Addition predominates over polymerization in the presence of those reagents.

Problem 12.17. Methane, octane, and polyethylene are members of the alkane family. Combustion of methane (gas heating) and octane (liquid-burning engines) are used as sources of energy. In contrast polyethylene is used for producing objects such as milk, food, and detergent bottles. What macroscopic property is responsible for this difference? What molecular-level property determines this difference?

Ans. The physical states and physical properties of members of the alkane or any other family depend on molecular mass. Polyethylene is a solid while methane is a gas and octane is a liquid. A solid can be fabricated into objects that have physical integrity (strength). You cannot make a milk bottle out of a gas (methane) or liquid (octane). The difference in physical strength is, like boiling and melting points and solubility, determined by the level of intermolecular forces. The intermolecular forces in polyethylene are much higher than in methane and octane because of the much higher molecular mass.

12.6.4 Oxidation

Alkenes, like all hydrocarbons, undergo combustion to yield CO_2 and H_2O. Combustion is the most extreme form of oxidation possible. Each of the carbon atoms of the alkene is oxidized to its highest oxidation state, i.e., each carbon ends up with the maximum number of bonds to oxygen—four—in forming CO_2. Mild oxidations of alkene are possible using certain oxidizing agents. A *mild oxidation* is an oxidation in which only some of the carbons of the compound undergo oxidation. For alkenes, mild oxidation involves only the carbons of the double bond. Oxidation at ambient or lower temperature with dilute potassium permanganate in basic solution results in the addition of hydroxyl groups to both carbons of the double bond:

$$CH_3-CH=CH-CH_2CH_3 \xrightarrow[\text{cold, base}]{\text{dil } KMnO_4} CH_3-\underset{}{\overset{OH}{\overset{|}{C}H}}-\overset{OH}{\overset{|}{C}H}-CH_2CH_3$$

Oxidation at elevated temperature with either basic permanganate or acidic dichromate brings about cleavage of the molecule at the double bond, without cleavage of other carbon-carbon bonds. The products are carboxylic acids:

$$CH_3-CH=CH-CH_2CH_3 \xrightarrow[\text{hot}]{KMnO_4 \text{ or } K_2Cr_2O_7} CH_3-\overset{OH}{\overset{|}{C}}=O + O=\overset{OH}{\overset{|}{C}H}-CH_2CH_3$$

The oxidations by permanganate and dichromate can be used as *simple chemical tests* for alkenes since there is a visible change upon reaction—either a color change or a precipitation. $Cr_2O_7^{2-}$ (orange solution) is converted to Cr^{3+} (green solution) when dichromate is the oxidizing agent. MnO_4^- (purple solution) is converted to MnO_2 (brown precipitate) when permanganate is the oxidizing agent.

Problem 12.18. Complete each of the following reactions by writing the structure of the organic product. If no reaction occurs, write "No Reaction." If there is more than one product, indicate which are the major and which the minor products.

(*a*) $CH_3-CH=CH_2 \xrightarrow{HCl} ?$ (*b*) $CH_3-CH=CH_2 \xrightarrow{NaOH} ?$

(*c*) $CH_3-CH=CH_2 \xrightarrow{\text{combustion}} ?$ (*d*) $CH_3-CH=CH_2 \xrightarrow{H_2O,\ H^+} ?$

(*e*) $nCH_2=\overset{CH_3}{\overset{|}{\underset{\underset{OCH_3}{\underset{|}{C=O}}}{\underset{|}{C}}}} \xrightarrow{\text{polymerization}} ?$ (*f*) $CH_3-CH=CH_2 \xrightarrow[\text{cold, base}]{\text{dil } KMnO_4} ?$

(*g*) $CH_3-CH=CH-CH_3 \xrightarrow{H_2SO_4} ?$ (*h*) $CH_3-CH=CH-CH_3 \xrightarrow[\text{hot, acid}]{K_2Cr_2O_7} ?$

(*i*) $CH_2=CH-CH=CH_2 \xrightarrow{Br_2} ?$

Ans. There is No Reaction for (*b*). All others undergo reaction. Markovnikoff's rule is applicable to (*a*) and (*d*) since both alkene and reagent are unsymmetrical.

(*a*) $$CH_3-CH=CH_2 \xrightarrow{HCl} \underset{\text{Major}}{CH_3-\overset{Cl}{\overset{|}{C}H}-CH_3} + \underset{\text{Minor}}{CH_3-CH_2-CH_2-Cl}$$

(c) $2\,CH_3{-}CH{=}CH_2 + 9\,O_2 \xrightarrow{\text{combustion}} 3\,CO_2 + 3\,H_2O$

(d) $CH_3{-}CH{=}CH_2 \xrightarrow{H_2O,\,H^+} CH_3{-}\underset{}{\overset{OH}{\overset{|}{CH}}}{-}CH_3 + CH_3{-}CH_2{-}CH_2{-}OH$

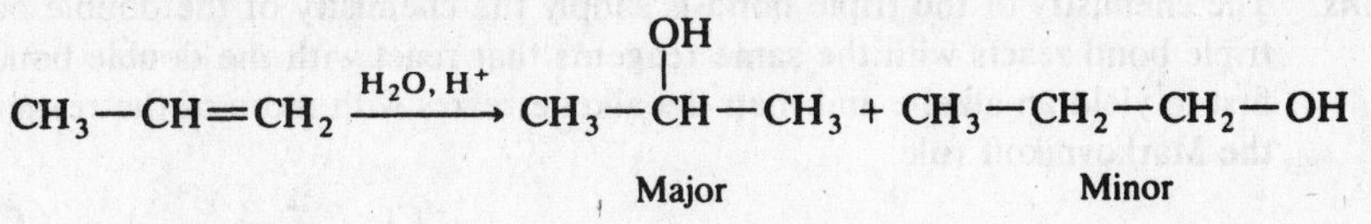

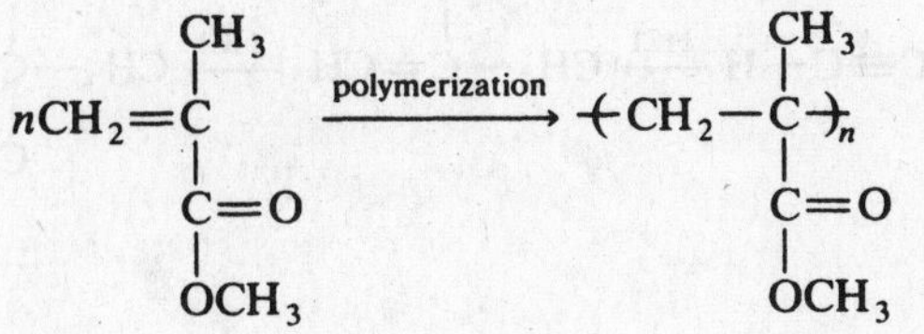

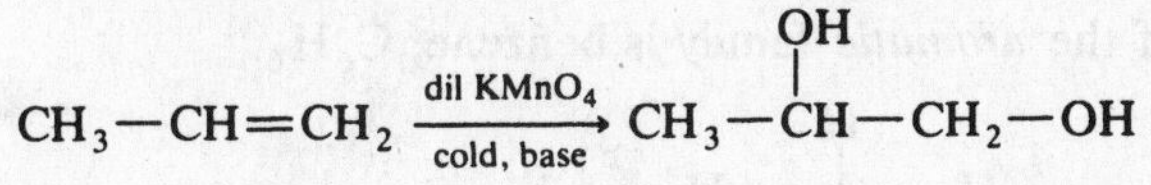

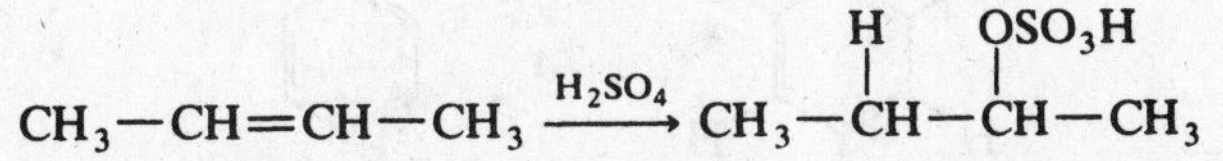

Major Minor

(e) $n\,CH_2{=}\overset{CH_3}{\overset{|}{C}}{-}C({=}O){-}OCH_3 \xrightarrow{\text{polymerization}} {+}CH_2{-}\overset{CH_3}{\overset{|}{C}}{+}_n$ (with $C{=}O$, OCH_3 on the polymer carbon)

(f) $CH_3{-}CH{=}CH_2 \xrightarrow[\text{cold, base}]{\text{dil }KMnO_4} CH_3{-}\overset{OH}{\overset{|}{CH}}{-}CH_2{-}OH$

(g) $CH_3{-}CH{=}CH{-}CH_3 \xrightarrow{H_2SO_4} CH_3{-}\overset{H}{\overset{|}{CH}}{-}\overset{OSO_3H}{\overset{|}{CH}}{-}CH_3$

(h) $CH_3{-}CH{=}CH{-}CH_3 \xrightarrow[\text{hot, acid}]{K_2Cr_2O_7} 2\,CH_3{-}\overset{OH}{\overset{|}{C}}{=}O$

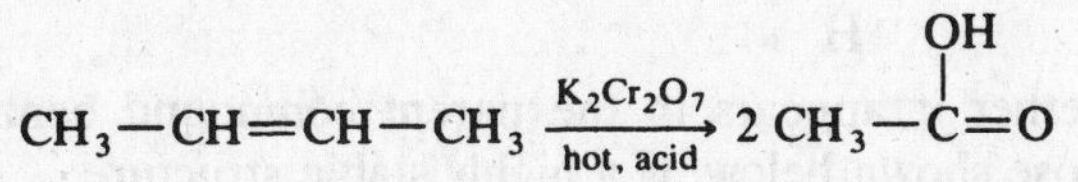

Part (*i*) involves reaction of a diene, 1,3-butadiene, a molecule with two double bonds. There is nothing special here as each double bond undergoes the reaction of a double bond:

(i) $CH_2{=}CH{-}CH{=}CH_2 \xrightarrow{Br_2} \overset{Br}{\overset{|}{CH_2}}{-}\overset{Br}{\overset{|}{CH}}{-}\overset{Br}{\overset{|}{CH}}{-}\overset{Br}{\overset{|}{CH_2}}$

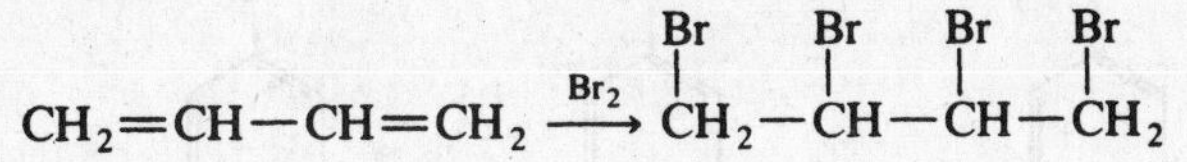

12.7 ALKYNES

Alkynes are unsaturated hydrocarbons that contain a carbon-carbon triple bond, i.e., two adjacent carbon atoms are joined by three bonds. Alkynes have the general formula C_nH_{2n-2} where n is an integer greater than 1. The simplest member of the alkyne family is C_2H_2 (IUPAC name: ethyne; common name: acetylene):

$$H{-}C{\equiv}C{-}H$$

Unlike the alkene double bond, there is no possibility of *cis* and *trans* isomers at the triple bond since atoms attached at the carbons of a triple bond lie in a straight line with the triple bond.

There is very little new to learn here with regard to nomenclature and properties. If you know the alkenes, you essentially know the alkynes.

Problem 12.19. Name the following by the IUPAC system:

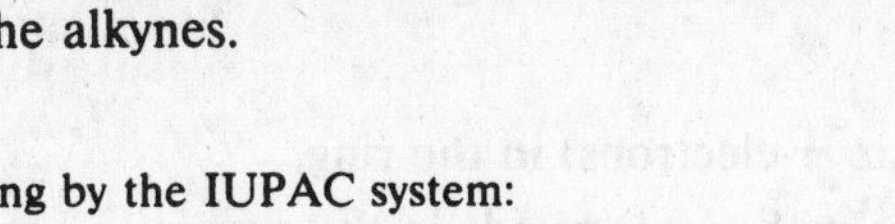

$$CH_3{-}CH_2{-}CH_2{-}\underset{CH(CH_3)_2}{\underset{|}{CH}}{-}C{\equiv}C{-}CH_3$$

Ans. The IUPAC rules for alkynes are exactly the same as those for alkenes except that the ending *-ene* is changed to *-yne*. The compound is 4-isopropyl-2-heptyne with the numbering of the carbon chain shown below:

$$\overset{7}{CH_3}{-}\overset{6}{CH_2}{-}\overset{5}{CH_2}{-}\underset{CH(CH_3)_2}{\underset{|}{\overset{4}{CH}}}{-}\overset{3}{C}{\equiv}\overset{2}{C}{-}\overset{1}{CH_3}$$

Problem 12.20. Show the product formed when propyne reacts with excess HCl.

Ans. The chemistry of the triple bond is simply the chemistry of the double bond multiplied by 2. The triple bond reacts with the same reagents that react with the double bond. The reaction proceeds first to yield an alkene and then the alkene reacts with more of the reagent. Each addition follows the Markovnikoff rule.

$$CH_3-C\equiv C-H \xrightarrow{HCl} CH_3-\underset{}{\overset{Cl}{\overset{|}{C}}}=CH_2 \xrightarrow{HCl} CH_3-\underset{\underset{Cl}{|}}{\overset{Cl}{\overset{|}{C}}}-CH_3$$

12.8 AROMATICS

The simplest member of the *aromatic* family is benzene, C_6H_6:

C_6H_6 or ⌬

The *benzene structure*, whether it appears in the parent compound benzene or in substituted benzene compounds such as those shown below, is a highly stable structure.

Cl	CH_3	OH	$\overset{O}{\overset{\|}{C}}-OH$	NH_2
Chlorobenzene	Toluene Methylbenzene	Phenol Hydroxybenzene	Benzoic acid Carboxybenzene	Aniline Aminobenzene

Aromatic compounds behave very differently from what is expected of alkenes. The double bonds in the benzene ring do not act like the double bonds in alkenes. Benzene is highly resistant to the reagents that add to alkene double bonds. Benzene will add hydrogen but much higher reaction temperatures are required (200 °C versus ambient temperatures for alkenes). Benzene will not undergo addition with halogen, hydrogen halide, sulfuric acid, or water under any conditions. The high stability of benzene also results in its combustion reaction being much less exothermic than that of alkanes and alkenes. These properties of the benzene structure define the behavior referred to as *aromaticity*.

Aromaticity requires all three of the following structural features:

1. Cyclic structure.
2. Three double bonds (six π-electrons) in the ring.
3. Double bonds alternate with single bonds in the ring.

Problem 12.21. Which of the following compounds are aromatic?

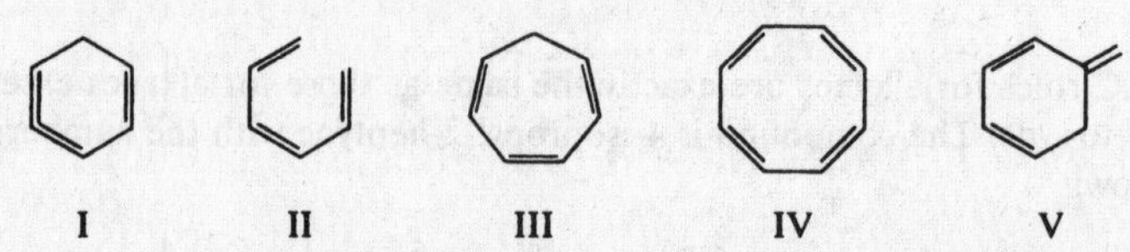

Ans. None of these compounds are aromatic. None of the compounds possess all three structural features necessary for aromatic behavior. Two of the double bonds in compound I do not alternate

with a single bond. Compound II has three double bonds alternating with single bonds but is not cyclic. There are two single bonds, instead of one, between two of the double bonds in compound III. Compound IV has four double bonds, instead of three, alternating with single bonds in a cyclic structure. Compound V has only two double bonds in the ring.

Problem 12.22. **Why do three double bonds alternating with single bonds in a cyclic arrangement have properties very different from alkenes?**

Ans. **In alkenes, *p* orbitals on adjacent sp^2 carbons overlap and share two electrons to form a π-bond. Each ring carbon in benzene is sp^2-hybridized. Every *p* orbital overlaps, not just with one other *p* orbital, but with *p* orbitals on each side. This results in a highly stable molecular orbital because there is a continuous and circular overlap of six successive *p* orbitals. The six π-electrons are *delocalized* over the whole molecule instead of being *localized* into three separate double bonds.**

Problem 12.23. **What do each of the following structural drawings represent?**

Ans. **A superficial look indicates the two structures are different. There is a double bond between the carbons holding the chlorines in the left structure, but only a single bond in the right structure. However, there is no difference in reality between any of the carbons in the ring and the bonding between any pair of adjacent carbons is identical. This is the meaning of the description in the answer to Problem 12.22. The two structures represent the same compound. Chemists sometimes draw the two structures separated by a two-headed arrow (as shown above) and describe the compound as being a *resonance hybrid* of the two structures. Chemists often (but not always) use an alternative structural drawing for the benzene ring to avoid this ambiguity. The benzene ring is shown as a hexagon with a circle inside it:**

12.9 NOMENCLATURE OF AROMATIC COMPOUNDS

Monosubstituted benzene compounds are named in the IUPAC system by placing the name of the substituent before *-benzene*—for example, the compounds chlorobenzene, methylbenzene, hydroxybenzene, carboxybenzene, and aminobenzene shown in Sec. 12.8. Some of these compounds have common names, e.g., toluene for methylbenzene, phenol for hydroxybenzene, benzoic acid for carboxybenzene, and aniline for aminobenzene. IUPAC names for disubstituted benzenes (which are constitutional isomers) follow the nomenclature for the corresponding cyclohexane compounds except there is an alternative to the numbering system for designating the positions of substituents on the ring. The 1,2-, 1,3-, and 1,4- are alternatively described by the prefixes *o*- for *ortho*-, *m*- for *meta*-, and *p*- for *para*-. There is no corresponding alternative to the numbering system for benzene compounds containing more than two substituents.

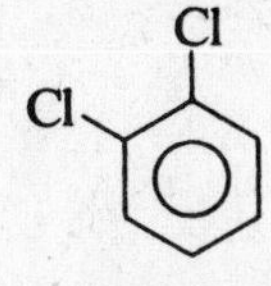

1,2-Dichlorobenzene
***o*-Dichlorobenzene**

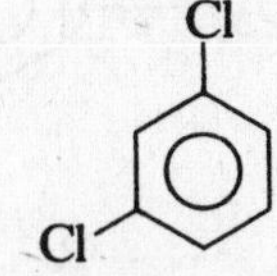

1,3-Dichlorobenzene
***m*-Dichlorobenzene**

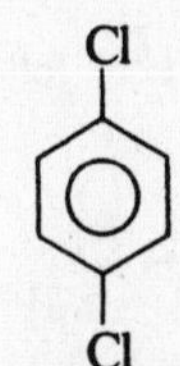

1,4-Dichlorobenzene
***p*-Dichlorobenzene**

Problem 12.24. Name each of the following by the IUPAC system:

(*a*) HO, $CH(CH_3)_2$ on benzene ring (*b*) CH_3, CH_2CH_3, Cl on benzene ring (*c*) $CH{=}CH{-}CH_2CH_3$ on benzene ring

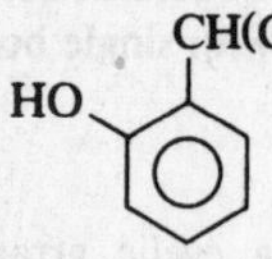

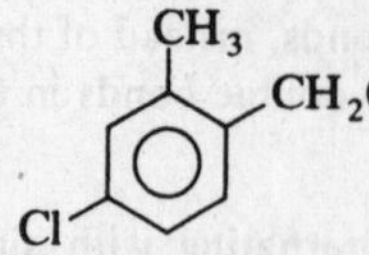

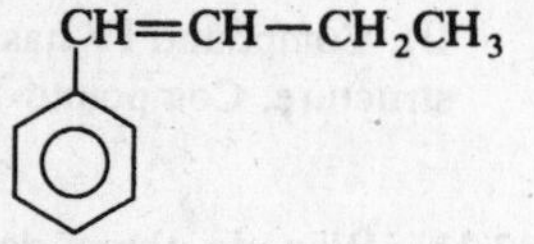

Ans. (*a*) 1-hydroxy-2-isopropylbenzene or *o*-hydroxy-isopropylbenzene. An alternative is to use *phenol* as the base name for hydroxybenzene, which yields the name 2-isopropylphenol or *o*-isopropylphenol.

(*b*) The smallest set of numbers (1,2,4) is obtained by starting at the ethyl group and proceeding counterclockwise. The name is 4-chloro-1-ethyl-2-methylbenzene.

(*c*) It is difficult to name this compound as a substituted benzene since the nonaromatic portion is not one of the simple alkyl groups. The compound is more easily named as a derivative of the alkene with the benzene ring noted as a substituent, referred to as *phenyl*. The name is 1-phenyl-1-butene.

The benzene ring in a compound such as 1-phenyl-1-butene is often abbreviated by C_6H_5, Ph or the Greek letter ϕ:

$$C_6H_5{-}CH{=}CH{-}CH_2CH_3 \quad \text{or} \quad Ph{-}CH{=}CH{-}CH_2CH_3 \quad \text{or} \quad \phi{-}CH{=}CH{-}CH_2CH_3$$

A more general abbreviation for aromatic systems is the use of Ar. For example, Ar—Cl represents any aromatic compound with a chlorine substituted on it.

12.10 REACTIONS OF BENZENE

Benzene undergoes many substitution reactions in which some group such as halogen, sulfonic acid (SO_3H), nitro (NO_2), or alkyl replaces hydrogen:

Halogenation

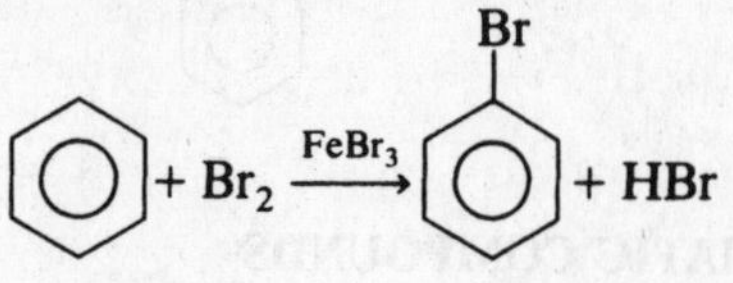

$$C_6H_6 + Br_2 \xrightarrow{FeBr_3} C_6H_5Br + HBr$$

Sulfonation

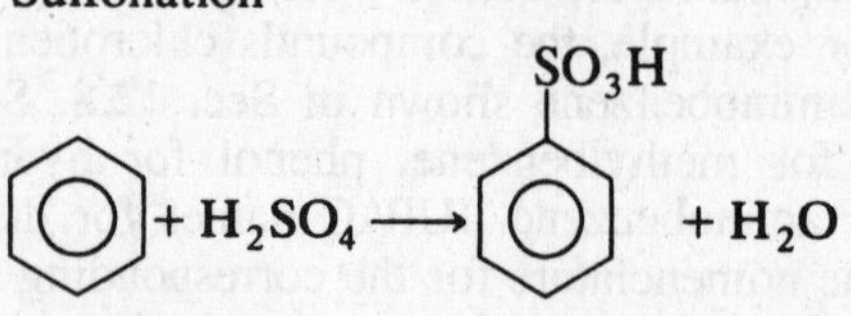

$$C_6H_6 + H_2SO_4 \longrightarrow C_6H_5SO_3H + H_2O$$

Nitration

$$C_6H_6 + HNO_3 \xrightarrow{H_2SO_4} C_6H_5NO_2 + H_2O$$

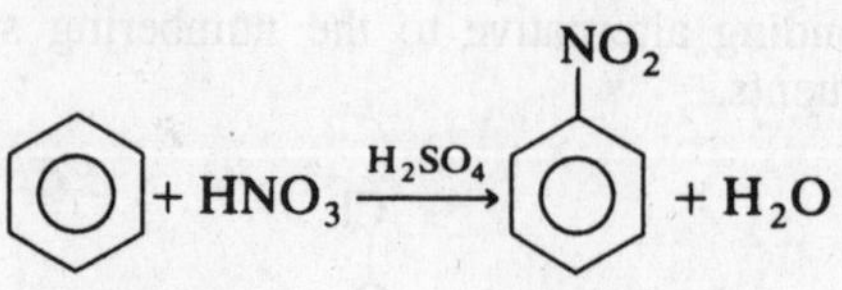

Alkylation

$$C_6H_6 + RCl \xrightarrow{AlCl_3} C_6H_5R + HCl$$

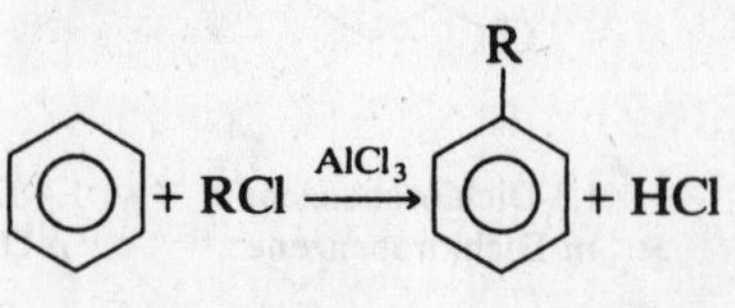

Note that metal halides such as $FeBr_3$ and $AlCl_3$ are catalysts for halogenation and alkylation, and sulfuric acid is a catalyst for nitration. These reactions do not contradict the earlier discussion on the high stability of the aromatic structure. The aromatic structure is intact after these reactions. All of the reagents in these reactions bring about addition to alkenes (including the nitration and alkylation reagents which were not previously discussed in Chap. 11), but there is no addition to the double bonds in benzene, only substitution.

Problem 12.25. What is substituting for what in the sulfonation of benzene?

Ans. The reaction is more clearly seen by using expanded formulas for both benzene and sulfuric acid:

Bonds broken — Bonds formed

$C_6H_6 + HO\text{—}SO_3H \longrightarrow C_6H_5SO_3H + H\text{—}OH$

The aromatic C—H and O—S bonds are broken. The SO_3H fragment bonds to the aromatic C to yield benzenesulfonic acid while the H and OH fragments form water.

Problem 12.26. Is the high stability of the aromatic structure evident in other reactions?

Ans. Alkylbenzenes undergo oxidation with strong oxidizing agents such as hot $KMnO_4$ or $K_2Cr_2O_7$ under acidic conditions. The benzene ring is unchanged but the alkyl group is oxidized.

$$C_6H_5\text{—}CH_2CH_2CH_3 \xrightarrow{(O)} C_6H_5\text{—}C(=O)\text{—}OH$$

All carbons of the alkyl group except that attached to the ring are oxidized to CO_2. The carbon attached to the ring is oxidized to the highest oxidation state possible (COOH group) without breakage of the bond to the aromatic ring. Breaking that bond results subsequently in destroying the aromatic system by oxidation of the ring carbons. Note the use of (O) to indicate reaction conditions for oxidations. This is often done but is not as informative as giving the specific reagents and reaction conditions. Complete oxidation of all carbons, including the aromatic carbons, can be accomplished by more powerful oxidizing agents (including combustion).

Benzene undergoes addition of hydrogen to each of its double bonds to yield cyclohexane. This is the only addition reaction of benzene. The resistance of the aromatic system to reaction is evident because a temperature of 200 °C is required for reaction compared to ambient temperature for hydrogenation of alkenes.

$$C_6H_6 + 3H_2 \xrightarrow[200\ ^\circ C]{Pt} C_6H_{12}$$

ADDITIONAL SOLVED PROBLEMS

ALKENES

Problem 12.27. Which of the following are correct molecular formulas? (*a*) C_5H_{10}, (*b*) C_4H_8Br, (*c*) C_4H_7Br.

Ans. Compare each formula to the general formulas C_nH_{2n+2} (alkanes) and C_nH_{2n} (alkenes and cycloalkanes). C_5H_{10} is a correct molecular formula since it fits the general formula for an alkene or cycloalkane.

Halogens in organic compounds are "counted" as hydrogens with regard to a comparison with general formulas for different families. Halogen and hydrogen are each monovalent, which means that they are one-to-one replacements for each other. C_4H_8Br is equivalent to C_4H_9, which does not fit either general formula. C_4H_8Br is not a correct molecular formula.

C_4H_7Br is equivalent to C_4H_8, which fits the general formula for an alkene or cycloalkane, and is a correct molecular formula.

THE CARBON-CARBON DOUBLE BOND

Problem 12.28. For each of the following compounds, what is the hybridization of the carbons indicated by arrows? What are the bond angles about those carbons?

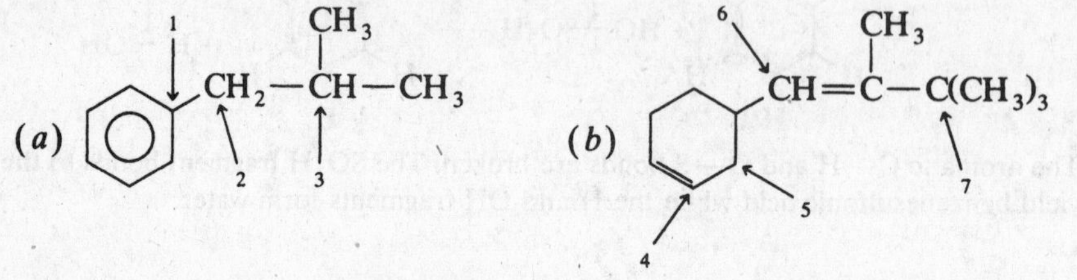

(a) C_6H_5–CH_2–CH(CH_3)–CH_3 (arrows: 1 at ring carbon, 2 at CH_2, 3 at CH)

(b) cyclohexenyl–CH=C(CH_3)–C(CH_3)$_3$ (arrows: 4 and 5 at ring carbons, 6 at CH=, 7 at $C(CH_3)_3$)

Ans. Carbons 1, 4, and 6 are carbons of a double bond, which means they are sp^2-hybridized with 120° bond angles. Carbons 2, 3, 5, and 7 are in carbon-carbon single bonds, which means they are sp^3-hybridized with 109.5° bond angles.

CONSTITUTIONAL ISOMERISM IN ALKENES

Problem 12.29. How many alkenes exist for the molecular formula C_5H_{10}? Show only one structural formula for each constitutional isomer.

Ans. There are three different carbon skeletons possible for a C_5 compound:

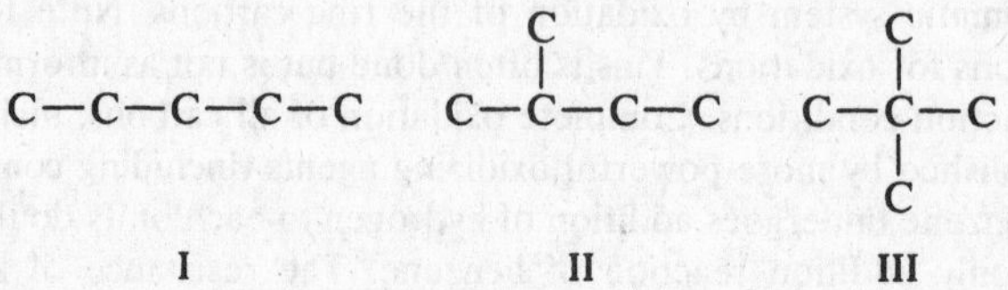

I: C–C–C–C–C

II: C–C(C)–C–C

III: C–C(C)(C)–C

There are two possible locations for the double bond within skeleton I—between the first two carbons or between the second and third carbon:

I(a): CH_2=CH–CH_2–CH_2–CH_3

I(b): CH_3–CH=CH–CH_2–CH_3

As always one must guard against drawing duplicate structural formulas of the same isomer. Location of the double bond between the far right pair of carbons gives a duplicate of I(*a*) while location of the double bond between the second and third carbons from the right-hand side gives a duplicate of I(*b*). There are three constitutional isomers with skeleton II:

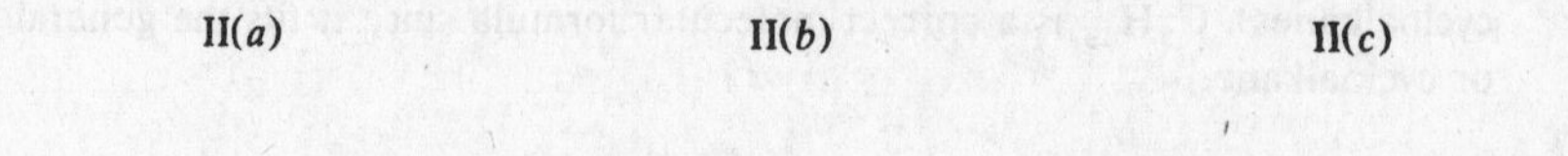

II(a): CH_2=C(CH_3)–CH_2–CH_3

II(b): CH_3–C(CH_3)=CH–CH_3

II(c): CH_3–CH(CH_3)–CH=CH_2

There are no alkenes with skeleton III as it is not possible to have a double bond in that structure because carbon is tetravalent. Structure III(*a*) has five bonds to the central carbon.

$$\begin{array}{c} CH_3 \\ | \\ CH_3{-}C{=}CH_2 \\ | \\ CH_3 \end{array}$$

III(*a*)

There are five C_5 alkenes [I(*a*), I(*b*), II(*a*), II(*b*), II(*c*)]. This compares to only three C_5 alkanes (corresponding to skeletons I, II, III).

Problem 12.30. How many constitutional isomers are possible for the molecular formula C_5H_{10}? Show only one structural formula for each isomer.

Ans. This question is broader than Problem 12.29 since it is not limited to alkenes. Alkenes and cycloalkanes are isomers since both follow the same general molecular formula C_nH_{2n}. The answer to this question is the sum of the answers to Problems 11.11 (C_5H_{10} cycloalkanes) and 12.29 (C_5H_{10} alkenes).

NOMENCLATURE OF ALKENES

Problem 12.31. Name the following compound by the IUPAC system:

$$\begin{array}{l} \quad\;\; Cl \\ \quad\;\; | \\ CH_3{-}C{-}CH_2{-}C{-}CH_3 \\ \quad\;\; | \qquad\quad\; \| \\ \quad CH_3 \qquad CH{-}CH_3 \end{array}$$

Ans. The longest continuous chain contains six carbons and is numbered as shown:

$$\begin{array}{l} \quad\;\; Cl \\ \overset{6}{CH_3}{-}\overset{5}{C}{-}\overset{4}{CH_2}{-}\overset{3}{C}{-}CH_3 \\ \quad\;\; | \qquad\quad\; \| \\ \quad CH_3 \qquad \underset{2}{CH}{-}\underset{1}{CH_3} \end{array}$$

The base name is 2-hexene and this leaves three substituents unaccounted for—the methyl groups at C-3 and C-5, and the Cl at C-5. The IUPAC name is 5-chloro-3,5-dimethyl-2-hexene.

CIS-TRANS ISOMERS

Problem 12.32. Which of the alkenes in Problem 12.29 exist as separate *cis* and *trans* isomers? Show structural formulas for the *cis* and *trans* isomers that exist.

Ans. Inspect each carbon of the double bond. If either carbon has two of the same substituents, the structure cannot have *cis* and *trans* isomers. Only I(*b*) exists as a pair of *cis* and *trans* isomers:

$$\begin{array}{ccc} H & & H \\ & C{=}C & \\ CH_3 & & CH_2CH_3 \end{array} \qquad \begin{array}{ccc} H & & CH_2CH_3 \\ & C{=}C & \\ CH_3 & & H \end{array}$$

cis-2-Pentene *trans*-2-Pentene

I(*a*) (1-pentene), II(*a*) (2-methyl-1-butene), and II(*c*) (3-methyl-1-butene) cannot exist as *cis* and *trans* isomers since each has a doubly bonded carbon with two hydrogens. II(*b*) (2-methyl-2-butene) cannot do so either because it has two methyl groups on one of the doubly bonded carbons.

Problem 12.33. Which of the following exist as separate *cis* and *trans* isomers?

(*a*) CH_3 CH_3 1,1-Dimethylcyclohexane

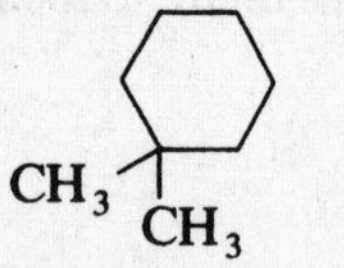

(*b*) CH_3 CH_3 1,2-Dimethylcyclohexane

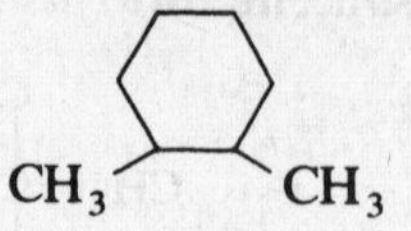

(*c*) CH_3 CH_3 1,2-Dimethylcyclohexene

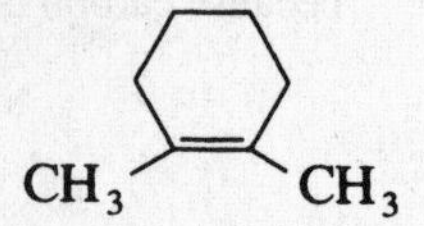

Ans. For *cis* and *trans* isomers to exist, two ring carbons must each contain two different substituents. (*a*) 1,1-Dimethylcyclohexane does not exist as a pair of *cis* and *trans* isomers, since each of the ring carbons has two of the same substituents. (*b*) 1,2-Dimethylcyclohexane exists as *cis* and *trans* isomers since C-1 and C-2 of 1,2-dimethylcyclohexane each have two different substituents (H and CH_3).

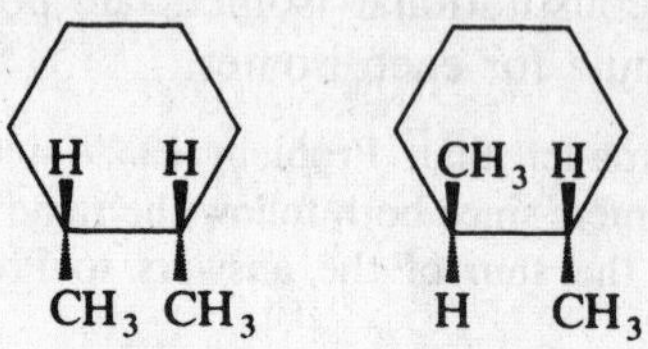

H H / CH_3 CH_3 — *cis*

CH_3 H / H CH_3 — *trans*

(*c*) 1,2-Dimethylcyclohexene does not exist as a pair of *cis* and *trans* isomers. Although neither carbon of the double bond has two of the same substituents, *cis* and *trans* isomers are not possible at the double bond. The geometry at both the double bond and the ring carbons is flat, which means there is only one placement possible for each methyl group—in the plane of the ring. Geometrical isomerism at the other four ring carbons is also not possible since each carbon has two of the same substituents.

Problem 12.34. For each of the following pairs of structural drawings, indicate whether the pair represents (1) the same compound, or (2) different compounds which are constitutional isomers, or (3) different compounds which are *cis-trans* isomers, or (4) different compounds which are not isomers:

(*a*) $CH_3(Cl)C{=}C(CH_3)(Cl)$ and $CH_3(Cl)C{=}C(Cl)(CH_3)$

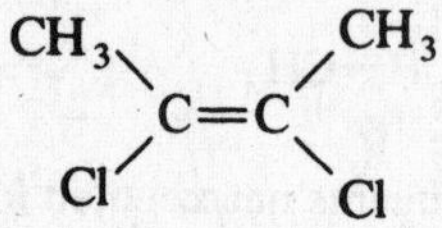

(*b*) $CH_2{=}CH$–cyclopropyl–Cl and cyclopentene–CH_2Cl

(*c*) $CH_3CH_2(H)C{=}C(CH_3)(H)$ and $H(H)C{=}C(H)(CH_2CH_2CH_3)$

(*d*) CH_3, H, H, CH_3 (cyclobutane) and H, H, CH_3, CH_3 (cyclobutane)

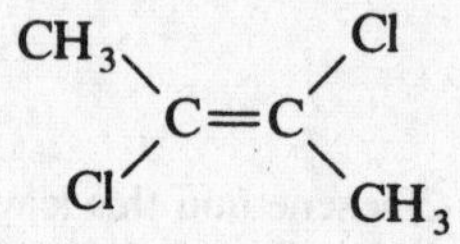

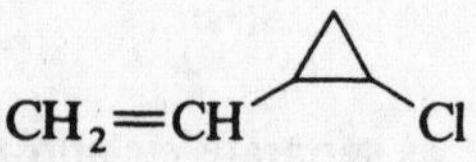

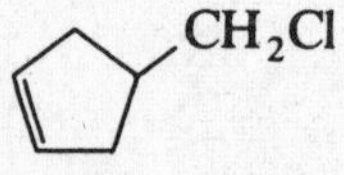

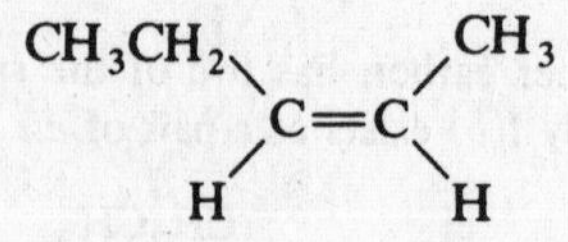

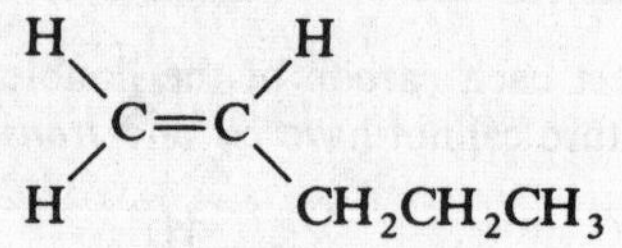

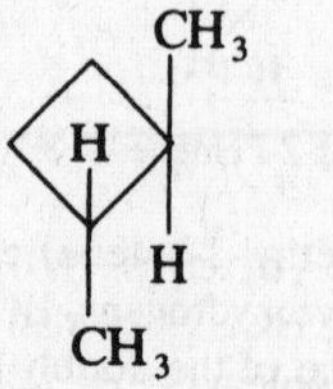

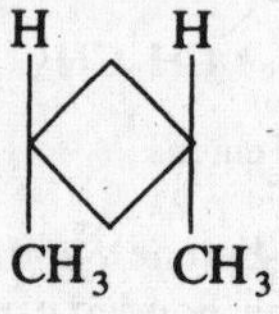

Ans. (*a*) *Cis-trans* isomers of 1-bromo-1,2-dichloropropene. The left-hand structure has the two methyl groups *cis* while the right-hand structure has the methyls *trans*.
(*b*) Different compounds which are not isomers. The left-hand structure is C_5H_7Cl while the right-hand structure is C_6H_9Cl.
(*c*) Constitutional isomers. The left- and right-hand structures are *cis*-2-pentene and 1-pentene, respectively.
(*d*) Constitutional isomers. The left- and right-hand structures are *trans*-1,2-dimethylcyclobutane and *cis*-1,3-dimethylcyclobutane, respectively.

CHEMICAL REACTIONS OF ALKENES

Problem 12.35. Give the equation for each of the following reactions. If there is more than one product, indicate which is the major product(s) and which the minor. If there is no reaction, state "No Reaction." (*a*) 1-Butene + NaOH, (*b*) 1-methylcyclohexene + HCl, (*c*) 1,3-butadiene + excess Cl_2, (*d*) polymerization of 2-methylpropene, (*e*) oxidation of 2-pentene with hot, acidic $K_2Cr_2O_7$, (*f*) combustion of 2-pentene, (*g*) 1-methylcyclohexene + H_2O in the presence of an acid catalyst.

Ans. (*a*) No Reaction.

(*b*) 1-methylcyclohexene (ring with CH_3) $\xrightarrow{HCl}$ ring with CH_3 and Cl on same carbon, H on adjacent carbon (Major) + ring with CH_3 and H on same carbon, Cl on adjacent carbon (Minor)

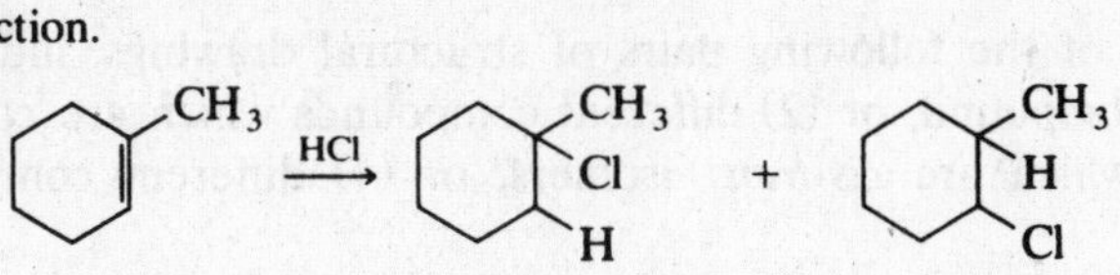

(*c*)
$$CH_2{=}CH{-}CH{=}CH_2 \xrightarrow{Cl_2} \underset{Cl}{CH_2}{-}\underset{Cl}{CH}{-}\underset{Cl}{CH}{-}\underset{Cl}{CH_2}$$

(*d*)
$$nCH_2{=}\overset{CH_3}{\underset{CH_3}{C}} \xrightarrow{\text{polymerization}} {+}CH_2{-}\overset{CH_3}{\underset{CH_3}{C}}{+}_n$$

(*e*)
$$CH_3{-}CH{=}CH{-}CH_2CH_3 \xrightarrow[\text{hot, acid}]{K_2Cr_2O_7} CH_3{-}\overset{OH}{C}{=}O + O{=}\overset{OH}{C}{-}CH_2CH_3$$

(*f*)
$$2\ CH_3{-}CH{=}CH{-}CH_2CH_3 + 15\ O_2 \longrightarrow 10\ CO_2 + 10\ H_2O$$

(*g*) 1-methylcyclohexene (ring with CH_3) $\xrightarrow[H^+]{H_2O}$ ring with CH_3 and OH on same carbon, H on adjacent carbon (Major) + ring with CH_3 and H on same carbon, OH on adjacent carbon (Minor)

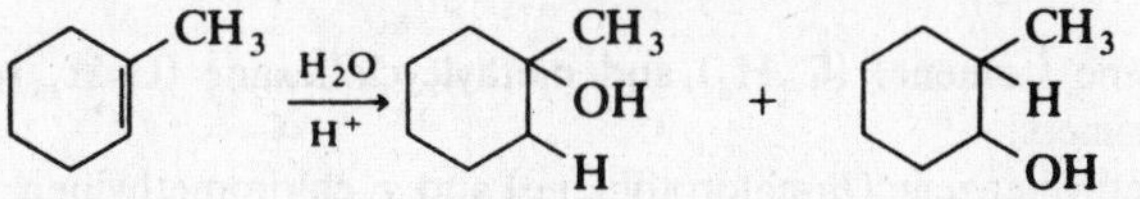

ALKYNES

Problem 12.36. Give the equation for the reaction of 2-butyne with HCl. If there is more than one product, indicate which is the major product(s) and which the minor. If there is no reaction, state "No Reaction."

Ans.

$$CH_3{-}C{\equiv}C{-}CH_3 \xrightarrow{HCl} CH_3{-}\overset{Cl}{C}{=}CH{-}CH_3 \xrightarrow{HCl} CH_3{-}\overset{Cl}{\underset{Cl}{C}}{-}CH_2{-}CH_3$$

2,2-dichlorobutane is the major product in the presence of excess HCl. Varying amounts of 2-chloro-2-butene are formed with less than an excess of HCl. Note that the orientation of the second addition follows Markovnikoff's rule. There is no problem of orientation in the first addition since the alkyne is symmetrical.

AROMATICS

Problem 12.37. Which of the following are aromatic?

I II III

Ans. All three compounds are aromatic, but compound II is different from compounds I and III. We analyze each ring independent of the others and note whether the three structural requirements for aromaticity are met by that ring. Only the left-hand ring of II is aromatic. The right-hand ring of II is not aromatic. Each ring in I and III is aromatic.

Problem 12.38. For each of the following pairs of structural drawings, indicate whether the pair represents (1) the same compound, or (2) different compounds which are constitutional isomers, or (3) different compounds which are *cis-trans* isomers, or (4) different compounds which are not isomers:

(*a*) CH_3 and CH_3 (*b*) CH_3, Cl and Cl, CH_3

(*c*) CH_3, Cl and Cl, CH_3

Ans. (*a*) Methylbenzene (toluene) (C_7H_8) and methylcyclohexane (C_7H_{16}) are different compounds which are not isomers.

(*b*) *m*-Chloromethylbenzene (*m*-chlorotoluene) and *p*-chloromethylbenzene (*p*-chlorotoluene) are constitutional isomers.

(*c*) The two structures represent the same compound, *p*-chloromethylbenzene (*p*-chlorotoluene).

NOMENCLATURE OF AROMATIC COMPOUNDS

Problem 12.39. Name each of the following by the IUPAC system:

(*a*) $CH_2CH_2CH_3$, Br, NO_2 (*b*) $C(CH_3)_3$, CH_3

Ans. (*a*) 2-bromo-1-propyl-3-nitrobenzene, (*b*) 1-*t*-butyl-3-methylbenzene

REACTIONS OF BENZENE

Problem 12.40. Write equations to show the products formed in each of the following reactions. If no reaction occurs, write "No Reaction." If more than one product is formed, indicate the major and minor products. (*a*) Benzene + NaOH, (*b*) benzene + Cl_2, (*c*) benzene + Cl_2 + $FeCl_3$, (*d*) methylbenzene (toluene) + Cl_2 + light or heat, (*e*) methylbenzene (toluene) + Cl_2 + $FeCl_3$.

Ans. (*a*) No Reaction. (*b*) No Reaction. Substitution does not occur in the absence of $FeCl_3$ or some other metal halide as a catalyst.

(*c*) benzene $\xrightarrow[FeCl_3]{Cl_2}$ benzene ring with Cl

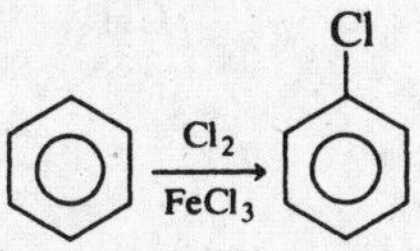

(*d*) benzene ring with CH_3 $\xrightarrow[\text{heat or light}]{Cl_2}$ benzene ring with CH_2Cl

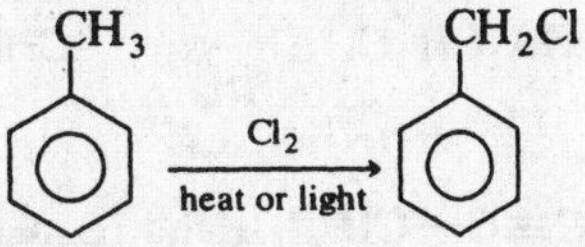

Methylbenzene has two sites where reaction with Cl_2 is possible—the aromatic ring or the alkane portion (methyl group). In the absence of a metal halide catalyst, aromatic substitution is not possible. In the absence of the catalyst and at high temperature or in the presence of light, substitution occurs at the alkane C—H bond.

(*e*) In the presence of the catalyst, aromatic substitution occurs. The reaction is complicated since substitution at different positions of the ring yields three different products. The para product is the major product for complex reasons.

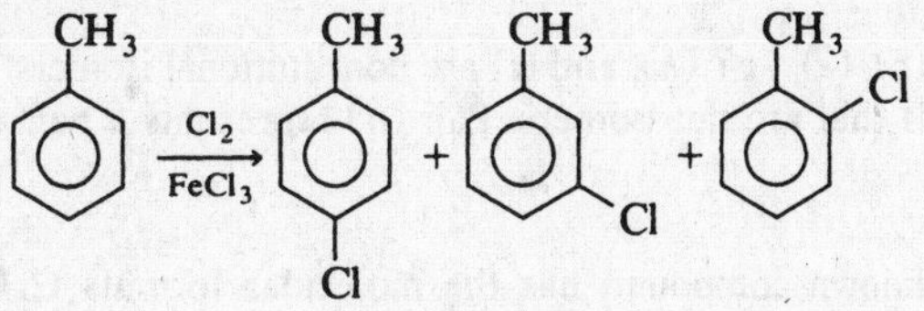

benzene ring with CH_3 $\xrightarrow[FeCl_3]{Cl_2}$ ring with CH_3 and Cl (para) + ring with CH_3 and Cl (meta) + ring with CH_3 and Cl (ortho)

SUPPLEMENTARY PROBLEMS

Problem 12.41. For each of the following pairs of structural drawings, indicate whether the pair represents (1) the same compound, or (2) different compounds which are constitutional isomers, or (3) different compounds which are *cis-trans* isomers, or (4) different compounds which are not isomers:

(*a*) CH_3CH_2(Cl)C=C(Br)(H) and CH_3(Cl)C=C(H)(CH_2Br)

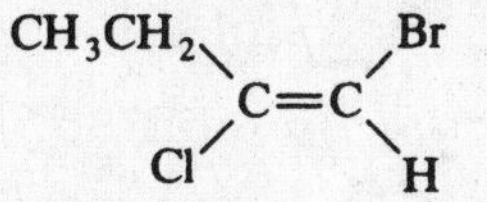

(*b*) $(CH_3)_3C$(H)C=C(CH_3)(H) and H(H)C=C(H)($CH_2CH_2CH_3$)

(*c*) CH_3CH_2(H)C=C(CH_3)(H) and H(H)C=C(H)($CH_2CH_2CH_3$)

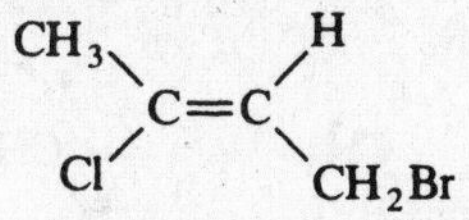

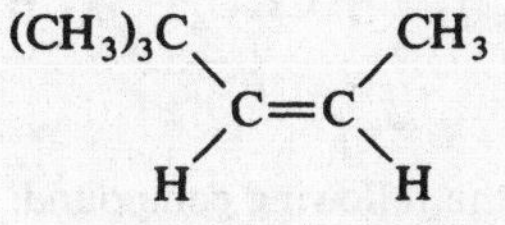

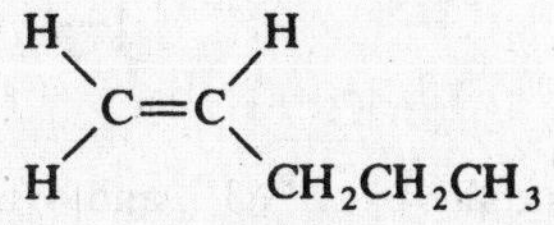

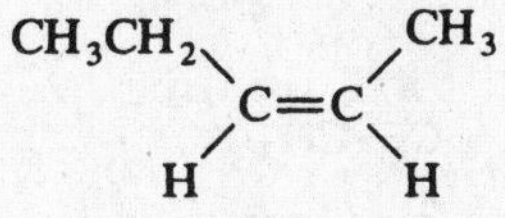

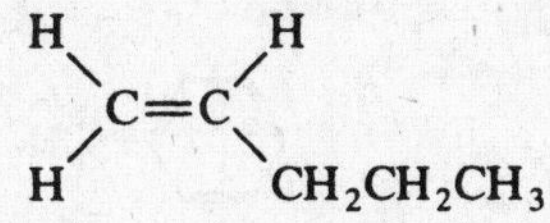

(*d*) $(CH_3)_3C(H)C{=}C(CH_3)(H)$ and $(CH_3)(H)C{=}C(H)(C(CH_3)_3)$

(*e*) $C_6H_5CH(CH_3)_2$ and $C_6H_3(CH_3)_3$

(*f*) Cl, CH_2CH_2Cl, CH_3-substituted benzene and Cl, Cl, CH_3, CH_3-substituted benzene

(*g*) $CH_3{-}C{\equiv}C{-}CH_3$ and $CH_2{=}CH{-}CH{=}CH_2$

(*h*) $C_6H_5CH_2CH_3$ and vinylcyclohexadiene

(*i*) cyclopentene and $CH_2{=}CH{-}CH{=}CH{-}CH_3$

Ans. Pairs (*a*), (*c*), (*e*), (*g*), (*h*), and (*i*) are constitutional isomers. Pairs (*b*) and (*f*) represent different compounds that are not isomers. Pair (*d*) represents a pair of *cis* and *trans* isomers.

Problem 12.42. An unknown compound has the molecular formula C_5H_8. Based on the molecular formula, what family (families) is (are) possible for the compound? Give an example of each family.

Ans. The unknown has four less hydrogens than the corresponding alkane, i.e., the alkane of the same number of carbons (the five-carbon alkane is C_5H_{12}). For each deficiency of two hydrogens, the unknown must have a ring or an extra bond between adjacent carbons. For a deficiency of four hydrogens, the unknown has some combination of two rings and/or extra bonds. The unknown C_5H_8 is a compound with two double bonds or one triple bond or two rings or one double bond + one ring.

$CH_2{=}CH{-}CH{=}CH{-}CH_3$ $\quad HC{\equiv}C{-}CH_2{-}CH_2{-}CH_3$

bicyclo[2.1.0]pentane, cyclopentene

methylcyclobutene (CH_3), methylenecyclobutane-type (CH_3), dimethylcyclopropene (CH_3, CH_3), dimethylcyclopropene (CH_3, CH_3)

Problem 12.43. Indicate the 1°, 2°, 3°, and 4° carbons in the following compound:

$$C_6H_5{-}CH_2{-}C(CH_3)_2{-}CH_2{-}C(CH_3){=}CH{-}CH_3$$

Ans. Recall from Sec. 11.7.3 that 1°, 2°, 3°, and 4° carbons, respectively, have a total of one, two, three, and four bonds to other carbons. Thus

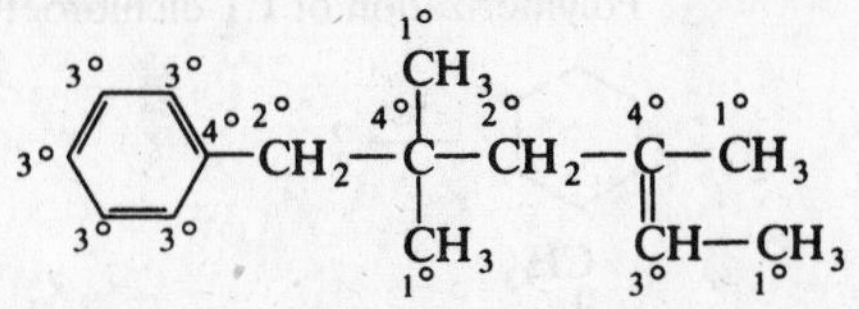

3° 3° 1° CH_3 4° 2° 4° 2° 4° 1°
3° [benzene ring] $-CH_2-C-CH_2-C-CH_3$
3° 3° CH_3 1° $CH-CH_3$ 3° 1°

Problem 12.44. What is the molecular formula of each of the following compounds?

(*a*) 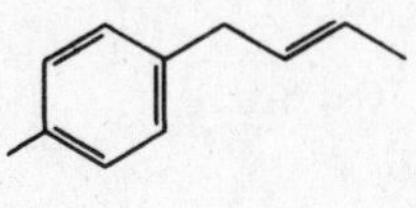 (*b*) C—C=C—C with C branch

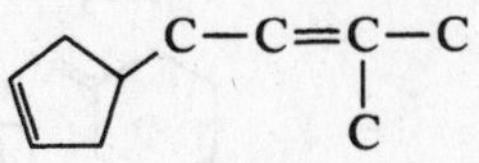

Ans. (*a*) $C_{11}H_{14}$, (*b*) $C_{10}H_{16}$

Problem 12.45. An unknown compound is either benzene, cyclohexene, or cyclohexane. Describe how you would determine the identity of the unknown by using simple chemical tests.

Ans. Add a drop of Br_2 to the unknown. If the red color disappears, the unknown is cyclohexene. If the red color does not disappear, add some $FeBr_3$ to the test tube containing the unknown plus unreacted Br_2. If the red color disappears, the unknown is benzene. If the red color does not disappear, the unknown is cyclohexane.

Problem 12.46. Which of the following exist as a pair of *cis* and *trans* isomers? Show structures of the cis and trans isomers. (*a*) 4-Methyl-2-pentene, (*b*) 2,4-dimethyl-2-pentene, (*c*) 4-methyl-2-pentyne, (*d*) 1,1-dichlorocyclopropane, (*e*) 1,2-dichlorocyclopropane.

Ans. The compounds in (*b*), (*c*), and (*d*) cannot exist as *cis* and *trans* isomers. Those in (*a*) and (*e*) can exist as *cis* and *trans* isomers.

(*a*)

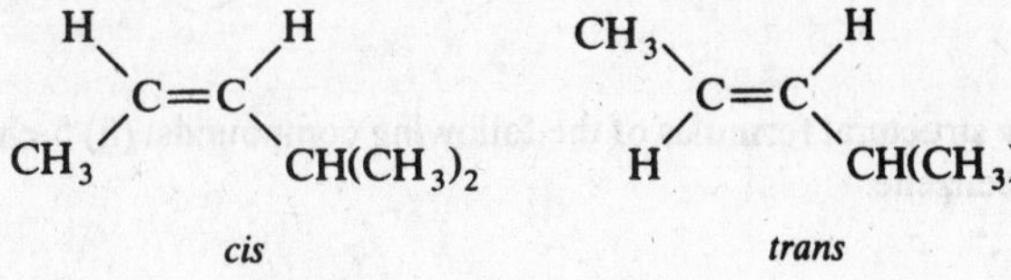

H H / C=C / CH_3 $CH(CH_3)_2$ — *cis*

CH_3 H / C=C / H $CH(CH_3)_2$ — *trans*

(*e*)

H H / [cyclopropane] / Cl Cl — *cis*

H Cl / [cyclopropane] / Cl H — *trans*

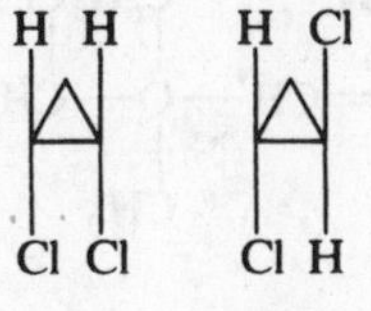

Problem 12.47. Place cyclohexane, cyclohexene, and benzene in order of increasing boiling points.

Ans. The boiling points are very nearly the same since there is little difference in polarity among alkanes, alkenes, and aromatics. Any differences in boiling or melting points among different hydrocarbons arises almost exclusively from differences in molecular mass, branching, or ring structure. The three hydrocarbons have the same ring size and very nearly the same molecular mass.

Problem 12.48. Write equations to show the products formed in each of the following reactions. If no reaction occurs, write "No Reaction." If more than one product is formed, indicate the major and minor products.

(*a*) Polymerization of 1,1-dichloroethene

(*b*) methylenecyclohexane $\xrightarrow{HCl}$?

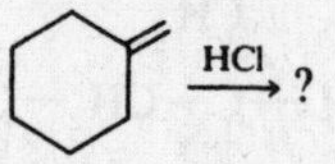

(*c*) p-xylene (CH_3, CH_3) $\xrightarrow{H_2SO_4}$?

(*d*) methylenecyclohexane $\xrightarrow[Pt]{H_2}$?

Ans.

(*a*) $nCH_2{=}CCl_2 \longrightarrow \left(CH_2{-}CCl_2 \right)_n$

(*b*) methylenecyclohexane $\xrightarrow{HCl}$ 1-chloro-1-methylcyclohexane (CH_3, Cl)

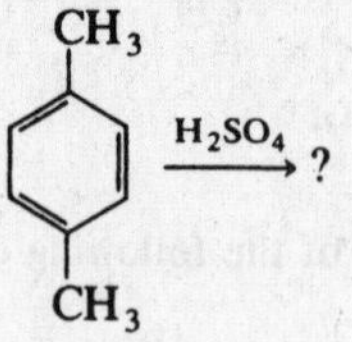

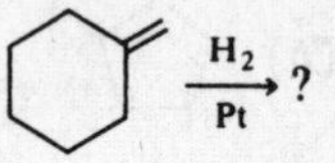

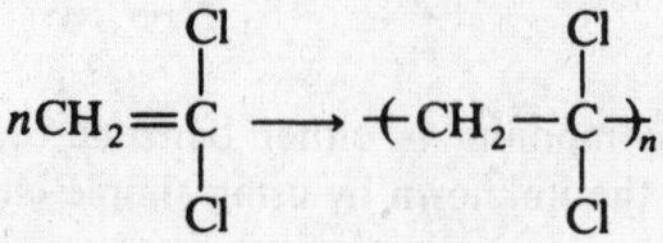

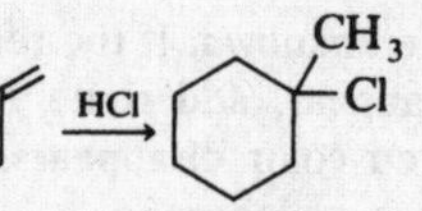

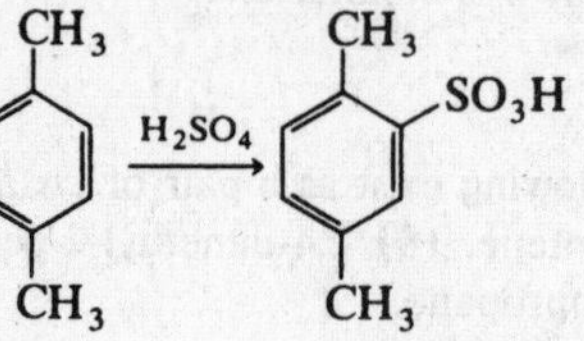

(*c*) p-xylene (CH_3, CH_3) $\xrightarrow{H_2SO_4}$ 2,5-dimethylbenzenesulfonic acid (CH_3, SO_3H, CH_3)

(*d*) methylenecyclohexane $\xrightarrow[Pt]{H_2}$ methylcyclohexane (CH_3, H)

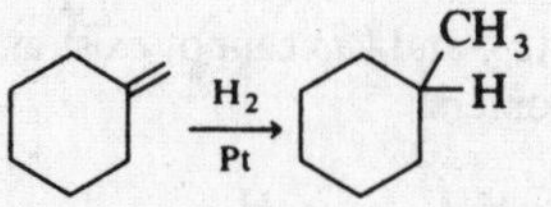

Problem 12.49. Draw structural formulas of the following compounds: (*a*) 5-chloro-2,4-dimethyl-4-phenyl-2-hexene, (*b*) 1-bromo-4-*t*-butylbenzene.

Ans.

(*a*) $CH_3{-}\overset{CH_3}{\overset{|}{C}}{=}CH{-}\underset{CH_3}{\underset{|}{\overset{C_6H_5}{\overset{|}{C}}}}{-}\overset{Cl}{\overset{|}{CH}}{-}CH_3$

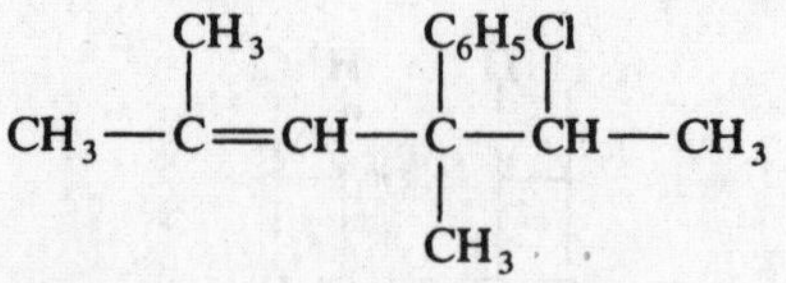

(*b*) benzene ring with Br and $C(CH_3)_3$ para

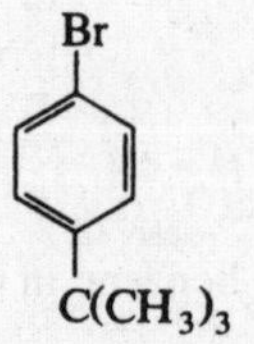

Problem 12.50. The acid-catalyzed hydration of propene, i.e., addition of water to propene, yields a mixture of two different products. What are the two products? Write the step-by-step mechanism for the reaction. (*Hint:* The first step is addition of a proton from sulfuric acid to propene.) Which is the major product and why?

Ans. **The major and minor products are 2-hydroxypropane and 1-hydroxypropane, respectively. (The IUPAC names are 2-propanol and 1-propanol, as will be discussed in Chap. 13.) The mechanism for formation of each product is shown below. 2-Propanol is the major product, as predicted by the Markovnikoff rule, because it is formed by the pathway that proceeds via formation of the most stable carbocation (2° instead of 1°).**

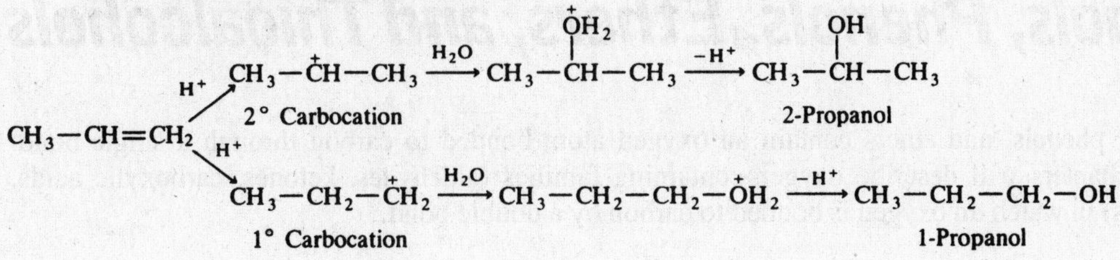

$$CH_3-CH=CH_2 \xrightarrow{H^+} \underset{\text{2° Carbocation}}{CH_3-\overset{+}{C}H-CH_3} \xrightarrow{H_2O} CH_3-\underset{}{\overset{\overset{+}{O}H_2}{\overset{|}{C}H}}-CH_3 \xrightarrow{-H^+} \underset{\text{2-Propanol}}{CH_3-\overset{OH}{\overset{|}{C}H}-CH_3}$$

$$CH_3-CH=CH_2 \xrightarrow{H^+} \underset{\text{1° Carbocation}}{CH_3-CH_2-\overset{+}{C}H_2} \xrightarrow{H_2O} CH_3-CH_2-CH_2-\overset{+}{O}H_2 \xrightarrow{-H^+} \underset{\text{1-Propanol}}{CH_3-CH_2-CH_2-OH}$$

Alcohols, Phenols, Ethers, and Thioalcohols

Alcohols, phenols, and ethers contain an oxygen atom bonded to carbon through a single bond. Subsequent chapters will describe oxygen-containing families (aldehydes, ketones, carboxylic acids, esters, amides) in which an oxygen is bonded to carbon by a double bond.

13.1 ALCOHOLS

Alcohols are organic compounds that contain a *hydroxyl group* (OH) attached to a saturated carbon. A *saturated carbon* is an sp^3-hybridized carbon, i.e., a carbon that is bonded to other atoms only through single bonds, no double or triple bonds. The alcohol family has the general formula R—OH and the first member is CH_3OH; IUPAC name: methanol; common name: methyl alcohol. Other alcohols are

$$CH_3{-}CH_2{-}OH \qquad CH_3{-}CH_2{-}CH_2{-}OH \qquad CH_3{-}\underset{\displaystyle CH_3}{\underset{|}{CH}}{-}OH$$

Ethanol / **Ethyl alcohol** — **1-Propanol** / **Propyl alcohol** — **2-Propanol** / **Isopropyl alcohol**

Ethanol is drinking alcohol, produced by fermentation of sugars in rye and corn (whiskey), grapes (wine), and barley (beer). It is also used as a solvent for perfumes, hairsprays, and antiseptic preparations. 2-Propanol is rubbing alcohol. Other alcohols are used as antifreeze for automobile engines and as ingredients in skin moisturizing creams. Carbohydrates such as glucose and starch belong to the alcohol family. Many alcohols are used as raw materials for producing other important organic compounds.

Problem 13.1. What is the structural relationship between an alcohol and other compounds we are familiar with?

Ans. An alcohol is usually considered to be derived from some other family of organic compound by replacing an H by OH. For example, ethanol is obtained by replacing an H of ethane by an OH:

$$CH_3{-}CH_3 \longrightarrow CH_3{-}CH_2{-}O{-}H$$

Replace any H by OH

An alternative view is that an alcohol is an organic derivative of water, whereby an H of water is replaced by an organic group. For example, ethanol is obtained if one replaces an H of water by an ethyl group (C_2H_5):

$$H{-}O{-}H \longrightarrow CH_3{-}CH_2{-}O{-}H$$

Replace either H by CH_3CH_2

Problem 13.2. Which of the following are alcohols and which are not alcohols?

$$\underset{\textbf{I}}{CH_3{-}CH_2{-}\overset{\displaystyle OH}{\overset{|}{\underset{\displaystyle CH_3}{\underset{|}{C}}}}{-}CH_3} \qquad \underset{\textbf{II}}{C_6H_5{-}OH} \qquad \underset{\textbf{III}}{C_6H_5{-}CH_2{-}OH} \qquad \underset{\textbf{IV}}{CH_3{-}\overset{\displaystyle O}{\overset{\|}{C}}{-}OH} \qquad \underset{\textbf{V}}{CH_3{-}O{-}CH_3}$$

Ans. To be classified as an alcohol, a compound must contain an OH group and that OH group must be bonded to a saturated carbon. Compound V contains an oxygen but the oxygen is not part of an OH group. Compound V is not an alcohol; it is an ether (Sec. 13.7). Compounds II and IV each contain an OH group but the OH group is bonded to an unsaturated carbon (a carbon that is doubly bonded to some atom). Compound II is a phenol (Sec. 13.6) while compound IV is a carboxylic acid (Chap. 15). Compounds I and III are alcohols since each contains an OH group attached to a saturated carbon.

Compound III illustrates that an alcohol can possess unsaturated carbons. It belongs both to the alcohol and aromatic families. The key requirement for an alcohol is that the OH not be bonded to an unsaturated carbon. In compound III, the OH is connected to a saturated carbon (the CH_2 carbon).

Oxygen uses sp^3 orbitals in forming bonds, similar to carbon. The difference between oxygen and carbon is that oxygen in its ground state has two electrons more than carbon. The result is that oxygen is divalent instead of tetravalent. Two of the four sp^3 orbitals of oxygen are filled, each containing a pair of *nonbonding electrons*.

Problem 13.3. What is the bond angle for the C—O—H angle in an alcohol?

Ans. The bond angles about any sp^3-hybridized atom are close to 109.5 °, the tetrahedral bond angle. This is the case for oxygen as well as carbon. Figure 13-1 shows the bond angles about oxygen in CH_3OH. Each of the two "bonds" from O that is not connected to another atom represents a nonbonded pair of electrons. The more usual representation of such electrons is structure I. Often, the nonbonded electrons are not shown, but understood to be present, as in structure II.

H—:Ö:—CH_3 (I) H—O—CH_3 (II)

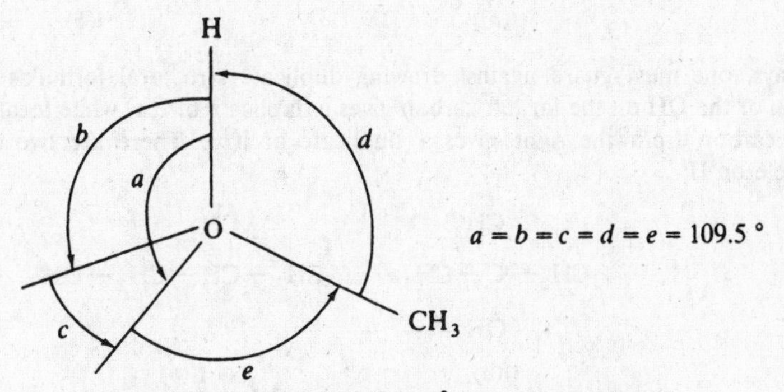

Fig. 13-1 Bond angles for sp^3 oxygen in methanol.

13.2 CONSTITUTIONAL ISOMERISM IN ALCOHOLS

Analogous to the alkenes and other families of compounds (except for alkanes), there are two sources of constitutional isomerism in alcohols—different carbon skeletons and different locations of the OH group within any carbon skeleton.

Problem 13.4. Compare the C:H ratio in the two compounds in each of the following pairs. What important lesson results from this problem?

(*a*) $CH_3CH_2CH_2OH$ and $CH_3CH_2CH_3$

(*b*) $CH_2{=}CHCH_2OH$ and $CH_3CH{=}CH_2$

(*c*) cyclopropyl–OH and cyclopropane

Ans. The compounds in each pair have the same C:H ratio. Pair (*a*) fits the general formula C_nH_{2n+2} for alkanes. Pairs (*b*) and (*c*) fit the general formula for an alkene or cycloalkane (C_nH_{2n}). The presence of oxygen in a compound does not change the C:H ratio compared to the corresponding compound without oxygen. The presence or absence of extra bonds and/or rings for an oxygen-containing formula can be analyzed in exactly the same manner as previously done for a formula without oxygen.

Problem 13.5. How many alcohols exist for the molecular formula $C_4H_{10}O$? Show only one structural formula for each constitutional isomer.

Ans. $C_4H_{10}O$ fits the general formula C_nH_{2n+2} for alkanes. There are no extra bonds and/or rings for these alcohols. Proceed in two steps. First, draw the carbon skeletons possible for molecules containing four carbons. Second, for each carbon skeleton, examine how many different compounds result from placement of the OH in different locations. There are two different carbon skeletons possible for a C_4 compound:

```
                 C
                 |
C—C—C—C      C—C—C
   I            II
```

There are two possible locations for the OH group within skeleton I—on an end carbon or next-to-end carbon of the carbon chain:

```
CH3—CH2—CH2—CH2—OH      CH3—CH—CH2—CH3
                            |
                            OH
      I(a)                  I(b)
```

As always, one must guard against drawing duplicate structural formulas of the same isomer. Location of the OH on the far left carbon gives a duplicate of I(*a*) while location of the OH on the second carbon from the right gives a duplicate of I(*b*). There are two constitutional isomers with skeleton II:

```
      CH3                 CH3
      |                   |
CH3—C—CH3           CH3—CH—CH2—OH
      |
      OH
     II(a)               II(b)
```

13.3 NOMENCLATURE OF ALCOHOLS

The IUPAC rules follow those for the alkanes with a few modifications:

1. The longest continuous chain must contain the carbon holding the OH group.
2. The ending of the name is changed from *-ane* to *-anol*.
3. The longest continuous chain is numbered from the end which gives the smallest number for the carbon holding the OH group.

Aside from IUPAC names, the simple alcohols (one to four carbons) are also known by common names. The alcohol is named by naming the alkyl substituent attached to the OH group followed by the word *alcohol*.

Problem 13.6. Give the common and IUPAC names for the alcohols in Problem 13.5.

Ans. Refer to Sec. 11.7.1 for the structures and names of the various butyl groups. The common names of alcohols I(*a*), I(*b*), II(*a*), and II(*b*) are butyl alcohol, *s*-butyl alcohol, *t*-butyl alcohol, and isobutyl alcohol, respectively.

The longest continuous carbon chain containing the OH for alcohols I(*a*) and I(*b*) is four carbons. The OH group is at carbon-1 and carbon-2, respectively, and the IUPAC names are 1-butanol and 2-butanol. Note that the carbon chain in each case is numbered from the end that gives the lowest number for the carbon holding the OH group. The longest continuous carbon chain holding the OH for both alcohols II(*a*) and II(*b*) is three carbons. Alcohol II(*a*) is 2-methyl-2-propanol and alcohol II(*b*) is 2-methyl-1-propanol.

$$\overset{4}{CH_3}-\overset{3}{CH_2}-\overset{2}{CH_2}-\overset{1}{CH_2}-OH \qquad \overset{1}{CH_3}-\overset{2}{CH}(OH)-\overset{3}{CH_2}-\overset{4}{CH_3}$$

I(*a*) I(*b*)

$$\overset{1}{CH_3}-\overset{2}{C}(CH_3)(OH)-\overset{3}{CH_3} \qquad \overset{3}{CH_3}-\overset{2}{CH}(CH_3)-\overset{1}{CH_2}-OH$$

II(*a*) II(*b*)

Problem 13.7. Identify the alcohols in Problem 13.6 as 1°, 2°, or 3°.

Ans. Primary, secondary, and tertiary alcohols are alcohols in which the OH is attached to a primary, secondary, and tertiary carbon, respectively. Recall that 1°, 2°, and 3° carbons are carbons having a total of one, two, and three bonds to other carbons. I(*a*) and II(*b*) are primary alcohols, I(*b*) is a secondary alcohol, and II(*a*) is a tertiary alcohol.

Problem 13.8. Name the following compounds by the IUPAC system:

$$HO-CH_2-CH_2-OH \qquad HO-CH_2-CH_2-CH_2-OH \qquad HO-CH_2-CH(OH)-CH_2-OH$$

I II III

Ans. Compounds with two and three OH groups per molecule are referred to as *diols* and *triols*, respectively. Diols are also referred to as *glycols* or *dihydric alcohols*, and triols as *trihydric alcohols*. The IUPAC system uses the endings *-anediol* and *-anetriol* for diols and triols, respectively. One needs to specify one number in the name for each of the positions of the OH groups. Alcohol I is 1,2-ethanediol, alcohol II is 1,3-propanediol, and alcohol III is 1,2,3-propanetriol. 1,2-Ethanediol (common name: ethylene glycol) and 1,2-propanediol (common name: propylene glycol) are the major components of antifreezes used in automobile coolant systems. 1,2,3-Propanetriol (common names: glycerol, glycerine) is used in a variety of applications (cosmetic lotions, rectal suppositories, foods) for its moisturizing properties.

13.4 PHYSICAL PROPERTIES OF ALCOHOLS

The OH bond of an alcohol, like each of the two OH bonds of water, participates in hydrogen bonding. The OH bond is highly polarized and the resulting intermolecular attractions are the strongest of all intermolecular forces. Alcohols have the strongest intermolecular attractive forces of all families of organic compounds except for carboxylic acids.

Problem 13.9. Which of the following correctly represents hydrogen bonding?

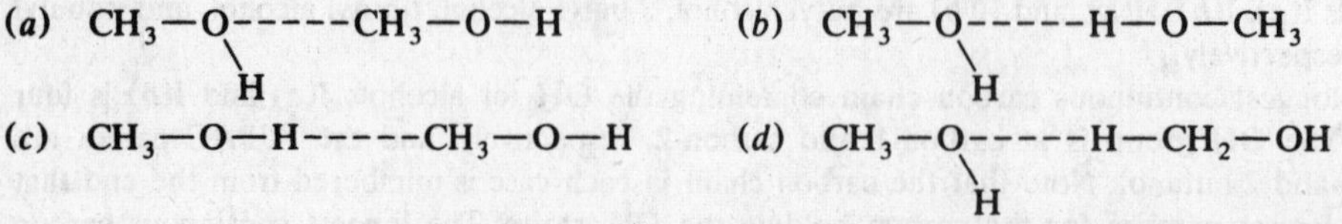

(*a*) $CH_3—O(—H)$- - - -$CH_3—O—H$ (*b*) $CH_3—O(—H)$- - - -$H—O—CH_3$

(*c*) $CH_3—O—H$- - - -$CH_3—O—H$ (*d*) $CH_3—O(—H)$- - - -$H—CH_2—OH$

Ans. The OH bond of an alcohol is polarized with the H end as $\delta+$ and O end as $\delta-$. Hydrogen bonding is the attraction of the $H^{\delta+}$ of one alcohol molecule for the $O^{\delta-}$ of another alcohol molecule. Formula (*a*) is incorrect because the C of a methyl group does not carry significant charge; there is no strong attraction of O from the OH of one molecule for that C of another molecule. Formula (*c*) is incorrect because the C of a methyl group does not carry significant charge and is not attracted to the H of an OH bond. Formula (*d*) is incorrect since an H attached to a C carries no significant charge and is not attracted to the O of an OH bond. Hydrogen bonding is represented only by formula (*b*). A more informative version of formula (*b*) uses $\delta+$ and $\delta-$ notations to show more clearly the source of the attractive forces:

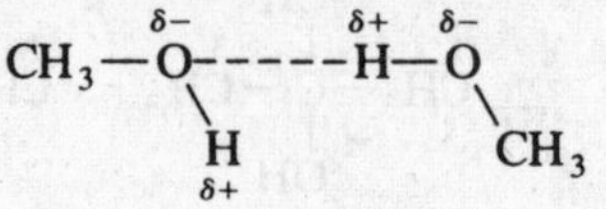

$$CH_3—\overset{\delta-}{O}(—\underset{\delta+}{H})\text{- - - -}\overset{\delta+}{H}—\overset{\delta-}{O}—CH_3$$

Note also that this representation more correctly notes the tetrahedral bond angles present in the alcohol.

Problem 13.10. Explain the difference in boiling points in each of the following pairs of compounds: (*a*) CH_3CH_2OH (78 °C) versus CH_3OH (65 °C), (*b*) CH_3CH_2OH (78 °C) versus $CH_3CH_2CH_3$ (−42 °C), (*c*) CH_3OH (65 °C) versus H_2O (100 °C), (*d*) $CH_3CH_2CH_2CH_2OH$ (117 °C) versus H_2O (100 °C), (*e*) $HOCH_2CH_2OH$ (197 °C) versus $CH_3CH_2CH_2OH$ (97 °C).

Ans. (*a*) Intermolecular attractions, boiling point, and melting point increase with increasing molecular mass for compounds within the same family.

(*b*) The intermolecular attractions in ethanol are much larger since they are the result of hydrogen bonding. Propane has the weakest intermolecular forces—those due to dispersion forces.

(*c*) There are two opposing factors. Methanol is the larger molecule but water has more hydrogen bonding because of its two OH bonds. The higher boiling point of water shows that the second factor is more important than the first.

(*d*) The same two opposing factors (molecular mass and number of OH bonds/molecule) are operative here. The greater number of OH bonds/molecule in water is insufficient to overcome the much larger molecular mass of 1-butanol.

(*e*) Molecular mass is about the same for 1,2-ethanediol and propanol but 1,2-ethanediol has more hydrogen bonding since it has twice the number of OH bonds per molecule.

Problem 13.11. Place the following compounds in order of increasing solubility in water. Explain.

$CH_3CH_2CH_2CH_3$	CH_3OH	$CH_2{=}CHCH_2CH_3$	$CH_3CH_2CH_2CH_2OH$
Butane	Methanol	1-Butene	1-Butanol

Ans. The order of increasing solubility is methanol > 1-butanol > butane = 1-butene according to the "like dissolves like" rule. Solubility involves breaking the hydrogen bonding interactions of water molecules for each other. This occurs only if a compound replaces the water-water interactions by forming strong intermolecular attractions with water. Alcohols have high solubility in water because they can hydrogen-bond with water. The solubility of an alcohol in water decreases as the size of the organic group increases. 1-Butanol is less soluble than methanol since the nonpolar portion of the molecule does not strongly interact with water. There is little difference between the solubility of butane and 1-butene in water since alkanes and alkenes have very nearly the same polarity.

13.5 CHEMICAL REACTIONS OF ALCOHOLS

13.5.1 Acid-Base Properties

Alcohols are very weak acids and bases, slightly weaker than water. The extremely weak acidity of alcohol is evidenced by the negligible degree of reaction with a strong base such as NaOH. The equilibrium for the reaction

$$\text{R—O—H} + \text{OH}^- \rightleftharpoons \text{R—O}^- + \text{H}_2\text{O}$$

is far to the left.

The very weak basicity of an alcohol is expressed by its ability to directly accept a proton from a strong acid such as sulfuric or phosphoric acid to form the alcohol equivalent of a hydronium ion (H_3O^+).

$$\text{R—O—H} + \text{H}^+ \rightleftharpoons \text{R—}\overset{\text{H}}{\underset{+}{\text{O}}}\text{—H}$$

Here also the equilibrium is far to the left. Although alcohols are very weak acids and bases, their acidity and basicity like that of water far exceeds the acidity and basicity of hydrocarbons.

13.5.2 Dehydration

Alcohols undergo *dehydration* (loss of H_2O) to form alkenes when heated in the presence of a strong acid such as H_2SO_4 or H_3PO_4. The reaction involves the loss of the OH group along with an H from a carbon adjacent to the carbon holding the OH. The reaction is referred to as an *elimination reaction*.

$$-\underset{\text{H}}{\overset{|}{\underset{|}{\text{C}}}}-\underset{\text{OH}}{\overset{|}{\underset{|}{\text{C}}}}- \xrightarrow[\text{heat}]{\text{H}_2\text{SO}_4} -\overset{|}{\text{C}}=\overset{|}{\text{C}}- + \text{H}_2\text{O}$$

Eliminated from alcohol

Problem 13.12. Give the product obtained when each of the following is heated in the presence of a strong acid: (*a*) 1-butanol, (*b*) cyclohexanol, (*c*) 2,2-dimethyl-1-propanol.

Ans.

(*a*) $CH_3CH_2CH_2—CH_2—OH \xrightarrow[\text{heat}]{H_3PO_4} CH_3CH_2CH{=}CH_2 + H_2O$

(*b*) cyclohexanol (cyclohexane ring with OH) $\xrightarrow[\text{heat}]{H_3PO_4}$ cyclohexene $+ H_2O$

(*c*) $CH_3—\overset{CH_3}{\underset{CH_3}{\overset{|}{\underset{|}{C}}}}—CH_2—OH \xrightarrow[\text{heat}]{H_3PO_4}$ No reaction

Adjacent C has no H

2,2-Dimethyl-1-propanol cannot undergo dehydration because there is no H available for elimination on the carbon adjacent to the carbon holding the OH.

Problem 13.13. We have described acid-catalyzed *dehydration* (loss of water) of an alcohol to yield an alkene. However, Sec. 12.6.1 described the opposite reaction—acid-catalyzed *hydration* (addition of water) of an alkene to yield an alcohol. Which is correct?

Ans. Both are correct. There is an equilibrium between the dehydration and hydration reactions. Reaction conditions determine whether the equilibrium lies on the hydration or dehydration side. If hydration of an alkene is the desired reaction, we add a large amount of water to the alkene in order to push the equilibrium toward alcohol. If dehydration of an alcohol is the desired reaction, we allow the alkene to distill out of the reaction vessel as it is formed (since the alkene has a lower boiling point than the alcohol). This pushes the equilibrium toward alkene.

Problem 13.14. Give the mechanism for the dehydration of an alcohol.

Ans. The first step is protonation of oxygen of the OH group. This weakens the bond between carbon and oxygen and H_2O is eliminated to form a carbocation. The carbocation eliminates H^+ from the adjacent carbon. This makes available a pair of electrons to form the π-bond by neutralizing the positive charge:

$$\underset{\substack{|\quad\ \ |\\ \mathrm{H}\quad \mathrm{OH}}}{-\overset{|}{\mathrm{C}}-\overset{|}{\mathrm{C}}-} \xrightleftharpoons{\mathrm{H}^+} \underset{\substack{|\quad\ \ |\\ \mathrm{H}\quad \overset{+}{\mathrm{OH}_2}}}{-\overset{|}{\mathrm{C}}-\overset{|}{\mathrm{C}}-} \xrightleftharpoons{-\mathrm{H_2O}} \underset{\substack{|\\ \mathrm{H}}}{-\overset{|}{\mathrm{C}}-\overset{|}{\underset{+}{\mathrm{C}}}-} \xrightleftharpoons{-\mathrm{H}^+} -\overset{|}{\mathrm{C}}=\overset{|}{\mathrm{C}}-$$

e^- pair forms π-bond

O is protonated H_2O is eliminated H^+ is eliminated

Each of the steps in the mechanism is written as an equilibrium since the whole process is reversible as described in Problem 13.13. The mechanism for dehydration is the reverse of the mechanism of hydration. The hydration begins by addition of H^+ to the double bond, followed by addition of water, and then loss of H^+.

The dehydration of certain secondary and tertiary alcohols is complicated by the formation of more than one alkene because there is more than one adjacent carbon with a hydrogen for elimination. An example is 2-butanol. 1-Butene is formed by the loss of H from CH_3 while 2-butene is formed by loss of H from CH_2. 1-Butene and 2-butene are not formed in equal amounts. 2-Butene is the major product because it is the more stable alkene compared to 1-butene.

$$\mathrm{CH_3}-\overset{\substack{\mathrm{OH}\\|}}{\mathrm{CH}}-\mathrm{CH_2}-\mathrm{CH_3} \xrightarrow[-\mathrm{H_2O}]{\mathrm{H_2SO_4}} \underset{\substack{\text{2-Butene}\\ \text{Major}}}{\mathrm{CH_3}-\mathrm{CH}=\mathrm{CH}-\mathrm{CH_3}} + \underset{\substack{\text{1-Butene}\\ \text{Minor}}}{\mathrm{CH_2}=\mathrm{CH}-\mathrm{CH_2}-\mathrm{CH_3}}$$

H on adjacent C

The stability of alkenes increases with decreasing numbers of hydrogens attached to the carbons of the double bond. The more usual way of discussing this structural feature is the number of alkyl groups attached to the double bond, i.e., the degree of substitution of the double bond. Stability increases with the increasing substitution of alkyl groups on the double bond. The order of stability of alkenes is

$$\underset{\text{Tetrasubstituted}}{\mathrm{R_2C}=\mathrm{CR_2}} > \overset{\text{Trisubstituted}}{\mathrm{R_2C}=\mathrm{CHR}} > \mathrm{R_2C}=\mathrm{CH_2} \underset{\text{Disubstituted}}{=} \mathrm{RCH}=\mathrm{CHR} > \overset{\text{Monosubstituted}}{\mathrm{RCH}=\mathrm{CH_2}} > \underset{\text{Unsubstituted}}{\mathrm{H_2C}=\mathrm{CH_2}}$$

2-Butene is more stable than 1-butene since 2-butene is a disubstituted double bond while 1-butene is monosubstituted.

The dehydration of 2-butanol is somewhat more complicated than shown above since the 2-butene formed as the major product is a mixture of *cis*-2-butene and *trans*-2-butene. *Trans*-2-butene is formed in greater abundance than *cis*-2-butene. *Trans* isomers are more stable than *cis* isomers for steric reasons. Alkyl groups sterically interfere with each other when close together as in the *cis* configuration.

Problem 13.15. Give the equation for the dehydration of 2-methyl-2-butanol. If more than one product is formed, show which hydrogen is lost in the formation of each product and indicate the major product.

Ans.

Loss of H yields 2-methyl-1-butene

$$\underset{\text{OH}}{\overset{\text{CH}_3}{\text{CH}_3-\text{C}-\text{CH}_2-\text{CH}_3}} \xrightarrow[-\text{H}_2\text{O}]{\text{H}_2\text{SO}_4} \underset{\substack{\text{2-Methyl-1-butene} \\ \text{Minor}}}{\text{CH}_2=\overset{\text{CH}_3}{\text{C}}-\text{CH}_2-\text{CH}_3} + \underset{\substack{\text{2-Methyl-2-butene} \\ \text{Major}}}{\text{CH}_3-\overset{\text{CH}_3}{\text{C}}=\text{CH}-\text{CH}_3}$$

Loss of H yields 2-methyl-2-butene

Loss of an H from either methyl group yields the same product, 2-methyl-1-butene. Loss of an H from the methylene group yields 2-methyl-2-butene. 2-Methyl-2-butene is a trisubstituted alkene, more stable, and the major product compared to 2-methyl-1-butene, which is a disubstituted alkene.

13.5.3 Oxidation

Alcohols undergo combustion (burning in air) to CO_2 and H_2O as do all organic compounds.

Mild oxidations of alcohols are carried out in the laboratory with reagents such as $KMnO_4$ or $Na_2Cr_2O_7$. Oxidation involves the loss of H from both the OH group and the carbon to which the OH group is attached. This results in the formation of a *carbonyl group*—a double bond between the carbon and oxygen.

$$\underset{\text{H}}{\overset{\text{R}'}{\text{R}-\text{C}-\text{O}-\text{H}}} \xrightarrow[\text{or Cr}_2\text{O}_7^{2-},\,\text{H}^+]{\text{MnO}_4^-,\,\text{OH}^-} \overset{\text{R}'}{\text{R}-\text{C}=\text{O}}$$

Loss of 2H — Carbonyl group

The reaction is an oxidation since the carbon that previously held the OH group goes to a higher oxidation state, i.e., it has two bonds to oxygen instead of one. The product is referred to as *aldehyde* if either R or R′ or both are H and a *ketone* if neither R nor R′ is H. Primary and secondary alcohols yield aldehydes and ketones, respectively. This type of oxidation, also referred to as a *dehydrogenation*, is carried out in living systems under the control of a *dehydrogenase enzyme* such as nicotinamide adenine dinucleotide (NADH).

Tertiary alcohols do not undergo oxidation since there is no H attached to the carbon holding the OH group.

$$\underset{\text{R}''}{\overset{\text{R}'}{\text{R}-\text{C}-\text{O}-\text{H}}}$$

Not attached directly to H

The oxidation of primary alcohols is more complicated than that of secondary alcohols. The aldehyde formed by the oxidation of a primary alcohol is itself easily oxidized. The H attached to the carbonyl carbon (formerly the carbon holding the OH group in the alcohol) is oxidized to OH. The resulting COOH group is referred to as a *carboxyl group* and a compound containing this functional group is referred to as a *carboxylic acid*.

$$R{-}\overset{\displaystyle H}{\underset{\displaystyle H}{C}}{-}O{-}H \xrightarrow[\text{or } Cr_2O_7^{2-},\,H^+]{MnO_4^-,\ OH^-} R{-}\overset{\displaystyle H}{C}{=}O \longrightarrow R{-}\overset{\displaystyle OH}{C}{=}O$$

Loss of 2H; Oxidized to OH; Carbonyl group; Carboxyl group

The oxidation of a primary alcohol normally yields the carboxylic acid unless one carries out the reaction in a special manner. Stopping the oxidation at the aldehyde stage requires one to carry out the reaction at a temperature (and partial vacuum if necessary) such that the aldehyde distills out of the reaction vessel as quickly as it is formed. This is possible because the aldehyde has a lower boiling point than the alcohol.

Problem 13.16. Give the product obtained when each of the following is oxidized with permanganate or dichromate. If there is more than one product, indicate which is the major product and which the minor product. If there is no reaction, indicate "No Reaction." (*a*) 2-Butanol, (*b*) 2-methyl-1-propanol, (*c*) 2-methyl-2-butanol.

Ans.

(*a*) $$CH_3{-}\overset{\displaystyle OH}{CH}{-}CH_2{-}CH_3 \xrightarrow{(O)} CH_3{-}\overset{\displaystyle O}{\overset{\|}{C}}{-}CH_2{-}CH_3$$

(*b*) $$CH_3{-}\overset{\displaystyle CH_3}{CH}{-}CH_2{-}OH \xrightarrow{(O)} CH_3{-}\overset{\displaystyle CH_3}{CH}{-}\underset{\displaystyle H}{C}{=}O \xrightarrow{(O)} CH_3{-}\overset{\displaystyle CH_3}{CH}{-}\underset{\displaystyle OH}{C}{=}O$$

Unless the aldehyde is removed by distillation immediately as formed or a very mild oxidizing agent is used, the major product is the carboxylic acid.

(*c*) $$CH_3{-}\overset{\displaystyle CH_3}{\underset{\displaystyle OH}{C}}{-}CH_2{-}CH_3 \longrightarrow \text{No Reaction}$$

Note that the symbol (O) is used to indicate reaction conditions for oxidation. This is an alternative to specifying the dichromate or permanganate reagent.

The oxidations of primary and secondary alcohols by dichromate and permanganate are useful as *simple chemical tests* for identification purposes since there is a visible change upon reaction. $Cr_2O_7^{2-}$ (orange solution) is converted to Cr^{3+} (green solution) when dichromate is the oxidizing agent. MnO_4^- (purple solution) is converted to MnO_2 (brown precipitate) when permanganate is the oxidizing agent. Dichromate and permanganate distinguish primary and secondary alcohols from tertiary alcohols and also from alkanes, but not from alkenes since alkenes also react with these oxidizing agents (Sec. 12.6.4).

Problem 13.17. An unknown is either hexane or 1-hexanol. Shaking the unknown with a few drops of basic potassium permanganate solution (purple color) results in a colorless solution with the formation of a brown precipitate. Identify the unknown and explain how you arrived at your answer.

Ans. The loss of purple color and formation of brown MnO_2 precipitate shows that the unknown has been oxidized. The unknown is 1-hexanol since primary alcohols are oxidized by permanganate but alkanes such as hexane are not.

13.6 PHENOLS

Phenols contain an OH group attached to the sp^2 carbon of a benzene ring. Phenol itself has no other substituent on the benzene ring:

OH

Phenol

The phenol structure is widespread in nature as vitamin E, oil of wintergreen, vanillin, the amino acid tyrosine, adrenalin, and tetrahydrocannibinaol (the hallucinogen ingredient of marijuana). Phenol compounds are used as antiseptics and disinfectants, as antioxidants, and to manufacture plastics and aspirin.

Problem 13.18. Which of the following are phenols and which are not phenols?

OH, CH_3, CH_2CH_3 (I); OH, CH_3, CH_2CH_3 (II); $CH_2CH_2CH_2OH$ (III); OH (IV)

I II III IV

Ans. Only compound I is a phenol. A phenol has a benzene ring with an OH group directly attached to the benzene ring. Compounds II, III, and IV are alcohols. Compounds II and IV have no benzene ring. Compound III has a benzene ring and OH group but the OH is not attached to the benzene ring.

Problem 13.19. Name compound I in Problem 13.18 by the IUPAC system.

Ans. The preferred name uses phenol as the parent name. The name is 4-ethyl-3-methylphenol. Less preferred but acceptable is 4-ethyl-1-hydroxy-3-methylbenzene, based on benzene as the parent name (Sec. 12.9).

Problem 13.20. Compare the boiling points and solubilities in water of toluene (methylbenzene), phenol, and cyclohexanol.

Ans. Phenol and cyclohexanol have much higher boiling points and are more soluble in water compared to toluene. Toluene is nonpolar whereas phenol and cyclohexanol are polar and participate in hydrogen bonding. The OH bond of a phenol is somewhat more polar than the OH of an alcohol. This results in a somewhat higher boiling point and water solubility for phenol as compared to cyclohexanol. Note that the water solubility of both phenol and cyclohexanol is low compared to a compound such as CH_3CH_2OH because of the big difference in the number of carbons.

The reason for the more polar OH in phenols is the slightly electron-withdrawing property of a benzene ring compared to alkyl groups. This results in enhanced electron-withdrawal of the electrons in the OH bond toward O and away from hydrogen.

The more polar OH bond in phenols results in increased acidity compared to alcohols. Phenols are weak acids, much stronger than alcohols, but much weaker than strong acids such as HCl. For example, the pH of 0.1 *M* solutions of water, ethanol, phenol, and HCl are 7, 7, 5.5, and 1.0, respectively.

Problem 13.21. An unknown is one of the following compounds. The unknown is only slightly soluble in water but dissolves completely in aqueous NaOH. Identify the unknown.

OH — benzene ring — $CH_2CH_2CH_3$
3-Propylphenol

OH — cyclohexane ring — $CH_2CH_2CH_3$
3-Propylcyclohexanol

Ans. Both compounds would be only slightly soluble in water but only 3-propylphenol is sufficiently acidic to dissolve completely in sodium hydroxide solution.

Phenols do not undergo dehydration since dehydration would destroy the highly stable benzene ring by converting a double bond into a triple bond. Phenols undergo oxidations but the reactions are too complex to discuss here. Phenols are often added to products (foodstuffs, rubber) to protect against the damaging effects of oxidizing agents. Phenols act as antioxidants by their own sacrificial oxidation. Vitamin E has this role in the body.

13.7 ETHERS

Ethers are compounds with the general formula R—O—R′ where R and R′ can be aliphatic or aromatic. Ethers are characterized by an oxygen which is directly bonded to two different carbons, neither of which is part of a carbon-oxygen double bond.

Problem 13.22. Which of the following are ethers and which are not ethers?

I: $CH_3CH_2—O—CH_2CH_3$

II: benzene ring with OH

III: benzene ring with OCH_2CH_3

IV: $CH_3CH_2—O—C(=O)—CH_2CH_3$

Ans. Compounds I and III are ethers. Compound II is not an ether since the oxygen is not connected to two different carbons; it is a phenol. Compound IV is not an ether. One of the oxygens is twice bonded to the same carbon instead of being bonded to two different carbons. The other oxygen is bonded to two different carbons but one of the carbons is part of a carbon-oxygen double bond. Compound IV is an ester.

Problem 13.23. How many ethers exist for the molecular formula $C_4H_{10}O$? Show only one structural formula for each isomer.

Ans. There are no rings and/or extra bonds for $C_4H_{10}O$ since the formula follows the C:H ratio for an alkane (C_nH_{2n+2}). There are two ways for the four carbons to be divided among the R and R′ groups attached to the oxygen—either two carbons in each group or one and three carbons. There are a total of three ethers with the formula $C_4H_{10}O$:

$CH_3CH_2—O—CH_2CH_3$ $CH_3—O—CH_2CH_2CH_3$ $CH_3—O—CH(CH_3)_2$

Problem 13.24. How many isomers exist for the molecular formula $C_4H_{10}O$? Show only one structural formula for each isomer.

Ans. This problem is not the same as Problem 13.23. The latter specified ethers with the formula $C_4H_{10}O$ but this problem does not have that limitation. Alcohols and ethers are constitutional isomers of each other. The answer to this problem is the sum of the answers to Problems 13.5 ($C_4H_{10}O$ alcohols) and 13.23 ($C_4H_{10}O$ ethers).

Ethers with simple substituents such as methyl, ethyl, propyl, butyl, and phenyl are known by their common names. The common name is obtained by placing the names of the substituents before the word *ether*. For example, compounds I and III in Problem 13.22 are diethyl ether and ethyl phenyl ether, respectively.

The IUPAC system is used for an ether which contains a complex substituent. An ether is named as a derivative of some other family with one of the R substituents of the ether together with the ether oxygen being named as an alkoxy (RO) substituent.

Problem 13.25. Name the following ethers:

$$\underset{\text{I}}{(CH_3)_3C{-}O{-}CH(CH_3)_2} \qquad \underset{\text{II}}{CH_3CH_2{-}O{-}CH_2{-}\overset{\overset{\displaystyle Cl}{|}}{C}H{-}CH_2{-}CH_3}$$

Ans. The common name of compound I is *t*-butyl isopropyl ether. There is no need for an IUPAC name.

The IUPAC name of compound II is 2-chloro-1-ethoxybutane, i.e., the compound is named as a derivative of butane which contains ethoxy and chloro groups at C-1 and C-2. Giving a common name for compound II is very difficult and inappropriate because there is no simple name for the right-side C_4 substituent attached to the oxygen.

Problem 13.26. Explain the difference in boiling points in each of the following pairs of compounds: (*a*) CH_3OCH_3 (−23 °C) versus $CH_3CH_2CH_3$ (−42 °C), (*b*) $CH_3CH_2OCH_2CH_3$ (35 °C) versus $CH_3(CH_2)_3CH_3$ (36 °C), (*c*) CH_3OCH_3 (−23 °C) versus CH_3CH_2OH (78 °C).

Ans. The variable in each pair is the family of the compound since there is little difference in molecular weight.

(*a*) The two carbon-oxygen bonds in an ether are polarized $^{\delta+}C{-}O^{\delta-}$ since the electronegativity of oxygen is greater than that of carbon. The two bond polarities do not cancel because the C—O—C bond angle is the tetrahedral bond angle (109.5 °C) and this makes ethers more polar than hydrocarbons. The greater polarity results in higher intermolecular attractions, higher boiling points, and higher melting points for ethers.

(*b*) Note that there is a 19 °C difference in boiling points between dimethyl ether and propane but a negligible difference between diethyl ether and pentane. As one goes to larger members of the ether family, the hydrocarbon portion of the ether becomes larger and the polar portion becomes smaller. There would still be a significant difference in boiling points between an ether and an alkane if the ether group were very polar. The fact that the boiling points of diethyl ether and pentane are so close tells us that the ether group is only slightly polar.

(*c*) The big difference (101 °C) in the boiling points of dimethyl ether and ethanol indicates once again that intermolecular attractive forces due to hydrogen bonding (alcohol) are much stronger than those due to polarity (ether).

Problem 13.27. Explain why diethyl ether and its isomer 1-butanol have close to the same solubility in water.

Ans. Alcohols are high boiling liquids because they hydrogen-bond with themselves. Alcohols are also highly soluble in water due to hydrogen bonding between alcohol and water. An ether cannot hydrogen-bond with itself since there is no OH bond and this results in lower boiling points for ethers. However, the oxygen of an ether can hydrogen-bond with the H of an OH bond of water and this results in high solubility of ethers in water.

$$\begin{array}{l} CH_3CH_2{-}O{-}{-}{-}{-}{-}H{-}O \\ \qquad\qquad\quad \backslash \qquad\qquad\quad \backslash \\ \qquad\qquad CH_2CH_3 \qquad\quad H \end{array}$$

The dehydration of 1° (but not 2° and 3°) alcohols is a more complicated reaction than previously described (Sec.13.5.2). *Intramolecular dehydration* to yield an alkene is not the only reaction that occurs.

There is simultaneous *intermolecular dehydration* in which one molecule of water is eliminated between two molecules of alcohol to form a symmetrical ether:

$$CH_3CH_2CH_2CH_2{-}OH \xrightarrow{H_2SO_4} CH_3CH_2CH{=}CH_2$$

$$CH_3CH_2CH_2CH_2{-}OH \xrightarrow{H_2SO_4} CH_3CH_2CH_2CH_2{-}O{-}CH_2CH_2CH_2CH_3$$

The alkene is the major product when dehydration is carried out at higher temperatures while ether is the major product at lower temperatures. The specific higher and lower temperatures vary with the catalyst used. For H_2SO_4, higher and lower are about 180 °C and 140 °C, respectively.

Problem 13.28. Give the product when cyclohexanol is heated in the presence of a strong acid.

Ans. This is the same question as Problem 13.12(*b*) and the answer is the same. Dehydration proceeds only intramolecularly to form an alkene, not intermolecularly to form an ether, because the alcohol is not a primary alcohol:

$$\text{cyclohexanol (OH)} \xrightarrow[\text{heat}]{H_2SO_4} \text{cyclohexene}$$

Intermolecular dehydration to form an ether competes with intramolecular dehydration to form an alkene only for primary alcohols.

Ethers are unreactive, being stable to a variety of acids, bases, and oxidizing agents. This lack of reactivity makes ethers useful as solvents for carrying out reactions of other organic compounds. Diethyl ether has been used in the past as an anesthetic. Diethyl ether has a tendency to explode and other compounds, such as 1-bromo-1-chloro-2,2,2-trifluoroethane, are now used.

13.8 THIOALCOHOLS

Thioalcohols, general formula R—SH, are the sulfur analogs of alcohols. Sulfur is below oxygen in the Periodic Table. Both are divalent and there are similarities in their behavior. Thioalcohols are also referred to as *mercaptans*. The SH group is referred to as the *thiol* or *mercaptan* or *sulfhydryl* group.

Thioalcohols are named by the IUPAC system in the same manner as alcohols are named except that the ending of the name is *-anethiol* instead of *-anol*.

Thioalcohols undergo oxidation to disulfides:

$$2\,R{-}SH \xrightarrow{(O)} \underset{\textit{Disulfide}}{R{-}S{-}S{-}R}$$

Disulfides undergo reduction back to thioalcohols:

$$R{-}S{-}S{-}R \xrightarrow{(H)} 2\,R{-}SH$$

Note that (H) is used to indicate conditions for reduction; the term *reduction* is also used. Oxidation of thiols and reduction of disulfides are important in the functioning of many proteins.

Problem 13.29. Name the following compounds by the IUPAC system:

(*a*) $CH_3—CH_2—CH_2—CH_2—SH$ (b) 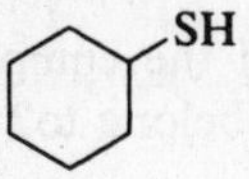 cyclohexane ring bearing SH

Ans. (*a*) 1-Butanethiol, (*b*) cyclohexanethiol.

Problem 13.30. Give the product obtained in each of the following reactions:

(*a*) $CH_3—CH(SH)—CH_2—CH_3 \xrightarrow{(O)} ?$

(*b*) $CH_3—CH_2—S—S—CH_2—CH_3 \xrightarrow{\text{reduction}} ?$

Ans.

(*a*) $CH_3—CH(SH)—CH_2—CH_3 \xrightarrow{(O)} CH_3—CH(—S—S—CH(CH_3)—CH_2—CH_3)—CH_2—CH_3$

(*b*) $CH_3—CH_2—S—S—CH_2—CH_3 \xrightarrow{\text{reduction}} 2\ CH_3—CH_2—SH$

ADDITIONAL SOLVED PROBLEMS

ALCOHOLS

Problem 13.31. Which of the following are alcohols and which are not alcohols?

I	II	III	IV
$HO—CH{=}CH—CH_3$	$CH_2{=}CH—CH_2—OH$	cyclopentane ring bearing OH	$HO—CH_2—CH_2—OH$

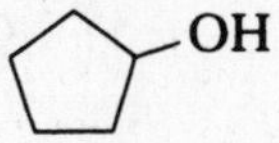

Ans. Compounds II, III, and IV are alcohols since each of the OH groups in these compounds is attached to a saturated carbon. Compound I is not an alcohol since the OH is attached to an unsaturated carbon. Compound I is referred to as *enol* (which distinguishes it from a phenol where the OH is attached to an aromatic unsaturated carbon).

Problem 13.32. Give the bond angles for *A*, *B*, *C*, and *D*:

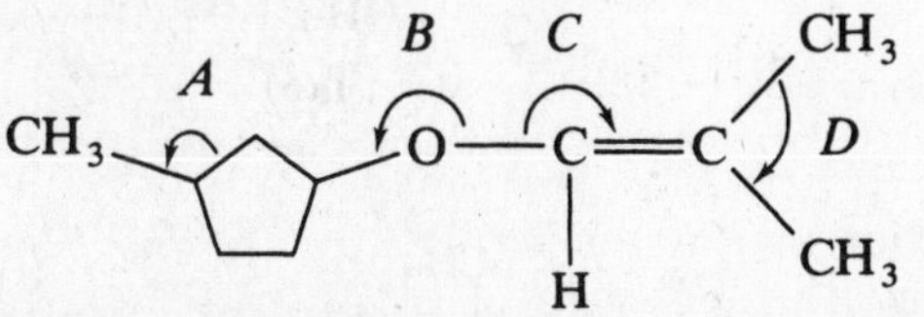

CH_3–(cyclopentane ring, angle *A*)–O(angle *B*)–$C(H)$(angle *C*)$=C(CH_3)_2$ (angle *D*)

Ans. A bond angle is defined by the angle between two bonds sharing a common or central atom. The bond angle is determined by the hybridization of the central atom. *A* and *B* are 109.5° since the central atom (carbon and oxygen, respectively) is sp^3-hybridized. *C* and *D* are 120° since the central atom (carbon) is sp^2-hybridized.

CONSTITUTIONAL ISOMERISM IN ALCOHOLS

Problem 13.33. Calculate the C:H ratios of the compounds in Problem 13.31. What family other than the alcohol family does each compound belong to?

Ans. Compounds I and II are C_3H_6 and fit the general formula for an alkene or cycloalkane, C_nH_{2n}. Compound III is C_5H_{10} and also fits the general formula for an alkene or cycloalkane, C_nH_{2n}. Compound IV is C_2H_6 and fits the general formula for an alkane, C_nH_{2n+2}. Compatible with these general formulas, Compounds I and II are alkenes, Compound III is a cycloalkane, and Compound IV is an alkane. The presence of oxygen in a compound does not change the C:H ratio.

Problem 13.34. How many alcohols exist for the molecular formula $C_5H_{12}O$? Show only one structural formula for each constitutional isomer.

Ans. $C_5H_{12}O$ contains no rings or double or triple bonds since the formula fits that for an alkane, C_nH_{2n+2}. There are three different carbon skeletons possible for a C_5 compound:

```
                      C               C
                      |               |
C—C—C—C—C         C—C—C—C         C—C—C
                                      |
                                      C

    I                 II              III
```

There are three constitutional isomers with skeleton I as there are three possible locations for the OH group within skeleton I—at an end carbon, a next-to-end carbon, or the middle carbon:

```
                                          OH
                                          |
CH3—CH2—CH2—CH2—CH2—OH               CH3—CH—CH2—CH2—CH3
         I(a)                              I(b)

                     OH
                     |
            CH3—CH2—CH—CH2—CH3
                   I(c)
```

As always one must guard against drawing duplicate structural formulas of the same isomer. Location of the OH on the far left carbon gives a duplicate of compound I(*a*) while location of the OH on the second carbon from the right gives a duplicate of compound I(*b*). There are four constitutional isomers with skeleton II:

```
     CH2OH                 CH3                   CH3
     |                     |                     |
CH3—CH—CH2—CH3        CH3—C—CH2—CH3         CH3—CH—CH—CH3
                           |                        |
                           OH                       OH
     II(a)                 II(b)                 II(c)

                  CH3
                  |
             CH3—CH—CH2—CH2—OH
                  II(d)
```

(Placing the OH on the far left carbon gives a duplicate of compound II(*a*).) There is only one constitutional isomer with skeleton III:

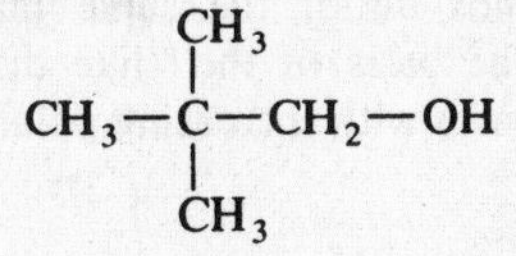

$$CH_3-\underset{CH_3}{\overset{CH_3}{\underset{|}{\overset{|}{C}}}}-CH_2-OH$$

III(*a*)

Problem 13.35. Which of the following are correct molecular formulas? (*a*) C_5H_9O, (*b*) C_5H_9ClO, (*c*) $C_5H_{10}O$?

Ans. C_5H_9ClO and $C_5H_{10}O$ are correct but C_5H_9O is not a correct molecular formula. Each formula must be analyzed to determine if it is compatible with the general formula for one of the families. Keep in mind that oxygen is not counted while halogen is counted like a hydrogen. C_5H_9O is equivalent to C_5H_9 which cannot be a correct formula. A consideration of the general formulas for different families (C_nH_{2n+2} for alkanes, C_nH_{2n} for cycloalkanes and alkenes) shows that no compound can have an odd number of hydrogens. C_5H_9ClO and $C_5H_{10}O$ are each equivalent to C_5H_{10} which is the general formula of an alkene or cycloalkane, C_nH_{2n}.

NOMENCLATURE OF ALCOHOLS

Problem 13.36. Name the following compounds by the IUPAC system:

(a) 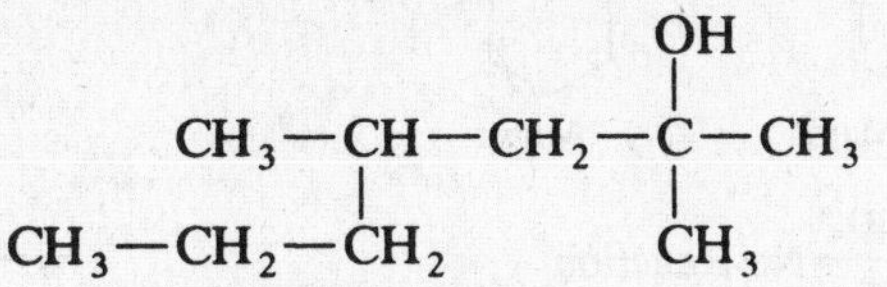 $CH_3-\underset{CH_3-CH_2-CH_2}{\underset{|}{CH}}-CH_2-\underset{CH_3}{\overset{OH}{\underset{|}{\overset{|}{C}}}}-CH_3$

(b) 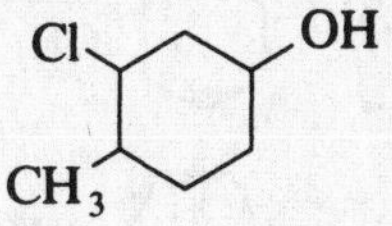 cyclohexane ring bearing OH, Cl, and CH_3 substituents

(c) $H_2C{=}CH-CH_2CH_2OH$

Ans. (*a*) The longest continuous chain that contains the carbon holding the OH is the seven-carbon chain numbered as shown below:

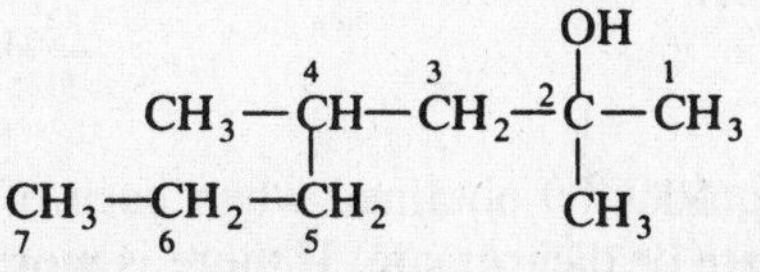

$$\overset{7}{CH_3}-\overset{6}{CH_2}-\overset{5}{CH_2}-\overset{4}{CH}-\overset{3}{CH_2}-\overset{2}{\underset{CH_3}{\overset{OH}{C}}}-\overset{1}{CH_3}$$

The chain is numbered in the direction that yields the smallest number for the carbon holding the OH group. The base name is 2-heptanol and this leaves two methyl groups unaccounted for—the methyls at C-2 and C-4. The IUPAC name is 2,4-dimethyl-2-heptanol.

(*b*) The base name is cyclohexanol. By convention a "1-" in front of the base name is not used as it is understood that the OH is at C-1. (*Note:* This convention does not apply to acyclic alcohols where the number "1-" is used, except for methanol and ethanol.) The ring carbons are numbered by proceeding counterclockwise from C-1 as this yields the lowest numbers for the carbons holding the Cl and CH_3 groups. The IUPAC name is 3-chloro-4-methylcyclohexanol.

(*c*) The compound is both an alkene and alcohol. The numbering of the longest carbon chain is determined by the direction that gives the lowest number for the OH. The name is 3-butene-1-ol.

PHYSICAL PROPERTIES OF ALCOHOLS

Problem 13.37. What is the order of solubility of the compounds in Problem 13.11 in a nonpolar solvent such as hexane?

Ans. The order of solubility in hexane (a nonpolar compound) is the reverse of the order in water: butane = 1-butene > 1-butanol > methanol. Solubility requires that the intermolecular attractions between two compounds be of the same magnitude as those within each of the unmixed compounds. (This is the basis of the "like dissolves like" rule.) Solubility of compounds in a nonpolar solvent decreases with increasing polarity.

Problem 13.38. Draw a representation of the hydrogen bonding present in 1-propanol.

Ans.

$$CH_3CH_2CH_2-\overset{\delta-}{\underset{\underset{\delta+}{H}}{O}}----\overset{\delta+}{H}-\overset{\delta-}{\underset{CH_2CH_2CH_3}{O}}$$

CHEMICAL REACTIONS OF ALCOHOLS

Problem 13.39. Give the product(s) obtained when each of the following is heated in the presence of a strong acid at high temperature. If there is more than one product, indicate which is the major product and which the minor product. If there is no reaction, indicate "No Reaction." (*a*) 1-Methylcyclohexanol, (*b*) phenylmethanol, (*c*) 3-methyl-2-butanol.

Ans.

(*a*) 1-methylcyclohexanol (CH_3, OH) $\xrightarrow[\text{heat}]{H_2SO_4}$ 1-methylcyclohexene (CH_3) (Major) + methylenecyclohexane ($=CH_2$) (Minor)

(*b*) $C_6H_5-CH_2-OH \xrightarrow[\text{heat}]{H_2SO_4}$ No Reaction

No H on adjacent C

(*c*)

$$CH_3-\overset{OH}{\overset{|}{CH}}-\overset{CH_3}{\overset{|}{CH}}-CH_3 \xrightarrow[\text{heat}]{H_2SO_4} \underset{\text{Major}}{CH_3-CH=\overset{CH_3}{\overset{|}{C}}-CH_3} + \underset{\text{Minor}}{CH_2=CH-\overset{CH_3}{\overset{|}{CH}}-CH_3}$$

Problem 13.40. Give the product(s) obtained when each of the following undergoes oxidation by a reagent such as permanganate or dichromate. If there is more than one product, indicate which is the major product and which the minor product. If there is no reaction, indicate "No Reaction." (*a*) Phenylmethanol, (*b*) cyclohexanol, (*c*) 1-methylcyclohexanol.

Ans.

(*a*) $C_6H_5-CH_2-OH \xrightarrow{(O)} C_6H_5-CH=O \xrightarrow{(O)} C_6H_5-\overset{OH}{\overset{|}{C}}=O$

Unless the aldehyde is removed by distillation immediately as formed, the major product is the carboxylic acid.

(*b*) cyclohexanol (OH) $\xrightarrow{(O)}$ cyclohexanone (=O)

(*c*) 1-methylcyclohexanol (CH_3, OH) $\xrightarrow{(O)}$ No Reaction

Problem 13.41. An unknown is either hexene or hexanol. Shaking the unknown with a few drops of acidic dichromate solution (orange color) results in a green solution. Identify the unknown and explain how you arrive at your answer.

Ans. The color change from orange to green indicates that dichromate has oxidized the unknown and has itself been converted to the green-colored Cr^{3+}. However, this test does not distinguish hexene from hexanol since alkenes as well as primary and secondary alcohols are oxidized by dichromate.

PHENOLS

Problem 13.42. Complete each of the following reactions by writing the structure of the organic product. If there is more than one product, indicate which is the major product(s) and which the minor. If there is no reaction, state "No Reaction."

(*a*) $C_6H_5CH_2CH_2CH_2OH \xrightarrow{NaOH}$? (benzene ring bearing $CH_2CH_2CH_2OH$)

(*b*) m-$HOC_6H_4CH_2CH_2CH_3 \xrightarrow{NaOH}$? (benzene ring bearing OH and, meta, $CH_2CH_2CH_3$)

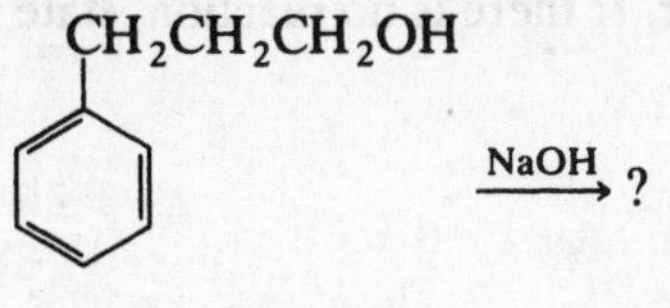

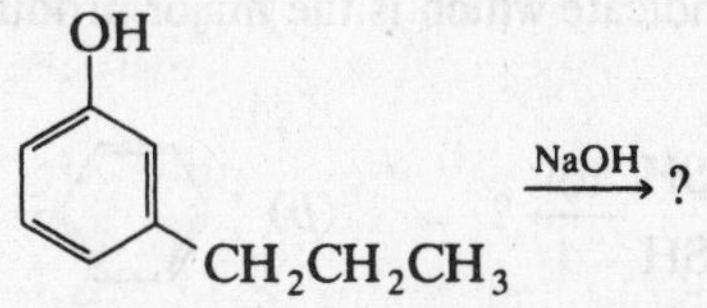

Ans. (*a*) There is No (or negligible) Reaction because alcohols are only very weakly acidic.

(*b*) m-$HOC_6H_4CH_2CH_2CH_3 \xrightarrow[-H_2O]{NaOH} m$-$Na^+\,{}^-OC_6H_4CH_2CH_2CH_3$

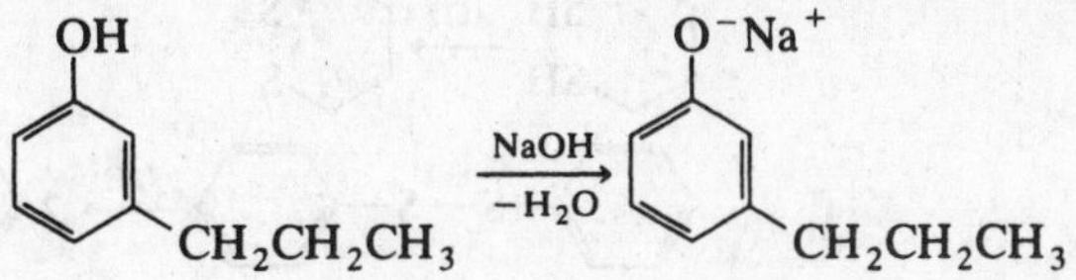

ETHERS

Problem 13.43. Draw the structures for each of the following compounds: (*a*) butyl isopropyl ether, (*b*) *m*-ethylisopropoxybenzene.

Ans.

(*a*) $CH_3CH_2CH_2CH_2—O—CH(CH_3)_2$

(*b*) benzene ring bearing $OCH(CH_3)_2$ and, meta, CH_2CH_3

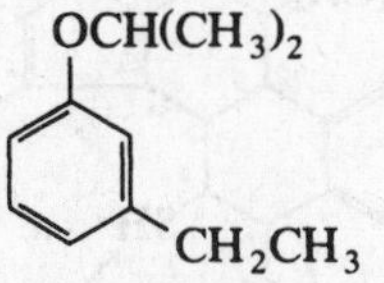

Problem 13.44. Draw a representation of the hydrogen bonding present when dimethyl ether dissolves in 1-propanol.

Ans. $CH_3—O$- - - - -$H—O$ (with CH_3 on the ether O and $CH_2CH_2CH_3$ on the alcohol O)

Problem 13.45. Complete each of the following reactions by writing the structure of the organic product. If there is more than one product, indicate which is the major product(s) and which the minor. If there is no reaction, state "No Reaction."

(*a*) $CH_3CH_2CH_2CH_2—O—CH(CH_3)_2 \xrightarrow{NaOH}$?

(*b*) $CH_3CH_2CH_2CH_2—O—CH(CH_3)_2 \xrightarrow{(O)}$?

Ans. The answer is No Reaction to both (*a*) and (*b*). Ethers are unreactive.

Problem 13.46. Place the following compounds in order of increasing boiling points:

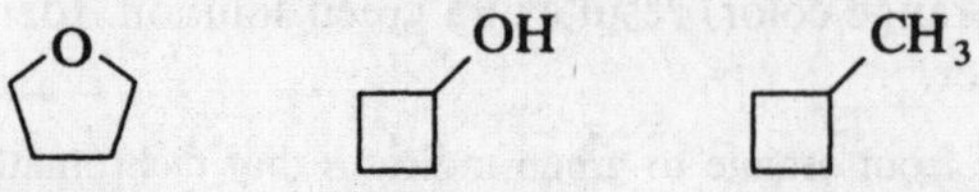

Tetrahydrofuran   Cyclobutanol   Methylcyclobutane

Ans. Cyclobutanol has a much higher boiling point than tetrahydrofuran and methylcyclobutane. Tetrahydrofuran and methylcyclobutane are close in boiling point. The boiling point of tetrahydrofuran is somewhat higher.

THIOALCOHOLS

Problem 13.47. Give the equation for each of the following reactions. If there is more than one product, indicate which is the major product(s) and which the minor. If there is no reaction, state "No Reaction."

(*a*) $HS(CH_2)_3SH \xrightarrow{(O)} ?$ (*b*) $C_6H_5{-}S{-}S{-}C_6H_5 \xrightarrow{(H)} ?$

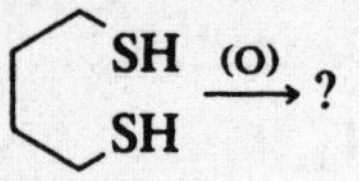

Ans.

(*a*) $HS(CH_2)_3SH \xrightarrow{(O)}$ cyclic disulfide (ring with S–S)

(*b*) $C_6H_5{-}S{-}S{-}C_6H_5 \xrightarrow{(H)} 2\ C_6H_5{-}SH$

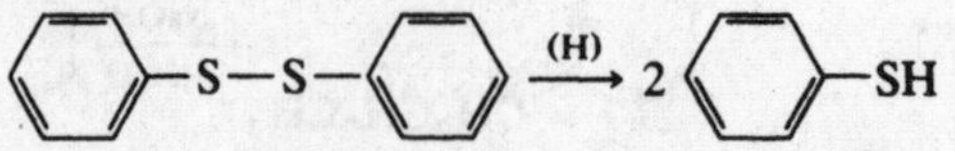

SUPPLEMENTARY PROBLEMS

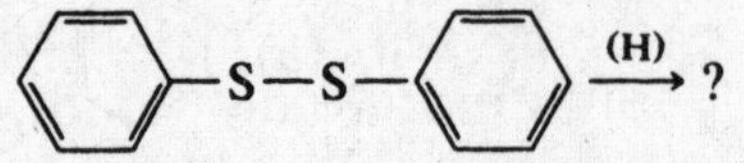

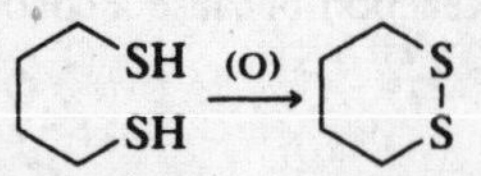

Problem 13.48. Name the functional groups to which the arrows point:

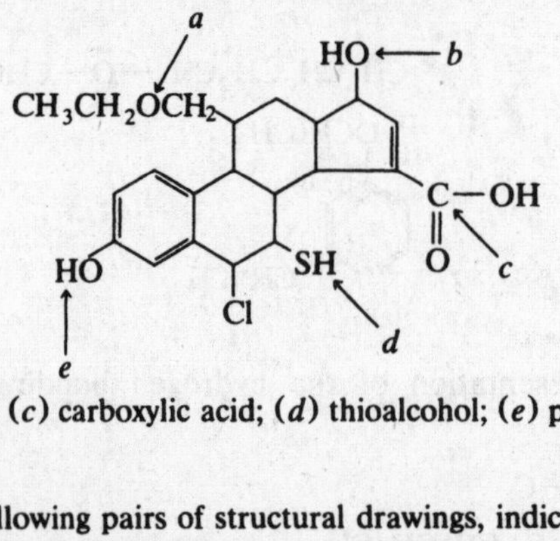

(Structure with labeled groups: *a* → $CH_3CH_2OCH_2$–; *b* → HO–; *c* → –C(=O)–OH; *d* → SH; *e* → HO on aromatic ring; also Cl.)

Ans. (*a*) ether; (*b*) alcohol; (*c*) carboxylic acid; (*d*) thioalcohol; (*e*) phenol.

Problem 13.49. For each of the following pairs of structural drawings, indicate whether the pair represents (1) the same compound, or (2) different compounds which are constitutional isomers, or (3) different compounds which are *cis* and *trans* isomers, or (4) different compounds which are not isomers.

(*a*) $CH_3CH_2CH_2OCH(CH_3)_2$ and $CH_3{-}CH(CH_3){-}CH_2{-}CH_2{-}O{-}CH_3$

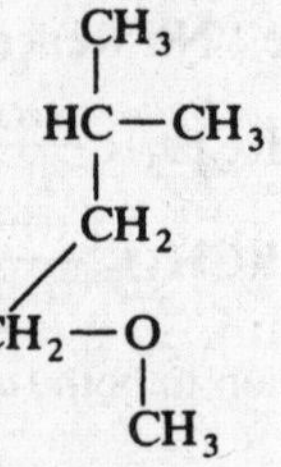

(*b*) $CH_3CH_2CH_2OCH(CH_3)_2$ and $CH_3CH_2CHCH(CH_3)_2$ (with OH on the CH)

(*c*) cyclohexanol (OH) and phenol (OH)

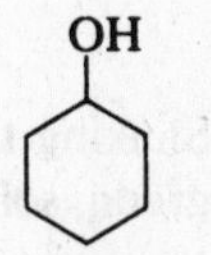

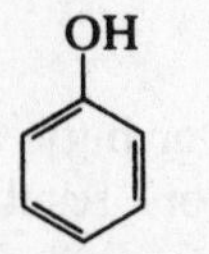

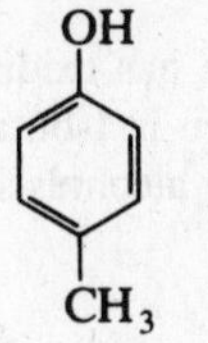

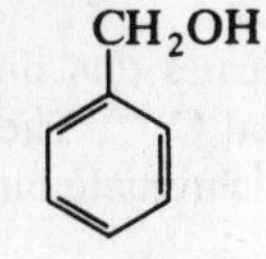

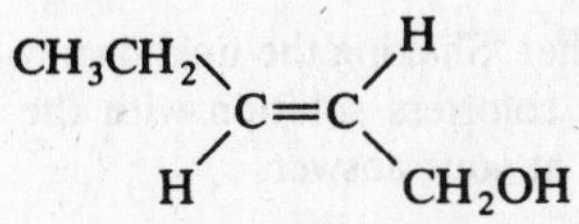

(*d*) 4-methylphenol (OH, CH_3) and benzyl alcohol (CH_2OH)

(*e*) CH_3CH_2(H)C=C(H)(CH_2OH) and CH_3CH_2(H)C=C(CH_2OH)(H)

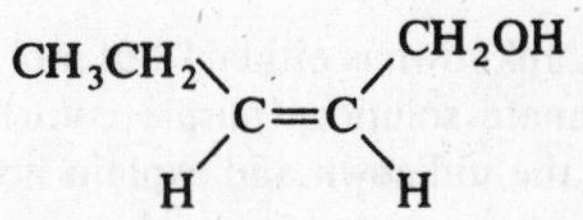

Ans. **Pairs (*a*), (*b*), and (*d*) are constitutional isomers. Pair (*c*) represents a pair of different compounds that are not isomers. Pair (*e*) represents a pair of *cis* and *trans* isomers.**

Problem 13.50. How many constitutional isomers exist for the molecular formula C_3H_6O? Show only one structural formula for each isomer.

Ans. **C_3H_6O fits the formula C_nH_{2n}, which means that it has a deficiency of two hydrogens compared to the corresponding alkane (C_nH_{2n+2}). C_3H_6O fits either a cycloalkane or alkene that also belongs to either the ether or alcohol family. There are a total of six isomers:**

I: $CH_2{=}CH{-}CH_2OH$

II: $CH_2{=}C(OH){-}CH_3$

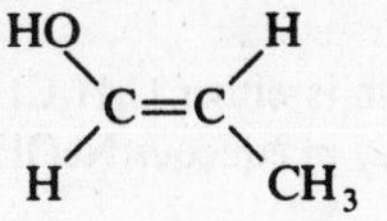

III: (HO)(H)C=C(H)(CH_3)

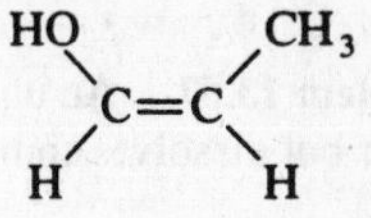

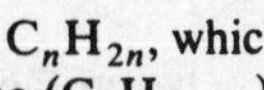

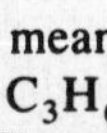

IV: (HO)(H)C=C(CH_3)(H)

V: cyclopropyl–OH

VI: oxetane (four-membered ring with O)

VII: $CH_2{=}CH{-}OCH_3$

Problem 13.51. Which of the compounds in Problem 13.50 are ethers and which are alcohols?

Ans. **Compounds VI and VII are ethers. Compounds I and V are alcohols. Compounds II, III, and IV are not alcohols since the OH group is not attached to a saturated carbon; they are enols.**

Problem 13.52. Draw structural formulas of the following compounds: (*a*) 3-chloro-2-methyl-4-ethoxyhexane, (*b*) 3-methyl-4-hexene-2-ol.

Ans. (*a*) $CH_3{-}CH(CH_3){-}CH(Cl){-}CH(OCH_2CH_3){-}CH_2{-}CH_3$ (*b*) $CH_3{-}CH(OH){-}CH(CH_3){-}CH{=}CH{-}CH_3$

Problem 13.53. Give the order of increasing water solubility for the compounds methyl ethyl ether, 1,2-butanediol, butane, and 1-butanol.

Ans. 1,2-butanediol > 1-butanol = methyl ethyl ether > butane.

Problem 13.54. An unknown is either *t*-butyl alcohol or butyl alcohol. Shaking the unknown with a few drops of acidic dichromate solution (orange color) results in a green solution. Identify the unknown and explain how you arrive at your answer.

Ans. The color change from orange to green indicates that dichromate has oxidized the unknown and has itself been converted to the green-colored Cr^{3+}. The unknown is 1-butanol since primary (as well as secondary) alcohols are oxidized by dichromate but tertiary alcohols such as *t*-butyl alcohol are not oxidized by dichromate.

Problem 13.55. An unknown is either 1-butanol or diethyl ether. Shaking the unknown with a few drops of basic potassium permanganate solution (purple color) results in a colorless solution with the formation of a brown precipitate. Identify the unknown and explain how you arrive at your answer.

Ans. The loss of purple color and formation of brown MnO_2 precipitate shows that the unknown has been oxidized. The unknown is 1-butanol since primary (as well as secondary) alcohols are oxidized by permanganate but ethers are not.

Problem 13.56. An unknown is either 2-methyl-2-propanol or diethyl ether. Shaking the unknown with a few drops of basic potassium permanganate solution (purple color) does not result in a color change or brown precipitate. Identify the unknown and explain how you arrive at your answer.

Ans. The test with permanganate cannot distinguish between the two possibilities since neither ethers nor tertiary alcohols are oxidized by permanganate (or dichromate).

Problem 13.57. An unknown is either $C_6H_5CH_2OH$ or *p*-$CH_3C_6H_4OH$. The unknown is only slightly soluble in water but dissolves completely in aqueous NaOH. Identify the unknown.

Ans. Both compounds would be only slightly soluble in water but only *p*-$CH_3C_6H_4OH$, since it is a phenol, is sufficiently acidic to dissolve completely in a sodium hydroxide solution.

Problem 13.58. Complete each of the following reactions by writing the structure of the organic product. If no reaction occurs, write "No Reaction." If there is more than one product, indicate which is the major product and which the minor product.

(*a*) $CH_3-C(CH_3)_2-OH \xrightarrow[180\ ^\circ C]{H_2SO_4} ?$

(*b*) $CH_3-C(CH_3)_2-OH \xrightarrow[140\ ^\circ C]{H_2SO_4} ?$

(*c*) 1-methylcyclopentanol (cyclopentane ring bearing OH and CH_3 on the same carbon) $\xrightarrow{(O)} ?$

(*d*) cyclopentanol (cyclopentane ring bearing OH) $\xrightarrow{(O)} ?$

(*e*) cyclopentanol (cyclopentane ring bearing OH) $\xrightarrow{(H)} ?$

(*f*) 2-methyltetrahydrofuran (five-membered ring containing O, with CH_3 on the carbon next to O) $\xrightarrow{\text{combustion}} ?$

(*g*) $CH_3CH_2CH_2-S-S-CH_2CH_2CH_3 \xrightarrow{(O)} ?$

(*h*) cyclohexyl$-CH_2-OH \xrightarrow{(O)} ?$

(*i*) $CH_3CH_2CH_2—S—S—CH_2CH_2CH_3 \xrightarrow{(H)} ?$

(*j*) $C_6H_{11}—CH_2—OH \xrightarrow{(H)} ?$

(*k*) $C_6H_{11}—CH_2—OH \xrightarrow{NaOH} ?$

(*l*) $C_6H_{11}—CH_2—SH \xrightarrow{(O)} ?$

(*m*) $C_6H_{11}—CH_2—SH \xrightarrow{(H)} ?$

Ans.

(*a*) $CH_3—C(CH_3)_2—OH \xrightarrow[180\ ^\circ C]{H_2SO_4} CH_3—C(=CH_2)—CH_3$

(*b*) $CH_3—C(CH_3)_2—OH \xrightarrow[140\ ^\circ C]{H_2SO_4} CH_3—C(=CH_2)—CH_3$.

(*c*) No Reaction.

(*d*) cyclopentanol $\xrightarrow{(O)}$ cyclopentanone

(*e*) No Reaction.

(*f*) $C_5H_{10}O + 7\ O_2 \longrightarrow 5\ CO_2 + 5\ H_2O$

(*g*) No Reaction.

(*h*) $C_6H_{11}—CH_2—OH \xrightarrow{(O)} C_6H_{11}—CH{=}O \xrightarrow{(O)} C_6H_{11}—C(OH){=}O$

Unless the aldehyde is removed by distillation immediately as formed, the major product is the carboxylic acid.

(*i*) $CH_3CH_2CH_2—S—S—CH_2CH_2CH_3 \xrightarrow{(H)} 2\ CH_3CH_2CH_2—SH$

(*j*) No Reaction.
(*k*) No (or negligible) Reaction.

(*l*) $2\ C_6H_{11}—CH_2—SH \xrightarrow{(O)} C_6H_{11}—CH_2S—SCH_2—C_6H_{11}$

(*m*) No Reaction.

Problem 13.59. Give an equation to show the starting alcohol needed to obtain each of the indicated products. Indicate other required reagents and/or reaction conditions.

(*a*) $CH_3CH_2—CH{=}CH_2$ (*b*) $CH_3CH_2CH_2—O—CH_2CH_2CH_3$

(*c*) $CH_3CH_2—C(=O)—CH_3$ (*d*) $CH_3CH_2—C(=O)—OH$

Ans.

(*a*) $$CH_3CH_2-CH_2-CH_2-OH \xrightarrow[180\ °C]{H_2SO_4} CH_3CH_2-CH=CH_2$$

(*b*) $$2\ CH_3CH_2CH_2-OH \xrightarrow[140\ °C]{H_2SO_4} CH_3CH_2CH_2-O-CH_2CH_2CH_3$$

(*c*) $$CH_3CH_2-\underset{}{\overset{OH}{\overset{|}{C}H}}-CH_3 \xrightarrow{(O)} CH_3CH_2-\overset{O}{\overset{\|}{C}}-CH_3$$

(*d*) $$CH_3CH_2-CH_2-OH \xrightarrow{(O)} \underset{\text{Aldehyde}}{CH_3CH_2-\overset{O}{\overset{\|}{C}}-H} \xrightarrow{(O)} \underset{\text{Carboxylic acid}}{CH_3CH_2-\overset{O}{\overset{\|}{C}}-OH}$$

Carry out the reaction without allowing the aldehyde to distill out of the reaction prior to its oxidation to the carboxylic acid.

Problem 13.60. Write the step-by-step mechanism for the acid-catalyzed dehydration of 1-propanol to yield propene.

Ans.

$$CH_3-CH_2-CH_2-OH \xrightarrow{H^+} CH_3-CH_2-CH_2-\overset{+}{O}H_2 \xrightarrow{-H_2O}$$

$$CH_3-CH_2-\overset{+}{C}H_2 \xrightarrow{-H^+} CH_3-CH=CH_2$$

Aldehydes and Ketones

14.1 STRUCTURE OF ALDEHYDES AND KETONES

Aldehydes and ketones contain a *carbonyl group* ($C{=}O$) in which the carbonyl carbon is not directly attached to any atoms other than hydrogen or carbon. *Aldehydes* have either one or two hydrogens directly attached to the carbonyl carbon. *Ketones* contain only carbons directly attached to the carbonyl carbon. Methanal (common name: formaldehyde) is the only aldehyde with two hydrogens attached to the carbonyl carbon. The simplest aldehyde with one hydrogen attached to the carbonyl carbon is ethanal (common name: acetaldehyde). The simplest ketone is propanone (common names: dimethyl ketone, acetone).

$$H{-}\overset{\overset{O}{\|}}{C}{-}H \qquad CH_3{-}\overset{\overset{O}{\|}}{C}{-}H \qquad CH_3{-}\overset{\overset{O}{\|}}{C}{-}CH_3$$

Methanal, **Formaldehyde** — **Ethanal**, **Acetaldehyde** — **Propanone**, **Dimethyl ketone**, **Acetone**

Formaldehyde has been used for preserving biological specimens and manufacturing plastics. Acetone is a useful organic solvent. Aldehydes and ketones are used by organic chemists for synthesizing other organic compounds. A variety of biological molecules are aldehydes or ketones, e.g., lipids (progesterone, testosterone) and carbohydrates.

Problem 14.1. Which of the following are aldehydes or ketones and which are not?

$$H{-}\overset{\overset{O}{\|}}{C}{-}CH_2CH_2CH(CH_3)_2 \qquad CH_3{-}\overset{\overset{O}{\|}}{C}{-}OCH_3 \qquad \text{(cyclopentanone)} \qquad CH_3{-}\overset{\overset{O}{\|}}{C}{-}NH_2$$

I — II — III — IV

Ans. Compound I is an aldehyde; compound III is a ketone. Compounds II and IV have carbonyl groups but the carbonyl groups are not those of an aldehyde or ketone. Compound II is an ester and compound IV is an amide since the carbonyl carbons are bonded to OR and NH_2, respectively.

Problem 14.2. Compare the bonding in $C{=}C$ and $C{=}O$ double bonds.

Ans. Oxygen is similar to carbon, sp^3-hybridized for single bonds and sp^2-hybridized for double bonds (Sec. 12.2). Both types of double bonds have a strong σ-bond (formed from sp^2 orbitals) coupled with a weak π-bond (formed from p orbitals). Oxygen differs from carbon in that two of the three sp^2 orbitals contain nonbonding pairs of electrons instead of forming bonds to H or C.

Problem 14.3. Give the bond angles for *A*, *B*, and *C*:

$$CH_3{-}\overset{\overset{O}{\|}}{C}{-}H$$

(*A*: angle CH_3–C=O; *B*: angle CH_3–C–H; *C*: angle O=C–H)

Ans. $A = B = C = 120^\circ$. The bond angles about any sp^2-hybridized atom, whether carbon or oxygen, are 120° with the atoms bonded to the C or O through sp^2 orbitals lying in a flat plane.

14.2 CONSTITUTIONAL ISOMERISM IN ALDEHYDES AND KETONES

The presence of oxygen is ignored in evaluating the C:H ratio of an unknown compound. Each pair of missing hydrogens compared to C_nH_{2n+2} corresponds to the presence of a ring or extra bond. For the possibility where there is an extra bond, the C:H ratio does not distinguish between a C=C and C=O.

Problem 14.4. How many aldehydes and ketones exist for the molecular formula C_3H_6O?

Ans. There is one aldehyde and one ketone:

$$CH_3-CH_2-\overset{\displaystyle O}{\overset{\|}{C}}H \quad \text{and} \quad CH_3-\overset{\displaystyle O}{\overset{\|}{C}}-CH_3$$

Problem 14.5. How many compounds exist for the molecular formula C_3H_6O?

Ans. This is a broader question than Problem 14.4. The molecular formula fits the general formula C_nH_{2n}, which is one pair of hydrogens less than an alkane. There is one ring or one extra bond present in C_3H_6O. We have the aldehyde and ketone shown in Problem 14.4 if the extra bond is in a C=O group. If the extra bond is in a C=C group, we have three different alkene-alcohols and one alkene-ether. If there is a ring instead of an extra bond, we have two possibilities—a cycloalkane-alcohol and a cyclic ether.

$CH_2=CHCH_2-OH$ $\quad$ (CH₃)(H)C=C(OH)(H) $\quad$ (CH₃)(H)C=C(H)(OH) $\quad$ $CH_2=CH-OCH_3$

(cyclopropanol) $\quad$ (oxetane)

Problem 14.6. How many aldehydes and ketones exist for the molecular formula $C_5H_{10}O$?

Ans. The molecular formula contains one pair of hydrogens less than the general formula for an alkane. Since the question specifies aldehydes and ketones, other possibilities for one extra bond or ring are excluded. There are four aldehydes and three ketones for $C_5H_{10}O$:

$$\underset{\text{I}}{CH_3CH_2CH_2CH_2-\overset{\displaystyle O}{\overset{\|}{C}}-H} \qquad \underset{\text{II}}{CH_3CH_2\overset{\displaystyle CH_3}{\overset{|}{C}H}-\overset{\displaystyle O}{\overset{\|}{C}}-H} \qquad \underset{\text{III}}{CH_3\overset{\displaystyle CH_3}{\overset{|}{C}H}CH_2-\overset{\displaystyle O}{\overset{\|}{C}}-H}$$

$$\underset{\text{IV}}{(CH_3)_3C-\overset{\displaystyle O}{\overset{\|}{C}}-H} \qquad \underset{\text{V}}{CH_3-\overset{\displaystyle O}{\overset{\|}{C}}-CH_2CH_2CH_3} \qquad \underset{\text{VI}}{CH_3CH_2-\overset{\displaystyle O}{\overset{\|}{C}}-CH_2CH_3}$$

$$\underset{\text{VII}}{CH_3-\overset{\displaystyle O}{\overset{\|}{C}}-CH(CH_3)_2}$$

14.3 NOMENCLATURE OF ALDEHYDES AND KETONES

The IUPAC rules follow those for the alkanes with a few modifications:

1. The longest continuous chain must contain the carbonyl carbon.
2. The ending of the name is changed from *-ane* to *-anal* for aldehydes and from *-ane* to *-anone* for ketones.
3. The longest continuous chain is numbered from the end that gives the smallest number for the carbonyl carbon. For aldehydes the carbonyl carbon is understood to be carbon-1 and the number 1 is not used in the name. For the three and four carbon ketones, a number is not needed because only one ketone is possible in each case.

Cyclic compounds containing an attached aldehyde group HC=O are named by adding *-carbaldehyde* to the name of the cyclic compound. The ring carbon holding the HC=O is C-1.

Common names for acyclic aldehydes containing 1 to 4 carbons are obtained by combining the prefixes *form-*, *acet-*, *propion-*, and *butyr-*with the word *-aldehyde,* i.e., formaldehyde, acetaldehyde, propionaldehyde, and butyraldehyde. C_6H_5-CHO is benzaldehyde.

Common names for some ketones are obtained by naming the groups attached to the carbonyl carbon followed by the word *ketone*. Dimethyl ketone is also known as acetone.

Problem 14.7. Name the compounds in Problem 14.6 by the IUPAC system. Indicate common names where appropriate.

Ans. The longest continuous chain containing the carbonyl carbon is five carbons for compounds I, V, and VI, four carbons for compounds II, III, and VII, and three carbons for compound IV. IUPAC names: pentanal for compound I, 2-methylbutanal for compound II, 3-methylbutanal for compound III, 2,2-dimethylpropanal for compound IV, 2-pentanone for compound V, 3-pentanone for compound VI, and 3-methylbutanone for compound VII.

The common names are methyl propyl ketone for compound V, diethyl ketone for compound VI, and isopropyl methyl ketone for compound VII.

Problem 14.8. Name the following compounds:

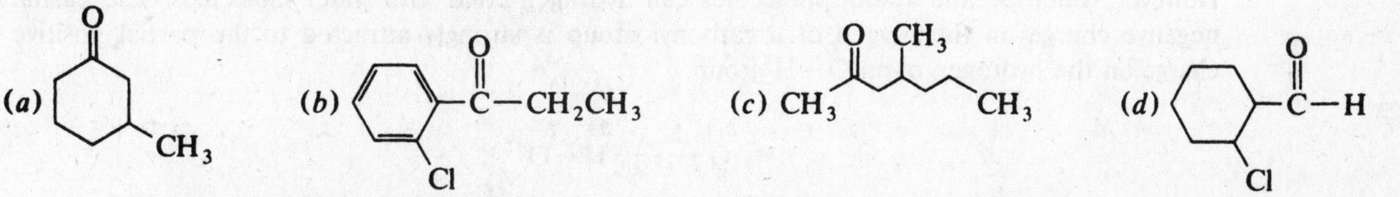

(*a*) cyclohexanone ring (O at C-1) with CH_3 (*b*) benzene ring with Cl, –C(=O)–CH_2CH_3 (*c*) CH_3–C(=O)–CH$_2$–CH(CH_3)–CH$_2$–CH$_2$–CH_3 (*d*) cyclohexane ring with Cl, –C(=O)–H

Ans. (*a*) 3-Methylcyclohexanone. (*b*) Common names are used for ketones where one (or both) substituent(s) attached to the carbonyl carbon is (are) cyclic (aromatic or nonaromatic): 2-chlorophenyl ethyl ketone or *o*-chlorophenyl ethyl ketone. (*c*) 4-Methyl-2-heptanone. (*d*) 2-Chlorocyclohexanecarbaldehyde.

14.4 PHYSICAL PROPERTIES OF ALDEHYDES AND KETONES

The carbonyl group is very different from the C=C. The weakly held electrons of the π-bond are closer to oxygen than carbon since oxygen is more electronegative than carbon. This polarization of the π-bond, resulting in a partial positive charge on carbon and partial negative charge on oxygen, is often expressed by using the structural representation

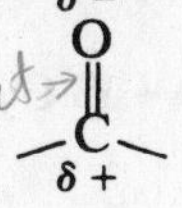

$$\overset{\delta-}{O}=\underset{\delta+}{C}<$$

Chemists also express the same idea by showing the carbonyl group as a resonance hybrid of the two structures

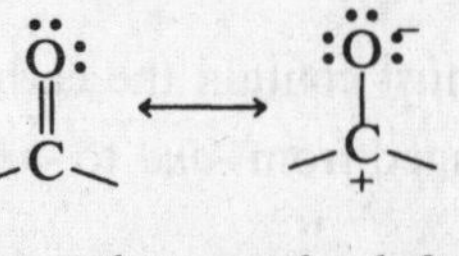

$$\begin{array}{c} \ddot{\text{O}}\text{:} \\ \| \\ \diagup\text{C}\diagdown \end{array} \longleftrightarrow \begin{array}{c} \text{:}\ddot{\text{O}}\text{:}^{-} \\ | \\ \diagup\text{C}^{+}\diagdown \end{array}$$

The structure on the right is obtained from that on the left by moving the two electrons from the π-bond onto oxygen. This results in + and − charges on C and O.

Problem 14.9. **Explain the order of boiling points in the following compounds:**

$CH_3CH_2CH_2CH_2CH_3$	36 °C	$CH_3CH_2CH_2\overset{\overset{\text{O}}{\|}}{C}H$	76 °C
$CH_3\overset{\overset{\text{O}}{\|}}{C}CH_2CH_3$	80 °C	$CH_3CH_2CH_2CH_2OH$	118 °C

Ans. **The molecular masses of the compounds are comparable. The differences in boiling point are due to differences in the intermolecular attractive forces. Boiling point increases with increasing intermolecular attractive forces. The lowest intermolecular forces are dispersion forces which are present in alkanes. Aldehydes and ketones have the higher attractive forces—those due to dipole-dipole interactions—because of the polar carbonyl group. Alcohols have higher intermolecular attractive forces because of the hydrogen bonding which results from the presence of OH groups.**

Problem 14.10. **Although the boiling and melting points of aldehydes and ketones are considerably lower than those of alcohols, there is little difference in their solubilities in water. Explain why aldehydes and ketones are as soluble in water as are alcohols.**

Ans. **Molecules of an aldehyde or ketone cannot hydrogen-bond together since there is no O—H group. However, aldehyde and ketone molecules can hydrogen-bond with water molecules. The partial negative charge on the oxygen of a carbonyl group is strongly attracted to the partial positive charge on the hydrogen of an O—H group.**

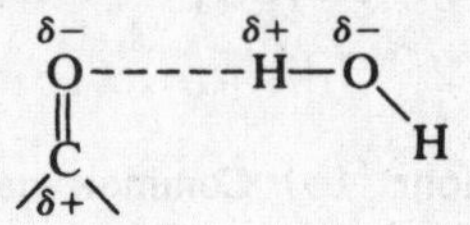

$$\begin{array}{l} \overset{\delta-}{\text{O}}\text{-----}\overset{\delta+}{\text{H}}\text{—}\overset{\delta-}{\text{O}} \\ \| \qquad\qquad\quad \diagdown \\ \text{C} \qquad\qquad\quad\ \text{H} \\ \diagup^{\delta+}\diagdown \end{array}$$

14.5 OXIDATION OF ALDEHYDES AND KETONES

Aldehydes and ketones undergo combustion to CO_2 and H_2O as do all organic compounds.

Problem 14.11. **Write the balanced equations for the combustion of the following compounds: (*a*) butanone, (*b*) butanal.**

Ans. The same equation describes the combustion of butanone and butanal since the two compounds are constitutional isomers:

$$2\,C_5H_8O + 13\,O_2 \longrightarrow 10\,CO_2 + 8\,H_2O$$

$C_4H_8O + 6O_2 \rightarrow 4CO_2 + 4H_2O$

Problem 14.12. Write an equation to show the product(s) of mild oxidation of the following compounds: (*a*) butanone, (*b*) butanal.

Ans.

(*a*) $$CH_3\overset{O}{\overset{\|}{C}}CH_2CH_3 \xrightarrow{(O)} \text{No Reaction}$$

(*b*) $$CH_3CH_2CH_2\overset{O}{\overset{\|}{C}}H \xrightarrow{(O)} CH_3CH_2CH_2\overset{O}{\overset{\|}{C}}{-}OH$$

This is not new information. We learned, during our study of the mild oxidation of alcohols (Sec. 13.5.3), that aldehydes undergo mild oxidation to carboxylic acids but ketones are resistant to mild oxidation. Mild oxidation of a primary alcohol with permanganate or dichromate proceeds initially to yield an aldehyde and then the aldehyde is oxidized to a carboxylic acid. Mild oxidation of a secondary alcohol yields a ketone which does not oxidize further.

Permanganate and dichromate distinguish an aldehyde from a ketone but not from primary or secondary alcohols. *Tollens' reagent* is a selective (mild) oxidizing reagent that forms the basis for a simple chemical test to distinguish aldehydes from ketones as well as from primary and secondary alcohols.

Tollens' reagent contains an alkaline solution of silver ion (Ag^+) in the form of its complex ion with ammonia, $[Ag(NH_3)_2]^+$. A positive test for an aldehyde is easily detected. Tollens' reagent is colorless. Silver ion is reduced to silver metal which precipitates from the solution as a gray solid. This usually coats the inside of the test tube and gives the appearance of a silver mirror.

Benedict's reagent is a specialized selective oxidizing reagent that is useful for distinguishing α-hydroxy aldehydes and α-hydroxy ketones from other aldehydes and ketones.

Benedict's reagent contains an alkaline solution of cupric ion (Cu^{2+}) complexed with citrate anion. A positive test is detected by the disappearance of the blue color of cupric ion and the formation of a red precipitate. The red precipitate is Cu_2O which forms when cupric ion is reduced to cuprous (Cu^+).

The designation α- refers to a carbon atom adjacent to a C=O. An α-hydroxy aldehyde or α-hydroxy ketone contains an OH group on a carbon adjacent to the C=O:

$$H{-}\overset{O}{\overset{\|}{C}}{-}\underset{}{\overset{OH}{\overset{|}{CH}}}{-}R \qquad\qquad R{-}\overset{O}{\overset{\|}{C}}{-}\overset{OH}{\overset{|}{CH}}{-}R'$$

α-Carbon (arrow to the CH carbon) — α-Carbon (arrow to the C=O carbon)

α-Hydroxy aldehyde — α-Hydroxy ketone

α-Hydroxy aldehydes and α-hydroxy ketones give positive tests not only with Benedict's reagent, but also with Tollens' reagent.

Problem 14.13. What structural feature in carbohydrates such as glucose and fructose is responsible for their positive tests with Tollens' and Benedict's reagents?

$$\underset{OH}{\underset{|}{CH_2}}{-}\underset{OH}{\underset{|}{CH}}{-}\underset{OH}{\underset{|}{CH}}{-}\underset{OH}{\underset{|}{CH}}{-}\underset{OH}{\underset{|}{CH}}{-}\overset{O}{\overset{\|}{CH}} \qquad\qquad \underset{OH}{\underset{|}{CH_2}}{-}\underset{OH}{\underset{|}{CH}}{-}\underset{OH}{\underset{|}{CH}}{-}\underset{OH}{\underset{|}{CH}}{-}\overset{O}{\overset{\|}{C}}{-}\underset{OH}{\underset{|}{CH_2}}$$

Glucose — Fructose

Ans. Glucose is an α-hydroxy aldehyde and fructose is an α-hydroxy ketone. Benedict's reagent has served as the basis of a clinical test for glucose and other carbohydrates in the urine and blood.

Problem 14.14. For each of the following compounds, indicate whether it gives a positive test with Benedict's or Tollens' reagents:

$$\underset{\text{I}}{C_6H_5-\overset{O}{\overset{\|}{C}}-\overset{OH}{\overset{|}{C}H}-CH_3} \qquad \underset{\text{II}}{HO-\overset{O}{\overset{\|}{C}}-CH_2-\overset{OH}{\overset{|}{C}H}-CH_3} \qquad \underset{\text{III}}{CH_3-\overset{O}{\overset{\|}{C}}-CH_2-\overset{O}{\overset{\|}{C}}-H} \qquad \underset{\text{IV}}{CH_3-\overset{O}{\overset{\|}{C}}-H}$$

Ans. Compounds III and IV give positive tests with Tollens' reagent because they are aldehydes, but do not react with Benedict's reagent because they are not α-hydroxy aldehydes or α-hydroxy ketones. Compound I gives positive tests with both Benedict's and Tollens' reagents because it is an α-hydroxy aldehyde. Compound II does not react with either reagent because it is not an aldehyde, α-hydroxy aldehyde, or α-hydroxy ketone.

14.6 REDUCTION OF ALDEHYDES AND KETONES

Reduction of the carbonyl group converts an aldehyde or ketone to the corresponding alcohol. The overall result involves the addition of hydrogen atoms to the C and O of the carbonyl group:

$$-\overset{O}{\overset{\|}{C}}- \longrightarrow -\underset{H}{\underset{|}{\overset{OH}{\overset{|}{C}}}}-$$

Problem 14.15. Why is the conversion of an aldehyde or ketone to an alcohol referred to as reduction?

Ans. Ignoring the carbons attached to it, the carbonyl carbon has an O:H ratio of 1:0. The corresponding carbon in the alcohol has an O:H ratio of 1:2. The decrease in O content (increase in H content) is defined as reduction.

Reduction of the carbonyl group is accomplished in the laboratory by two methods:

1. *Catalytic hydrogenation:* An aldehyde or ketone is reacted with H_2 in the presence of a metal catalyst such as platinum, nickel or palladium:

$$-\overset{O}{\overset{\|}{C}}- \xrightarrow[\text{Pt}]{H_2} -\underset{H}{\underset{|}{\overset{OH}{\overset{|}{C}}}}-$$

   This is analogous to the catalytic reduction of the alkene double bond (Sec. 12.6.1).

2. *Hydride reduction:* An aldehyde or ketone is first treated with a metal hydride and then with water. Examples of metal hydrides are NaH, $LiAlH_4$, and $NaBH_4$.

Problem 14.16. How do metal hydrides differ from other compounds containing hydrogen?

Ans. In all ionic or highly polar compounds of hydrogen (e.g., HCl and H_2O) other than metal hydrides, hydrogen is bonded to an atom that is more electronegative than itself. The H in such bonds acts as H^+ with the other atom carrying a negative charge, e.g., Cl^- and HO^-. In metal hydrides, H is bonded to a metal which is less electronegative than itself. The H in such bonds acts as H:⁻ while the other atom carries a positive charge. H:⁻ is referred to as a *hydride ion*.

Problem 14.17. What is the mechanism of hydride reduction of the carbonyl group?

Ans. The reaction is visualized in terms of the ionic structure (Sec. 14.4) for the carbonyl group. Hydride reduction proceeds in two steps. First, the metal hydride supplies a hydride ion which adds to the C^+ of the carbonyl group. Second, H_2O supplies H^+ which adds to the O^- of the carbonyl group:

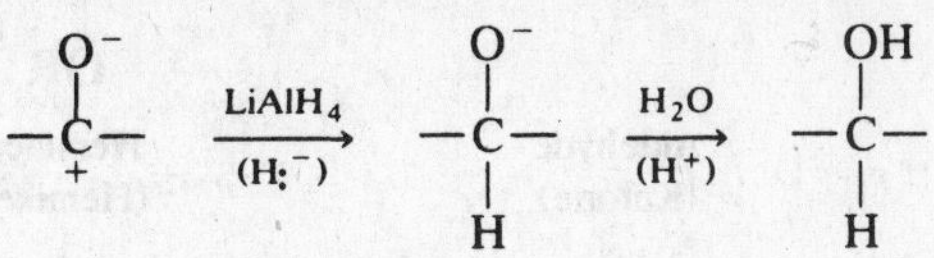

$$\underset{+}{-\overset{O^-}{\overset{|}{C}}-} \xrightarrow[(H:^-)]{LiAlH_4} -\overset{O^-}{\overset{|}{\underset{\underset{H}{|}}{C}}}- \xrightarrow[(H^+)]{H_2O} -\overset{OH}{\overset{|}{\underset{\underset{H}{|}}{C}}}-$$

Problem 14.18. Complete the following reactions by writing the structure of the organic product:

(*a*) cyclohexanone $\xrightarrow[Pt]{H_2}$? (*b*) $CH_3CH_2CH_2\overset{O}{\overset{\|}{C}}H \xrightarrow[2.\ H_2O]{1.\ LiAlH_4}$?

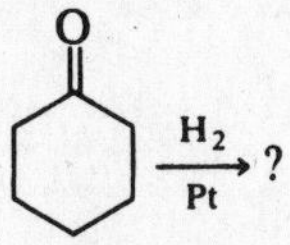

Ans.

(*a*) cyclohexanone $\xrightarrow[Pt]{H_2}$ cyclohexanol (OH)

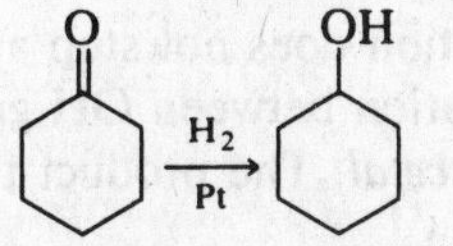

(*b*) $CH_3CH_2CH_2\overset{O}{\overset{\|}{C}}H \xrightarrow[2.\ H_2O]{1.\ LiAlH_4} CH_3CH_2CH_2CH_2OH$

Problem 14.19. In part (*b*) of Problem 14.18, is it necessary to specify "1." and "2." before the reagents $LiAlH_4$ and H_2O?

Ans. Yes. Specifying "1." and "2." gives the required sequence for carrying out the reaction. One adds $LiAlH_4$ to the carbonyl compound, waits some period of time, and then adds H_2O. This is how this reaction must be carried out to successfully convert an aldehyde or ketone to alcohol. Leaving "1." and "2." out means that you are instructing someone to simultaneously add $LiAlH_4$ and H_2O to the carbonyl compound. This would not result in reduction of the carbonyl group. $LiAlH_4$ reacts faster with H_2O than with C=O. The only reaction would be that between the H:⁻ from $LiAlH_4$ and H^+ from H_2O to form H_2. To obtain reduction of C=O, the addition of H_2O must be delayed until after $LiAlH_4$ has reacted with C=O.

Some reactions require a sequence for addition of reagents. Other reactions do not.

Nicotinamide adenine dinucleotide (NADH) supplies a hydride ion for reductions of carbonyl groups in biological systems. NADH is converted to NAD^+ in the process. An example is the reduction of pyruvic acid to lactic acid, which occurs during the metabolism of glucose:

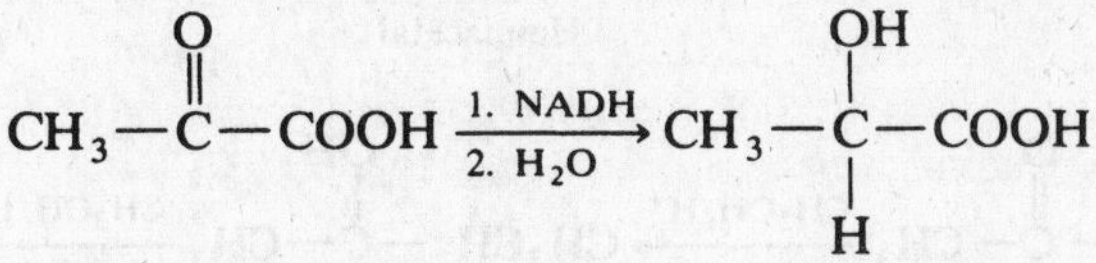

$$\underset{\text{Pyruvic acid}}{CH_3-\overset{O}{\overset{\|}{C}}-COOH} \xrightarrow[2.\ H_2O]{1.\ NADH} \underset{\text{Lactic acid}}{CH_3-\overset{OH}{\overset{|}{\underset{\underset{H}{|}}{C}}}-COOH}$$

14.7 REACTION OF ALDEHYDES AND KETONES WITH ALCOHOL

Alcohols add to the carbonyl groups of aldehydes and ketones to yield *hemiacetals* and *hemiketals*, respectively. The reaction requires an acid catalyst in the laboratory or an enzyme in biological systems.

$$\underset{\text{Aldehyde (Ketone)}}{R'{-}\overset{O}{\overset{\|}{C}}{-}H(R'')} \xrightarrow{ROH,\ H^+} \underset{\text{Hemiacetal (Hemiketal)}}{R'{-}\underset{OR}{\underset{|}{\overset{OH}{\overset{|}{C}}}}{-}H(R'')}}$$

The reaction is visualized in terms of the ionic form of the carbonyl group. ROH is a source of RO^- and H^+ ions which add to the C^+ and O^- of C=O:

$$-\overset{O^-}{\overset{|}{\underset{+}{C}}}- \ \ \overset{H^+}{}\ \ \overset{}{RO^-} \longrightarrow -\underset{OR}{\underset{|}{\overset{OH}{\overset{|}{C}}}}-$$

If there is excess alcohol present, the reaction does not stop at the hemiacetal or hemiketal. The hemicompound reacts with alcohol via dehydration between OH groups of the alcohol and hemicompound. The product from a hemiacetal is an *acetal*. The product from a hemiketal is a *ketal*.

$$\underset{\text{Hemiacetal (Hemiketal)}}{R'{-}\underset{OR}{\underset{|}{\overset{OH}{\overset{|}{C}}}}{-}H(R'')} \xrightarrow[-H_2O]{ROH,\ H^+} \underset{\text{Acetal (Ketal)}}{R'{-}\underset{OR}{\underset{|}{\overset{OR}{\overset{|}{C}}}}{-}H(R'')}}$$

The reaction is similar to the reaction by which an ether is formed via dehydration between two molecules of alcohol (Sec. 13.7).

Some texts do not use the terms hemiketal and ketal. Those texts use hemiacetal for both hemiacetals and hemiketals, and acetal for both acetals and ketals.

Problem 14.20. Write equations to show the reactions of benzaldehyde and butanone with excess methanol. Show both the hemiacetal (hemiketal) and acetal (ketal) compounds.

Ans.

$$C_6H_5{-}\overset{O}{\overset{\|}{C}}{-}H \xrightarrow{CH_3OH,\ H^+} \underset{\text{Hemiacetal}}{C_6H_5{-}\underset{OCH_3}{\underset{|}{\overset{OH}{\overset{|}{C}}}}{-}H} \xrightarrow[-H_2O]{CH_3OH,\ H^+} \underset{\text{Acetal}}{C_6H_5{-}\underset{OCH_3}{\underset{|}{\overset{OCH_3}{\overset{|}{C}}}}{-}H}$$

$$CH_3CH_2{-}\overset{O}{\overset{\|}{C}}{-}CH_3 \xrightarrow{CH_3OH,\ H^+} \underset{\text{Hemiketal}}{CH_3CH_2{-}\underset{OCH_3}{\underset{|}{\overset{OH}{\overset{|}{C}}}}{-}CH_3} \xrightarrow[-H_2O]{CH_3OH,\ H^+} \underset{\text{Ketal}}{CH_3CH_2{-}\underset{OCH_3}{\underset{|}{\overset{OCH_3}{\overset{|}{C}}}}{-}CH_3}$$

Problem 14.21. Show the *intramolecular* formation of a hemiacetal from 5-hydroxypentanal.

Ans.

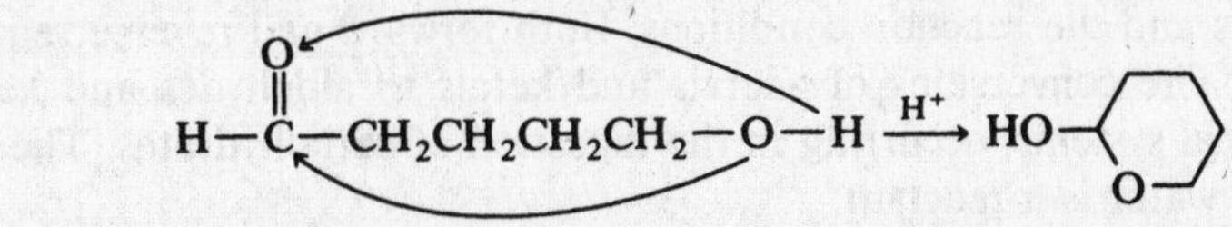

$$H-\overset{O}{\overset{\|}{C}}-CH_2CH_2CH_2CH_2-O-H \xrightarrow{H^+} HO-\langle\text{tetrahydropyran ring}\rangle$$

The curved arrows show the bonds that form in the process of hemiacetal formation. The hydrogen of the OH group becomes bonded to the carbonyl oxygen. The oxygen of the OH group becomes bonded to the carbonyl carbon. The carbonyl oxygen ends up as an OH and the OH oxygen ends up as an ether.

Problem 14.22. Show the formation of an acetal from ethanol and the hemiacetal of Problem 14.21.

Ans.

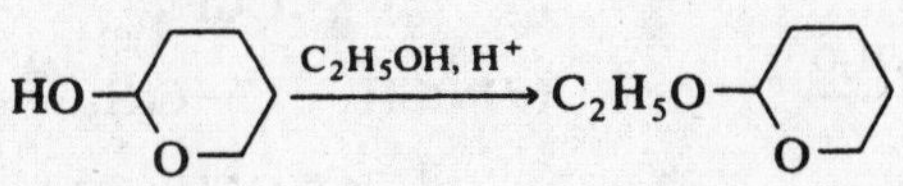

$$HO-\langle\text{ring O}\rangle \xrightarrow{C_2H_5OH,\ H^+} C_2H_5O-\langle\text{ring O}\rangle$$

Hemiacetal → Acetal

Intramolecular hemiacetal and hemiketal formation is critical to understanding the structure and properties of carbohydrates (Chap. 18).

Problem 14.23. For each of the following, indicate whether the compound is an acetal, hemiacetal, ketal, hemiketal, or something else:

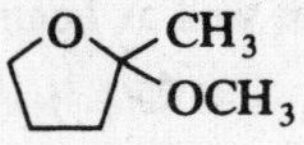

I: five-membered ring (O) with C bearing CH_3 and OCH_3

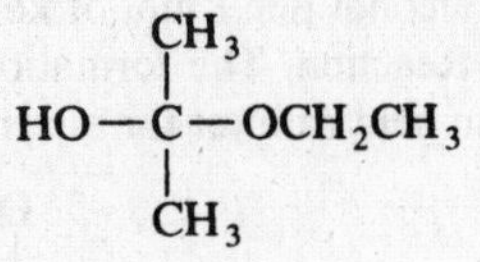

II: $HOCH_2OCH_3$

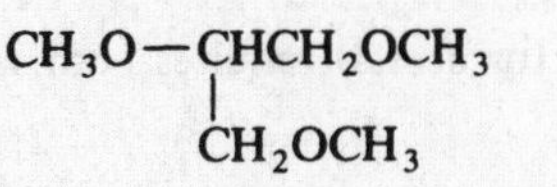

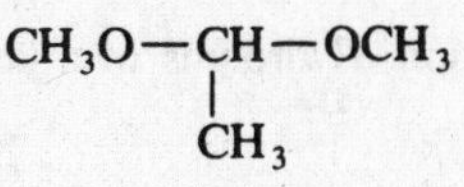

III: $HO-C(CH_3)_2-OCH_2CH_3$

IV: $CH_3O-CHCH_2OCH_3$ with CH_2OCH_3 on CH

V: $CH_3O-CH(CH_3)-OCH_3$

Ans. Acetals, hemiacetals, ketals, and hemiketals have two different oxygens bonded to the same carbon. One of the two oxygens is an OH group in hemiacetals and hemiketals. Neither of the two oxygens is an OH group in acetals and ketals:

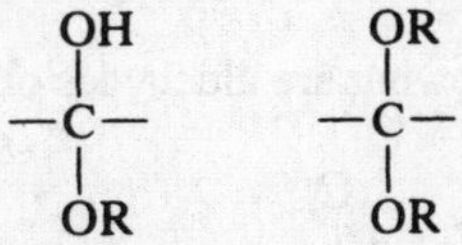

$$-\overset{OH}{\underset{OR}{C}}- \qquad -\overset{OR}{\underset{OR}{C}}-$$

Hemiacetal / Hemiketal — Acetal / Ketal

Aside from carbon having two bonds to two different oxygens, the remaining two bonds to the carbon are both to other carbons in hemiketals and ketals. For hemiacetals and acetals, no more than one of the remaining bonds is to carbon. At least one of the bonds is to hydrogen.

Compound I is a ketal, compound II is a hemiacetal, compound III is a hemiketal, compound IV is something else (triether), and compound V is an acetal.

The formation of hemiacetals and hemiketals from aldehydes and ketones and their subsequent conversions to acetals and ketals are reversible reactions. The position of equilibrium depends on the specific compounds and the reaction conditions. Both forward and reverse reactions are catalyzed by acids or enzymes. The conversions of acetals and ketals to aldehydes and ketones is an important reaction in biological systems, occurring in the digestion of carbohydrates. The reaction is referred to as *hydrolysis* since water is a reactant.

Problem 14.24. Complete each of the following reactions by writing the structure of the organic product. If no reaction occurs, write "No Reaction." If there is more than one product, indicate which are the major and which the minor products.

(*a*) $CH_3-\overset{\overset{\large O}{\|}}{C}-CH_2CH_3 \xrightarrow[H^+]{CH_3OH} ?$ (*b*) $CH_3-\overset{\overset{\large O}{\|}}{C}-CH_2CH_3 \xrightarrow{CH_3OH} ?$

(*c*) $CH_3-\underset{OCH_3}{\overset{OCH_3}{C}}-C_6H_5 \xrightarrow{H^+,\,H_2O} ?$ (*d*) $CH_3-\underset{OCH_3}{\overset{OCH_3}{C}}-C_6H_5 \xrightarrow{H^+} ?$

Ans.

(*a*) $CH_3-\overset{\overset{\large O}{\|}}{C}-CH_2CH_3 \xrightarrow[H^+]{CH_3OH} CH_3-\underset{OH}{\overset{OCH_3}{C}}-CH_2CH_3 \xrightarrow[H^+]{CH_3OH} CH_3-\underset{OCH_3}{\overset{OCH_3}{C}}-CH_2CH_3$

Ketal is the major product since the question is written in the manner corresponding to the presence of excess alcohol. If the question did not specify an excess of alcohol (i.e., having only 1 mol of alcohol per 1 mol of ketone), the major product would have been the hemiketal.

(*b*) No Reaction. The formation of hemiketal and ketal (as well as hemiacetal and acetal) both require an acid catalyst (or enzyme).

(*c*) $CH_3-\underset{OCH_3}{\overset{OCH_3}{C}}-C_6H_5 \xrightarrow{H^+,\,H_2O} CH_3-\overset{\overset{\large O}{\|}}{C}-C_6H_5 + 2\,CH_3OH$

(*d*) No Reaction. Hydrolysis of a ketal (or acetal) requires both water and an acid catalyst (or enzyme).

ADDITIONAL SOLVED PROBLEMS

STRUCTURE OF ALDEHYDES AND KETONES

Problem 14.25. Which of the following are aldehydes or ketones and which are not?

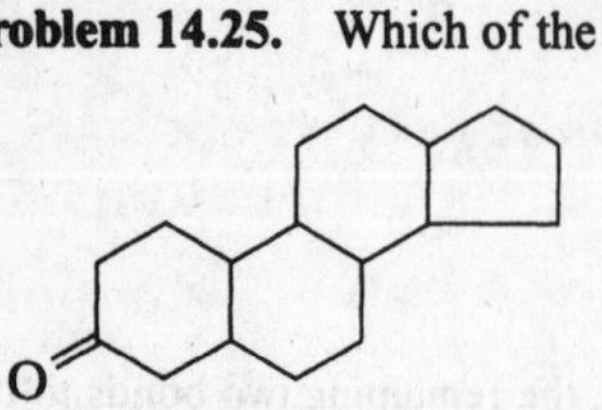

I (steroid ring skeleton with C=O) II $CH_3-\overset{\overset{\large O}{\|}}{C}-OCH_3$ III $CH_3O-CH_2-\overset{\overset{\large O}{\|}}{C}-H$ IV (cyclopentyl)$-\overset{\overset{\large O}{\|}}{C}-C_6H_5$

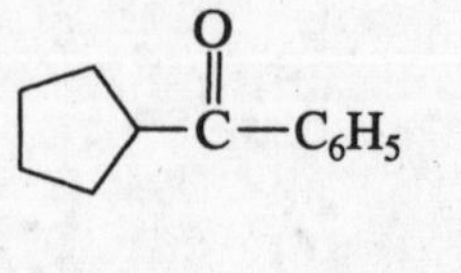

Ans. Compounds I and IV are ketones. Compound III is an aldehyde as well as an ether. Compound II is an ester.

Problem 14.26. What is the hybridization of the carbon and oxygen atoms in compounds III and IV of Problem 14.25?

Ans.

$$\underset{sp^3}{CH_3O}-CH_2-\overset{O\ (sp^2)}{\overset{\|}{C}}-H \qquad (sp^3\ \text{ring})-\overset{O\ (sp^2)}{\overset{\|}{C}}-\phi$$

CONSTITUTIONAL ISOMERISM IN ALDEHYDES AND KETONES

Problem 14.27. For each of the following structural drawings, indicate whether the pair represents (1) the same compound, (2) different compounds which are constitutional isomers, or (3) different compounds which are not isomers:

(*a*) $CH_3CH_2CH_2-\overset{O}{\overset{\|}{C}}-CH_3$ and $CH_2{=}CHCH_2CH_2CH_2-OH$

(*b*) $CH_3-\overset{O}{\overset{\|}{C}}-CH_2CH_3$ and $CH_3CH_2-\overset{O}{\overset{\|}{C}}-CH_3$

(*c*) $CH_3-\overset{O}{\overset{\|}{C}}-CH_3$ and cyclopropanone (ring$=$O)

(*d*) $CH_3-CH{=}CH-C_6H_4-OH$ and $CH_3CH_2-\overset{O}{\overset{\|}{C}}-C_6H_5$

(*e*) $CH_3-\overset{O}{\overset{\|}{C}}-CH_2CH_3$ and $CH_3CH_2-\overset{O}{\overset{\|}{C}}-OCH_3$

(*f*) $CH_3-\overset{O}{\overset{\|}{C}}-CH_2CH_2CH_3$ and $CH_3CH_2-\overset{O}{\overset{\|}{C}}-CH_2CH_3$

Ans. (*a*) Constitutional isomers of $C_5H_{10}O$
(*b*) Same compound
(*c*) Different compounds which are not isomers
(*d*) Constitutional isomers of $C_9H_{10}O$
(*e*) Different compounds which are not isomers
(*f*) Constitutional isomers of $C_5H_{10}O$

NOMENCLATURE OF ALDEHYDES AND KETONES

Problem 14.28. Draw structural formulas for each of the following compounds: (*a*) 5-methyl-3-hexanone, (*b*) 4-methylpentanal, (*c*) 2-bromo-4-methylbenzaldehyde.

Ans. (*a*) $CH_3CH_2-\overset{O}{\overset{\|}{C}}-CH_2\overset{CH_3}{\overset{|}{C}H}CH_3$ (*b*) $(CH_3)_2CHCH_2CH_2-\overset{O}{\overset{\|}{C}}H$

(*c*) $CH_3-C_6H_3(Br)-CHO$ (Br ortho to CHO)

Problem 14.29. Name the following compounds:

(*a*) $(CH_3)_2CH-\overset{O}{\overset{\|}{C}}-C_6H_{11}$ (cyclohexyl) (*b*) $CH_3\overset{Cl}{\overset{|}{C}H}-\overset{CH_3}{\overset{|}{C}H}CH_2-\overset{O}{\overset{\|}{C}}-H$

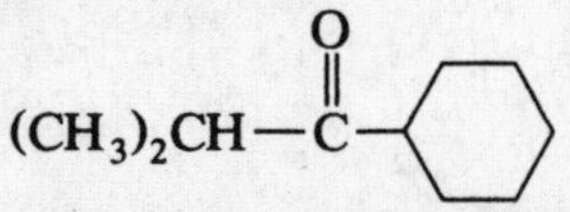

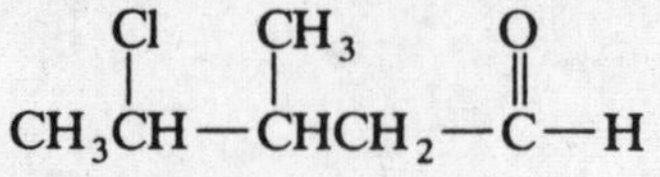

Ans. (*a*) cyclohexyl isopropyl ketone, (*b*) 4-chloro-3-methylpentanal.

PHYSICAL PROPERTIES OF ALDEHYDES AND KETONES

Problem 14.30. For each of the following pairs of compounds, indicate which compound has the higher value of the property. Explain, (*a*) Boiling point: methyl propyl ether versus butanone, (*b*) solubility in water: methyl propyl ether versus butanone, (*c*) solubility in water: propanal versus butanal, (*d*) boiling point: butanone versus butanal.

Ans. (*a*) Butanone has the higher boiling point because ketones (and aldehydes) are more polar than ethers.
(*b*) Methyl propyl ether and butanone have the same solubility in water because ethers and ketones (as well as aldehydes) have the same ability to hydrogen-bond with water.
(*c*) Propanal has the higher solubility in water. Water solubility goes down the hydrophobic portion of the molecule becomes larger.
(*d*) Butanone and butanal have the same boiling point. There is little difference in the boiling points of aldehydes and ketones.

Problem 14.31. Draw a representation of the hydrogen bonding present when acetone (propanone) dissolves in methanol.

Ans.

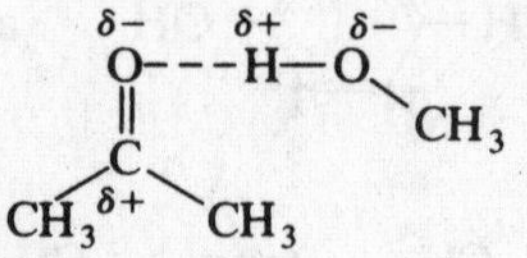

$$\overset{\delta-}{O}\text{- - -}\overset{\delta+}{H}-\overset{\delta-}{O}-CH_3$$
$$\|$$
$$CH_3-\overset{\delta+}{C}-CH_3$$

OXIDATION OF ALDEHYDES AND KETONES

Problem 14.32. Why is the conversion of an aldehyde to a carboxylic acid referred to as oxidation?

Ans. Ignoring the carbons attached to it, the carbonyl carbon has an O:H ratio of 1:1 in the aldehyde and 2:1 in the carboxylic acid. An increase in O content (decrease in H content) is defined as oxidation.

Problem 14.33. For each of the following compounds, indicate whether it gives a positive test with Benedict's reagent:

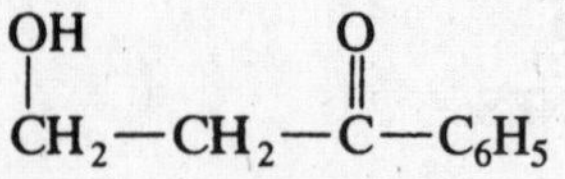

$$\underset{\text{I}}{\overset{OH}{\overset{|}{C}H_2}-CH_2-\overset{O}{\overset{\|}{C}}-C_6H_5} \qquad \underset{\text{II}}{H-\overset{O}{\overset{\|}{C}}-CH_2-\overset{OH}{\overset{|}{C}H}-C_6H_5}$$

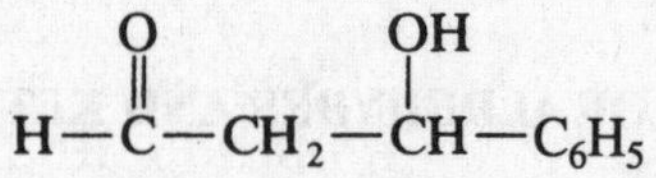

$$\underset{\text{III}}{CH_3-\overset{O}{\overset{\|}{C}}-CH_2-\overset{O}{\overset{\|}{C}}-CH_3} \qquad \underset{\text{IV}}{CH_3-\overset{O}{\overset{\|}{C}}-\overset{OH}{\overset{|}{C}H}-CH_3}$$

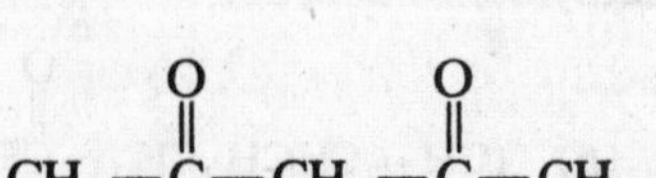

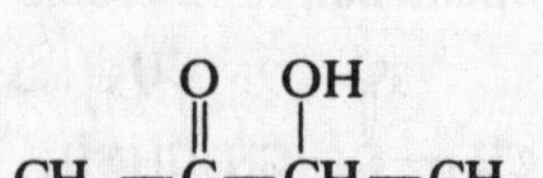

Ans. Only compound IV (α-hydroxy ketone) gives a positive test. None of the other compounds is an α-hydroxy aldehyde or α-hydroxy ketone.

Problem 14.34. An unknown is either 1-butanol or butanal. Shaking the unknown with a few drops of basic potassium permanganate solution (purple color) results in a colorless solution with the formation of a brown precipitate. Identify the unknown and explain.

Ans. The test with permanganate does not differentiate between 1-butanol and butanal since primary (and secondary) alcohols as well as aldehydes undergo oxidation with permanganate. Permanganate is not a selective enough mild oxidizing agent to distinguish between the possibilities.

Problem 14.35. What test would you perform to distinguish whether the unknown is 1-butanol or butanal?

Ans. Add a few drops of Tollens' reagent. A positive test is indicated by the formation of a silver mirror on the inside of the test tube.

Problem 14.36. An unknown is either 1-butanol or butanone. Shaking the unknown with a few drops of basic potassium permanganate solution (purple color) does not result in a loss of color and the formation of a brown precipitate. Identify the unknown and explain.

Ans. The unknown is butanone since ketones are not oxidized by permanganate while primary alcohols are oxidized.

REDUCTION OF ALDEHYDES AND KETONES

Problem 14.37. Write the structure of the aldehyde or ketone that would yield the following compounds upon reduction:

(*a*) cyclopentyl–CH_2OH (*b*) $C_6H_5-CH(OH)-CH_3$

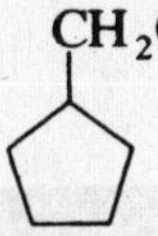

Ans. (*a*) cyclopentyl–CHO (*b*) $C_6H_5-C(=O)-CH_3$

REACTION OF ALDEHYDES AND KETONES WITH ALCOHOL

Problem 14.38. For each of the following, indicate whether the compound is an acetal, hemiacetal, ketal, hemiketal, or something else:

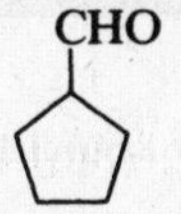

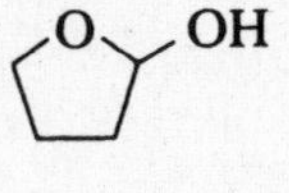

I: five-membered ring containing O, with OH on the carbon adjacent to O

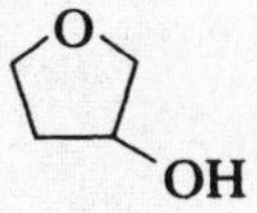

II: five-membered ring containing O, with OH on a carbon not adjacent to O

III: $HO-C(CH_3)_2-OCH_2CH_3$

IV: $CH_3O-CH(CH_3)CH_2OCH_3$

V: $CH_3-C(=O)CH_2OH$

Ans. Compound I is a hemiacetal, compound II is something else (ether-alcohol), compound III is a hemiketal, compound IV is something else (diether), compound V is something else (alcohol-ketone).

Problem 14.39. Write equations to show the formation of hemiacetal and acetal from the reaction of butanal with ethanol.

Ans.

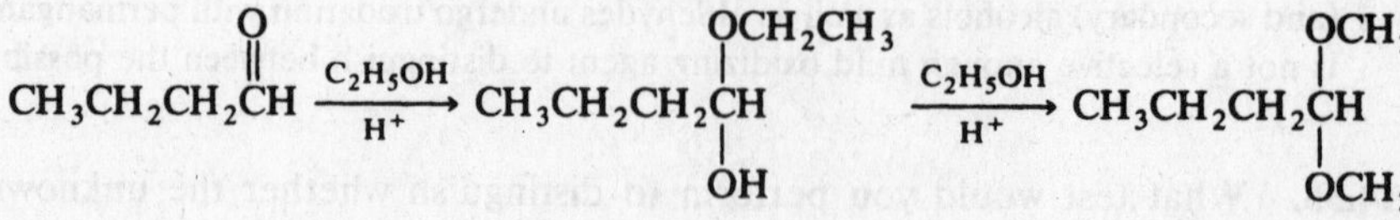

$$CH_3CH_2CH_2\overset{O}{\overset{\|}{C}}H \xrightarrow[H^+]{C_2H_5OH} CH_3CH_2CH_2\underset{OH}{\underset{|}{\overset{OCH_2CH_3}{\overset{|}{C}H}}} \xrightarrow[H^+]{C_2H_5OH} CH_3CH_2CH_2\underset{OCH_2CH_3}{\underset{|}{\overset{OCH_2CH_3}{\overset{|}{C}H}}}$$

Problem 14.40. Write a balanced equation for the hydrolysis of each of the following compounds:

$$(a)\ C_6H_5-\underset{OCH_3}{\underset{|}{\overset{CH_2CH_3}{\overset{|}{C}}}}-OCH_3 \qquad (b)\ CH_3O-\overset{CH_3}{\overset{|}{C}}HOCH_3$$

Ans.

$$(a)\ C_6H_5-\underset{OCH_3}{\underset{|}{\overset{CH_2CH_3}{\overset{|}{C}}}}-OCH_3 + 2\,H_2O \xrightarrow{H^+} C_6H_5-\overset{CH_2CH_3}{\overset{|}{C}}=O + 2\,CH_3OH$$

$$(b)\ CH_3O-\overset{CH_3}{\overset{|}{C}}HOCH_3 + 2\,H_2O \xrightarrow{H^+} O=\overset{CH_3}{\overset{|}{C}}H + 2\,CH_3OH$$

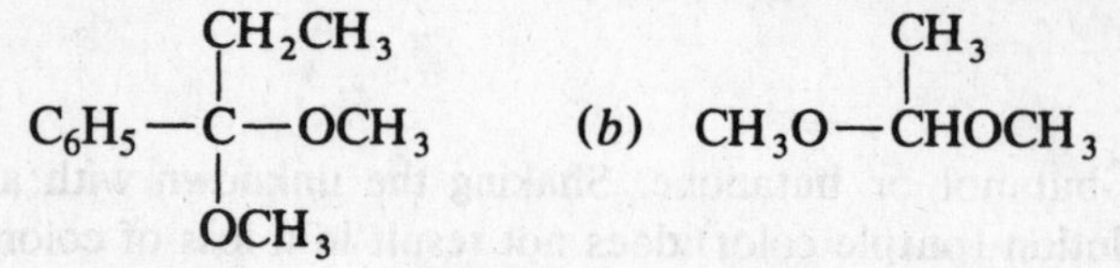

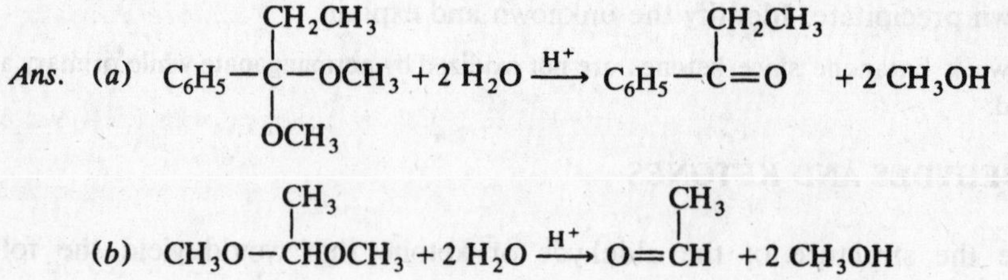

SUPPLEMENTARY PROBLEMS

Problem 14.41. Draw structural formulas of the following compounds: (*a*) phenyl *t*-butyl ketone, (*b*) 3-ethylpentanal, (*c*) 4-isopropylbenzaldehyde.

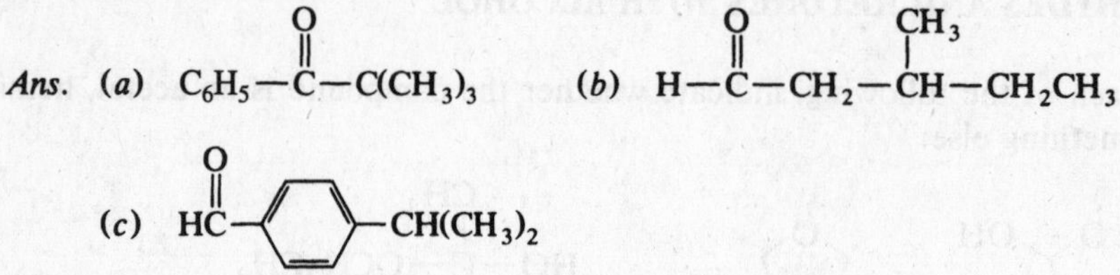

Ans.

$$(a)\ C_6H_5-\overset{O}{\overset{\|}{C}}-C(CH_3)_3 \qquad (b)\ H-\overset{O}{\overset{\|}{C}}-CH_2-\overset{CH_3}{\overset{|}{C}H}-CH_2CH_3$$

$$(c)\ H\overset{O}{\overset{\|}{C}}-C_6H_4-CH(CH_3)_2$$

Problem 14.42. An unknown has the molecular formula $C_5H_{12}O$. Is it an ether, alcohol, aldehyde, or ketone? Explain.

Ans. The unknown is either an alcohol or ether since the C:H ratio fits the general formula for an alkane.

Problem 14.43. An unknown has the molecular formula C_4H_8O. Is it an ether, alcohol, aldehyde, or ketone? Explain.

Ans. The C:H ratio indicates there is a ring or double bond. The possibilities are an aldehyde or ketone, alkene-ether, alkene-alcohol, cyclic alcohol, and cyclic ether.

Problem 14.44. An unknown is either 2-pentanone or pentanal. Shaking the unknown with a few drops of Tollens' reagent results in the formation of a silver mirror. Identify the unknown and explain.

Ans. Pentanal. Tollens' reagent gives positive tests with aldehydes but negative tests with ketones.

Problem 14.45. An unknown is either butanone or *t*-butyl alcohol. Shaking the unknown with a few drops of Tollens' reagent does not result in the formation of a silver mirror. Identify the unknown and explain.

Ans. This test does not differentiate between butanone and *t*-butyl alcohol since neither ketones nor tertiary alcohols are oxidized by Tollens' reagent.

Problem 14.46. For each of the following pairs of compounds, indicate which compound has the higher value of the property. Explain. (*a*) Boiling point: 1-butanol versus butanal; (*b*) solubility in water: 1-butanol versus butanal; (*c*) solubility in hexane: methyl propyl ether versus butanone; (*d*) solubility in hexane: propanal versus butanal.

Ans. (*a*) 1-Butanol has the higher boiling point since it can hydrogen-bond with itself but butanal cannot.
(*b*) 1-Butanol and butanal have the same solubility in water since both can hydrogen-bond with water.
(*c*) Methyl propyl ether, being less polar than butanone, is more soluble in the nonpolar hexane.
(*d*) Butanal is more soluble in hexane since a larger portion of butanal is nonpolar compared to propanal.

Problem 14.47. Which of the following compounds can be oxidized to an aldehyde or ketone? If a compound can, write the structure of the aldehyde or ketone which is formed.

(*a*) [cyclopentane ring bearing OH and, on the adjacent carbon, $-CH_3$] (*b*) [cyclopentane ring bearing CH_3 and OH on the same carbon] (*c*) $CH_3-C_6H_4-CH_2OH$ (*para*)

Ans. (*a*) [cyclopentanone ring (C=O) with $-CH_3$ on the adjacent carbon] (*b*) No oxidation (*c*) $CH_3-C_6H_4-CHO$ (*para*)

Problem 14.48. For each of the following, indicate whether the compound is an acetal, hemiacetal, ketal, hemiketal, or something else:

I: [cyclohexane ring bearing CH_3O and OH on the same carbon]

II: [five-membered ring containing O, with OCH_2CH_3 and CH_3 on the carbon next to the ring O]

III: $C_6H_5-CH(OH)(OCH_3)$

IV: $CH_3-CH(OH)-CH_2OCH_3$

V: $C_6H_5-C(OCH_3)_2-CH_3$

Ans. Compound I is a hemiketal, compound II is a ketal, compound III is a hemiacetal, compound IV is something else (ether-alcohol), and compound V is a ketal.

Problem 14.49. For each of the following compounds, indicate whether it gives a positive test with Benedict's reagent:

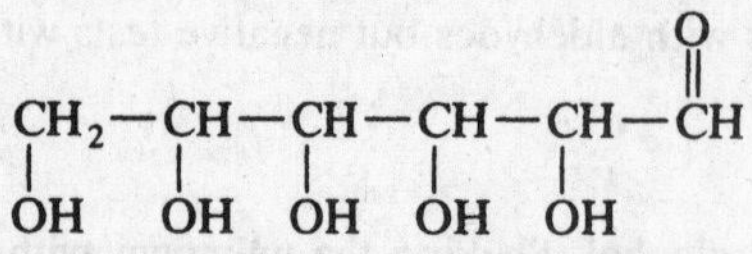

I: $CH_2(OH)-CH(OH)-CH(OH)-CH(OH)-CH(OH)-CH{=}O$

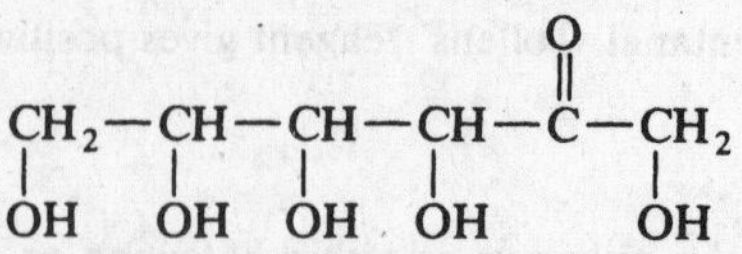

II: $CH_2(OH)-CH(OH)-CH(OH)-CH(OH)-C(=O)-CH_2OH$

III: 2-hydroxycyclopentanone (cyclopentane ring with OH and =O on adjacent carbons)

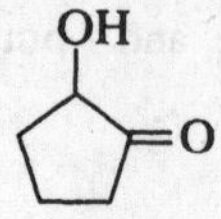

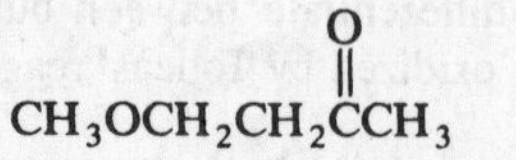

IV: $CH_3OCH_2CH_2\overset{O}{\overset{\|}{C}}CH_3$

Ans. **Positive tests are given by all except compound IV. Compound IV is neither an α-hydroxy aldehyde nor an α-hydroxy ketone; it is an ether-ketone. Compound I is an α-hydroxy aldehyde; compounds II and III are α-hydroxy ketones.**

Problem 14.50. Complete each of the following reactions by writing the structure of the organic product. If no reaction occurs, write "No Reaction." If there is more than one product, indicate which are the major and which the minor products.

(*a*) $CH_3-\overset{O}{\overset{\|}{C}}-CH_2CH_3 \xrightarrow[Ni]{H_2} ?$

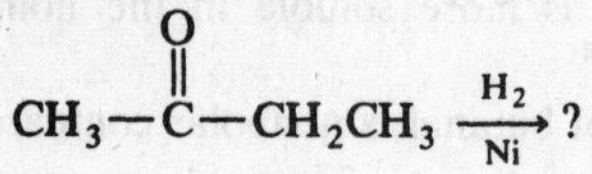

(*b*) $CH_3-C(OCH_3)_2-C_6H_5 \xrightarrow[Ni]{H_2} ?$

(*c*) $H-\overset{O}{\overset{\|}{C}}-CH_2C_6H_5 \xrightarrow[2.\ H_2O]{1.\ NADH} ?$

(*d*) 3-methylcyclopentanone $\xrightarrow[2.\ H_2O]{1.\ LiAlH_4} ?$

(*e*) 3-methylcyclopentanone $\xrightarrow[H^+]{CH_3OH} ?$

(*f*) 2-methoxytetrahydrofuran (OCH_3) $\xrightarrow[H^+]{H_2O} ?$

Ans.

(*a*) $CH_3-\overset{O}{\overset{\|}{C}}-CH_2CH_3 \xrightarrow[Ni]{H_2} CH_3-CH(OH)-CH_2CH_3$

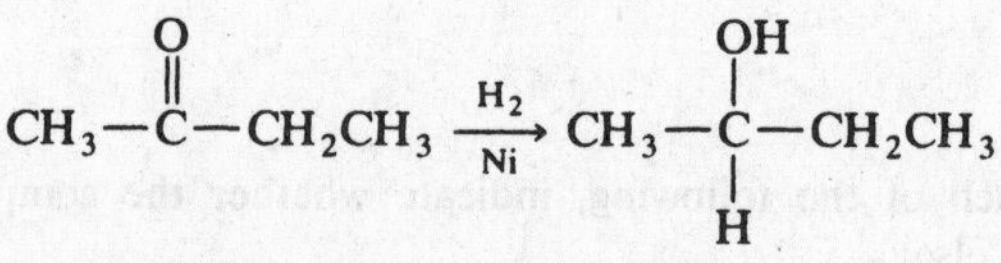

(*b*) No Reaction.

(*c*) $H-\overset{O}{\overset{\|}{C}}-CH_2C_6H_5 \xrightarrow[2.\ H_2O]{1.\ NADH} HOCH_2CH_2C_6H_5$

(*d*) 3-methylcyclopentanone $\xrightarrow[2.\ H_2O]{1.\ LiAlH_4}$ 3-methylcyclopentanol (OH, CH_3)

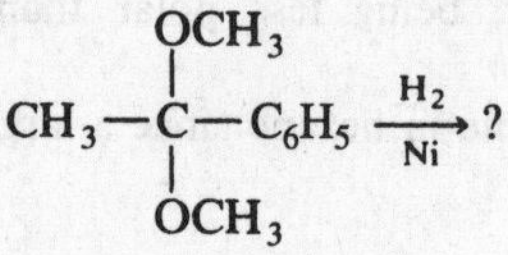

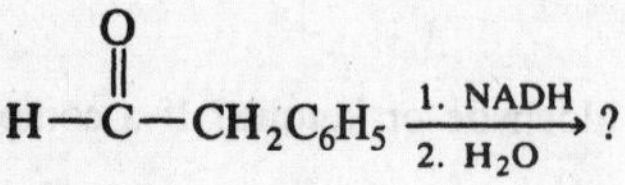

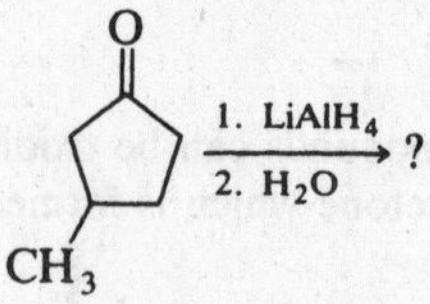

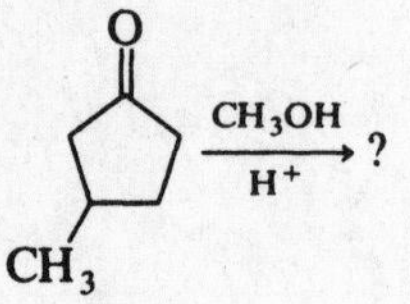

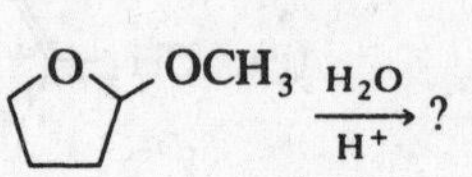

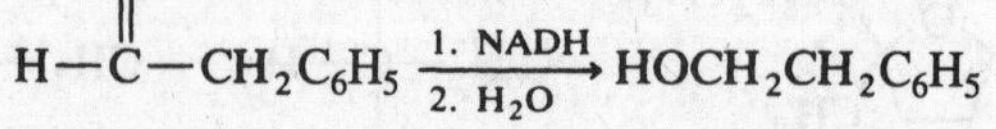

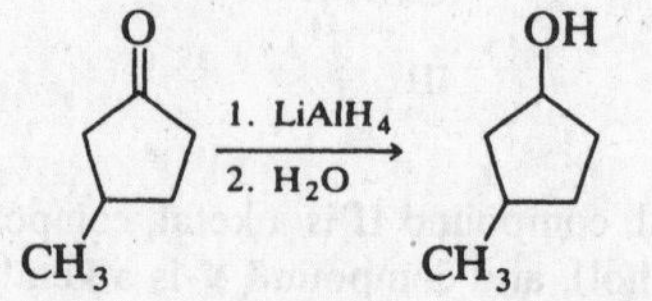

(*e*) 3-Methylcyclopentanone $\xrightarrow[H^+]{CH_3OH}$ 1,1-dimethoxy-3-methylcyclopentane (CH_3O, OCH_3; CH_3)

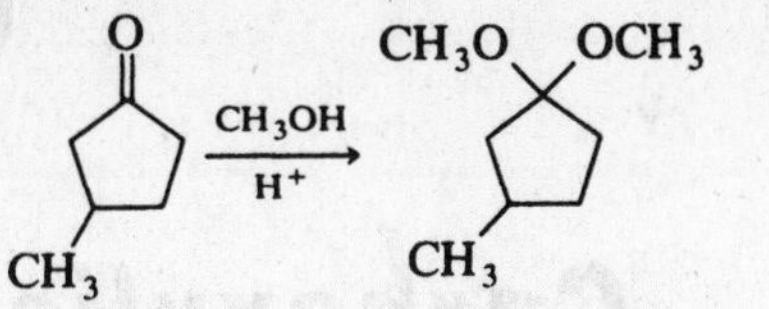

(*f*) 2-Methoxytetrahydropyran (OCH_3) $\xrightarrow[H^+]{H_2O}$ 5-hydroxypentanal (OH O, H) $+ CH_3OH$

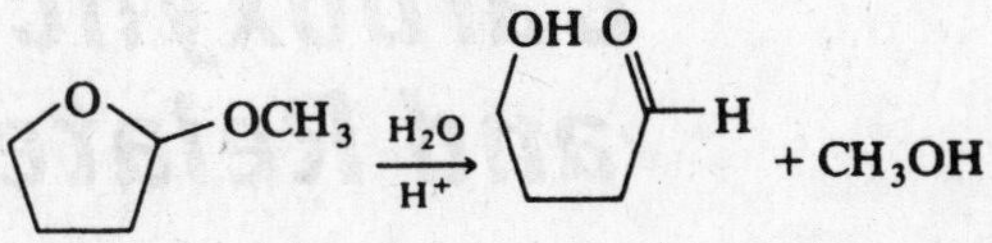

Problem 14.51. An unknown compound has the molecular formula C_3H_6O and gives a positive test with Benedict's reagent. Identify the compound.

Ans. Propanal

Problem 14.52. An unknown compound has the molecular formula C_3H_6O and gives a negative test with Tollens' reagent. Identify the compound.

Ans. The unknown is one of the following compounds, none of which are oxidized by Tollens' reagent:

$CH_3-\overset{O}{\overset{\|}{C}}-CH_3$ $\quad CH_2{=}CHCH_2OH$ $\quad CH_2{=}\overset{OH}{\overset{|}{C}}CH_3$ $\quad HOCH{=}CHCH_3$ (*cis + trans*) $\quad$ oxetane (four-membered ring with O)

Carboxylic Acids, Esters, and Related Compounds

15.1 STRUCTURE OF CARBOXYLIC ACIDS

Carboxylic acids (often referred to simply as *organic acids* or *acids*) contain the *carboxyl* group, a carbonyl group to which is attached a hydroxyl group:

$$-\overset{\displaystyle O}{\overset{\|}{C}}-OH$$

Carboxyl group

The carboxyl group is abbreviated as — COOH or — CO_2H. The carbon of the carboxyl group can be attached to a hydrogen as in methanoic acid (common name: formic acid) or to a carbon. The carbon can be part of an alkyl group as in ethanoic acid (common name: acetic acid) or an aryl group as in benzoic acid.

$$H-\overset{\displaystyle O}{\overset{\|}{C}}-OH \qquad CH_3-\overset{\displaystyle O}{\overset{\|}{C}}-OH \qquad C_6H_5-\overset{\displaystyle O}{\overset{\|}{C}}-OH$$

Methanoic acid / Formic acid — Ethanoic acid / Acetic acid — Benzoic acid

Formic acid is the irritant in the sting of certain ants. Vinegar is a 5 percent solution of acetic acid in water. Benzoic acid is used to synthesize aspirin. Carboxylic acids are found extensively in nature. The tartness of many fruits as well as sour milk is due to carboxylic acids. Some acids have very unpleasant odors. Derivatives of carboxylic acids are important in lipids (fats), synthetic plastics and fibers, aspirin, and soaps.

Problem 15.1. Which of the following are carboxylic acids and which are not?

$$\underset{I}{CH_3-\overset{\displaystyle O}{\overset{\|}{C}}-OH} \qquad \underset{II}{H-\overset{\displaystyle O}{\overset{\|}{C}}-CH_2CH_3} \qquad \underset{III}{H-\overset{\displaystyle O}{\overset{\|}{C}}-CH_2-OH} \qquad \underset{IV}{HO-\overset{\displaystyle O}{\overset{\|}{C}}-C_5H_9}$$

Ans. Compounds I and IV are carboxylic acids. Compound II is an aldehyde. Compound III is not a carboxylic acid because the carbonyl and hydroxyl groups are not connected to each other as in the carboxyl group. Compound III is both an aldehyde and an alcohol.

Problem 15.2. Give the bond angles for A, B, C, and D:

(Structure: acetic acid, $CH_3-C(=O)-O-H$, with angle A between C=O and C–CH_3, angle B between C–CH_3 and C–O, angle C between C=O and C–O, and angle D at C–O–H.)

Ans. $A = B = C = 120°$. The bond angles about any sp^2 atom, whether carbon or oxygen, are $120°$ with the atoms bonded to the C or O through sp^2 orbitals lying in a flat plane. $D = 109.5°$ since the OH oxygen is sp^3-hybridized.

Problem 15.3. What is the relationship between compounds I and III in Problem 15.1?

Ans. Compounds I and III are constitutional isomers of $C_2H_4O_2$ but not in the same family. Compound I is a carboxylic acid and compound III is an aldehyde-alcohol.

Problem 15.4. Draw the isomeric carboxylic acids that exist for the molecular formula $C_5H_{10}O_2$.

Ans. The molecular formula fits the general formula C_nH_{2n}, which is one pair of hydrogens less than the general formula for an alkane. One extra bond or ring is present. Since the question specifies carboxylic acids, the extra bond or ring is accounted for by the double bond of the carboxyl group. There is no other extra bond or ring. Four carboxylic acids exist for $C_5H_{10}O_2$:

$$CH_3CH_2CH_2CH_2{-}\overset{O}{\overset{\|}{C}}{-}OH \quad \text{I}$$

$$CH_3CH_2\overset{CH_3}{\overset{|}{C}H}{-}\overset{O}{\overset{\|}{C}}{-}OH \quad \text{II}$$

$$CH_3\overset{CH_3}{\overset{|}{C}H}CH_2{-}\overset{O}{\overset{\|}{C}}{-}OH \quad \text{III}$$

$$(CH_3)_3C{-}\overset{O}{\overset{\|}{C}}{-}OH \quad \text{IV}$$

15.2 NOMENCLATURE OF CARBOXYLIC ACIDS

The IUPAC rules follow those for the alkanes with a few modifications:

1. The longest continuous chain must contain the carboxyl carbon.
2. The ending of the name is changed from *-ane* to *-anoic acid*.
3. The carboxyl carbon is C-1.

Cyclic compounds containing an attached COOH group are named by adding *-carboxylic acid* to the name of the cyclic compound. The ring carbon holding the COOH is C-1.

C_6H_5-COOH is named as benzoic acid.

Common names for acyclic acids containing one to four carbons are obtained by combining the prefixes *form-*, *acet-*, *propion-*, *butyr-* with *-ic acid*, i.e., formic acid, acetic acid, propionic acid, and butyric acid. Common names for acids with more carbons follow in a similar manner.

Problem 15.5. Name the compounds in Problem 15.4 by the IUPAC system.

Ans. The longest continuous chain containing the carboxyl carbon is 5, 4, 4, and 3 carbons, respectively, for compounds I, II, III, and IV. Compound I is pentanoic acid, compound II is 2-methylbutanoic acid, compound III is 3-methylbutanoic acid, and compound IV is 2,2-dimethylpropanoic acid.

Problem 15.6. Name the following compounds by the IUPAC system:

(*a*) benzene ring bearing $-\overset{O}{\overset{\|}{C}}-OH$, with Cl and CH_3 substituents

(*b*) cyclohexane ring bearing $-\overset{O}{\overset{\|}{C}}-OH$, with Cl and CH_3 substituents

(*c*) HO–C(=O)–chain with CH_3, Cl, and terminal CH_3

Ans. (*a*) 5-chloro-2-methylbenzoic acid, (*b*) 5-chloro-2-methylcyclohexanecarboxylic acid, (*c*) 4-chloro-3-methylhexanoic acid.

15.3 PHYSICAL PROPERTIES OF CARBOXYLIC ACIDS

The carboxyl group contains two polar groups—the hydroxyl as in alcohols and the carbonyl as in aldehydes and ketones:

Problem 15.7. Prior to studying acids, we had learned that alcohols have the highest boiling and melting points of any family. Carboxylic acids have even higher boiling and melting points than alcohols. Explain.

Ans. Boiling and melting points increase with increasing intermolecular attractive forces. Alcohols have high intermolecular attractive forces because of the hydrogen bonding which results from the presence of OH groups. Carboxylic acids have even higher intermolecular forces because of the presence of both carbonyl and hydroxyl groups.

Problem 15.8. What is special about the hydrogen bonding in carboxylic acids?

Ans. Pairs of carboxylic acid molecules form very strong hydrogen bonding because the carbonyl group of one molecule hydrogen bonds with the hydroxyl group of another molecule:

The molecular mass of a carboxylic acid in the liquid state and in solution is twice the formula mass because of this strong hydrogen bonding. The molecular mass equals the formula mass only at higher temperature or in dilute solutions in highly polar solvents where the hydrogen bonding is broken.

Problem 15.9. Compare the solubilities in water of alcohols and carboxylic acids.

Ans. Carboxylic acids are somewhat more soluble in water than alcohols since acids have two groups (C = O and OH) for interaction with water while alcohols have only the OH.

Problem 15.10. Which of the following correctly describe(s) the hydrogen bonding present when a carboxylic acid dissolves in water?

(*a*) (*b*) (*c*)

Ans. The hydrogen bonding present between water and a carboxylic acid involves all three interactions. The hydrogens of water hydrogen bond to the oxygens of both the carbonyl and hydroxyl groups of the acid. The oxygen of water hydrogen bonds with the hydroxyl hydrogen of the carboxylic acid.

15.4 ACIDITY OF CARBOXYLIC ACIDS

Carboxylic acids are the most acidic of all organic compounds—much more acidic than phenols which are in turn much more acidic than alcohols. However, carboxylic acids are considerably weaker acids than strong acids such as HCl, H_2SO_4, H_3PO_4, $HClO_4$.

Problem 15.11. Write an equation to show acetic (ethanoic) acid acting as an acid when added to water.

Ans. $CH_3COOH + H_2O \rightleftharpoons CH_3COO^- + H_3O^+$

Acidity in the Brønsted definition involves donation of H^+ to a base, water in this case. The reaction of acetic acid with water is shown as an equilibrium reaction. The equilibrium is only partially to the right-hand side because carboxylic acids are weak acids.

Problem 15.12. How does one experimentally determine the relative acidities of compounds such as ethanol, phenol, acetic acid, and HCl?

Ans. Compare the pH of solutions of the compounds at the same molar concentration. For example, the pH of 0.1 *M* solutions of ethanol, phenol, acetic acid, and HCl is 7.0, 5.5, 2.9, and 1.0, respectively.

Problem 15.13. Explain the order of acidities: $HCl > RCOOH > C_6H_5OH > ROH$.

Ans. Acidity increases with the increasing ability of a compound to donate protons to a base. Acidity for a series of compounds containing the OH group depends on the strength of the O — H bond. The O — H bond is weakened when there are electron-withdrawing groups attached to the oxygen. C_6H_5OH is more acidic than ROH because C_6H_5 is more electron-withdrawing than R. RCOOH is even more acidic because the C=O is a stronger electron-withdrawing group than C_6H_5. HC1 is the strongest acid because Cl is more electron-withdrawing than C=O. (Acidity can also be explained by considering the stability of the anion formed by ionization of a proton. Higher stability for the anion means greater acidity for the compound. Anion stability increases when electron-withdrawing groups are present.)

Problem 15.14. Write an equation to show acetic acid acting as an acid when added to aqueous NaOH solution.

Ans. $CH_3COOH + NaOH \longrightarrow CH_3COONa + H_2O$

The reaction of acetic acid with NaOH goes completely to the right side in spite of acetic acid being a weak acid because NaOH is a very strong base.

The product of the reaction of a carboxylic acid with a strong base such as NaOH as described in Problem 15.14 is referred to as a *carboxylate salt*.

The rules for naming carboxylate salts are the following:

1. Name the carboxylic acid (either IUPAC or common name) from which the carboxylate salt is derived.
2. Change the ending of the name from *-ic acid* to *-ate*.
3. Place the name of the inorganic ion before the name of the carboxylic acid.

Carboxylate salts are used as additives for inhibiting or stopping the growth of bacteria and fungi, e.g., in foods, athlete's feet and diaper rash powders, and cervical creams.

Problem 15.15. Name the following carboxylate salts:

(*a*) $CH_3\overset{O}{\overset{\|}{C}}—ONa$ (*b*) $CH_3CH_2CH_2CH_2—\overset{O}{\overset{\|}{C}}—OK$

Ans. (*a*) IUPAC name: sodium ethanoate; common name: sodium acetate. (*b*) IUPAC name: potassium pentanoate.

Problem 15.16. Carboxylate salts are weak bases. Write equations to show CH_3COOK acting as a base toward (*a*) water and (*b*) HCl.

Ans.

(*a*) $$CH_3COOK + H_2O \rightleftharpoons CH_3COOH + KOH$$

(*b*) $$CH_3COOK + HCl \longrightarrow CH_3COOH + KCl$$

Carboxylate salts are bases because the carboxylate anion $RCOO^-$ accepts a proton from a base, water in (*a*) and HCl in (*b*). Note that the equilibrium is only partially to the right-hand side for reaction with the weak base water but completely to the right-hand side for the strong acid HCl.

Problem 15.17. Compare the water solubilities of carboxylic acids with their sodium carboxylate salts.

Ans. There is no difference in water solubilities for acids and salts containing up to four carbons. Both are miscible with water in all proportions. However, there is a large difference for acids and salts of five or more carbons. Acids of five or more carbons are much less soluble than acids of less than five carbons. The polar portion of the molecule is water-compatible while the nonpolar portion is not. Water solubility occurs unless the nonpolar portion is too large. "Too large" is five or more carbons for all families except for salts such as the sodium carboxylate salts. The metal-carboxylate bond is ionic and much more water-compatible than the polar covalent COOH. Carboxylate salts show water solubility even when there are more than five carbons.

Problem 15.18. Compare the melting points of carboxylic acids with their sodium carboxylate salts.

Ans. The carboxylate salts have considerably higher boiling and melting points since they are ionic compounds while the carboxylic acids are polar compounds. Intermolecular attractive forces in ionic compounds are greater than those in the most polar compounds, even those with hydrogen bonding forces.

The high water solubility of carboxylate salts is important in biological systems. The body's bicarbonate-carbonic acid and other buffer systems solubilize carboxylic acids by conversion to carboxylate salts. Intravenous drugs containing the COOH group are usually administered in the form of the carboxylate salt instead of the carboxylic acid to achieve faster absorption into the body.

15.5 SOAPS AND DETERGENTS

Sodium (Na^+) and potassium (K^+) carboxylate salts of fatty acids are used as *soaps*. *Fatty acids*, the carboxylic acids derived from fats, are mixtures of unbranched carboxylic acids of 14 to 20 carbons (Chap. 19).

Problem 15.19. What is the purpose of using a soap? Explain how a soap works on a molecular level.

Ans. Soaps are used to cleanse our skins (hands, bodies, feet) of greasy dirt. The typical greasy dirt consists of hydrocarbon molecules or molecules whose hydrocarbon portion is much larger than the polar portion. Such molecules are not washed away by water since they have no affinity for water. Recall that only like dissolves like.

A soap molecule consists of a long nonpolar hydrocarbon chain attached to a sodium or potassium carboxylate group. Such molecules and their cleansing action are shown in Fig. 15-1. The wiggly line represents the nonpolar *tail* of the soap and the circle represents the smaller ionic carboxylate *head*. The nonpolar tail has a strong affinity for grease and is absorbed into the grease particle. The carboxylate head of soap has a strong affinity for water. Water pulls on the soap heads which in turn pull on the soap tails which in turn pull on the grease. The result is that grease on the skin is broken up into smaller particles, pulled away from our skin, and solubilized in the water.

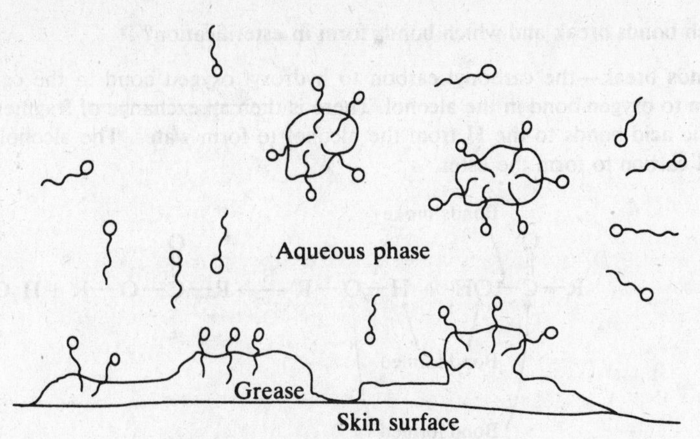

Fig. 15-1 Cleansing action of soap.

Problem 15.20. Why does a soap not work well in hard water?

Ans. Hard water contains Ca^{2+}, Mg^{2+}, Fe^{2+}, and Fe^{3+}. (Soft water does not contain these ions.) These ions exchange with the sodium and potassium ions of a soap. The calcium, magnesium, and iron carboxylate salts of soaps have extremely low water solubility. Precipitation of the soap interferes with its cleaning action.

Problem 15.21. Detergents, unlike soaps, work well in hard water. Describe the structure of a detergent and why detergents work well in hard water.

Ans. Alkyl benzene sulfonates are one of common types of detergents. They have nonpolar hydrocarbon tails and ionic heads as in soaps:

$$CH_3(CH_2)_{10}CH_2-C_6H_4-\overset{\overset{O}{\|}}{\underset{\underset{O}{\|}}{S}}-O^-K^+$$

A detergent has a sulfonate ionic head while a soap has a carboxylate ionic head. This difference is responsible for the difference in utilization in hard water. Ca^{2+}, Mg^{2+}, Fe^{2+}, and Fe^{3+} sulfonates are water-soluble whereas the carboxylates are water-insoluble. The detergent molecules stay in solution in hard water and exert their cleansing action. Soap molecules cannot exert their cleansing action in hard water because they precipitate out in the form of their Ca^{2+}, Mg^{2+} Fe^{2+}, and Fe^{3+} carboxylates.

15.6 CONVERSION OF CARBOXYLIC ACIDS TO ESTERS

Carboxylic acids react with alcohols or phenols (R′ = alkyl or aryl) by dehydration in the presence of an acid catalyst or enzyme to form *esters*:

$$\underset{\text{Acid}}{R-\overset{\overset{O}{\|}}{C}-OH} + \underset{\text{Alcohol}}{HO-R'} \overset{H^+}{\rightleftharpoons} \underset{\text{Ester}}{R-\overset{\overset{O}{\|}}{C}-O-R'} + H_2O$$

The reaction, referred to as *esterification*, is a reversible reaction. Ester formation is maximized by forcing the equilibrium to the right. This is accomplished by the removal of water from the reaction mixture by running the reaction above the boiling point of water.

The lower-molecular-mass esters are responsible for the odors of many fruits. Synthetic esters are used as food flavorings. Many lipids are esters.

Problem 15.22. Which bonds break and which bonds form in esterification?

Ans. Two bonds break—the carbonyl carbon to hydroxyl oxygen bond in the carboxylic acid and the hydrogen to oxygen bond in the alcohol. There is then an exchange of fragments. The OH from the carboxylic acid bonds to the H from the alcohol to form water. The alcohol oxygen bonds to the carbonyl carbon to form the ester.

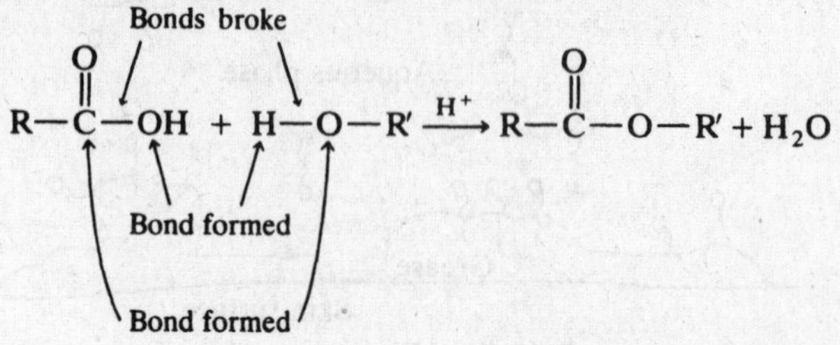

Bonds broke

$$R-\overset{O}{\overset{\|}{C}}-OH + H-O-R' \xrightarrow{H^+} R-\overset{O}{\overset{\|}{C}}-O-R' + H_2O$$

Bond formed

Bond formed

Problem 15.23. Write the equation for esterification of benzoic acid with propanol.

Ans. $$C_6H_5-\overset{O}{\overset{\|}{C}}-OH + HO-CH_2CH_2CH_3 \xrightarrow{H^+} C_6H_5-\overset{O}{\overset{\|}{C}}-O-CH_2CH_2CH_3 + H_2O$$

Problem 15.24. What structural features distinguish esters from carboxylic acids?

Ans. Both have an oxygen to which is attached a carbonyl group. Acids and esters differ in the other atom bonded to the oxygen, hydrogen for acids and carbon for esters.

$$-\overset{O}{\overset{\|}{C}}-O-H \qquad -\overset{O}{\overset{\|}{C}}-O-\overset{|}{\underset{|}{C}}-$$

Acid Ester

Problem 15.25. Which carboxylic acid and alcohol are required to produce each of the following esters?

(*a*) $CH_3CH_2CH_2\overset{O}{\overset{\|}{C}}-OCH(CH_3)_2$ (*b*) $CH_3CH_2\overset{O}{\overset{\|}{C}}-OC_6H_5$

Ans. In the ester structure R—CO—OR′, the R—CO— fragment, often referred to as the *acyl* group, comes from the carboxylic acid. The —OR′ fragment comes from the alcohol.

(*a*) $CH_3CH_2CH_2\overset{O}{\overset{\|}{C}}-OH$ and $HOCH(CH_3)_2$ (*b*) $CH_3CH_2\overset{O}{\overset{\|}{C}}-OH$ and HOC_6H_5

Esterification yields a polymer if one uses *bifunctional* reactants—a dicarboxylic acid and a diol.

HOOC—R—COOH + HO—R′—OH + HOOC—R—COOH + HO—R′—OH

+ HOOC—R—COOH + HO—R′—OH, etc.

⟶ ~~OC—R—COO—R′—OOC—R—COO—R′—OOC—R—COO—R′—O~~

The more usual (abbreviated) way of depicting polymerization is

$$n\text{HOOC}-R-\text{COOH} + n\text{HO}-R'-\text{OH} \longrightarrow \left[\overset{O}{\overset{\|}{C}}-R-\overset{O}{\overset{\|}{C}}-O-R'-O\right]_n$$

Problem 15.26. Write the equation for polymerization of terephthalic acid, *p*-HOOC—C_6H_4—COOH, with ethylene glycol, $HOCH_2CH_2OH$.

Ans.

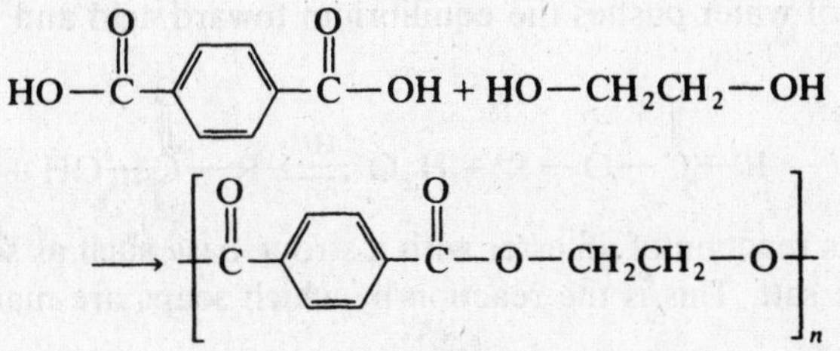

$$HO-\overset{O}{\overset{\|}{C}}-C_6H_4-\overset{O}{\overset{\|}{C}}-OH + HO-CH_2CH_2-OH \longrightarrow \left[\overset{O}{\overset{\|}{C}}-C_6H_4-\overset{O}{\overset{\|}{C}}-O-CH_2CH_2-O\right]_n$$

This polymer is produced commercially in large volumes and used widely as *polyester* or *Dacron fiber* (wearing apparel, upholstery, tire cord) and *Mylar plastic* (photographic, magnetic, and x-ray films and tapes).

15.7 NOMENCLATURE AND PHYSICAL PROPERTIES OF ESTERS

The rules for naming esters are the following:

1. Name the carboxylic acid (either IUPAC or common name) from which the ester is derived.
2. Change the ending of the name from *-ic acid* to *-ate*.
3. Place the name of the substituent in the alcohol portion of the ester at the front of the name.

Problem 15.27. Name the esters in Problem 15.25.

Ans. (*a*) Isopropyl butanoate (IUPAC); isopropyl butyrate (common), (*b*) phenyl propanoate (IUPAC); phenyl propionate (common).

Problem 15.28. Draw the structures of the following compounds: (*a*) *t*-butyl 4-methylpentanoate, (*b*) propyl benzoate.

Ans. (*a*) $(CH_3)_2CHCH_2CH_2-\overset{O}{\overset{\|}{C}}-OC(CH_3)_3$ (*b*) $C_6H_5-\overset{O}{\overset{\|}{C}}-OCH_2CH_2CH_3$

Esters are the least polar of the families of carbonyl-containing compounds, less polar than aldehydes and ketones.

Problem 15.29. Explain the order of boiling points:

$$CH_3-\overset{O}{\overset{\|}{C}}-OCH_2CH_3 \quad CH_3-\overset{O}{\overset{\|}{C}}-CH_2CH_3 \quad CH_3-\overset{O}{\overset{\|}{C}}-CH_2CH_2CH_3$$

77 °C 80 °C 102 °C

Ans. A comparison of ethyl acetate and methyl propyl ketone involves compounds of comparable molecular mass. The more polar compound (ketone) has the higher boiling point since it has higher intermolecular attractive forces. When comparing ethyl acetate and ethyl methyl ketone, we need to note that the former has a higher molecular mass which counteracts its lower polarity. Ethyl acetate boils only a few degrees lower than ethyl methyl ketone.

15.8 CHEMICAL REACTIONS OF ESTERS

Esters undergo two important reactions—hydrolysis and saponification. Hydrolysis, the reverse of esterification, is the acid-catalyzed reaction of an ester with water to yield the carboxylic acid and alcohol. An excess of water pushes the equilibrium toward acid and alcohol.

$$R-\overset{\overset{\displaystyle O}{\|}}{C}-O-R' + H_2O \overset{H^+}{\rightleftharpoons} R-\overset{\overset{\displaystyle O}{\|}}{C}-OH + HO-R'$$

Saponification is reaction of an ester with a strong base such as sodium hydroxide to form alcohol and the carboxylate salt. This is the reaction by which soaps are manufactured from fats (Chap. 19).

$$R-\overset{\overset{\displaystyle O}{\|}}{C}-O-R' + NaOH \overset{H_2O}{\rightleftharpoons} R-\overset{\overset{\displaystyle O}{\|}}{C}-ONa + HO-R'$$

Problem 15.30. Write the equation for the acid-catalyzed hydrolysis of butyl propanoate.

Ans.

$$CH_3CH_2-\overset{\overset{\displaystyle O}{\|}}{C}-OCH_2CH_2CH_2CH_3 \xrightarrow[H^+]{H_2O} CH_3CH_2-\overset{\overset{\displaystyle O}{\|}}{C}-OH + HOCH_2CH_2CH_2CH_3$$

Problem 15.31. Write the equation for the saponification of butyl propanoate.

Ans.

$$CH_3CH_2-\overset{\overset{\displaystyle O}{\|}}{C}-OCH_2CH_2CH_2CH_3 \xrightarrow[H_2O]{KOH} CH_3CH_2-\overset{\overset{\displaystyle O}{\|}}{C}-OK + HOCH_2CH_2CH_2CH_3$$

15.9 CARBOXYLIC ACID ANHYDRIDES, HALIDES, AND AMIDES

The term *carboxylic acid derivatives* is used to encompass various compounds that can be synthesized from and hydrolyzed back to carboxylic acids. In addition to esters, these include *acid halides*, *anhydrides*, and *amides*:

$$R-\overset{\overset{\displaystyle O}{\|}}{C}-Cl \qquad R-\overset{\overset{\displaystyle O}{\|}}{C}-O-\overset{\overset{\displaystyle O}{\|}}{C}-R \qquad R-\overset{\overset{\displaystyle O}{\|}}{C}-NH_2$$

Acid halide — Acid anhydride — Amide

Acid halides and anhydrides are not found in biological systems because they react rapidly with water and alcohol to form carboxylic acids and esters, respectively. Amides, very important in proteins, are described in Chap. 16.

Acid halides are named by changing the ending of the name of the corresponding acid from *-ic acid* to *-yl halide*; acid anhydrides are named by replacing the word *acid* by *anhydride*.

Problem 15.32. Show the structures of acetyl chloride and propanoic anhydride.

Ans. (*a*) $CH_3-\overset{\overset{\displaystyle O}{\|}}{C}-Cl$ (*b*) $CH_3CH_2-\overset{\overset{\displaystyle O}{\|}}{C}-O-\overset{\overset{\displaystyle O}{\|}}{C}-CH_2CH_3$

15.10 PHOSPHORIC ACID ANHYDRIDES AND ESTERS

Phosphoric acid, diphosphoric acid, and triphosphoric acid and their organic derivatives are important in carbohydrates and nucleic acids (Chapters 18, 21).

$$\underset{\text{Phosphoric acid}}{\mathrm{HO{-}P(=O)(OH){-}OH}} \qquad \underset{\text{Diphosphoric acid}}{\mathrm{HO{-}P(=O)(OH){-}O{-}P(=O)(OH){-}OH}} \qquad \underset{\text{Triphosphoric acid}}{\mathrm{HO{-}P(=O)(OH){-}O{-}P(=O)(OH){-}O{-}P(=O)(OH){-}OH}}$$

Diphosphoric acid is formed by dehydration between pairs of phosphoric acid molecules. The oxygen linkage between the two phosphate units is an *anhydride* linkage:

$$2\,\mathrm{HO{-}P(=O)(OH){-}OH} \xrightarrow{-H_2O} \mathrm{HO{-}P(=O)(OH){-}\underset{\text{Anhydride}}{O}{-}P(=O)(OH){-}OH}$$

Triphosphoric acid, formed by dehydration between phosphoric acid and diphosphoric acid, contains two anhydride linkages:

$$\mathrm{HO{-}P(=O)(OH){-}OH} + \mathrm{HO{-}P(=O)(OH){-}O{-}P(=O)(OH){-}OH} \xrightarrow{-H_2O} \mathrm{HO{-}P(=O)(OH){-}O{-}P(=O)(OH){-}O{-}P(=O)(OH){-}OH}$$

(Anhydride: both P–O–P linkages)

Each of the OH groups attached to phosphorus in these compounds has properties exactly analogous to the OH group in a carboxylic acid except that acidity is higher for phosphoric acid OH groups.

Problem 15.33. Write an equation for the reaction of diphosphoric acid with NaOH.

Ans. Each acidic H^+ is replaced by Na^+ until all four have been replaced:

$$\mathrm{HO{-}P(=O)(OH){-}O{-}P(=O)(OH){-}OH} \xrightarrow{NaOH} \mathrm{HO{-}P(=O)(OH){-}O{-}P(=O)(OH){-}ONa} \xrightarrow{NaOH} \mathrm{NaO{-}P(=O)(ONa){-}O{-}P(=O)(ONa){-}ONa}$$

Problem 15.34. Write an equation to show esterification of one of the HO groups of triphosphoric acid with 1-propanol.

Ans.

$$\mathrm{HO{-}P(=O)(OH){-}O{-}P(=O)(OH){-}O{-}P(=O)(OH){-}OH} + \mathrm{HOCH_2CH_2CH_3} \xrightarrow{-H_2O} \mathrm{HO{-}P(=O)(OH){-}O{-}P(=O)(OH){-}O{-}P(=O)(OH){-}OCH_2CH_2CH_3}$$

The alcohol adenosine forms esters with diphosphoric and triphosphoric acid (ADP and ATP, respectively), which are used for energy production in biological systems.

ADDITIONAL SOLVED PROBLEMS

STRUCTURE OF CARBOXYLIC ACIDS

Problem 15.35. For each of the following structural drawings, indicate whether the pair represents (1) the same compound, (2) different compounds which are constitutional isomers, or (3) different compounds which are not isomers:

(*a*) $CH_3-\overset{O}{\overset{\|}{C}}-CH(CH_3)_2$ and $CH_3CH_2CH_2CH_2-\overset{O}{\overset{\|}{C}}-OH$

(*b*) $CH_3CH_2CH_2-\overset{O}{\overset{\|}{C}}-OCH_3$ and $CH_3CH_2CH_2CH_2-\overset{O}{\overset{\|}{C}}-OH$

(*c*) $CH_3(CH_2)_3COOH$ and $CH_3CH_2CH_2CH_2-\overset{O}{\overset{\|}{C}}-OH$

(*d*) $(CH_3)_3C-COOH$ and $CH_3CH_2CH_2CH_2-\overset{O}{\overset{\|}{C}}-OH$

(*e*) cyclohexane ring with COOH and Cl (1,3-positions) and cyclohexane ring with COOH and Cl (1,2-positions)

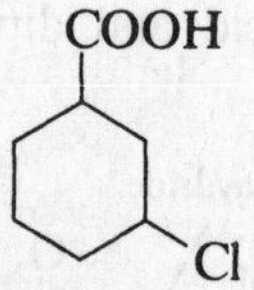

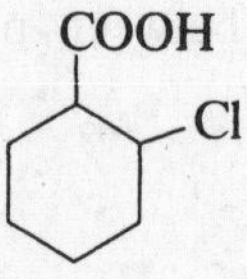

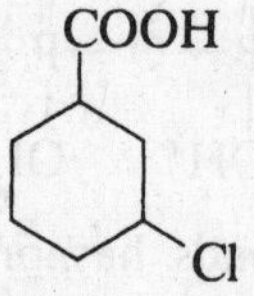

(*f*) cyclohexane ring with COOH and Cl and benzene ring with COOH and Cl

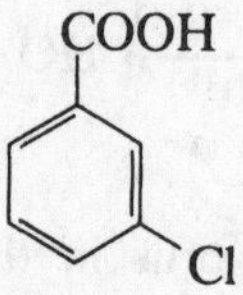

Ans. Different compounds which are not isomers: (*a*) and (*f*). Constitutional isomers: (*b*), (*d*), and (*e*). Same compound: (*c*).

NOMENCLATURE OF CARBOXYLIC ACIDS

Problem 15.36. Draw a structural formula for each of the following compounds: (*a*) 3-chlorobutanoic acid, (*b*) 4-methylpentanoic acid, (*c*) 2-bromo-4-methylbenzoic acid, (*d*) 3-methylcyclopentanecarboxylic acid.

Ans. (*a*) $CH_3-\underset{}{\overset{Cl}{\overset{|}{C}H}}-CH_2-\overset{O}{\overset{\|}{C}}-OH$ (*b*) $(CH_3)_2CHCH_2CH_2-\overset{O}{\overset{\|}{C}}-OH$

(*c*) CH_3–(benzene ring, Br ortho to COOH)–$COOH$ (*d*) CH_3–(cyclopentane ring)–$COOH$

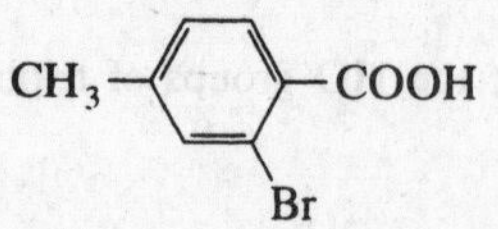

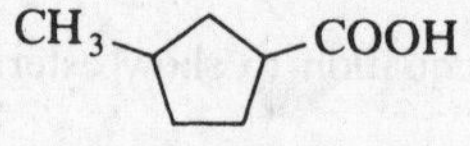

Problem 15.37. Name the following compounds by the IUPAC system:

(*a*) $\overset{CH_3}{\underset{H}{}}\!>C=C<\!\overset{H}{\underset{CH_2COOH}{}}$ (*b*) $HO-\overset{O}{\overset{\|}{C}}-\underset{CH_3}{\underset{|}{C}H}CH_2CH_2-\overset{O}{\overset{\|}{C}}-OH$

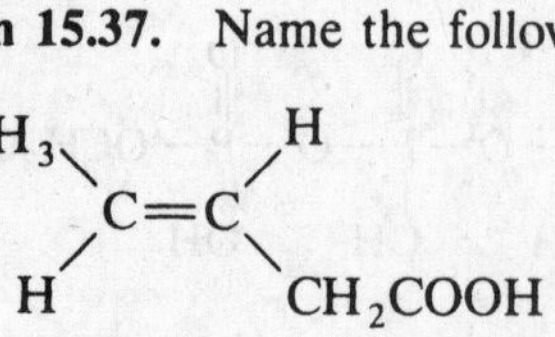

Ans. (*a*) *trans*-3-Pentenoic acid. For a carboxylic acid with a C$=$C, the ending is *-enoic acid* instead of *-anoic acid*. Numbering of the continuous chain is based on giving the lowest number to the carboxyl carbon instead of the double bond carbon.

(*b*) 2-Methylpentanedioic acid. There is no need to use numbers for the positions of the carboxyl groups as the correct longest continuous chain must contain both carboxyl groups.

PHYSICAL PROPERTIES OF CARBOXYLIC ACIDS

Problem 15.38. Explain the order of boiling points in the following compounds:

$CH_3CH_2CH_2CH_2CH_3$	$CH_3CH_2CH_2\overset{O}{\overset{\|}{C}}H$	$CH_3CH_2CH_2CH_2OH$	$CH_3CH_2\overset{O}{\overset{\|}{C}}OH$
36 °C	76 °C	118 °C	141 °C

Ans. The molecular masses of the compounds are comparable. The differences in boiling point are due to differences in the intermolecular attractive forces. Boiling point increases with increasing intermolecular forces. The lowest intermolecular forces are dispersion forces which are present in alkanes. Aldehydes (and ketones) have the higher attractive forces—those due to dipole-dipole interactions—because of the polar carbonyl group. Alcohols have higher intermolecular attractive forces because of the hydrogen bonding which results from the presence of OH groups. Carboxylic acids have even higher intermolecular forces because of the presence of both carbonyl and hydroxyl groups.

Problem 15.39. Compare the solubilities in water of the compounds in Problem 15.38.

Ans. Propanoic acid > 1-butanol = butanal ≫ pentane. Pentane has negligible water solubility since there is negligible intermolecular attraction between the nonpolar alkane and the highly polar water. Butanal and 1-butanol have significant solubilities in water since each can hydrogen-bond with water. Propanoic acid is somewhat more soluble in water than either the aldehyde or alcohol since it has two sites (carbonyl and hydroxyl) for interaction with water.

ACIDITY OF CARBOXYLIC ACIDS

Problem 15.40. Write an equation for the reaction of benzoic acid with $Ca(OH)_2$.

Ans.

$$2\ C_6H_5\text{—}COOH + Ca(OH)_2 \longrightarrow (C_6H_5\text{—}COO)_2Ca + 2\ H_2O$$

Problem 15.41. Which of the following compounds is more acidic? Explain.

$$CH_3\overset{O}{\overset{\|}{C}}CH_2OH \quad \text{or} \quad CH_3CH_2\overset{O}{\overset{\|}{C}}OH$$

Ans. Propanoic acid is enormously more acidic than hydroxyacetone. The latter is an alcohol and ketone. The alcohol group is only very slightly acidic while the ketone carbonyl shows no acidity.

Problem 15.42. Compare the solubility in hexane of sodium heptanoate and heptanoic acid.

Ans. Heptanoic acid is more soluble in the nonpolar hexane since it is a polar compound while sodium heptanoate is ionic.

SOAPS AND DETERGENTS

Problem 15.43. Which of the following compounds is a soap?

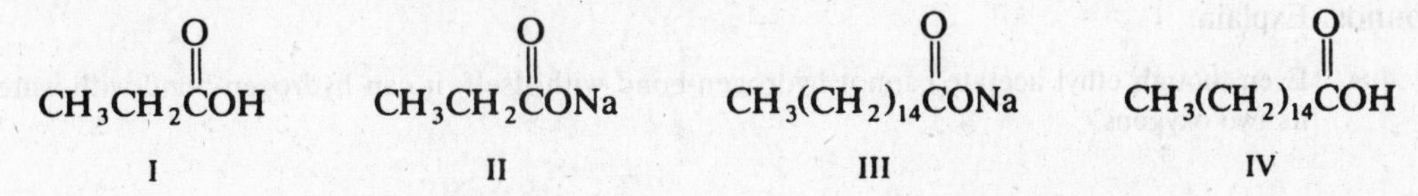

$CH_3CH_2\overset{O}{\overset{\|}{C}}OH$	$CH_3CH_2\overset{O}{\overset{\|}{C}}ONa$	$CH_3(CH_2)_{14}\overset{O}{\overset{\|}{C}}ONa$	$CH_3(CH_2)_{14}\overset{O}{\overset{\|}{C}}OH$
I	II	III	IV

Ans. Only compound III is a soap. A soap must have both an ionic portion and a long hydrophobic portion. Compound II has only the ionic portion, compound IV has only the long hydrophobic portion, and compound I has neither.

CONVERSION OF CARBOXYLIC ACIDS TO ESTERS

Problem 15.44. Complete the following equations. If no reaction occurs, write "No Reaction."

(*a*) $(CH_3)_2CHCH_2\overset{\overset{\displaystyle O}{\|}}{C}{-}OH \xrightarrow{CH_3OH} ?$ (*b*) $(CH_3)_2CHCH_2\overset{\overset{\displaystyle O}{\|}}{C}{-}OH \xrightarrow[H^+]{CH_3OH} ?$

Ans. (*a*) No Reaction since there is no acid catalyst present.

(*b*) $(CH_3)_2CHCH_2\overset{\overset{\displaystyle O}{\|}}{C}{-}OH \xrightarrow[H^+]{CH_3OH} (CH_3)_2CHCH_2\overset{\overset{\displaystyle O}{\|}}{C}{-}OCH_3$

Problem 15.45. Which carboxylic acid and alcohol are required to produce phenylmethyl hexanoate?

Ans. $C_6H_5CH_2OH$ and $CH_3(CH_2)_4COOH$

Problem 15.46. Which of the following combinations of reactants yields polymer upon esterification? Write equations to show the product formed in each system. (*a*) $HOOCC_6H_4COOH + CH_3OH$, (*b*) $C_6H_5COOH + HOCH_2CH_2OH$, (*c*) $HOOCC_6H_4COOH + HOCH_2CH_2OH$.

Ans. Combinations (*a*) and (*b*) cannot give polymer. Polymer is formed only when both reactants are bifunctional.

(*a*) $HOOCC_6H_4COOH + 2\,CH_3OH \xrightarrow{-H_2O} CH_3O{-}OCC_6H_4CO{-}OCH_3$

(*b*) $2C_6H_5COOH + HOCH_2CH_2OH \xrightarrow{-H_2O} C_6H_5CO{-}OCH_2CH_2O{-}OCC_6H_5$

(*c*) $HOOCC_6H_4COOH + HOCH_2CH_2OH \xrightarrow{-H_2O} \left(OCC_6H_4CO{-}OCH_2CH_2O\right)_n$

NOMENCLATURE AND PHYSICAL PROPERTIES OF ESTERS

Problem 15.47. Give the IUPAC name for each of the following compounds:

(*a*) $Cl{-}C_6H_4{-}\overset{\overset{\displaystyle O}{\|}}{C}{-}O{-}\overset{\overset{\displaystyle CH_3}{|}}{C}HCH_2CH_3$ (*b*) $Cl{-}C_6H_4{-}O{-}\overset{\overset{\displaystyle O}{\|}}{C}{-}\overset{\overset{\displaystyle CH_3}{|}}{C}HCH_2CH_3$

Ans. (*a*) *s*-Butyl *p*-chlorobenzoate, (*b*) *p*-chlorophenyl 2-methyl butanoate

Problem 15.48. Explain why the boiling point of ethyl acetate is 40 °C below that of 1-butanol (77 °C versus 117 °C).

Ans. The intermolecular attractive forces in 1-butanol are much higher since 1-butanol hydrogen-bonds with itself. Ethyl acetate cannot hydrogen-bond with itself since there is no O—H group.

Problem 15.49. Ethyl acetate is less soluble in water than 1-butanol. However, the difference in water solubilities is considerably less than the difference in the boiling (or melting) points of the two compounds. Explain.

Ans. Even though ethyl acetate cannot hydrogen-bond with itself, it can hydrogen-bond with water via its two oxygens.

CHEMICAL REACTIONS OF ESTERS

Problem 15.50. Write the equation for the acid-catalyzed hydrolysis of propyl benzoate.

Ans.

$$C_6H_5-\overset{O}{\overset{\|}{C}}-OCH_2CH_2CH_3 \xrightarrow[H^+]{H_2O} C_6H_5-\overset{O}{\overset{\|}{C}}-OH + HOCH_2CH_2CH_3$$

Problem 15.51. Write the equation for the saponification of phenyl butanoate by NaOH.

Ans.

$$CH_3CH_2CH_2-\overset{O}{\overset{\|}{C}}-OC_6H_5 \xrightarrow[H_2O]{NaOH} CH_3CH_2CH_2-\overset{O}{\overset{\|}{C}}-ONa + HOC_6H_5$$

CARBOXYLIC ACID ANHYDRIDES AND HALIDES

Problem 15.52. Give equations for the following reactions: (*a*) acetic anhydride + water, (*b*) Benzoyl bromide + ethanol.

Ans. (*a*) $CH_3-\overset{O}{\overset{\|}{C}}-O-\overset{O}{\overset{\|}{C}}-CH_3 + H_2O \longrightarrow 2\,CH_3-\overset{O}{\overset{\|}{C}}-OH$

(*b*) $C_6H_5-\overset{O}{\overset{\|}{C}}-Br + CH_3CH_2OH \xrightarrow{-HBr} C_6H_5-\overset{O}{\overset{\|}{C}}-OCH_2CH_3$

PHOSPHORIC ACID ANHYDRIDES AND ESTERS

Problem 15.53. Give the equation for hydrolysis of the methyl ester of diphosphoric acid.

Ans.

$$HO-\underset{OH}{\underset{|}{\overset{O}{\overset{\|}{P}}}}-O-\underset{OH}{\underset{|}{\overset{O}{\overset{\|}{P}}}}-OCH_3 \xrightarrow{H_2O} HO-\underset{OH}{\underset{|}{\overset{O}{\overset{\|}{P}}}}-O-\underset{OH}{\underset{|}{\overset{O}{\overset{\|}{P}}}}-OH + CH_3OH$$

SUPPLEMENTARY PROBLEMS

Problem 15.54. Identify the family of each of the following compounds: (*a*) $(CH_3)_2CHCOOH$, (*b*) $CH_3COCH_2CH_3$, (*c*) $C_6H_5COOCOC_6H_5$, (*d*) $CH_3CH(OCH_3)_2$, (*e*) $CH_3OCH_2COCH_3$, (*f*) $CH_3CH_2COOCH_3$, (*g*) $C_6H_5COCH_2OH$, (*h*) $C_6H_5CH_2COOH$, (*i*) CH_3CH_2COCl

(*j*) [five-membered ring: O in ring, C=O at ring carbon not adjacent to O] (*k*) [five-membered ring: O in ring, C=O adjacent to ring O] (*l*) $HO-\underset{OH}{\underset{|}{\overset{O}{\overset{\|}{P}}}}-OCH(CH_3)_2$

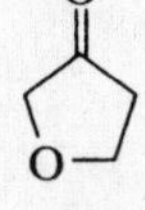

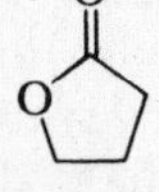

Ans. (*a*) carboxylic acid, (*b*) ketone, (*c*) acid anhydride, (*d*) acetal, (*e*) ether-ketone, (*f*) ester, (*g*) alcohol-ketone, (*h*) carboxylic acid, (*i*) acid chloride, (*j*) ether-ketone, (*k*) ester (cyclic esters are commonly referred to as *lactones*), (*l*) phosphate ester.

Problem 15.55. Which of the compounds in Problem 15.54 are isomers of each other?

Ans. The following pairs are isomers: (*e*) and (*f*); (*g*) and (*h*); (*j*) and (*k*).

Problem 15.56. Name compounds (*a*), (*b*), (*c*), (*e*), (*f*), (*g*), (*h*), and (*i*) in Problem 15.54.

Ans. (*a*) 2-Methylpropanoic acid, (*b*) butanone or ethyl methyl ketone, (*c*) benzoic anhydride, (*e*) 1-methoxypropanone or methoxyacetone, (*f*) methyl propanoate, (*g*) hydroxymethyl phenyl ketone, (*h*) phenylacetic acid, and (*i*) propanoyl chloride.

Problem 15.57. What is the order of boiling points for hexane, pentanal, butanoic acid, 1-pentanol, and methyl propanoate?

Ans. Butanoic acid > 1-pentanol > pentanal > methyl propanoate > hexane. The order is based on differences in intermolecular forces since the compounds have very nearly the same molecular mass. Butanoic acid and 1-pentanol have hydrogen bonding but that in the acid is more extensive because of the special interaction between the carbonyl and HO groups. Aldehydes boil higher than esters because the polarity of the carbonyl group is greater than that of an ester group. The attractive forces in alkanes, the dispersion forces, are the weakest.

Problem 15.58. Compare the water solubilities of the compounds in Problem 15.57.

Ans. Butanoic acid > 1-pentanol = pentanal = methyl propanoate > hexane. The order is similar to the order of boiling points except that the alcohol, aldehyde, and ester have the same solubility. Although neither the aldehyde nor the ester can hydrogen-bond with itself, each can hydrogen-bond with water.

Problem 15.59. Carboxylic acids have distinct and disagreeable odors. The odor of a 1 percent sodium acetate aqueous solution is similar but much less intense than that of a 1 percent acetic acid aqueous solution. Explain.

Ans. Sodium acetate reacts with water to produce acetic acid and sodium hydroxide. The concentration of acetic acid is very low because acetate is a weak base and the equilibrium is only slightly to the right-hand side.

$$CH_3COONa + H_2O \rightleftharpoons CH_3COOH + NaOH$$

Problem 15.60. What is the order of acidities for water, phenol, cyclohexanol, H_2SO_4, and propanoic acid?

Ans. H_2SO_4 > propanoic acid > phenol > cyclohexanol = water.

Problem 15.61. Calculate the pH of a 0.100 M aqueous solution of acetic acid. $K_a = 1.80 \times 10^{-5}$. (Refer to Problem 9.7.)

Ans.

$$CH_3COOH + H_2O \overset{K}{\rightleftharpoons} H_3O^+ + CH_3COO^-$$

$$K_a = K[H_2O] = \frac{[H_3O^+][CH_3COO^-]}{[CH_3COOH]}$$

Let $$x = [H_3O^+] = [CH_3COO^-]$$

$$1.80 \times 10^{-5} = \frac{x^2}{(0.100 - x)}$$

Assume $(0.100 - x) = 0.100$ and solve for x:

$$x = 1.34 \times 10^{-3}$$

$$pH = -\log [H_3O^+] = -\log 1.34 \times 10^{-3} = 2.87$$

Problem 15.62. Complete the following reactions. If no reaction occurs, write "No Reaction."

(*a*) (four-membered cyclic lactone, C=O) $\xrightarrow[H^+]{H_2O}$? (*b*) (five-membered cyclic lactone, C=O) $\xrightarrow[H^+]{H_2O}$?

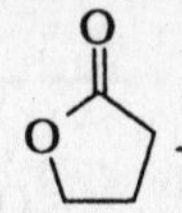

(*c*) $HO-\overset{O}{\overset{\|}{P}}(OH)-OH \xrightarrow{CH_3OH}$? (*d*) $CH_3CH_2\overset{O}{\overset{\|}{C}}H \xrightarrow{(O)}$?

(*e*) $CH_3CH_2COOH \xrightarrow{(O)}$? (*f*) $CH_3CH_2COOH \xrightarrow[H^+]{C_6H_5OH}$?

(*g*) $HCOOH \xrightarrow{NaOH}$? (*h*) $C_6H_5COOC_6H_5 \xrightarrow[H^+]{H_2O}$?

(*i*) $HO(CH_2)_4COOH \xrightarrow{H^+}$?

Ans. (*a*) and (*e*) No Reaction

(*b*) (five-membered cyclic lactone) $\xrightarrow[H^+]{H_2O} HO(CH_2)_3COOH$

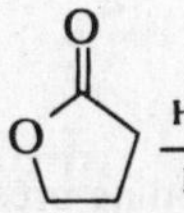

(*c*) $HO-\overset{O}{\overset{\|}{P}}(OH)-OH \xrightarrow{CH_3OH} HO-\overset{O}{\overset{\|}{P}}(OH)-OCH_3$

(*d*) $CH_3CH_2\overset{O}{\overset{\|}{C}}H \xrightarrow{(O)} CH_3CH_2\overset{O}{\overset{\|}{C}}OH$

(*f*) $CH_3CH_2COOH \xrightarrow[H^+]{C_6H_5OH} CH_3CH_2COOC_6H_5$

(*g*) $HCOOH \xrightarrow{NaOH} HCOONa$

(*h*) $C_6H_5COOC_6H_5 \xrightarrow[H^+]{H_2O} C_6H_5COOH + HOC_6H_5$

(*i*) $HO(CH_2)_4COOH \xrightarrow{H^+}$ (six-membered cyclic lactone, C=O)

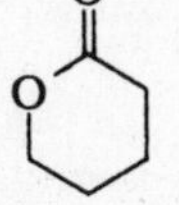

Problem 15.63. Thioalcohols behave like alcohols in reacting with carboxylic acids to form esters. Write the equation for esterification of benzoic acid with propanethiol.

Ans.

$$C_6H_5-\overset{O}{\overset{\|}{C}}-OH + HS-CH_2CH_2CH_3 \xrightarrow{H^+} C_6H_5-\overset{O}{\overset{\|}{C}}-S-CH_2CH_2CH_3 + H_2O$$

Problem 15.64. An ester $C_4H_8O_2$ yields acid X and alcohol Y upon hydrolysis. Oxidation of Y yields X. Identify the ester and show the reactions involved.

Ans. The ester is ethyl acetate:

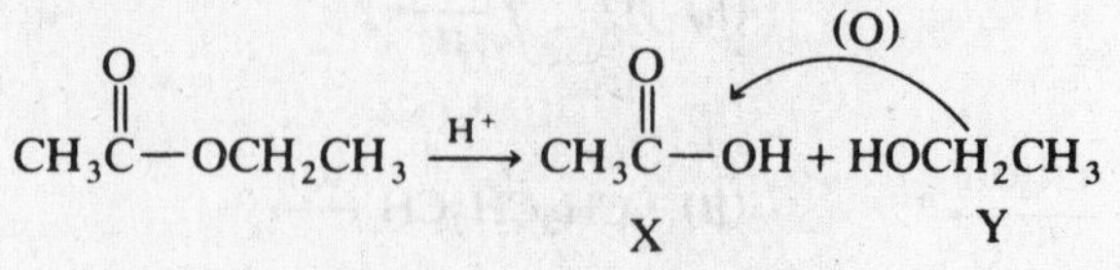

$$CH_3\overset{\overset{\displaystyle O}{\|}}{C}-OCH_2CH_3 \xrightarrow{H^+} \underset{X}{CH_3\overset{\overset{\displaystyle O}{\|}}{C}-OH} + \underset{Y}{HOCH_2CH_3} \quad (Y \xrightarrow{(O)} X)$$

Problem 15.65. Draw a representation of the hydrogen bonding present when acetic acid dissolves in water.

Ans.

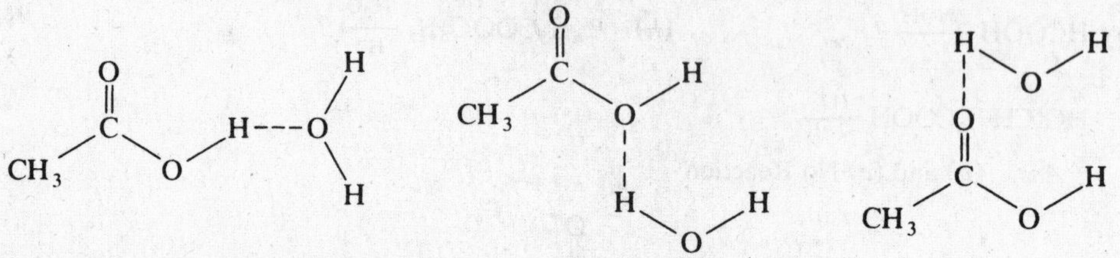

$$CH_3-\overset{\overset{\displaystyle O}{\|}}{C}-O\cdots H-O-H \qquad CH_3-\overset{\overset{\displaystyle O}{\|}}{C}-\underset{\underset{\displaystyle H-O-H}{\vdots}}{O}-H \qquad CH_3-\overset{\overset{\displaystyle O\cdots H-O-H}{\|}}{C}-O-H$$

Problem 15.66. Draw a representation of the hydrogen bonding present when methyl acetate dissolves in water.

Ans.

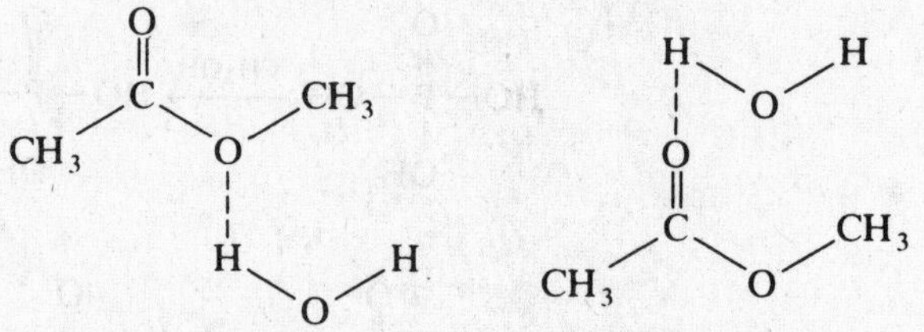

$$CH_3-\overset{\overset{\displaystyle O}{\|}}{C}-\underset{\underset{\displaystyle H-O-H}{\vdots}}{O}-CH_3 \qquad CH_3-\overset{\overset{\displaystyle O\cdots H-O-H}{\|}}{C}-O-CH_3$$

Problem 15.67. The normal metabolic activity of cells results in the production of carboxylic acids, which would alter the pH of blood and other biological fluids if unchecked. This is a potential danger to proteins and other biologically important compounds whose structures and physiological functions depend on pH (Sec. 20.3.4). The pH of blood and other biological fluids is maintained within very narrow limits to avoid such changes in structure and function. How is pH control achieved?

Ans. The body contains buffer systems, e.g., bicarbonate-carbonic acid, which prevent changes in pH by neutralization of carboxylic acids (and other acids) (Sec. 9.8). (The buffer systems also neutralize basic compounds such as amines (Refer to Chapter 16).

CHAPTER 16

Amines and Amides

Amines and *amides* contain a nitrogen atom bonded to one, two, or three carbon atoms. Amides differ from amines in having a carbonyl carbon bonded to the nitrogen:

$$\underset{\text{Amine}}{-\overset{|}{\underset{|}{N}}-} \qquad \underset{\text{Amide}}{-\underset{|}{N}-\overset{\overset{\displaystyle O}{\|}}{C}-}$$

A variety of naturally occurring and synthetic amines and amides are physiologically active. The amines include pain killers (heroin, codeine, morphine) and stimulants and decongestants (adrenalin, cocaine, dopamine, amphetamines, nicotine). The amides include pain relievers (acetaminophen), tranquilizers (phenobarbital, diazepam), and local anesthetics (novocaine). Caffeine is both an amine and an amide. Polypeptides, proteins, and synthetic nylon are amides.

16.1 AMINES

Amines are organic derivatives of ammonia in which one, two, or all three of the hydrogens have been replaced by an alkyl or aryl group. The amines are classified as *primary*, *secondary*, and *tertiary amines*, respectively.

$$\underset{\text{Ammonia}}{H-\underset{\underset{\displaystyle H}{|}}{N}-H} \qquad \underset{1^\circ\ \text{Amine}}{R^1-\underset{\underset{\displaystyle H}{|}}{N}-H} \qquad \underset{2^\circ\ \text{Amine}}{R^1-\underset{\underset{\displaystyle H}{|}}{N}-R^2} \qquad \underset{3^\circ\ \text{Amine}}{R^1-\underset{\underset{\displaystyle R^3}{|}}{N}-R^2}$$

The nitrogen atom in 1°, 2°, and 3° amines has one, two, and three bonds, respectively, to carbon.

Amines are also classified as aromatic or aliphatic. An *aromatic amine* is an amine in which an aromatic carbon is directly bonded to the nitrogen. An *aliphatic amine* is an amine in which the carbon attached to the amine nitrogen is not part of an aromatic ring.

Problem 16.1. Which of the following are amines and which are not amines? Classify amines as primary, secondary, or tertiary.

$$\underset{\text{I}}{CH_3CH_2-\underset{\underset{\displaystyle CH_3}{|}}{N}-CH_3} \qquad \underset{\text{II}}{C_6H_5-NH_2} \qquad \underset{\text{III}}{C_6H_5-CH_2-NH_2} \qquad \underset{\text{IV}}{CH_3-\overset{\overset{\displaystyle O}{\|}}{C}-NH_2} \qquad \underset{\text{V}}{CH_3-\underset{\underset{\displaystyle H}{|}}{N}-CH_3}$$

Ans. All except compound IV are amines. Compound IV is an amide since there is a carbonyl carbon bonded to the nitrogen. Compounds II and III are primary amines, compound V is a secondary amine, and compound I is a tertiary amine.

Problem 16.2. Compounds II and III in Problem 16.1 contain the benzene ring. Are these amines classified as aromatic amines?

Ans. Compound II is an aromatic amine since an aromatic carbon is directly bonded to nitrogen. Compound III is an aliphatic amine since the carbon bonded to nitrogen is an aliphatic carbon, not an aromatic carbon.

Problem 16.3. Describe the orbitals used by nitrogen in the amines in Problem 16.1.

Ans. Nitrogen uses sp^3-hybrid orbitals in forming single bonds, similar to oxygen in alcohols and ethers and similar to carbon in alkanes and cycloalkanes. The difference between nitrogen and carbon is that the ground state of N contains one electron more than carbon. The result is that nitrogen is trivalent instead of tetravalent. One of the four sp^3-hybrid orbitals of nitrogen is filled, containing a pair of nonbonding electrons.

Problem 16.4. What are the bond angles about the N atom in compound V in Problem 16.1?

Ans. The bond angles about any sp^3-hybridized atom are close to 109.5 °, the tetrahedral bond angle. This is the case for nitrogen as well as for carbon and oxygen. Figure 16-1 shows the bond angles about nitrogen in $(CH_3)_2NH$. The "bond" from N that is not connected to another atom represents a nonbonded pair of electrons. The more usual representation of such electrons is structure I, shown below. Often, the nonbonded electrons are not shown, but understood to be present, as in structure II:

```
      CH3            CH3
       |              |
       N:             N
     /   \          /   \
    H     CH3      H     CH3
        I              II
```

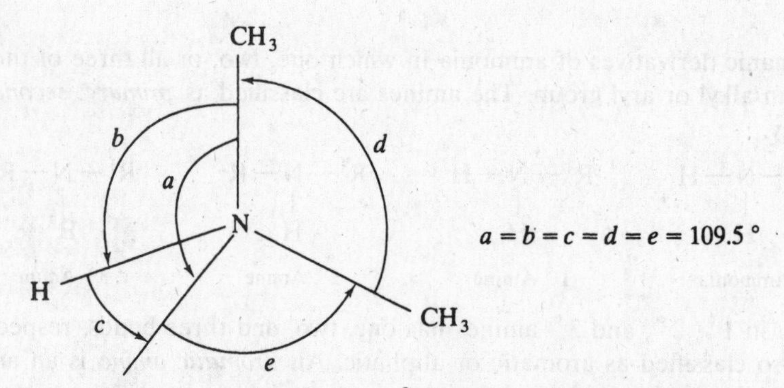

Fig. 16-1 Bond angles for sp^3 nitrogen in dimethylamine.

16.2 CONSTITUTIONAL ISOMERISM IN AMINES

Analogous to all families except the alkanes, there are two sources of constitutional isomerism in amines—different carbon skeletons and different locations of the nitrogen.

Problem 16.5. Draw structural formulas of isomeric amines of molecular formula C_3H_9N. Show only one structural formula for each isomer.

Ans.

```
CH3—N—CH3     CH3CH2—N—CH3     CH3CH2CH2—N—H     (CH3)2CH—N—H
    |                |                   |                 |
    CH3              H                   H                 H

    I                II                  III               IV
```

Problem 16.6. Compare the C:H ratio in the two compounds in each of the following pairs. What important lesson results from this problem?

(*a*) $CH_3CH_2CH_2NH_2$ and $CH_3CH_2CH_3$

(*b*) $CH_2{=}CHCH_2NH_2$ and $CH_3CH{=}CH_2$

(*c*) cyclopropyl–NH_2 and cyclopropane

Ans. The compounds in each pair do not have the same C:H ratio even though each compound has the same number of extra bonds and/or rings. The presence of nitrogen in a compound changes the C:H ratio relative to what it is in the corresponding compound without nitrogen. The C:H ratio in compounds containing nitrogen can be analyzed in the same manner as done for compounds without nitrogen if the nitrogen and one hydrogen are ignored. The amine in pair a has a C:H ratio of 3:8 after ignoring one hydrogen, which means pair a fits the general formula (C_nH_{2n+2}) for alkanes. Pairs b and c fit the general formula for an alkene or cycloalkane (C_nH_{2n}).

16.3 NOMENCLATURE OF AMINES

There is more than one system for naming amines. For amines with simple substituents, the preferred nomenclature system names the substituents, without spaces between them, followed by the suffix *-amine*. This is often referred to as the *common nomenclature system.*

Primary amines are also named as derivatives of hydrocarbons by changing the ending of the name from *-e* to *-amine*. For secondary and tertiary amines, the NHR and NRR′ substituents are named as *N-substituted* and *N,N-disubstituted amino* substituents, respectively.

An amine that is also a member of another family can be named based on the other family with the NH_2 named as an *amino* substituent.

Compound II in Problem 16.1 is almost always referred to as aniline, although phenylamine and benzenamine are also correct.

Problem 16.7. Name the following amines:

I: $CH_3{-}CH_2{-}NH_2$

II: $CH_3{-}N(CH_3){-}CH_3$

III: $CH_3{-}CH_2{-}CH_2{-}NH{-}CH_3$

IV: $CH_3{-}CH_2{-}CH(N(CH_3)CH_2CH_3){-}CH_3$

Ans. Compound I: ethylamine or ethanamine. Compound II: trimethylamine or *N,N*-Dimethylaminomethane. Compound III: methylpropylamine or 1-(*N*-methylamino)propane. Compound IV: *s*-butylethylmethylamine or 2-(*N*-ethyl-*N*-methylamino)butane. (Parentheses are used in the second names for compounds III and IV to avoid ambiguity.)

Problem 16.8. Name the following compounds:

I: benzene ring bearing $N(CH_3)_2$

II: benzene ring bearing $HN{-}CH_3$ and, para to it, CH_3

III: benzene ring bearing NH_2, with CH_3 ortho and CH_3 para to it

Ans. Compound I: *N,N*-dimethylaniline. Compound II: *p*-methyl-*N*-methylaniline. Compound III: 2,4-dimethylaniline.

Problem 16.9. Name the following compounds:

$$H_2N-CH_2-CH_2-CH_2-CH_2-NH_2 \qquad CH_2{=}CH-CH_2-CH_2-NH_2$$
I II

$$\underset{\displaystyle NH_2}{CH_3-\underset{|}{CH}-CH_2-CH_2-OH}$$
III

Ans. Compound I: 1,4-butanediamine; compound II: 3-butene-1-amine or 4-amino-1-butene; compound III: 3-amino-1-butanol

A *heterocyclic amine* is an amine in which the nitrogen is part of a cyclic structure. Heterocyclic amine structures are found in a number of compounds of biochemical interest. Some important heterocyclic structures are shown below along with their names:

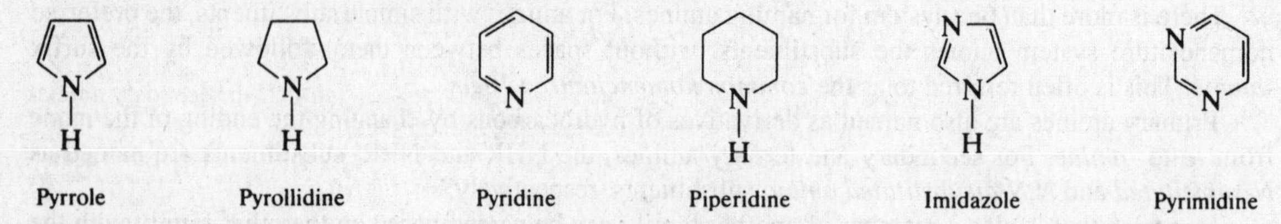

Pyridine and pyrimidine are nitrogen analogues of benzene and possess aromatic properties. (Pyrrole and imidazole also possess aromatic properties but the reasons for this are beyond the scope of this text.)

Problem 16.10. Classify the heterocyclic amines shown above as primary, secondary, or tertiary.

Ans. Pyrrole, pyrollidine, and piperidine are secondary amines since there are two bonds from nitrogen to carbon. Pyridine and pyrimidine are tertiary amines since there are three bonds from nitrogen to carbon. Imidazole is both a secondary and a tertiary amine. The nitrogen which has an attached hydrogen is the secondary amine nitrogen.

16.4 PHYSICAL PROPERTIES OF AMINES

The N—H bond in primary and secondary amines, like the O—H bond of an alcohol, participates in hydrogen bonding. The N—H bond is not as polarized as the O—H bond and this results in weaker hydrogen-bonding intermolecular attractions in amines compared to alcohols. Primary and secondary amines have higher boiling and melting points than hydrocarbons and ethers but lower than alcohols. The solubility of amines in water is comparable to that of alcohols and ethers.

Problem 16.11. Which of the following correctly represents hydrogen bonding?

(*a*) $(CH_3)_2N$ with H bonded to N: $(CH_3)_2N{-}H \cdots CH_3-\underset{|}{\overset{CH_3}{N}}-H$

(*b*) $(CH_3)_2\underset{\backslash H}{N} \cdots H-N(CH_3)_2$

(*c*) $(CH_3)_2N-H \cdots CH_3-\underset{CH_3}{N}-H$

(*d*) $(CH_3)_2\underset{\backslash H}{N} \cdots H-CH_2-\underset{CH_3}{NH}$

Ans. The NH bond of an amine is polarized with the H end as $\delta+$ and N end as $\delta-$. Hydrogen bonding is the attraction of the $H^{\delta+}$ of one amine molecule for the $N^{\delta-}$ of another amine molecule. Formulas (*a*) and (*c*), which are equivalent, are incorrect because the C of a methyl group does not carry significant charge; there is no strong attraction of N from the NH of one molecule for that C of another molecule. Formula (*d*) is incorrect since an H attached to a C carries no significant charge and is not attracted to the N of an NH bond. Hydrogen bonding is represented only by formula (*b*). A more informative version of (*b*) uses $\delta+$ and $\delta-$ notations to more clearly show the source of the attractive forces:

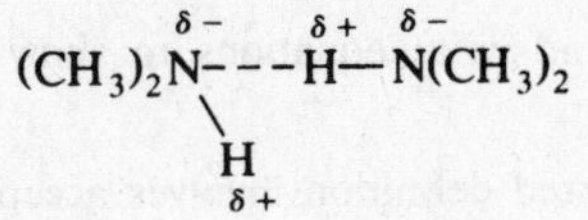

$$\overset{\delta-}{(CH_3)_2N}\text{---}\overset{\delta+}{H}\text{—}\overset{\delta-}{N}(CH_3)_2$$
$$\underset{\delta+}{H}$$

Problem 16.12. Explain the difference in boiling points in each of the following pairs of compounds: (*a*) $CH_3OCH_2CH_3$ (10.8 °C) versus $CH_3NHCH_2CH_3$ (37 °C), (*b*) $CH_3CH_2CH_2OH$ (97 °C) versus $CH_3CH_2CH_2NH_2$ (48 °C), (*c*) $CH_3NHCH_2CH_3$ (37 °C) versus $CH_3CH_2CH_2NH_2$ (48 °C), (*d*) $CH_3NHCH_2CH_3$ (37 °C) versus $(CH_3)_3N$ (3.5 °C), (*e*) $CH_3CH_2CH_3NH_2$ (48 °C) versus CH_3CH_2CHO (49 °C).

Ans. (*a*) Hydrogen bonding attractions among amine molecules are greater than the polar attractions among ether molecules.

(*b*) The hydrogen bonding attractions among alcohol molecules are greater than those among amine molecules since the O—H bond is more polar than the N—H bond.

(*c*) The amount of hydrogen bonding in primary amines is greater than that in secondary amines since there are twice as many N—H bonds in primary amines.

(*d*) Tertiary amines do not possess hydrogen bonding since there is no N—H bond. Tertiary amines possess only weak polar attractive forces since the C—N is only slightly polar.

(*e*) The hydrogen-bonding forces in amines are comparable to the polar forces in aldehydes. The hydrogen-bonding forces in amines are the weaker of the hydrogen-bonding forces (the alcohols have the stronger hydrogen-bonding forces) while the polar forces in aldehydes and ketones are the strongest of the polar forces.

Problem 16.13. Hydrogen bonding is responsible for the solubility of primary and secondary amines in water. Give structural representations of the hydrogen bonding between dimethylamine and water.

Ans.

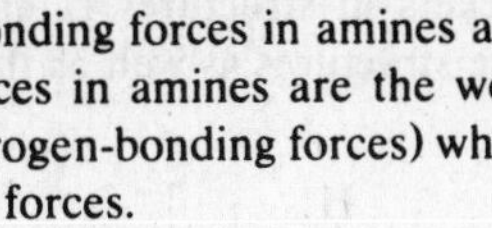

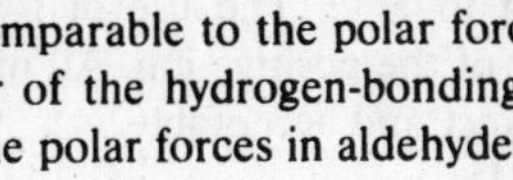

$$\overset{\delta-}{(CH_3)_2N}(\underset{\delta+}{H}\text{---}O^{\delta-}(\overset{\delta+}{H})(H)) \qquad \overset{\delta-}{(CH_3)_2N}(\underset{\delta+}{H})\text{---}\overset{\delta+}{H}\text{—}\overset{\delta-}{O}\text{—}H$$

Problem 16.14. Explain why tertiary amines are almost as water-soluble as primary and secondary amines.

Ans. Although a tertiary amine cannot hydrogen-bond with itself, it can hydrogen-bond with water due to the presence of the nonbonded electron pair of nitrogen. There is an attractive force between the nonbonded electron pair of N and the $H^{\delta+}$ of water, as shown below for trimethylamine and water:

$$(CH_3)_3N\text{---}\overset{\delta+}{H}\text{—}\overset{\delta-}{O}\text{—}H$$

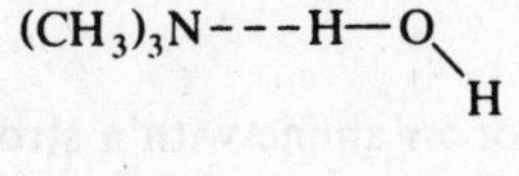

16.5 CHEMICAL REACTIONS OF AMINES

16.5.1 Basicity

Amines are the most basic of all organic compounds—much more basic than alcohols and water. However, amines are considerably weaker bases than strong bases such as NaOH and KOH. The basicity of amines, like that of ammonia, is due to the ability of the nonbonded pair of electrons on nitrogen to bond to a proton donated by some acid HA:

$$\diagdown\!\!\!-\text{N}: + \text{HA} \rightleftharpoons -\overset{+}{\text{N}}:\text{H} + \text{A}^-$$

Problem 16.15. Define basicity and write equations to show methylamine acting as a base when added to (*a*) water and (*b*) HCl.

Ans. Basicity in the Brønsted definition involves acceptance by the base (amine) of a proton donated from an acid.

(*a*) $CH_3NH_2 + H_2O \rightleftharpoons CH_3NH_3{}^+OH^-$

(*b*) $CH_3NH_2 + HCl \longrightarrow CH_3NH_3{}^+Cl^-$

Water and HCl are the acids in a and b, respectively. The reaction of methylamine with water is shown as an equilibrium reaction with the equilibrium only partially to the right-hand side because amines are weak bases and water is a weak acid. The reaction with HCl goes completely to the right-hand side because HCl is a strong acid.

Problem 16.16. Draw an expanded structure for $CH_3NH_3{}^+Cl^-$ to more clearly indicate what atoms are bonded to nitrogen and where the positive charge resides.

Ans. Several variations are equally acceptable. Structures I and II emphasize that the positive charge resides primarily on nitrogen (because a neutral N in the amine has donated electron density in bonding to H^+). The use of [ ] brackets in structures II and III emphasizes more clearly the identity of the positive ion. All of these structures as well as the condensed version $CH_3NH_3{}^+Cl^-$ are correct and acceptable.

$$\underset{\text{I}}{CH_3-\overset{\overset{\displaystyle H}{|}}{\underset{\underset{\displaystyle H}{|}}{N^+}}-H\ \ Cl^-} \qquad \underset{\text{II}}{\left[CH_3-\overset{\overset{\displaystyle H}{|}}{\underset{\underset{\displaystyle H}{|}}{N^+}}-H\right]Cl^-} \qquad \underset{\text{III}}{\left[CH_3-\overset{\overset{\displaystyle H}{|}}{\underset{\underset{\displaystyle H}{|}}{N}}-H\right]^+Cl^-}$$

Problem 16.17. How does one experimentally determine the relative basicities of compounds such as ethanol, methylamine, and KOH?

Ans. Compare the pH of solutions of the compounds at the same molar concentrations. For example, the pH of 0.1 *M* solutions of ethanol, methylamine, and KOH is 7.0, 11.8 and 13.0, respectively.

Problem 16.18. Explain the order of basicities: aliphatic amines > ammonia > aromatic amines.

Ans. Basicity increases with increasing electron density on nitrogen, i.e., with increased availability of the nonbonded electron pair. Aliphatic amines are more basic than ammonia because alkyl substituents are electron-releasing (relative to H in ammonia) and this increases electron density on nitrogen. Aromatic amines such as $C_6H_5-NH_2$ are less basic than ammonia because aryl substituents are electron-withdrawing and this decreases electron density on nitrogen.

The product of the reaction of an amine with a strong acid such as HCl as described in Problem 16.15(*b*) is an *amine salt*.

The rules for naming simple amine salts are as follows:

1. Name the amine from which the amine salt is derived.
2. Change the ending of the name from *-amine* to *-ammonium*. For amine salts of aniline, change the ending to *-anilinium*.
3. Add the name of the anion at the end of the name.

Problem 16.19. Name the following amine salts:

$$(a)\ \underset{\displaystyle CH_3NH_2{}^+Cl^-}{CH_3\underset{|}{C}HCH_2CH_3} \qquad (b)\ \left[\begin{array}{c}(CH_3CH_2)_2NH^+\\ |\\ C_6H_5\end{array}\right]_2 SO_4{}^{2-}$$

Ans. (*a*) *s*-butylmethylammonium chloride, (*b*) *N*,*N*-diethylanilinium sulfate.

Problem 16.20. Amine salts are weak acids. Write equations to show dimethylammonium chloride acting as an acid toward (*a*) water and (*b*) KOH.

Ans.

$$(a)\ (CH_3)_2NH_2{}^+Cl^- + H_2O \rightleftharpoons (CH_3)_2NH + H_3O^+ + Cl^-$$

$$(b)\ (CH_3)_2NH_2{}^+Cl^- + KOH \longrightarrow (CH_3)_2NH + H_2O + KCl$$

Water and KOH are the bases in (*a*) and (*b*), respectively. The reaction of dimethylammonium chloride with water is shown as an equilibrium reaction with the equilibrium only partially to the right-hand side because ammonium salts are weak acids and water is a weak base. The reaction with KOH goes completely to the right-hand side because KOH is a strong base.

Problem 16.21. Compare the water solubilities of amines with their amine salts.

Ans. There is no difference in water solubilities for amines and amine salts containing up to four carbons. Both are miscible with water in all proportions. However, there is a large difference for amines and amine salts of five or more carbons. Amines of five or more carbons are much less soluble than amines of less than five carbons. The polar portion of the molecule is water-compatible while the nonpolar portion is not. Water solubility occurs unless the nonpolar portion is too large. "Too large" is five or more carbons for all families except for salts such as the amine salts. Amine salts are ionic and much more water-compatible than the polar amines. Amine salts show water solubility even when there are more than five carbons.

Problem 16.22. Compare the melting points of amines with their amine salts.

Ans. The amine salts have considerably higher boiling and melting points since they are ionic compounds while the amines are polar compounds. Intermolecular attractive forces in ionic compounds are greater than those in the most polar compounds, even those with hydrogen-bonding forces.

The high water solubility of amine salts is important in biological systems. The body's buffer systems neutralize and solubilize amines by conversion to amine salts. Intravenous drugs containing the amino group are usually administered in the form of the amine salt instead of the amine to achieve faster absorption into the body.

16.5.2 Nucleophilic Substitution on Alkyl Halides

Tertiary amines react with alkyl halides by *nucleophilic substitution* to form *quaternary ammonium salts*:

$$R_3N\!: + R'Cl \longrightarrow R_3\overset{+}{N}\!:\!R' + Cl^-$$

The nonbonded electron pair of nitrogen bonds to the carbon of an alkyl halide and this forces breakage of the carbon-halogen bond. The reaction is analogous to basicity, the difference being that nitrogen bonds to a carbon instead of to a hydrogen. The term *nucleophilic substitution* describes the two aspects of the reaction. *Nucleophilic* refers to the attacking species (nitrogen of amine) being an electron donor. *Substitution* refers to one group (amine nitrogen) substituting for another (halogen). The reaction is also referred to as *alkylation* of the amine.

Problem 16.23. Write equations to show ethyl chloride reacting with (*a*) *N*,*N*-dimethylaniline and (*b*) trimethylamine.

Ans.

$$(a)\quad C_6H_5-N(CH_3)_2 + CH_3CH_2Cl \longrightarrow \left[C_6H_5-N^+(CH_3)_2-CH_2CH_3\right]Cl^-$$

$$(b)\quad CH_3-N(CH_3)_2 + CH_3CH_2Cl \longrightarrow \left[CH_3-N^+(CH_3)_2-CH_2CH_3\right]Cl^-$$

Quaternary ammonium salts in which one of the alkyl groups is a long alkyl group, e.g., $C_{16}H_{33}$, possess cleansing properties and are referred to as *cationic soaps* or *detergents*. These are cationic analogs of the sodium and potassium carboxylate soaps discussed in Sec. 15.5. There is a cationic group instead of an anionic group attached to a long hydrophobic chain.

16.6 CONVERSION OF AMINES TO AMIDES

Problem 16.24. Give the equation describing the acid-base reaction between RCOOH and $R'NH_2$.

Ans.

$$RCOOH + R'NH_2 \longrightarrow RCOO^-[R'NH_3^+]$$

The acid-base reaction occurs at ambient or moderate temperatures (20–50 °C). A different reaction occurs when the amine and carboxylic acid are reacted at moderately high temperatures (>100 °C). Dehydration takes place between carboxylic acid and amine to produce an *amide*:

$$\underset{\text{Acid}}{R-\overset{O}{\overset{\|}{C}}-OH} + \underset{\text{Amine}}{H-\overset{H}{\overset{|}{N}}-R'} \xrightarrow{\text{heat}} \underset{\text{Amide}}{R-\overset{O}{\overset{\|}{C}}-\overset{H}{\overset{|}{N}}-R'} + H_2O$$

The reaction, referred to as *amidation*, is a reversible reaction. Amide formation is maximized by forcing the equilibrium to the right by removal of water via running the reaction above the boiling point of water. Amide formation is the reaction by which polypeptides and proteins are synthesized in biological systems.

Problem 16.25. Which bonds break and which bonds form in amidation?

Ans. Two bonds break—the carbonyl carbon to hydroxyl oxygen bond in the carboxylic acid and the hydrogen to nitrogen bond in the amine. There is then an exchange of fragments. The OH from the carboxylic acid bonds to the H from the amine to form water. The amine nitrogen bonds to the carbonyl carbon to form the amide:

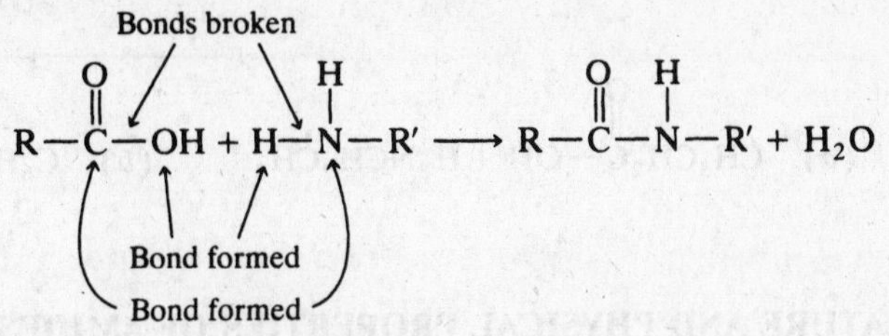

$$\mathrm{R{-}\overset{\overset{\large O}{\|}}{C}{-}OH + H{-}\underset{}{\overset{\overset{\large H}{|}}{N}}{-}R' \longrightarrow R{-}\overset{\overset{\large O}{\|}}{C}{-}\overset{\overset{\large H}{|}}{N}{-}R' + H_2O}$$

Bonds broken

Bond formed

Bond formed

Problem 16.26. Write equations for amidation of benzoic acid with (*a*) methylamine and (*b*) dimethylamine.

Ans.

(*a*) $$\mathrm{C_6H_5{-}\overset{\overset{\large O}{\|}}{C}{-}OH + H\overset{\overset{\large H}{|}}{N}{-}CH_3 \longrightarrow C_6H_5{-}\overset{\overset{\large O}{\|}}{C}{-}\overset{\overset{\large H}{|}}{N}{-}CH_3 + H_2O}$$

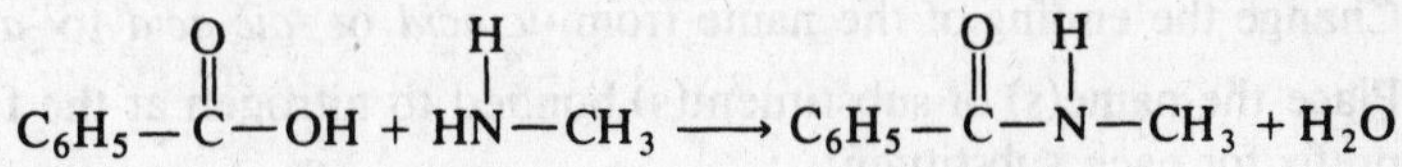

(*b*) $$\mathrm{C_6H_5{-}\overset{\overset{\large O}{\|}}{C}{-}OH + H\overset{\overset{\large CH_3}{|}}{N}{-}CH_3 \longrightarrow C_6H_5{-}\overset{\overset{\large O}{\|}}{C}{-}\overset{\overset{\large CH_3}{|}}{N}{-}CH_3 + H_2O}$$

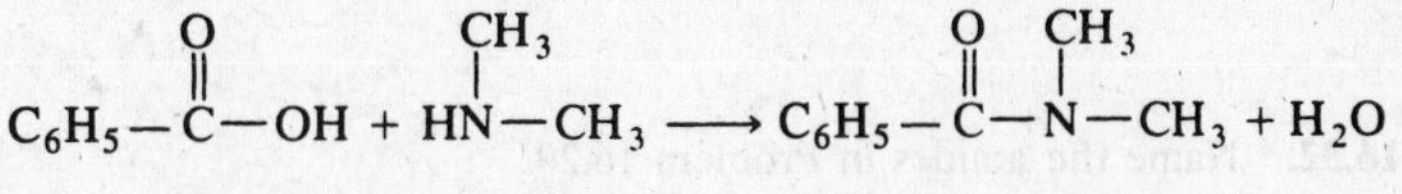

Problem 16.27. What structural feature distinguishes an amide from an amine?

Ans. Amines and amides contain a nitrogen atom bonded to one or more carbon atoms. Amides differ from amines in having a carbonyl carbon bonded to the nitrogen:

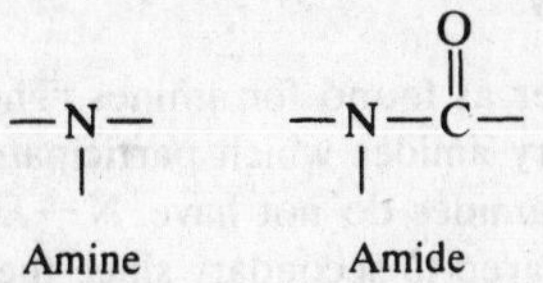

$$\underset{\text{Amine}}{\mathrm{{-}\underset{|}{N}{-}}} \qquad \underset{\text{Amide}}{\mathrm{{-}\underset{|}{N}{-}\overset{\overset{\large O}{\|}}{C}{-}}}$$

Problem 16.28. What structural features distinguish primary, secondary, and tertiary amides?

Ans. The difference is the same as for amines. The nitrogen in 1°, 2°, and 3° amides has one, two, and three bonds to carbon, respectively:

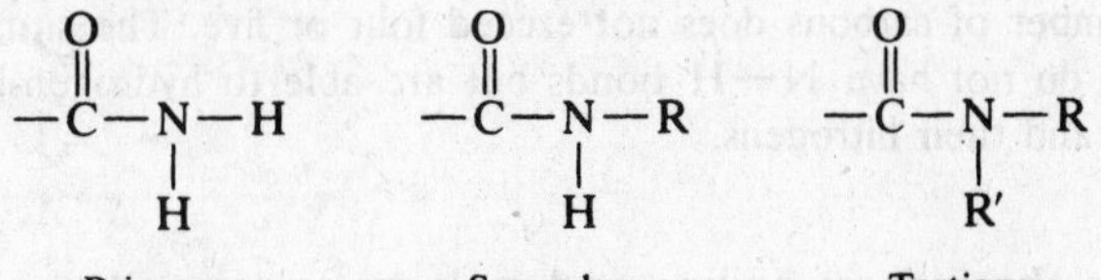

$$\underset{\text{Primary}}{\mathrm{{-}\overset{\overset{\large O}{\|}}{C}{-}\underset{\underset{\large H}{|}}{N}{-}H}} \qquad \underset{\text{Secondary}}{\mathrm{{-}\overset{\overset{\large O}{\|}}{C}{-}\underset{\underset{\large H}{|}}{N}{-}R}} \qquad \underset{\text{Tertiary}}{\mathrm{{-}\overset{\overset{\large O}{\|}}{C}{-}\underset{\underset{\large R'}{|}}{N}{-}R}}$$

Problem 16.29. Classify the following compounds as primary, secondary, or tertiary amides:

(*a*) $\mathrm{CH_3CH_2\overset{\overset{O}{\|}}{C}{-}\overset{\overset{H}{|}}{N}CH_2CH_3}$ (*b*) $\mathrm{C_6H_5\overset{\overset{O}{\|}}{C}{-}\overset{\overset{CH_3}{|}}{N}CH(CH_3)_2}$

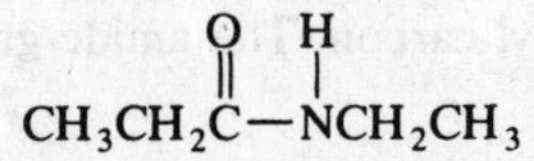

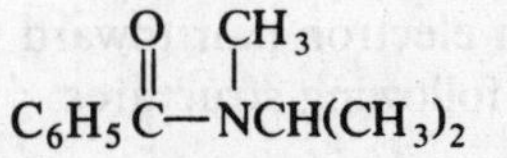

Ans. (*a*) Secondary amide since there are two carbons and only one H attached to N. (*b*) Tertiary amide since there are three carbons and no H attached to N.

Problem 16.30. Why do tertiary amines not form amides?

Ans. Tertiary amines have no hydrogen attached to nitrogen. Amidation requires the presence of a hydrogen on the nitrogen. There must be a hydrogen lost from the nitrogen to couple with the OH that is to be lost from the carboxylic acid.

Problem 16.31. Which carboxylic acid and amine are required to produce each of the amides in Problem 16.29?

Ans.

$$(a)\ CH_3CH_2\overset{\overset{\displaystyle O}{\|}}{C}{-}OH + H_2NCH_2CH_3 \qquad (b)\ C_6H_5\overset{\overset{\displaystyle O}{\|}}{C}{-}OH + H\overset{\overset{\displaystyle CH_3}{|}}{N}CH(CH_3)_2$$

16.7 NOMENCLATURE AND PHYSICAL PROPERTIES OF AMIDES

The rules for naming amides derived from simple amines are as follows:

1. Name the carboxylic acid (either IUPAC or common name) from which the amide is derived.
2. Change the ending of the name from *-ic acid* or *-oic acid* to *-amide*.
3. Place the name(s) of substituent(s) bonded to nitrogen at the front of the name. Use *N-* as a prefix for each substituent.

Problem 16.32. Name the amides in Problem 16.29.

Ans. (*a*) *N*-Ethylpropanamide or *N*-ethylpropionamide, (*b*) *N*-isopropyl-*N*-methyl-benzamide.

Problem 16.33. Explain the order of increasing boiling and melting points for amides of comparable molecular mass: primary > secondary > tertiary.

Ans. This is the same order as found for amines. The level of intermolecular attractions is greater for primary and secondary amides which participate in hydrogen bonding since they possess N—H bonds while tertiary amides do not have N—H bonds. Hydrogen bonding is more extensive for primary amides compared to secondary since there are twice as many N—H bonds in the former.

Problem 16.34. Compare the water solubilities of primary, secondary, and tertiary amides of comparable molecular weight.

Ans. Unlike their boiling and melting point differences, all amides show high water solubility as long as the number of carbons does not exceed four or five. The situation is similar to amines. Tertiary amides do not have N—H bonds but are able to hydrogen-bond with water via their carbonyl groups and their nitrogens.

Amides possess the highest boiling and melting points and water solubilities of any previously studied family of organic compounds, somewhat higher even than the carboxylic acids. This is attributed to their exceptionally high polarity which results from interaction between the carbonyl group and the nonbonded electron pair on nitrogen. The carbonyl group is polarized $^{+}C{-}O^{-}$ and this results in withdrawal of the nitrogen electron pair toward the carbonyl carbon. The amide group is pictured as a resonance hybrid of the following structures:

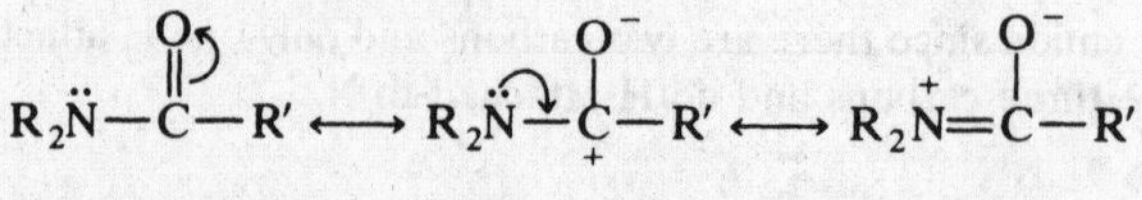

$$R_2\ddot{N}{-}\overset{\overset{\displaystyle O}{\|}}{C}{-}R' \longleftrightarrow R_2\ddot{N}{-}\overset{\overset{\displaystyle O^-}{|}}{\underset{+}{C}}{-}R' \longleftrightarrow R_2\overset{+}{N}{=}\overset{\overset{\displaystyle O^-}{|}}{C}{-}R'$$

The far right structural representation shows the + and − charges separated by three atoms instead of the usual two atoms (e.g., for the C=O in aldehydes and ketones). Larger charge separation results in increased polarity and increased intermolecular forces, even greater than those due to hydrogen bonding.

16.8 CHEMICAL REACTIONS OF AMIDES

Problem 16.35. Explain why amides, unlike amines, are neutral.

Ans. Basicity of amines is due to the availability of the nonbonded electron pair on nitrogen for bonding to a proton. The nitrogen electron pair in amides is not available for bonding to a proton. The carbonyl group pulls electron density toward it from the nitrogen as described in the previous paragraph.

Amides undergo hydrolysis in the laboratory in the presence of either an acid or base catalyst to produce the corresponding carboxylic acid and amine. This is the same reaction responsible for the enzyme-controlled digestion of proteins in biological systems.

Problem 16.36. Write the equation for hydrolysis of *N*-butyl-*N*-methylpropanamide.

Ans.

$$CH_3CH_2-\overset{\overset{O}{\|}}{C}-\overset{\overset{CH_3}{|}}{N}CH_2CH_2CH_2CH_3 \xrightarrow[OH^-]{H_2O} CH_3CH_2-\overset{\overset{O}{\|}}{C}-O^- + H\overset{\overset{CH_3}{|}}{N}CH_2CH_2CH_2CH_3$$

$$CH_3CH_2-\overset{\overset{O}{\|}}{C}-\overset{\overset{CH_3}{|}}{N}CH_2CH_2CH_2CH_3 \xrightarrow[H^+]{H_2O} CH_3CH_2-\overset{\overset{O}{\|}}{C}-OH + H\underset{H}{\overset{\overset{CH_3}{+|}}{N}}CH_2CH_2CH_2CH_3$$

The products are different depending on whether hydrolysis is carried out under basic or acidic conditions. The carboxylate salt and amine are obtained in base while the carboxylic acid and ammonium salt are obtained in acid.

ADDITIONAL SOLVED PROBLEMS

AMINES

Problem 16.37. Classify the nitrogen atoms in the following compounds as being part of amine or amide groups. Classify amines as primary, secondary, or tertiary. Indicate whether an amine is heterocyclic.

Caffeine — Amphetamine ($C_6H_5CH_2CH(CH_3)-NH_2$)

Ans. Both nitrogen atoms in the 6-membered ring of caffeine are part of tertiary amide groups. Both nitrogens in the 5-membered ring of caffeine are part of heterocyclic tertiary amine groups. Amphetamine contains a primary amine group.

Problem 16.38. Give the bond angles for *A, B, C, D, E,* and *F*:

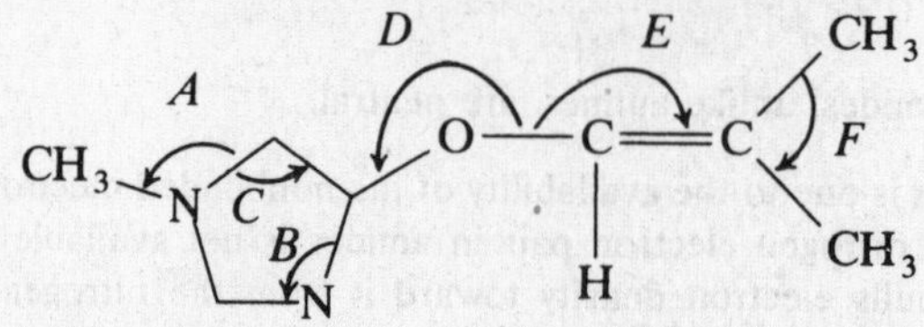

D E CH_3
A
CH_3 N C O—C=C F
B CH_3
N H

Ans. A bond angle is defined by the angle between two bonds sharing a common or central atom. The bond angle is determined by the hybridization of the central atom. *A, B, C,* and *D* are 109.5 ° since the central atoms (nitrogen, carbon, and oxygen) are sp^3-hybridized. *E* and *F* are 120 ° since the central atom (carbon) is sp^2-hybridized.

CONSTITUTIONAL ISOMERISM IN AMINES

Problem 16.39. Draw structural formulas of isomeric amines of molecular formula $C_4H_{11}N$. Show only one structural formula for each isomer.

Ans.

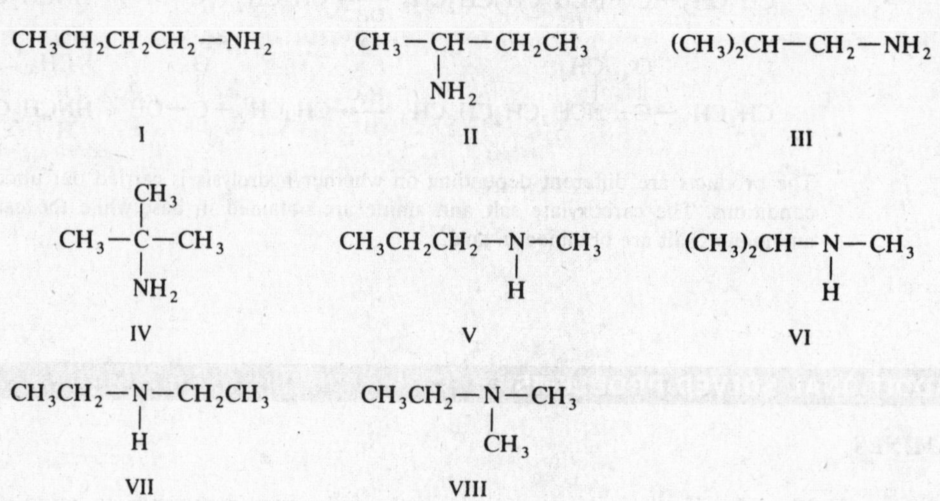

I: $CH_3CH_2CH_2CH_2—NH_2$

II: $CH_3—CH(NH_2)—CH_2CH_3$

III: $(CH_3)_2CH—CH_2—NH_2$

IV: $CH_3—C(CH_3)(NH_2)—CH_3$

V: $CH_3CH_2CH_2—N(H)—CH_3$

VI: $(CH_3)_2CH—N(H)—CH_3$

VII: $CH_3CH_2—N(H)—CH_2CH_3$

VIII: $CH_3CH_2—N(CH_3)—CH_3$

Problem 16.40. Identify the amines in Problem 16.39 as primary, secondary, or tertiary amines.

Ans. Compounds I, II, III, and IV are the primary amines. Compounds V, VI, and VII are secondary amines. Compound VIII is a tertiary amine.

NOMENCLATURE OF AMINES

Problem 16.41. Name the amines in Problem 16.39.

Ans. Compound I: Butylamine or 1-butanamine, compound II: *s*-butylamine or 2-butanamine, compound III: isobutylamine or 2-methyl-1-butanamine, compound IV: *t*-butylamine or 2-methyl-2-propanamine, compound V: methylpropylamine or 1-(*N*-methylamino)propane, compound VI: methylisopropylamine or 2-(*N*-methylamino)propane, compound VII: diethylamine or *N*-ethylaminoethane, compound VIII: ethyldimethylamine or *N,N*-dimethylaminoethane.

Problem 16.42. Draw the structure of each of the following compounds: (*a*) cyclohexyldimethylamine, (*b*) 4-methyl-*N*-methylaminocyclohexane, (*c*) cyclopentylammonium chloride.

Ans.

(*a*) $N(CH_3)_2$ on cyclohexane ring (*b*) $HN-CH_3$ and CH_3 on cyclohexane ring (*c*) $NH_3{}^+Cl^-$ on cyclopentane ring

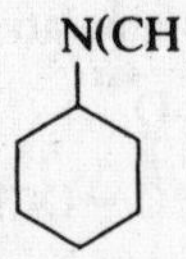

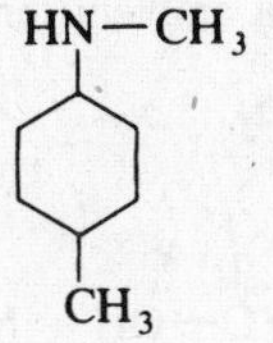

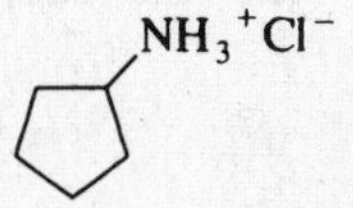

PHYSICAL PROPERTIES OF AMINES

Problem 16.43. For each of the following pairs of compounds, indicate which compound has the higher value of the property. Explain. (*a*) Melting point: hexylamine or dipropyl ether. (*b*) Solubility in water: methylamine or octylamine. (*c*) Solubility in hexane: methylamine or octylamine. (*d*) Boiling point: nonylamine or 1,8-octanediamine.

Ans. (*a*) Hexylamine has the higher boiling point due to higher intermolecular attractions from hydrogen bonding. Dipropyl ether has weaker intermolecular attractions arising from moderate polarity.

(*b*) Methylamine is much more soluble in water compared to octylamine. Both molecules possess the hydrogen-bonding amine group. However, the major portion of the octylamine molecule is hydrophobic and this negates the effect of the hydrophilic portion.

(*c*) Octylamine is much more soluble in hexane since most of the molecule is nonpolar like hexane.

(*d*) 1,8-Octanediamine has the higher boiling point. The total amount of hydrogen bonding is much greater in 1,8-octanediamine since there are two amino groups compared to one in nonylamine.

CHEMICAL REACTIONS OF AMINES

Problem 16.44. Give the equation describing the acid-base reaction between trimethylamine and benzoic acid.

Ans.

$$(CH_3)_3N + C_6H_5COOH \longrightarrow \left[(CH_3)_3\overset{+}{N}H\right] C_6H_5COO^-$$

Problem 16.45. Write the equation to show the nucleophilic substitution reaction between *N*-ethyl-*N*-methylaniline and isopropyl iodide.

Ans.

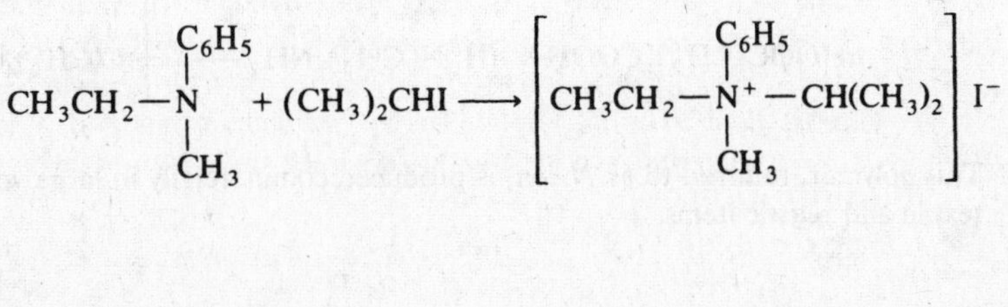

$$CH_3CH_2-\underset{\underset{CH_3}{|}}{\overset{\overset{C_6H_5}{|}}{N}} + (CH_3)_2CHI \longrightarrow \left[CH_3CH_2-\underset{\underset{CH_3}{|}}{\overset{\overset{C_6H_5}{|}}{N^+}}-CH(CH_3)_2\right] I^-$$

CONVERSION OF AMINES TO AMIDES

Problem 16.46. Give equations showing the reaction of formic (methanoic) acid with ammonia (NH_3) at (*a*) ambient temperature and (*b*) higher temperature (>100 °C).

Ans. Acid-base reaction occurs at ambient temperature while amidation occurs at higher temperature:

$$\mathrm{H{-}\overset{O}{\overset{\|}{C}}{-}OH + NH_3 \longrightarrow H{-}\overset{O}{\overset{\|}{C}}{-}O^-NH_4^+} \quad \text{Acid-base reaction}$$

$$\mathrm{H{-}\overset{O}{\overset{\|}{C}}{-}OH + NH_3 \longrightarrow H{-}\overset{O}{\overset{\|}{C}}{-}NH_2 + H_2O} \quad \text{Amidation}$$

Problem 16.47. Amides can also be synthesized by using the acid chloride or acid anhydride instead of the carboxylic acid. Give equations for the synthesis of formamide from formyl chloride and formic anhydride.

Ans.

$$(a)\quad \mathrm{H{-}\overset{O}{\overset{\|}{C}}{-}Cl + NH_3 \longrightarrow H{-}\overset{O}{\overset{\|}{C}}{-}NH_2 + HCl}$$

$$(b)\quad \mathrm{H{-}\overset{O}{\overset{\|}{C}}{-}O{-}\overset{O}{\overset{\|}{C}}{-}H + NH_3 \longrightarrow H{-}\overset{O}{\overset{\|}{C}}{-}NH_2 + HO{-}\overset{O}{\overset{\|}{C}}{-}H}$$

These equations are oversimplified. In actual practice, the HCl and HCOOH byproducts in equations (*a*) and (*b*) would react with excess NH_3 to form $NH_4^+Cl^-$ and $HCOO^-NH_4^+$, respectively.

Problem 16.48. Which of the following compounds are amides?

$$\mathrm{CH_3CH_2{-}\overset{O}{\overset{\|}{C}}{-}NH_2} \qquad\qquad \mathrm{CH_3\underset{NH_2}{\underset{|}{CH}}{-}\overset{O}{\overset{\|}{C}}H}$$

I   II

Ans. Compound I is an amide while II is both an aldehyde and an amine. Both compounds are structural isomers of C_3H_7NO.

Problem 16.49. Amidation yields a polymer if one uses bifunctional reactants. Write the equation for polymerization of 1,6-hexanedioic acid, $HOOC(CH_2)_4COOH$, and 1,6-diaminohexane, $H_2N(CH_2)_6NH_2$.

Ans.

$$n\mathrm{HOOC(CH_2)_4COOH} + n\mathrm{H_2N(CH_2)_6NH_2} \longrightarrow \left[\mathrm{\overset{O}{\overset{\|}{C}}(CH_2)_4\overset{O}{\overset{\|}{C}}{-}\overset{H}{\overset{|}{N}}(CH_2)_6\overset{H}{\overset{|}{N}}}\right]_n$$

This polymer, referred to as *Nylon*, is produced commercially in large volume and used to produce textile and plastic items.

NOMENCLATURE AND PHYSICAL PROPERTIES OF AMIDES

Problem 16.50. Draw the structure of each of the following compounds: (*a*) *N*-phenylbutanamide, (*b*) *N*-ethyl-*N*-methyl-2-methylpropanamide.

Ans.

$$(a)\ CH_3CH_2CH_2\overset{\overset{O}{\|}}{C}-NHC_6H_5 \qquad (b)\ CH_3\underset{\underset{CH_3}{|}}{C}H\overset{\overset{O}{\|}}{C}-\overset{\overset{CH_3}{|}}{N}-CH_2CH_3$$

Problem 16.51. Place the following compounds in order of increasing boiling point and solubility in water:

$$\underset{I}{CH_3CH_2CH_2CH_2CH_2-NH_2} \qquad \underset{II}{CH_3CH_2CH_2\overset{\overset{O}{\|}}{C}-NH_2} \qquad \underset{III}{CH_3\overset{\overset{O}{\|}}{C}-N(CH_3)_2}$$

Ans. **Compound II > compound III > compound I. All amides, including tertiary amides which have no hydrogen bonding, have higher melting and boiling points and water solubilities than amines because the amide group is very highly polar. The intermolecular attractions due to very high polarity in amides exceeds the intermolecular attractions due to hydrogen bonding in amines. Within the amides, intermolecular forces and melting and boiling points and water solubility increase in the order 1° > 2° > 3° due to the additive effect of hydrogen bonding (1° > 2°; none for 3°).**

CHEMICAL REACTIONS OF AMIDES

Problem 16.52. Which is more basic, compound I or II in Problem 16.48?

Ans. **Compound II is more basic since it is an amine. Amides such as compound I are neutral.**

Problem 16.53. Write the equation for hydrolysis of *N*-methyl-*N*-phenyl-2,2-dimethylpropanamide.

Ans.

$$(CH_3)_3C-\overset{\overset{O}{\|}}{C}-\overset{\overset{CH_3}{|}}{N}-C_6H_5 \xrightarrow{H_2O} (CH_3)_3C-\overset{\overset{O}{\|}}{C}-OH + H\overset{\overset{CH_3}{|}}{N}-C_6H_5$$

SUPPLEMENTARY PROBLEMS

Problem 16.54. Identify each of the following as an amine or amide. Classify the amine or amide as 1°, 2°, or 3°.

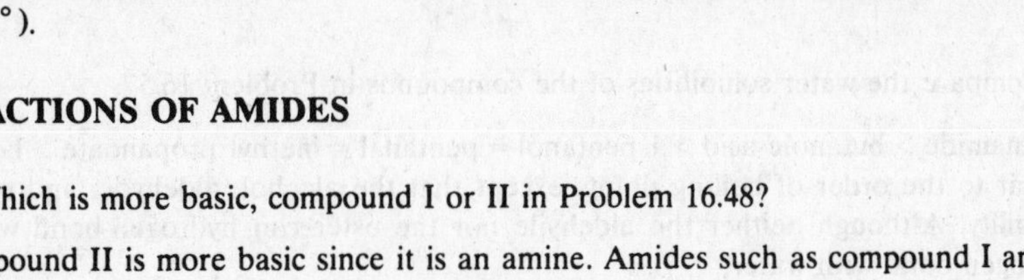

[Structures I–IV: I, a ring ketone with N–CH₃; II, a ring amide (O=C–N) with N–CH₃; III, a ring ketone with NH and a CH₃ substituent; IV, a six-membered ring with C=O next to NH]

Ans. **Compound I: 3° amine + ketone, compound II: 3° amide, compound III: 2° amine + ketone, compound IV: 2° amide.**

Problem 16.55. Which of the compounds in Problem 16.54 are isomers of each other?

Ans. All of the compounds are isomers of C_5H_9NO.

Problem 16.56. Draw structural formulas of the following compounds: (*a*) *N*-isopropylaniline, (*b*) *N*-propyl-3-methylpentanamide, (*c*) cyclopentyltrimethylammonium chloride.

Ans.

(*a*) $C_6H_5-N(H)-CH(CH_3)_2$ (*b*) $CH_3CH_2CH(CH_3)CH_2-C(=O)-N(H)-CH_2CH_2CH_3$ (*c*) $C_5H_9-\overset{+}{N}(CH_3)_3\ Cl^-$ (cyclopentyl ring)

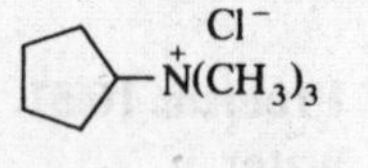

Problem 16.57. What is the order of boiling points for hexane, pentanal, butanoic acid, butanamide, 1-pentanol, and methyl propanoate?

Ans. Butanamide > butanoic acid > 1-pentanol > pentanal > methyl propanoate > hexane. The order is based on differences in intermolecular forces since the compounds have very nearly the same molecular mass. Butanoic acid and 1-pentanol have hydrogen bonding but that in the acid is more extensive because of the special interaction between carbonyl and HO groups. Butanamide has a higher boiling than butanoic acid because of the very high polarity of the amide group, the result of resonance interaction between the carbonyl group and the nonbonded electron pair of nitrogen. Aldehydes boil higher than esters because the polarity of the carbonyl group is greater than that of an ester group. The attractive forces in alkanes, the dispersion forces, are the weakest.

Problem 16.58. Compare the water solubilities of the compounds in Problem 16.57.

Ans. Butanamide > butanoic acid > 1-pentanol = pentanal = methyl propanoate > hexane. The order is similar to the order of boiling points except that the alcohol, aldehyde, and ester have the same solubility. Although neither the aldehyde nor the ester can hydrogen-bond with itself, each can hydrogen-bond with water.

Problem 16.59. The disagreeable odor of fish is due to methylamine, formed by decomposition of proteins. Lemon juice, which contains the carboxylic acid citric acid, is often added to fish to mask the fishy odor. Explain.

Ans. The carboxylic acid reacts with methylamine to convert it to the ammonium salt. The ammonium salt does not have an odor since it is not volatile, i.e., it has a much higher boiling point.

Problem 16.60. Many drugs which are amines (e.g., morphine and codeine) are administered not as amines but as the amine salts. Explain.

Ans. The amines with medicinal properties generally show low solubility in water because they contain too many carbons. Conversion to the amine salt is advantageous for intravenous injection and also for absorption after oral administration since the amine salt is much more water-soluble.

Problem 16.61. What is the order of basicity for water, NaOH, propylamine, acetamide, and aniline?

Ans. NaOH > propylamine > aniline > acetamide = water.

Problem 16.62. Complete the following reactions. If no reaction occurs, write "No Reaction."

(*a*) $CH_3CH_2-C(=O)-NH_2 \xrightarrow[H^+ \text{ or } OH^-]{H_2O} ?$ (*b*) $CH_3CH(NH_2)-C(=O)H \xrightarrow[H^+ \text{ or } OH^-]{H_2O} ?$

(*c*) N-methylpyrrolidine $\xrightarrow{CH_3CH_2Cl}$?

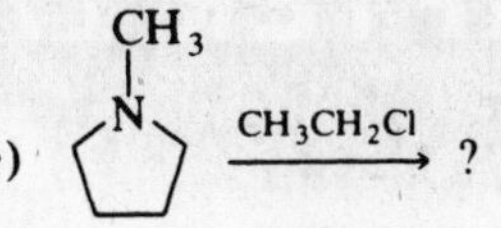

(*d*) $H_2N-C_6H_4-NH_2 + HOOC(CH_2)_2COOH \rightarrow$?

Ans.

(*a*) $$CH_3CH_2-\overset{O}{\overset{\|}{C}}-NH_2 \xrightarrow[H^+]{H_2O} CH_3CH_2-\overset{O}{\overset{\|}{C}}-OH + NH_4^+$$

(*b*) The compound is an amine and does not hydrolyze. However, reaction with acid does occur:

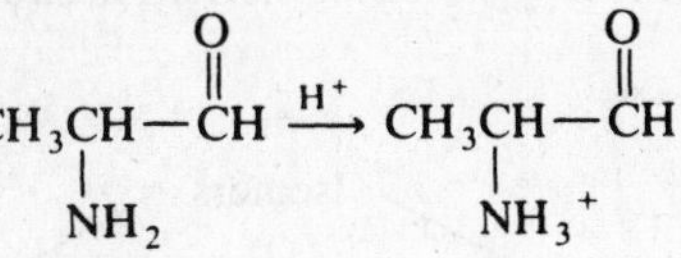

$$CH_3CH(NH_2)-\overset{O}{\overset{\|}{C}}H \xrightarrow{H^+} CH_3CH(NH_3^+)-\overset{O}{\overset{\|}{C}}H$$

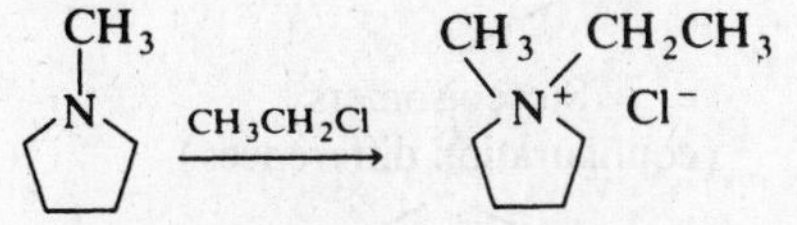

(*c*) N-methylpyrrolidine $\xrightarrow{CH_3CH_2Cl}$ N-ethyl-N-methylpyrrolidinium (CH_3, CH_2CH_3, N^+) Cl^-

(*d*) $$H_2N-C_6H_4-NH_2 + HOOC(CH_2)_2COOH \longrightarrow \left[-\overset{H}{N}-C_6H_4-\overset{H}{N}-\overset{O}{\overset{\|}{C}}(CH_2)_2\overset{O}{\overset{\|}{C}}-\right]_n$$

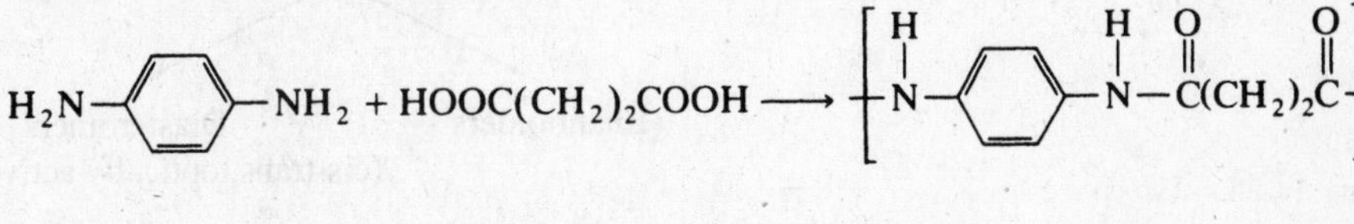

Problem 16.63. Draw a representation of the hydrogen bonding present when acetamide dissolves in water.

Ans. (Structures showing acetamide, $CH_3-\overset{O}{\overset{\|}{C}}-NH_2$, hydrogen-bonded to water molecules: the amide N–H hydrogen to the water oxygen, the water hydrogen to the carbonyl oxygen, and water O–H to the amide nitrogen.)

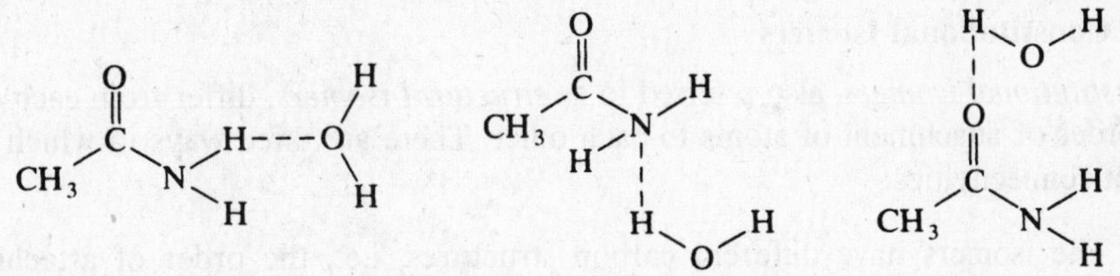

Problem 16.64. Calculate the pH of a 0.100 *M* aqueous solution of methyl amine. $K_b = 4.40 \times 10^{-4}$. (Refer to Problem 9.42.)

Ans.

$$CH_3NH_2 + H_2O \overset{K}{\rightleftharpoons} OH^- + CH_3NH_3^+$$

$$K_b = K[H_2O] = \frac{[OH^-][CH_3NH_3^+]}{[CH_3NH_2]}$$

Let $$x = [OH^-] = [CH_3NH_3^+]$$

$$4.40 \times 10^{-4} = \frac{x^2}{(0.1 - x)}$$

Assume $(0.100 - x) = 0.100$ and solve for x:

$$x = 6.60 \times 10^{-3}$$

$$[H_3O^+][OH^-] = 10^{-14}$$

$$[H_3O^+] = \frac{10^{-14}}{6.60 \times 10^{-3}} = 1.52 \times 10^{-12}$$

$$pH = -\log[H_3O^+] = -\log 1.52 \times 10^{-12} = 11.80$$

11.81

Stereoisomerism

17.1 REVIEW OF ISOMERISM

Constitutional isomers were discussed in Sections 11.6 and 12.3 and geometric isomers, a class of diastereomers, in Section 12.5. The discussion of stereoisomers is completed in this chapter with a discussion of enantiomers and two other classes of diastereomers – optically active and meso compounds. The relationship between different types of isomerism is shown in Fig. 17-1.

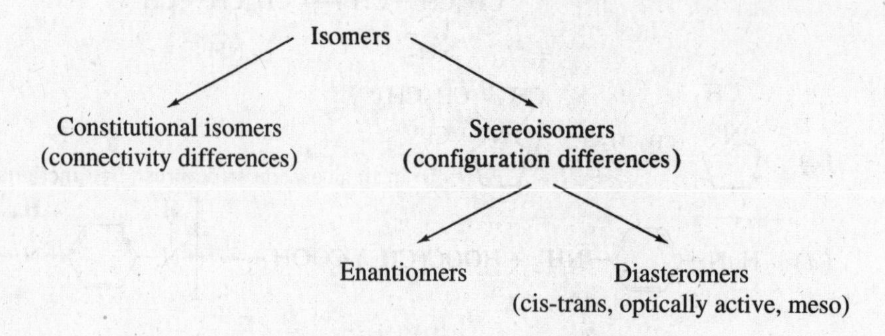

Fig. 17-1 Relationship of different types of isomers.

17.1.1 Constitutional Isomers

Constitutional isomers, also referred to as *structural isomers*, differ from each other in *connectivity* —the order of attachment of atoms to each other. There are three ways in which molecules can have different connectivities:

1. The isomers have different carbon structures, i.e., the order of attachment of the carbon atoms is different.
2. The isomers have the same carbon structure but differ in the position of a functional group such as Cl, OH, C = O, C = C.
3. The isomers have different functional groups, e.g., ethers and alcohols.

Some isomers differ from each other in more than one way.

Problem 17.1. Draw the constitutional isomers of C_4H_{10} and describe how they differ.

Ans. Butane and 2-methylpropane differ from each other in their carbon structures:

$$CH_3-CH_2-CH_2-CH_3 \qquad \begin{array}{c} CH_3-CH-CH_3 \\ | \\ CH_3 \end{array}$$

Problem 17.2. Draw the constitutional isomers of C_3H_7Cl and describe how they differ.

Ans. 1-Chloropropane and 2-chloropropane have the same carbon structure but differ from each other in the placement of the chlorine atom:

$$CH_3-CH_2-CH_2-Cl \qquad \begin{array}{c} Cl \\ | \\ CH_3-CH-CH_3 \end{array}$$

Problem 17.3. Draw the constitutional isomers of C_4H_9Cl and describe how they differ.

Ans. 1-Chlorobutane and 2-chlorobutane have the same carbon structure but differ from each other in the placement of the chlorine atom. The same is true in comparing 2-chloro-2-methylpropane and 1-chloro-2-methylpropane. Both the carbon structure and placement of Cl differ when comparing either 1-chlorobutane or 2-chlorobutane with either 2-chloro-2-methylpropane or 1-chloro-2-methylpropane.

$CH_3{-}CH_2{-}CH_2{-}CH_2Cl$

1-Chlorobutane

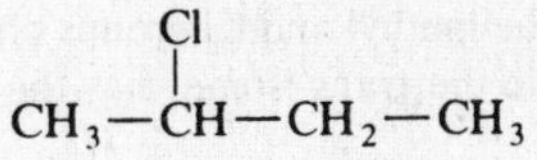

$CH_3{-}CH(Cl){-}CH_2{-}CH_3$

2-Chlorobutane

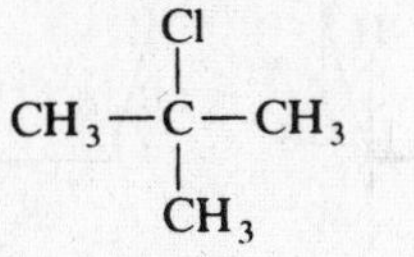

$CH_3{-}C(Cl)(CH_3){-}CH_3$

2-Chloro-2-methylpropane

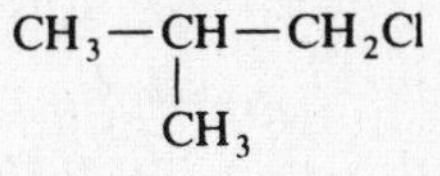

$CH_3{-}CH(CH_3){-}CH_2Cl$

1-Chloro-2-methylpropane

Problem 17.4. Draw the constitutional isomers of C_2H_6O and describe how they differ.

Ans. Ethanol and dimethyl ether differ both in their carbon structures and functional groups.

$CH_3-CH_2-OH \qquad CH_3-O-CH_3$

17.1.2 Geometrical Isomers

Cis-trans isomers, also referred to as *geometrical isomers*, are a type of *stereoisomer*. Other types are discussed in this chapter. Stereoisomers do not differ from each other in connectivity (the order of attachment of atoms to each other). Stereoisomers differ in *configuration*, i.e., the arrangement of their atoms in space.

Cis-trans isomers differ in configuration relative to a C = C or ring. More specifically, *cis* and *trans* isomers differ in the spatial arrangement of atoms (or groups of atoms) on one stereocenter relative to the spatial arrangement of atoms (or groups of atoms) on a second stereocenter. A *stereocenter* is an atom bearing other atoms or groups of atoms whose identities are such that an interchange of two of the groups produces a stereoisomer. Appropriately substituted carbons of a double bond or ring are stereocenters.

Problem 17.5. Are *cis-trans* isomers possible for (*a*) 1-butene and (*b*) 2-butene? If yes, draw the *cis-trans* isomers and describe how they differ.

Ans. (*a*) *Cis-trans* isomers are not possible for 1-butene since only one of the carbons of the double bond has two different substituents.

(*b*) *Cis-trans* isomers are possible for 2-butene since both carbons of the double bond have two different substituents.

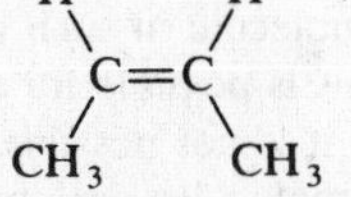

H		H		H		CH₃

$(H)(CH_3)C{=}C(H)(CH_3)$ — *cis*-2-Butene (H and H on top; CH_3 and CH_3 on bottom)

$(H)(CH_3)C{=}C(CH_3)(H)$ — *trans*-2-Butene (H and CH_3 on top; CH_3 and H on bottom)

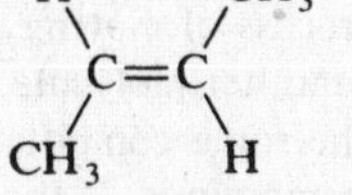

The *cis* and *trans* isomers differ in the spatial positions of the methyl groups on the double bond. *Cis*-2-butene has the methyl groups on the same side of the double bond while *trans*-2-butene has the methyl groups on opposite sides of the double bond.

Problem 17.6. Are *cis-trans* isomers possible for (*a*) chlorocyclopentane and (*b*) 1-chloro-2-methylcyclopentane? If yes, draw the *cis* and *trans* isomers and describe how they differ.

Ans. (*a*) *Cis-trans* isomers are not possible for chlorocyclopentane since only one of the carbons of the ring has two different substituents.

(*b*) *Cis-trans* isomers are possible for 1-chloro-2-methylcyclopentane since there are two carbons of the ring with two different substituents. The *cis* and *trans* isomers differ in the spatial positions of the methyl and Cl groups on the ring. The *cis* isomer has the groups on the same side of the ring while the trans isomer has the groups on opposite sides of the ring.

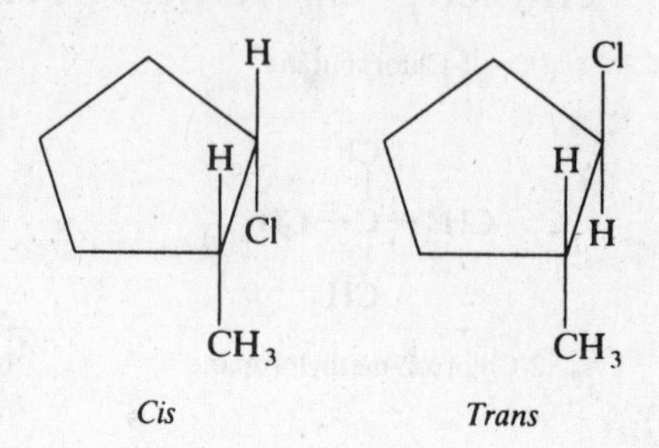

Cis *Trans*

17.2 ENANTIOMERS

Enantiomers arise from differences in the configuration at a *stereocenter*, more precisely called a *tetrahedral stereocenter*, which is a carbon with four different substituents attached to it. Such stereocenters have also been called *chiral carbons, asymmetric carbons,* and *stereogenic centers.*

When this occurs, there is not one compound but two different compounds, referred to as *enantiomers*. Enantiomers are optical isomers which are alike in all respects except that they possess molecular *chirality*, i.e., the molecules are *nonsuperimposable* mirror images of each other. Two molecules that are superimposable mirror images of each other are two molecules of the same compound. A molecule that is nonsuperimposable on its mirror image is referred to as *chiral.* A molecule that is superimposable on its mirror image is referred to as *achiral.*

GENERAL RULE. The absence of a stereocenter with four different substituents precludes the existence of enantiomers.

Problem 17.7. What is meant by two molecules being mirror images of each other?

Ans. Imagine putting a two-sided mirror in between two molecules. Each molecule looking in its side of the mirror sees as its mirror image the other molecule.

Problem 17.8. What is meant by two molecules being superimposable or nonsuperimposable?

Ans. Imagine the process of moving one molecule in such a manner as to make it go into and become identical with another molecule. If this is possible for two molecules, they are superimposable and molecules of the same compound. If it is not possible, the two molecules are nonsuperimposable and different compounds. If the two molecules are nonsuperimposable and mirror images of each other, they are enantiomers.

Problem 17.9. For each of the following compounds, indicate whether there is a stereocenter. If yes, label the stereocenter with an asterisk, list the four different substituents attached to the stereocenter, and draw the other enantiomer.

(a) H, Cl (top), CH_3, Cl (bottom) (b) H, Cl (top), CH_3, C_2H_5 (bottom)

Ans. (*a*) There is only one 1,1-dichloroethane, not a pair of enantiomers, since there is no stereocenter.
(*b*) 2-Chlorobutane exists as a pair of enantiomers since there is a stereocenter (noted by *). The stereocenter has four different substituents: H, Cl, CH_3, C_2H_5. The other enantiomer is drawn as the mirror image of the first structure.

(b) H, Cl (top), * CH_3, C_2H_5 (bottom) — I; CH_3, Cl (top), * H, C_2H_5 (bottom) — II

Problem 17.10. Objects as well as molecules are either chiral or achiral. Which of the following are chiral and which achiral? (*a*) sock, (*b*) shoe, (*c*) foot, (*d*) hand.

Ans. Achiral: sock; chiral: shoe, foot, hand

Problem 17.11. What is the relationship of the following structural drawings to those in the answer to Problem 17.9(*b*)?

CH_3, H (top), * Cl, C_2H_5 (bottom) — III; Cl, H (top), * CH_3, C_2H_5 (bottom) — IV

Ans. This pair (III, IV) represents the same pair of enantiomers as pair I and II. Pair III and IV are drawn from a different observation point than pair I and II. If you visualize yourself positioned midway behind the CH_3 and Cl groups of I and II, you would then draw the molecules as III and IV, respectively. This imaginary visualization is not easy for the beginning student. Building and manipulation of molecular models of structures I, II, III, and IV is the surest way to verify that structure I = structure III and structure II = structure IV.

Problem 17.11 points out the visualization difficulties in discussing enantiomers. One must always understand that there is only one pair of enantiomers for a compound with one stereocenter. It is correct to use either the pair of drawings in Problem 17.9(*b*) or the pair in Problem 17.11 to represent the enantiomers of 2-chlorobutane. It is incorrect to draw all four structures to represent 2-chlorobutane.

The spatial relationships among the groups attached to a stereocenter are best understood by using molecular models. Unfortunately, books have the limitation of two dimensions. Drawings such as structures I, II, III, and IV in Problems 17.9 and 17.11 attempt to overcome this limitation by incorporating some three-dimensional aspects. Brevity pushes chemists and publishers to use drawings with less three-dimensional aspects such as structures III(*b*) and III(*c*) below:

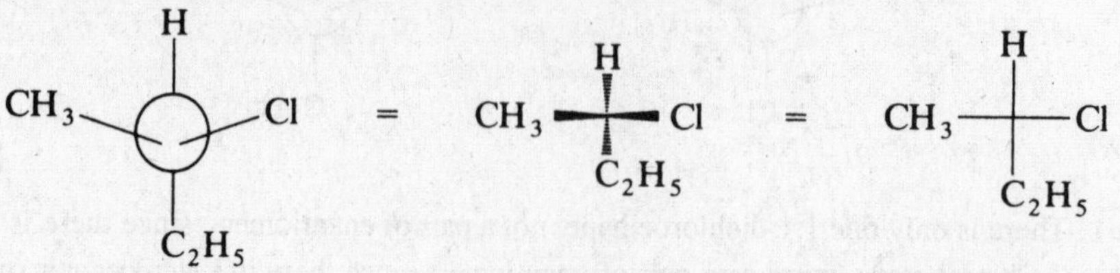

H; CH_3; Cl; C_2H_5 — III(*a*) = H; CH_3; Cl; C_2H_5 — III(*b*) = H; CH_3; Cl; C_2H_5 — III(*c*)

A "flat" drawing such as III(*c*), referred to as a *Fischer projection*, shows no obvious three-dimensional aspect of the molecule except that such drawings are used with the convention that horizontal bonds come forward and vertical bonds go backward. This convention must not be forgotten.

One often needs to compare two isomeric structural drawings of the type illustrated by drawing III(*c*) and conclude whether the structures represent the same compound or a pair of enantiomers. This can be done with the following rules:

RULE 1. The structures represent the same compound if they have the same connectivity (i.e., the compounds are not constitutional isomers) and no stereocenter. If there is a stereocenter, proceed to Rule 2.

RULE 2. Note whether the two structures are superimposable (same compound) or nonsuperimposable mirror images (enantiomers) before or after rotating either structure 180° in the plane of the paper. Rotation out of the plane of the paper is not allowed. If this does not allow one to distinguish between the same compound and enantiomers, go to Rule 3. Rotations of 90° or 270° are not allowed.

RULE 3. Exchange positions of any two of the groups attached to the stereocenter of one of the structures. Note whether the two structures are now identical (superimposable). Repeat the process of exchanging any two groups attached to the stereocenter until the two structures become identical. If an even number of exchanges is needed to make the two structures identical, the two structures represent the same enantiomer. If an odd number of exchanges is needed to make the two structures identical, the two structures represent enantiomers.

Problem 17.12. What is the rationale behind Rule 2 which allows rotation only 180° in the plane of the paper?

Ans. Only rotation 180° in the plane of the paper preserves the correct spatial arrangements of bonds—horizontal bonds coming forward and vertical bonds going backward. Rotation 180° out of the plane of the paper as well as rotation 90° or 270° in the plane of the paper reverses those spatial arrangements and this does not allow a valid comparison with the other structural drawing.

Problem 17.13. What is the rationale behind Rule 3?

Ans. An odd number of exchanges reverses the configuration at the stereocenter while an even number of exchanges maintains the configuration of the stereocenter.

Problem 17.14. What do the following structures represent, enantiomers or the same compound?

CH_2OH / H, OH / CHO   CHO / HO, H / CH_2OH

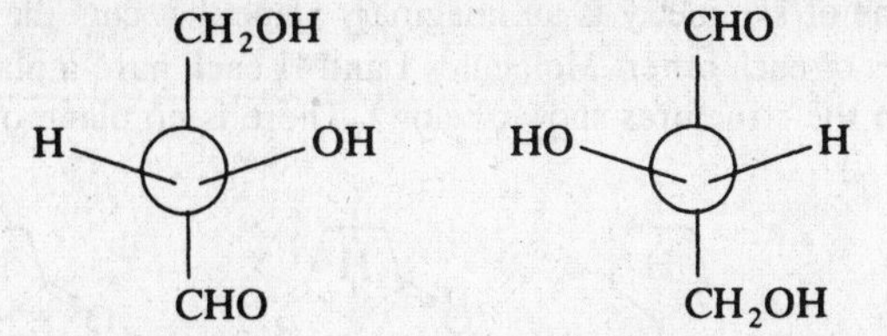

Ans. The same compound. Rotating either structure 180° in the plane of the paper allows it to be superimposed on the other structure.

Problem 17.15. What do structures I and II represent, enantiomers or the same compound?

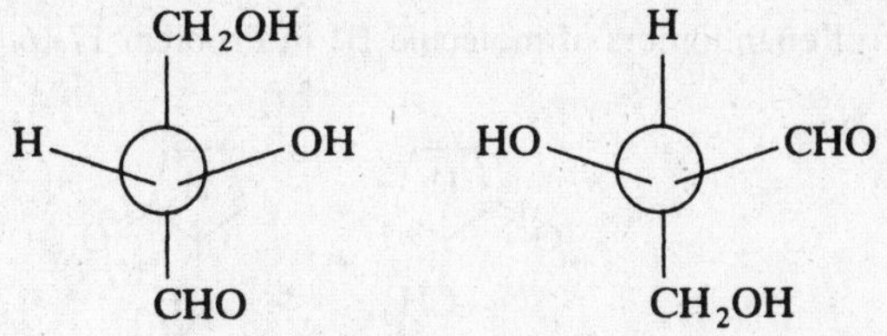

I: CH_2OH / H, OH / CHO   II: H / HO, CHO / CH_2OH

Ans. Rotating either structure 180° in the plane of the paper does not show it to be either an enantiomer or the same compound. We use Rule 3 and make the following exchanges on structure II: CHO and CH_2OH, CH_2OH and H, HO and H. This converts structure II to III to IV to I:

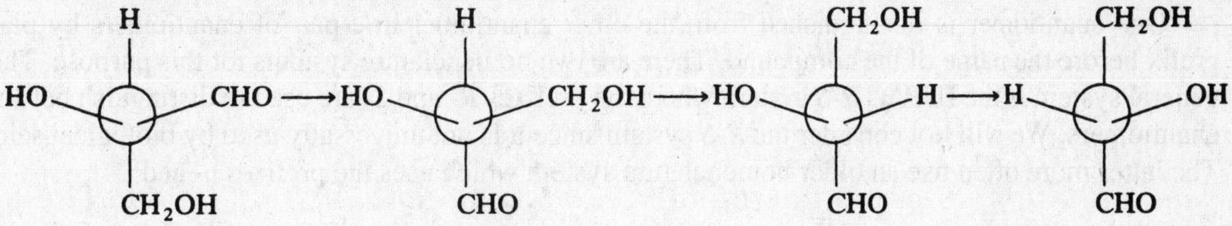

II (H / HO, CHO / CH_2OH) ⟶ III (H / HO, CH_2OH / CHO) ⟶ IV (CH_2OH / HO, H / CHO) ⟶ I (CH_2OH / H, OH / CHO)

Three exchanges—an odd number—converted II to I, which tells us that I and II are enantiomers.

Problem 17.16. Which of the following molecules are chiral? For each chiral molecule, label the stereocenter with an asterisk.

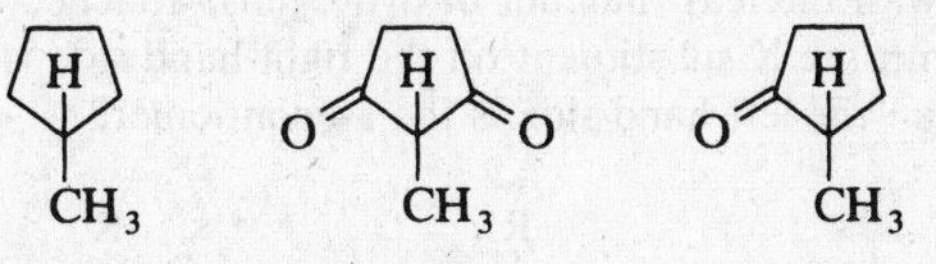

I (H, CH_3)   II (O, O, H, CH_3)   III (O, H, CH_3)

Ans. A carbon atom in a ring structure is a stereocenter if its two nonring substituents are different and if the ring is not symmetrical about that carbon. Neither molecule I nor molecule II is chiral. The only possible candidate as a stereocenter in either molecule I or molecule II is the carbon holding the H and CH_3 substituents. (The other ring carbons each have two H substituents in molecule I and either two H substituents or two bonds to the same oxygen in molecule II.) Referenced to the carbon with H and CH_3 substituents, each ring has two halves which are the same (two CH_2CH_2 halves in molecule I and two $O{=}C{-}CH_2$ halves in molecule II). Molecule III is chiral since the two halves of the ring to the left and right of the carbon with H and CH_3 substituents are not the same: One half is $O{=}C{-}CH_2$ and the other half is CH_2CH_2.

An alternative way of detecting a stereocenter in a ring is to note if the molecule has a *plane of symmetry*. A plane of symmetry is an imaginary plane that cuts the molecule into two halves that are mirror images of each other. Molecules I and II each have a plane of symmetry (symbolized by the dotted line in the structures shown below). There is no plane of symmetry for molecule III.

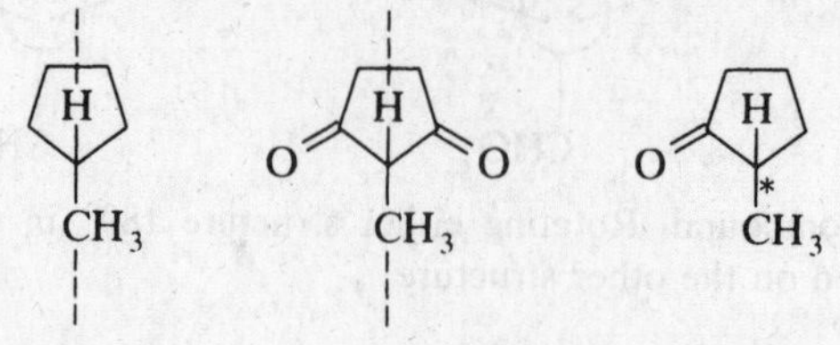

H, CH_3 — I; H, O, O, CH_3 — II; H, O, *, CH_3 — III

Problem 17.17. Draw the pair of enantiomers of molecule III in Problem 17.16.

Ans.

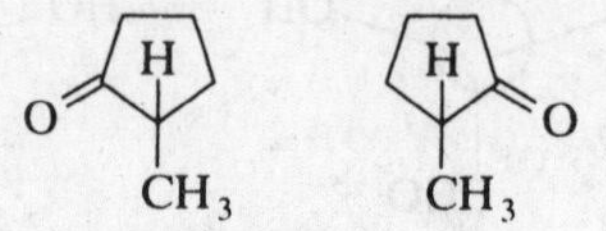

H, O, CH_3 — H, O, CH_3

17.3 NOMENCLATURE AND PROPERTIES OF ENANTIOMERS

17.3.1 Nomenclature

One enantiomer is distinguished from the other enantiomer in a pair of enantiomers by placing a prefix before the name of the compound. There are two nomenclature systems for this purpose. The most general system is the IUPAC *R-S* system where the prefixes *R*- and *S*- are used to distinguish between the enantiomers. We will not consider the *R-S* system since it is not universally used by biological scientists. The latter more often use an older nomenclature system which uses the prefixes D- and L-.

Problem 17.18. Describe how the D-L nomenclature system assigns the prefixes D- and L-.

Ans. The D-L nomenclature system applies only for the case of a stereocenter attached to two different carbon substituents R and R′, a hydrogen, and a hetero atom substituent X (e.g., OH or NH_2). The molecule is drawn so that the R and R′ substituents are vertical while the H and hetero atom substituents are horizontal. R, the carbon substituent which has the most highly oxidized carbon atom (the carbon with the least number of hydrogens) attached to the stereocenter, is placed up. The enantiomer with the X substituent on the right-hand side is the D-enantiomer while that with the X substituent on the left-hand side is the L-enantiomer:

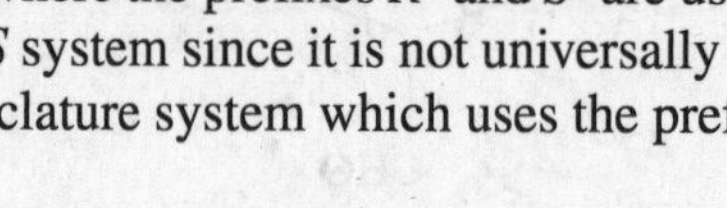

R; H—X; R′ — D   R; X—H; R′ — L

Problem 17.19. The D-L nomenclature system was originally used for the two glyceraldehydes. Show that the two structures below are correctly named as D and L:

HC=O; H—OH; CH_2OH HC=O; HO—H; CH_2OH

D-Glyceraldehyde   L-Glyceraldehyde

Ans. HC = O and CH_2OH, the two carbon substituents, are placed vertically. The more highly oxidized substituent HC = O is placed up and the least highly oxidized substituent CH_2OH is placed down. The enantiomers with the OH on the right- and left-hand sides, respectively, are D-glyceraldehyde and L-glyceraldehyde.

17.3.2 Physical Properties

Chiral molecules differ from achiral molecules in their interaction with plane-polarized light. Chiral molecules such as D-glyceraldehyde and L-glyceraldehyde possess *optical activity*, i.e., they rotate the plane of plane-polarized light. Achiral molecules do not rotate the plane of plane-polarized light.

Although each enantiomer of a pair of enantiomers rotates the plane of plane-polarized light by the same number of degrees, they differ in the direction of that rotation. One enantiomer rotates clockwise and the other counterclockwise. This is noted in the name of the compound by placing (+)- and (−)-, respectively, before the name of each enantiomer. The designations *dextrorotatory* and *levorotatory* are also used.

Optical activity is measured by a *polarimeter* and expressed as the *specific rotation*, $[\alpha]$, defined by the expression

$$[\alpha] = \frac{\alpha}{cl}$$

where α is the observed rotation in degrees for a sample concentration of c grams per milliliter and sample length of l decimeters (1 dm = 10 cm).

Enantiomers do not differ in any other physical property except the direction of rotation of plane-polarized light.

Compounds that rotate the plane of plane-polarized light are said to be *optically active* and often called *optical isomers*.

Problem 17.20. Calculate $[\alpha]$ for D-glyceraldehyde. The observed rotation is 5.4° in the clockwise direction when a solution containing 0.80 g/mL is placed in a sample holder of length 0.50 dm in a polarimeter.

Ans.

$$[\alpha] = \frac{\alpha}{cl} = \frac{+5.4^\circ}{0.80 \times 0.50} = +13.5^\circ$$

Problem 17.21. What is the more complete name of D-glyceraldehyde given that $[\alpha] = +13.5^\circ$.

Ans. D-(+)-Glyceraldehyde

Problem 17.22. Which is the correct designation for (+)-tartaric acid, D- or L-?

Ans. There is no relationship between the direction of optical rotation and the D- or L-configuration. D- or L-refers to the actual configuration at the stereocenter and can be determined only by x-ray crystallography. Only after a determination of the configuration by X-ray crystallography can we use the convention described in Problem 17.18 to designate one enantiomer as D- and the other as L-. Optical rotation is determined by using a polarimeter and its direction is unrelated to the D- or L-nomenclature.

Problem 17.23. What is the specific rotation of L-glyceraldehyde given that $[\alpha] = +13.5^\circ$ for D-glyceraldehyde?

Ans. -13.5°

Problem 17.24. What is the observed rotation for a solution containing equal amounts of D-glyceraldehyde and L-glyceraldehyde?

Ans. There will be no observed rotation. Rotation of plane-polarized light is taking place but the two enantiomers rotate in opposite directions. There is complete cancellation of optical rotation since the enantiomers are present in equal amounts. Such solutions are referred to as *racemic mixtures* and are said to be *optically inactive*.

17.3.3 Chemical Properties

Enantiomers show no difference in their reactions with achiral molecules, e.g., NaOH, HCl, and Br_2. The situation is very different for reactions with chiral molecules or for reactions in which chiral enzymes are involved. The typical biological process involves some chiral molecule undergoing reaction under the influence of a chiral enzyme. One enantiomer reacts rapidly while the other reacts at a much slower rate or does not react at all. The body can use one enantiomer efficiently but not the other. For example, D-glucose is utilized by the body but L-glucose cannot be utilized. L-Dopa is used to treat Parkinson's disease; the D-enantiomer has no effect. Taste and odor also involve highly selective biochemical reactions. For example, (−)-carvone has the odor of caraway seeds while (+)-carvone has the odor of spearmint.

Problem 17.25. What is the mechanism of the high selectivity of biochemical reactions involving chiral enzymes with chiral molecules?

Ans. A portion of the enzyme, referred to as the *active site*, is responsible for binding to and holding the chiral molecule while it undergoes reaction with some reagent(s). The three-dimensional shape of the active site is such that it only matches the configuration of one of the enantiomers, e.g., D-glucose, but not the other enantiomer, L-glucose. One of the enantiomers undergoes reaction but not the other. This mechanism is often referred to as the *lock-and-key* or *chiral recognition* mechanism. Only one enantiomer (key) fits a particular active site (lock).

17.4 COMPOUNDS WITH MORE THAN ONE STEREOCENTER

A compound with n stereocenters has a maximum of 2^n stereoisomers. For two stereocenters, there is a maximum of 2^2 or 4 stereoisomers. Whether or not the maximum number of stereoisomers occurs depends on whether or not each stereocenter has the same or a different set of four different substituents.

Problem 17.26. Draw the different stereoisomers of 2,3-dichloropentane.

Ans.

I	II	III	IV
CH_3	CH_3	CH_3	CH_3
H—C—Cl	Cl—C—H	H—C—Cl	Cl—C—H
Cl—C—H	H—C—Cl	H—C—Cl	Cl—C—H
C_2H_5	C_2H_5	C_2H_5	C_2H_5

There are two stereocenters (C-2 and C-3) and a total of four stereoisomers. The four stereoisomers are organized into two pairs of enantiomers—structures I and II are one pair and III and IV are another pair. A new term, *diastereomer*, is used to indicate the relationship of either compound in one pair with either compound in the other pair. Neither structure I nor structure II is superimposable or the mirror image of structure III or structure IV.

Problem 17.27. Draw the different stereoisomers of 2,3-dichlorobutane.

Ans.

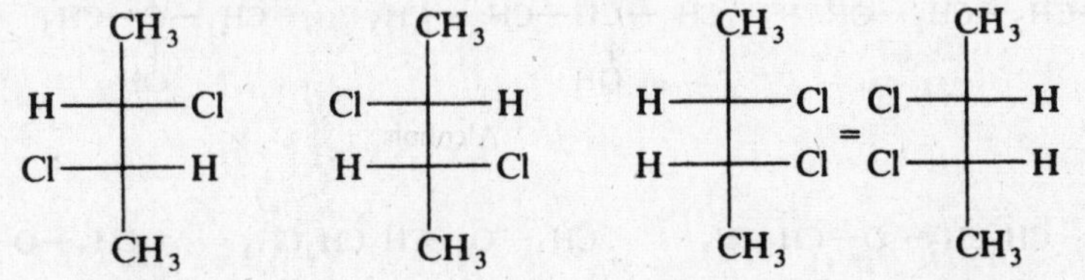

$$\begin{array}{ccccccc} \begin{array}{c} CH_3 \\ H-\!\!\!+\!\!\!-Cl \\ Cl-\!\!\!+\!\!\!-H \\ CH_3 \end{array} & \quad & \begin{array}{c} CH_3 \\ Cl-\!\!\!+\!\!\!-H \\ H-\!\!\!+\!\!\!-Cl \\ CH_3 \end{array} & \quad & \begin{array}{c} CH_3 \\ H-\!\!\!+\!\!\!-Cl \\ H-\!\!\!+\!\!\!-Cl \\ CH_3 \end{array} & = & \begin{array}{c} CH_3 \\ Cl-\!\!\!+\!\!\!-H \\ Cl-\!\!\!+\!\!\!-H \\ CH_3 \end{array} \\ \text{I} & & \text{II} & & \text{III} & & \text{IV} \end{array}$$

There are two stereocenters for 2,3-dichlorobutane but there are only three stereoisomers. There is a pair of enantiomers (I and II). Unlike the situation with 2,3-dichloropentane, structures III and IV are not a pair of enantiomers. Structures III and IV are superimposable by rotating either one 180° in the plane of the paper. This occurs because the two stereocenters (C-2 and C-3) have the same set of four different substituents (H, Cl, CH_3, $CHCl-CH_3$). A correct answer for this problem must show either the equal sign between structures III and IV or only one structure from III and IV. The lone compound, whether shown as III or IV, is referred to as a *meso* compound. The meso compound has a plane of symmetry (symbolized by a dotted line in the structure shown below). The meso compound is achiral and optically inactive even though it possesses two stereocenters.

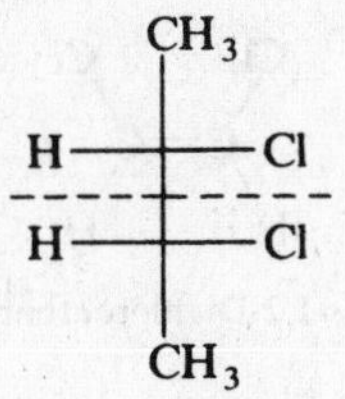

$$\begin{array}{c} CH_3 \\ H-\!\!\!+\!\!\!-Cl \\ \text{- - - - - - - - -} \\ H-\!\!\!+\!\!\!-Cl \\ CH_3 \\ \text{III} \end{array}$$

ADDITIONAL SOLVED PROBLEMS

REVIEW OF ISOMERISM

Problem 17.28. How many constitutional isomers exist for the molecular formula C_6H_{14}? Show one structural formula for each isomer. Do not show more than one structural formula for each isomer.

Ans. There are five different compounds (constitutional isomers) of C_6H_{14}:

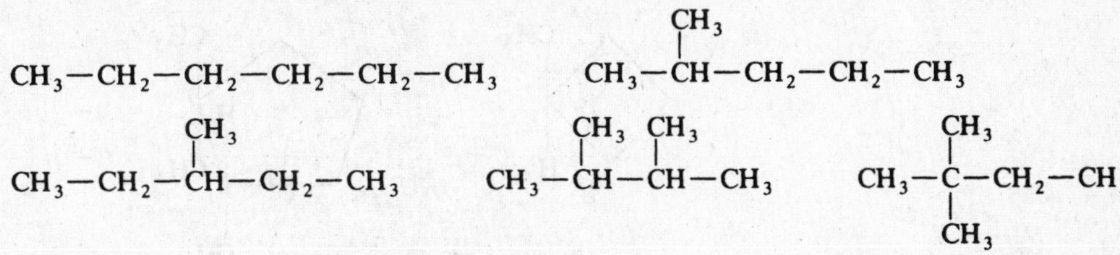

$$CH_3-CH_2-CH_2-CH_2-CH_2-CH_3 \qquad CH_3-\underset{}{\overset{CH_3}{\overset{|}{C}H}}-CH_2-CH_2-CH_3$$

$$CH_3-CH_2-\overset{CH_3}{\overset{|}{C}H}-CH_2-CH_3 \qquad CH_3-\overset{CH_3}{\overset{|}{C}H}-\overset{CH_3}{\overset{|}{C}H}-CH_3 \qquad CH_3-\underset{CH_3}{\underset{|}{\overset{CH_3}{\overset{|}{C}}}}-CH_2-CH_3$$

Problem 17.29. How many constitutional isomers exist for the molecular formula $C_4H_{10}O$? Show only one structural formula for each isomer.

Ans. There are no rings and/or extra bonds for $C_4H_{10}O$ since the formula follows the C:H ratio for an alkane (C_nH_{2n+2}). Both alcohols and ethers fulfill the molecular formula $C_4H_{10}O$.

There are four alcohols and three ethers:

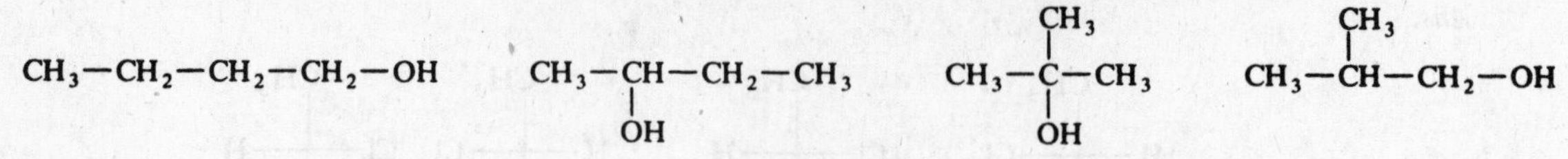

$CH_3-CH_2-CH_2-CH_2-OH$ $CH_3-CH(OH)-CH_2-CH_3$ $CH_3-C(CH_3)(OH)-CH_3$ $CH_3-CH(CH_3)-CH_2-OH$

Alcohols

$CH_3CH_2-O-CH_2CH_3$ $CH_3-O-CH_2CH_2CH_3$ $CH_3-O-CH(CH_3)_2$

Ethers

Problem 17.30. Which of the following exist as a pair of *cis* and *trans* isomers? Show structures of the *cis* and *trans* isomers. (*a*) 1,1-Dichloro-1-butene, (*b*) 1,2-dichloroethene, (*c*) 3-Hexene.

Ans. (*a*) 1,1-Dichloro-1-butene does not exist as *cis* and *trans* isomers since one of the double-bonded carbons has two of a kind (two chlorines):

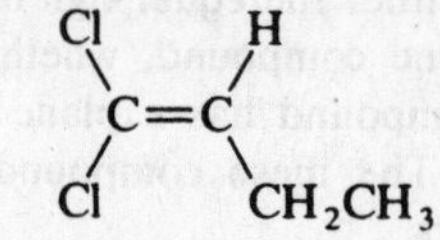

$Cl_2C=CH(CH_2CH_3)$

(*b*)

cis: Cl, Cl on top; H, H on bottom — $C=C$ trans: Cl, H on top; H, Cl on bottom — $C=C$

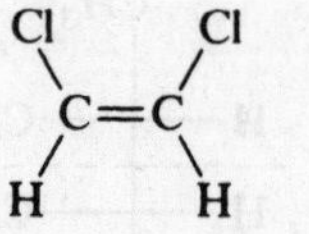

cis-1,2-Dichloroethane 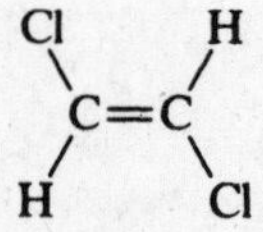   *trans*-1,2-Dichloroethene

(*c*)

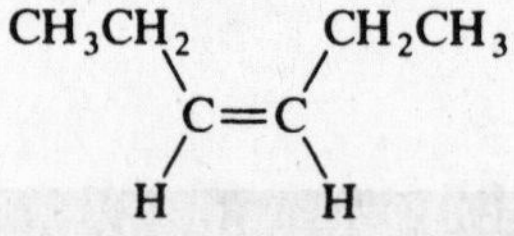

cis: CH_3CH_2, CH_2CH_3 on top; H, H on bottom — $C=C$ trans: CH_3CH_2, H on top; H, CH_2CH_3 on bottom — $C=C$

cis-3-Hexene 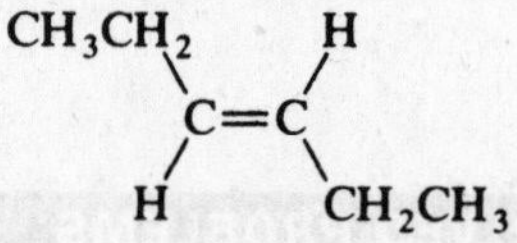   *trans*-3-Hexene

Problem 17.31. Which of the following exist as *cis* and *trans* isomers? (*a*) 1-methylcyclobutane, (*b*) 1,1,2-trimethylcyclopentane, (*c*) 1,2-dimethylcyclopentane.

Ans. 1-Methylcyclobutane and 1,1,2-trimethylcyclopentane do not exist as *cis* and trans *isomers* since each has only one ring carbon with two different substituents:

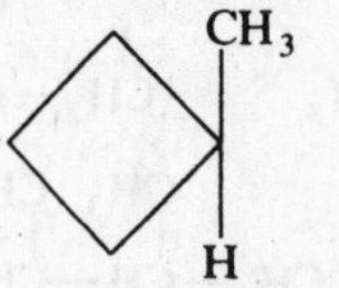

(cyclobutane ring bearing CH_3 and H)   (cyclopentane ring bearing H, CH_3 on one carbon and CH_3, CH_3 on the adjacent carbon)

1-Methylcyclobutane 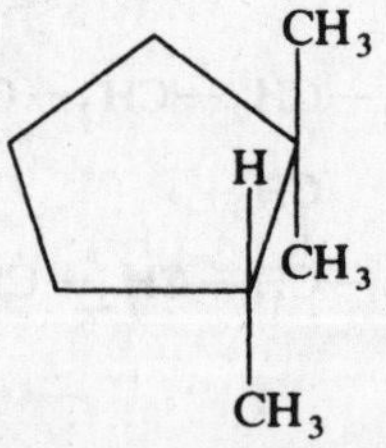   1,1,2-Trimethylcyclopentane

Two stereocenters are required for the existence of *cis-trans* isomerism. Three of the four ring carbons of 1-methylcyclobutane have two of the same substituents (two hydrogens). Four of the five ring carbons of 1,1,2-trimethylcyclopentane have two of the same substituents—C-1 has two methyl groups while C-3, C-4, and C-5 each have two hydrogens.

1,2-Dimethylcyclopentane exists as two different compounds since there are two stereocenters:

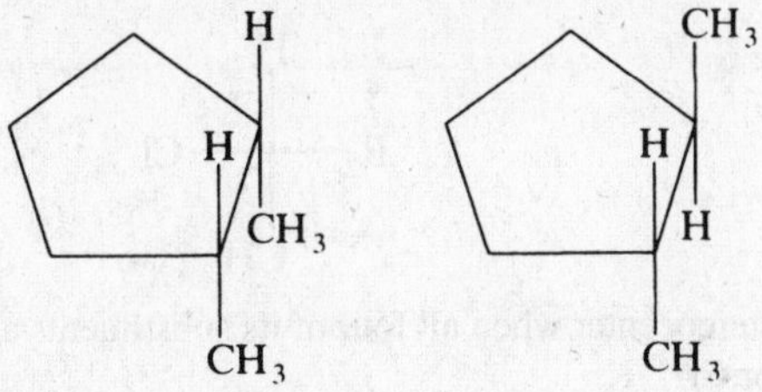

Cis — Trans

Ring carbons C-1 and C-2 each have two different groups (one hydrogen and one methyl).

Note: There are other problems in previous chapters for reviewing constitutional and *cis-trans* isomers:

Constitutional isomers: Problems 11.11, 11.24, 11.25, 11.35 through 11.37, 12.5, 12.6, 12.29, 12.30, 12.38, 13.34, 13.47, 13.48, 14.4, 14.5, 14.27, 15.4, 15.35, 16.5, 16.39

Cis-trans isomers: Problems 12.32 through 12.34

ENANTIOMERS

Problem 17.32. Which of the following exist as enantiomers? Place an asterisk near each stereocenter.

(*a*) $CH_3CH{=}CHCH_2Cl$ (*b*) $CH_3CH(Cl){-}CH{=}CH_2$ (*c*) $HO{-}CH_2CH(NH_2){-}C_6H_5$

(*d*) $HO{-}C(CH_2COOH)_2{-}COOH$ (*e*) 1-chloro-2-methylbenzene (ring with Cl and CH_3) (*f*) 1-chloro-2-methylcyclohexane (ring with Cl and CH_3)

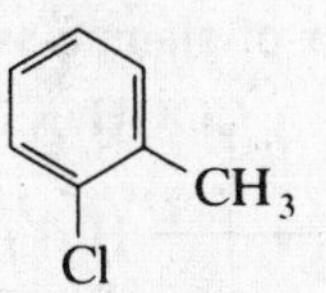

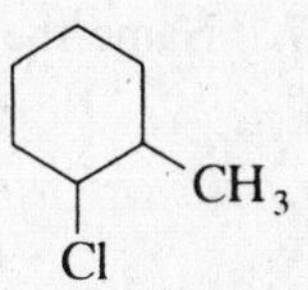

Ans. Molecules (*b*), (*c*), and (*f*) exist as enantiomers:

(*b*) $CH_3\overset{*}{C}H(Cl){-}CH{=}CH_2$ (*c*) $HO{-}CH_2\overset{*}{C}H(NH_2){-}C_6H_5$ (*f*) cyclohexane ring with * at the C bearing CH_3 and * at the C bearing Cl

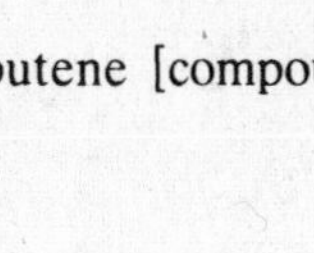

Problem 17.33. Which of the molecules in Problem 17.32 are chiral molecules?

Ans. This is basically the same question as in Problem 17.32. Molecules that contain stereocenters are chiral molecules, which is synonymous with saying they exist as enantiomers. Molecules (*b*), (*c*), and (*f*) are chiral.

Problem 17.34. Draw structures for the two enantiomers of 3-chloro-1-butene [compound (*b*) in Problem 17.32].

Ans.

$CH_3{-}C(H)(Cl){-}CH{=}CH_2$ $CH_2{=}CH{-}C(H)(Cl){-}CH_3$

Problem 17.35. What must R be in the structure below for the compound to exist as enantiomers?

H
R—+—Cl
CH_3

Ans. A carbon is a stereocenter when all four of its substituents are different. R can be any substituent other than H, CH_3, or Cl.

Problem 17.36. What is the relationship of structures I through V?

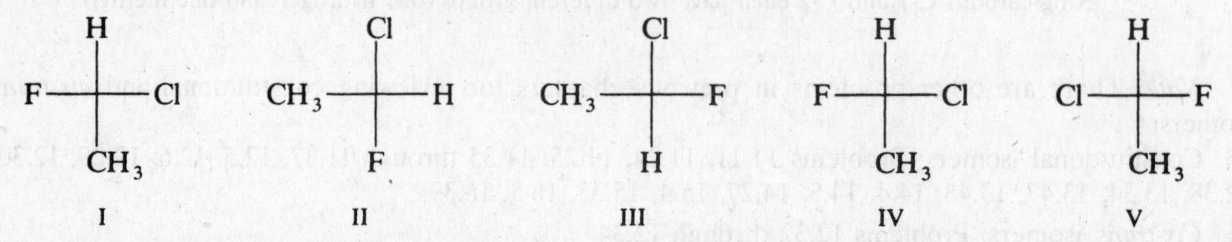

Ans. The relationship of structures I, IV, and V is relatively easy to see. Structure I = structure IV since the two are superimposable by moving either structure I to the right or structure IV to the left. Structures I and V are enantiomers since the two are nonsuperimposable mirror images. The relationships of structures II and III to the other structures is less obvious. Structure I = structure II since they are interconvertible by two exchanges (H and Cl, F and CH_3), an even number of exchanges. Structure III = structure V since they are interconvertible by two exchanges (H and Cl, H and CH_3).

NOMENCLATURE AND PROPERTIES OF ENANTIOMERS

Problem 17.37. Name the following enantiomer of alanine as D- or L-:

COOH
NH_2—+—H
CH_3

Ans. The two carbon substituents are vertical, with the more oxidized carbon up. This is the L-enantiomer since the hetero substituent (NH_2) is on the left-hand side.

Problem 17.38. A solution of morphine at a concentration of 0.20 g/mL is placed in a sample tube of length 0.50 dm. The sample tube is placed in a polarimeter and the optical rotation is found to be −13.2°. What will be the optical rotation if the solution concentration is decreased to 0.050 g/mL and a sample tube of 1.00 dm is used?

Ans.

$$[\alpha] = \frac{\alpha}{cl} = \frac{-13.2^\circ}{0.20 \times 0.50} = -132^\circ$$

$$\alpha = [\alpha]cl = -132^\circ \times 0.050 \times 1.00 = -6.6^\circ$$

Problem 17.39. The properties of (+)-tartaric acid are: $[\alpha] = +12.0^\circ$, melting point = 169°C, solubility in water = 1.4 g/mL. What are the corresponding properties of (−)-tartaric acid?

Ans. $[\alpha] = -12.0^\circ$ instead of $+12.0^\circ$; melting point and solubility are exactly the same as for (+)-tartaric acid.

COMPOUNDS WITH MORE THAN ONE STEREOCENTER

Problem 17.40. Draw the different stereoisomers of 1-chloro-2-methylcyclopentane.

Ans. There are four stereoisomers since the two stereocenters (C-1 and C-2) do not have the same four different substituents. There is a *cis* pair of enantiomers and a *trans* pair of enantiomers. Each of the *cis* enantiomers is a diastereomer of each of the *trans* enantiomers and vice versa.

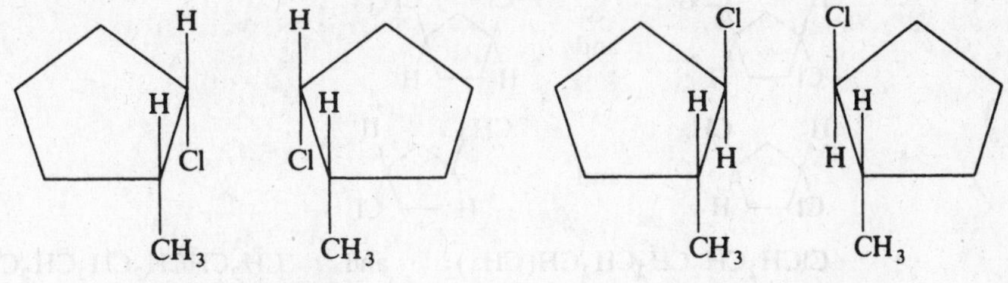

Cis pair of enantiomers *Trans* pair of enantiomers

Problem 17.41. Draw the different stereoisomers of 1,2-dichlorocyclopentane.

Ans. There are three stereoisomers since the two stereocenters have the same four different substituents. There is a *trans* pair of enantiomers and a *cis* meso compound. Each of the *trans* enantiomers is a diastereomer of the meso and vice versa.

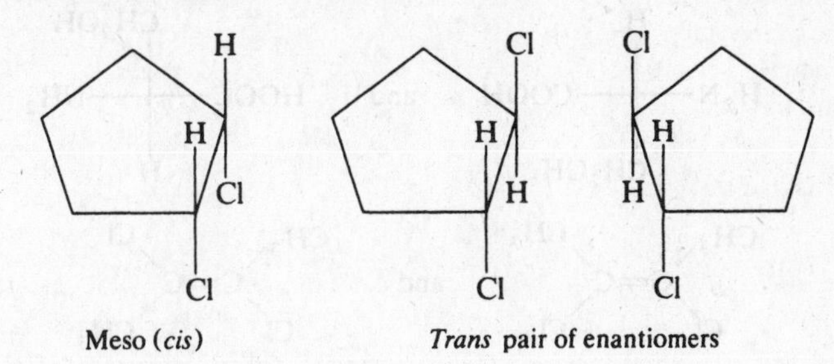

Meso (*cis*) *Trans* pair of enantiomers

Problem 17.42. Which of the stereoisomers in Problems 17.40 and 17.41 are optically active?

Ans. All except the meso compound in Problem 17.37

Problem 17.43. How many stereoisomers are possible for compounds (*b*), (*c*), and (*f*) in Problem 17.32?

Ans. The maximum number of stereoisomers is 2^n where n is the number of stereocenters. There are two stereoisomers each for (*b*) and (*c*) and four for (*f*). The maximum number is found for (*f*) since neither stereocenter has the same set of four different substituents.

SUPPLEMENTARY PROBLEMS

Problem 17.44. The relationship between the two structures in each pair of structural drawings shown below fits one of the following categories: (1) the same compound, or (2) different compounds which are constitutional isomers, or (3) different compounds which are diastereomers, or (4) different compounds which are enantiomers, or (5) different compounds which are not isomers.

For each of the following paris, indicate the relationship between the structures. If the compounds are diasteromers, indicate whether they are *cis-trans* isomers.

(*a*) [3-methyltetrahydrofuran ring, substituent CH_3] and $CH_3CH{=}CH{-}CH_2CH_2OH$

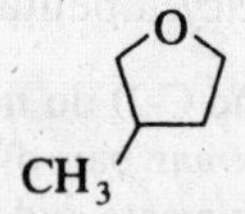

(*b*) [cyclopentane with H, CH_3 on one carbon and Cl, H on adjacent carbon] and [cyclopentane with Cl, CH_3 up and H, H down]

(*c*) [cyclopentane with H, CH_3 up and Cl, H down] and [cyclopentane with CH_3, H up and H, Cl down]

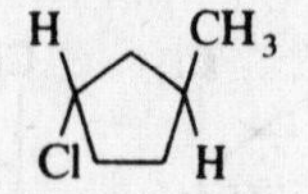

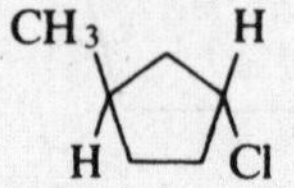

(*d*) $ClCH_2CH_2CH_2CH_2CH(CH_3)_2$ and $CH_3CH(CH_3)CH_2CH_2CH_2Cl$

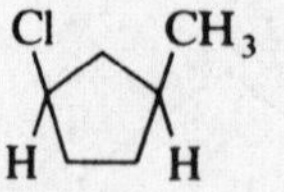

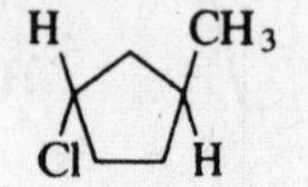

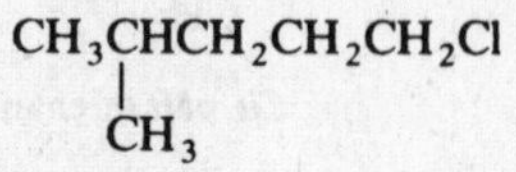

(*e*) Fischer projection: top H, left H_2N, right COOH, bottom CH_2OH and Fischer projection: top CH_2OH, left H_2N, right COOH, bottom H

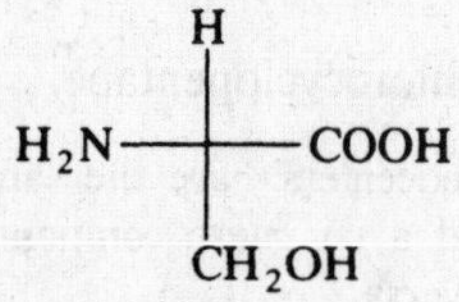

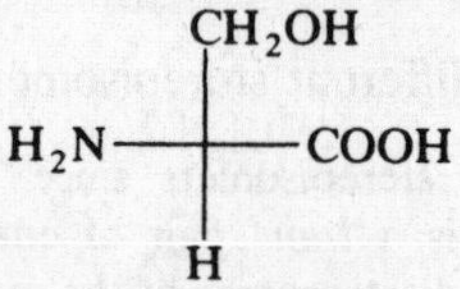

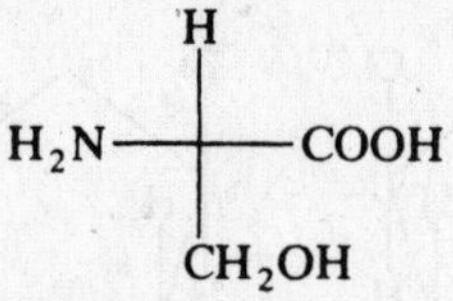

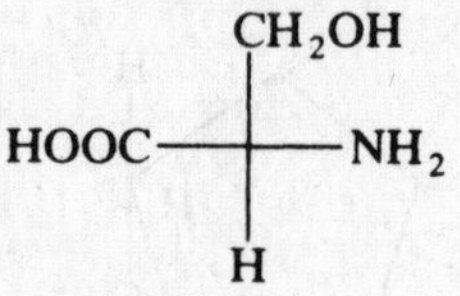

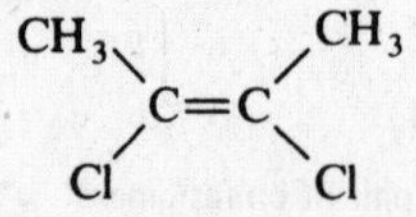

(*f*) Fischer projection: top H, left H_2N, right COOH, bottom CH_2OH and Fischer projection: top CH_2OH, left HOOC, right NH_2, bottom H

(*g*) $(CH_3)(Cl)C{=}C(CH_3)(Cl)$ [CH_3 groups on same side, Cl groups on same side] and $(CH_3)(Cl)C{=}C(Cl)(CH_3)$ [CH_3 groups on opposite sides]

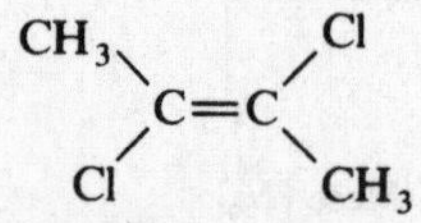

Ans. (*a*) Constitutional isomers; (*b*) *cis-trans* isomers which are also categorized as optical isomers (diastereomers); (*c*) enantiomers; (*d*) different compounds which are not isomers; (*e*) enantiomers; (*f*) same compound; (*g*) *cis-trans* isomers (diastereomers).

Problem 17.45. Which of the following exist as stereoisomers? Place an asterisk near each stereocenter.

(*a*)
$HC{=}O$
$HC{-}OH$
$HC{-}OH$
$HC{-}OH$
$H_2C{-}OH$

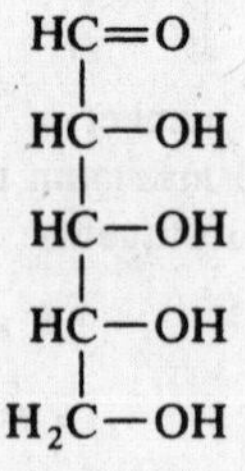

(*b*) $CH_3CH(CH_3){-}CH_2CH_3$

(*c*)
$H_2C{-}OH$
$HC{=}O$
$H_2C{-}OH$

(*d*) [steroid skeleton with $CO{-}CH_3$ side chain, two angular H_3C groups, and ketone O on ring A]

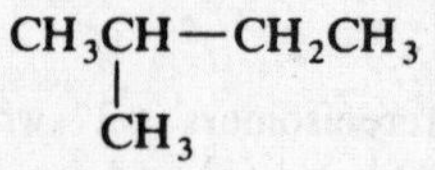

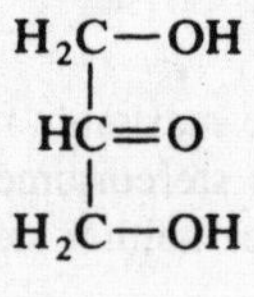

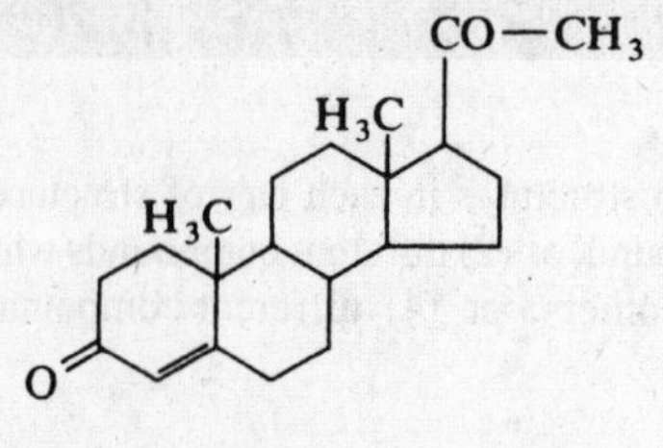

Ans. Only (*a*) and (*d*) have stereocenters and exist as stereoisomers:

(*a*)

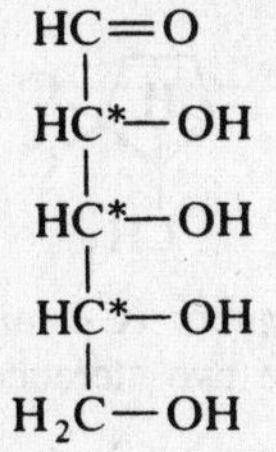

HC=O
|
HC*—OH
|
HC*—OH
|
HC*—OH
|
H_2C—OH

(*d*)

CH—CH_3
H_3C
H_3C
O

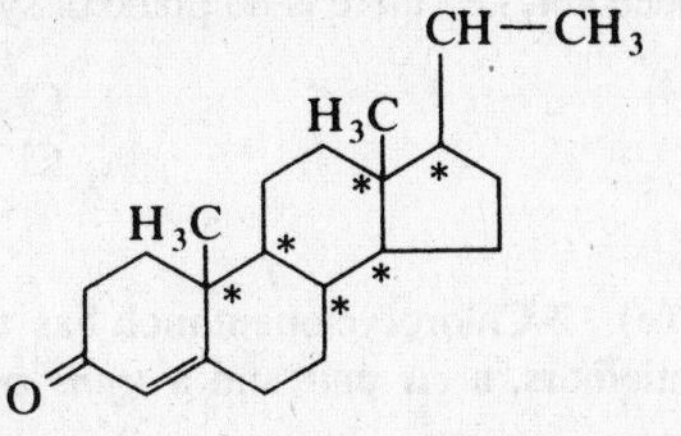

Problem 17.46. How many stereoisomers are there for compounds (*a*) and (*d*) in Problem 17.45?

Ans. The maximum of 2^n stereoisomers occurs since none of the stereocenters has the same set of four different substituents in either compound. There are 8 stereoisomers for (*a*) and 64 for (*d*).

Problem 17.47. Draw the stereoisomers, if any, of each of the following: (*a*) 1,1-dimethyl-1-propanol, (*b*) 2-methyl-1-propanol, (*c*) 1-propanol, (*d*) 1-methyl-1-propanol, (*e*) 3-chloro-2-butanol, (*f*) 2,3-butanediol.

Ans. There are no stereoisomers for (*a*), (*b*), or (*c*) since no stereocenter is present.

(*d*) 1-Methyl-1-propanol has a stereocenter (C-1) and exists as a pair of enantiomers:

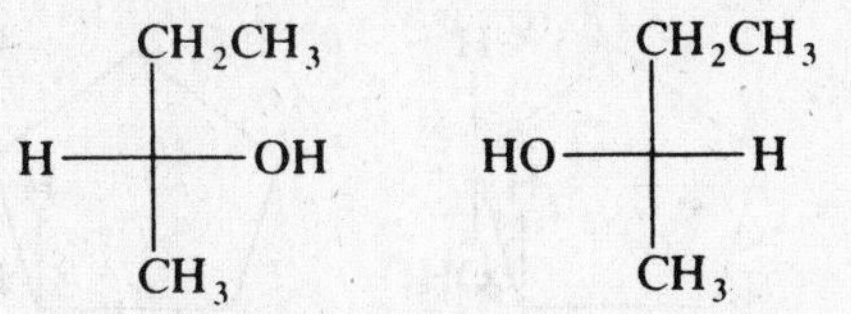

CH_2CH_3	CH_2CH_3
H—+—OH	HO—+—H
CH_3	CH_3

(*e*) 3-Chloro-2-butanol has two stereocenters (C-2 and C-3). There are two pairs of enantiomers since the two stereocenters do not have the same four different substituents:

CH_3	CH_3	CH_3	CH_3
H—+—OH	HO—+—H	H—+—OH	HO—+—H
Cl—+—H	H—+—Cl	H—+—Cl	Cl—+—H
CH_3	CH_3	CH_3	CH_3

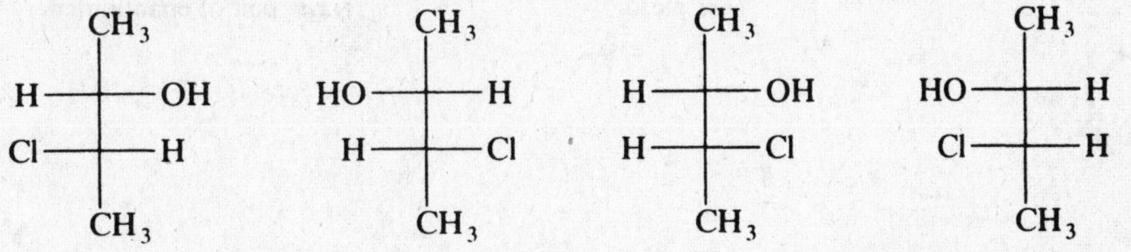

(*f*) 2,3-Butanediol has two stereocenters (C-2 and C-3). There are only three stereoisomers, one pair of enantiomers and one meso compound, since the two stereocenters have the same four different substituents:

CH_3	CH_3	CH_3
H—+—OH	HO—+—H	H—+—OH
HO—+—H	H—+—OH	H—+—OH
CH_3	CH_3	CH_3
Pair of enantiomers		Meso

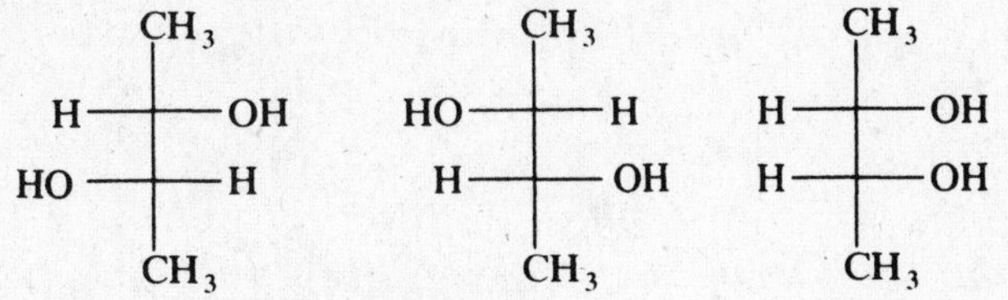

Problem 17.48. Draw the enantiomers, if any, of each of the following: (*a*) 1-chlorocyclopentane, (*b*) 1,1-dichlorocyclopentane, (*c*) 1,1,3,3-tetrachloro-2-methylcyclopentane, (*d*) 1,1-dichloro-2-methylcyclopentane, (*e*) 2-chlorocyclopentanol, (*f*) 1,2-cyclopentanediol.

Ans. Compounds (*a*), (*b*), and (*c*) do not exist as stereoisomers since no stereocenter is present. (Each of the rings has a plane of symmetry.)

(*d*) 1,1-Dichloro-2-methylcyclopentane exists as a pair of enantiomers since there is a stereocenter, i.e., there is no plane of symmetry:

(*e*) 2-Chlorocyclopentanol has two stereocenters (C-1 and C-2). There are two pairs of enantiomers, a *cis* pair and a *trans* pair, since the two stereocenters have different substituents:

Cis pair of enantiomers *Trans* pair of enantiomers

(*f*) 1,2-Cyclopentanediol has two stereocenters (C-1 and C-2). There are three stereoisomers, a ***trans*** pair of enantiomers and a *cis* meso compound, since the *cis* compound has a plane of symmetry:

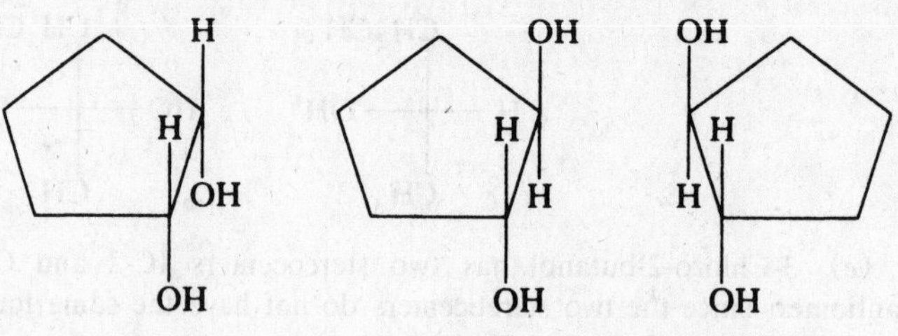

Cis meso *Trans* pair of enantiomers

Carbohydrates

Biochemistry is the study of (a) the structure of *biomolecules*, the molecules of living organisms, both organic (*carbohydrates*, *lipids*, *proteins*, and *nucleic acids*) and inorganic (*water and various inorganic salts*) and (b) their *physiological functions* (the processes by which an organism maintains its living state), that include *heredity*, the molecular basis by which genetic information is stored and transmitted (Chapters 18-21), and *metabolism*, the physical and chemical processes which allow organisms to use materials and energy from their environment, (Chapters 22-23).

Carbohydrates, also referred to as *saccharides*, are polyhydroxyaldehydes or polyhydroxyketones or compounds that can be hydrolyzed to or derived from polyhydroxyaldehydes or polyhydroxyketones. Carbohydrates have many important biological functions—to store and generate energy in animals and plants (glucose, starch, glycogen), as supportive structures in plants (cellulose) and crustaceans (chitin), and as components of cell membranes and nucleic acids.

The simplest carbohydrates, referred to as *monosaccharides*, have the following structures:

```
  H—C=O                   CH2OH
    |                       |
 [H—C—OH]a                 C=O
    |                       |
  CH2OH                 [H—C—OH]b
                            |
                          CH2OH

Polyhydroxyaldehyde     Polyhydroxyketone
```

The numbering of the carbon chain in monosaccharides involves giving the smallest number for the carbonyl carbon, C-1 for the polyhydroxyaldehydes and C-2 for the polyhydroxyketones.

The monosaccharides of biological importance are those of three to six carbons, i.e., $a = 1$ to 4 and $b = 0$ to 3. Monosaccharides are the basic building blocks for forming larger-sized saccharides.

Disaccharides contain two monosaccharide units per molecule. *Polysaccharides* contain large numbers of monosaccharide units, usually in the thousands. The term *oligosaccharide* is used to describe saccharides intermediate between disaccharides and polysaccharides, although more specific terms such as *trisaccharide* and *tetrasaccharide* are also applicable. Higher saccharides yield monosaccharides upon hydrolysis. Monosaccharides do not undergo hydrolysis.

Monosaccharides and disaccharides are often referred to as *sugars*.

IUPAC names are not used for carbohydrates because common names are too strongly entrenched in the biological literature. Most common names have the ending *-ose*, e.g., the monosaccharides D-glucose and D-fructose, the disaccharides lactose and sucrose, and the polysaccharides amylose and cellulose.

18.1 MONOSACCHARIDES

Monosaccharides are classified as *aldoses* or *ketoses*, respectively, depending on whether the carbonyl group is an aldehyde or a ketone. Classification into *trioses*, *tetroses*, *pentoses*, and *hexoses*, respectively, is used to indicate there are three, four, five, and six carbon atoms in the monosaccharide. Both types of prefixes are also used together, e.g., aldohexose for a six-carbon monosaccharide with an aldehyde group.

Problem 18.1. Assign a name to each of the following monosaccharides that indicates both the type of carbonyl group and the number of carbon atoms:

I	II	III	IV
H—C=O	CH_2OH	CH_2OH	H—C=O
H—C—OH	C=O	C=O	H—C—OH
CH_2OH	H—C—OH	H—C—OH	H—C—OH
	CH_2OH	H—C—OH	H—C—OH
		H—C—OH	CH_2OH
		CH_2OH	

Ans. Compound I = aldotriose, compound II = ketotetrose, compound III = ketohexose, compound IV = aldopentose

Figures 18-1 and 18-2 show the aldoses and ketoses of biological interest. D-Glucose, D-fructose, and D-galactose are the most abundant hexoses. D-Glucose, also known as *dextrose* or *blood sugar*, is the most abundant of all organic compounds. It is the building block for starch and cellulose.

D-Glyceraldehyde
CHO
H—OH
CH_2OH

D-Erythrose	D-Threose
CHO	CHO
H—OH	HO—H
H—OH	H—OH
CH_2OH	CH_2OH

D-Ribose	D-Arabinose	D-Xylose	D-Lyxose
CHO	CHO	CHO	CHO
H—OH	HO—H	H—OH	HO—H
H—OH	H—OH	HO—H	HO—H
H—OH	H—OH	H—OH	H—OH
CH_2OH	CH_2OH	CH_2OH	CH_2OH

D-Allose	D-Altrose	D-Glucose	D-Mannose	D-Gulose	D-Iodose	D-Galactose	D-Talose
CHO	CHO	CHO	CHO	CHO	CHO	CHO	CHO
H—OH	HO—H	H—OH	HO—H	H—OH	HO—H	H—OH	HO—H
H—OH	H—OH	HO—H	HO—H	H—OH	H—OH	HO—H	HO—H
H—OH	H—OH	H—OH	H—OH	HO—H	HO—H	HO—H	HO—H
H—OH	H—OH	H—OH	H—OH	H—OH	H—OH	H—OH	H—OH
CH_2OH	CH_2OH	CH_2OH	CH_2OH	CH_2OH	CH_2OH	CH_2OH	CH_2OH

Fig. 18-1 D-Family of aldoses.

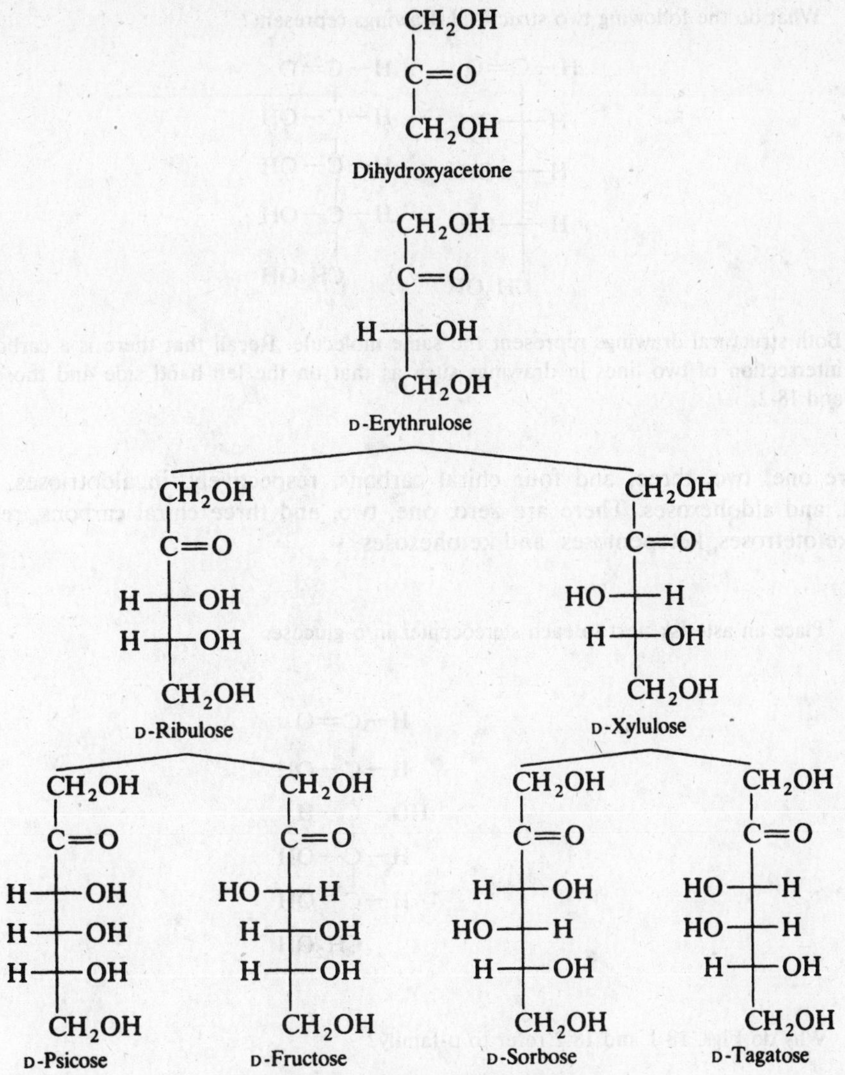

$$\begin{array}{c} CH_2OH \\ | \\ C{=}O \\ | \\ CH_2OH \end{array}$$

Dihydroxyacetone

$$\begin{array}{c} CH_2OH \\ | \\ C{=}O \\ H\text{—}\!\!+\!\!\text{—}OH \\ CH_2OH \end{array}$$

D-Erythrulose

$$\begin{array}{cc} \begin{array}{c} CH_2OH \\ C{=}O \\ H\text{—}\!\!+\!\!\text{—}OH \\ H\text{—}\!\!+\!\!\text{—}OH \\ CH_2OH \end{array} & \begin{array}{c} CH_2OH \\ C{=}O \\ HO\text{—}\!\!+\!\!\text{—}H \\ H\text{—}\!\!+\!\!\text{—}OH \\ CH_2OH \end{array} \\ \text{D-Ribulose} & \text{D-Xylulose} \end{array}$$

$$\begin{array}{cccc} \begin{array}{c} CH_2OH \\ C{=}O \\ H\text{—}\!\!+\!\!\text{—}OH \\ H\text{—}\!\!+\!\!\text{—}OH \\ H\text{—}\!\!+\!\!\text{—}OH \\ CH_2OH \end{array} & \begin{array}{c} CH_2OH \\ C{=}O \\ HO\text{—}\!\!+\!\!\text{—}H \\ H\text{—}\!\!+\!\!\text{—}OH \\ H\text{—}\!\!+\!\!\text{—}OH \\ CH_2OH \end{array} & \begin{array}{c} CH_2OH \\ C{=}O \\ H\text{—}\!\!+\!\!\text{—}OH \\ HO\text{—}\!\!+\!\!\text{—}H \\ H\text{—}\!\!+\!\!\text{—}OH \\ CH_2OH \end{array} & \begin{array}{c} CH_2OH \\ C{=}O \\ HO\text{—}\!\!+\!\!\text{—}H \\ HO\text{—}\!\!+\!\!\text{—}H \\ H\text{—}\!\!+\!\!\text{—}OH \\ CH_2OH \end{array} \\ \text{D-Psicose} & \text{D-Fructose} & \text{D-Sorbose} & \text{D-Tagatose} \end{array}$$

Fig. 18-2 D-Family of ketoses.

D-Fructose, also referred to as *levulose*, is present in equal amounts with D-glucose in fruits and honey. D-Galactose is combined with D-glucose in lactose. Trioses are present as intermediates in metabolic processes.

A *deoxycarbonhydrate* has one fewer oxygen than the corresponding carbohydrate. For example, 2-deoxy-D-ribose has two hydrogens, but no OH, at C-2 whereas D-ribose has one H and one OH at C-2. D-Ribose and 2-deoxy-D-ribose, the most abundant pentoses, are used in forming ribonucleic (RNA) and deoxyribonucleic (DNA) acids:

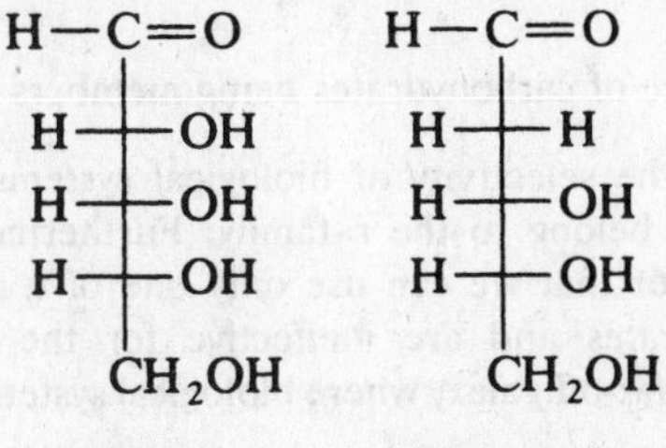

$$\begin{array}{cc} \begin{array}{c} H\text{—}C{=}O \\ H\text{—}\!\!+\!\!\text{—}OH \\ H\text{—}\!\!+\!\!\text{—}OH \\ H\text{—}\!\!+\!\!\text{—}OH \\ CH_2OH \end{array} & \begin{array}{c} H\text{—}C{=}O \\ H\text{—}\!\!+\!\!\text{—}H \\ H\text{—}\!\!+\!\!\text{—}OH \\ H\text{—}\!\!+\!\!\text{—}OH \\ CH_2OH \end{array} \\ \text{D-Ribose} & \text{2-Deoxy-D-ribose} \end{array}$$

Problem 18.2. What do the following two structural drawings represent?

$$
\begin{array}{ccc}
 & H-C=O & \\
H & + & OH \\
H & + & OH \\
H & + & OH \\
 & CH_2OH &
\end{array}
\qquad
\begin{array}{c}
H-C=O \\
| \\
H-C-OH \\
| \\
H-C-OH \\
| \\
H-C-OH \\
| \\
CH_2OH
\end{array}
$$

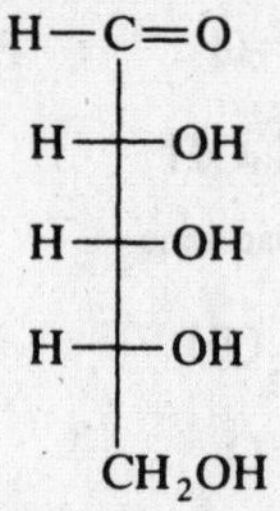

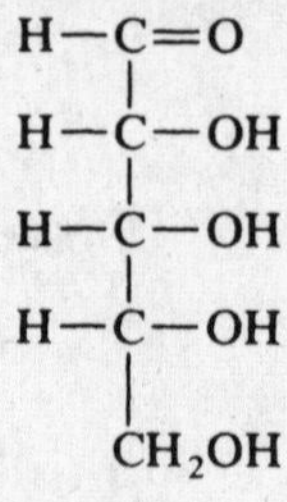

Ans. Both structural drawings represent the same molecule. Recall that there is a carbon atom at the intersection of two lines in drawings such as that on the left-hand side and those in Figs. 18-1 and 18-2.

There are one, two, three, and four chiral carbons, respectively, in aldotrioses, aldotetroses, aldopentoses, and aldohexoses. There are zero, one, two, and three chiral carbons, respectively, in ketotrioses, ketotetroses, ketopentoses, and ketohexoses.

Problem 18.3. Place an asterisk next to each stereocenter in D-glucose.

Ans.

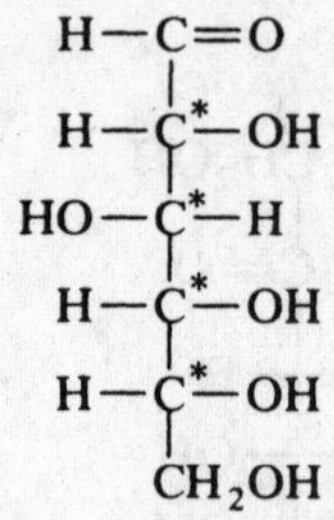

$$
\begin{array}{c}
H-C=O \\
| \\
H-C^{*}-OH \\
| \\
HO-C^{*}-H \\
| \\
H-C^{*}-OH \\
| \\
H-C^{*}-OH \\
| \\
CH_2OH
\end{array}
$$

Problem 18.4. Why do Figs. 18-1 and 18-2 refer to D-family?

Ans. All of the monosaccharides are related to D-glyceraldehyde. The configuration of the stereocenter furthest from the carbonyl group is the same as the stereocenter in D-glyceraldehyde, i.e., the OH attached to the stereocenter is on the right-hand side:

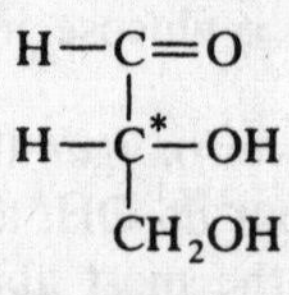

$$
\begin{array}{c}
H-C=O \\
| \\
H-C^{*}-OH \\
| \\
CH_2OH
\end{array}
$$

D-Glyceraldehyde

Problem 18.5. What is the significance of carbohydrates being members of the D-family?

Ans. This is an example of the selectivity of biological systems. With very few exceptions, naturally occurring carbohydrates belong to the D-family. Furthermore, organisms such as ourselves have evolved in such a manner that we can use only one of a pair of enantiomers. Our enzymes are specific for D-carbohydrates and are ineffective for the L-carbohydrates. Perhaps, there is a mirror-image planet in a far-off galaxy where biological systems have evolved to use L-carbohydrates?

The names of carbohydrates are often written without the prefix D-. Whether D- is included or not, it is almost always the D-enantiomer that is being discussed.

Problem 18.6. How many stereoisomers are there for aldohexoses?

Ans. There are 2^4 or 16 stereoisomers since the aldohexose structure (see Problem 18.3) has four stereocenters, none of which have the same set of four different substituents. Eight of the stereoisomers are shown as the bottom horizontal row of structures in Fig. 18-1. The other eight stereoisomers are the mirror images of the D-family stereoisomers.

Problem 18.7. Draw and name the enantiomer of D-glucose.

Ans. This is the mirror image of D-glucose:

```
      H—C=O
        |
     HO—C—H
        |
      H—C—OH
        |
     HO—C—H
        |
     HO—C—H
        |
       CH₂OH
```

Problem 18.8. What is the relationship of the four D-ketohexoses (Fig. 18-2) to each other?

Ans. D-Psicose, D-fructose, D-sorbose, and D-tagatose are stereoisomers because all four have the same connectivity. Since each is neither superimposable nor the mirror image of the others, the four D-ketohexoses are diastereomers.

Problem 18.9. What is the relationship of the four D-ketohexoses (Fig. 18-2) to the eight D-aldohexoses (Fig. 18-1)?

Ans. Constitutional isomers. Ketohexoses differ from aldohexoses in connectivity.

A pair of monosaccharides that differ in their configurations at only one stereocenter are referred to as *epimers*.

Problem 18.10. Which of the following pairs are epimers? (*a*) D-glucose and D-gulose, (*b*) D-glucose and D-mannose.

Ans. D-Glucose and D-mannose are epimers, differing in configuration only at C-2. D-Glucose and D-gulose differ in configurations at more than one stereocenter; they differ at C-3 and C-4.

Problem 18.11. Why are the monosaccharides in Figs. 18-1 and 18-2 arranged in pairs (as noted by the ∧)?

Ans. **Such pairs are *epimers*, which differ only in the configuration at C-2 for aldoses, only at C-3 for ketoses. The pair can be considered as being derived from the monosaccharide above them by insertion of another stereocenter (i.e., CHOH unit). Two different monosaccharides are obtained since the new stereocenter can be added in two different configurations. For example, D-glucose and D-mannose are derived from D-arabinose, by addition of the new unit as H—C—OH and HO—C—H, respectively.**

18.2 CYCLIC HEMIACETAL AND HEMIKETAL STRUCTURES

The actual structure of monosaccharides such as D-glucose is not what has been shown up to this point. We previously learned that aldehydes and ketones react reversibly with alcohols to form hemiacetals and hemiketals, respectively. (Refer back to Sec. 14.7 to review this reaction.) This reaction takes place intramolecularly with monosacchardes since both OH and C=O are present in the same molecule. The structures of aldoses and ketoses are cyclic hemiacetals and hemiketals, respectively.

Hemiacetal formation in D-glucose involves addition of the OH at C-5 to the C=O group to form a 6-membered ring structure, referred to as a *pyranose* ring. To follow this process we first turn the left-hand structure in Fig. 18-3 on its side. Next, the bond between C-4 and C-5 is rotated to bring the OH on C-5 in close proximity to the carbonyl group. The OH adds to the C=O as indicated by the arrows. The H of the OH adds to the O of the C=O while the O of the OH adds to the C of C=O. This addition converts C-1 to a chiral carbon with the result that two compounds are formed, α-D-glucose and β-D-glucose.

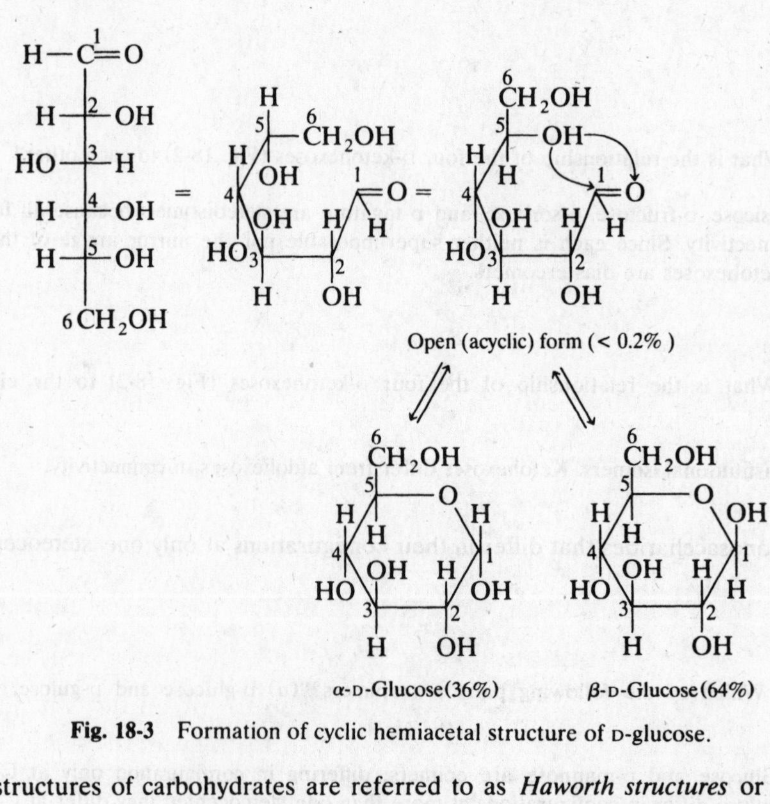

Open (acyclic) form ($< 0.2\%$)

α-D-Glucose (36%) β-D-Glucose (64%)

Fig. 18-3 Formation of cyclic hemiacetal structure of D-glucose.

The cyclic structures of carbohydrates are referred to as *Haworth structures* or *Haworth projections.*

Problem 18.12. What is the distinguishing feature of the hemiacetal structure in glucose?

Ans. There are one H, one C, and two different oxygen substituents attached to the same carbon. This carbon (C-1) (referred to as the *hemiacetal carbon*) corresponds to the carbonyl carbon of the open-chain form prior to hemiacetal formation. One of the oxygen substituents is an OH while the other is an OR.

Problem 18.13. Why are two hemiacetals, α-D-glucose and β-D-glucose, formed?

Ans. Hemiacetal formation converts C-1 to a stereocenter with two possible configurations, OH up or OH down. Neither is excluded from forming upon ring formation and the result is two hemiacetals.

Problem 18.14. What is the relationship of α-D-glucose and β-D-glucose? Are they enantiomers or diastereomers?

Ans. Although classified as diastereomers, they are a unique type of diastereomers. α-D-Glucose and β-D-glucose have the same configuration at all stereocenters except the hemiacetal carbon (C-1). Such diastereomeric pairs are referred to as *anomers*.

Ketohexoses such as D-fructose also exist as cyclic compounds. Addition of the OH at C-5 to the carbonyl group forms the 5-membered cyclic hemiketal (Fig. 18-4). The 5-membered ring is often referred to as a *furanose* ring. The α- and β-notations are the same as for D-glucose. The OH groups at C-2 in α-D-fructose and β-D-fructose are *trans* and *cis*, respectively, to the CH_2OH at C-5.

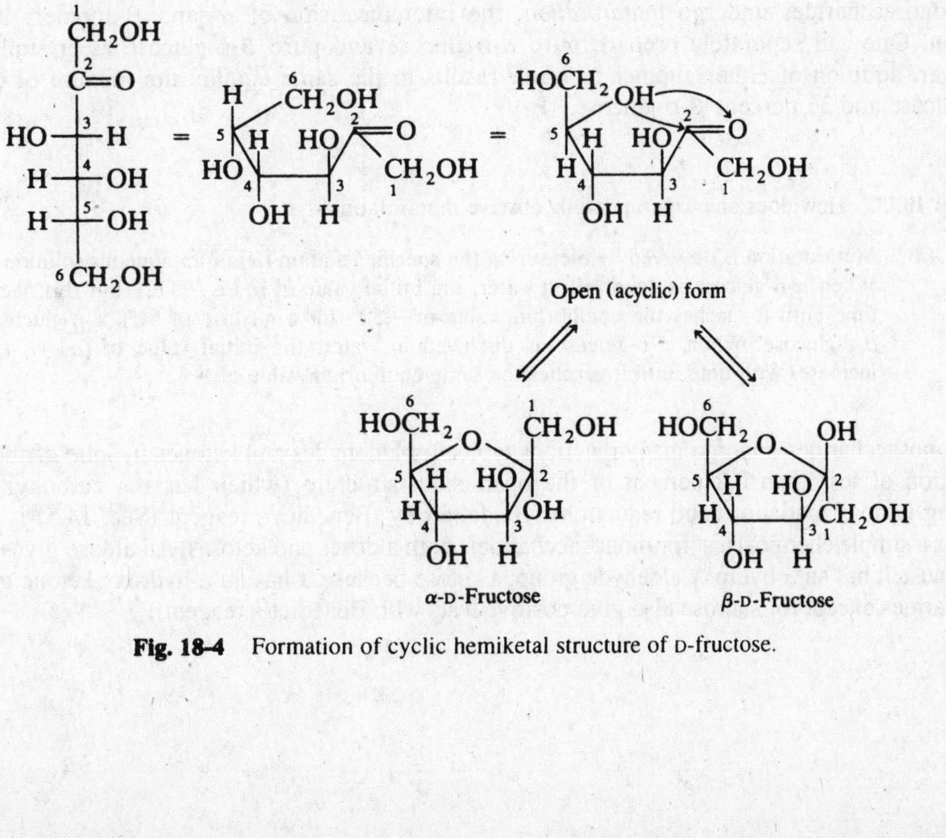

Fig. 18-4 Formation of cyclic hemiketal structure of D-fructose.

Problem 18.15. What is the distinguishing feature of the hemiketal structure in fructose?

Ans. There are two different carbon substituents and two different oxygen substituents attached to the same carbon. This carbon (C-2) (referred to as the ***hemiketal carbon***) corresponds to the carbonyl carbon of the open-chain form prior to hemiketal formation. One of the oxygen substituents is an OH while the other is an OR.

Problem 18.16. Given the open-chain structures of D-mannose and D-glucose (Fig. 18-1) and the pyranose structure of β-D-glucose (Fig. 18-3), draw the pyranose structure of β-D-mannose.

Ans. β-D-Mannose differs from β-D-glucose only in the configuration at C-2. Take the pyranose structure of β-D-glucose and reverse the positions of OH and H at C-2:

6
CH_2OH
H 5 O H
H
4 OH HO 1
HO 3 2 OH
H H

β-D-Mannose

18.3 PROPERTIES AND REACTIONS OF MONOSACCHARIDES

Monosaccharides are very soluble in water because of the hydrogen-bonding ability of the OH groups. They are only slightly soluble in alcohol and insoluble in nonpolar solvents such as ethers and hydrocarbons.

Monosaccharides undergo *mutarotation*, the interconversion of α- and β-anomers in aqueous solution. One can separately prepare pure α-D-glucose and pure β-D-glucose as crystalline solids. However, addition of either anomer to water results in the same equilibrium mixture of 64 percent α-D-glucose and 36 percent β-D-glucose.

Problem 18.17. How does one experimentally observe mutarotation?

Ans. Mutarotation is observed by measuring the specific rotation $[\alpha]$ of an aqueous solution of glucose. When α-D-glucose is dissolved in water, the initial value of $[\alpha]$ is $+112°$ but this decreases with time until it reaches the equilibrium value of $+52°$ for a mixture of 64% α-D-glucose and 36% β-D-glucose. When β-D-glucose is dissolved in water, the initial value of $[\alpha]$ is $+19°$ which increases with time until it reaches the same equilibrium value of $+52°$.

Monosaccharides undergo many reactions typical of aldehydes and ketones in spite of the presence in solution of less than 0.2 percent of the open-chain structure (which has the carbonyl function), including various oxidations and reductions. Oxidation by Benedict's reagent (Sec. 14.5) is especially useful as a simple chemical test for monosaccharides, both aldoses and ketoses. An aldose gives a positive test because it has an α-hydroxy aldehyde group; a ketose because it has an α-hydroxy ketone group. (All disaccharides except for sucrose also give positive tests with Benedict's reagent.)

Problem 18.18. Benedict's reagent is the basis for a clinical test for monosaccharides. Describe what occurs when D-glucose is reacted with Benedict's reagent.

Ans. Benedict's reagent is an alkaline solution of Cu^{2+} complexed with citric anion to keep it soluble. A positive test occurs in the presence of a monosaccharide and is detected by the disappearance of the blue color of Cu^{2+} with the simultaneous formation of a red precipitate. The red precipitate is Cu_2O which forms due to reduction of Cu^{2+} to Cu^{+}. The monosaccharide is referred to as a *reducing carbohydrate* or *reducing sugar* since it brings about the reduction. Oxidation of D-glucose involves conversion of the aldehyde group to COOH.

$$\begin{array}{ccc} \text{H—C=O} & & \text{HO—C=O} \\ \text{H—}\!\!\!\!\!\!\!\!\!\!\!\!\quad\text{H}\!\!\!+\!\!\!\text{OH} & & \text{H}\!\!\!+\!\!\!\text{OH} \\ \text{HO}\!\!\!+\!\!\!\text{H} & \xrightarrow{Cu^{2+}} & \text{HO}\!\!\!+\!\!\!\text{H} \\ \text{H}\!\!\!+\!\!\!\text{OH} & & \text{H}\!\!\!+\!\!\!\text{OH} \\ \text{H}\!\!\!+\!\!\!\text{OH} & & \text{H}\!\!\!+\!\!\!\text{OH} \\ CH_2OH & & CH_2OH \end{array}$$

Problem 18.19. The carbonyl group of a monosaccharide can be reduced to an alcohol group by various reducing agents, e.g., H_2 in the presence of a metal catalyst or $NaBH_4$ (Sec. 14.6). Show the product(s) of the reduction of D-fructose by $NaBH_4$.

Ans.

$$\begin{array}{ccccc} CH_2OH & & CH_2OH & & CH_2OH \\ \text{C=O} & & \text{H}\!\!\!+\!\!\!\text{OH} & & \text{HO}\!\!\!+\!\!\!\text{H} \\ \text{HO}\!\!\!+\!\!\!\text{H} & \xrightarrow[\text{2. } H_2O]{\text{1. } NaBH_4} & \text{HO}\!\!\!+\!\!\!\text{H} & + & \text{HO}\!\!\!+\!\!\!\text{H} \\ \text{H}\!\!\!+\!\!\!\text{OH} & & \text{H}\!\!\!+\!\!\!\text{OH} & & \text{H}\!\!\!+\!\!\!\text{OH} \\ \text{H}\!\!\!+\!\!\!\text{OH} & & \text{H}\!\!\!+\!\!\!\text{OH} & & \text{H}\!\!\!+\!\!\!\text{OH} \\ CH_2OH & & CH_2OH & & CH_2OH \end{array}$$

There are two products because the carbonyl carbon is converted to a stereocenter upon reduction. There are two different configurations for placement of groups at that stereocenter—OH on the left- and right-hand sides.

Problem 18.20. The yields in reactions of the carbonyl group in monosaccharides (e.g., the reactions described in Problems 18.18 and 18.19) are very high even though the amount of the open-chain carbonyl structure is extremely low (< 0.2 percent). How can this occur?

Ans. The cyclic hemiacetal and hemiketal structures have no carbonyl function but are in equilibrium with the open-chain carbonyl structure. As the small amount of the open-chain carbonyl structure reacts, some of the α- and β-anomers undergo ring-opening to replenish the equilibrium concentration of the open-chain structure. The newly formed open-chain structure undergoes reaction. More α- and β-anomers open up. There is a continuous shifting of α- and β-anomers to the open-chain carbonyl structure until all of the monosaccharide initially present has reacted.

The hemiacetal and hemiketal groups of monosaccharides undergo dehydration with alcohol in the presence of an acid catalyst to form acetal and ketal groups, respectively. For example, reaction of glucose with methanol yields a mixture of methyl α- and β-acetals:

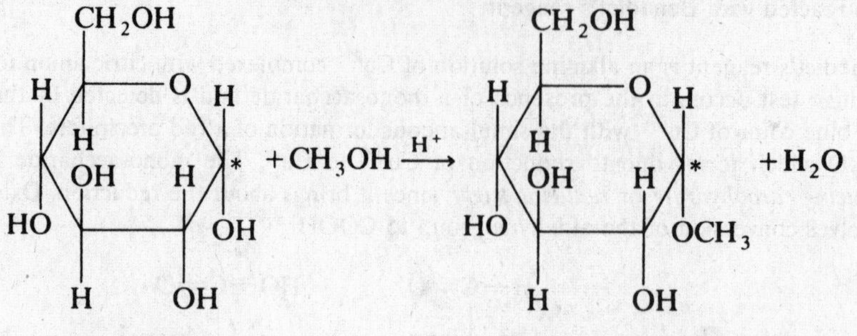

$$\text{α-D-Glucose} + CH_3OH \xrightarrow{H^+} \text{Methyl α-D-glucoside} + H_2O$$

α-D-Glucose: CH_2OH, H, O, H, H, OH, H, *, HO, OH, H, OH

Methyl α-D-glucoside: CH_2OH, H, O, H, H, OH, H, *, HO, OCH_3, H, OH

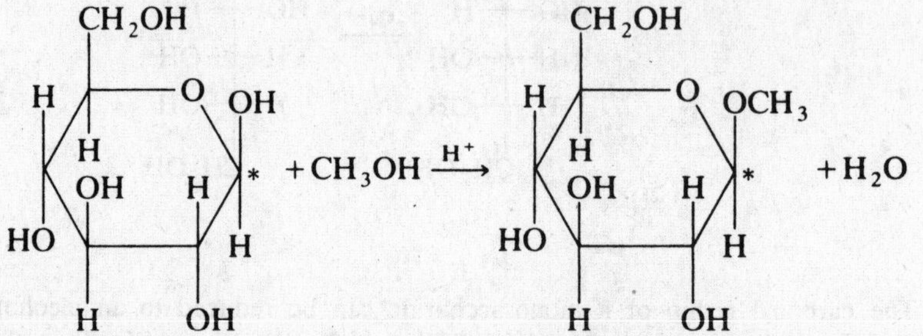

$$\text{β-D-Glucose} + CH_3OH \xrightarrow{H^+} \text{Methyl β-D-glucoside} + H_2O$$

β-D-Glucose: CH_2OH, H, O, OH, H, OH, H, *, HO, H, H, OH

Methyl β-D-glucoside: CH_2OH, H, O, OCH_3, H, OH, H, *, HO, H, H, OH

The hemiacetal carbons and the acetal carbons to which they are converted are marked with asterisks. Carbohydrate acetals (and ketals) are referred to as *glycosides*. The acetal (or ketal) carbon and its two OR substituents are referred to as a *glycosidic linkage*. To name the glycoside of a specific carbohydrate, the *-ose* ending of the carbohydrate's name is replaced by *-oside*, as shown above for methyl α- and β-D-glucosides.

Glycosides are not reducing sugars like the corresponding hemiacetals and hemiketals. A glycoside is not in equilibrium with an open-chain form of the carbohydrate and, thus, there is no aldehyde or α-hydroxy ketone compound available for reaction with an oxidizing agent such as Benedict's reagent. Glycosides also do not undergo mutarotation.

Glycosides undergo hydrolysis back to the corresponding hemiacetal or hemiketal under acidic or enzymatic conditions.

Problem 18.21. What is the distinguishing feature of acetals and ketals in comparison to hemiacetals and hemiketals, respectively?

Ans. Neither of the two oxygen substituents in acetals and ketals is an OH; both are OR substituents.

Problem 18.22. Give equations for the conversion of α- and β-D-fructoses to the corresponding methyl α- and β-D-fructosides. Place asterisks next to the hemiketal and ketal carbons.

Ans.

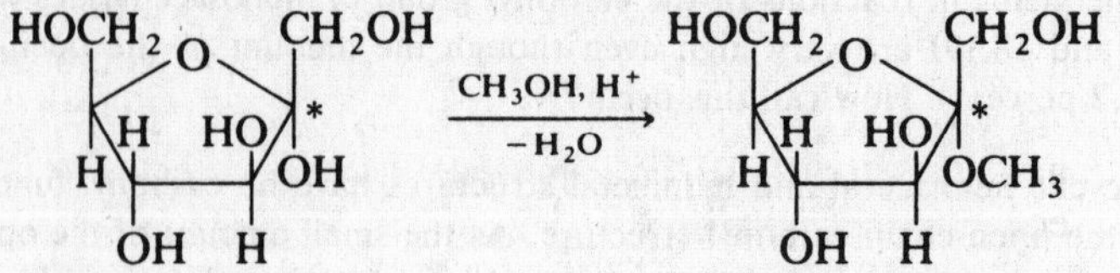

$$\text{α-D-Fructose} \xrightarrow[-H_2O]{CH_3OH,\ H^+} \text{Methyl α-D-fructoside}$$

α-D-Fructose: $HOCH_2$, O, CH_2OH, *, H, HO, H, OH, OH, H

Methyl α-D-fructoside: $HOCH_2$, O, CH_2OH, *, H, HO, H, OCH_3, OH, H

$$\text{β-D-Fructose} \xrightarrow[-H_2O]{CH_3OH,\ H^+} \text{Methyl β-D-fructoside}$$

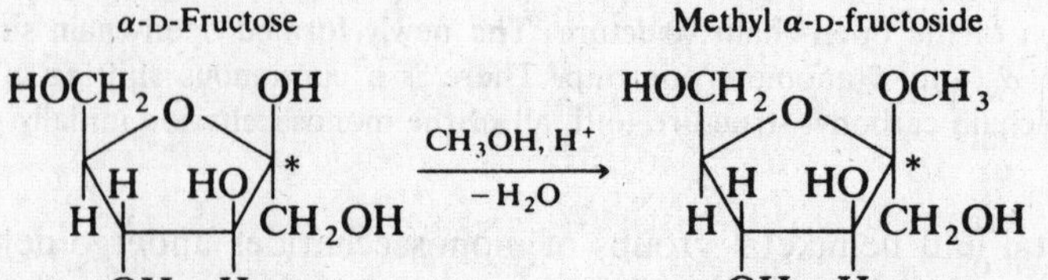

β-D-Fructose: $HOCH_2$, O, OH, *, H, HO, H, CH_2OH, OH, H

Methyl β-D-fructoside: $HOCH_2$, O, OCH_3, *, H, HO, H, CH_2OH, OH, H

Phosphate esters of monosaccharides are important intermediates in the metabolism of carbohydrates and in the formation of nucleic acids.

Problem 18.23. Generate the structure of β-D-glucose-1-phosphate by showing the loss of water between the OH group on C-1 of β-D-glucose and one of the OH groups of phosphoric acid.

Ans.

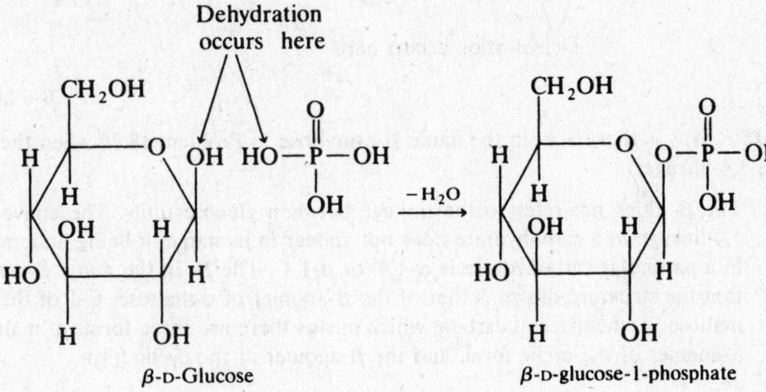

Dehydration occurs here

CH_2OH ... O OH HO—P(=O)(OH)—OH $\xrightarrow{-H_2O}$ CH_2OH ... O O—P(=O)(OH)—OH

β-D-Glucose β-D-glucose-1-phosphate

18.4 DISACCHARIDES

Disaccharides are glycosides formed from two molecules of monosaccharide, either the same monosaccharide or different monosaccharides, by dehydration between OH substituents. To identify a disaccharide, the following structural features must be specified:

1. Which monosaccharides are present?
2. Is the linkage α- or β-? The linkage between the two monosaccharide molecules is always a *glycosidic linkage*, i.e., one of the monosaccharide molecules is always linked through its hemiacetal (or hemiketal) OH. Naturally occurring disaccharides are either the α- or the β-glycoside, not a mixture of α- and β- as formed in the laboratory.
3. Which of the alcohol OH groups is involved in the linkage? The OH from the other second monosaccharide is usually not the hemiacetal (hemiketal) OH.

Maltose, cellobiose, lactose, and sucrose are the important disaccharides. Maltose, also referred to as *malt sugar* and *corn sugar*, is not present as such in any plant. It is the product of partial hydrolysis of starch and is found in processed corn syrup and germinating grains. Hydrolysis of maltose yields only glucose. Maltose consists of two glucose units linked together by an α-1,4-glycosidic linkage, i.e., one glucose unit is linked through its hemiacetal carbon (C-1), with its oxygen in the α-position, to the second glucose molecule's C-4.

Problem 18.24. Show the loss of water between OH groups of two glucose molecules to form maltose. Mark each hemiacetal carbon with an asterisk.

Ans. First, draw two glucose molecules side by side so that the OH on C-1 of one glucose is near the OH on C-4 of the other glucose. The glucose on the left-hand side must be α-glucose since the linkage in maltose is an α-1,4-linkage. Second, dehydrate water from between those two OH groups to link together the glucose units by an oxygen.

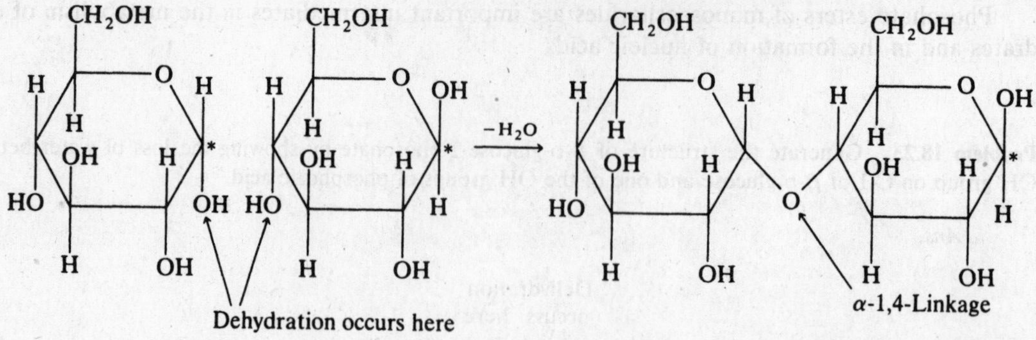

CH₂OH CH₂OH → −H₂O → CH₂OH CH₂OH

Dehydration occurs here

α-1,4-Linkage

β-D-Maltose

Problem 18.25. Why is there a β- in the name β-D-maltose in Problem 18.24 when the linkage between glucose units is an α-1,4-linkage?

Ans. The β- does not refer to the linkage between glucose units. The convention is that the type of 1,4-linkage in a carbohydrate does not appear in its name, it being understood whether the linkage in a particular carbohydrate is α-1,4- or β-1,4-. The β- in the name β-D-maltose refers to the fact that the structure drawn is that of the β-anomer of D-maltose. C-1 of the far right glucose unit of maltose is a hemiacetal carbon, which means there are three forms of maltose—an open-chain, the α-anomer of the cyclic form, and the β-anomer of the cyclic form.

Problem 18.26. Is maltose a reducing sugar and does it undergo mutarotation? Why?

Ans. Yes. Maltose contains a hemiacetal group.

Cellobiose is a disaccharide obtained by partial hydrolysis of cellulose. The only difference between cellobiose and maltose is that there is a β-1,4-linkage between glucose units instead of an α-1,4-linkage.

Problem 18.27. Show the loss of water between OH groups of two glucose molecules to form cellobiose. Mark each hemiacetal carbon with an asterisk. Indicate whether cellobiose is a reducing sugar and undergoes mutarotation.

Ans. Follow the same procedure as in Problem 18.24 except the glucose on the left-hand side must be β-glucose since the linkage in cellobiose is a β-1,4-linkage:

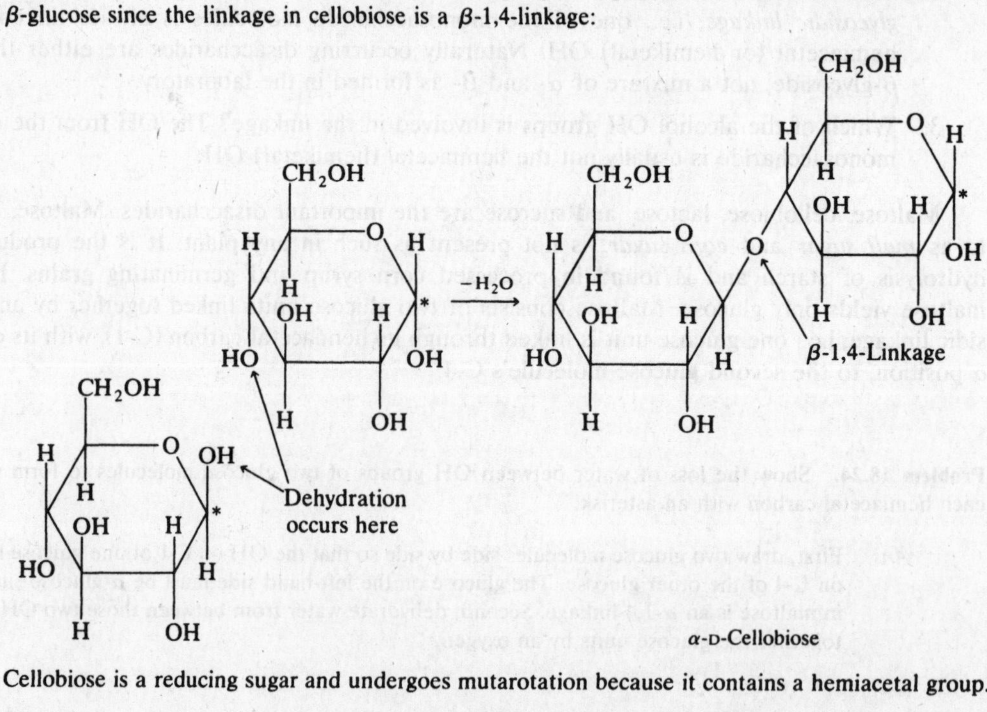

−H₂O →

Dehydration occurs here

β-1,4-Linkage

α-D-Cellobiose

Cellobiose is a reducing sugar and undergoes mutarotation because it contains a hemiacetal group.

Lactose or *milk sugar*, a disaccharide found in mammalian milk, consists of glucose and galactose. The linkage between the monosaccharide units is a β-1,4-linkage from C-1 of galactose to C-4 of glucose.

Problem 18.28. Show the formation of lactose from β-galactose and glucose. Mark each hemiacetal carbon with an asterisk. Is lactose a reducing sugar and does it mutarotate. (Galactose differs from glucose only in having the opposite configuration at C-4.)

Ans. Line up β-galactose to the left of glucose and then dehydrate between the OH groups at C-1 of galactose and C-4 of glucose:

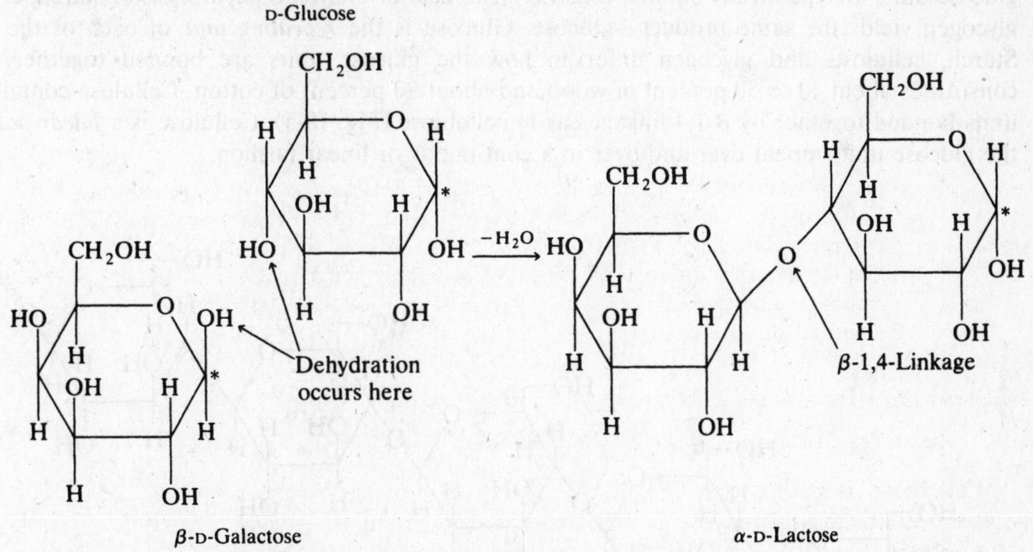

D-Glucose

Dehydration occurs here

$-H_2O$

β-1,4-Linkage

β-D-Galactose

α-D-Lactose

Lactose is a reducing sugar and undergoes mutarotation since it contains a hemiacetal group.

Sucrose, or *table sugar*, obtained from sugar cane and sugar beets, is a disaccharide of glucose and fructose. The linkage between the monosaccharides is a 1,2-linkage from C-1 of α-D-glucose to C-2 of β-D-fructose. Since linkage occurs through the hemiacetal OH of glucose and the hemiketal OH of fructose, sucrose has no hemiacetal or hemiketal group. Unlike maltose, cellobiose, and lactose, sucrose is not a reducing sugar and does not undergo mutarotation.

Problem 18.29. Show the formation of sucrose from α-glucose and β-fructose. Mark each hemiacetal and hemiketal carbon with an asterisk. Label acetal and ketal groups.

Ans. α-D-Glucose and β-D-fructose must be lined up so that the hemiacetal and hemiketal OH groups are near each other. This requires that β-D-fructose be rotated in the plane of the paper so that C-2 is on the left-hand side while C-5 is on the right-hand side (relative to the representation of fructose in Fig. 18.4):

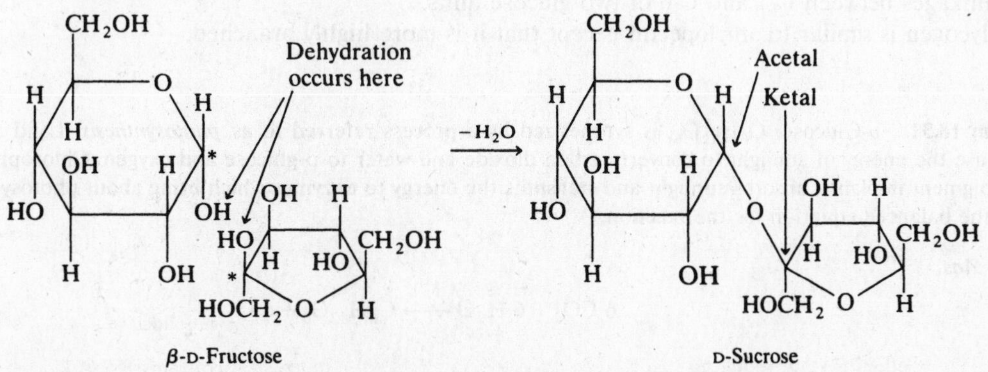

α-D-Glucose

Dehydration occurs here

$-H_2O$

Acetal

Ketal

β-D-Fructose

D-Sucrose

Problem 18.30. Is sucrose an α- or β-glycoside?

Ans. Both. It is an α-glycoside from the perspective of glucose but a β-glycoside from the perspective of fructose.

18.5 POLYSACCHARIDES

Starch, cellulose, and glycogen are polymers of glucose, i.e., large numbers (many thousands) of glucose units are chemically bonded together. The acid or enzymatic hydrolysis of starch, cellulose, or glycogen yields the same product—glucose. Glucose is the *repeating unit* of each of the polymers. Starch, cellulose, and glycogen differ in how the glucose units are bonded together. Cellulose constitutes about 40 to 50 percent of wood and about 90 percent of cotton. Cellulose contains glucose units bonded together by β-1,4-linkages as in cellobiose (Fig. 18-5). Cellulose is a *linear polymer*, i.e., the glucose units repeat over and over in a continuous or linear fashion.

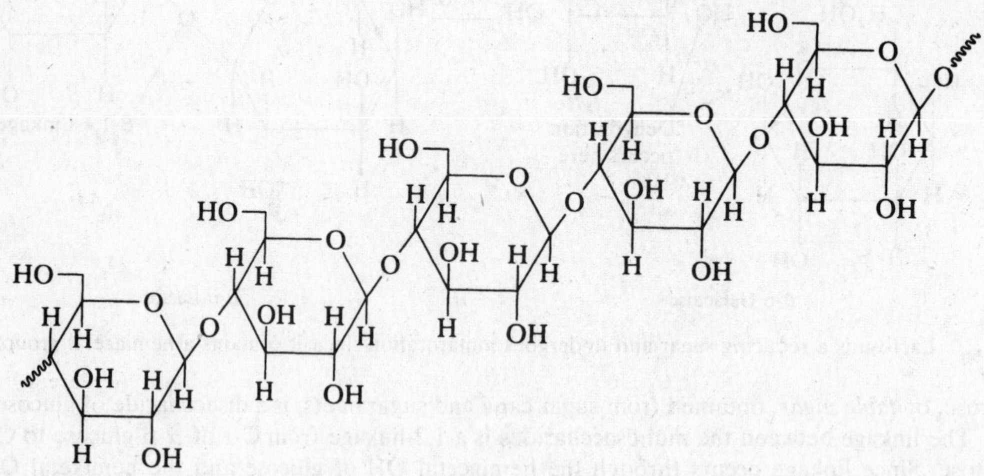

Fig. 18-5 Cellulose is a linear polymer of glucose units bonded together by β-1,4-linkages.

Starch is composed of 10 to 20 percent amylose and 80 to 90 percent amylopectin. Amylose is a linear polymer with glucose units bonded together by α-1,4-linkages as in maltose (Fig. 18-6). Amylopectin also has glucose units bonded together by α-1,4-linkages but there is an additional structural feature not present in amylose (Fig. 18-7). Amylopectin is a *branched polymer*, i.e., there are polyglucose chains which branch off of other polyglucose chains. The branching occurs through α-1,6-linkages between C-1 and C-6 of two glucose units.

Glycogen is similar to amylopectin except that it is more highly branched.

Problem 18.31. D-Glucose, $C_6H_{12}O_6$, is synthesized by a process referred to as *photosynthesis*. Land and sea plants use the energy of sunlight to convert carbon dioxide and water to D-glucose and oxygen. Chlorophyll, the green pigment in plants, absorbs sunlight and transmits the energy to enzymes which bring about photosynthesis. Write the balanced equation for the reaction.

Ans.

$$6\ CO_2 + 6\ H_2O \longrightarrow C_6H_{12}O_6 + 6\ O_2$$

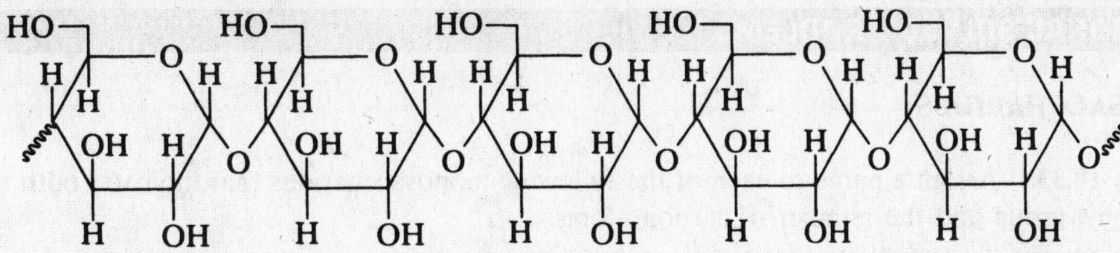

Fig. 18-6 Amylose is a linear polymer of glucose units bonded together by α-1,4-linkages.

α-1,6-Linkage

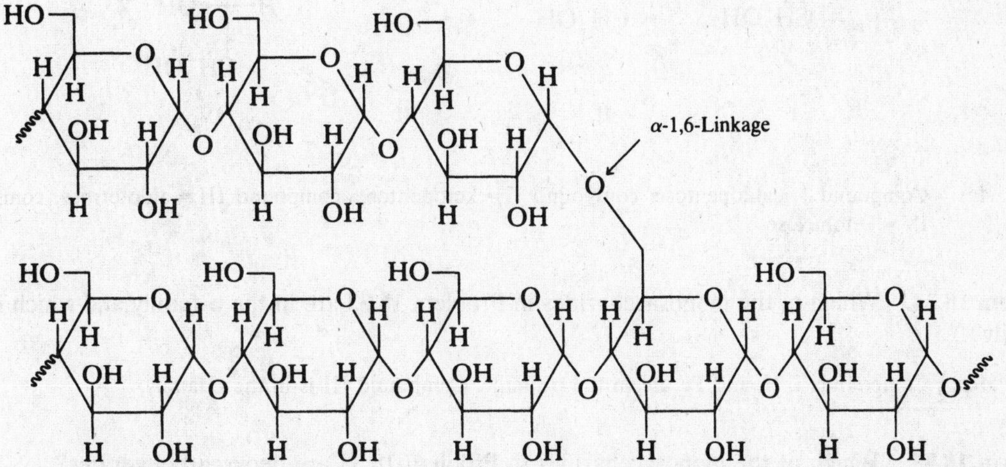

Fig. 18-7 Amylopectin contains glucose units bonded together by α-1,4-linkages plus branching through α-1,6-linkages.

Problem 18.32. Discuss the importance of photosynthesis for life processes.

Ans. Most life forms on earth owe their existence to photosynthetic processes. Photosynthetic organisms capture the energy from the sun in the form of oxidizable organic compounds which most other organisms require for their metabolic processes. Photosynthesis is critical from many other viewpoints. It is responsible for using up the carbon dioxide produced by our metabolic processes and the burning of wood and petroleum fuels and for regenerating oxygen needed for metabolic processes. Plants use the D-glucose produced by photosynthesis to synthesize cellulose and starch. Cellulose is used by plants as their structural material. Starch is used as the storage form of glucose. Humans as well as other animals eat plants. Starch is digested, i.e., it is enzymatically hydrolyzed to glucose. Whatever glucose is needed immediately for energy production is metabolized. (Carbon dioxide and water are byproducts of this metabolism.) Excess glucose is stored in two different forms: (*a*) It is polymerized to glycogen which is stored in the liver and muscle tissue, and (*b*) it is metabolically converted to fat and deposited in fat tissue. When needed, glycogen is hydrolyzed to glucose. Humans and most mammals cannot digest and utilize cellulose since we do not possess the enzyme for hydrolyzing β-1,4-glycosidic linkages. However, cellulose does serve the useful function of roughage in our digestive tract. Cattle, sheep, and termites utilize cellulose because microorganisms in their digestive tracts possess the necessary enzyme for hydrolyzing β-1,4-glycosidic linkages.

ADDITIONAL SOLVED PROBLEMS

MONOSACCHARIDES

Problem 18.33. Assign a name to each of the following monosaccharides that indicates both the type of carbonyl group and the number of carbon atoms:

I	II	III	IV
H—C=O	CH_2OH	H—C=O	CH_2OH
H—OH	C=O	H—OH	C=O
H—OH	H—OH	HO—H	H—H
H—OH	H—OH	CH_2OH	H—OH
CH_2OH	CH_2OH		H—OH
			CH_2OH

Ans. Compound I = aldopentose, compound II = ketopentose, compound III = aldotetrose, compound IV = ketohexose

Problem 18.34. Which of the monosaccharides in Problem 18.33 are in the D-family and which in the L-family?

Ans. Compounds I, II, and IV are in the D-family. Compound III is in the L-family.

Problem 18.35. Which of the monosaccharides in Problem 18.33 are deoxycarbohydrates?

Ans. Compound IV is deoxy at C-3.

Problem 18.36. How many stereoisomers are there for each of the monosaccharides in Problem 18.33?

Ans. Compound I has three stereocenters and a total of eight stereoisomers. Compounds II, III, and IV each have two stereocenters and four stereoisomers.

Problem 18.37. Draw the stereoisomers of compound IV of Problem 18.33 and indicate the relationship of the various compounds.

Ans.

IV	V	VI	VII
CH_2OH	CH_2OH	CH_2OH	CH_2OH
C=O	C=O	C=O	C=O
H—H	H—H	H—H	H—H
H—OH	HO—H	HO—H	H—OH
H—OH	HO—H	H—OH	HO—H
CH_2OH	CH_2OH	CH_2OH	CH_2OH

There are two pairs of enantiomers: One pair is compounds IV and V and the other is compounds VI and VII. Compounds IV and VI are diastereomers as well as epimers. Compounds V and VII are diastereomers as well as epimers.

CYCLIC HEMIACETAL AND HEMIKETAL STRUCTURES

Problem 18.38. Why does a solution of D-glucose contain more β-D-glucose than α-D-glucose?

Ans. There is a difference in stability between the two structures. α-D-Glucose is less stable because of steric and electrostatic repulsions between the large groups (OH) on C-1 and C-2. The OH groups are further from each other in β-D-Glucose.

Problem 18.39. α-D-Fructose and β-D-fructose are also formed in unequal amounts, and like the glucose system, β-D-fructose is more abundant than α-D-fructose. Why?

Ans. The reason for the unequal amounts of α- and β-anomers is exactly the same in both the glucose and fructose systems. For fructose, the β-anomer has the large substituents at C-2 (CH_2OH) and C-3 (OH) furthest apart.

Problem 18.40. Distinguish anomers from epimers.

Ans. A pair of anomers like a pair of epimers differ in the configuration at only one stereocenter. A pair of epimers differ in the configuration at any single stereocenter. Anomers are a specific type of epimers, differing in the configuration only at the hemiacetal or hemiketal carbon.

Problem 18.41. Why does hemiacetal formation in D-glucose involve the OH at C-5 instead of one of the other OH groups?

Ans. Different-sized rings form depending on which OH reacts. Reactions at C-2, C-3, C-4, C-5, and C-6 yield rings of size 3, 4, 5, 6, and 7, respectively. Six-membered rings, being the most stable, are usually formed in preference to rings of other sizes. (The overall order of stability is 6 > 5, 7 ≫ 3, 4.)

Problem 18.42. What do the following structures represent?

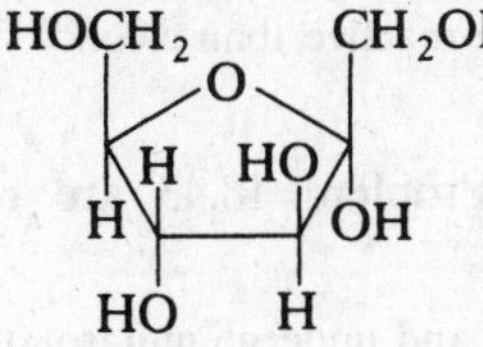

$HOCH_2$ O CH_2OH / H HO / H OH / HO H — I

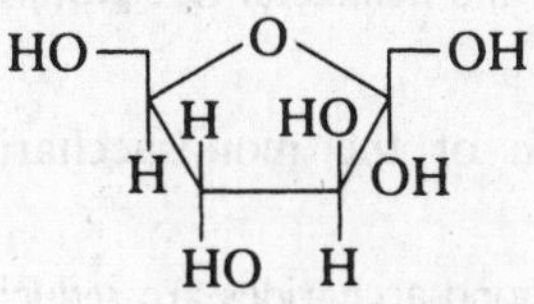

HO— O —OH / H HO / H OH / HO H — II

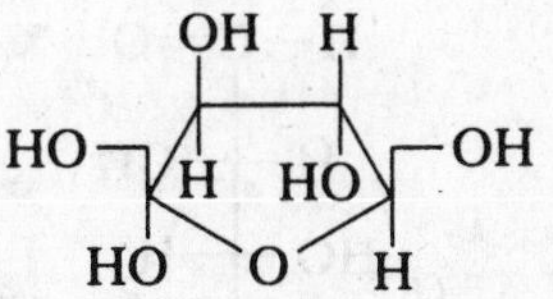

OH H / HO— H HO —OH / HO O H — III

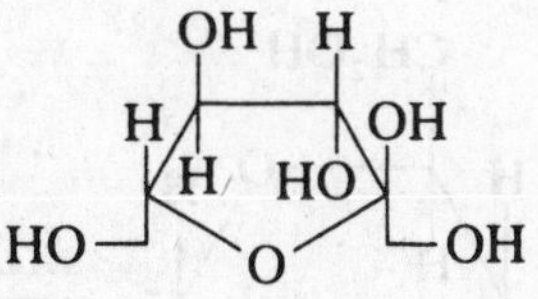

OH H / H OH / H HO / HO— O —OH — IV

Ans. All four structures represent the same compound, α-D-fructose. Structures I and II use slightly different ways of representing the CH_2 carbons. Structure III is rotated 180° in the plane of the paper relative to structure II. Structure IV is rotated 180° out of the plane of the paper relative to structure III.

Problem 18.43. Answer the following questions for the compounds shown below: (*a*) Which are aldoses and which are ketoses? (*b*) Which are pentoses and which are hexoses? (*c*) Which are α-anomers and which are β-anomers? (*d*) Which are deoxy sugars?

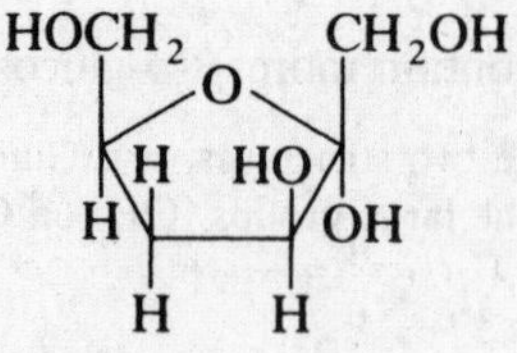

I 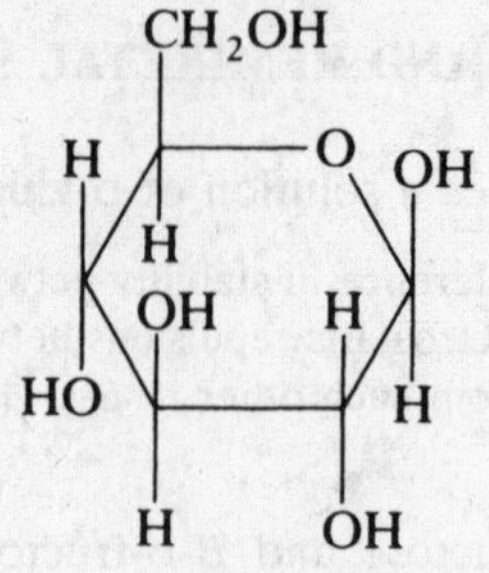 II

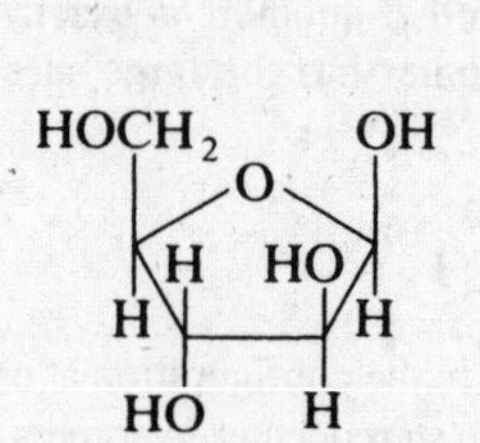

III 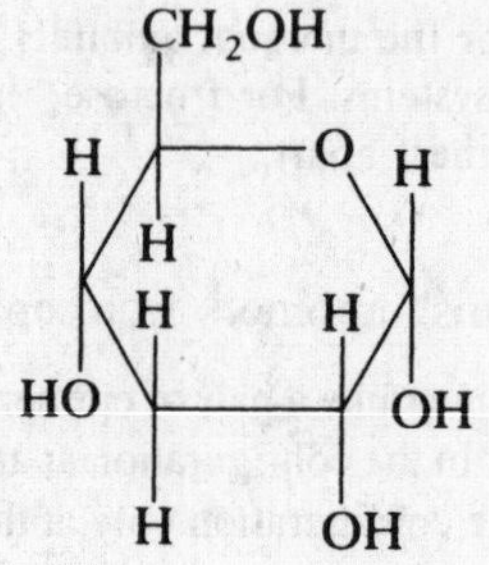 IV

Ans. (*a*) Compounds II, III, and IV are aldoses; compound I is a ketose. (*b*) Compounds I, II, and IV are hexoses; compound III is a pentose. (*c*) Compounds II and III are β-anomers; compounds I and IV are α-anomers. (*d*) Compounds I and IV are deoxy sugars.

PROPERTIES AND REACTIONS OF MONOSACCHARIDES

Problem 18.44. In the reaction of methanol with glucose and fructose, why does reaction occur only at the hemiacetal and hemiketal OH groups instead of also occurring at the alcohol OH groups?

Ans. Hemiacetal and hemiketal OH groups are more reactive than alcohol OH groups.

Problem 18.45. Which of the monosaccharides in Problem 18.43 are reducing sugars? Which undergo mutarotation?

Ans. All of the monosaccharides are reducing sugars and undergo mutarotation.

Problem 18.46. Give the product formed in each of the following reduction reactions:

(*a*) D-Glucose $\xrightarrow[\text{2. } H_2O]{\text{1. } LiAlH_4}$?

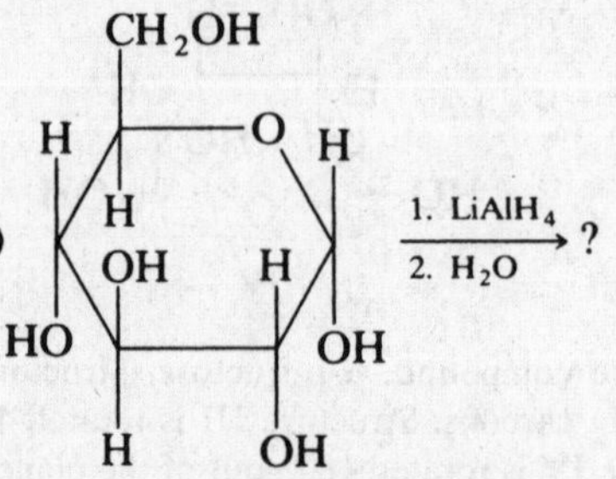

(*b*) $\xrightarrow[\text{2. } H_2O]{\text{1. } LiAlH_4}$?

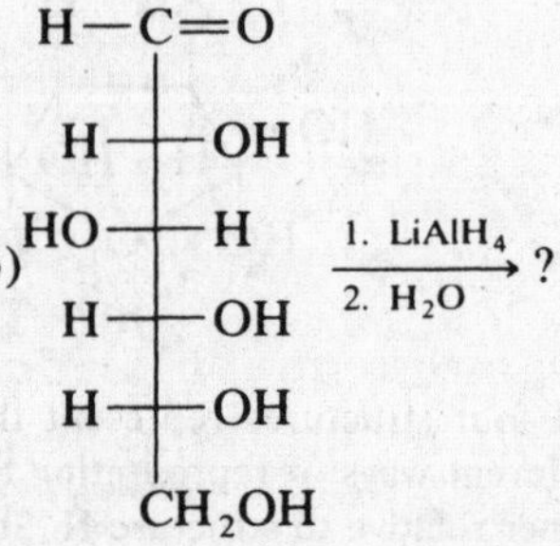

Ans. The compound undergoing reaction is the same in both (*a*) and (*b*): D-glucose. The product is the same, D-sorbitol, which can be represented by any of the following structural drawings:

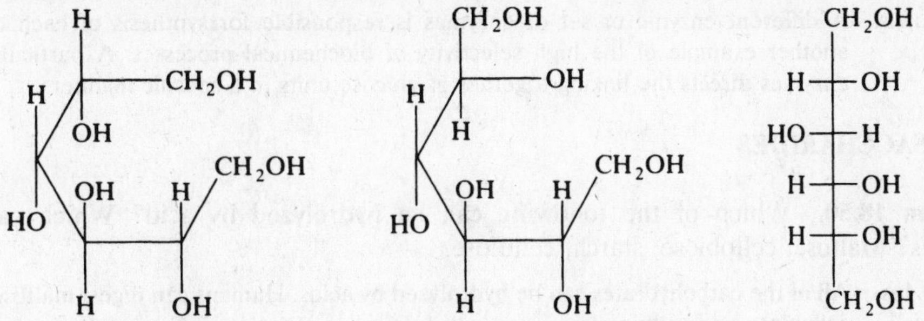

CHO$_2$OH	
H—C—OH	
HO—C—H	
H—C—OH	
H—C—OH	
CH$_2$OH	

Problem 18.47. Show equations for the formation of the following compounds from the corresponding monosaccharides: (*a*) ethyl β-glucoside, (*b*) β-2-deoxyribose-5-phosphate.

Ans.

(*a*) $\xrightarrow[H^+]{CH_3CH_2OH}$ (OCH$_2$CH$_3$)

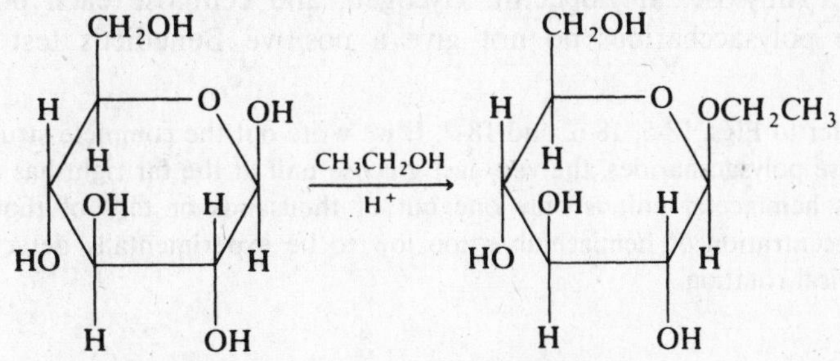

(*b*) $\xrightarrow{H_3PO_4}$ (HO—P(OH)(OH)—OCH$_2$)

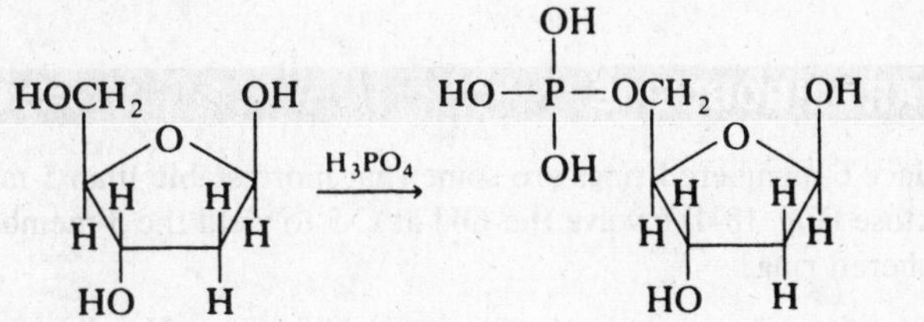

DISACCHARIDES

Problem 18.48. Gentiobiose consists of two glucose units linked together by an β-1,6-glycosidic linkage, i.e., one glucose unit is linked through its hemiacetal carbon (C-1), with its oxygen in the β-position, to the second glucose molecule's C-6. Show the formation of gentiobiose from glucose. Label acetal and hemiacetal groups in gentiobiose, if present. Does gentiobiose give a positive Benedict's test? Does it mutarotate?

Ans. Gentiobiose gives a positive Benedict's test and undergoes mutarotation since it has a hemiacetal group.

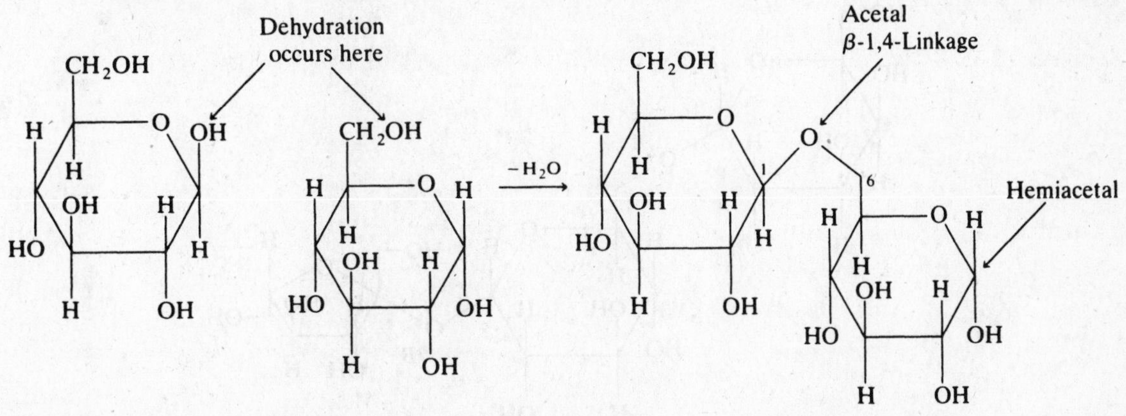

Dehydration occurs here

$\xrightarrow{-H_2O}$

Acetal β-1,4-Linkage

Hemiacetal

α-D-Gentiobiose

Problem 18.49. Maltose, cellobiose, and gentiobiose are disaccharides of glucose in which the two glucose units are linked together differently. How does this occur?

Ans. A different enzyme or set of enzymes is responsible for synthesis of each disaccharide. This is another example of the high selectivity of biochemical processes. A particular enzyme or set of enzymes directs the linking together of glucose units in a specific manner.

POLYSACCHARIDES

Problem 18.50. Which of the following can be hydrolyzed by acid? Which can be digested by humans? Maltose, cellobiose, starch, cellulose.

Ans. All of the carbohydrates can be hydrolyzed by acids. Humans can digest maltose and starch, but not cellobiose and cellulose.

Problem 18.51. Amylose, amylopectin, glycogen, and cellulose each possess a hemiacetal group. However, these polysaccharides do not give a positive Benedict's test or undergo mutarotation. Explain.

Ans. Refer to Figs. 18-5, 18-6, and 18-7. If we write out the complete structure for a molecule of any of these polysaccharides, the very last glucose unit at the far right has a hemiacetal group. However, this hemiacetal unit is only one out of thousands or tens of thousands of glucose units. The concentration of hemiacetal is too low to be experimentally detected by the Benedict's test or optical rotation.

SUPPLEMENTARY PROBLEMS

Problem 18.52. Since 6-membered rings are somewhat more stable than 5-membered rings, why does hemiketal formation in D-fructose (Fig. 18-4) involve the OH at C-5 to yield the 5-membered ring instead of reaction at C-6 to yield the 6-membered ring?

Ans. The previous discussion was an oversimplification. Hemiketal formation in D-fructose and other ketohexoses results in reaction at both C-5 and C-6 to give a mixture of 5- and 6-membered rings. Pure fructose in solution consists of a mixture of the α- and β-anomers of both furanose and pyranose structures, as well as a very small amount (<0.2 percent) of the open-chain form. However, all combined forms of D-fructose, e.g., fructose combined with glucose in sucrose, involve only the 5-membered ring. For this reason, most texts simplify matters for the student by ignoring the pyranose structures. The exclusive use of the 5-membered ring in combined forms of D-fructose is not understood.

Problem 18.53. Raffinose, a trisaccharide found in small amounts in some plants, has the structure shown below:

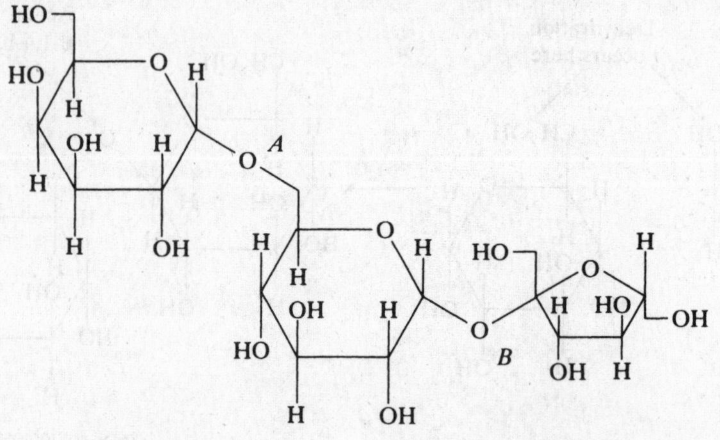

Answer the following questions regarding raffinose: (*a*) What are the monosaccharides obtained upon hydrolysis? (*b*) Identify linkages *A* and *B*. (*c*) Is raffinose a reducing sugar?

Ans. (*a*) From left to right: galactose, glucose, fructose.
(*b*) *A* is an acetal α-1,6-linkage. *B* is 1,2-linkage (C-1 of glucose and C-2 of fructose). *B* is an acetal and α-anomer from the perspective of glucose but a ketal and α-anomer from the perspective of fructose.
(*c*) Raffinose is not a reducing sugar since there is no hemiacetal or hemiketal group.

Problem 18.54. In the Fischer projection (Figs. 18-1 and 18-2), D-carbohydrates are those where the OH on the stereocenter furthest from the C = O is placed on the right-hand side. What is common to all D-carbohydrates in the Haworth structures?

Ans. Refer to the cyclic structures of D-glucose and D-fructose in Figs. 18-3 and 18-4. When the cyclic structures are placed in this manner so that the oxygen in the ring is up and the hemiacetal or hemiketal carbon is on the right-hand side, all D-carbohydrates have the CH_2OH substituent on the stereocenter furthest from the hemiacetal or hemiketal carbon in the up position (above the plane of the ring).

Problem 18.55. Which of the following monosaccharides belong to the D-family?

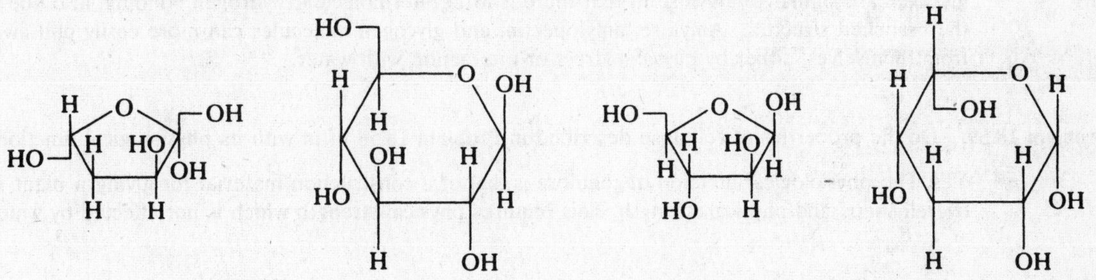

[Structures I, II, III, IV]

Ans. Compounds II and III are in the D-family while compounds I and IV are in the L-family.

Problem 18.56. All of the six-membered cyclic structures are incorrect from the viewpoint of conveying the correct bond angles for the atoms of the ring. Six-membered rings are not flat (Sec. 11.8). They are nonplanar, puckered chair structures which allow the bond angles to be the stable tetrahedral bond angle (109.5°). Draw the α- and β-anomers of D-glucose in the chair structure.

Ans.

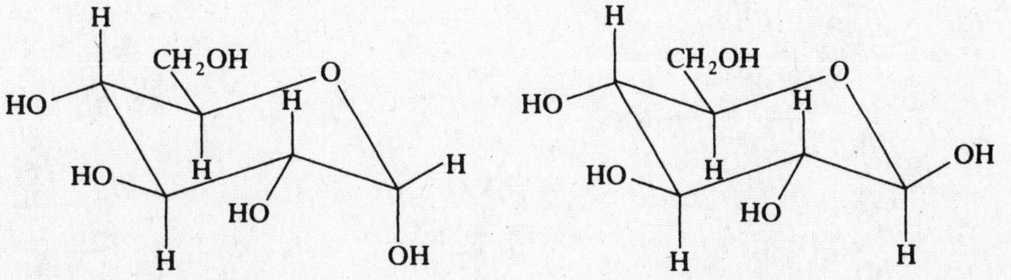

α-Glucose

β-Glucose

Problem 18.57. Draw the α-anomer of lactose in the chair structure.

Ans.

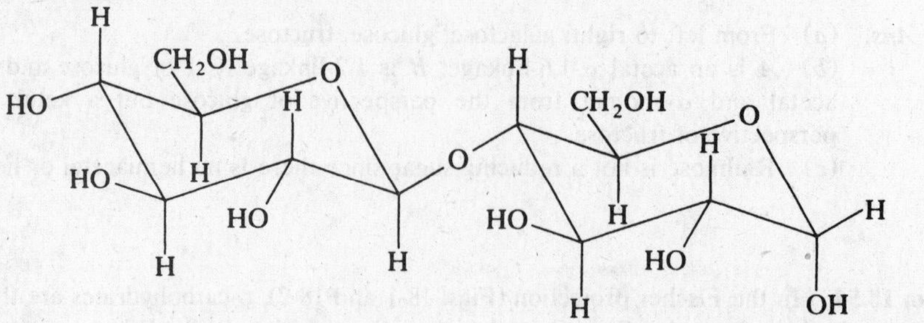

H
CH_2OH H O
HO H CH_2OH O
O H
HO H
HO HO H H
H H HO
H OH

Problem 18.58. Cellulose differs markedly from amylose, amylopectin, and glycogen. Cellulose has high physical strength and is not highly wetted by water. Amylose, amylopectin, and glycogen have much lower strengths and absorb large quantities of water (so much that they form pastes). Explain the origin of the differences.

Ans. The property differences arise from the differences in chemical structure (β-1,4- versus α-1,4-linkage and linear versus branched). Cellulose molecules exist in extended or linear conformations and this results in strong intermolecular hydrogen bonding. Cellulose molecules do not pull away from each other easily, either by physical stress or interaction with water. Amylose molecules exist in helical conformations where there is considerable intramolecular hydrogen bonding but much less intermolecular hydrogen bonding with other amylose molecules. Amylopectin and glycogen do not exist in helical conformations because of their branched structures. However, amylopectin and glycogen are similar to amylose in that there is little intermolecular hydrogen bonding, also due to the branched structure. Amylose, amylopectin, and glycogen molecules can more easily pull away from themselves, either by physical stress or interaction with water.

Problem 18.59. Do the properties of cellulose described in Problem 18.58 fit in with its physiological function?

Ans. Yes. The physiological function of cellulose is that of a construction material for giving a plant its overall shape and physical strength. This requires physical strength which is not affected by water.

Problem 18.60. Why do α-hydroxy ketones such as the ketoses give positive Benedict's tests? *Hint*: Fructose gives exactly the same product as glucose (Problem 18.18) upon oxidation by Benedict's reagents.

Ans. Fructose undergoes rearrangement (a process referred to as *isomerization*) to glucose under the alkaline conditions of these reagents.

CHAPTER 19

Lipids

19.1 INTRODUCTION

Lipids are not a family of compounds in the same sense as previously discussed families, i.e., they do not possess one characteristic functional group. Lipids are a class of organic compounds in plants and animals that share a common physical property—solubility in nonpolar and low-polarity solvents and insolubility in water.

Problem 19.1. Describe the solubility experiment by which compounds are categorized as lipids.

Ans. Undecomposed plant or animal matter is crushed into small particles or powder using a blender or other mechanical device and then a nonpolar or low-polarity solvent such as diethyl ether, hexane, benzene, or carbon tetrachloride is added. Lipids dissolve in the nonpolar solvent while other compounds (proteins, carbohydrates, nucleic acids, and inorganics) do not. The solution is filtered from the insoluble matter and then the solvent evaporated to yield the lipids.

Problem 19.2. What functional groups are present in compounds belonging to the class of compounds called lipids?

Ans. A number of functional groups are found in lipids—carboxylic acid and ester, phosphate ester, amide, alcohol, ether, ketone. Many lipids contain more than one kind of functional group. Some lipids contain none of these functional groups.

Problem 19.3. Why do lipids with polar functional groups such as ester and amide have the same solubility behavior as lipids without those functional groups?

Ans. The molecules of a lipid with a polar functional group contain large portions that are nonpolar. The polar group comprises a very small portion of the molecule, too small to impart significant polarity to the molecule as a whole.

Lipids are categorized into *hydrolyzable* (*saponifiable*) and nonhydrolyzable (*nonsaponifiable*) *lipids*. Hydrolyzable lipids contain at least one ester group, which undergoes hydrolysis in the presence of an acid, a base, or an enzyme. Hydrolysis by base is referred to as *saponification*. Hydrolysis cleaves a saponifiable lipid into two or more smaller molecules. Nonhydrolyzable lipids do not undergo hydrolytic cleavage into smaller molecules.

Figure 19-1 gives an overview of the different types of lipids.

Lipids perform a variety of physiological functions. Triacylglycerols are used for energy storage and metabolic fuel. Phospholipids, sphingolipids, and cholesterol (a steroid) are structural components of cell membranes. Nonhydrolyzable lipids perform a variety of regulatory functions (hormones, vitamins).

19.2 FATTY ACIDS

Fatty acids are the carboxylic acids used as building blocks for saponifiable lipids. Only very small amounts of fatty acids are found in uncombined form in nature. Several dozen different fatty acids have been isolated from various plant and animal lipids. Table 19-1 lists the most important fatty acids, all of which have a long hydrocarbon chain attached to the carboxyl (COOH) group.

Fatty acids without double bonds are referred to as *saturated fatty acids*. Fatty acids with double bonds are referred to as *unsaturated fatty acids*. Fatty acids with one double bond are *monounsatu-*

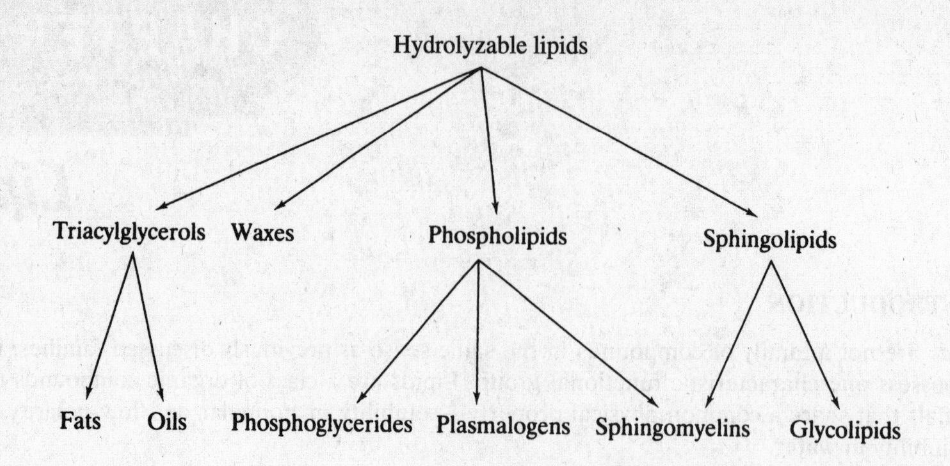

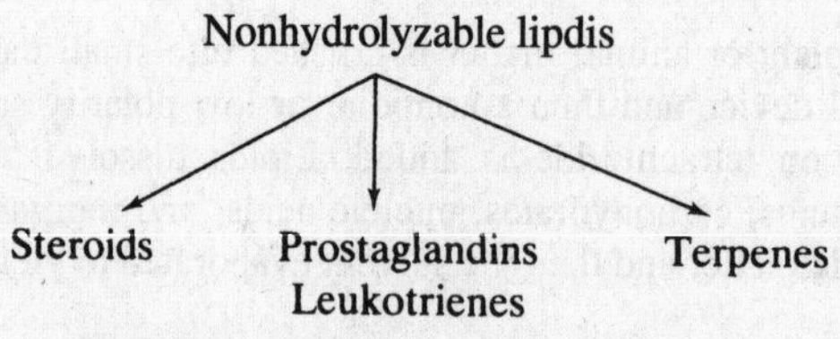

Fig. 19-1 Classification of lipids.

Table 19-1. Common Saturated and Unsaturated Fatty Acids

Name of acid	Number of double bonds	Number of carbons	Structure	Melting point, °C
Saturated fatty acids				
Lauric	0	12	$CH_3(CH_2)_{10}COOH$	44
Myristic	0	14	$CH_3(CH_2)_{12}COOH$	54
Palmitic	0	16	$CH_3(CH_2)_{14}COOH$	63
Stearic	0	18	$CH_3(CH_2)_{16}COOH$	69
Arachidic	0	20	$CH_3(CH_2)_{18}COOH$	77
Unsaturated fatty acids				
Palmitoleic	1	16	$CH_3(CH_2)_5CH{=}CH(CH_2)_7COOH$	1
Oleic	1	18	$CH_3(CH_2)_7CH{=}CH(CH_2)_7COOH$	13
Linoleic	2	18	$CH_3(CH_2)_4(CH{=}CHCH_2)_2(CH_2)_6COOH$	−5
Linolenic	3	18	$CH_3CH_2(CH{=}CHCH_2)_3(CH_2)_6COOH$	−11
Arachidonic	4	20	$CH_3(CH_2)_4(CH{=}CHCH_2)_4(CH_2)_2COOH$	−49
Eicosapentaenoic	5	20	$CH_3CH_2(CH{=}CHCH_2)_5(CH_2)_2COOH$	−54

rated fatty acids; those with two or more double bonds are *polyunsaturated fatty acids*. The double bonds in fatty acids are exclusively *cis* double bonds, not *trans* double bonds. The most abundant saturated fatty acids are palmitic and stearic acids. The most abundant unsaturated fatty acids are oleic and linoleic. Linoleic and linolenic acids are *essential fatty acids*, i.e., they cannot be synthesized in the human body. All other fatty acids are *nonessential fatty acids* since they can be synthesized either from other fatty acids or from carbohydrates and proteins.

Problem 19.4. What are the common structural features of the fatty acids used to synthesize saponifiable lipids in plants and animals (refer to Table 19-1)?

Ans. All are monocarboxylic acids, R—COOH, where the R group is unbranched and the total number of carbons is an even number. (Exceptions are found in bacterial lipids.)

Problem 19.5. Which of the following carboxylic acids are incorporated into plant and animal lipids?

$$\underset{\text{I}}{CH_3CH_2\overset{\overset{\large CH_3}{|}}{C}H(CH_2)_{10}COOH} \qquad \underset{\text{II}}{CH_3(CH_2)_{17}COOH} \qquad \underset{\text{III}}{CH_3(CH_2)_4(CH{=}CHCH_2)_2(CH_2)_5COOH}$$

Ans. None of these acids are used. Compound I is branched while compounds II and III each contain an odd number of carbons. Only unbranched acids with an even number of carbons are incorporated into lipids.

Unsaturated fatty acids are often referred to by an *omega*-(ω-) number to indicate the location of the double bond nearest the methyl end of the carbon chain. A number is placed after omega- or ω- to indicate the number of the first carbon of the double bond, counting from the methyl carbon, not the COOH carbon. For example, palmitoleic and oleic acids are omega-7 and omega-9 fatty acids, respectively.

Problem 19.6. Give the ω- designation for linoleic and linolenic acids (refer to Table 19-1).

Ans. Linoleic and linolenic acids are ω-6 and ω-3, respectively.

The melting points of unsaturated fatty acids are lower than those of saturated fatty acids with the same number of carbons (compare stearic, oleic, linoleic, and linolenic acids in Table 19-1). The more double bonds present, the lower the melting point of the fatty acid. The effect of a double bond in lowering the melting point is a consequence of its presence in nature in the *cis* geometry instead of *trans*. Unsaturated fatty acids are liquids at temperatures where saturated fatty acids are solids.

Problem 19.7. Why do *cis* double bonds lower the melting points of fatty acids? *Trans* double bonds do not have the same effect as *cis* double bonds. Why? *Hint*: Use appropriate drawings to show the effect of double bonds on molecular shape and indicate how molecular shape affects melting point.

Ans. The molecular shapes of stearic, oleic, and linoleic acids are shown in Fig. 19-2. Saturated fatty acid molecules are elongated. The intermolecular attractive forces between such molecules in the solid state are relatively high since the molecules can pack together tightly in an orderly fashion. Higher temperatures are needed during melting to overcome the higher intermolecular forces. *Cis* double bonds introduce rigid kinks or bends into the carbon chain. Packing of molecules in the solid state is looser and less orderly, intermolecular attractive forces are lower, and high temperatures are not needed to separate molecules during melting. The more double bonds, the more kinks in the chain, the less orderly the solid state packing, and the lower the melting temperatures. *Trans* double bonds do not affect the melting point in the same manner as *cis* double bonds. *Trans* double bonds, although rigid, do not introduce significant bends into the chain. The melting point of the *trans* isomer of oleic acid is much higher than that of the *cis* isomer.

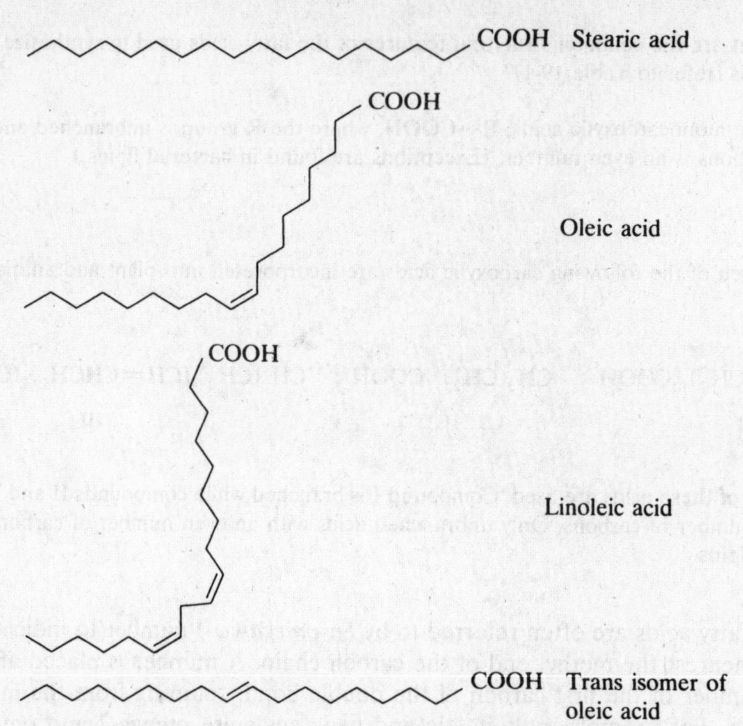

Fig. 19-2 Molecular shapes of saturated and unsaturated fatty acids.

19.3 TRIACYLGLYCEROLS

19.3.1 Structure and Physical Properties

Animal fats such as butter, beef, pork, and poultry fats and *vegetable oils* such as corn, peanut, safflower, and olive oils are *triacylglycerols* or *triglycerides*. These are triesters of glycerol (IUPAC name: 1,2,3-propanetriol). Each of the three OH groups of glycerol forms an ester group by reaction with the COOH group of a fatty acid to form the triacylglycerol:

$$
\begin{array}{lcl}
CH_2-OH & HO-\overset{\overset{\large O}{\|}}{C}-R^1 & CH_2-O-\overset{\overset{\large O}{\|}}{C}-R^1 \\
| & & | \\
CH-OH & +\ HO-\overset{\overset{\large O}{\|}}{C}-R^2 \xrightarrow{-3H_2O} & CH-O-\overset{\overset{\large O}{\|}}{C}-R^2 \\
| & & | \\
CH_2-OH & HO-\overset{\overset{\large O}{\|}}{C}-R^3 & CH_2-O-\overset{\overset{\large O}{\|}}{C}-R^3 \\
\text{Glycerol} & & \text{Triglyceride}
\end{array}
$$

A variety of triacylglycerols are possible. *Simple triacylglycerols* are those in which R^1, R^2, and R^3 are the same, i.e., three molecules of the same fatty acid react with glycerol. *Complex triacylglycerols* are those in which R^1, R^2, and R^3 are different. Naturally occurring triacylglycerols are complex triacylglycerols.

Problem 19.8. Write the equation showing triacylglycerol formation from 1 mol of glycerol and 1 mol each of palmitic, stearic, and oleic acids.

Ans.

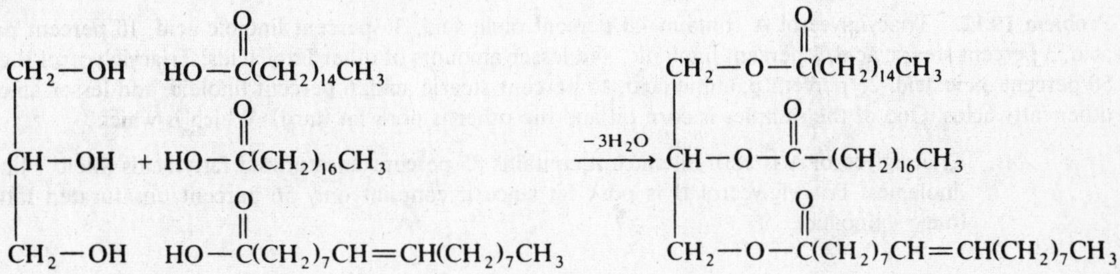

$$\begin{array}{l} CH_2-OH \quad HO-\overset{O}{\overset{\|}{C}}(CH_2)_{14}CH_3 \\ | \\ CH-OH \;+\; HO-\overset{O}{\overset{\|}{C}}(CH_2)_{16}CH_3 \\ | \\ CH_2-OH \quad HO-\overset{O}{\overset{\|}{C}}(CH_2)_7CH=CH(CH_2)_7CH_3 \end{array} \xrightarrow{-3H_2O} \begin{array}{l} CH_2-O-\overset{O}{\overset{\|}{C}}(CH_2)_{14}CH_3 \\ | \\ CH-O-\overset{O}{\overset{\|}{C}}-(CH_2)_{16}CH_3 \\ | \\ CH_2-O-\overset{O}{\overset{\|}{C}}(CH_2)_7CH=CH(CH_2)_7CH_3 \end{array}$$

Problem 19.9. The answer to Problem 19.8 is an oversimplification. What is the relationship of the following triacylglycerols to that in Problem 19.8?

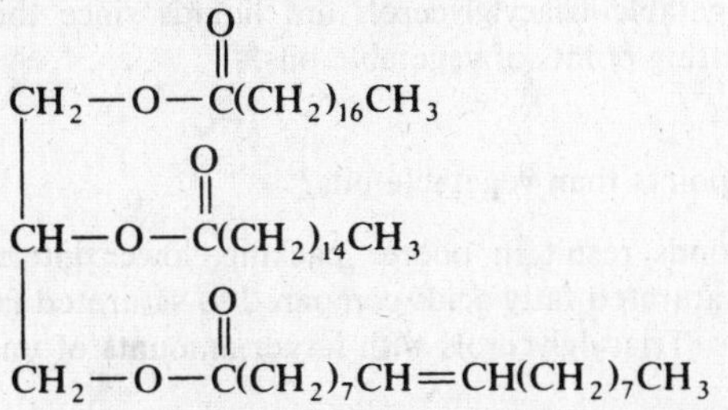

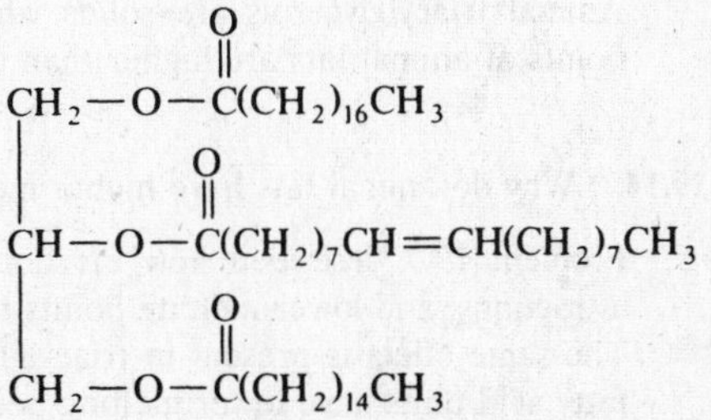

$$\begin{array}{l} CH_2-O-\overset{O}{\overset{\|}{C}}(CH_2)_{16}CH_3 \\ | \\ CH-O-\overset{O}{\overset{\|}{C}}(CH_2)_{14}CH_3 \\ | \\ CH_2-O-\overset{O}{\overset{\|}{C}}(CH_2)_7CH=CH(CH_2)_7CH_3 \end{array} \qquad \begin{array}{l} CH_2-O-\overset{O}{\overset{\|}{C}}(CH_2)_{16}CH_3 \\ | \\ CH-O-\overset{O}{\overset{\|}{C}}(CH_2)_7CH=CH(CH_2)_7CH_3 \\ | \\ CH_2-O-\overset{O}{\overset{\|}{C}}(CH_2)_{14}CH_3 \end{array}$$

Ans. The three triacylglycerols are constitutional isomers. A mixture of the three triacylglycerols is formed by the reaction of 1 mol of glycerol with 1 mol each of palmitic, stearic, and oleic acids. Many textbooks gloss over this point and show only one of the isomers. Now that we understand the situation, we will do the same unless the problem specifically asks otherwise. Any one of the three constitutional isomers is acceptable as the answer to Problem 19.8.

Problem 19.10. Explain why the answers to the Problems 19.8 and 19.9 are still oversimplifications.

Ans. The product is a mixture of different complex triacylglycerol molecules. Some triacylglycerol molecules contain one each of palmitic, stearic, and oleic units, i.e., the three constitutional isomers shown in Problems 19.8 and 19.9. But there are other triacylglycerol molecules present. Some triacylglycerol molecules contain two stearic units and one palmitic unit, some contain one stearic and two palmitics, some contain one stearic and two oleics, some contain two stearics and one oleic, some contain one palmitic and two oleics, and some contain two palmitics and one oleic. Any of the triacylglycerol structures in Problems 19.8 and 19.9 give the average structure in the triacylglycerol mixture. Most textbooks gloss over this situation and give the average composition without indicating that it is an average composition. Now that we understand the situation, we will do the same unless the problem specifically asks otherwise.

Problem 19.11. Chemical analysis of a typical animal fat or vegetable oil shows the presence of five or more fatty acids. For example, corn oil contains 55 percent linoleic acid, 31 percent oleic acid, 10 percent palmitic acid, 3 percent stearic acid, and 1 percent myristic acid. Explain the meaning of this composition in view of the fact that any triacylglycerol molecule contains only three fatty acid units.

Ans. The composition is an average composition. Corn oil like all fats and oils is a mixture of many different complex triacylglycerols. There are some triacylglycerol molecules that contain two linoleic units and one oleic unit, some that contain one each of linoleic, oleic, and palmitic, some that contain one each of linoleic, oleic, and stearic, and so on.

The most abundant fatty acid units, both saturated and unsaturated, in fats and oils are the C_{16} and C_{18} acids. Vegetable oils differ from animal fats in the relative amounts of saturated and unsaturated fatty acid units. Animal fats generally contain less than 50 to 60 percent unsaturated fatty acid units. Vegetable oils generally contain more than 80 percent unsaturated fatty acid units.

Problem 19.12. Triacylglycerol A contains 48 percent oleic acid, 36 percent linoleic acid, 10 percent palmitcic acid, 3 percent stearic acid, 1 percent linolenic, and lesser amounts of other fatty acids. Triacylglycerol B contains 50 percent oleic acid, 26 percent palmitic acid, 15 percent stearic acid, 6 percent linoleic, and lesser amounts of other fatty acids. One of the samples is corn oil and the other is pork fat (lard). Which is which?

Ans. Triacylglycerol A is corn oil since it contains 85 percent unsaturated fatty acids (oleic + linoleic + linolenic). Triacylglycerol B is pork fat since it contains only 56 percent unsaturated fatty acids (oleic + linoleic).

Problem 19.13. Why are animal and vegetable triacylglycerols referred to as *fats* and *oils*, respectively?

Ans. The terms *fat* and *oil* convey the physical state of the triacylglycerol at ambient temperatures. Animal triacylglycerols are solids while vegetable triacylglycerols are liquids since the melting points of animal fats are higher than the melting points of vegetable oils.

Problem 19.14. Why do animal fats have higher melting points than vegetable oils?

Ans. Problem 19.7 described how *cis* double bonds result in poorer packing, lower intermolecular attractions, and lower melting points for unsaturated fatty acids compared to saturated fatty acids. The same effect is present in triacylglycerols. Triacylglycerols with larger amounts of unsaturated fatty acid units have lower melting points.

Problem 19.15. There are some notable exceptions to the generalization that vegetable triacylglycerols are liquids while animal triacylglycerols are solid. Coconut and palm oils are solids while fish oils are liquids. Explain.

Ans. Coconut and palm oils contain smaller amounts of unsaturated fatty acid units than other vegetable oils. Fish oils contain larger amounts of unsaturated fatty acid units than other animal fats. Fish oils approximate the fatty acid content of vegetable oils while coconut and palm oils approximate the fatty acid content of animal fats.

19.3.2 Chemical Reactions

The chemical reactions of triacylglycerols are those expected of molecules containing ester and C=C functional groups. The ester group undergoes hydrolysis while the C=C undergoes addition reactions. Hydrolysis of a triacylglycerol is the reverse of triacylglycerol formation. It yields glycerol and fatty acids. The reaction requires an acid or base in the laboratory. Biological hydrolysis of a triacylglycerol occurs during digestion and requires the appropriate enzymes, referred to as *lipases*.

Problem 19.16. Write the equation for the acid-catalyzed hydrolysis of the triacylglycerol containing equimolar amounts of palmitic, stearic, and oleic acids.

Ans.

$$\begin{array}{lcll}
CH_2-O-\overset{O}{\overset{\|}{C}}(CH_2)_{14}CH_3 & & CH_2-OH & HO-\overset{O}{\overset{\|}{C}}(CH_2)_{14}CH_3 \\
| & & | & \\
CH-O-\overset{O}{\overset{\|}{C}}(CH_2)_{16}CH_3 & \xrightarrow[H^+]{H_2O} & CH-OH & +\ HO-\overset{O}{\overset{\|}{C}}(CH_2)_{16}CH_3 \\
| & & | & \\
CH_2-O-\overset{O}{\overset{\|}{C}}(CH_2)_7CH=CH(CH_2)_7CH_3 & & CH_2-OH & HO-\overset{O}{\overset{\|}{C}}(CH_2)_7CH=CH(CH_2)_7CH_3
\end{array}$$

Problem 19.17. Write the equation for the saponification of the triacylglycerol in Problem 19.16 by NaOH.

Ans. The answer is essentially the same as in Problem 19.16 with one difference. The fatty acids are not obtained as the fatty acids but in the form of their sodium salts:

$$\begin{array}{l} CH_2-O-\overset{O}{\overset{\|}{C}}(CH_2)_{14}CH_3 \\ | \\ CH-O-\overset{O}{\overset{\|}{C}}(CH_2)_{16}CH_3 \\ | \\ CH_2-O-\overset{O}{\overset{\|}{C}}(CH_2)_7CH{=}CH(CH_2)_7CH_3 \end{array} \xrightarrow[NaOH]{H_2O} \begin{array}{l} CH_2-OH \\ | \\ CH-OH \\ | \\ CH_2-OH \end{array} + \begin{array}{l} NaO-\overset{O}{\overset{\|}{C}}(CH_2)_{14}CH_3 \\ \\ NaO-\overset{O}{\overset{\|}{C}}(CH_2)_{16}CH_3 \\ \\ NaO-\overset{O}{\overset{\|}{C}}(CH_2)_7CH{=}CH(CH_2)_7CH_3 \end{array}$$

Hydrolysis by base is referred to as *saponification* because this is the reaction for manufacturing soaps. The mixture of the sodium (or potassium) salts of fatty acids is a soap. Section 15.5, including Problems 15.19 through 15.21 and 15.43, should be reviewed for the chemistry of soaps.

The C = C double bonds present in triacylglycerols undergo addition reactions with H_2 and halogen (Cl_2, Br_2, I_2). Hydrogen addition, referred to as *hydrogenation*, requires the presence of a metallic catalyst such as palladium or platinum. Partial hydrogenation of vegetable oils such as corn oil is used to produce margarine, but results in converting a portion of the cis double bonds to trans double bonds--which have negative cardiovascular consequences.

Problem 19.18. Write the equation for hydrogenation of the triacylglycerol that contains one each of palmitic, oleic, and linoleic acid units.

Ans.

$$\begin{array}{l} CH_2-O-\overset{O}{\overset{\|}{C}}(CH_2)_{14}CH_3 \\ | \\ CH-O-\overset{O}{\overset{\|}{C}}-(CH_2)_6(CH_2CH{=}CH)_2(CH_2)_4CH_3 \\ | \\ CH_2-O-\overset{O}{\overset{\|}{C}}(CH_2)_7CH{=}CH(CH_2)_7CH_3 \end{array} \xrightarrow[Pt]{H_2} \begin{array}{l} CH_2-O-\overset{O}{\overset{\|}{C}}(CH_2)_{14}CH_3 \\ | \\ CH-O-\overset{O}{\overset{\|}{C}}(CH_2)_{16}CH_3 \\ | \\ CH_2-O-\overset{O}{\overset{\|}{C}}(CH_2)_{16}CH_3 \end{array}$$

Problem 19.19. Write the equation for addition of iodine to the triacylglycerol that contains one each of palmitic, oleic, and linoleic acid units.

Ans.

$$\begin{array}{l} CH_2-O-\overset{O}{\overset{\|}{C}}(CH_2)_{14}CH_3 \\ | \\ CH-O-\overset{O}{\overset{\|}{C}}(CH_2)_6(CH_2CH{=}CH)_2(CH_2)_4CH_3 \\ | \\ CH_2-O-\overset{O}{\overset{\|}{C}}(CH_2)_7CH{=}CH(CH_2)_7CH_3 \end{array} \xrightarrow{I_2} \begin{array}{l} CH_2-O-\overset{O}{\overset{\|}{C}}(CH_2)_{14}CH_3 \\ | \\ CH-O-\overset{O}{\overset{\|}{C}}(CH_2)_6(CH_2\overset{I}{C}H\overset{I}{C}H)_2(CH_2)_4CH_3 \\ | \\ CH_2-O-\overset{O}{\overset{\|}{C}}(CH_2)_7\overset{I}{C}H\overset{I}{C}H(CH_2)_7CH_3 \end{array}$$

Fats and oils become *rancid* on standing, i.e., they develop disagreeable odors and tastes. Two reactions are involved in this process—hydrolysis and oxidation. Hydrolysis is important as a source of rancid odors only for certain triacylglycerols that contain C_4 and C_6 carboxylic acids. Butter is an example—its triacylglycerols contain about 3 to 4 percent butanoic acid and about 1 to 2 percent hexanoic acid. Hydrolysis due to bacteria from the air releases these volatile and malodorous acids. Air oxidation of the double bonds of unsaturated fatty acids released by bacterial hydrolysis also contributes to rancidity. Oxidation cleaves the double bond, with each carbon of the double bond being converted to a COOH group.

Problem 19.20. Write the equation for air oxidation of palmitoleic acid.

Ans.

$$CH_3(CH_2)_5CH=CH(CH_2)_7COOH \xrightarrow{(O)} CH_3(CH_2)_5COOH + HOOC(CH_2)_7COOH$$

19.4 WAXES

Waxes form natural coatings on fruits, leaves, furs, feathers and skin. These coatings have protective functions against physical damage as well as water-repellency functions (waterproofing of birds, control of water loss by transpiration in plants). Waxes derived from natural sources are used commercially in polishes, cosmetics, and ointments. Most waxes are mixtures of esters although some are hydrocarbons, alcohols, or ketones. The ester waxes are esters of a monohydric alcohol (an alcohol containing one OH) and a fatty acid. The alcohols and fatty acids are those of 16 to 36 carbons with unbranched structures and an even number of carbons.

Problem 19.21. Which of the following are found in waxes?

$$\underset{\text{I}}{CH_3(CH_2)_{27}COO(CH_2)_{21}CH_3} \qquad \underset{\text{II}}{CH_3(CH_2)_{26}COO(CH_2)_{22}CH_3} \qquad \underset{\text{III}}{CH_3(CH_2)_{26}COO(CH_2)_{21}CH_3}$$

Ans. Compounds I and II are not waxes. The acyl (R — CO —) portion of compound I contains an odd number of carbons. The alcohol (— OR) portion of compound II contains an odd number of carbons. Compound III is a wax since both the acyl and alcohol portions contain an even number of carbons (28 and 22 carbons, respectively).

19.5 PHOSPHOLIPIDS

Phospholipids contain a phosphate ester functional group and ionic charges. *Phosphoglycerides* are phospholipids based on glycerol. Two of the three OH groups of glycerol are esterified as in the triacylglycerols but the third OH is a phosphate diester. A phosphoglyceride forms by reaction of glycerol with two fatty acids and one phosphoric acid to form *phosphatidic acid*. The latter is then further esterified with a small amino alcohol such as ethanolamine, $HOCH_2CH_2NH_2$. The reactions are dehydrations between OH groups—between glycerol OH and fatty acid OH, glycerol OH and phosphoric acid OH, phosphoric acid OH and amino alcohol OH:

$$
\begin{array}{lll}
CH_2\text{—}OH & HO\text{—}\overset{O}{\overset{\|}{C}}\text{—}R^1 & \\
| & & \\
CH\text{—}OH & +\ HO\text{—}\overset{O}{\overset{\|}{C}}\text{—}R^2 & \xrightarrow{-3H_2O} \\
| & & \\
CH_2\text{—}OH & HO\text{—}\overset{O}{\overset{\|}{P}}\text{—}OH & \\
 & \quad\ \ |\ OH &
\end{array}
\begin{array}{l}
CH_2\text{—}O\text{—}\overset{O}{\overset{\|}{C}}\text{—}R^1 \\
| \\
CH\text{—}O\text{—}\overset{O}{\overset{\|}{C}}\text{—}R^2 \\
| \\
CH_2\text{—}O\text{—}\overset{O}{\overset{\|}{P}}\text{—}OH \\
\qquad\quad\ \ |\ OH \\
\text{Phosphatidic acid}
\end{array}
\xrightarrow[-H_2O]{HOCH_2CH_2NH_2}
\begin{array}{l}
CH_2\text{—}O\text{—}\overset{O}{\overset{\|}{C}}\text{—}R^1 \\
| \\
CH\text{—}O\text{—}\overset{O}{\overset{\|}{C}}\text{—}R^2 \\
| \\
CH_2\text{—}O\text{—}\overset{O}{\overset{\|}{P}}\text{—}OCH_2CH_2NH_2 \\
\qquad\quad\ \ |\ OH \\
\text{Phosphoglyceride}
\end{array}
$$

Problem 19.22. The structure of a phosphoglyceride is usually shown as compound II below instead of compound I. Why? What is the difference?

$$
\begin{array}{l}
CH_2\text{—}O\text{—}\overset{O}{\overset{\|}{C}}\text{—}R^1 \\
| \\
CH\text{—}O\text{—}\overset{O}{\overset{\|}{C}}\text{—}R^2 \\
| \\
CH_2\text{—}O\text{—}\overset{O}{\overset{\|}{P}}\text{—}OCH_2CH_2NH_2 \\
\qquad\quad\ \ |\ OH \\
\qquad\quad \text{I}
\end{array}
\qquad\qquad
\begin{array}{l}
CH_2\text{—}O\text{—}\overset{O}{\overset{\|}{C}}\text{—}R^1 \\
| \\
CH\text{—}O\text{—}\overset{O}{\overset{\|}{C}}\text{—}R^2 \\
| \\
CH_2\text{—}O\text{—}\overset{O}{\overset{\|}{P}}\text{—}OCH_2CH_2NH_3^+ \\
\qquad\quad\ \ |\ O^- \\
\qquad\quad \text{II}
\end{array}
$$

Ans. Phosphoglycerides actually exist as compound II, not compound I. Compound II differs from compound I in having an H^+ moved from the OH of the phosphate group to the nitrogen of the amino group, resulting in negative and positive charges on O and N, respectively. The pH of a living system (which is buffered) is such that both the phosphate OH and amine nitrogen are in ionic form. (The presence of ionic charges in the phospholipid is very important when we discuss cell membranes—Sec. 19.8).

Problem 19.23. Write the structure of the phosphoglyceride formed from 1 mol each of glycerol, palmitic and stearic acids, phosphoric acid, and choline, $HOCH_2CH_2\overset{+}{N}(CH_3)_3$.

Ans.

$$
\begin{array}{ll}
CH_2\text{—}OH & HO\text{—}\overset{O}{\overset{\|}{C}}(CH_2)_{14}CH_3 \\
| & \\
CH\text{—}OH & HO\text{—}\overset{O}{\overset{\|}{C}}(CH_2)_{16}CH_3 \\
| & \\
CH_2\text{—}OH & HO\text{—}\overset{O}{\overset{\|}{P}}\text{—}OH\quad HOCH_2CH_2\overset{+}{N}(CH_3)_3 \\
 & \qquad\ |\ O^-
\end{array}
\xrightarrow{-4H_2O}
\begin{array}{l}
CH_2\text{—}O\text{—}\overset{O}{\overset{\|}{C}}(CH_2)_{14}CH_3 \\
| \\
CH\text{—}O\text{—}\overset{O}{\overset{\|}{C}}(CH_2)_{16}CH_3 \\
| \\
CH_2\text{—}O\text{—}\overset{O}{\overset{\|}{P}}\text{—}OCH_2CH_2\overset{+}{N}(CH_3)_3 \\
\qquad\quad\ \ |\ O^-
\end{array}
$$

Phosphoglycerides are often named by placing *phosphatidyl-* before the name of the amino alcohol as in phosphatidylethanolamine and phosphatidylcholine, although the common names cephalin and lecithin, respectively, are more often used.

Plasmalogens are similar to phosphoglycerides except that the fatty ester unit at C-1 is replaced by a fatty vinyl ether unit (— O — CH = CH — R).

Problem 19.24. Draw the plasmalogen which contains a stearic acid unit, choline, and a C_{18} vinyl ether unit.

Ans.

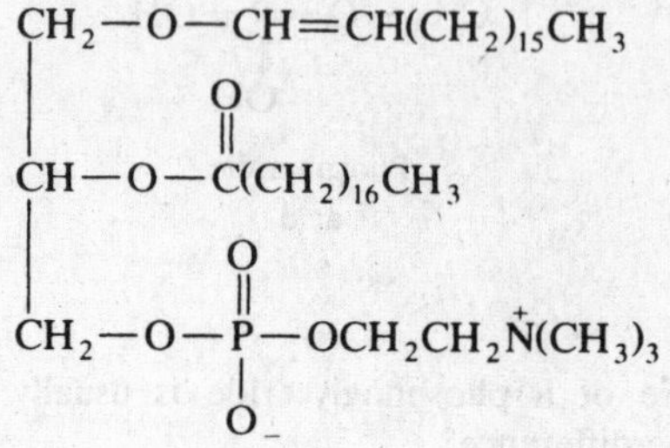

$$
\begin{array}{l}
CH_2-O-CH{=}CH(CH_2)_{15}CH_3 \\
|\qquad\qquad\quad O \\
|\qquad\qquad\quad \| \\
CH-O-C(CH_2)_{16}CH_3 \\
|\qquad\qquad\quad O \\
|\qquad\qquad\quad \| \\
CH_2-O-P-OCH_2CH_2\overset{+}{N}(CH_3)_3 \\
\qquad\qquad\quad | \\
\qquad\qquad\quad O_-
\end{array}
$$

19.6 SPHINGOLIPIDS

Sphingolipids are based on sphingosine instead of glycerol:

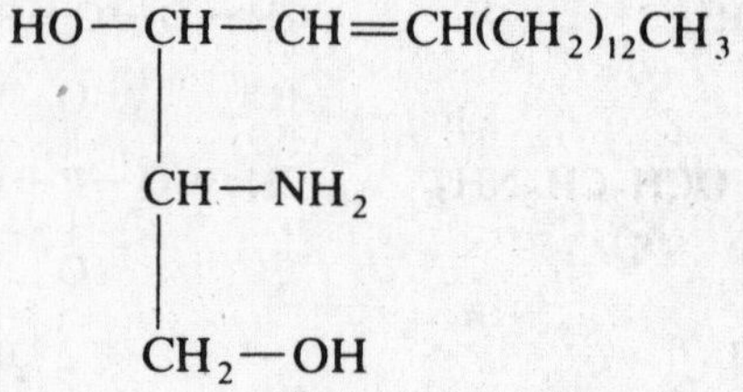

$$
\begin{array}{l}
HO-CH-CH{=}CH(CH_2)_{12}CH_3 \\
\qquad\;| \\
\qquad CH-NH_2 \\
\qquad\;| \\
\qquad CH_2-OH
\end{array}
$$

Sphingosine

Other structural units are incorporated into sphingosine through reactions at the NH_2 and OH groups to form *sphingomyelins* or *cerebrosides*. Both sphingomyelins and cerebrosides incorporate a fatty acid unit by formation of an amide by reaction with the NH_2 group. Sphingomyelins have a phosphate diester group with an amino alcohol incorporated as in phosphoglycerides and plasmalogens. (Sphingomyelins are categorized as phospholipids, as well as sphingolipids, since they contain the phosphate diester group present in other phospholipids.) Cerebrosides, also referred to as *glycolipids*, contain a saccharide unit attached by acetal formation at the OH of sphingosine.

Problem 19.25. Show the formation of the sphingomyelin containing oleic acid, phosphoric acid, and choline.

Ans.

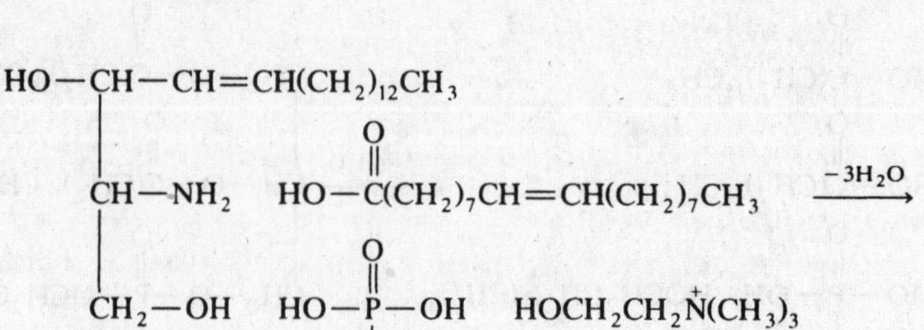

$$
\begin{array}{llll}
HO-CH-CH{=}CH(CH_2)_{12}CH_3 & & & \\
\qquad\;| & \qquad O & & \\
\qquad\;| & \qquad \| & & \\
\qquad CH-NH_2 & HO-C(CH_2)_7CH{=}CH(CH_2)_7CH_3 & & \xrightarrow{-3H_2O} \\
\qquad\;| & \qquad O & & \\
\qquad\;| & \qquad \| & & \\
\qquad CH_2-OH & HO-P-OH & HOCH_2CH_2\overset{+}{N}(CH_3)_3 & \\
 & \qquad | & & \\
 & \qquad O_- & &
\end{array}
$$

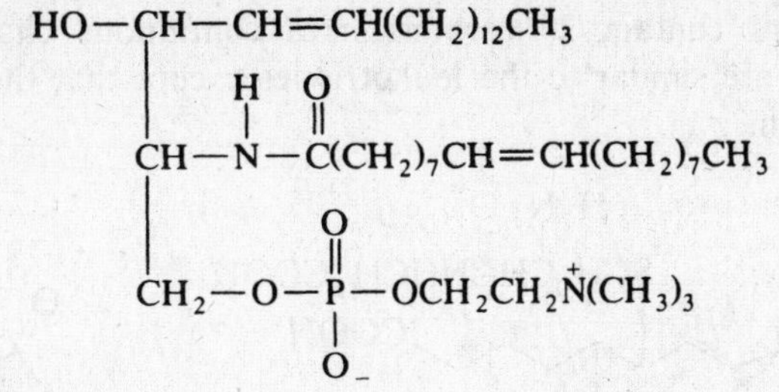

$HO-CH-CH=CH(CH_2)_{12}CH_3$

$CH-N(H)-C(=O)(CH_2)_7CH=CH(CH_2)_7CH_3$

$CH_2-O-P(=O)(O^-)-OCH_2CH_2\overset{+}{N}(CH_3)_3$

Problem 19.26. Show the formation of the cerebroside containing oleic acid and β-D-glucose.

Ans.

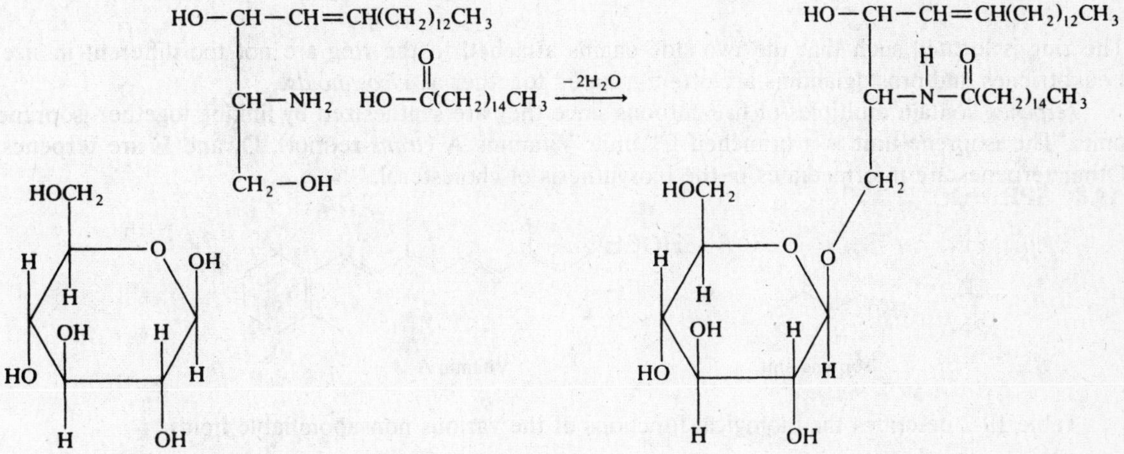

$HO-CH-CH=CH(CH_2)_{12}CH_3$

$CH-NH_2 \quad HO-C(=O)(CH_2)_{14}CH_3 \xrightarrow{-2H_2O}$

CH_2-OH

β-D-glucose ring: $HOCH_2$, O, OH, H, H, OH, H, HO, H, H, OH

$HO-CH-CH=CH(CH_2)_{12}CH_3$

$CH-N(H)-C(=O)(CH_2)_{14}CH_3$

CH_2 — O (glycosidic link to glucose ring: $HOCH_2$, O, H, H, OH, H, HO, H, H, OH)

19.7 NONHYDROLYZABLE LIPIDS

Nonhydrolyzable lipids are not cleaved into smaller molecules by hydrolysis (including acid, saponification, and digestion) because of the absence of ester groups. Three classes of compounds comprise the nonsaponifiable lipids, as shown in the bottom of Fig. 19-1. *Steroids* contain a four-ring system of three 6-membered rings and one 5-membered ring:

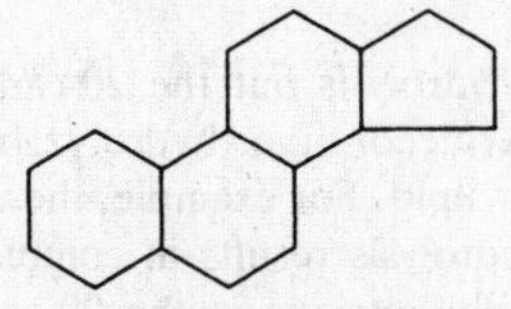

CH_3

$CH(CH_2)_3CH(CH_3)_2$

H_3C

H_3C

HO

Steroid ring system

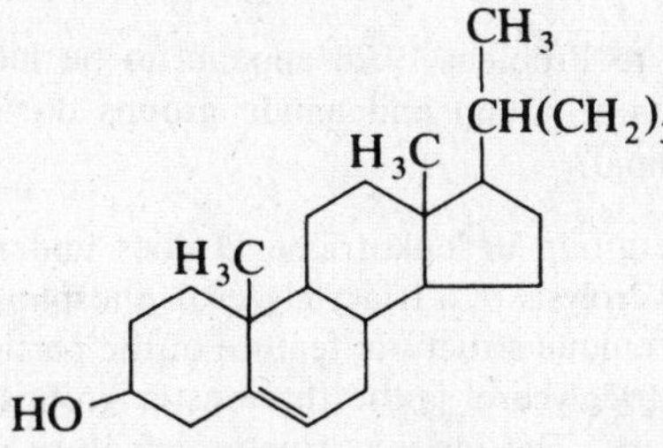

Cholesterol

Cholesterol, the most abundant steroid, is an important constituent of cell membranes and the starting material for biosynthesis of other steroids. This includes the *sex hormones*, such as progesterone, estradiol, testosterone, and androsterone, *adrenocortical hormones* such as aldosterone and cortisone, *bile salts*, and vitamin D.

Leukotrienes contain 20 carbons in a continuous chain with a COOH group at one end. *Prostaglandins* are similar to the leukotrienes except that there is a 5-membered ring as part of the 20-carbon chain.

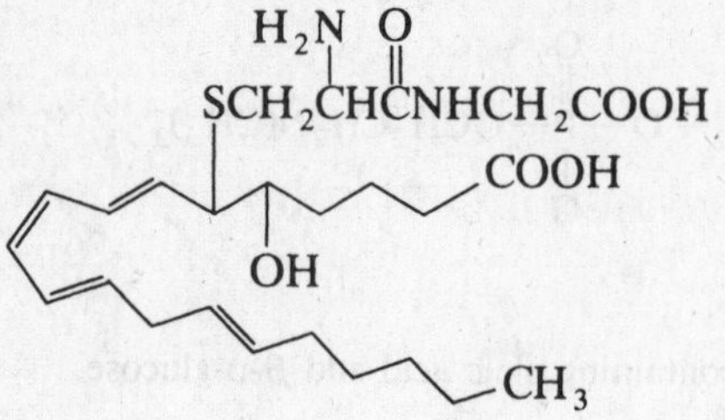

H_2N O
$SCH_2CHCNHCH_2COOH$
COOH
OH
CH_3
O
COOH
HO
CH_3
OH

Leukotriene D

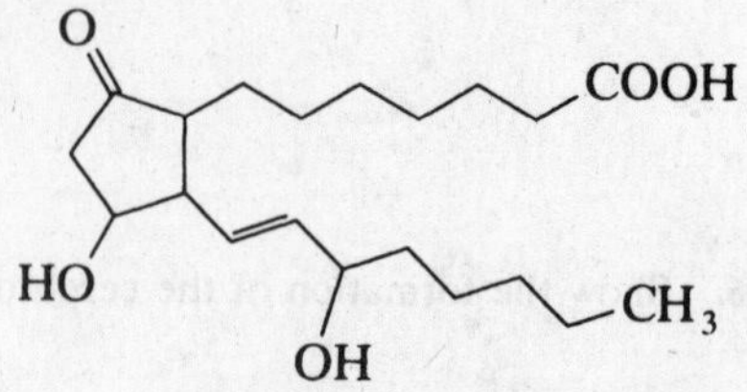

Prostaglandin E_1

The ring is located such that the two side chains attached to the ring are not too different in size. Leukotrienes and prostaglandins are often grouped together as *eicosanoids*.

Terpenes contain multiples of five carbons since they are synthesized by linking together isoprene units. The isoprene unit is a branched C_5 unit. Vitamins A (*trans*-retinol), D, and E are terpenes. Other terpenes are intermediates in the biosynthesis of cholesterol.

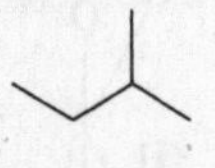

$HOCH_2$

Isoprene unit

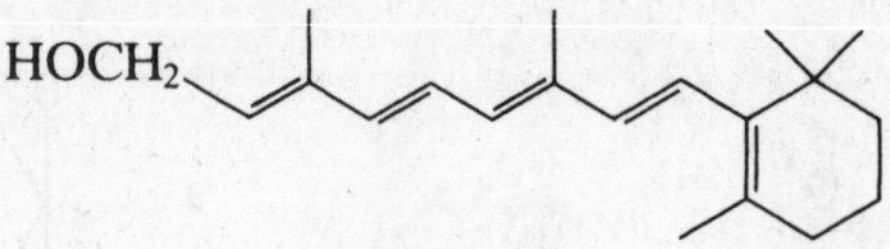

Vitamin A

Table 19-2 describes the biological functions of the various nonsaponifiable lipids.

Problem 19.27. Why are steroids, prostaglandins, leukotrienes, and terpenes classified as lipids?

Ans. They are soluble in nonpolar and low-polarity solvents and insoluble in water.

Problem 19.28. Why are steroids, prostaglandins, leukotrienes, and terpenes classified as nonhydrolyzable lipids?

Ans. They do not undergo hydrolysis.

Problem 19.29. The answer to Problem 19.28 appears to be incorrect for leukotriene D. The side group in leukotriene D contains an amide group and amide groups do undergo hydrolysis. Should leukotriene D be classified as an hydrolyzable lipid?

Ans. The amide side group in leukotriene D does undergo hydrolysis but the 20-carbon chain is not affected. The hydrolysis of a triacylglycerol, phosphoglyceride, or other hydrolyzable lipid results in a change in the unique structural feature of the particular lipid. For example, the unique structural feature of a triacylglycerol is the three ester groups. Hydrolysis results in complete loss of those three ester groups. The unique structural feature of a leukotriene is the 20-carbon continuous chain. That chain remains intact after hydrolysis of leukotriene D.

Problem 19.30. Why is vitamin A not classified as a prostaglandin?

Ans. Vitamin A has a 6-membered ring, not a 5-membered ring, and the side chains are of very unequal size.

Table 19-2. Biological Functions of Nonhydrolyzable Lipids

Compound	Function
Steroids	
Cholesterol	Component of cell membranes
Androgens*	Development of male reproductive organs; maintenance of secondary sex characteristics
Synthetic androgens†	Muscle development for athletes
Progesterone, Estrogens‡	Development of female reproductive organs; maintenance of secondary sex characteristics; control of menstrual cycle
Synthetic estrogens	Oral contraceptives
Aldosterone	Controls water and electrolyte balances
Cortisone	Controls metabolism of proteins, carbohydrates, lipids; controls water and electrolyte balances; controls inflammation
Bile salts	Facilitate digestion of saponifiable lipids; facilitate absorption of fat-soluble vitamins (A, D, E, K) in intestinal tract
Vitamin D	Controls calcium absorption in intestinal tract and deposition in bone
Prostaglandins	Lower or raise blood pressure; involved in blood clotting; control gastric secretions; cause inflammation; induce labor
Leukotrienes	Trigger responses to inflammation, allergy, asthma
Terpenes	
Vitamin A	Facilitates vision in dim light (night vision)
Vitamin E	Antioxidant; maintains cell membrane integrity by preventing oxidation of unsaturated fatty acid units
Vitamin K	Essential for blood clotting

*Androsterone, testosterone.
†Referred to as anabolic steroids.
‡Estrone, estradiol.

Problem 19.31. Why is vitamin A classified as a terpene?

Ans. Vitamin A does not fit into any of the other classes and one might classify it as a terpene simply by default. However, the key to its classification as a terpene is that we can examine its structure and note that it is made of isoprene units linked to each other. The dotted lines in the structure below show where isoprene units have been linked together:

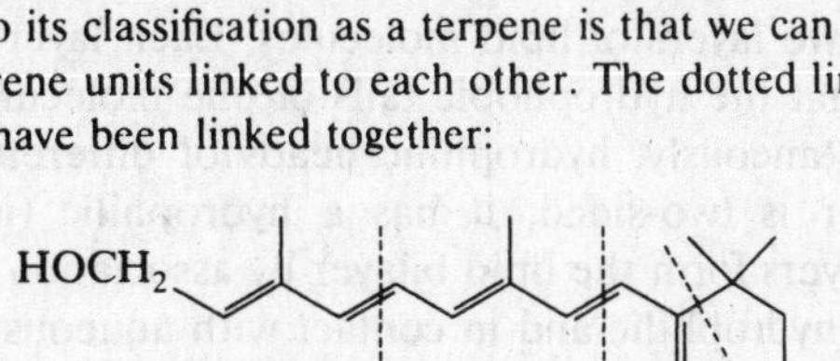

$HOCH_2$

19.8 CELL MEMBRANES

Cell membranes provide the mechanical barrier that separates cells from their environment. Their ability to control the entry of various materials into cells and the exit of other materials from cells are critical to the functioning of cells. Metabolic fuels to generate energy and building blocks for synthesizing proteins, nucleic acids, and other compounds are extracted from the environment and concentrated in cells. Waste materials from cell activities are expelled into the environment. Enzymes which are structural components of membranes maintain the intracellular pH and ionic composition within narrow limits to regulate intracellular enzyme activity. They also generate appropriate ionic concentration gradients across cell membranes, essential for nerve and muscle action. (Similar membranes separate intracellular organelles such as the nucleus and mitachondria from the cytoplasm.)

Cell membranes are constructed principally of phospholipids, sphingolipids, and proteins, together with small amounts of cholesterol. The common feature of the membrane lipids is their similarity to soap and detergent molecules—the presence of both hydrophilic and hydrophobic portions in the same molecule. Such molecules are referred to as *amphipathic molecules.*

Problem 19.32. Show the hydrophobic and hydrophilic portions of the phosphoglyceride from Problem 19.23 (contains one each of palmitic and stearic acids and choline).

Ans. The palmitate and stearate units constitute the hydrophobic portions while the phosphate O^- and choline N^+ centers constitute the hydrophilic portions. The hydrophobic-hydrophilic aspects of phosphoglyceride molecules are best represented by redrawing the molecule as in structure I below. The hydrophobic and hydrophilic portions repell each other and are directed away from each other. Structure II below is a schematic representation of such molecules. The spherical *head* and long wiggly *tails* represent the hydrophilic and hydrophobic portions, respectively.

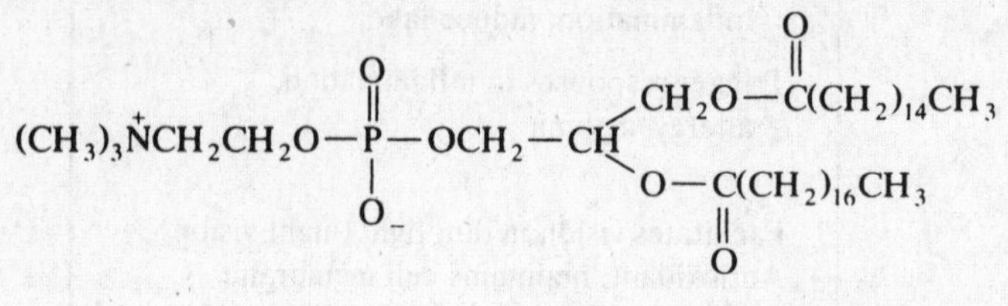

$$(CH_3)_3\overset{+}{N}CH_2CH_2O-\underset{\underset{O}{|}}{\overset{\overset{O}{\|}}{P}}-OCH_2-CH\begin{cases}CH_2O-\overset{\overset{O}{\|}}{C}(CH_2)_{14}CH_3\\ O-\underset{\underset{O}{\|}}{C}(CH_2)_{16}CH_3\end{cases}$$

I

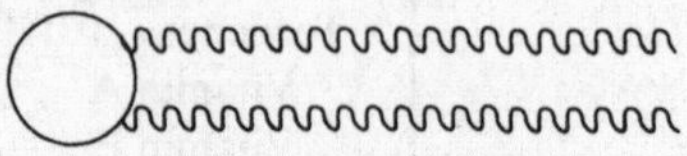

II

Problem 19.33. Ionic head portions are not present in all membrane lipids. The other phospholipids, the plasmalogens and sphingomyelins, are like the phosphoglycerides in having an ionic head and two nonpolar tails, but the glycolipids and cholesterol do not possess ionic heads. What is the structural feature common to all membrane lipids?

Ans. The head portions are hydrophilic while the tail portions are hydrophobic. The head portion need not be ionic to be hydrophilic. Glycolipids have a head portion, the saccharide unit, which is hydrophilic because it has several OH groups. The hydrophilic head of cholesterol is its OH group.

Cell membranes result from the association of membrane lipid molecules to form a *lipid bilayer* (Fig. 19-3). There are two layers of lipid molecules. Each layer consists of lipid molecules associated with each other such that the hydrophobic tails of one molecule are adjacent to hydrophobic tails of other molecules. Simultaneously, hydrophilic heads of different lipid molecules are aligned next to each other. Each layer is two-sided, it has a hydrophilic (ionic/polar) side and a hydrophobic (nonpolar) side. Two layers form the lipid bilayer by association of their hydrophobic sides. Both sides of the lipid bilayer are hydrophilic and in contact with aqueous solutions. The lipid bilayer is the cell membrane, surrounding the aqueous cell contents and isolating them from the aqueous environment outside the cell.

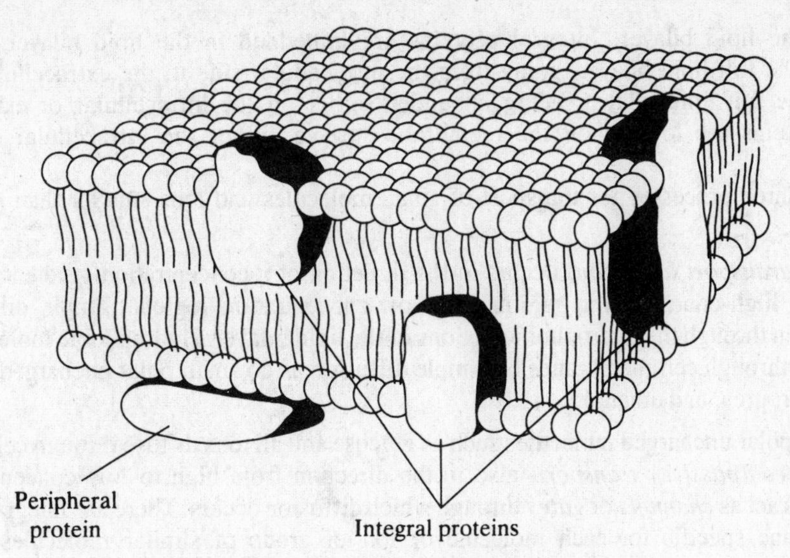

Fig. 19-3 Fluid mosaic model of cell membrane.

Problem 19.34. Be more specific in describing what holds the lipid molecules together in each layer of the bilayer and what holds the two layers together.

Ans. Intermolecular attractive and repulsive forces. These are the same forces responsible for the "like dissolves in like" rule of solubility. The hydrophobic portions of each lipid layer are repulsed by water while the hydrophilic portions are attracted to water. The hydrophobic portions of the two lipid layers are attracted to each other.

Cell membranes have a *fluid mosaic structure*. The term *fluid* conveys the fluid (flexible) or liquid-like nature of the cell membrane. The term *mosaic* conveys the complex nature of cell membranes in terms of their structure and composition. Not only does the lipid bilayer contain different lipids (phosphoglycerides, plasmalogens, sphingomyelins, glycolipids, cholesterol), it also contains a variety of proteins essential to cell function. The physiological function of a membrane depends on its fluidity, which varies with the relative amounts of saturated fatty acids, unsaturated fatty acids, and cholesterol.

Problem 19.35. How does the fatty acid composition of a cell membrane affect its flexibility?

Ans. Flexibility increases with the increasing amount of unsaturated fatty acid units present in the lipids. The kinks in the *cis* double bonds prevent close packing of lipid molecules in the lipid bilayer. This decreases intermolecular attractions and increases flexibility.

Problem 19.36. What is the function of cholesterol in cell membranes?

Ans. The hydrophobic portion (steroid ring system) of cholesterol is more rigid than that of other membrane lipids. Cholesterol is a regulator of membrane fluidity, keeping membranes from being too fluid. However, excess cholesterol would make the membrane too rigid.

Cell membranes contain anywhere from 20 to 75 percent protein depending on the type of cell. Proteins are either *integral proteins* or *peripheral proteins* depending on whether or not they are imbedded in the lipid bilayer. Integral proteins are imbedded in the lipid bilayer. Some integral proteins span the full lipid bilayer, going from the intracellular side to the extracellular side. Others do not span the full lipid bilayer, being imbedded either on the intracellular or extracellular side. Peripheral proteins are located on the membrane surface, either the intracellular or extracellular surface.

There are three processes for transport of small molecules and ions smaller than polymers across cell membranes:

1. *Simple transport (diffusion)* occurs in the direction of a concentration gradient, i.e., movement from a high-concentration region to a low-concentration region. Simple diffusion involves diffusion through the hydrophobic regions of the lipid bilayer. Hydrophobic molecules such as O_2 diffuse through cell membranes by simple diffusion as do small polar uncharged molecules such as water, urea, and ethanol.
2. Larger polar uncharged molecules such as glucose and all ions diffuse through cell membranes by *facilitated (passive) transport*, also in the direction from high to low concentration. Integral proteins act as *channels* or *gates* through which diffusion occurs. There are integral proteins in the membrane specific for each molecule or ion or group of similar molecules or ions to be transported.
3. The normal functioning of cells requires that the concentrations of some solutes be different inside and outside cells. This sometimes requires transport against the normal concentration gradient, referred to as *active transport*. Active transport, like facilitated diffusion, occurs through integral proteins but requires an expenditure of energy (typically ATP). Active transport systems are often referred to as *pumps*.

Polymers (e.g., proteins and polysaccharides) and large particles (e.g., lipoproteins) are too large to pass through membranes by any of the three transport processes. Such materials pass into and out of cells by specialized processes called *endocytosis* and *exocytosis*, respectively. In the cytosis processes the materials to be transported are packaged into membrane-surrounded *vesicles* that move to the cell membrane. The vesicle and cell membranes merge together. This results in a temporary opening in the cell membranes through which the polymers and large particles move into or out of the cell.

Problem 19.37. For red blood cells, the extracellular concentration of Na^+ is about 140 millimolar (mM) while the intracellular concentration is about 10 mM. Active transport is needed for the proper functioning of red blood cells. In which direction does active transport of Na^+ take place?

Ans. Active transport of Na^+ occurs from inside red blood cells to outside (plasma), i.e., transport from 10 mM concentration to 140 mM concentration.

19.9 LIPIDS AND HEALTH

Problem 19.38. What is atherosclerosis (cardiovascular disease)?

Ans. Cholesterol separates from blood and deposits on the inner walls of arteries to form *plaque*. This narrowing of arteries results in the heart working harder to pump blood, blood pressure increasing, and the heart enlarging over time. Plaque formation also converts the smooth inner walls of arteries to rough surfaces and this promotes *thrombosis*, the formation of blood clots which block arteries. A *heart attack* involves the blockage of arteries feeding the heart and results in the death of heart tissue. A *stroke* involves the blockage and usually also the rupture of arteries in the brain and results in damage to brain tissue.

There are statistical links between our dietary lipids and cardiovascular disease. Medical practioners generally make the following dietery recommendations to decrease the incidence of cardiovascular disease: (1) Eat less total meat, especially red meat. (2) Eat more vegetables. (3) Eat more fish.

Problem 19.39. What do these recommendations mean in terms of the chemical composition of our food intake?

Ans. Decrease the amount of lipid compared to protein and carbohydrate. Decrease the amount of lipid that contains saturated fatty acids and cholesterol. Meat contains much more lipid and a much greater saturated fatty acid content than vegetables. Meat also has a high cholesterol content while vegetables contain no cholesterol. Red meat tends to contain much more lipid than poultry (if one avoids the skin) or fish (if one avoids certain fatty fish). Fish contains a higher unsaturated fatty acid content than red meat or poultry.

Problem 19.40. What is the health benefit of decreasing the amount of lipid in the diet?

Ans. More than twice as much energy is generated per gram of lipid metabolized compared to protein or carbohydrate. A high lipid intake results in excess (unmetabolized) lipid which is then stored in fat (*adipose*) tissue. Weight increase results in a higher load on the heart.

Problem 19.41. What is the health benefit of decreasing the amount of cholesterol in the diet?

Ans. The major fraction of the cholesterol in the blood is synthesized in the liver. Decreasing dietary cholesterol has some effect in decreasing the cholesterol level in the blood. The effect is measurable but not huge unless the person's normal diet contains excessive animal fat. However, dietary control is prudent for most people since any decrease in blood cholesterol level decreases the tendency for plaque formation.

Plaque deposition is a very complicated process and not well understood. Cholesterol and triacylglycerols are insoluble in blood. They are carried in the blood stream by complexation with phospholipids and special proteins to form *lipoprotein* complexes. (The mechanism for solubilization of cholesterol and triacylglycerols is the same as the cleansing action of soap molecules on oily nonpolar dirt.) The lipoproteins are classified according to density—*chylomicrons*, *very-low-density lipoproteins* (*VLDL*), *low-density lipoproteins* (*LDL*), and *high-density lipoproteins* (*HDL*). Chylomicrons transport exogenous (i.e., dietary) triacylglycerols and cholesterol from the intestines to other tissues. VLDL and LDL transport endogenous (internally synthesized) triacylglycerols and cholesterol from the liver to other tissues. HDL transports cholesterol back to the liver from other tissues. Management of cholesterol results from the interplay of HDL with LDL, VLDL, and chylomicrons. Too low an HDL level or too high a level of the other lipoproteins results in higher blood cholesterol levels and an increased tendency for plaque formation. This brings us to another medical recommendation: Exercise more. Exercise, especially anaerobic exercise, increases the HDL level.

ADDITIONAL SOLVED PROBLEMS

INTRODUCTION

Problem 19.42. Which of the following are hydrolyzable lipids? (*a*) triacylglycerols, (*b*) prostaglandins, (*c*) waxes, (*d*) leukotrienes, (*e*) sphingomyelins, (*f*) steroids, (*g*) glycolipids, (*h*) cerebrosides, (*i*) terpenes, (*j*) plasmalogens, (*k*) phosphoglycerides?

Ans. (*a*), (*c*), (*e*), (*g*), (*h*), (*j*), (*k*).

FATTY ACIDS

Problem 19.43. Which of the following carboxylic acids are incorporated into lipids?

$$\underset{\text{I}}{CH_3CH_2\overset{\overset{\large CH_3}{|}}{C}H(CH_2)_9COOH} \qquad \underset{\text{II}}{CH_3(CH_2)_{18}COOH} \qquad \underset{\text{III}}{CH_3(CH_2)_4(CH{=}CHCH_2)_2(CH_2)_6COOH}$$

Ans. Compounds II and III, fatty acids with even numbers of carbon atoms, are utilized. Compound I is not utilized since it is branched.

TRIACYLGLYCEROLS

Problem 19.44. Reaction with iodine is used to quantify the degree of unsaturation of lipids that contain fatty acid units. The amount of iodine that reacts with a triacylglycerol is referred to as the *iodine number* of the lipid. The greater the unsaturated fatty acid content of a lipid, the higher the iodine number. Consider triacylglycerols A and B. Triacylglycerol A contains one each of palmitic, stearic, and oleic acids. Triacylglycerol B contains one each of palmitic, oleic, and linoleic acids. Which triacylglycerol has the higher iodine number?

Ans. Triacylglycerol B has the higher iodine number. Triacylglycerols A and B have one and three double bonds per molecule, respectively. Triacylglycerol B reacts with 3 times as much iodine as does triacylglycerol A.

Problem 19.45. Write the equation for the saponification of the triacylglycerol containing one each of palmitic, oleic, and linoleic acids.

Ans.

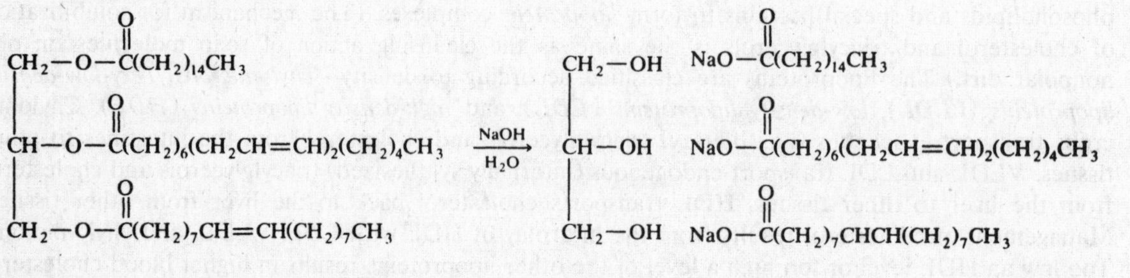

$$\begin{array}{l} CH_2-O-\overset{O}{\overset{\|}{C}}(CH_2)_{14}CH_3 \\ | \\ CH-O-\overset{O}{\overset{\|}{C}}(CH_2)_6(CH_2CH{=}CH)_2(CH_2)_4CH_3 \\ | \\ CH_2-O-\overset{O}{\overset{\|}{C}}(CH_2)_7CH{=}CH(CH_2)_7CH_3 \end{array} \xrightarrow[H_2O]{NaOH} \begin{array}{l} CH_2-OH \quad NaO-\overset{O}{\overset{\|}{C}}(CH_2)_{14}CH_3 \\ | \\ CH-OH \quad NaO-\overset{O}{\overset{\|}{C}}(CH_2)_6(CH_2CH{=}CH)_2(CH_2)_4CH_3 \\ | \\ CH_2-OH \quad NaO-\overset{O}{\overset{\|}{C}}(CH_2)_7CHCH(CH_2)_7CH_3 \end{array}$$

Problem 19.46. Medical practioners advise that we increase the proportion of our triacylglycerol intake that is unsaturated. The manufacture of margarine by partial hydrogenation of corn or other vegetable oils decreases the unsaturated content of the product. Explain.

Ans. Corn oil is not palatable to the consumer as a substitute for butter. How many of you would spread corn oil on your toast, bagel, or muffin? Hydrogenation is carried out to the extent required to convert the liquid corn oil into a solid or semisolid product acceptable to the consumer's palate. The process is oten referred to as *hardening*. Only partial hydrogenation is carried out so as to retain a substantial amount of the unsaturated content.

WAXES

Problem 19.47. Draw the structure of the wax which yields $C_{28}H_{56}O_2$ and $C_{20}H_{42}O$ upon hydrolysis.

Ans.

$$CH_3(CH_2)_{26}\overset{O}{\overset{\|}{C}}-O(CH_2)_{19}CH_3$$

PHOSPHOLIPIDS

Problem 19.48. Which of the following are phospholipids? (*a*) triacylglycerols, (*b*) prostaglandins, (*c*) waxes, (*d*) leukotrienes, (*e*) sphingomyelins, (*f*) steroids, (*g*) glycolipids, (*h*) cerebrosides, (*i*) terpenes, (*j*) plasmalogens, (*k*) phosphoglycerides.

Ans. (*e*), (*h*), (*j*), (*k*).

Problem 19.49. Show the formation of a sphingolipid from one each of sphingosine, myristic and phosphoric acids, and serine.

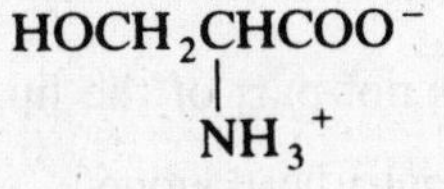

$$\underset{\text{Serine}}{HOCH_2\underset{\displaystyle NH_3^+}{\underset{|}{C}}HCOO^-}$$

Ans.

$$\begin{array}{l} HO-CH-CH=CH(CH_2)_{12}CH_3 \\ \quad\;\;| \\ \quad CH-NH_2 \quad HO-\overset{O}{\overset{\|}{C}}(CH_2)_{12}CH_3 \\ \quad\;\;| \\ \quad CH_2-OH \quad HO-\underset{O^-}{\underset{|}{\overset{O}{\overset{\|}{P}}}}-OH \quad HOCH_2\underset{NH_3^+}{\underset{|}{C}}HCOO^- \end{array} \xrightarrow{-3H_2O} \begin{array}{l} HO-CH-CH=CH(CH_2)_{12}CH_3 \\ \quad\;\;| \\ \quad CH-\overset{H}{\overset{|}{N}}-\overset{O}{\overset{\|}{C}}(CH_2)_{12}CH_3 \\ \quad\;\;| \\ \quad CH_2-O-\underset{O^-}{\underset{|}{\overset{O}{\overset{\|}{P}}}}-OCH_2\underset{NH_3^+}{\underset{|}{C}}HCOO^- \end{array}$$

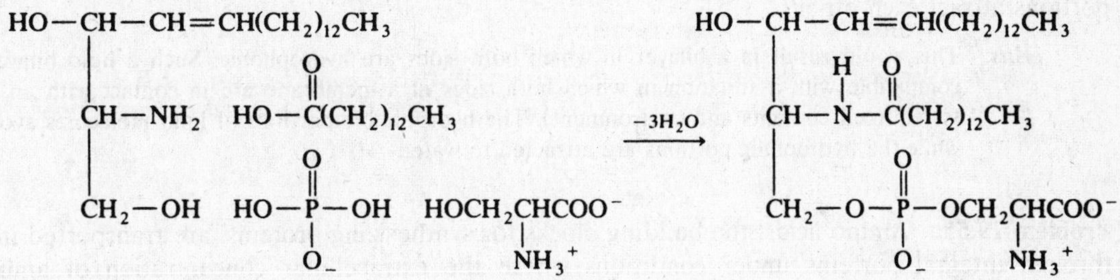

NONSAPONIFIABLE LIPIDS

Problem 19.50. Which of the following are nonhydrolyzable lipids? (*a*) triacylglycerols, (*b*) prostaglandins, (*c*) waxes, (*d*) leukotrienes, (*e*) sphingomyelins, (*f*) steroids, (*g*) glycolipids, (*h*) cerebrosides, (*i*) terpenes, (*j*) plasmalogens, (*k*) phosphoglycerides.

Ans. (*b*), (*d*), (*f*), (*i*).

Problem 19.51. Why is vitamin D_3 (shown below) classified as a steroid when it does not have four rings?

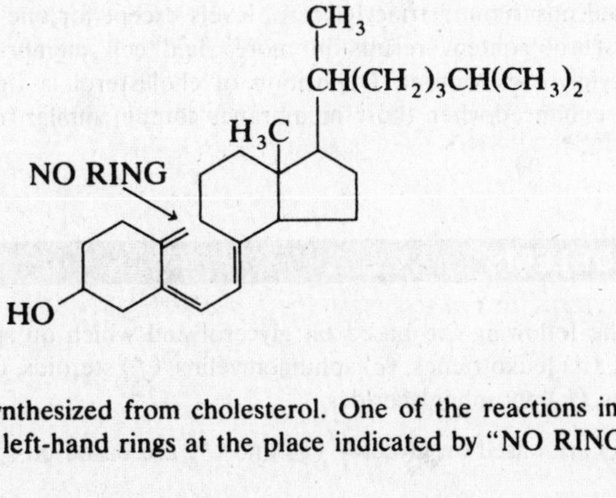

CH_3

$CH(CH_2)_3CH(CH_3)_2$

H_3C

NO RING

HO

Ans. Vitamin D_3 is synthesized from cholesterol. One of the reactions involves breakage of the bond between the two left-hand rings at the place indicated by "NO RING."

Problem 19.52. The terms *hormone* and *vitamin* have been used but not defined. What is a hormone? What is a vitamin?

Ans. *Hormones* such as the sex and adrenocortical hormones are chemical messengers that are synthesized in endocrine (ductless) glands and carried to a target organ or cells by the bloodstream. Hormones initiate a chemical response at the target and alter its activity. Prostaglandins and leukotrienes are referred to as *local hormones* because they are short-lived. Local hormones are not carried in the bloodstream. They alter the activity of the cells in which they are synthesized and/or adjoining cells. *Vitamins* are organic molecules which cannot be synthesized by an organism but are required for some biological function.

CELL MEMBRANES

Problem 19.53. Why are triacylglycerols not part of the lipid bilayer?

Ans. Triacylglycerols do not have a polar head group.

Problem 19.54. Why do the two lipid layers not form a reverse bilayer, one in which the polar portions attract each other?

Ans. This would result in a bilayer in which both sides are hydrophobic. Such a lipid bilayer is not compatible with a situation in which both sides of a membrane are in contact with an aqueous system (cell contents and environment). The hydrophobic portions of lipid molecules avoid water while the hydrophilic portions are attracted to water.

Problem 19.55. Amino acids, the building blocks for synthesizing proteins, are transported into cells through integral porteins under conditions where the extracellular concentration of amino acid exceeds the intracellular concentration. What type of transport is involved, simple transport, facilitated transport, or active transport?

Ans. Facilitated transport

LIPIDS AND HEALTH

Problem 19.56. What is the cardiovascular benefit of increasing the amount of unsaturated triacylglycerols relative to saturated triacylglycerols in the diet?

Ans. It is not clear that increasing the unsaturated fatty acid content of the diet has a direct effect on health. Studies clearly show that diets low in unsaturated triacylglycerols are beneficial. However, those diets (low in red meat, high in vegetables) are simultaneously low in cholesterol and total lipids. The benefits of lowered cholesterol and total lipids are understood—see Problems 19.40 and 19.41. The literature is generally mute on a mechanism for improved cardiovascular performance due to increased unsaturated triacylglycerol levels except for one speculation. Decreased unsaturated triacylglycerol content results in more rigid cell membranes because of the increased saturated triacylglycerol content. Deposition of cholesterol, a rigid molecule, on arterial membranes may be enhanced when those membranes contain similar (more rigid) molecules.

SUPPLEMENTARY PROBLEMS

Problem 19.57. Which of the following are based on glycerol and which on sphingosine? (*a*) triacylglycerols, (*b*) prostaglandins, (*c*) waxes, (*d*) leukotrienes, (*e*) sphingomyelins, (*f*) steroids, (*g*) glycolipids, (*h*) cerebrosides, (*i*) terpenes, (*j*) plasmalogens, (*k*) phosphoglycerides.

Ans. (*a*), (*j*), and (*k*) are based on glycerol. (*g*) and (*h*) are based on sphingosine.

Problem 19.58. Identify the class of each of the following lipids:

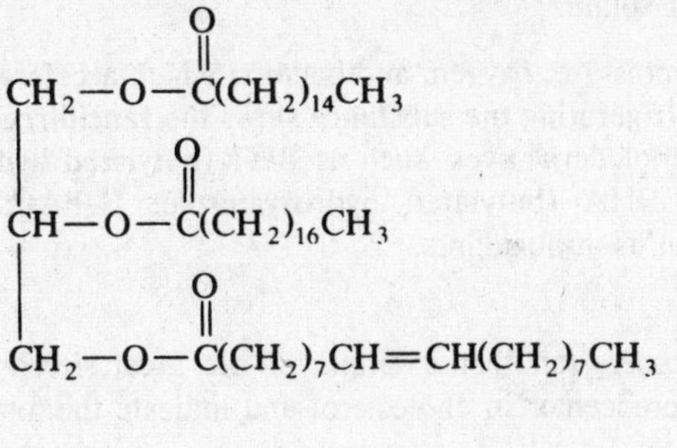

$CH_2-O-\overset{O}{\overset{\|}{C}}(CH_2)_{14}CH_3$

$CH-O-\overset{O}{\overset{\|}{C}}(CH_2)_{16}CH_3$

$CH_2-O-\overset{O}{\overset{\|}{C}}(CH_2)_7CH=CH(CH_2)_7CH_3$

I

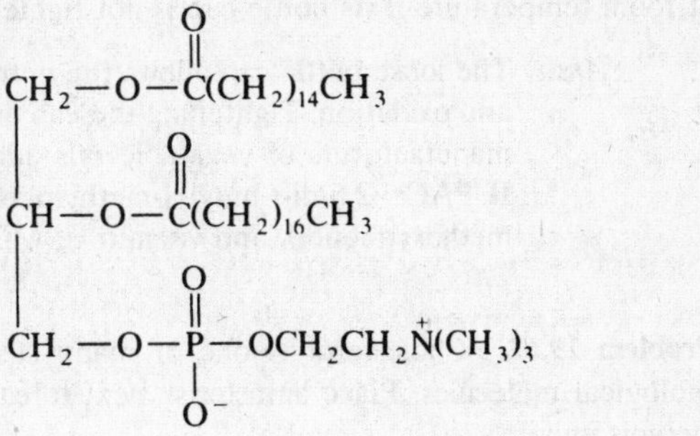

$CH_2-O-\overset{O}{\overset{\|}{C}}(CH_2)_{14}CH_3$

$CH-O-\overset{O}{\overset{\|}{C}}(CH_2)_{16}CH_3$

$CH_2-O-\overset{O}{\overset{\|}{P}}(O^-)-OCH_2CH_2\overset{+}{N}(CH_3)_3$

II

HO COOH

HO CH$_3$

OH

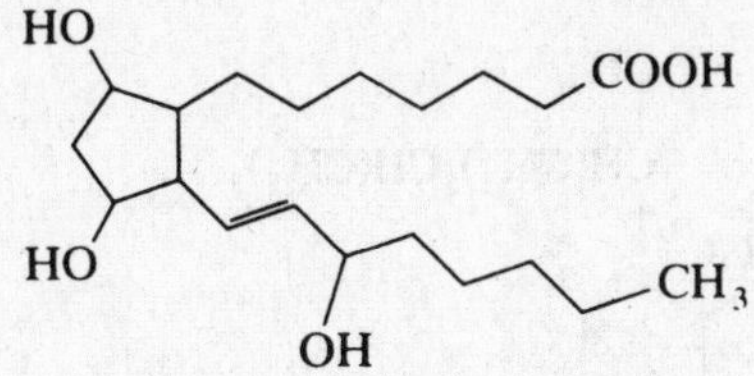

III

Ans. Compound I = triacylglycerol, compound II = phosphoglyceride, compound III = prostaglandin.

Problem 19.59. **The sodium salt of cholic acid, one of the bile salts, aids the digestion (i.e., accelerates the rate) of saponifiable lipids. What mechanism is responsible for this action of the bile salts?**

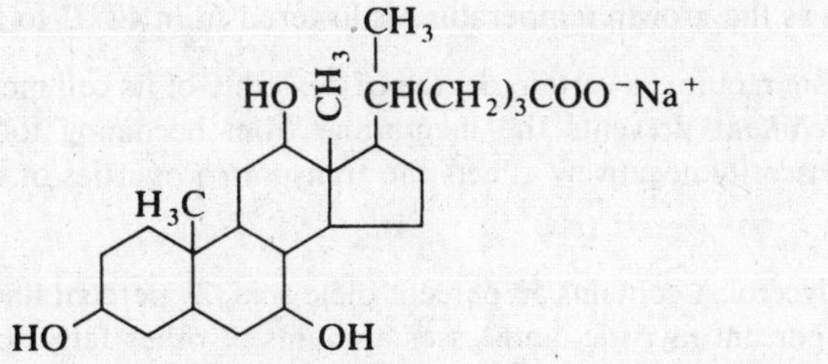

CH$_3$

HO CH$_3$ $CH(CH_2)_2COO^-Na^+$

H$_3$C

HO OH

Sodium cholate

Ans. Bile salts act like soaps or detergents. Lipids are insoluble in the contents of the duodenum where lipid digestion occurs. Without bile salts, lipid digestion would occur only at the surfaces of the insoluble fat globules—a slow process. Bile salts break up the fat globules into a larger number of smaller globules. The mechanism for this is the same as when oily dirt is washed off your hands with soap. The total surface area of the fat is increased and this accelerates the rate of digestion.

Problem 19.60. Draw dotted lines to show where isoprene units have been linked together to synthesize limonene (oil of lemon):

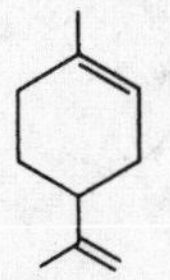

Limonene

Ans.

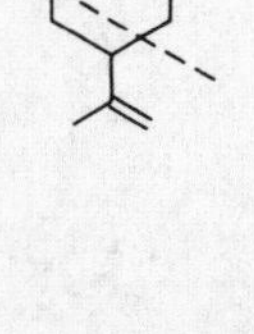

Problem 19.61. A salad or other cooking oil can develop an unpleasant odor and taste (become rancid) if stored at room temperature if its bottle cap is not tightened. Explain.

Ans. The loose bottle cap allows the entry of moisture, oxygen, and bacteria which accelerate hydrolysis and oxidation. Tightening the cap and refrigerating the substance slows the rancid reactions. Some manufacturers of vegetable oils add phenol derivatives, such as BHT (butylated hydroxytoluene; IUPAC: 2,6-di-*t*-butyl-4-methylphenol), BHA (butylated hydroxyanisole; IUPAC: 2-*t*-butyl-4-methoxyphenol), and vitamin E, which act as antioxidants.

Problem 19.62. Cholesterol is one of many stereoisomers, another example of the great stereoselectivity of biological molecules. Place an asterisk next to each stereocenter in cholesterol and indicate the total number of stereoisomers.

Ans.

CH_3
*$CH(CH_2)_3CH(CH_3)_2$
H_3C
H_3C
HO

There are 2^8 or 256 optical isomers.

Problem 19.63. The ratio of saturated to unsaturated fatty acid units in the *E. coli* bacterial membrane decreases from 1.6 to 1.0 as the growth temperature is lowered from 42 °C to 27 °C. Why?

Ans. The organism requires a certain degree of flexibility of its cell membrane. Decreasing the saturated fatty acid content prevents the membrane from becoming too viscous at lower temperatures. Increased viscosity negatively effects the transport properties of the membrane.

Problem 19.64. Triacylglycerol A contains 53 percent oleic acid, 29 percent linoleic acid, 9 percent palmitic acid, 4 percent stearic acid, 1 percent myristic, and lesser amounts of other fatty acids. Triacylglycerol B contains 39 percent oleic acid, 29 percent palmitic acid, 24 percent stearic acid, 3 percent myristic, 2 percent linoleic, and lesser amounts of other fatty acids. One of the samples is peanut oil and the other is beef fat (tallow). Which is which?

Ans. Triacylglycerol A is peanut oil since it contains 84 percent unsaturated fatty acids (oleic + linoleic). Triacylglycerol B is beef fat since it contains only 41 percent unsaturated fatty acids (oleic + linoleic).

Problem 19.65. Both prostaglandins and leukotrienes are synthesized in the body from arachidonic acid. In very general terms, what chemical changes occur during these syntheses?

COOH
8
12
CH_3

Arachidonic acid

Ans. Leukotriene synthesis from arachidonic acid involves alterations in the positions of some of the double bonds and the introduction of various functional groups (OH and the sulfur-containing group). For prostaglandin synthesis, there is also bond formation between C-8 and C-12 to generate the 5-membered ring.

CHAPTER 20

Proteins

Proteins constitute about half of the body's dry weight. They perform the widest range of physiological functions of all the biological molecules (Table 20-1). There are many thousands of individual proteins, each with a specialized function, which together direct and carry out the range of physiological functions required for life.

Table 20-1. Classification of Proteins by Physiological Functions

Class	Physiological Functions
Contractile protein	Contracts and expands to give motion Muscle (actin, myosin)
Defense protein	Protection against viruses, bacteria, other foreign bodies Antibodies (immunoglobulins); blood clotting (thrombin and fibrinogen); toxic proteins (snake venom)
Enzyme	Catalysis of virtually all reactions in living organisms Peptide and lipid digestion (peptidase, esterase)
Regulatory protein	Regulation and control of cell processes Metabolism (insulin, parathyroid hormone); skeletal growth (growth hormone); neurotransmitters (enkephalin, dopamine)
Storage protein	Store nutrients for organism Seed proteins of green plants; egg white protein (ovalbumin); milk protein (casein); iron storage in spleen (ferritin)
Structural protein	Mechanical support for organism's structure Feathers, hair, hoof, horn, skin, wool (α-keratin); bone, cornea, tendons and other connective tissue (collagen); blood vessels, ligaments (elastin); blood clots (fibrin)
Transport protein	Transports chemicals in bloodstream and cells Oxygen (hemoglobin); lipids (lipoproteins); active transport by integral proteins

Polypeptides are copolymers of amino acids, i.e., different amino acids are joined to each other in a specific sequence. Proteins are naturally occurring polypeptides, usually in aggregation with one or more other polypeptides and/or with other types of molecules or ions.

20.1 AMINO ACIDS

Amino acids, the building blocks for polypeptides, contain both an amine and carboxyl group in the same molecule. The amino acids used to synthesize polypeptides in nature are the *α-amino acids*, those in which the amine group is located on the α-carbon. The α-carbon is the first carbon attached to the COOH carbon, the second carbon from the COOH is the β-carbon, the third carbon from the COOH is the γ-carbon, and so on.

Problem 20.1. Which of the following is an α-amino acid?

$$\underset{\text{I}}{\underset{|\atop NH_2}{CH_3CH_2CHCOOH}} \qquad \underset{\text{II}}{\underset{|\atop NH_2}{CH_3CHCH_2COOH}} \qquad \underset{\text{III}}{NH_2CH_2CH_2CH_2COOH}$$

Ans. Compound I = α-amino acid, compound II = β-amino acid, and compound III = γ-amino acid.

Although a variety of amino acids are known in nature, the only amino acids used in polypeptide synthesis are the 20 α-amino acids listed in Table 20-2. The same 20 α-amino acids are used for all species of plants and animals. Ten amino acids are *essential amino acids* for humans, i.e., they cannot be synthesized in the body but must be provided by dietary intake. All 20 α-amino acids have the general formula $NH_2—CHR—COOH$. *Note:* The term *amino acid* refers to α-amino acid unless otherwise noted; the prefix α- is often left out.

The amino acids are separated into four categories depending on the identity of the R group (often referred to as the *side group*). Nonpolar neutral amino acids have hydrophobic R groups. Polar neutral amino acids have hydrophilic but neutral R groups. The side group of neutral amino acids, both nonpolar and polar, does not contain a COOH or basic nitrogen group. Polar acidic amino acids have hydrophilic acidic R groups, i.e., the R group contains a COOH. Polar basic amino acids have R groups that contain a basic nitrogen group.

Common names, shorter and more convenient than the IUPAC names, are used for the α-amino acids. Three-letter abbreviations (symbols) are also used (Table 20-2). For all, except four of the α-amino acids, the three letters are the first three letters of the common name. For the four α-amino acids, it is a combination of the first letter of the name with two other letters of the name. A one-letter abbreviation system has been introduced recently. For 11 α-amino acids, it is the first letter of the common name. For the other nine α-amino acids, it is some letter assigned by IUPAC convention. We will use only the three-letter abbreviations.

Problem 20.2. Which are the α-amino acids that have a three-letter symbol different from the first three letters of the name? What are their three-letter symbols?

Ans. Isoleucine = Ile, tryptophan = Trp, asparagine = Asn, glutamine = Gln.

Problem 20.3. How does proline differ from the other 19 α-amino acids?

Ans. Proline has a secondary amine group whereas all other α-amino acids have a primary amine group.

An α-amino acid does not exist as the neutral uncharged molecule $NH_2—CHR—COOH$. The COOH group is acidic while the NH_2 group is basic. An acid-base reaction occurs where an H^+ is transferred from COOH to the amine nitrogen to form a *dipolar ion*, also referred to as a *zwitterion*:

$$\underset{\text{Nonexistent}}{NH_2—\overset{R\atop|}{CH}—COOH} \longrightarrow \underset{\text{Zwitterion}}{{}^+NH_3—\overset{R\atop|}{CH}—COO^-}$$

The zwitterion contains carboxlate and ammonium ionic centers but has a net charge of zero. α-Amino acids exist exclusively as zwitterions in the solid state. Even when drawn as the uncharged molecule, we must understand that the actual structure is that of the dipolar ion. Amino acids are referred to as *amphoteric* compounds because both acidic and basic groups are contained in the same molecule.

The situation is more complicated in aqueous solution where three different species coexist in equilibrium with each other. The zwitterion is converted into the positive ion ${}^+NH_3—CHR—COOH$

Table 20-2. Amino Acids $NH_2-\underset{}{\overset{R}{\overset{|}{C}}}H-COOH$

Name	Symbols		Side chain, R	pI*
Nonpolar neutral				
Glycine	Gly	G	$H-$	5.97
Alanine	Ala	A	CH_3-	6.01
Valine†	Val	V	$(CH_3)_2CH-$	5.96
Leucine†	Leu	L	$CH_3CH(CH_3)CH_2-$	5.98
Isoleucine†	Ile	I	$CH_3CH_2CH(CH_3)-$	6.02
Phenylalanine†	Phe	F	$C_6H_5-CH_2-$ (benzene ring)	5.48
Methionine†	Met	M	$CH_3SCH_2CH_2-$	5.74
Proline‡	Pro	P	(pyrrolidine ring: HN–CH–COOH)	6.30
Tryptophan†	Trp	W	(indole ring, N–H)$-CH_2-$	5.88
Polar neutral				
Cysteine	Cys	C	$HSCH_2-$	5.05
Serine	Ser	S	$HOCH_2-$	5.68
Threonine†	Thr	T	$CH_3CH(OH)-$	5.60
Asparagine	Asn	N	$H_2NC(=O)CH_2-$	5.41
Glutamine	Gln	Q	$H_2NC(=O)CH_2CH_2-$	5.65
Tyrosine	Tyr	Y	$HO-C_6H_4-CH_2-$ (benzene ring)	5.66
Polar acidic				
Aspartic acid	Asp	D	$HOOCCH_2-$	2.77
Glutamic acid	Glu	E	$HOOCCH_2CH_2-$	3.22
Polar basic				
Lysine†	Lys	K	$H_2NCH_2CH_2CH_2CH_2-$	9.74
Arginine†	Arg	R	$H_2NC(=NH)NHCH_2CH_2CH_2-$	10.76
Histidine†	His	H	(imidazole ring, N, N–H)$-CH_2-$	7.59

* pI = isoelectric point.
† Essential for humans.
‡ Complete structure of proline is shown.

by protonation of the COO^- group. The zwitterion is converted into the negative ion $NH_2—CHR—COO^-$ by the loss of a proton from the $^+NH_3$ group:

$$\underset{\substack{\text{Positive ion}\\ \text{Acidic media}\\ \text{High }[H^+]\\ \text{Low pH}}}{{}^+NH_3—\overset{\overset{\displaystyle R}{|}}{C}H—COOH} \underset{HO^-}{\overset{H^+}{\rightleftharpoons}} \underset{\text{Zwitterion}}{{}^+NH_3—\overset{\overset{\displaystyle R}{|}}{C}H—COO^-} \underset{H^+}{\overset{HO^-}{\rightleftharpoons}} \underset{\substack{\text{Negative ion}\\ \text{Basic media}\\ \text{Low }[H^+]\\ \text{High pH}}}{NH_2—\overset{\overset{\displaystyle R}{|}}{C}H—COO^-}$$

The relative amounts of the three species depends on pH. This is quantified by the *isoelectric point*, pI, of an α-amino acid. Table 20-2 lists pI values for the α-amino acids. The isoelectric point is the pH at which essentially all of the α-amino acid is in the zwitterion form, with very low and equal concentrations of the positive and negative ions. In more acidic media (pH < pI), the concentration of positive ion increases while the concentration of zwitterion decreases, i.e., the equilibria shift to the left. In more basic media (pH > pI), the concentration of negative ion increases while the concentration of zwitterion decreases, i.e., the equilibria shift to the right. These structural changes resist pH changes for the neutral α-amino acids since their solutions act as buffers. More than 98 percent of the amino acid is in the zwitterion form over a pH range of 2 pH units either side of pI. However, in highly acidic media (pH ~ 1), the neutral α-amino acids are present almost exclusively (> 90 percent) as the positive ion. In highly basic media (pH ~ 12), the neutral α-amino acids are present almost exclusively (> 90 percent) as the negative ion.

The structure of an acidic or basic α-amino acid changes much more rapidly with pH compared to a neutral α-amino acid. At pH = pI, the amino acid is present as the zwitterion with one amine or carboxyl group in uncharged form. An acidic α-amino acid has an uncharged carboxyl group which undergoes dissociation to COO^- with an increase in pH. A basic α-amino acid has an uncharged amine group which undergoes protonation with a decrease in pH. The conversions to charged groups are greater than 90 percent complete when pH is changed by 2 units above or below pI.

An important generalization to remember is that amino and carboxyl groups on all amino acids, neutral, acid, and basic, are close to fully charged at physiological pH (pH ~ 7).

Problem 20.4. Alanine and lactic acid have very nearly the same molecular mass but their melting points differ greatly, 314 and 53 °C, respectively. Explain.

$$\underset{\text{Alanine}}{NH_2—\overset{\overset{\displaystyle CH_3}{|}}{C}H—COOH} \qquad \underset{\text{Lactic acid}}{HO—\overset{\overset{\displaystyle CH_3}{|}}{C}H—COOH}$$

Ans. Lactic acid is a polar compound whereas alanine is an ionic compound. The intermolecular forces are much higher in alanine because of its dipolar structure. The melting point is sufficiently high that alanine undergoes chemical decomposition during the melting transition.

Problem 20.5. Draw the structure of alanine at pH = 1, 6.02, and 12.

Ans.

$$\underset{\substack{\text{III}\\ \text{pH} = 1}}{{}^+NH_3—\overset{\overset{\displaystyle CH_3}{|}}{C}H—COOH} \underset{}{\overset{H^+}{\rightleftharpoons}} \underset{\substack{\text{I}\\ \text{pH} = 6.02}}{{}^+NH_3—\overset{\overset{\displaystyle CH_3}{|}}{C}H—COO^-} \overset{HO^-}{\rightleftharpoons} \underset{\substack{\text{II}\\ \text{pH} = 12}}{NH_2—\overset{\overset{\displaystyle CH_3}{|}}{C}H—COO^-}$$

Problem 20.6. The pI values for the nonpolar neutral and polar neutral α-amino acids fall in the narrow range 5.05 to 6.30. Why?

Ans. pI depends on the relative acidity and basicity of the COOH and NH_2 groups which in turn depends on the electron-withdrawing or electron-releasing character of the R group. The various R groups in the neutral amino acids are similar in their electron-withdrawing or electron-releasing characters.

Problem 20.7. Physiological pH is near neutral (pH = 7): Blood pH is 7.35 and the pH in most other cells is in the range 6.8 to 7.1. (The exception is the stomach, whose contents are much more acidic.) What is the structure and net charge of the neutral α-amino acids at pH = pI? At neutral pH? At pH = pI $\pm$ 2? Explain.

Ans. The zwitterion will predominate (98 percent or higher) under all these pH conditions and the net charge is zero. The zwitterion predominates over a pH range 2 units on each side of pI (i.e., pH = pI $\pm$ 2). This is especially important from a biological viewpoint as it means that all neutral amino acids are in the zwitterionic form at physiological pH.

Problem 20.8. What is the net charge on acidic and basic α-amino acids at physiological pH?

Ans. All acidic and basic α-amino acids are charged at physiological pH because all carboxyl and amino groups are present in their ionic forms. There is a net charge on the molecule since the number of carboxyl and amine groups are not the same. The acidic α-amino acids are negatively charged and the basic α-amino acids are positively charged.

Problem 20.9. Explain why an acidic amino acid such as aspartic acid (pI = 2.77) has a net charge at physiological pH?

Ans. Aspartic acid is present as the zwitterion (structure I below) at pH = 2.77 since that is the pI of aspartic acid. At neutral pH (which is much more basic than pI), the side chain COOH loses its proton to yield structure II, which has three charge centers and a net charge of 1– :

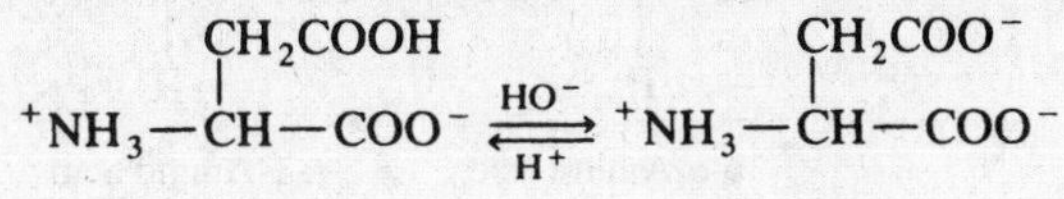

$$\underset{\substack{\text{I} \\ \text{pH} = 2.77}}{{}^{+}NH_3-\underset{\displaystyle |}{\overset{CH_2COOH}{CH}}-COO^-} \underset{H^+}{\overset{HO^-}{\rightleftharpoons}} \underset{\substack{\text{II} \\ \text{pH} = 7}}{{}^{+}NH_3-\overset{CH_2COO^-}{CH}-COO^-}$$

Problem 20.10. Explain why a basic amino acid such as lysine (pI = 9.74) has a net charge at physiological pH.

Ans. The zwitterion (structure I below) exists at pH = 9.74 since that is the pI of lysine. At neutral pH (which is much more acidic than pI), the side chain NH_2 becomes protonated to yield structure II, which has three charge centers and a net charge of 1+ :

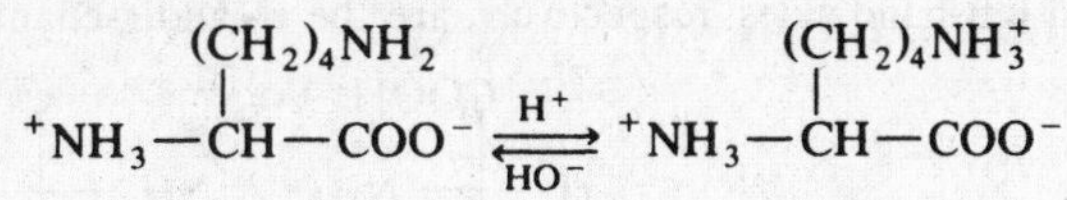

$$\underset{\substack{\text{I} \\ \text{pH} = 9.74}}{{}^{+}NH_3-\overset{(CH_2)_4NH_2}{CH}-COO^-} \underset{HO^-}{\overset{H^+}{\rightleftharpoons}} \underset{\substack{\text{II} \\ \text{pH} = 7}}{{}^{+}NH_3-\overset{(CH_2)_4NH_3^+}{CH}-COO^-}$$

Electrophoresis is an analytical method for identifying amino acids by observing their migration as a function of pH under an applied electric field gradient. The technique is referred to as *paper* or *gel electrophoresis* depending on whether the support used is paper or a polyacrylamide gel. In paper electrophoresis, a strip of paper is saturated with a buffer solution and then a solution of unknown amino acids is placed at the center of the paper. Electrodes are connected to the ends of the paper and an electric current passed through the solution. The electric current is stopped and a solution of ninhydrin sprayed onto the paper. Ninhydrin reacts with amino acids to yield colored products (usually blue-purple) and this gives the location of the amino acids. At its pI, an amino acid is present as the zwitterion with no net charge and will not migrate in electrophoresis. The amino acid carries a positive charge at pH < pI and migrates to the negative electrode (*cathode*). The amino acid carries a negative charge at pH > pI and migrates to the positive electrode (*anode*).

Problem 20.11. Referring to Problem 20.5, to which electrode will alanine migrate in electrophoresis at pH = 1, 6.02, and 12?

Ans. There is no migration at pH = 6.02 since structure I has zero net charge. There is migration to the positive and negative electrodes at pH = 12 and 1, respectively, since structures II and III have net charges of 1− and 1+.

Problem 20.12. An unknown, containing some combination of alanine, lysine, or aspartic acid, is subjected to paper electrophoresis at pH = 7. Ninhydrin treatment shows some amino acid at the negative electrode and some amino acid has not moved from the center. No amino acid is found at the positive electrode. Which amino acid(s) is (are) in the unknown?

Ans. Alanine, aspartic acid, and lysine if present would be present as zwitterion, negative ion, and positive ion, respectively, at pH = 7. Alanine would not migrate while aspartic acid and lysine would migrate to the positive and negative electrodes, respectively. The unknown contains alanine and lysine but no aspartic acid.

Except for glycine, the α-carbon of all α-amino acids is a stereocenter and a pair of enantiomers is possible:

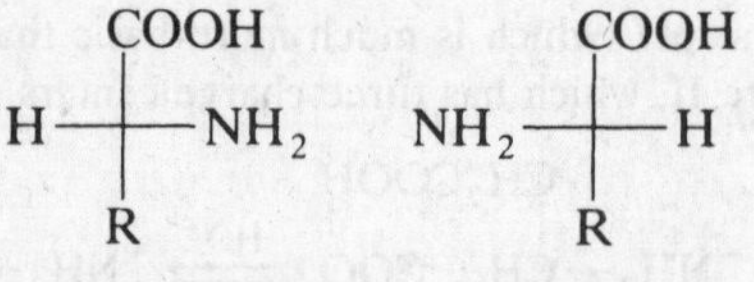

```
      COOH              COOH
  H ──┼── NH2     NH2 ──┼── H
       R                 R
 D-α-Amino acid    L-α-Amino acid
```

The α-amino acids used in synthesizing naturally occurring polypeptides are the L-enantiomers. This is another example of the stereoselectivity of biological systems.

Problem 20.13. Draw D-α-alanine and L-α-alanine using the convention described in Problem 17.18.

Ans. COOH and CH_3, the two carbon substituents, are placed vertically. The more highly oxidized substituent COOH is placed up and the least oxidized substituent is placed down. The enantiomers with the heteroatom substituent (NH_2 for the amino acid instead of OH for glyceraldehye) on the right- and left-hand sides, respectively, are the D- and L-enantiomers:

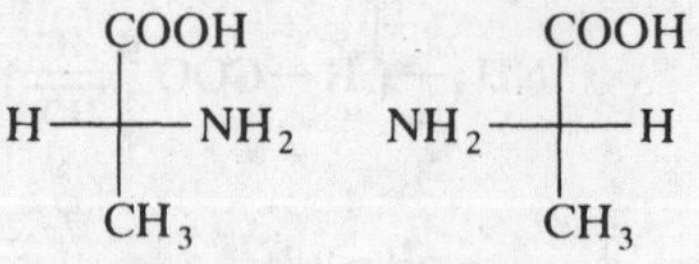

```
      COOH              COOH
  H ──┼── NH2     NH2 ──┼── H
      CH3               CH3
  D-α-Alanine       L-α-Alanine
```

Problem 20.14. Show that the following structure is that of the L-α-alanine:

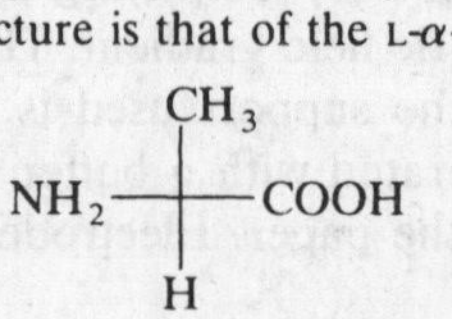

```
        CH3
  NH2 ──┼── COOH
         H
```

Ans. Follow Rule 2 in Sec. 17.2: Exchange the positions of any two substituents attached to the stereocenter. If an even number of exchanges is needed to make two structures identical, the two structures are the same compound. If an odd number of exchanges is needed to make two structures identical, the two structures represent a pair of enantiomers. The given structure is L-α-alanine since two exchanges are needed to convert it to the L-α-alanine structure of Problem 20.13:

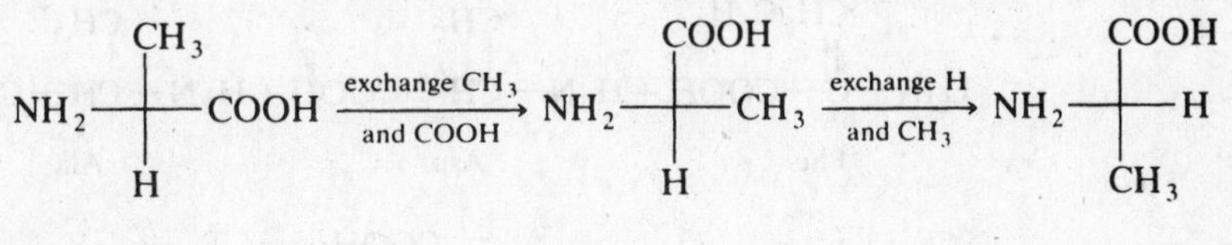

$$\begin{array}{c}CH_3\\ NH_2-\!\!\!+\!\!\!-COOH\\ H\end{array} \xrightarrow[\text{and COOH}]{\text{exchange }CH_3} \begin{array}{c}COOH\\ NH_2-\!\!\!+\!\!\!-CH_3\\ H\end{array} \xrightarrow[\text{and }CH_3]{\text{exchange H}} \begin{array}{c}COOH\\ NH_2-\!\!\!+\!\!\!-H\\ CH_3\end{array}$$

20.2 PEPTIDE FORMATION

Amino acids are linked together to form *peptides* by amide formation between the COOH of one amino acid and the amine group of another amino acid. Although the mechanism is not simple, the reaction can be considered as a dehydration between COOH and amine groups:

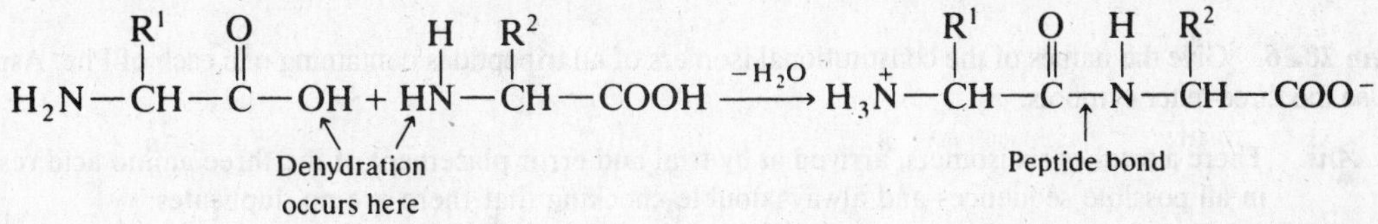

$$H_2N-\overset{R^1}{\overset{|}{C}H}-\overset{O}{\overset{\|}{C}}-OH + \overset{H}{\overset{|}{H}N}-\overset{R^2}{\overset{|}{C}H}-COOH \xrightarrow{-H_2O} H_3\overset{+}{N}-\overset{R^1}{\overset{|}{C}H}-\overset{O}{\overset{\|}{C}}-\overset{H}{\overset{|}{N}}-\overset{R^2}{\overset{|}{C}H}-COO^-$$

Dehydration occurs here (between OH and H of HN); Peptide bond (C–N)

Amino acids in the peptide, referred to as *amino acid residues*, are linked together by the *peptide bond*—the bond between the carbonyl carbon and the nitrogen. The functional group formed is the *peptide group*, —CO—NH—, which is an amide group. Peptides are *dipeptides*, *tripeptides*, *tetrapeptides*, and so on, depending on the number of amino acid residues. The term *oligopeptide* usually refers to peptides of less than 10–20 amino acid residues; larger peptides are termed *polypeptides*. (The demarcation between oligopeptides and polypeptides is arbitrary.)

Note that the dipeptide is written in the zwitterionic form. All peptides, no matter their size, are similar to amino acids in having an amino group at one end and a carboxyl at the other end. Like amino acids, the amine and carboxyl groups react to yield the zwitterion structure for peptides.

Peptide formation can yield different constitutional isomers, e.g., the reaction of Gly with Ala can proceed in two ways depending on which amine group reacts with which carboxyl group:

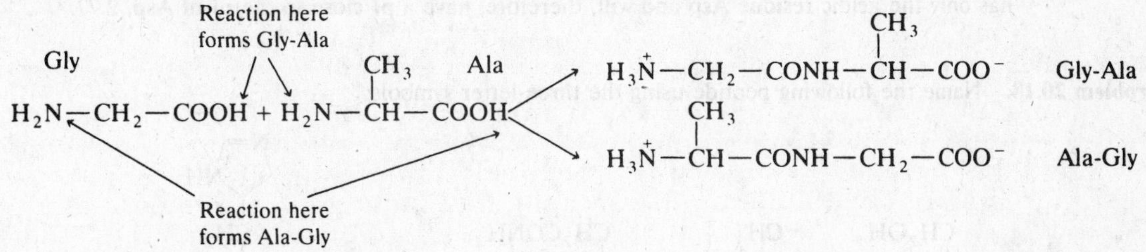

$$\underset{\text{Gly}}{H_2N-CH_2-COOH} + \underset{\text{Ala}}{H_2N-\overset{CH_3}{\overset{|}{C}H}-COOH} \longrightarrow \begin{cases} H_3\overset{+}{N}-CH_2-CONH-\overset{CH_3}{\overset{|}{C}H}-COO^- & \text{Gly-Ala} \\ H_3\overset{+}{N}-\overset{CH_3}{\overset{|}{C}H}-CONH-CH_2-COO^- & \text{Ala-Gly} \end{cases}$$

Reaction here forms Gly-Ala (Gly COOH with Ala H_2N); Reaction here forms Ala-Gly (Gly H_2N with Ala COOH)

The two dipeptides, like all peptides (and amino acids), contain ammonium and carboxylate end groups, referred to as *N-terminal and C-terminal residues* or *units*, respectively. The convention for drawing peptides is to place the N-terminal residue at the left and the C-terminal residue at the right. Names of peptides are obtained by naming the amino acid residues from left to right, usually with abbreviated symbols, e.g., Gly-Ala and Ala-Gly, as shown above. More expanded names use the full names of each amino acid. The far right amino acid residue retains the name of the amino acid. For all other amino acids except tryptophan, the *-ine* or *-ic acid* of the name is replaced by *-yl*. For tryptophan, *-yl* is added to the name.

The order of placement of amino acid residues in a peptide proceeding left to right is referred to as the *amino acid sequence* or simply *sequence*. Different amino acid sequences equal different constitutional isomers.

Problem 20.15. Show the formation of Phe-Asp-Ala using complete structural formulas.

Ans. Line up the amino acids in the order Phe Asp Ala left to right and then carry out dehydrations between adjacent carboxyl and amino groups. Show the product as the zwitterion:

$$\underset{\text{Phe}}{H_2N-\overset{\overset{\displaystyle CH_2C_6H_5}{|}}{C}-COOH} + \underset{\text{Asp}}{H_2N-\overset{\overset{\overset{\displaystyle COOH}{|}}{CH_2}}{\overset{|}{CH}}-COOH} + \underset{\text{Ala}}{H_2N-\overset{\overset{\displaystyle CH_3}{|}}{CH}-COOH} \longrightarrow$$

$$\underset{\text{Phe-Asp-Ala}}{H_3\overset{+}{N}-\overset{\overset{\displaystyle CH_2C_6H_5}{|}}{CH}-CONH-\overset{\overset{\overset{\displaystyle COOH}{|}}{CH_2}}{\overset{|}{CH}}-CONH-\overset{\overset{\displaystyle CH_3}{|}}{CH}-COO^-}$$

Problem 20.16. Give the names of the constitutional isomers of all tripeptides containing one each of Phe, Asp, and Ala. Use the three-letter symbols.

Ans. There a total of six isomers, arrived at by trial and error placement of the three amino acid residues in all possible sequences and always double-checking that there are no duplicates:

Phe-Asp-Ala	Asp-Phe-Ala	Ala-Asp-Phe
Phe-Ala-Asp	Asp-Ala-Phe	Ala-Phe-Asp

Keep in mind that reverse sequences, e.g., Phe-Asp-Ala and Ala-Asp-Phe, are not the same compound because the N- and C-terminal residues are at the left- and right-hand ends, respectively.

Problem 20.17. Estimate the pI value of Phe-Asp-Ala.

Ans. The pI of a peptide containing only neutral amino acid residues is in the range of pI values of the neutral amino acids, 5.05 to 6.30. For peptides with acidic and basic amino acid residues, the pI is approximately the average of the pI values of the acidic and basic amino acid residues. Phe-Asp-Ala has only the acidic residue Asp and will, therefore, have a pI close to the pI of Asp, 2.77.

Problem 20.18. Name the following peptide using the three-letter symbols:

$$H_3\overset{+}{N}-\overset{\overset{\displaystyle CH_2OH}{|}}{CH}-CONH-\overset{\overset{\displaystyle CH_3}{|}}{CH}-CONH-\overset{\overset{\displaystyle CH_2CONH_2}{|}}{CH}-CONH-CH_2-CONH-\overset{\overset{\displaystyle \text{imidazolyl (N, NH)}}{\overset{|}{CH_2}}}{\overset{|}{CH}}-COO^-$$

Ans. Ser-Ala-Asn-Gly-His

Problem 20.19. The amide bond in peptides does not have the properties of a single bond. There is no free rotation about the carbon-nitrogen bond and the bond length is shorter than expected for a single bond. The CO — N bond is said to have considerable double-bond character. Describe how resonance theory rationalizes the double-bond properties of the amide bond.

Ans. The carbonyl group is polarized $^+C-O^-$ and this leads to resonance interaction with the nonbonded pair of electrons of nitrogen. The amide group is a resonance hybrid of the following structures, which explains not only the lack of free rotation and shortened bond length for the carbon-nitrogen bond but also the highly polar nature of the amide group (Problem 16.34):

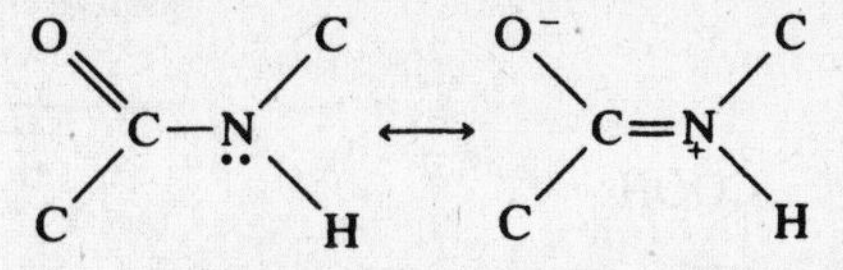

$$\text{O}{=}\text{C}(\text{C}){-}\ddot{\text{N}}(\text{C})(\text{H}) \longleftrightarrow \text{O}^{-}{-}\text{C}(\text{C}){=}\overset{+}{\text{N}}(\text{C})(\text{H})$$

Also, the peptide chain segments are located *trans* to each other on the C = N bond. The various atoms that are part of the double bond and those atoms directly attached to the double bond are coplanar, i.e., they lie in the same plane.

Problem 20.20. The number of constitutional isomers (i.e., different amino acid sequences) for peptides containing one each of n different amino acids is $n!$ (n factorial). What is the number of isomers for peptides containing one each of four different amino acids? Of 10 different amino acids?

Ans. For tetrapeptides, $n! = 4 \times 3 \times 2 \times 1 = 24$, i.e., there are 24 constitutional isomers. For decapeptides, $n! = 10 \times 9 \times 8 \times 7 \times 6 \times 5 \times 4 \times 3 \times 2 \times 1 = 3{,}628{,}800$, i.e., there are more than three million constitutional isomers.

Cysteine differs from the other α-amino acids in one important property because of the presence of the SH functional group, referred to as a *thiol* or *mercaptan* or *sulfhydryl* group. The thiol group undergoes oxidation in the presence of oxidizing agents (Sec. 13.8). This occurs in biological systems as well as laboratory conditions. Oxidation involves a reaction between the thiol groups of a pair of cysteine molecules. Hydrogen is removed from each SH group and two cysteine molecules are coupled together by a ***disulfide linkage*** or ***bridge***:

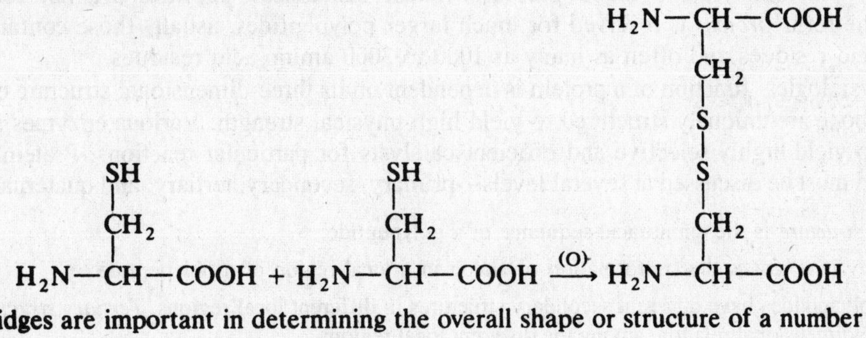

$$H_2N{-}CH(CH_2SH){-}COOH + H_2N{-}CH(CH_2SH){-}COOH \xrightarrow{(O)} H_2N{-}CH(COOH){-}CH_2{-}S{-}S{-}CH_2{-}CH(COOH){-}NH_2$$

Disulfide bridges are important in determining the overall shape or structure of a number of peptides and proteins.

Problem 20.21. Write the equation showing formation of a disulfide bridge in Phe-Cys-Ala.

Ans.

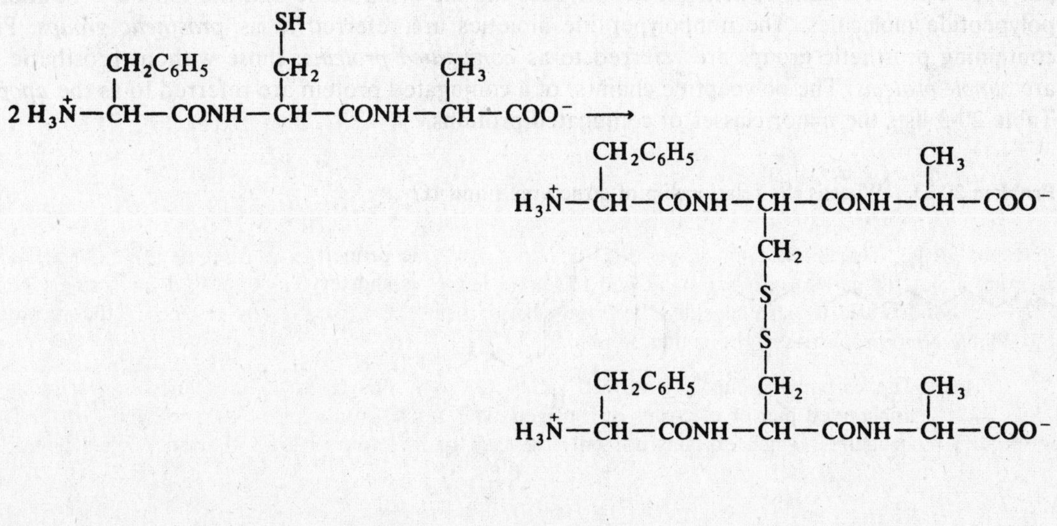

$$2\ H_3\overset{+}{N}{-}CH(CH_2C_6H_5){-}CONH{-}CH(CH_2SH){-}CONH{-}CH(CH_3){-}COO^- \longrightarrow$$

$$H_3\overset{+}{N}{-}CH(CH_2C_6H_5){-}CONH{-}CH{-}CONH{-}CH(CH_3){-}COO^-$$
$$|\ CH_2{-}S{-}S{-}CH_2\ |$$
$$H_3\overset{+}{N}{-}CH(CH_2C_6H_5){-}CONH{-}CH{-}CONH{-}CH(CH_3){-}COO^-$$

Problem 20.22. Write the equation for the digestion (hydrolysis) of Phe-Asp-Ala.

Ans.

$$\underset{\substack{|\\CH_2C_6H_5}}{H_3\overset{+}{N}-CH}-CONH-\underset{\substack{|\\CH_2\\|\\COOH}}{CH}-CONH-\underset{\substack{|\\CH_3}}{CH}-COO^- \longrightarrow$$

$$\underset{Phe}{H_2N-\underset{CH_2C_6H_5}{CH}-COOH} + \underset{Asp}{H_2N-\underset{CH_2-COOH}{CH}-COOH} + \underset{Ala}{H_2N-\underset{CH_3}{CH}-COOH}$$

20.3 PROTEIN STRUCTURE AND FUNCTION

20.3.1 Protein Shape

There are a number of small-sized peptides with physiological functions of the type associated with proteins. Glutathion, a tripeptide, is a scavenger for oxidizing agents. Enkephalins are pentapeptides which bind at receptor sites in the brain to reduce the sensation of pain. Oxytocin and vasopressin are nonapeptides. Oxytocin regulates uterine contractions and lactation. Vasopressin regulates excretion of water by the kidneys, and blood pressure. These small-sized peptides are not referred to as proteins. The term *protein* is reserved for much larger polypeptides, usually those containing at least 50 amino acid residues and often as many as 1000 to 3000 amino acid residues.

The physiological function of a protein is dependent on its three-dimensional structure or shape. The proteins of bone are uniquely structured to yield high physical strength. Various enzymes are uniquely structured to yield highly selective and efficient catalysts for particular reactions. Protein structure is complex and must be discussed at several levels—primary, secondary, tertiary, and quaternary:

Primary structure is the amino acid sequence of a polypeptide.

Secondary structure is the conformation or shape in a local region of a polypeptide.

Some polypeptides have different secondary structures in different local regions. *Tertiary structure* describes the three-dimensional relation among the different local regions.

Some proteins are assemblies of two or more polypeptide molecules. *Quaternary structure* refers to the conformational or spatial relationship between the different polypeptide molecules that comprise a protein.

Some proteins contain polypeptide molecule(s) associated with molecules or ions other than polypeptides. For example, hemoglobin includes the molecule heme and the ion Fe^{2+} in addition to polypeptide molecules. The nonpolypeptide moieties are referred to as *prosthetic groups*. Proteins containing prosthetic groups are referred to as *conjugated proteins*; those without prosthetic groups are *simple proteins*. The polypeptide chain(s) of a conjugated protein are referred to as the *apoprotein*. Table 20-3 lists the major classes of conjugated proteins.

Problem 20.23. What is the relationship of structures I and II?

I

II

Table 20-3. Classification of Conjugated Proteins

Class	Prosthetic group	Example
Glycoprotein	Saccharide	Gamma globulin (antibody); mucin (food lubricant in saliva); interferon (antiviral agent)
Hemoprotein	Heme	Hemoglobin (O_2 carrier in blood); myoglobin (O_2 storage in muscle)
Lipoprotein	Lipid	VLDL, LDL, HDL (lipid carriers)
Metalloprotein	Metal ion	Fe in hemoglobin and myoglobin; Fe in ferritin (Fe storage); Zn in alcohol dehydrogenase (enzyme for alcohol oxidation)
Nucleoprotein	Nucleic acid	RNA-bound protein (protein synthesis in ribosome)
Phosphoprotein	Phosphate ester	Caesin (milk protein)

Ans. Structures I and II are conformational isomers of dodecane. One conformational isomer differs from the other in the angles of rotation about successive C — C bonds. These are only two of the many possible conformational isomers of dodecane. Use your molecular models to build dodecane. Hold the terminal carbon atoms in your two hands. Move your hands with respect to each other. This results in different conformations due to rotations about the various C — C bonds. This process is often referred to as *folding* when applied to polypeptides. Different conformations of the same molecule have very different shapes.

Problem 20.24. What is the importance of the discussion on conformational isomers to our consideration of secondary, tertiary, and quaternary structures of proteins?

Ans. Different secondary, tertiary, or quaternary structures of a protein are different conformational isomers. There is a negligible difference in stability for the many conformational isomers of molecules such as dodecane. Dodecane molecules are continuously undergoing conversion from one conformational isomer to another. The situation is dramatically different for polypeptides. For the typical polypeptide, one of the many conformations is much more stable than the others and the polypeptide is present exclusively in that conformation with its specific three-dimensional shape. A complete description of this conformation or shape is what we refer to as the secondary, tertiary, and quaternary structures of a protein. Conformational shape directly affects the physiological function of a protein.

Various properties of a polypeptide chain are critical in determining why it folds into a specific conformation:

1. The coplanarity of the peptide linkage, the bond angles and bond lengths of the other bonds, and the sizes of various atoms in a polypeptide chain allow only certain conformations without introducing distortion and instability.
2. The major driving force for folding of a polypeptide chain into a specific conformation arises from its existence in an aqueous environment. A conformation is stabilized by:
   a. Shielding of the nonpolar amino acid residues (e.g., Ala, Phe, Leu, Ile) from contact with water.
   b. Hydrogen bonding of peptide linkages.

The peptide linkages of a polypeptide molecule can be stabilized either by hydrogen bonding to water or hydrogen bonding to other peptide linkages (either of the same molecule or of another polypeptide molecule). Hydrogen bonding exclusively with water is not important for naturally occurring polypeptides because it does not result in shielding of the nonpolar amino acid residues from contact with water. The conformation is destabilized as nonpolar residues are forced into contact with water. Hydrogen bonding of peptide linkages with each other is the much more important process because it results in conformations which minimize the contact of nonpolar amino acid residues with water. The *α-helix* and *β-pleated-sheet* conformations are the two most important regular (ordered) secondary structures observed in naturally occurring polypeptides.

The α-helix conformation (Fig. 20-1) involves the folding of the polypeptide in a helical manner such that the C=O oxygen of an amino acid residue is hydrogen-bonded to the N—H hydrogen of the fourth amino acid residing further down the polypeptide chain. The overall shape of the α-helix conformation is that of a coil spring. The α-helix is referred to as a *right-handed helix* because a clockwise movement is required to move along the length of the helix or coil. Note that the amino acid residues are represented by planar segments because of the double-bond character of the peptide linkage. The α-carbons are located at the junctions between successive planes. The R group at each α-carbon protrudes out at a right angle to the length (long axis) of the α-helix.

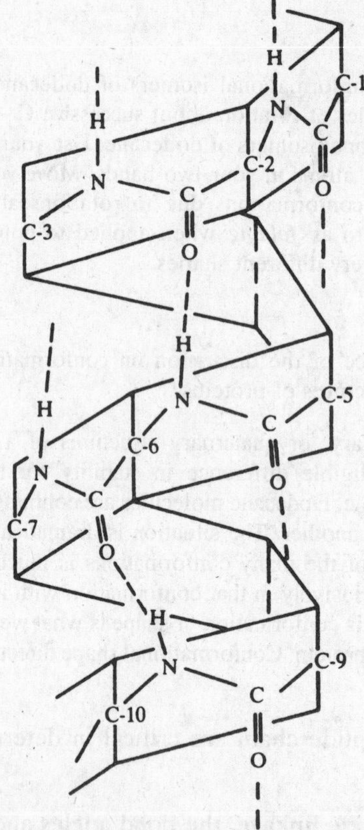

Fig. 20-1 α-Helix conformation. (*After P. W. Kuchel and G. B. Ralston, Theory and Problems of Biochemistry, Schaum's Outline Series, McGraw-Hill, New York, 1988.*)

Polypeptide chains are almost completely extended in the β-pleated-sheet conformation (Fig. 20-2). There is a side-by-side alignment of adjacent polypeptide chains. Hydrogen bonding occurs between adjacent polypeptide chains—more precisely, between the C=O oxygen of one chain and the N—H hydrogen of an adjacent chain. Each polypeptide chain is hydrogen-bonded to adjacent chains on each side of it, except for the chains at the two ends of the sheet. The sheets are pleated

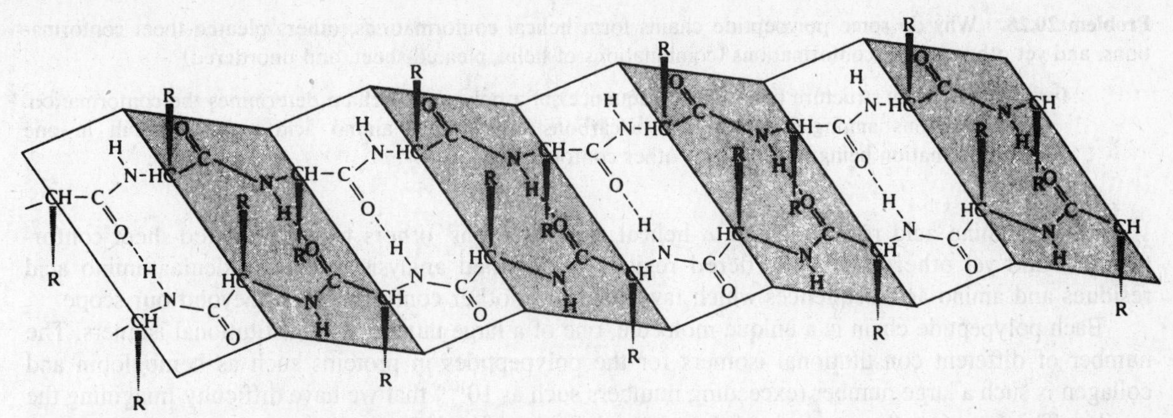

Fig. 20-2 β-Pleated sheet conformation. (*After W. H. Brown and E. P. Rogers, General, Organic, and Biochemistry, 3rd ed., Brooks/Cole, Monterey, CA, 1987.*)

sheets, not flat sheets, because of the planar structure of the peptide linkage. The R groups at successive α-carbons of a polypeptide chain protrude above and below the plane of the sheet. The sheet conformation permits stacking of sheets one on top of the other in addition to the side-by-side alignment.

Problem 20.25. Describe the classification of proteins as *fibrous* or *globular*. What is the relationship of this classification to physiological function?

Ans. This classification is based on solubility in the aqueous environment. Fibrous proteins are insoluble while globular proteins are soluble. Globular proteins are globe-shaped, i.e., more or less spherical. Fibrous proteins have shapes in which one dimension is much longer than the other dimensions, i.e., they are elongated. Fibrous proteins function as structural and contractile proteins (Table 20-1). The functions of globular proteins are those such as the transport, enzyme, defense, and regulatory functions which require solubility in the aqueous environment.

Problem 20.26. Fibrous proteins exist in conformations that are completely (or nearly completely) helical or pleated sheet. How does this fit the physiological functions of fibrous proteins?

Ans. Molecules in helical or pleated-sheet conformations pack together into tight assemblies with high intermolecular attractive forces. This yields assemblies with high physical strength and insolubility in an aqueous environment, the requirements that fit the physiological functions of a fibrous protein. Examples of fibrous proteins are fibrin (involved in blood clotting) and collagen in connective tissue.

Problem 20.27. The conformation of a globular protein is different in different parts of its polypeptide chain(s). Some have chains that contain a number of helical portions separated by *unordered regions* (i.e., regions which lack regularity in conformation). Others have chains that contain a number of different pleated-sheet regions separated by unordered regions. Yet other globular proteins have polypeptide chains with a number of both helical and pleated-sheet regions separated by unordered regions. How does this fit in with the physiological functions of globular proteins?

Ans. The unordered regions serve as *bends* (or *turns*) in the polypeptide chain. The bends allow the polypeptide chain to pack itself into a globular shape, in line with its physiological functions. The globular shape is not conducive for packing into a tight assembly with high intermolecular forces. This together with placement of hydrophilic amino acid residues on the outer surface of the protein results in solubility in the aqueous environment. Examples of globular proteins are immunoglobulins (antibodies) and hemoglobin (oxygen carrier).

Problem 20.28. Why do some polypeptide chains form helical conformations, others pleated-sheet conformations, and yet others mixed conformations (combinations of helix, pleated-sheet, and unordered)?

> *Ans.* The primary structure (amino acid sequence) of a polypeptide chain determines the conformation. Interactions among R groups on α-carbons of different amino acid residues result in one conformation being favored over other conformations.

Certain amino acid residues favor α-helical conformations, others favor β-pleated-sheet conformations, and yet others favor unordered regions. A detailed analysis of the particular amino acid residues and amino acid sequences which favor one or another conformation is beyond our scope.

Each polypeptide chain is a unique molecule, one of a large number of constitutional isomers. The number of different constitutional isomers for the polypeptides in proteins such as hemoglobin and collagen is such a large number (exceeding numbers such as 10^{100}) that we have difficulty imagining the number. This is yet another example of the great selectivity of biological systems.

In addition to hydrogen bonding among peptide linkages, interactions between R side groups of different amino acid residues are important in determining the conformation of a polypeptide chain. There are one covalent (*disulfide bridge*) and three noncovalent interactions (*hydrogen bonding*, *salt bridge*, *hydrophobic attraction*), each of which locks part of the polypeptide chain into a particular shape (conformation). A disulfide bridge forms by oxidation of mercaptan groups on a pair of cysteine residues (Fig. 20-3). The reaction is reversible under reducing conditions. (It is probably more appropriate to consider disulfide bridges as part of the primary structure of a polypeptide chain.) The three noncovalent interactions are shown in Fig. 20-4. Hydrophobic attractions are the dispersion or van der Waals attractions between hydrocarbon R groups of nonpolar amino acid residues. Hydrogen bonding occurs between a residue with an OH or NH and a residue with C$=$O, COOH, OH, or NH. Salt bridges (also referred to as *ionic attractions*) are the electrostatic attractions between COO^- and N^+ on acidic and basic side groups.

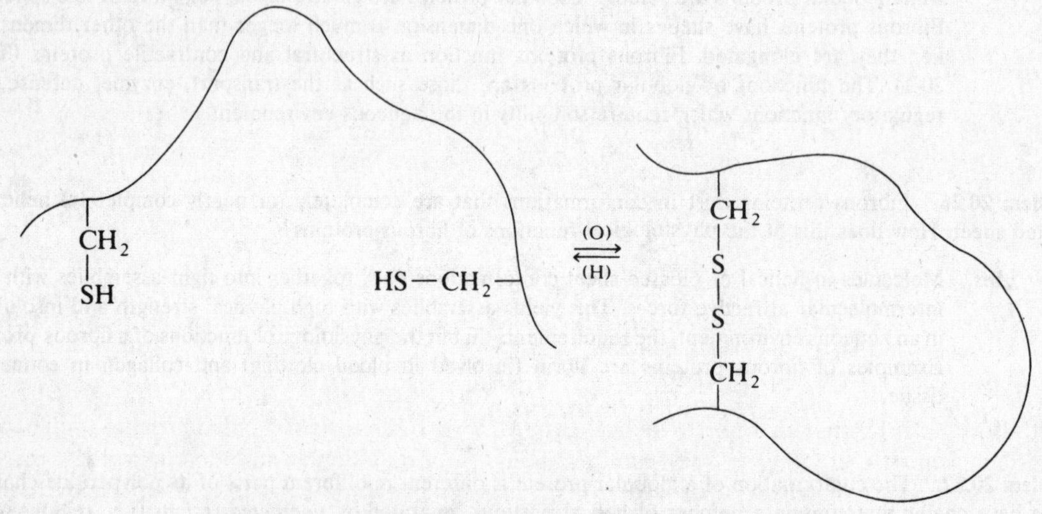

Fig. 20-3 Formation of disulfide bridge between Cys residues.

20.3.2 Fibrous Proteins

α-Keratin is the protein used in the construction of feathers, wool, nails, hair, cilia, horn, hoof, and skin. Electron microscopy and x-ray diffraction show a hierarchy of structures for these materials. Figure 20-5 shows a hair. The polypeptide chains in α-keratin are mostly α-helical. There are very

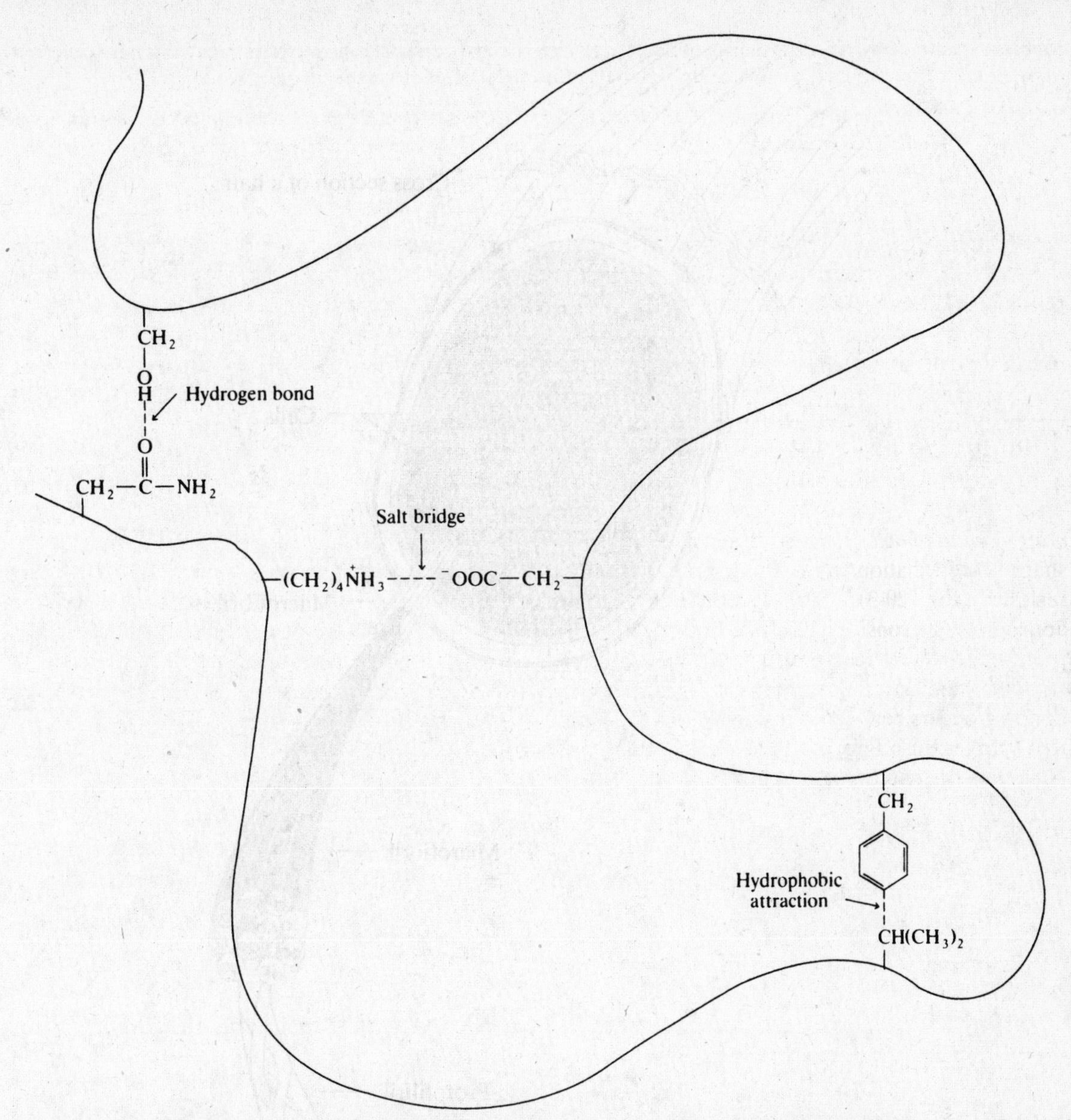

Fig. 20-4 Noncovalent attractive interactions of amino acid side groups: hydrogen bonding between Ser and Asn; salt bridge between Lys and Asp; hydrophobic attraction between Phe and Val.

short disordered regions separating α-helical regions. Two of these right-handed α-helical polypeptide chains coil around each other in a left-handed manner to form a double-stranded helical coil referred to as a *supercoil* or *superhelix*. Two supercoils coil around each other to form a left-handed helical *protofibril*. Protofibrils are coiled together to form *microfibrils* which are in turn coiled into *macrofibrils* which are further assembled into the hair. Coiling from one level to the next is always in the opposite direction. At each level—superhelices, protofibrils, microfibrils, and macrofibrils—structures are packed both side by side and lengthwise since the dimensions of a biological structure such as a hair are much greater than those of the superhelix, protofibril, microfibril, and macrofibril.

Problem 20.29. What keeps the two α-helices together in a superhelix? What keeps two superhelices together in a protofibril? What keeps the assembly of 11 protofibrils together in a microfibril? What keeps microfibrils together in a macrofibril?

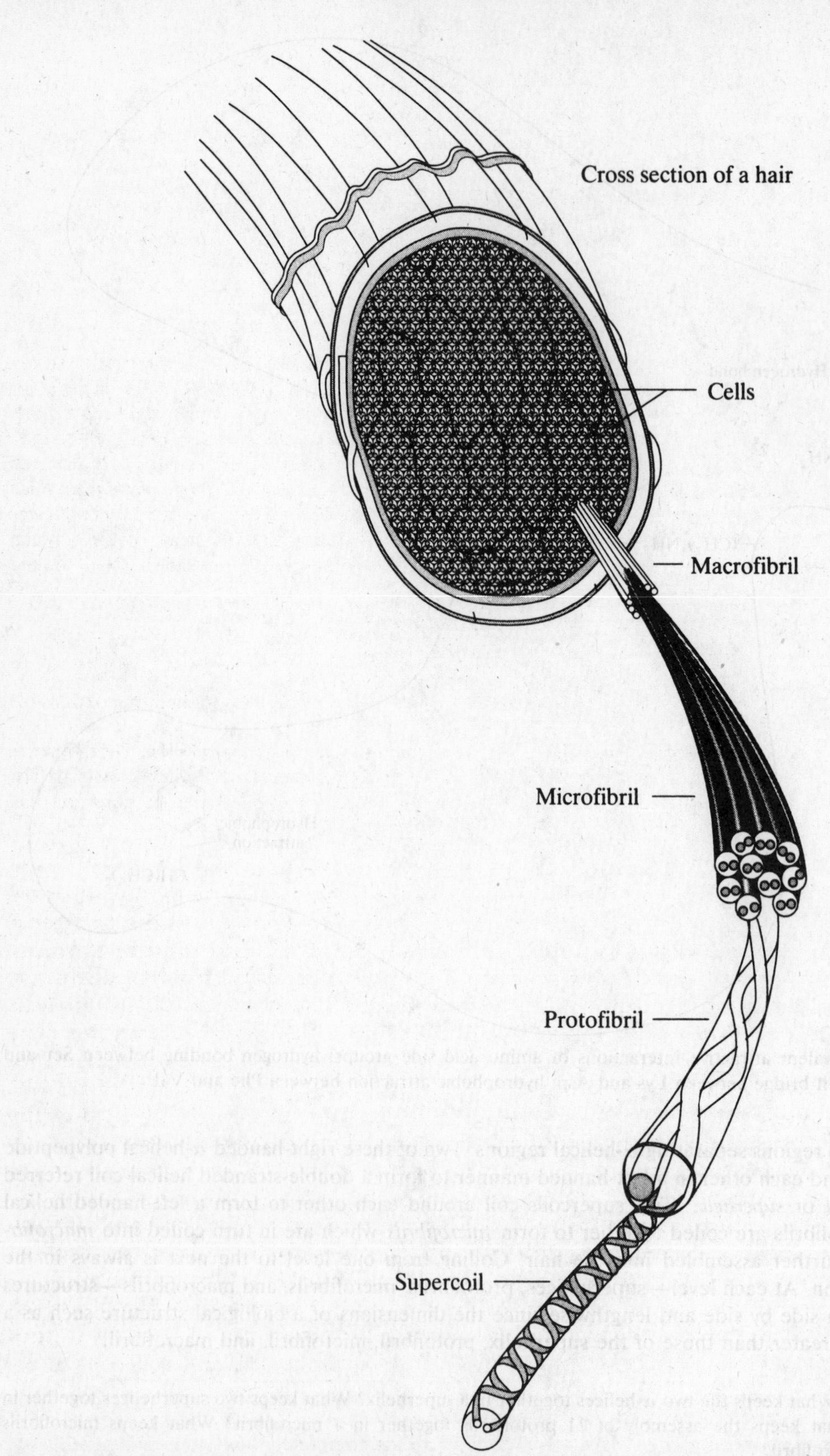

Fig. 20-5 Hierarchical structure of hair. (*After A. L. Lehninger, D. L. Nelson, and M. M. Cox, Principles of Biochemistry, 2nd ed., Worth Publishers, New York, 1993.*)

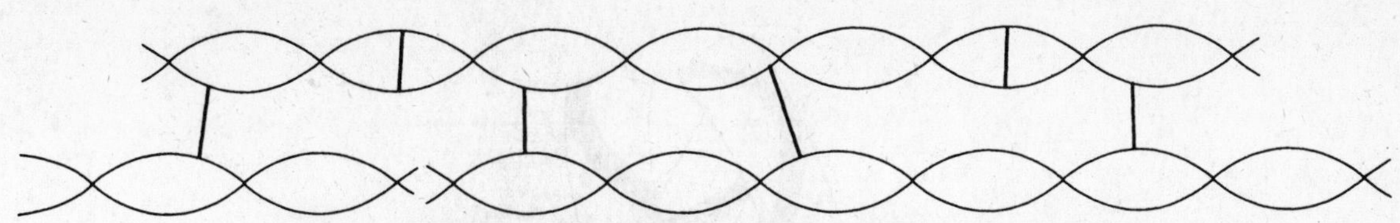

Fig. 20-6 Disulfide crosslinks between polypeptide chains in a protofibril.

Ans. Noncovalent interactions and disulfide bridges between different α-helices hold them together in a superhelix. Noncovalent interactions and disulfide bridges among polypeptide chains in different superhelices, in different protofibrils, and in different microfibrils hold together the superhelices, protofibrils, microfibrils, and macrofibrils. Disulfide bridges are more important than the noncovalent interactions in imparting insolubility, strength, and resistance to stretching. These disulfide bridges formed between different polypeptide chains are referred to as *intermolecular disulfide bridges* or *disulfide crosslinks*. (The disulfide bridges described in Fig. 20-3 are *intramolecular disulfide bridges*.) Figure 20-6 shows disulfide crosslinks (represented by heavy lines) between different polypeptide chains in a protofibril. Adjacent superhelices are not shown coiled around each other in order to simplify the representation of disulfide crosslinks.

Problem 20.30. Nail is harder than skin. How are these materials of different hardness formed from α-keratin?

Ans. The α-keratin polypeptide chains have similar but not exactly the same composition. The difference in hardness between nail and skin is a consequence of different amounts of disulfide crosslinks. The amount of cysteine residues is greater for the α-keratin polypeptides used to construct nail compared to the α-keratin polypeptides used to construct skin.

Collagen is the most abundant protein in mammals, comprising about one-third of the total protein. It is the major component of bone, cornea, tendons, and other connective tissues. Collagen is similar to keratin in having a hierarchical organization built up from helical polypeptide chains. However, there are substantial differences. The primary structure of the collagen polypeptide is very different compared to α-keratin, containing much more proline and glycine. This results in a helix different from the α-helix. The collagen polypeptide chain coils into a left-handed helix which is more elongated than the α-helix. This results in the formation of a right-handed superhelix, referred to as *tropocollagen* or *triple helix*, by the right-handed coiling of three polypeptide chains (Fig. 20-7). Tropocollagen coils are then organized into higher structures in a manner similar to α-keratin construction. Many other details are different for collagen compared to α-keratin. There is relatively little hydrogen bonding between peptide linkages in the same polypeptide chain. Extensive hydrogen bonding of peptide linkages occurs among the three polypeptide chains in a triple helix and also among polypeptide chains in different triple helices. Additional hydrogen bonding occurs through proline residues which are converted to 4-hydroxyproline (Hypro) residues during the biosynthesis of collagen. Crosslinking is critical to the properties of collagen but does not occur through cysteine residues as it does for most other proteins. Crosslinking of collagen occurs through a special reaction of lysine residues carried out after the formation of the collagen triple helix.

The structure of bone is even more complex. Bone consists of collagen fibrils imbedded in calcium phosphate polymer, $Ca_{10}(PO_4)_6(OH)_2$ (referred to as *hydroxyapatite*). The construction, analogous to that in reinforced materials, makes for very high physical strength.

There are very few proteins which are completely or nearly completely built from polypeptides of β-pleated-sheet conformations. *Silk fibroin*, produced by the silkworm, is an exception.

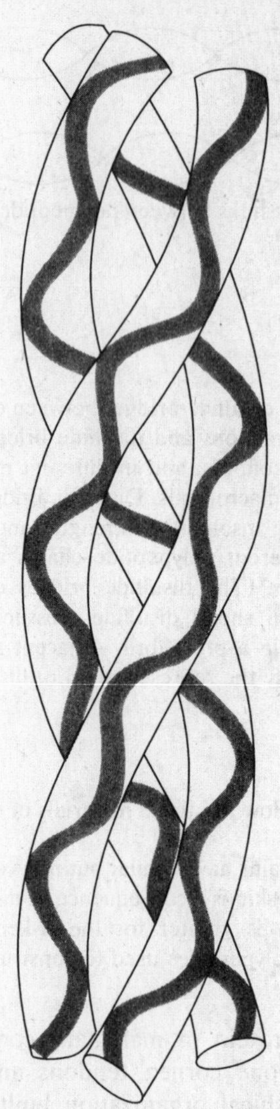

Fig. 20-7 Triple helix of collagen. (*After A. L. Lehninger, D. L. Nelson, and M. M. Cox, Principles of Biochemistry, 2nd ed., Worth Publishers, New York, 1993.*)

20.3.3 Globular Proteins

Hemoglobin binds molecular oxygen in the lungs and transports it to myoglobin in muscle. *Myoglobin* stores molecular oxygen until needed for metabolic oxidation. Myoglobin contains one polypeptide chain of 153 amino acid residues. More than three-fourths of the residues are contained in eight α-helical sections connected by short, disordered coils (referred to as *bends* or *turns*) (Fig. 20-8*a*). The bends contain mostly Pro, which is known to be an α-helix breaker, i.e., Pro does not fit into an α-helix. Nonpolar amino acid residues are located together in the interior where they are shielded from water. Hydrophobic attractions are important in folding of the polypeptide molecule. All of the hydrophilic side chains except for two His residues are located on the outer surfaces of the polypeptide. Hydrogen bonding with water results in the solubility of myoglobin in the aqueous environment. There is a cavity in the polypeptide molecule in which the prosthetic group *heme* is held by hydrophobic attractions (Fig. 20-8*b*). An Fe^{2+} ion is held in the center of heme by the four nitrogens of heme and the two His residues of myoglobin. It is the Fe^{2+} ion which holds and stores O_2.

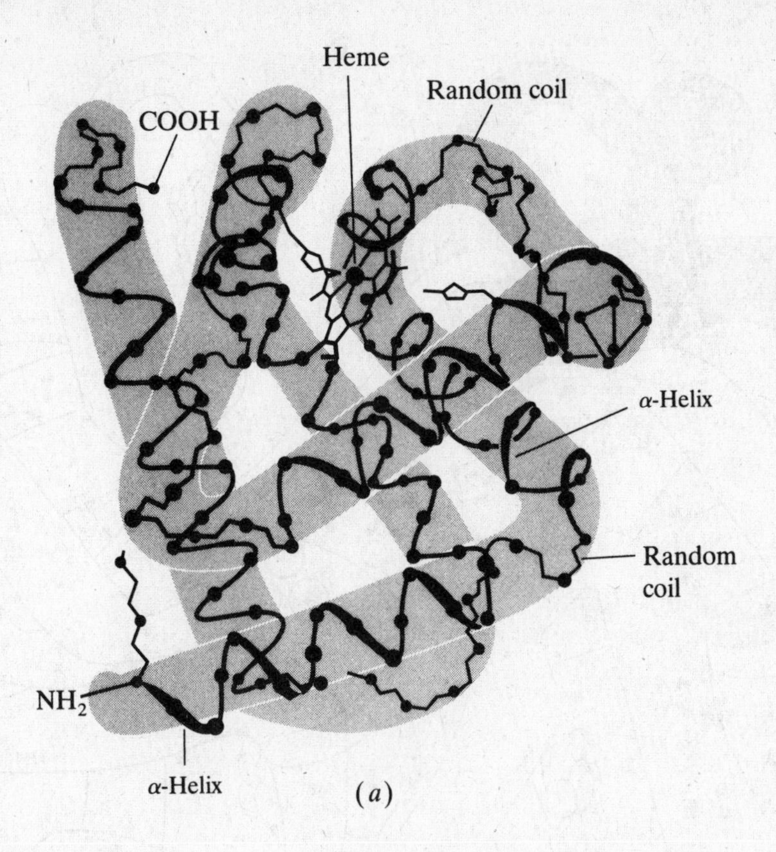

(*a*)

(*b*)

Fig. 20-8 (*a*) Myoglobin. (*After J. I. Kroschwitz and M. Winokur, Chemistry: General, Organic, Biological, McGraw-Hill, New York, 1990.* (*b*) Heme.

Problem 20.31. Which structural features in myoglobin are the primary, secondary, tertiary, and quaternary structures?

Ans. The primary structure is the particular amino acid sequence of myoglobin. There are two types of secondary structure in myoglobin—α-helical and bend conformations. The tertiary structure is the spatial relationship between the different α-helical and bend portions of the molecules and the placement of heme and Fe^{2+} ion. There is no quaternary structure for myoglobin. Quaternary structure relates only to proteins with two or more polypeptide chains.

Hemoglobin consists of an assembly of four separate polypeptide chains, two α-chains of 141 amino acid residues each and two β-chains of 146 amino acid residues each. The α- and β-chains are

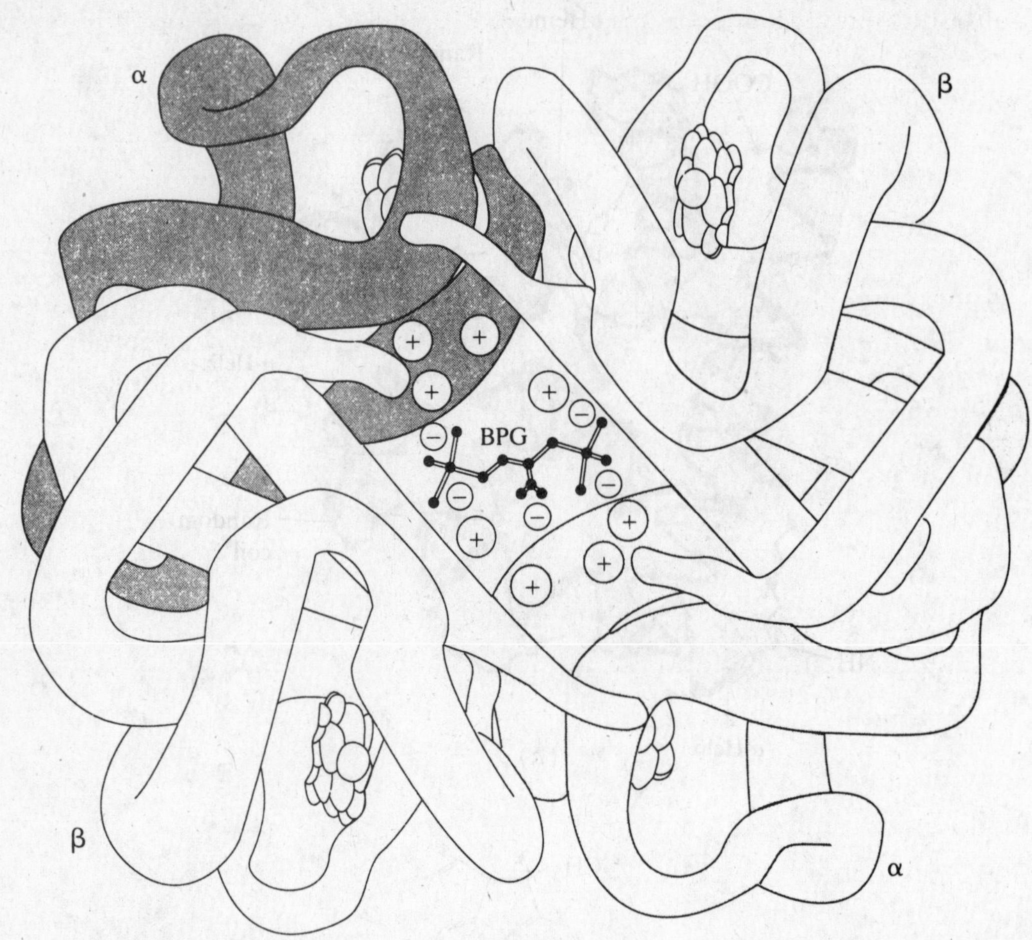

Fig. 20-9 Hemoglobin. (*After P. W. Kuchel and G. B. Ralston, Theory and Problems of Biochemistry, Schaum's Outline Series, McGraw-Hill, New York, 1988.*)

generally similar to the myoglobin polypeptide and this results in similar secondary and tertiary structures (α-helical portions separated by bends). The differences in the primary structures are such that there are "left over" hydrophobic surfaces after the α- and β-chains fold into their compact shapes. These hydrophobic surfaces are shielded from water by the aggregation of the two α- and two β-chains by the mutual attraction of their hydrophobic surfaces. The spatial relationship of the four polypeptide chains is shown in Fig. 20-9. Each α- and each β-chain has a heme and each heme has an Fe^{2+} ion. There is a central cavity in deoxygenated hemoglobin which binds 2,3-bisphosphoglycerate (BPG), also called 2,3-diphosphoglycerate (DPG).

Problem 20.32. What holds BPG in the central cavity of hemoglobin?

Ans. BPG is an anion with a 5− charge. There are basic amino acid residues in the portions of the β-chains in the vicinity of the central cavity and their positive charges bind BPG through ionic attractions:

$$\begin{array}{c} \quad\quad\quad O-PO_3^{2-} \\ \quad\quad\quad | \\ ^{-}OOC-CH-CH_2-O-PO_3^{2-} \end{array}$$

BPG

Problem 20.33. Oxygenated hemoglobin does not contain BPG, i.e., BPG leaves hemoglobin when hemoglobin picks up oxygen. Explain.

Ans. The conformational structure of the polypeptide chains changes somewhat when hemoglobin binds oxygen. The size of the central cavity decreases to the extent that it is too small to hold BPG.

Figure 20-10 shows the structure of hen egg lysozyme, a single polypeptide chain of 120 ammo acid residues with four disulfide linkages. It contains two α-helical regions, three β-pleated-sheet regions, and several regions of less regular conformation (*loops*). Here we see intramolecular side-by-side alignment of β-pleated-sheet regions, in comparison to the intermolecular alignment of β-pleated sheets previously described. The β-pleated-sheet regions are shown as arrows to indicate the direction from the N-terminal residue to the C-terminal residue. The head of the arrow is the C-terminal residue. The β-pleated-sheet regions are in antiparallel alignments in hen egg lysozyme. Proteins with both parallel and antiparallel alignment of β-pleated-sheet regions are known.

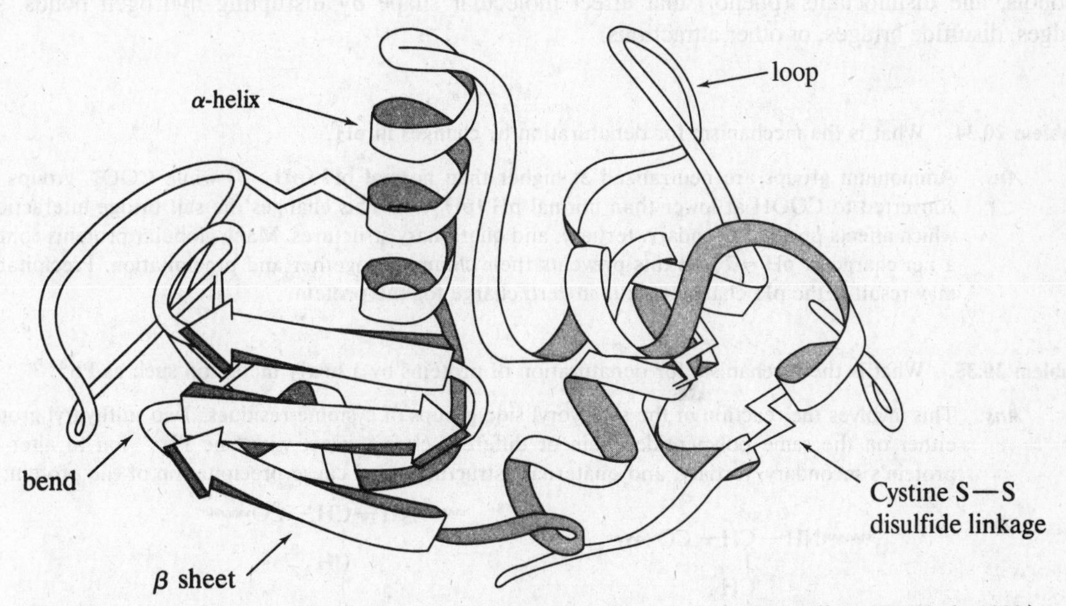

Fig. 20-10 Hen egg lysozyme. (*After G. Zubay, Biochemistry, 2nd ed., Macmillan, New York, 1988.*)

20.3.4 Denaturation

Denaturation is the change in physiological function of a protein under the influence of a *denaturing agent*. Denaturation is different than digestion. Digestion involves the hydrolysis of peptide linkages, i.e., the destruction of primary structure. Denaturation involves a change in secondary, tertiary, or quaternary structure, not a change in primary structure. These levels of structure are a consequence of the lower forces of noncovalent attractions which are disrupted by changes in the biological environment. Globular proteins tend to denature more easily than fibrous proteins because the noncovalent interactions are stronger in the latter. Proteins with extensive disulfide bridges are more resistant to denaturation since those bonds are stronger than the noncovalent interactions. Denaturation under mild conditions is generally reversible; denaturation under extreme conditions may not be reversible. An undenatured protein is often referred to as a *native protein*. Denaturation often results in the insolubility (precipitation) of polypeptide molecules due to an entangling of the chains after the unraveling of the secondary, tertiary, and quaternary structures.

Heat increases the thermal energy of the molecules, bonds rotate more easily, and hydrogen bonding of peptide linkages as well as other noncovalent interactions are disrupted. The physiological functions of a variety of proteins are affected, resulting in illness and even death. Temperatures above 106 °C are very dangerous for humans. On the other hand, we take advantage of the effect of heat on proteins for sterilizing surgical instruments, sealing small blood vessels in surgery by cauterization, and canning foods. Cooking meats, poultry, and fish not only kills microorganisms but makes the digestion of these foods easier by unraveling their conformational structure. Ultraviolet and ionizing radiations have effects similar to heat.

Mechanical energy (violent mixing) is used to whip (denature) egg white protein into a frothy solid in recipes for meringues and souffles. Mixing causes the unraveling of polypeptide chains from their native structure followed by entanglement with each other and precipitation.

Other denaturing agents include detergents, heavy metal ions (Ag^+, Hg^{2+}, Pb^{2+}), changes in pH, alcohols, and disinfectants (phenol) that affect molecular shape by disrupting hydrogen bonds, salt bridges, disulfide bridges, or other attractions.

Problem 20.34. What is the mechanism for denaturation by changes in pH?

Ans. Ammonium groups are neutralized at higher than normal pH (pH > 7) while COO^- groups are converted to COOH at lower than normal pH (pH < 7). This changes the salt bridge interactions, which affects protein secondary, tertiary, and quaternary structures. Many globular proteins contain a net charge at pH = 7 and this prevents their clumping together and precipitation. Precipitation may result if the pH change results in zero charge for the protein.

Problem 20.35. What is the mechanism for denaturation of proteins by a heavy metal ion such as Pb^{2+}?

Ans. This involves the reaction of the sulfhydryl side groups of cysteine residues. Two sulfhydryl groups, either on the same polypeptide chain or different chains, react with one Pb^{2+} ion to alter the protein's secondary, tertiary, and quaternary structures, and cause precipitation of the protein:

```
                                          ~~~~NH—CH—CO~~~~
~~~~NH—CH—CO~~~~                                  |
        |                                        CH2
       CH2                                        |
        |                                         S
       SH                                         |
                        Pb2+                      Pb
                      ------>                     |
       SH                                         S
        |                                         |
       CH2                                       CH2
        |                                         |
~~~~NH—CH—CO~~~~                          ~~~~NH—CH—CO~~~~
```

ADDITIONAL SOLVED PROBLEMS

AMINO ACIDS

Problem 20.36. What is the relationship of asparagine and glutamine to aspartic and glutamic acids?

Ans. Asparagine and glutamine are the amides derived from aspartic and glutamic acids, respectively.

Problem 20.37. What is the structural relationship of phenylalanine and tyrosine?

Ans. Everything is the same except that tyrosine has an OH group in the para position of the benzene ring.

Problem 20.38. Amino acids are water-soluble but their solubility varies with pH. Solubility is lowest at the isoelectric point. Why?

Ans. Zwitterions are less soluble than the positive and negative ions present at pH values below and above pI, respectively. Zwitterions associate together to form larger-sized aggregates with lower solubility.

Problem 20.39. Give the three-letter symbol for amino acid(s) whose side group contains (*a*) an aromatic group, (*b*) a primary alcohol, (*c*) a secondary alcohol, (*d*) sulfur, (*e*) phenol, (*f*) an amide.

Ans. (*a*) Phe, Trp, Tyr; (*b*) Ser; (*c*) Thr; (*d*) Met, Cys; (*e*) Tyr; (*f*) Asn, Gln.

Problem 20.40. Which has the more polar side group, alanine or serine?

Ans. Serine. Serine has CH_2OH while alanine has CH_3.

Problem 20.41. At what pH will Thr exist in the zwitterion form?

Ans. At its isoelectric point, 5.60.

Problem 20.42. Draw the structure of asparagine at pH = 1, 5.41, 7, and 12. To which electrode will asparagine migrate in electrophoresis at each pH?

Ans.

$$\overset{+}{N}H_3-\underset{}{CH}(CH_2\overset{O}{\overset{\|}{C}}NH_2)-COOH \qquad \text{pH} = 1$$

$$\overset{+}{N}H_3-CH(CH_2\overset{O}{\overset{\|}{C}}NH_2)-COO^- \qquad \text{pH} = 5.41, 7$$

$$NH_2-CH(CH_2\overset{O}{\overset{\|}{C}}NH_2)-COO^- \qquad \text{pH} = 12$$

There is migration to the negative electrode at pH = 1. No migration occurs at pH = 5.41 or 7. There is migration to the positive electrode at pH = 12.

Problem 20.43. An unknown, containing proline and/or glutamic acid, is subjected to paper electrophoresis at pH = 7. Ninhydrin treatment shows amino acid only at the positive electrode. Which amino acid(s) is (are) present in the unknown? Explain how you arrived at your answer.

Ans. Only glutamic acid is present. At pH = 7, all amino and carboxyl groups are ionized. Glutamic acid has a net charge of 1− and migrates to the positive electrode. Proline would be present as the zwitterion (no net charge) at this pH and would not migrate (i.e., it would be observed in the middle of the paper strip).

PEPTIDE FORMATION

Problem 20.44. Draw the structure of Leu-Pro-Ser.

Ans.

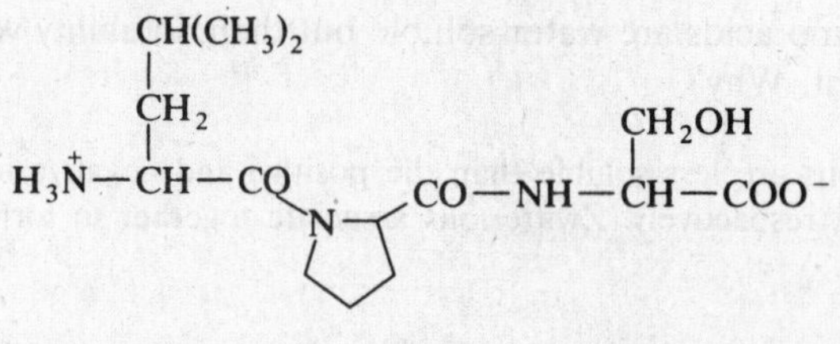

$$\begin{array}{l} CH(CH_3)_2 \\ | \\ CH_2 \qquad\qquad\qquad CH_2OH \\ | \qquad\qquad\qquad\qquad | \\ H_3\overset{+}{N}-CH-CO\text{—}N(\text{pyrrolidine ring})\text{—}CO-NH-CH-COO^- \end{array}$$

Problem 20.45. Draw the complete structures of Phe-Asp-Ala at pH = pI, 1, 7, and 12. Explain how you arrived at your answer.

Ans. The pI of Phe-Asp-Ala is approximately the same as the pI of Asp. The answer to this question is similar to that for the amino acid Asp (see Problem 20.9). Zwitterion (structure I below) exists at pH = 2.77 since that is the pI of the tripeptide. At neutral pH (more basic than pI), the side chain COOH loses its proton to yield structure II, which has three charge centers and a net charge of 1−. At pH = 12, the ammonium group loses its proton to yield structure III which has two charge centers and a net charge of 2−. At pH = 1 (more acidic than pI), the carboxyl group becomes protonated to yield structure IV, which has one charge center and a net change of 1+:

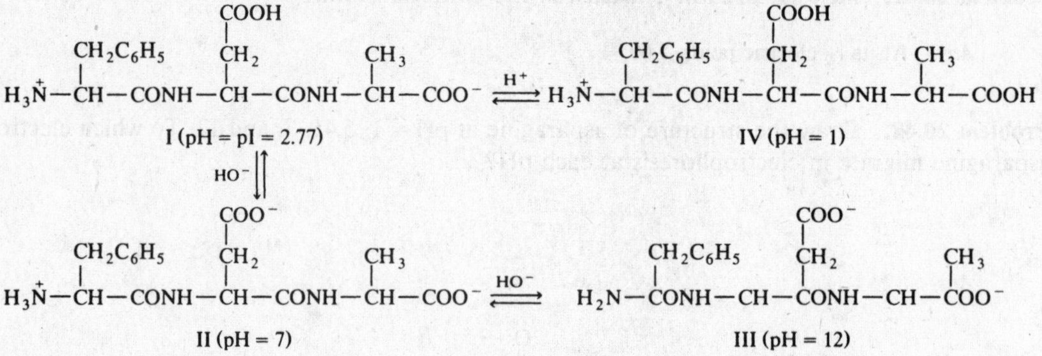

$$H_3\overset{+}{N}-\underset{CH_2C_6H_5}{CH}-CONH-\underset{CH_2COOH}{CH}-CONH-\underset{CH_3}{CH}-COO^- \underset{}{\overset{H^+}{\rightleftharpoons}} H_3\overset{+}{N}-\underset{CH_2C_6H_5}{CH}-CONH-\underset{CH_2COOH}{CH}-CONH-\underset{CH_3}{CH}-COOH$$

I (pH = pI = 2.77) IV (pH = 1)

$$\Big\downarrow HO^-$$

$$H_3\overset{+}{N}-\underset{CH_2C_6H_5}{CH}-CONH-\underset{CH_2COO^-}{CH}-CONH-\underset{CH_3}{CH}-COO^- \overset{HO^-}{\rightleftharpoons} H_2N-\underset{CH_2C_6H_5}{CONH}-CH-\underset{CH_2COO^-}{CONH}-CH-\underset{CH_3}{COO^-}$$

II (pH = 7) III (pH = 12)

PROTEIN STRUCTURE AND FUNCTION

Problem 20.46. Draw the structure of Met-Phe-Thr-Gly-Asp.

Ans.

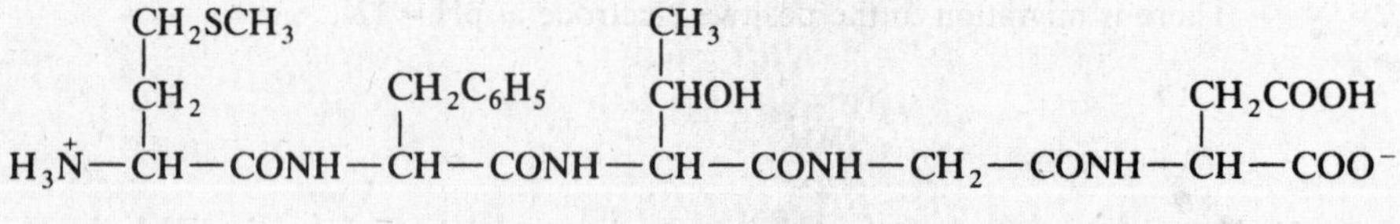

$$H_3\overset{+}{N}-\underset{CH_2CH_2SCH_3}{CH}-CONH-\underset{CH_2C_6H_5}{CH}-CONH-\underset{CHOH-CH_3}{CH}-CONH-CH_2-CONH-\underset{CH_2COOH}{CH}-COO^-$$

Problem 20.47. Which structural features in hemoglobin are the primary, secondary, tertiary, and quaternary structures?

Ans. The primary structure is the particular amino acid sequence in each of the α- and β-polypeptide chains. There are two types of secondary structure—α-helical and bend conformations. Tertiary structure is the spatial relationship between the different α-helical and bend portions of the molecules and the placement of heme and the Fe^{2+} ion relative to the polypeptide. Quaternary structure relates to the spatial relationship between the four polypeptide chains.

Problem 20.48. Explain the physiological function achieved by the coiling of polypeptide chains as well as the coiling of higher levels of structure for proteins such as α-keratin and collagen. Is there any advantage to having successive levels of structure coiled in opposite directions?

Ans. These structural features have the function of increased physical strength. Coiled chains carry a greater load than completely extended, uncoiled chains. Overall strength is increased by having successive levels coiled in opposite directions since the uncoiling of any level of structure is resisted not only by that level but also by the levels above and below. Cables for suspension bridges are constructed in the same manner.

Problem 20.49. What is the mechanism for the denaturation of proteins by a detergent? By alcohol?

Ans. The large nonpolar portions of detergent molecules interact with the hydrophobic portions of a polypeptide to alter the polypeptide's folding and conformational structure. Alcohol, being less polar than water, interacts differently than water (less hydrogen bonding) with the polar side groups of amino acid residues and this alters the polypeptide's folding and conformational structure.

Problem 20.50. Give the products of digestion (hydrolysis) of the following peptide:

$$\begin{array}{l}
\mathrm{H_3\overset{+}{N}-CH(CH_2C_6H_5)-CONH-CH-CONH-CH(CH_3)-COO^-} \\
\qquad\qquad\qquad\qquad\qquad\quad |\\
\qquad\qquad\qquad\qquad\qquad\mathrm{CH_2-S-S-CH_2} \\
\qquad\qquad\qquad\qquad\qquad\qquad\qquad |\\
\mathrm{H_3\overset{+}{N}-CH(CH_2C_6H_5)-CONH-CH-CONH-CH(CH_3)-COO^-}
\end{array}$$

Ans.

$$\begin{array}{c}
\mathrm{H_3\overset{+}{N}-CH-COO^-} \\
| \\
\mathrm{CH_2} \\
| \\
\mathrm{S} \\
| \\
\mathrm{S} \\
| \\
\mathrm{CH_2} \\
| \\
\mathrm{H_3\overset{+}{N}-CH-COO^-}
\end{array}$$

$$\mathrm{2\,H_3\overset{+}{N}-CH(CH_2C_6H_5)-COO^- \; + \; (\text{the above}) \; + \; 2\,H_3\overset{+}{N}-CH(CH_3)-COO^-}$$

Digestion cleaves the peptide bonds but not the disulfide bridge.

Problem 20.51. Disulfides present after digestion are reduced in the liver. Write the equation showing reduction of the disulfide in the answer to Problem 20.50.

Ans.

$$\begin{array}{c}
\mathrm{H_3\overset{+}{N}-CH-COO^-} \\
| \\
\mathrm{CH_2} \\
| \\
\mathrm{S} \\
| \\
\mathrm{S} \\
| \\
\mathrm{CH_2} \\
| \\
\mathrm{H_3\overset{+}{N}-CH-COO^-}
\end{array}
\xrightarrow{\text{(H)}}
\begin{array}{c}
\mathrm{SH} \\
| \\
\mathrm{CH_2} \\
| \\
\mathrm{2\,H_3\overset{+}{N}-CH-COO^-}
\end{array}$$

Problem 20.52. What is the most likely attractive interaction, if any, between the side chains in each of the following pairs of amino acid residues? (*a*) Phe and Leu, (*b*) Glu and Leu, (*c*) Glu and Lys, (*d*) Asp and Ser, (*e*) Cys and Cys, (*f*) Glu and Glu.

Ans. (*a*) Hydrophobic attractions; (*b*) none since Glu is polar and Leu is nonpolar; (*c*) salt bridge; (*d*) hydrogen bonding; (*e*) disulfide linkage; (*f*) none since COO^- groups repulse each other

Problem 20.53. Sickle cell anemia is an inherited abnormality of hemoglobin. The sixth residue in the β-polypeptide chain, normally Glu, is replaced by Val. This alters the shape of deoxygenated hemoglobin from biconcave disks to crescents, with the result that red blood cells change shape, tend to clump together, and reduce the flow capacity of capillaries. Why does the change of one residue have such a detrimental effect? Which amino acid, if substituted for Glu, might not have a large effect?

Ans. The side group in Glu is ionic (COO^-) and most likely participates in a salt bridge. Replacement of Glu by the hydrophobic Val destroys that salt bridge and results in a change in conformational structure. Replacement of Glu by Asp might not be a problem since both have a COO^- (although the size of the side groups are not exactly the same).

Problem 20.54. Humans with the disease diabetes mellitus do not produce sufficient insulin. Treatment involves subcutaneous injection of pig insulin. Interestingly, pig insulin differs from human insulin in only one amino acid residue—the polar Thr is replaced by the hydrophobic Ala. Compare this case with sickle cell anemia previously described in Problem 20.53. Explain the effectiveness of pig insulin.

Ans. The amino acid residue in the pig insulin chain must be in a portion of the chain whose conformation is not greatly affected by whether the residue is Thr or Ala, or that portion of the protein is not involved in the physiological functioning of insulin.

Problem 20.55. Consider a globular protein imbedded in a lipid bilayer. Which of the following amino acid residues will be in contact with the interior of the bilayer? Phe, Glu, Lys, Leu.

Ans. The nonpolar residues, Phe and Leu, are in contact with the nonpolar interior of the bilayer.

Problem 20.56. The isoelectric point of lysozyme is 11.0. Which of the following residues are present in large amounts? Asp, Glu, His, Lys.

Ans. The pI is on the basic side of neutral. This occurs when there is an excess of basic residues (His, Lys) compared to acidic residues (Asp, Glu).

SUPPLEMENTARY PROBLEMS

Problem 20.57. Consider a mixture of albumin (pI = 4.9), hemoglobin (pI = 6.8), and lysozyme (pI = 11.0). What occurs in the electrophoresis experiment at pH = 6.8?

Ans. At pH = 6.8, hemoglobin has no net charge, albumin has a negative charge, and lysozyme has a positive charge. Hemoglobin does not move from the center of the paper. Lysozyme and albumin migrate to the negative and positive electrodes, respectively.

Problem 20.58. What is the structure of aspartic acid at pH = 1, 2.77, 7, and 12? To which electrode does Asp migrate at each pH?

Ans. The zwitterion (structure I below) exists at pH = 2.77 since this is the pI of aspartic acid. At neutral pH (more basic than pI), the side chain COOH loses its proton to yield structure II, which has three charge centers and a net charge of 1 −. At pH = 12, the ammonium group loses its proton to yield structure III, which has two charge centers and a net charge of 2 −. At pH = 1 (more acidic than pI), the carboxyl group becomes protonated to yield structure IV, which has one charge center and a net charge of 1 +.

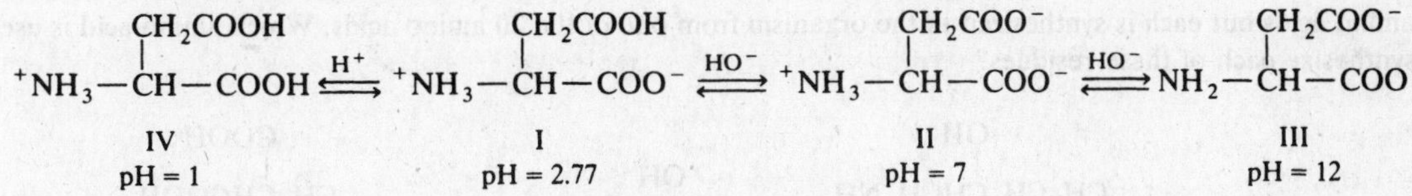

$$\underset{\substack{\text{IV}\\ \text{pH}=1}}{{}^{+}NH_3-\underset{}{\overset{CH_2COOH}{\overset{|}{C}H}}-COOH} \underset{H^+}{\rightleftharpoons} \underset{\substack{\text{I}\\ \text{pH}=2.77}}{{}^{+}NH_3-\overset{CH_2COOH}{\overset{|}{C}H}-COO^-} \underset{HO^-}{\rightleftharpoons} \underset{\substack{\text{II}\\ \text{pH}=7}}{{}^{+}NH_3-\overset{CH_2COO^-}{\overset{|}{C}H}-COO^-} \underset{HO^-}{\rightleftharpoons} \underset{\substack{\text{III}\\ \text{pH}=12}}{NH_2-\overset{CH_2COO^-}{\overset{|}{C}H}-COO^-}$$

No migration occurs at pH = 2.77. Migration occurs to the negative electrode at pH = 1. Migration occurs to the positive electrode at pH = 7 and 12.

Problem 20.59. Does the following form of arginine predominate at pH = 1, 7, 10.76, or 12?

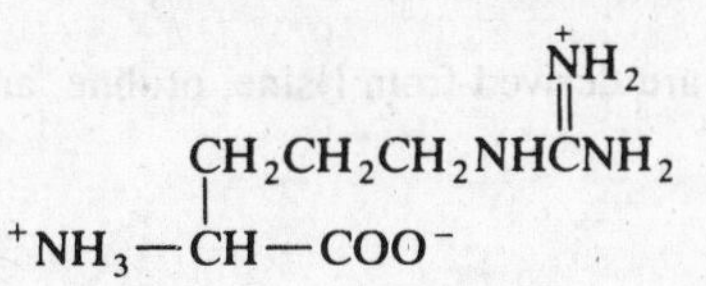

$${}^{+}NH_3-\overset{CH_2CH_2CH_2NH\overset{\overset{+}{N}H_2}{\overset{\|}{C}}NH_2}{\overset{|}{C}H}-COO^-$$

Ans. pH = 7

Problem 20.60. Draw L-α-tyrosine.

Ans.

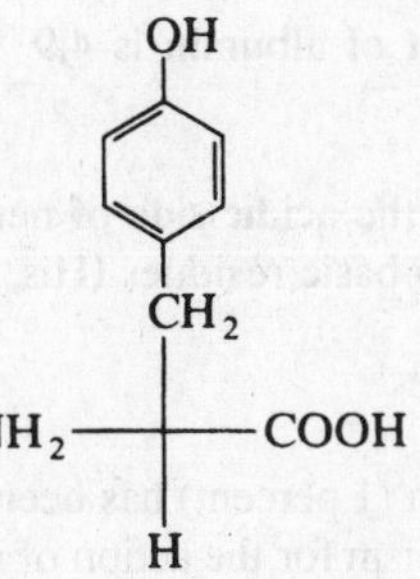

$$NH_2-\overset{\overset{p\text{-}HOC_6H_4}{|}\,CH_2}{\underset{H}{\overset{|}{\underset{|}{C}}}}-COOH$$

Problem 20.61. Show the synthesis of His-Ala-Asn-Gly-Ser from the five amino acids.

Ans.

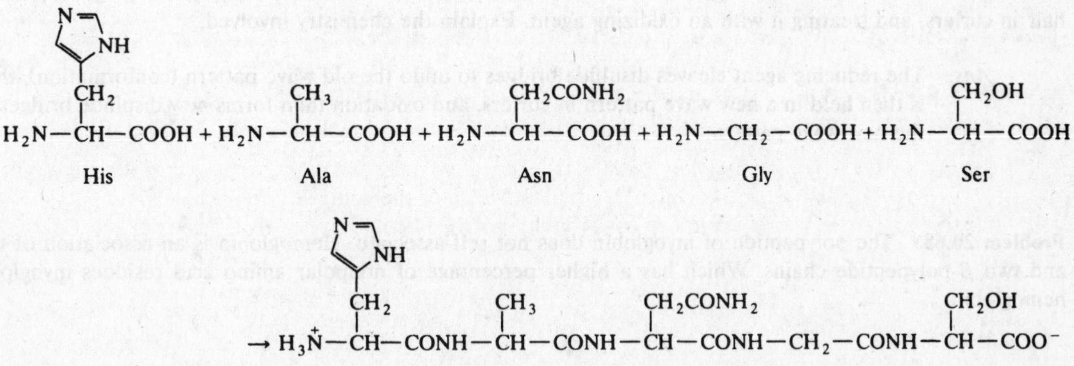

$$\underset{\text{His}}{H_2N-\overset{\overset{\text{imidazol-4-yl}}{|}CH_2}{\overset{|}{C}H}-COOH} + \underset{\text{Ala}}{H_2N-\overset{CH_3}{\overset{|}{C}H}-COOH} + \underset{\text{Asn}}{H_2N-\overset{CH_2CONH_2}{\overset{|}{C}H}-COOH} + \underset{\text{Gly}}{H_2N-CH_2-COOH} + \underset{\text{Ser}}{H_2N-\overset{CH_2OH}{\overset{|}{C}H}-COOH}$$

$$\rightarrow H_3\overset{+}{N}-\overset{\overset{\text{imidazol-4-yl}}{|}CH_2}{\overset{|}{C}H}-CONH-\overset{CH_3}{\overset{|}{C}H}-CONH-\overset{CH_2CONH_2}{\overset{|}{C}H}-CONH-CH_2-CONH-\overset{CH_2OH}{\overset{|}{C}H}-COO^-$$

Problem 20.62. What is the total number of constitutional isomers for the peptide of Problem 20.61?

Ans. $5! = 5 \times 4 \times 3 \times 2 \times 1 = 120$

Problem 20.63. The following amino acid residues are found in certain proteins. None of these are one of the 20 amino acids but each is synthesized in the organism from one of the 20 amino acids. Which amino acid is used to synthesize each of these residues?

I: $-NH-CH(CH_2CH_2CH(OH)CH_2NH_2)-CO-$

II: 4-hydroxyproline residue (ring N with OH on the ring, $CO-$)

III: $-NH-CH(CH_2CH(COOH)COOH)-CO-$

Ans. Structures I, II, and III are derived from lysine, proline, and glutamic acid, respectively.

Problem 20.64. Which of the following amino acid residues are likely to be in the interior of a globular protein? Ile, Phe, Glu, His, Lys, Asp, Trp.

Ans. Ile, Phe, Trp

Problem 20.65. The isoelectric point of albumin is 4.9. Which of the following residues are present in large amounts? Asp, Glu, His, Lys.

Ans. An isoelectric point on the acidic side of neutral occurs when there is an excess of acidic residues (Asp, Glu) compared to basic residues (His, Lys).

Problem 20.66. Silver nitrate solution (1 percent) has been placed in the eyes of newborn babies as a disinfectant against gonorrhea. What is the mechanism for the action of silver nitrate?

Ans. Ag^+ denatures gonorrhea proteins by reaction with sulfhydryl groups of Cys residues.

Problem 20.67. Permanent waving of hair involves the treatment of the hair with a reducing agent, placing the hair in curlers, and treating it with an oxidizing agent. Explain the chemistry involved.

Ans. The reducing agent cleaves disulfide bridges to undo the old wave pattern (conformation), the hair is then held in a new wave pattern in curlers, and oxidation then forms new disulfide bridges to set the new wave pattern.

Problem 20.68. The polypeptide of myoglobin does not self-associate. Hemoglobin is an association of two α- and two β-polypeptide chains. Which has a higher percentage of nonpolar amino acid residues myoglobin or hemoglobin.

Ans. Hemoglobin. The fact that the polypeptide of myoglobin does not self-associate indicates there are very few nonpolar amino acid residues on its exterior surface. There is no need for self-association to shield nonpolar residues from water. The folded α- and β-polypeptide chains of hemoglobin have more exterior nonpolar residues. The association of two α- and two β-polypeptide chains shields the nonpolar residues from water.

Problem 20.69. Certain amino acid residues favor or disfavor α-helical conformations while others favor or disfavor β-pleated-sheet conformations. The α-helical conformation is prevented from forming by Pro and by two or more consecutive residues with charged side groups of like charge. The β-pleated-sheet conformation is favored only for structures with small side groups, specifically, a high content of Gly with most other residues being Ser and Ala. The collagen superhelix requires a large content of Pro-X-Gly and/or Hypro-X-Gly sequences, where X is any amino acid residue and Hypro is 4-hydroxyproline. Table 20-4 gives the amino acid composition of three unknown proteins, compounds I, II, and III. One of the proteins is silk (β-pleated sheet), one is α-keratin (extensive α-helix), and the other is collagen. Which is which?

Table 20-4. Amino Acid Composition of Proteins I, II, and III in Problem 20.69

	Percentage in protein		
Residues	I	II	III
Ala	29.4	5.0	12.0
Arg	0.5	7.2	5.0
Asp + Asn	1.3	6.0	4.5
Cys	0	11.2	0
Glu + Gln	1.0	12.1	7.7
Gly	44.6	8.6	32.7
His	0.2	0.7	0.4
Hypro	0	0	8.6
Ile	0.7	2.8	0.9
Leu	0.5	6.9	2.1
Lys	0.3	2.3	3.7
Met	0	0.5	0.7
Phe	0.5	2.5	1.2
Pro	0.3	7.5	13.9
Ser	12.2	10.2	3.4
Trp	0.2	1.2	0
Tyr	5.2	4.2	0.4
Val	2.2	5.1	1.8

Ans. Compound I is silk since it has the very high Gly content along with large amounts of Ala and Ser. Gly, Ala, and Ser account for 86.2 percent of the amino acid residues. Compound III is collagen since it contains Hypro and the amount of Hypro + Pro is similar to the amount of Gly, indicating large amounts of Pro-X-Gly and Hypro-X-Gly sequences. Compound II is α-keratin since the composition is not conducive to β-pleated sheet (no large amounts of Gly, Ser, Ala) or collagen superhelix (no large amounts of Pro-X-Gly and Hypro-X-Gly sequences) and the amount of Pro is not high.

Problem 20.70. Which structural features in α-keratin are the primary, secondary, tertiary, and quaternary structures?

Ans. The primary structure is the particular amino acid sequence in the polypeptide chain. There are two types of secondary structure—mostly α-helical with small amounts of disordered conformations. Tertiary structure is the spatial relationship between the α-helical and disordered portions of the polypeptide chain. Quaternary structure describes the spatial relationship between the different polypeptide chains in forming superhelices, protofibrils, microfibrils, macrofibrils, and fibrils.

Problem 20.71. Write the equation for the hydrolysis of Ser-Ala-Asn-Gly-His.

Ans.

$$\overset{+}{H_3N}-\underset{CH_2OH}{CH}-CONH-\underset{CH_3}{CH}-CONH-\underset{CH_2CONH_2}{CH}-CONH-CH_2-CONH-\underset{CH_2(\text{imidazolyl})}{CH}-COO^-$$

$$\xrightarrow{H_2O} \underset{\text{Ser}}{H_2N-\underset{CH_2OH}{CH}-COOH} + \underset{\text{Ala}}{H_2N-\underset{CH_3}{CH}-COOH} + \underset{\text{Asn}}{H_2N-\underset{CH_2CONH_2}{CH}-COOH}$$

$$+ \underset{\text{Gly}}{H_2N-CH_2-COOH} + \underset{\text{His}}{H_2N-\underset{CH_2(\text{imidazolyl})}{CH}-COOH}$$

Problem 20.72. Some diseases are caused by the deposition of amyloid plaques. What are amyloid plaques, how do they form, and how do they cause loss of physiological function?

Ans. A native protein possesses a conformation (Sec. 20.3.4) that is critical to its normal physiological function. If denaturation (misfolding) of the protein occurs, either the native conformation is reestablished (by proteins called *chaperones*) or the denatured protein is degraded (by hydrolysis and disulfide reduction) and new native protein synthesized. Some diseases involve the formation of denatured proteins that are neither renatured nor degraded, but are instead deposited extracellularly in cells, tissues, and organs as insoluble material called *amyloid plaques.* The deposited amyloid plaque interferes with normal physiological functions with subsequent cell and organ failure and, eventually even death. Mad cow and Alzheimer's diseases are examples of neurodegenerative diseases that result from this process. Amyloid plaques are deposited in brain tissue, causing the death of nerve cells and evenually death of the organism.

Problem 20.73. What are prion diseases?

Ans. Scrapie disease in sheep and goats and mad cow disease in cattle are examples of prion diseases. The outer surfaces of normal neurons in these organisms contain a native protein called *normal prion protein*, whose conformation is highly α-helix with almost no β-pleated sheet. The diseases involve a change in conformation of the normal prion protein into *altered prion protein*, whose conformation has a much lower α-helix content but a high β-pleated sheet content. This conformation change occurs when the animals ingest nerve tissue from diseased animals (with scrapie or mad cow diseases). The altered prion protein is neither renatured nor degraded, but is deposited in brain tissue as amyloid plaque. The result is death of nerve cells and eventually death of the organism.

Nucleic Acids and Heredity

Chromosomes, located in the nuclei of cells, contain the genetic or hereditary information needed for directing and controlling the growth and reproduction of an organism. Each chromosome contains many *genes*, the fundamental unit of heredity. The differences between species, e.g., between humans and horses, and the differences between members of the same species, e.g., John Doe and Bill Smith, lie in the differences in their genes. Similarities between offspring of the same parents lie in similarities in their genes. Genes are responsible for characteristics of the species and the individual member of a species. Two types of *nucleic acids* are involved in carrying out the hereditary process—*ribonucleic acids* (*RNA*) and *deoxyribonucleic acids* (*DNA*). Regulation of growth and reproduction is achieved by regulating the synthesis of the wide variety of proteins responsible for growth and reproduction. Chromosomes contain DNA molecules and each gene is a portion of a DNA molecule. DNA contains the hereditary information and directs reproduction of itself and the synthesis of RNA. RNA molecules diffuse out of the cell nucleus and carry out the critical task of protein synthesis in ribosomes, located in the cytosol.

Nucleic acids, like carbohydrates and polypeptides, are polymers. Controlled hydrolysis of nucleic acids shows that *nucleotides* are the basic building blocks (*monomers*) of nucleic acids.

21.1 NUCLEOTIDES

Nucleotides are derived from three different types of molecules—phosphoric acid, pentose sugar, and heterocyclic nitrogen base (Fig. 21-1). Two different pentoses are used, D-ribose for RNA and 2-deoxy-D-ribose for DNA. Henceforth, the sugars will be referred to simply as ribose and deoxyribose. A total of five different heterocyclic bases are used, three pyrimidine and two purine bases. Each base is symbolized by the first letter of its name, C, T, U, A, and G for cytosine, thymine, uracil, adenine, and guanine, respectively. A, G, and C are used in synthesizing both DNA and RNA nucleotides. T is used only for DNA while U is used only for RNA.

Synthesis of a nucleotide from phosophoric acid, sugar, and base can be described as proceeding by two dehydrations, one between phosphoric acid and the sugar and the other between the sugar and the heterocyclic base. Synthesis of the ribonucleotide from phosphoric acid, ribose, and cytosine proceeds as follows:

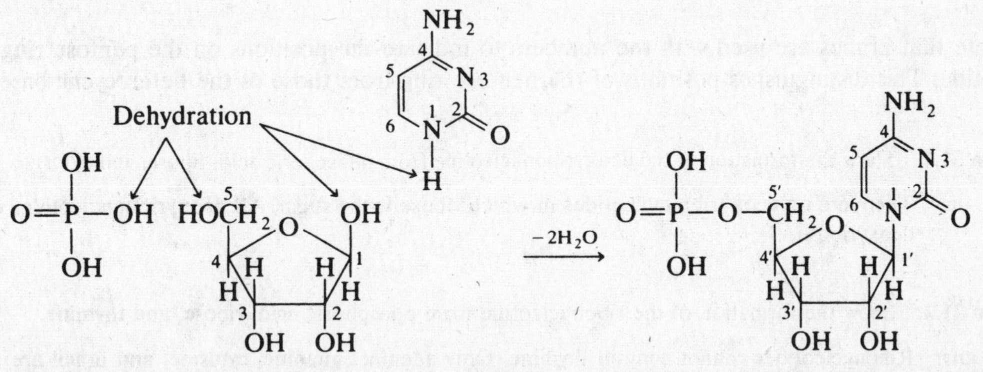

Dehydration

$$\xrightarrow{-2H_2O}$$

Dehydration between phosphoric acid and the sugar takes place at C-5 of the sugar. Dehydration between the sugar and the heterocyclic base takes place at C-1 of the sugar and an N—H bond of the heterocyclic base. The N—H hydrogens used in synthesizing nucleotides are enclosed in boxes in Fig. 21-1. These are the hydrogens at position-1 of the pyrimidine ring and position-9 of the purine ring.

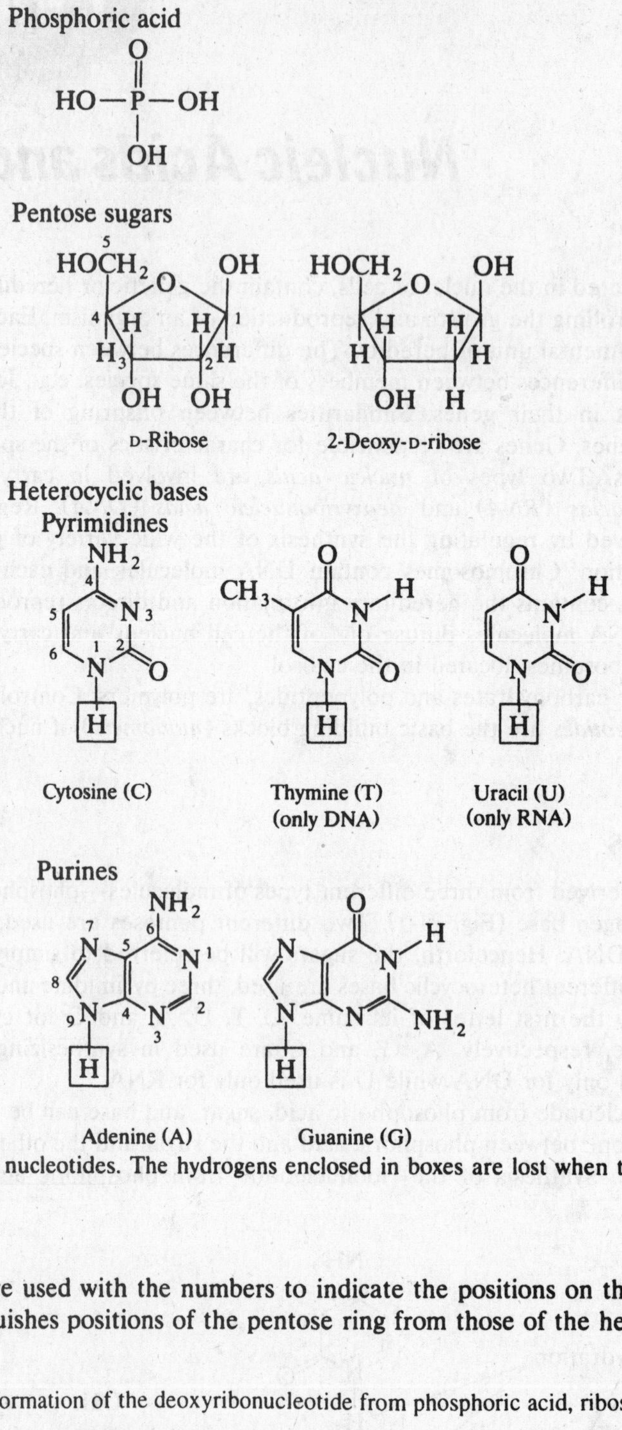

Phosphoric acid

O
‖
HO—P—OH
|
OH

Pentose sugars

5
$HOCH_2$ O OH $HOCH_2$ O OH
4 H H 1 H H
H_3 2 H H H
OH OH OH H

D-Ribose 2-Deoxy-D-ribose

Heterocyclic bases

Pyrimidines

NH_2 O O
CH_3 H H
N N N
H H H

Cytosine (C) Thymine (T) (only DNA) Uracil (U) (only RNA)

Purines

NH_2 O
H
N N N N
N N N N NH_2
H H

Adenine (A) Guanine (G)

Fig. 21-1 Components of nucleotides. The hydrogens enclosed in boxes are lost when the bases are bonded to the sugars.

Note that primes are used with the numbers to indicate the positions on the pentose ring in the nucleotide. This distinguishes positions of the pentose ring from those of the heterocyclic base.

Problem 21.1. Show the formation of the deoxyribonucleotide from phosphoric acid, ribose, and cytosine.

Ans. There are no deoxyribonucleotides in which ribose is the sugar. All deoxyribonucleotides contain deoxyribose.

Problem 21.2. Show the formation of the ribonucleotide from phosphoric acid, ribose, and thymine.

Ans. Ribonucleotides cannot contain thymine. Only adenine, guanine, cytosine, and uracil are used in ribonucleotides.

Problem 21.3. Show the formation of the deoxyribonucleotide from phosphoric acid, deoxyribose, and thymine.

Ans.

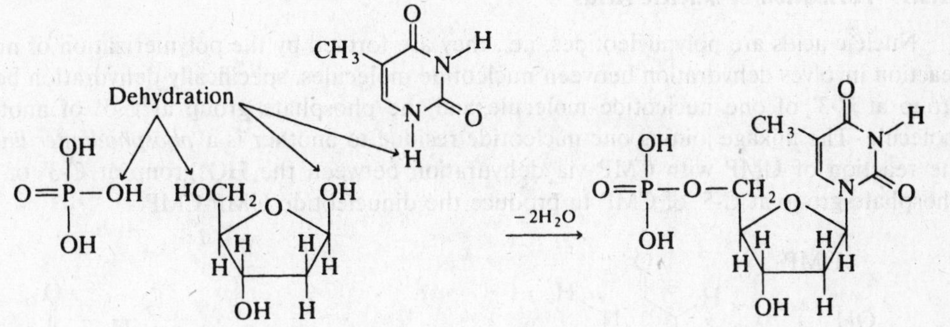

Dehydration

$$\text{O=P(OH)}_2\text{–OH} + \text{HOCH}_2\text{(deoxyribose)} + \text{thymine} \xrightarrow{-2H_2O} \text{O=P(OH)}_2\text{–O–CH}_2\text{(deoxyribose)–thymine}$$

Nucleotides are named by the following set of rules:

1. The prefix *deoxy-* is used to indicate deoxyribose as the sugar. No prefix means ribose.
2. The purine base combined with sugar is named by replacing the ending *-ine* of the base by *-osine*, i.e., adenosine = adenine + sugar and guanosine = guanine + sugar. The pyrimidine bases, cytosine, thymine, and uracil, combined with sugar are named as cytidine, thymidine, and uridine, respectively.
3. Monophosphate is the second word of the name.

An alternative nomenclature system for nucleotides emphasizes their acidic properties (a result of the phosphate group). The name of the base unit from the above nomenclature system forms the basis of the name. The ending *-osine monophosphate* or *-ine monophosphate* is replaced by *-ylic acid*. The prefix *deoxy-* is used as before to distinguish deoxyribose from ribose.

Abbreviations are also used. The prefix *d-* is used instead of *deoxy-*. This is followed by the one-letter symbol for the base and MP for monophosphate.

Problem 21.4. Name and give the abbreviation for the nucleotide in Problem 21.3.

Ans. Name: deoxythymidine monophosphate or deoxythymidylic acid; abbreviation: dTMP

Problem 21.5. Identify and name CMP.

Ans. This is the ribonucleotide that contains cytosine. Its name is cytidine monophosphate or cytidylic acid.

Problem 21.6. Draw the structure of deoxyguanosine monophosphate. Give the alternative name and abbreviation.

Ans.

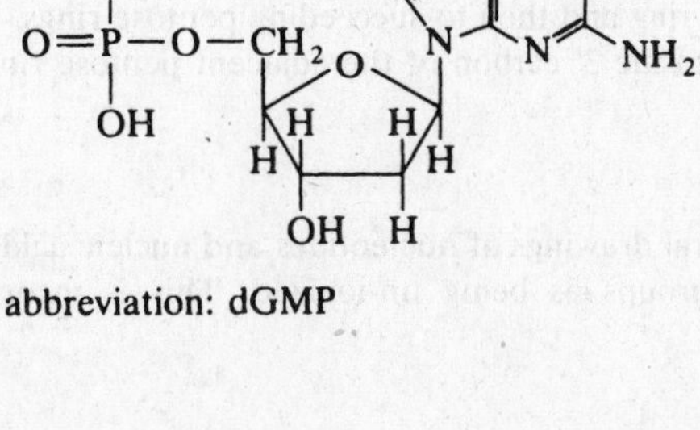

$$\text{O=P(OH)}_2\text{–O–CH}_2\text{(deoxyribose)–guanine}$$

Name: deoxyguanylic acid; abbreviation: dGMP

21.2 NUCLEIC ACIDS

21.2.1 Formation of Nucleic Acids

Nucleic acids are polynucleotides, i.e., they are formed by the polymerization of nucleotides. The reaction involves dehydration between nucleotide molecules, specifically dehydration between the OH group at C-3′ of one nucleotide molecule and the phosphate group at C-5′ of another nucleotide molecule. The linkage joining one nucleotide residue to another is a *phosphodiester linkage*. Consider the reaction of UMP with CMP via dehydration between the HO group at C-3′ of UMP and the phosphate group at C-5′ of CMP to produce the dinucleotide UMP-CMP.

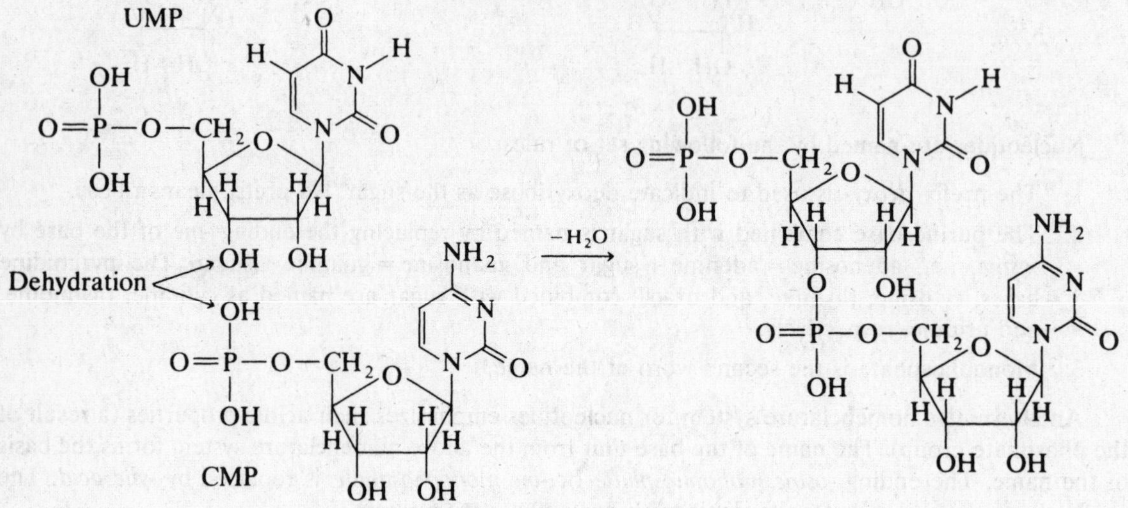

UMP

Dehydration

CMP

$\xrightarrow{-H_2O}$

Problem 21.7. What is the relation of UMP-CMP to the product from the reaction of UMP and CMP via dehydration between the HO group at C-3′of CMP and the phosphate group at C-5′ of UMP?

Ans. This would produce CMP-UMP, a constitutional isomer of UMP-CMP. CMP-UMP and UMP-CMP differ in the sequence of nucleotides, and even more specifically in the sequence of heterocyclic bases. The situation is analogous to the sequence of amino acid residues in a peptide.

Problem 21.8. What convention is used to name CMP-UMP and UMP-CMP?

Ans. The nucleotide residues are named in the 5′ ⟶ 3′ direction. There are two ways of stating the convention: (1) A polynucleotide has an unreacted phosphate group (a phosphate with two OH groups) at one end and an unreacted 3′ —OH group at the other end. The nucleotide residues are named by proceeding from the unreacted phosphate end to the unreacted 3′ —OH end. (2) Name the nucleotide residues by proceeding from the 5′ carbon of a pentose ring to the 3′ carbon of the same pentose ring and then to succeeding pentose rings. Do not proceed from the 5′ carbon of one pentose ring to the 3′ carbon of the adjacent pentose ring.

Problem 21.9. The structural drawings of nucleotides and nucleic acids in much of this chapter and other books show the phosphate OH groups as being un-ionized. This is incorrect but often done for simplicity. The phosphate hydroxyl groups are ionized at physiological pH (~ 7.0). Draw the structure of UMP-CMP showing ionized phosphate groups.

Ans.

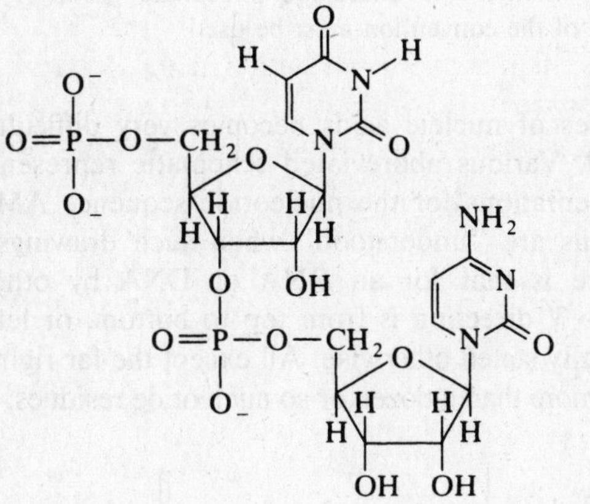

Problem 21.10. Show the formation of the sequence dAMP-dCMP-dTMP within a DNA molecule.

Ans. See Fig. 21-2.

Fig. 21-2 Synthesis of a dAMP-dCMP-dTMP sequence within a DNA molecule.

Problem 21.11. What convention is used for naming the sequence of three nucleotide residues in Problem 21.9?

Ans. The 5′ ⟶ 3′ convention is used. However, the first statement of the convention as described in Problem 21.8 cannot be used because the ends of the DNA molecule are not shown, i.e., one does not know whether the unreacted phosphate group is at the top or bottom end. The second statement of the convention must be used.

Drawing structures of nucleic acids becomes very difficult when more than a few nucleotide residues are involved. Various abbreviated schematic representations are used. Figure 21-3 shows some of these representations for the nucleotide sequence AMP-CMP-UMP-GMP-CMP. Note that many structural details are "understood" when such drawings are used. One knows whether the nucleic acid sequence is that for an RNA or DNA by other information that is supplied. By convention the 5′ ⟶ 3′ direction is from top to bottom, or left to right if drawn in the horizontal plane, unless specifically stated otherwise. All except the far right representation become too awkward to use when we have more than a dozen or so nucleotide residues.

```
|              |
Phosphate      P
|              |
Sugar—A        S—A      A     ACUGC
|              |        C
Phosphate      P        U
|              |        G
Sugar—C        S—C      C
|              |
Phosphate      P
|              |
Sugar—U        S—U
|              |
Phosphate      P
|              |
Sugar—G        S—G
|              |
Phosphate      P
|              |
Sugar—C        S—C
|              |
```

Fig. 21-3 Different representations of a nucleic acid containing the nucleotide sequence AMP-CMP-UMP-GMP-CMP.

Problem 21.12. What constitutes the backbone structure of nucleic acid chains?

Ans. Alternating pentose and phosphate groups

21.2.2 Secondary, Tertiary, and Quaternary Structures of DNA

DNA molecules do not exist as individual molecules. Two paired DNA molecules (often referred to as *strands* instead of molecules) fold around each other to form a right-handed *double helix*

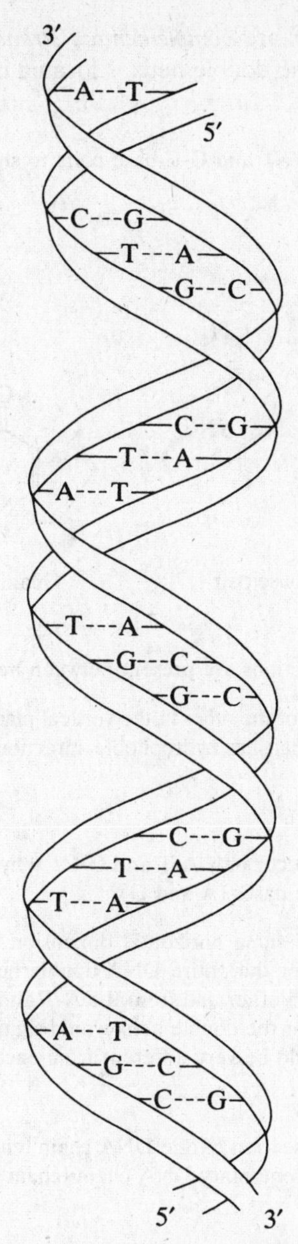

Fig. 21-4 DNA double helix.

(Fig. 21-4). The two DNA molecules in a DNA double helix run in opposite directions: one in the 5′ ⟶ 3′ direction and the other in the 3′ ⟶ 5′ direction. Phosphate groups are on the outer surface of the double helix while heterocyclic bases are inside the double helix. The placement of heterocyclic bases within the double helix is critical to its formation. The double helix is stabilized by a combination of hydrogen bonding and hydrophobic attractions between heterocyclic bases. The heterocyclic base on each nucleotide residue is in the opposite plane from that of the axis of the double helix, i.e., the bases are in the horizontal plane when the double helix is viewed in the vertical plane. Bases from opposite sides of the double helix are paired. This *base pairing* occurs in a *complementary* manner—adenine pairs only with thymine and cytosine pairs only with guanine. The two DNA strands in the double helix are *complementary strands* since the nucleotide sequences are such that each heterocyclic base in the double helix is located opposite its complementary base.

Problem 21.13. Draw structures of the A-T and G-C base pairs to show the hydrogen bonding that holds each base pair together.

Ans.

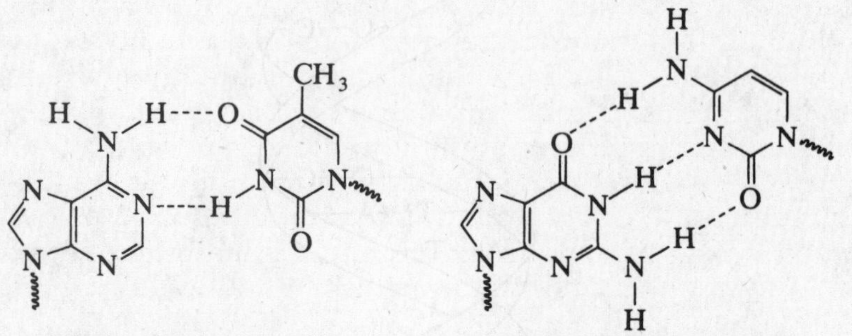

Adenine-thymine base pair Guanine-cytosine base pair

Problem 21.14. What hydrophobic attractions are present between heterocyclic bases in the double helix?

Ans. Base pairs stack one on top of the other in the vertical plane, like the steps in a spiral staircase. Each heterocyclic base in a base pair has hydrophobic attractions with the bases directly above and below it.

Problem 21.15. Why does base pairing occur only A-T and G-C? Why not A-C and G-T? Why not between two pyrimidine bases (T and C) or two purine bases (A and G)?

Ans. The base pair A-T has the same horizontal dimension as the G-C pair. This permits a uniform cross-sectional dimension for the entire DNA double helix. The dimensions of the A-C and G-T pairs are different from each other and from the A-T and G-C pairs. Any variation from a constant cross-sectional dimension for the double helix would significantly lower its stability. The dimensions for T-C and A-G pairs would be very different from each other and from A-T and G-C.

Problem 21.16. The base sequence in a section of one DNA chain (chain 1) is ACGTAG. What is the sequence in the corresponding section of the complementary DNA chain (chain 2)?

Ans.

5′ ACGTAG 3′ Chain 1

3′ TGCATC 5′ Chain 2

Since there is no statement to the contrary, the sequence in chain 1 must be assumed to be given in the 5′ ⟶ 3′ direction. This means that the complementary sequence if written as TGCATC must be identified as being in the 3′ ⟶ 5′ direction. The answer TGCATC without such identification is wrong since the reader assumes the 5′ ⟶ 3′ direction unless otherwise stated. If no identification is to be given, you must show the sequence as CTACGT, which is the sequence for chain 2 when read in the 5′ ⟶ 3′ direction.

Problem 21.17. What is the relationship between the amount of adenine in one DNA molecule and the amount of thymine in its complementary chain? Between cytosine and guanine in the two chains?

Ans. **The amount of A in one chain equals the amount of T in the other chain because A and T are complementary bases. The amount of C in one chain equals the amount of G in the other chain because C and G are complementary bases.**

The human cell nucleus has a diameter of about 5 micrometers (μm) (10^{-5} m) while the total DNA (referred to as the *genome*) contained in its 23 pairs of chromosomes has about 3.2×10^9 base pairs and a total length of 3 m if stretched out end to end. Placement of this enormous amount of DNA into the nucleus requires a high degree of compaction of the DNA double helix. The required compaction of DNA is achieved by its tertiary and quaternary structures (Fig. 21-5). DNA (*a*) is complexed with an equal amount of *histone proteins*. Two pairs each of four different histone proteins form a spool-like *nucleosome core*. Nucleosome cores act as templates for the compaction of the DNA double helix. The DNA double helix is wound around nucleosome cores to form a chain of *nucleosomes* (*b*), sometimes referred to as *chromatin fiber*. There are about 150–200 base pairs per nucleosome core. Large numbers (10^4–10^5) of nucleosome cores are needed to compact each DNA double helix. The chain of nucleosomes is further coiled to form a thick *solenoidal fiber* (*c*) which is even further coiled into a highly compact supercoiled fiber (*d*). Other levels of organization take place depending on the stage of cell division, e.g., Fig. 21-5e shows the structure of a chromosome during the later phases of cell division (mitosis).

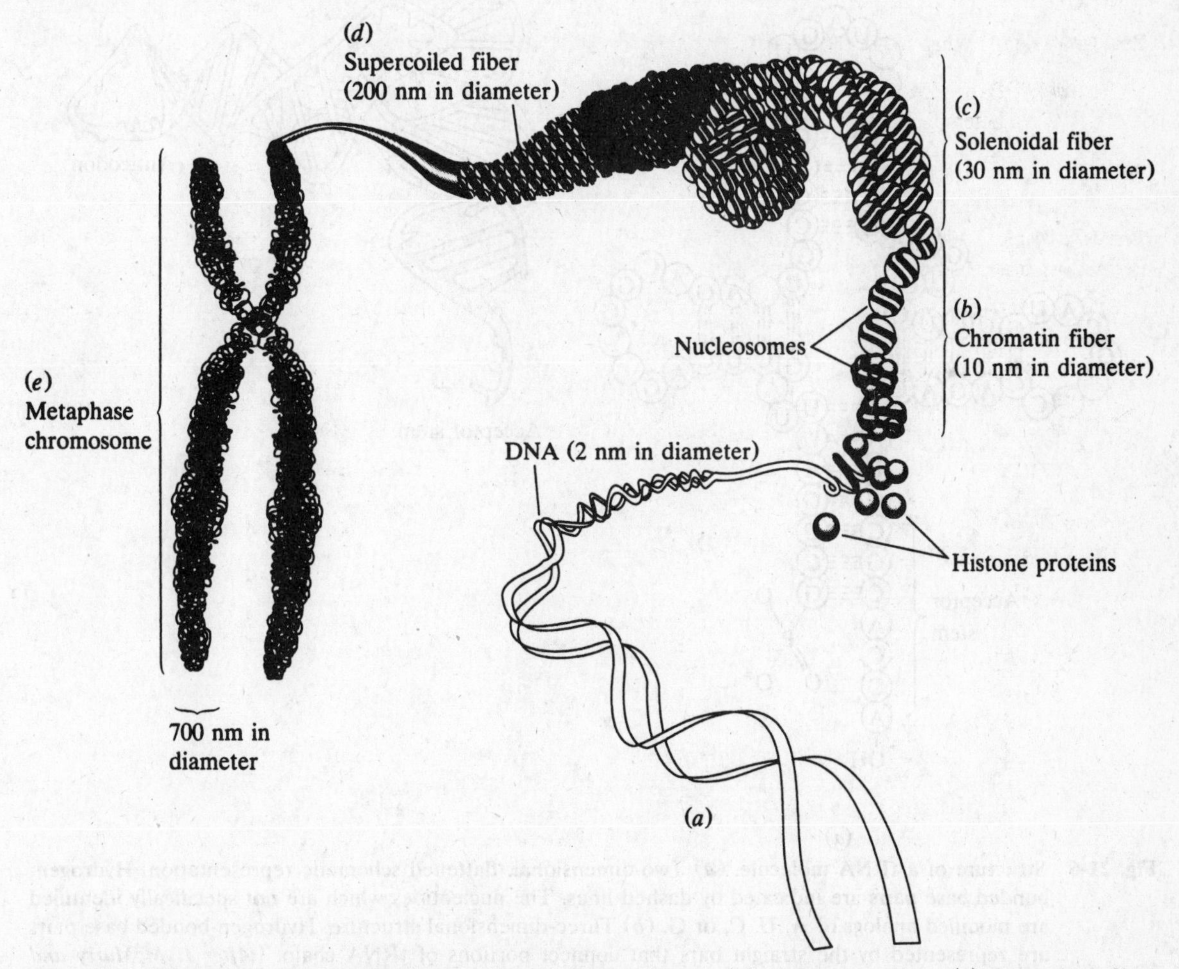

Fig. 21-5 Different levels of DNA structure. (*a*) Double helix. (*b*) Chromatin fiber. (*c*) Solenoidal fiber. (*d*) Highly supercoiled fiber. (*e*) Chromosome. (1 nm = 10^{-9} m) (*After C. K. Mathews and D. E. van Holde, Biochemistry, Benjamin/Cummings, Redwood City, CA, 1990.*)

21.2.3 Secondary, Tertiary, and Quaternary Structures of RNA

Ribonucleic acids exist as single-strand molecules, not double helices as do deoxynucleic acids. The presence of the OH at C-2′ of the ribose ring sterically interferes with the formation of long stretches of helical conformation. However, RNA molecules possess a wide variety of secondary, tertiary, and, quaternary structures. Figure 21-6 shows one type of RNA molecule referred to as *transfer RNA* or *tRNA*. The overall shape of tRNA resembles a cloverleaf, especially in the two-dimensional representation. There is considerable intramolecular base pairing of nucleotide residues. The base pairing scheme is the same as for DNA except that uracil is used in RNA instead of thymine. Adenine pairs with uracil and cytosine pairs with guanine. Roughly half of the nucleotide bases in an RNA molecule participate in base pairing. Base pairing does not occur in one long continuous run but in short runs. Sections of the RNA molecule with approximately helical conformation are separated by sections with no base pairing and without helical conformation. The latter sections, except for the linear section at the 3′ end of the RNA chain, are near circular and referred to as *loops*. The linear section at the 3′ end and the loop furthest from it are referred to as *acceptor stem* and *anticodon*, respectively.

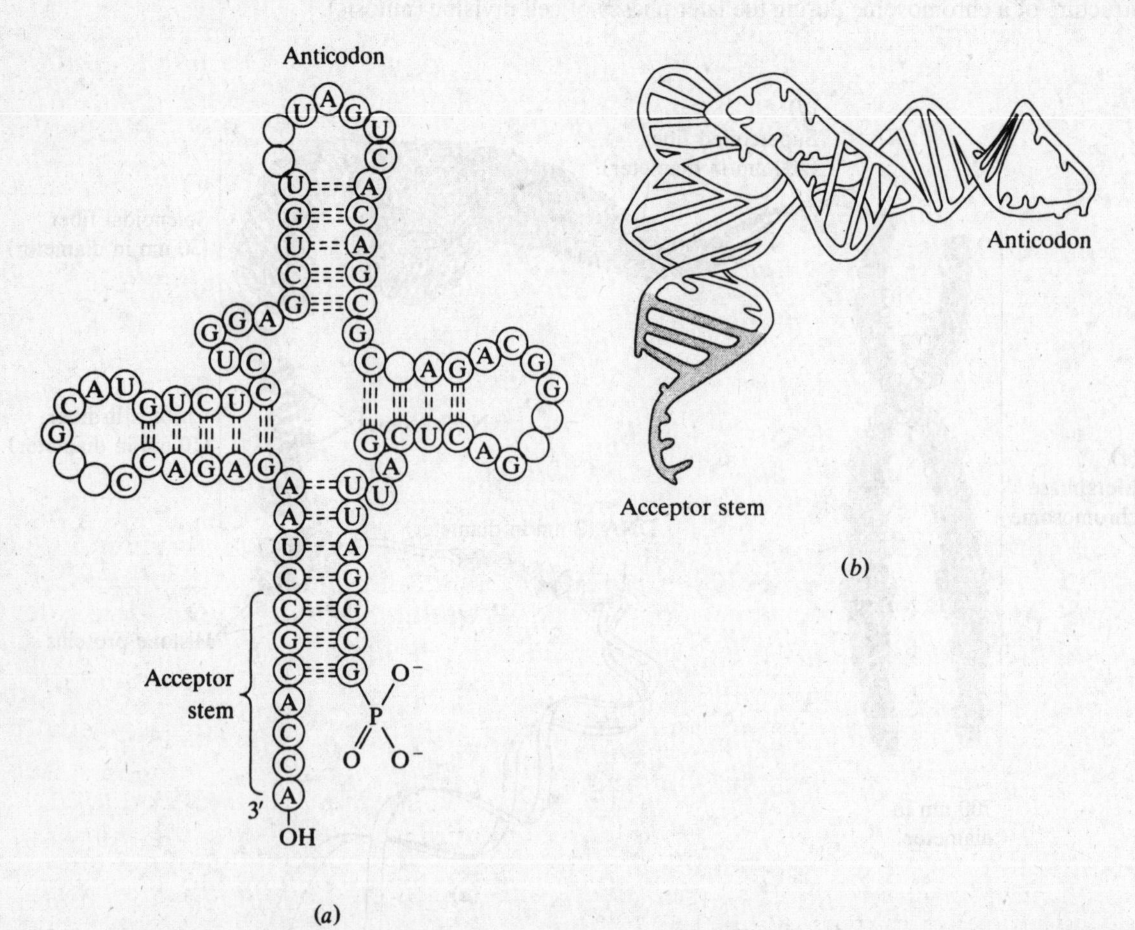

Fig. 21-6 Structure of a tRNA molecule. (*a*) Two-dimensional, flattened schematic representation. Hydrogen-bonded base pairs are indicated by dashed lines. The nucleotides which are not specifically identified are modified analogs of A, U, C, or G. (*b*) Three-dimensional structure. Hydrogen-bonded base pairs are represented by the straight bars that connect portions of tRNA chain. (*After J. McMurry and M. E. Castellion, Fundamentals of General, Organic, and Biological Chemistry, Prentice-Hall, Englewood Cliffs, NJ, 1992.*)

Problem 21.18. Why is there no base pairing in the loop sections?

Ans. The sequence of bases in the loop sections is such that complementary bases are not opposite each other or, if opposite each other, are not within hydrogen-bonding distance.

Problem 21.19. The above description of tRNA does not include quaternary structure. Does tRNA have quaternary structure?

Ans. Yes. The physiological functions of tRNA as well as other types of RNA are such that they are often associated with other molecules. This is discussed throughout Sec. 21.3.

21.3 FLOW OF GENETIC INFORMATION

The *central dogma of molecular genetics* describes how hereditary information is transmitted and used to synthesize proteins. The main features of this theory are shown in Fig. 21-7. DNA is the storage form of hereditary information. The hereditary information is the sequence of nucleotide residues in DNA which corresponds to a specific sequence of amino acid residues in a polypeptide. *Replication* is the process of copying DNA during cell division. *Transcription* is the synthesis of RNA from DNA, the first step in the transmission of hereditary information. Three types of RNA are produced, *ribosomal RNA (rRNA)*, *messenger RNA (mRNA)*, and *transfer RNA (tRNA)*. rRNA, mRNA, and tRNA participate in the process of *translation*—the synthesis of protein.

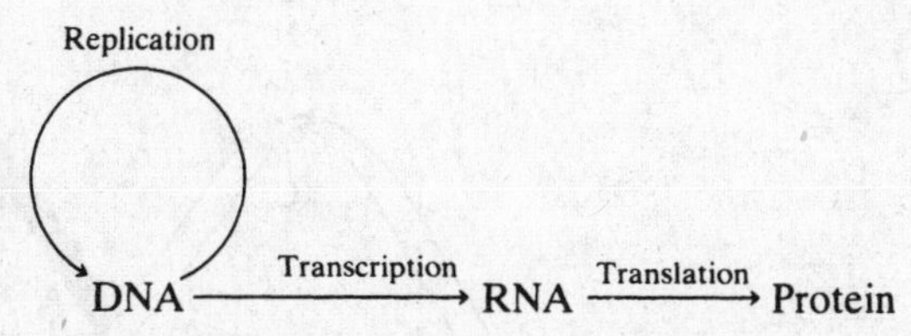

Fig. 21-7 Basic features of the central dogma of molecular genetics.

21.3.1 Replication

The two strands of a DNA double helix unwind over a short span (ca. 150 to 200 nucleotide residues) to form a *replication bubble* with a *replication fork* at each end (Fig. 21-8*a*). Each strand acts as a template for synthesizing a complementary strand. Over 20 enzymes are involved in directing and controlling replication. *Helicase* and *topoisomerase* catalyze the unwinding process, *single-strand DNA-binding proteins* stabilize the unwound DNA, and *DNA polymerase* catalyzes the addition of nucleotide to growing DNA chains. The two new strands grow in opposite directions since each grows in the $5' \longrightarrow 3'$ direction. The two strands start growing from replication forks at opposite sides of the bubble. The replication bubble expands with further unwinding of the DNA template double helix as DNA synthesis proceeds. One of the new strands, the *leading strand*, is synthesized in a continuous manner. The other strand, the *lagging strand*, is synthesized in a discontinuous manner to form a series of smaller DNA fragments, referred to as *Okazaki fragments*, which are linked together by the enzyme *DNA ligase*. Replication proceeds simultaneously at a series of replication bubbles along the DNA template chain. The DNA fragments from different replication bubbles are subsequently linked together by DNA ligase to produce the final new DNA chains.

DNA polymerase has a molecular mass of about one-half million. It directs the synthesis process by simultaneous complexation with the template DNA chain, growing new DNA chain, and nucleotide. Base pairing together with DNA polymerase control guarantees the accuracy of replication

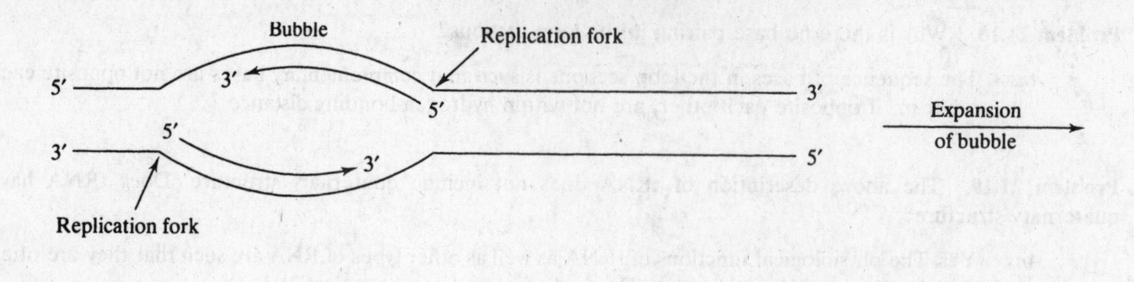

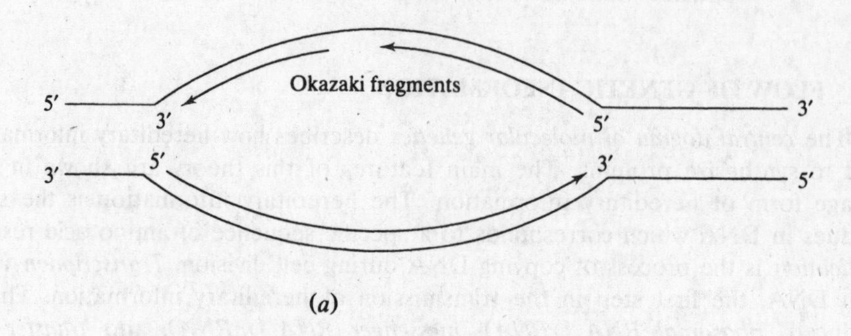

(*a*)

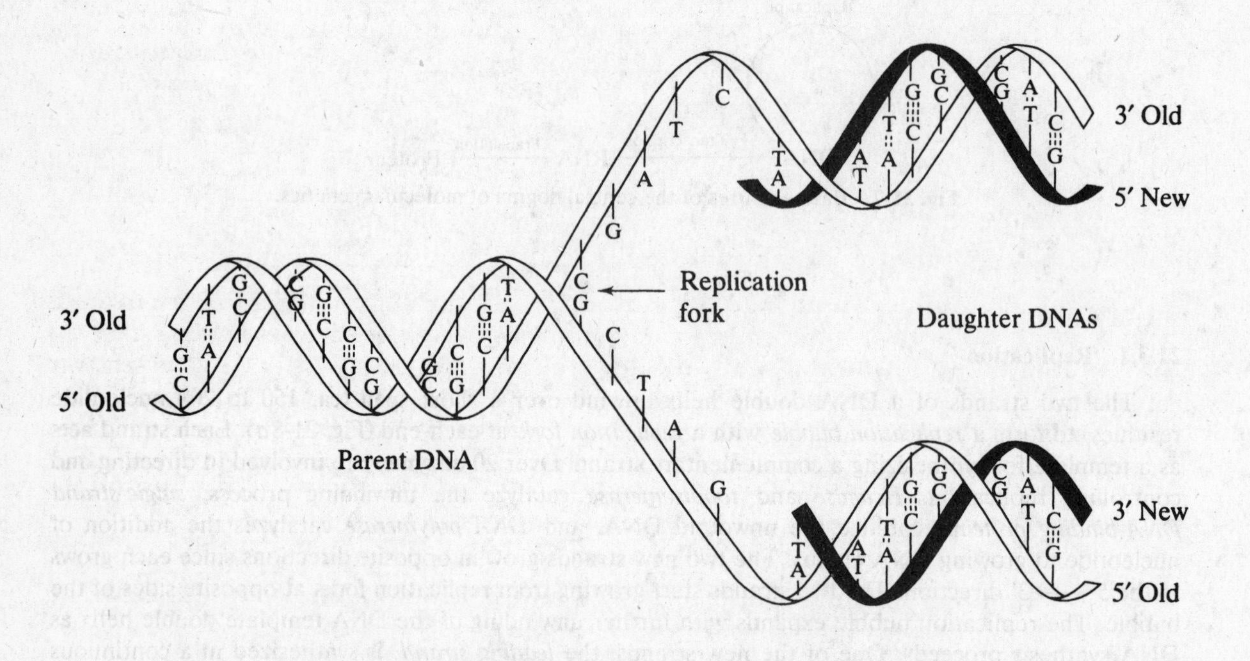

(*b*)

Fig. 21-8 Replication of DNA double helix. (*a*) DNA synthesis at a pair of replication forks with expansion of the replication bubble. DNA chains are not shown in helical form. (*b*) Complementary chains are synthesized via base pairing. (*After J. McMurry and M. E. Castellion, Fundamentals of General, Organic, and Biological Chemistry, Prentice-Hall, Englewood Cliffs, NJ, 1992.*) (*c*) Semiconservative replication. (*After J. D. DeLeo, Fundamentals of Chemistry: General, Organic, and Biological, Scott, Foresman, Glenview, IL, 1988.*)

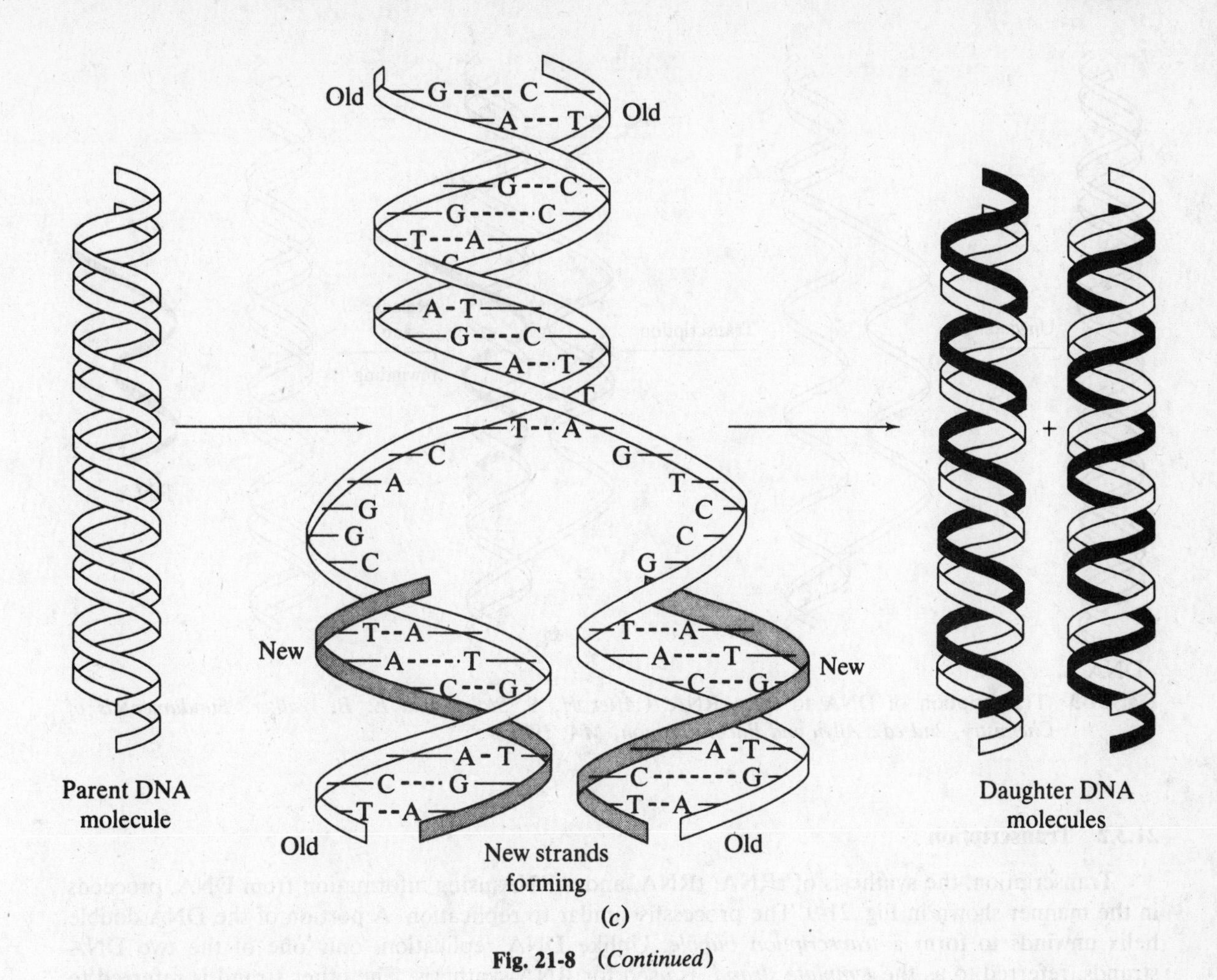

(c)

Fig. 21-8 (*Continued*)

(Fig. 21-8*b*). For example, if a template strand has A at a certain position, only dTMP nucleotide is allowed to hydrogen-bond at that point and become incorporated into the new strand. After the addition of a nucleotide, DNA polymerase moves to the next nucleotide residue in the DNA template strand and ensures addition of the complementary nucleotide in the new DNA chain. DNA polymerase not only catalyzes the addition of nucleotide to growing DNA chains but is also involved in proofreading and correcting errors. Replication is *semiconservative*, i.e., one strand of the *parent* DNA double helix is conserved intact in each of the *daughter* DNA double helices (Fig. 21-8*c*). Each daughter DNA double helix contains one *old* and one *new* DNA chain.

Problem 21.20. What is the relationship of the new DNA chain of a daughter DNA double helix and the old DNA chain in the other daughter DNA double helix?

Ans. They are identical.

Problem 21.21. The sequence ACGTGC (reading in the $5' \longrightarrow 3'$ direction) appears on a portion of one strand of a parent DNA double helix. What is the corresponding sequence on the complementary strand of the parent DNA double helix? What is the corresponding sequence on the new daughter strand made from this parent strand during replication?

Ans. The parent strand and the new daughter strand are identical and complementary to the ACGTGC sequence. The sequence is TGCACG written in the $3' \longrightarrow 5'$ and GCACGT written in the $5' \longrightarrow 3'$ direction.

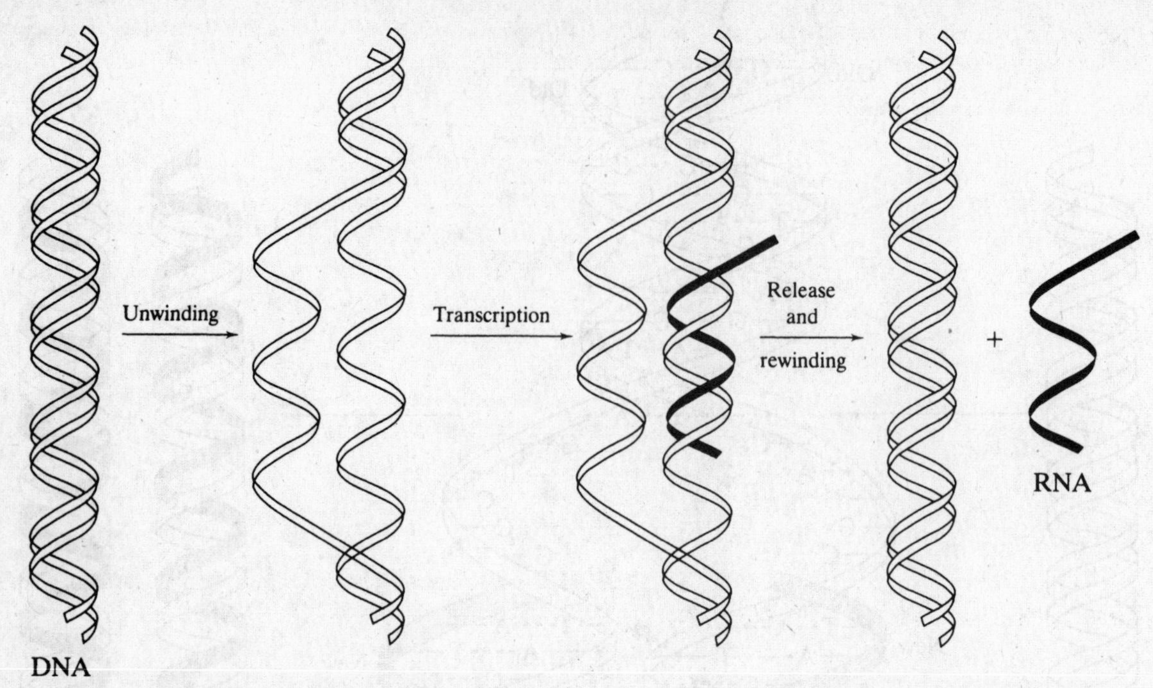

Fig. 21-9 Transcription of DNA to form RNA. (*After H. S. Stoker and E. B. Walker, Fundamentals of Chemistry, 2nd ed., Allyn and Bacon, Boston, MA, 1991.*)

21.3.2 Transcription

Transcription, the synthesis of rRNA, tRNA, and mRNA using information from DNA, proceeds in the manner shown in Fig. 21-9. The process is similar to replication. A portion of the DNA double helix unwinds to form a *transcription bubble*. Unlike DNA replication, only one of the two DNA strands, referred to as the *template strand*, is used for RNA synthesis. The other strand is referred to as the *information strand*. Transcription proceeds under the control of *RNA polymerase*, different RNA polymerases being responsible for rRNA, tRNA, and mRNA. Synthesis begins at an *initiation* (*promoter*) *site* whose base sequence is recognized by RNA polymerase as a start signal. Synthesis stops when RNA polymerase encounters a *termination site* whose base sequence is recognized as a stop signal. The RNA and RNA polymerase are released along with rewinding of the DNA into its double-helix conformation.

Problem 21.22. What base pairing occurs during transcription?

Ans. The same as in DNA replication except that uracil is used instead of thymine in the RNA chain. A, G, C, and T in the template strand of DNA result in the incorporation of U, C, G, and A, respectively, in the RNA strand being synthesized.

Problem 21.23. Compare the sizes of DNA, rRNA, mRNA, and tRNA.

Ans. DNA is enormously larger than any RNA. Molecular masses of DNA molecules range from a few billion to hundreds of billions. Most rRNAs have molecular masses in the range one-half million to one million although some have much lower molecular masses. mRNAs have molecular masses in the range one-quarter million to two million. tRNAs are much smaller, with molecular masses in the range 25,000 to 30,000.

Problem 21.24. What is the relationship of the RNA synthesized from a portion of a DNA template strand to the corresponding portion of the DNA information strand?

Ans. They are identical.

The initially-formed RNA, often referred to as a *primary transcript RNA* (*ptRNA*) is altered by chemical reactions to achieve biological activity. The *posttranscriptional processing* of ptRNAs includes capping the ends of an RNA molecule with certain nucleotide sequences, modification of certain nucleotides in the RNA chain (e.g., by methylation), and splicing operations. Some modifications are needed to stabilize RNAs to their environmental conditions. Other modifications optimize the physiological functions of RNAs by altering their secondary, tertiary, and quaternary structures. Splicing operations involve cutting out portions (*introns*) of RNA which do not appear to have any physiological function (Fig. 21-10). The portions (*exons*) with physiological function are linked together after excision of the introns. (There are also modifications after the translation (Sec. 21.3.3) process.)

Problem 21.25. Where do the posttranscriptional modifications occur, in the nucleus or in the cytosol?

Ans. Capping and splicing occur in the nucleus while most nucleotide modifications generally take place in the *cytosol* (the region outside the nucleus and other organelles).

21.3.3 Translation

Translation is the process by which rRNA, mRNA, and tRNA come together to synthesize polypeptides (Fig. 21-11). Polypeptide synthesis occurs at *ribosomes* located in the cytosol. Ribosomes are formed from rRNA and also contain about 40 percent protein, mostly the various enzymes required for protein synthesis. A ribosome is composed of two subunits, one large and one small. mRNA, carrying the message for which polypeptide is to be synthesized, binds to the ribosome. Synthesis proceeds when tRNAs transport various amino acids into the ribosome.

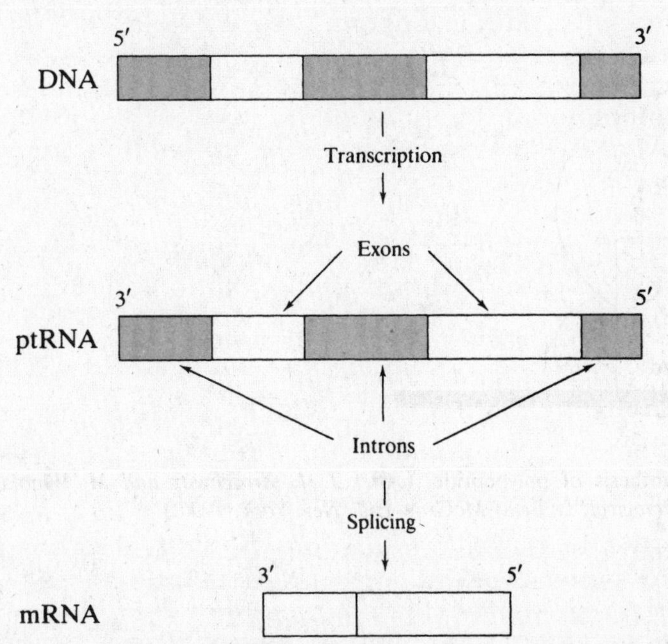

Fig. 21-10 Splicing of RNA by excision of introns and linking of exons.

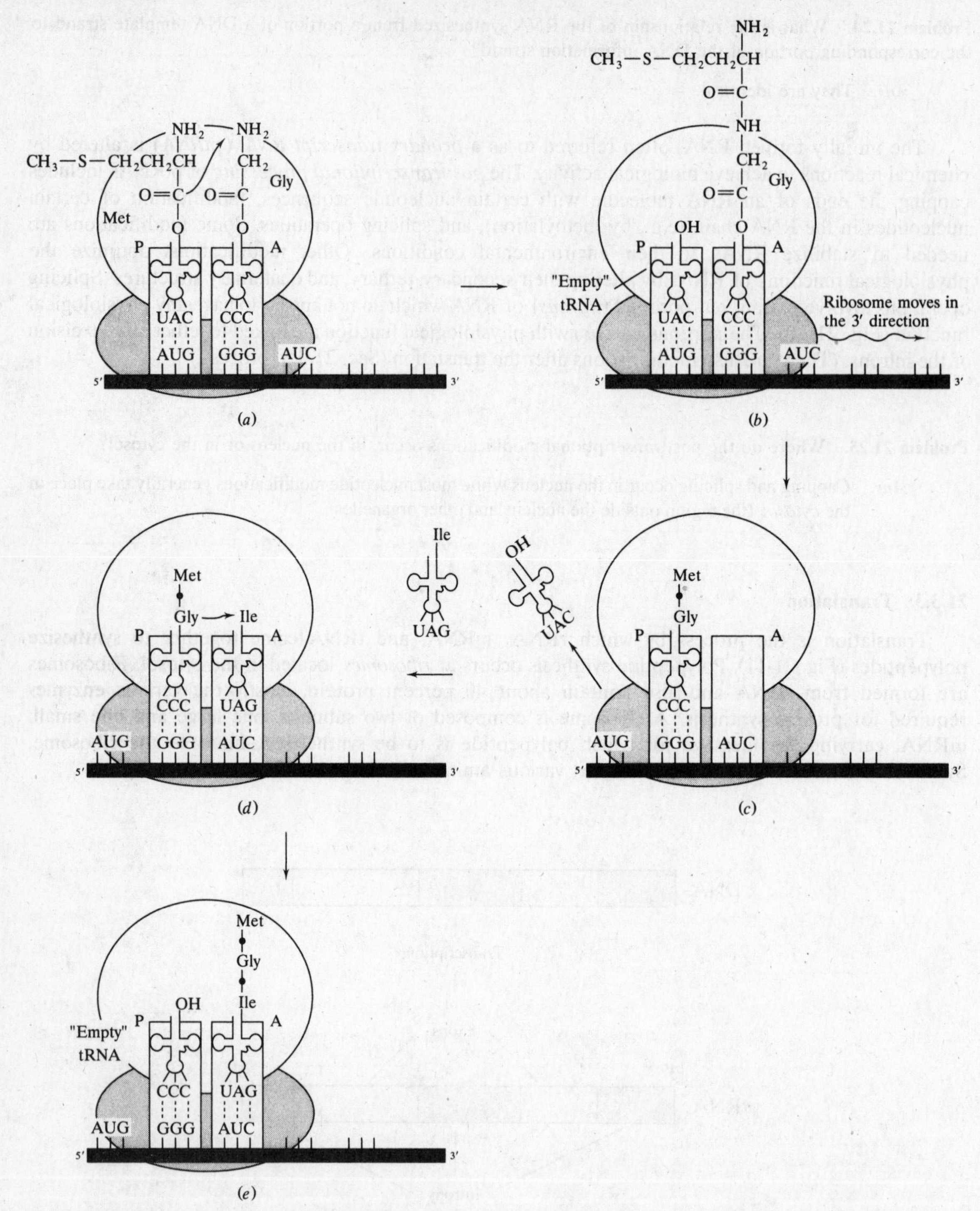

Fig. 21-11 (*a*)–(*e*) Synthesis of polypeptide. (*After J. I. Kroschwitz and M. Winokur, Chemistry: General, Organic, Biological, 2nd ed., McGraw-Hill, New York, 1990.*)

Table 21-1. Genetic Code: Codon Assignments

First Base (5′ end)	Second Base U	Second Base C	Second Base A	Second Base G	Third Base (3′ End)
U	Phe	Ser	Tyr	Cys	U
	Phe	Ser	Tyr	Cys	C
	Leu	Ser	Stop	Stop	A
	Leu	Ser	Stop	Trp	G
C	Leu	Pro	His	Arg	U
	Leu	Pro	His	Arg	C
	Leu	Pro	Gln	Arg	A
	Leu	Pro	Gln	Arg	G
A	Ile	Thr	Asn	Ser	U
	Ile	Thr	Asn	Ser	C
	Ile	Thr	Lys	Arg	A
	Met*	Thr	Lys	Arg	G
G	Val	Ala	Asp	Gly	U
	Val	Ala	Asp	Gly	C
	Val	Ala	Glu	Gly	A
	Val	Ala	Glu	Gly	G

*Codon AUG for Met also codes for the initiation of polypeptide synthesis.

Problem 21.26. What is the form of the message carried by mRNA?

Ans. The sequence of bases on the mRNA specifies the polypeptide to be synthesized. Each sequence of three consecutive bases (*base triplet*) comprises a *codon* which codes or specifies for a particular tRNA which carries a particular amino acid. The codon is written using the one-letter symbols for the bases, e.g., ACC designates the codon with the base sequence adenine-cytosine-cytosine.

Problem 21.27. How does the base triplet of mRNA recognize a particular tRNA?

Ans. Each tRNA has a particular base triplet referred to as an *anticodon* (Fig. 21-6). The three-dimensional structures of tRNA and mRNA are such that tRNA hydrogen bonds (i.e., base pairs) with mRNA only when its anticodon is complementary to the codon of mRNA. For example, only the tRNA with anticodon ACC (written in the $3' \longrightarrow 5'$ direction) base pairs to codon UGG (written in the $5' \longrightarrow 3'$ direction) of mRNA. That particular tRNA carries the amino acid Trp.

The *genetic code*, the complete description of the amino acids specified by various base triplets, is tabulated in terms of the codons on mRNA (Table 21-1). The base sequence in mRNA is identical with the base sequence in the information strand of DNA except that U replaces T. The information strand contains many messages (base sequences), each message corresponding to a particular polypeptide to be synthesized. (Each message corresponds to a gene.) The individual messages are separately sent out of the nucleus in the form of different mRNAs. The genetic code is degenerate, i.e., most of the amino acids are coded by more than one codon. One of the 64 codons (AUG) codes simultaneously for Met as well as for the initiation of synthesis. Three of the codons (UAA, UAG, UGA) code for the termination of polypeptide synthesis. Note that the codon sequence is always written in the $5' \longrightarrow 3'$ direction. The genetic code is universal, i.e., it applies to all organisms, plant and animal.

Problem 21.28. What amino acid is specified by each of the following? (*a*) UUU, (*b*) CUA, (*c*) UUA, (*d*) AAC, (*e*) GAC, (*f*) GCC.

Ans. (*a*) Phe, (*b*) Leu, (*c*) Leu, (*d*) Asn, (*e*) Asp, (*f*) Ala

Problem 21.29. Suppose that sections X, Y, and Z of the following hypothetical template DNA strand are the exons of a gene:

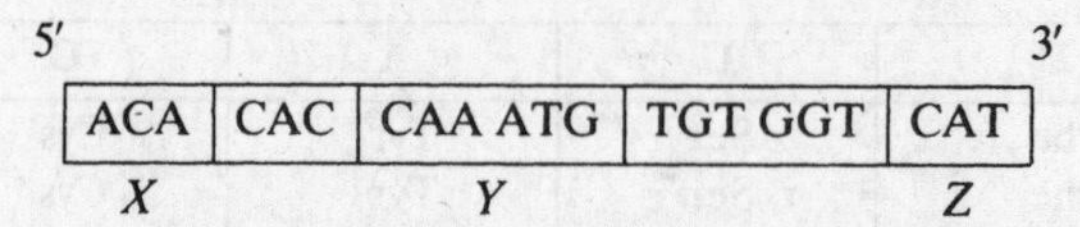

What are the structures of ptRNA and mRNA made under its direction? What polypeptide sequence will be synthesized?

Ans. The base sequence in ptRNA is the complement of that in template DNA (keep in mind that U is used in RNA instead of T). The direction of RNA synthesis proceeds in the opposite direction of the template DNA strand. In the splicing process, introns are removed and the results are mRNA. The codons in mRNA must be read in the $5' \longrightarrow 3'$ direction to correctly use the genetic code (Table 21-1) to determine the sequence of amino acids in the polypeptide. The mRNA is AUG-CAU-UUG-UGU, which codes for the polypeptide Met-His-Leu-Cys.

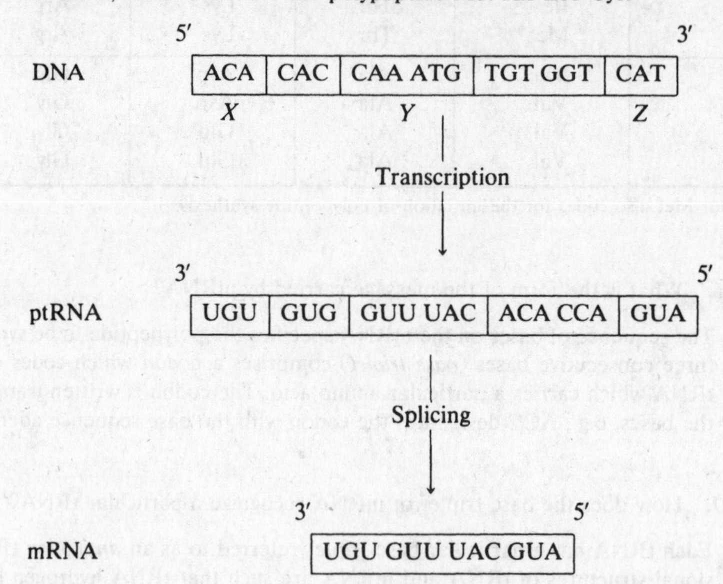

Problem 21.30. What is the mechanism by which a particular tRNA transports only a particular amino acid into the ribosome?

Ans. Pickup of amino acids by tRNAs is controlled by enzymes referred to as *aminoacyl-tRNA synthetases*. There is at least one aminoacyl-tRNA synthetase for each amino acid. The synthetase recognizes and allows only a particular amino acid to become bonded to a particular tRNA for transport to the ribosome. The recognition is a consequence of the three-dimensional structures of synthetase, tRNA, and amino acid. The amino acid is bonded to the acceptor stem (Fig. 21-6) at the 3′ end of tRNA, specifically by formation of an ester linkage between the carboxyl group of the amino acid and the HO group on C-3′ (or in some cases the C-2′) of ribose:

Polypeptide synthesis proceeds at *peptidyl (P)* and *aminoacyl (A) sites* in the ribosome (Fig. 21-11). The synthesis in Fig. 21-11 shows initiation followed by sequential addition of Gly and Ile. The 5′ end of mRNA, located in the smaller subunit of the ribosome, is prepared to receive the tRNAs for Met and Gly (by matching of codon of mRNA and anticodon of tRNA). The tRNAs for Met and Gly pick up Met and Gly from the cytosol, enter the ribosome, and line up at the P and A sites, respectively, by base pairing of codons of mRNA and anticodons of tRNA (Fig. 21-11*a*). The enzyme peptidyl transferase, contained in the large subunit of the ribosome, catalyzes the transfer and

NH_2
H—C—R
CO
O
OH
NH_2
H—C—R
COOH

bonding of Met to Gly. The reaction is an *acyl transfer reaction*. The result is a dipeptide at site A and an empty tRNA at site P (Fig. 21-11*b*). The empty tRNA is released from the ribosome (where it is available to transport another Met), the ribosome moves in the $5' \longrightarrow 3'$ direction, and the dipeptide is now at the P site (Fig. 21-11*c*). The tRNA with Ile base-pairs at the empty A site (Fig. 21-11*d*). The Met-Gly dipeptide is transferred and bonded to Ile to form the Met-Gly-Ile tripeptide at site A and an empty tRNA at site P (Fig. 21-11*e*). Synthesis proceeds in this manner over and over again with the ribosome moving to the end of the mRNA. Synthesis terminates and the polypeptide is released from the ribosome when a STOP codon is read on mRNA.

21.4 OTHER ASPECTS OF NUCLEIC ACIDS AND PROTEIN SYNTHESIS

21.4.1 Mutations

A *mutation* is a change in the sequence of bases in DNA. *Spontaneous mutations* result from errors in base pairing during replication. Replication errors are very rare, less than 1 base pair per 10^{10}, because of the proofreading function of DNA polymerase as well as the presence of various enzymes which have repair functions. Superimposed on spontaneous mutations are those produced by *mutagens* (agents that produce mutation). Mutagens include ultraviolet light (sunlight) and various high-energy radiations (x- and cosmic rays, radioactivity), and a variety of chemicals. An example of a chemical mutagen is sodium nitrite, which is used as a preservative for bologna and hot dogs. Mutagens react with nucleotide base pairs, either converting them to other bases or to other structures. Repair enzymes minimize the effects of mutagens; the overall effects of mutagens depend on their amounts and the duration of exposure.

Mutations occur by substitution, insertion, or deletion of bases. Substitution mutations are the most common types of mutation. A *substitution mutation*, involving the substitution of one base by another, changes one codon in mRNA. This may or may not alter the amino acid residue specified by that codon because of the degeneracy of the genetic code.

Problem 21.31. Consider the following substitution mutations for the first triplet (CAA) of the *Y* exon of the template DNA strand in Problem 21.29: (*a*) TAA instead of CAA, (*b*) AAA instead of CAA, (*c*) CTA instead of CAA. What change occurs in the codon on mRNA? What effect will there be on the amino acid sequence of the polypeptide being synthesized?

Ans. The normal triplet CAA results in the codon UUG in mRNA which codes for Leu. (*a*) There is no effect since the codon is changed to UUA which still codes for Leu. (*b*) The codon is changed to UUU. The amino acid sequence is changed since UUU codes for Phe instead of Leu. (*c*) The codon is changed to UAG. Polypeptide synthesis is terminated since UAG codes for STOP.

Problem 21.32. What is the physiological consequence of each substitution mutation in Problem 21.31?

Ans. (*a*) No effect since the same polypeptide is synthesized.
(*b*) The change of one amino acid residue will be significant only if the residue is near the active site of the protein and the change alters the protein conformation in that region. Protein conformation is changed, which usually changes physiological function, if the side group of the altered residue differs significantly from the original residue in size, polarity, or ionic charge. Changing from Leu to Phe, which differ in size, will probably have some effect but it may not be large.
(*c*) The prematurely terminated polypeptide will usually have no physiological function.
Mutations that have no significant negative biological consequence are referred to as *silent mutations*.

An *insertion mutation* involves the insertion of a base into the normal sequence. A *deletion mutation* involves the deletion of a base from the normal sequence. Insertion and deletion mutations, referred to as frame shift mutations, are rarer than substitution mutations but can cause much larger effects. A frame shift mutation changes all codons that follows an insertion or deletion in an exon or intron. There is no effect of a mutation (substitution or frame shift) in an intron. The mutation is constrained within the intron and has no effect on the codons in the exons because the introns are excised (Fig 21.10). But a frameshift mutation in an exon yields nonfunctional protein because all codons after the mutation are altered. The polypeptide sequence generally bears no resemblance to that coded in the original DNA.

Mutations that have negative biological consequences fall into two categories. Such mutations in *somatic cells* (cells other than egg and sperm cells) affect the physiological functioning of an individual organism. The individual may become ill and even die, depending on the level of the mutation and the particular physiological functions which are affected. Uncontrolled growth of cells (*cancer*) may result from mutation. A mutagen giving rise to cancer is referred to as a *carcinogen*. Most mutagens are carcinogens. Mutations in *germ cells* (egg and sperm cells) have much greater and far-reaching significance since they are passed onto offspring and subsequent generations. Such mutations are referred to as *genetic defects* or *hereditary diseases* when physiological functions are severely affected (Table 21-2).

Problem 21.33. Are there any positive effects of mutations?

Ans. Yes. Spontaneous mutations are the mechanism for the Darwinian survival of the fittest. A mutation occurs which makes an organism better adapted to survive under some set of altered environmental conditions. The population containing that mutation survives and passes on its altered genome to subsequent generations.

21.4.2 Antibiotics

Most *antibiotics* fight bacterial infections by interfering with some aspect of protein synthesis in the bacteria. We are most interested in those antibiotics that are specific for bacteria with minimal effect on animal protein synthesis. Table 21-3 lists some of the common antibiotics used in medicine. Large-scale use of antibiotics has been highly successful but some bacteria have "fought back" by spontaneous mutation and propagation of antibiotic-resistant strains. This requires that we keep ahead of the bacteria by synthesizing new antibiotics to kill the mutant strains.

Table 21-2. Hereditary Diseases

Disease	Affected Physiological Function
Albinism	Absence of pigment (melanin) which protects against UV damage in skin and iris of eye; skin cancer.
Cystic fibrosis	Inadequate lipase produced by pancreas; overproduction of mucus in lungs and digestive systems; breathing difficulties.
Diabetes mellitus	Defective insulin resulting in poor glucose metabolism and its accumulation in blood and urine; atherosclerosis, visual problems, circulatory problems in legs and feet.
Galactosemia	Defective transferase resulting in poor metabolism of galactose; accumulation of galactose in cells causes cataracts and mental retardation.
Hemophilia	Defective blood clotting factor(s); poor clotting, excess bleeding, internal hemorrhaging.
Phenylketonuria (PKU)	Lack of Phe hydroxylase for converting Phe to Tyr; Phe converted to phenylpyruvic acid which causes brain damage.
Sickle-cell anemia	Defective hemoglobin; sickled red blood cells aggregate to cause anemia, plugged capillaries, low oxygen pressure in tissues.
Tay-Sachs	Defective hexosaminidase A causing accumulation of glycolipids in brain and eyes; mental retardation and loss of motor control.

Table 21-3. Antibiotic Inhibition of Protein Synthesis in Bacteria

Antibiotic	Mechanism for Inhibition
Actinomycin	Binds to DNA and prevents it from acting as template for both DNA and RNA synthesis
Chloramphenicol	Inhibits peptide bond formation by interfering with peptidyl transferase
Erythromycin	Binds to large subunit of ribosome and prevents its movement along mRNA
Penicillin	Inhibits cell wall formation by inhibiting formation of needed enzyme
Puromycin	Binds to A site, becomes attached to growing polypeptide, and causes premature termination
Streptomycin	Inhibits initiation and causes incorrect reading of mRNA by binding to rRNA
Tetracyclin	Binds to rRNA and prevents binding of tRNA to mRNA

21.4.3 Viruses

Viruses are infectious, parasitic small particles containing either DNA or RNA (but not both) encapsulated by a protein overcoat. Viruses are at the borderline of life since they neither reproduce nor carry out metabolism. Viruses are unable to reproduce or synthesize proteins because they lack some or most of the apparatus (amino acids, nucleotides, and enzymes) for replication, transcription, and translation. Viruses are responsible for a wide range of diseases—influenza, poliomyelitis, leukemia, hepatitis, smallpox, chicken pox, tumors (including cancer), and AIDS (acquired immune deficiency syndrome).

Infection of a host cell by a virus particle occurs when the virus sticks to a host cell and then a portion of the cell membrane is dissolved by an enzyme present in the protein overcoat of the virus. This is followed by the entry of nucleic acid of the virus or the complete virus into the host cell. There is high specificity in this process as only certain types of hosts and host cells are invaded by any particular virus. For example, there are large numbers of plant viruses, none of which invade animal cells. The HIV virus (human immunodeficiency virus) which is responsible for AIDS infects only the T (lymphocyte) cells of the human immune system but the result is devastating. The immune system is inactivated and the person becomes easy prey to pneumonia, Kaposi's sarcoma (a form of cancer), and a variety of other opportunistic diseases.

The virus uses the apparatus of the host cell to reproduce itself, i.e., the host cell is fooled into reproducing the virus. Reproduction of the virus, requiring the synthesis of its nucleic acid and protein overcoat, proceeds differently depending on whether the virus contains DNA or RNA. DNA viruses are reproduced by the host cell using its enzymes and building blocks (nucleotides and amino acids) in the same manner that the host replicates its own DNA and directs its own protein synthesis.

Problem 21.34. Most RNA viruses reproduce by the synthesis of RNA from RNA. Since host cells have no enzymes for synthesizing RNA from RNA, how are such RNA viruses reproduced in host cells?

Ans. A portion of the viral RNA contains information for making an enzyme, *RNA replicase*, which directs RNA synthesis from RNA. RNA replicase, synthesized in the host cell from viral RNA, directs the subsequent replication and transcription of viral RNA.

Problem 21.35. What are *retroviruses* and how do they reproduce?

Ans. Retroviruses are RNA viruses that reproduce through the intermediate formation of DNA. The retrovirus contains genetic information for the synthesis of the enzyme *reverse transcriptase*. Reverse transcriptase, synthesized using the host cell's mechanism for protein synthesis, directs the synthesis of viral DNA from viral RNA. Subsequently, the host reproduces the viral RNA using the DNA as a template. The retroviruses include HIV and cancer-causing viruses (*oncogenic viruses*).

Problem 21.36. How does a virus harm the host cell? How does a bacterium harm the host cell?

Ans. Uncontrolled reproduction of the virus results in breakage of the cell membrane. Spillage of the cell contents leads to invasion of other host cells. Bacteria usually do not enter host cells but cause harm through the toxicity of their metabolic products.

21.4.4 Recombinant DNA Technology

Recombinant DNA technology (also referred to as *genetic engineering* or *cloning*) involves the alteration of the DNA of some organism with the objective of having that organism produce a desired polypeptide. The *E. coli* bacterium is most often used, although yeasts have also been employed. *E. coli* contains DNA in two forms: One form is its single chromosome and the other are more numerous circular double-stranded DNA, referred to as *plasmids*. *E. coli* are employed in genetic engineering because their plasmids can be easily isolated, altered, and then reintroduced into the *E. coli*. The key to the process (Fig. 21-12) is the splicing of a *donor DNA* fragment (corresponding to the gene containing the information for synthesizing the polypeptide of interest) into the plasmid (*vector DNA*) of *E. coli*. Plasmids are isolated from *E. coli* and cut at only one site using a *restriction enzyme*. The cut occurs in an uneven manner such that the overlapping single-strand ends are complementary to each other. The ends are TTAA and AATT when the restriction enzyme *EcoRI* is used. The donor DNA fragment must have the same uneven ends as the vector DNA. This is obtained

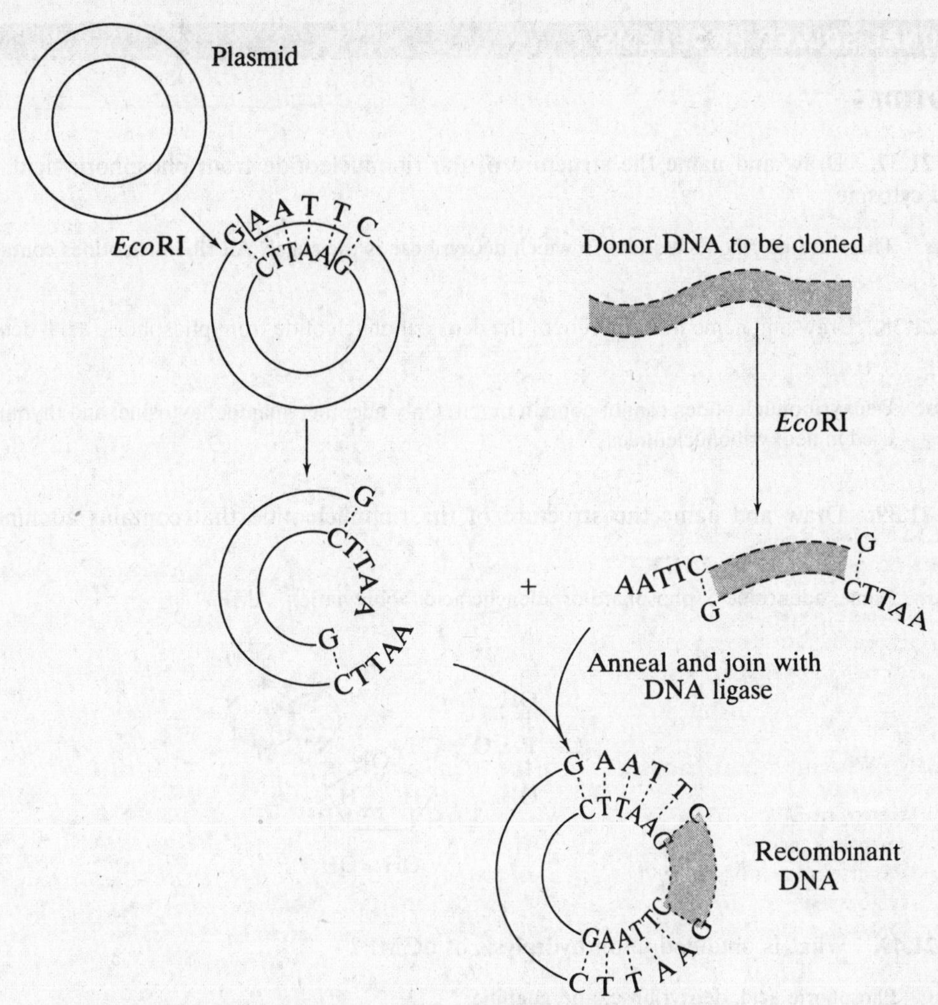

Fig. 21-12 Formation of recombinant DNA. (*After P. Kuchel and G. B. Ralston, Schaum's Outline of Theory and Problems of Biochemistry, McGraw-Hill, New York, 1988.*)

either by using the same restriction enzyme on a DNA from a source that contains the gene of interest or by *in-vitro DNA synthesis*. (Critical to the in-vitro synthesis of donor DNA is knowledge of the nucleotide sequence that codes for the polypeptide of interest.)

The vector and donor DNAs are joined together (*annealed*) in the presence of DNA ligase. Base pairing of the complementary uneven ends (referred to as *sticky ends*) of the vector and donor fragments ensures their proper alignment prior to covalent bonding of the terminal nucleotides by DNA ligase. The altered plasmids, referred to as *recombinant DNA*, are reintroduced into *E. coli*. This is accomplished by treatment of *E. coli* with calcium chloride at low temperature followed by incubation with the plasmids at higher temperature. The treatment makes the bacterial walls permeable to plasmids. *E. coli* subsequently performs replication, transcription, and translation functions. Replication reproduces many copies of the altered plasmid. These identical copies of the altered plasmid DNA are referred to as *clones*. Transcription and translation produce the normal *E. coli* polypeptides plus the polypeptide coded by the inserted donor DNA fragment. The desired polypeptide is separated and purified by the usual separation techniques. Human insulin, growth hormone, and interferons have been produced by this technique.

ADDITIONAL SOLVED PROBLEMS

NUCLEOTIDES

Problem 21.37. Draw and name the structure of the ribonucleotide from phosphoric acid, deoxyribose, and cytosine.

Ans. There are no ribonucleotides in which deoxyribose is the sugar. All ribonucleotides contain ribose.

Problem 21.38. Draw and name the structure of the deoxyribonucleotide from phosphoric acid, deoxyribose, and uracil.

Ans. Deoxyribonucleotides cannot contain uracil. Only adenine, guanine, cytosine, and thymine are used in deoxyribonucleotides.

Problem 21.39. Draw and name the structure of the ribonucleotide that contains adenine as the base.

Ans. Name: adenosine 5′-phosphate or adenylic acid; abbreviation: AMP.

NH_2 N N OH O=P—O—CH_2 O N N OH H H H H OH OH

Problem 21.40. What is obtained upon hydrolysis of dGMP?

Ans. Phosphoric acid, deoxyribose, and guanine

NUCLEIC ACIDS

Problem 21.41. Draw the structure of the sequence GMP-UMP in a ribonucleic acid. Indicate the 3′ and 5′ carbons and the 5′ ⟶ 3′ direction.

Ans.

O N H N 5′ O 5′ O=P—O—CH_2 O N N NH_2 OH H H H 3′ H O H 3′ O OH N 5′ O=P—O—CH_2 O N O OH H H H 3′ H CH_3 O OH

Problem 21.42. One DNA chain of a DNA double helix contains 18% A, 35% T, 26% C, and 21% G. What is the composition of the complementary chain?

Ans. 18% T, 35% A, 26% G, and 21% C

Problem 21.43. The total amount of purine bases (A and G) is equal to the total amount of pyrimidine bases (C and T) for the DNA double helix. Verify this relationship for Problem 21.42.

Ans. Use moles instead of percent for convenience. The total amount of purine bases in the two strands is $18 + 21 + 35 + 26 = 100$. The total amount of pyrimidine bases is $35 + 26 + 18 + 21 = 100$.

Problem 21.44. What is the relationship between the amounts of purine and pyrimidine bases in RNA?

Ans. There is no relationship because RNA exists as single strands without a requirement for complementary base pairing.

FLOW OF GENETIC INFORMATION

Problem 21.45. Why does DNA replication involve simultaneous synthesis at many replication bubbles?

Ans. The enormous size of DNA (molecular masses exceed many billions) requires this route in order to achieve the completion of replication in a reasonable time period (a few hours).

Problem 21.46. The genetic code, using three bases to specify an amino acid, is degenerate. A genetic code based on two bases specifying an amino acid would not be degenerate. Would it be adequate for protein synthesis based on 20 amino acids?

Ans. No. There are only 16 different codons possible for a two-base code and this is not enough to code for the 20 amino acids.

Problem 21.47. All cells in the body have a complete set of chromosomes, i.e., all cells have the complete message for producing all of the proteins required by all cells. How does the organism operate under these conditions?

Ans. Regulatory enzymes are present in each cell to prevent translation of most of the DNA messages in that particular cell. For example, heart cells do not need and must not synthesize the proteins found in brain cells or liver cells and so on. *Repressor enzymes* repress the synthesis of all proteins except those needed for the heart cell. Simultaneously, *inducer enzymes* induce the synthesis of proteins needed for the heart cell. Only about 2 percent of all the DNA in any cell is used for protein synthesis.

Problem 21.48. If exons *X* and *Y*, containing 280 and 260 nucleotide units, respectively, comprise one mRNA molecule, how many amino acid residues are contained in the polypeptide synthesized from that mRNA?

Ans. A total of 540 nucleotides, with three nucleotides coding for each amino acid residue, code for 180 amino acid residues in the polypeptide.

Problem 21.49. Consider the section TAA GTC CAA (written in the $3' \longrightarrow 5'$ direction) of a template strand of DNA. If this section is used to produce a messenger RNA, what amino acid sequence is produced?

Ans. mRNA = AUU CAG GUU, which codes for Ile-Gln-Val

Problem 21.50. Consider the section TAA GTC CAA (written in the $5' \longrightarrow 3'$ direction) of a template strand of DNA. If this section is used to produce a messenger RNA, what amino acid sequence is produced?

Ans. mRNA = UUG GAC UUA, which codes for Leu-Asp-Leu

Problem 21.51. The accuracy of transcription is 1 base pair per 10^5 compared to 1 base pair per 10^{10} for replication because RNA polymerase, unlike DNA polymerase, has no proofreading ability. Does this make biological sense?

Ans. Yes. The error in replication is not a problem since many correct copies (99,999 out of each 100,000) of the mRNA and polypeptide will be produced and this is sufficient for the proper functioning of the cell. The one incorrect (out of 100,000) copy of the mRNA and polypeptide will not be used by the cell. Furthermore, the error is not propagated to subsequent generations of cells or to offspring. A much higher accuracy is needed for replication because errors accumulate and are transmitted to subsequent generations of cells and offspring.

OTHER ASPECTS OF NUCLEIC ACIDS AND PROTEIN SYNTHESIS

Problem 21.52. For Problem 21.49, what amino acid sequence is produced if there occurs a mutation whereby the middle A in the TAA triplet of the DNA template strand is replaced by G?

Ans. mRNA = ACU CAG GUU, which codes for Thr-Gln-Val

Problem 21.53. For Problem 21.49, what amino acid sequence is produced if there occurs a mutation whereby A is inserted in front of the TAA triplet of the DNA template strand?

Ans. mRNA = UAU UCA GGU U, which codes for Tyr-Ser-Gly.

Problem 21.54. What is a reovirus and how does it reproduce?

Ans. A *reovirus* is a double-stranded RNA virus. Reproduction involves RNA replicase as in the case of single-stranded RNA viruses.

Problem 21.55. How are hereditary diseases treated?

Ans. There are a number of approaches: (1) Genetic counseling can give potential parents information on which to base a decision on procreation. (2) The patient can receive drug replacement therapy if the structure of the defective protein is known, can be synthesized in-vitro or obtained from a suitable organism, and can be appropriately introduced (e.g., ingestion or injection). For example, diabetics receive insulin from cattle or pigs or receive synthetic human insulin made by recombinant DNA technology. (3) Gene therapy, now in the experimental stage, involves modification of a patient's own genes to correct the genetic defect.

SUPPLEMENTARY PROBLEMS

Problem 21.56. What are the three types of biological polymers and what are the small-sized molecules (monomers) from which they are synthesized?

Ans. Nucleic acids, polypeptides, and polysaccharides are synthesized from nucleotides, amino acids, and glucose, respectively.

Problem 21.57. At high temperatures, deoxynucleic acids become denatured, i.e., they unwind from double helices into disordered single strands. Account for the fact that the higher the content of guanine-cytosine base pairs relative to adenine-thymine base pairs, the higher the temperature required to denature a DNA double helix.

Ans. There are three hydrogen bonds per G-C base pair but only two hydrogen bonds per A-T base pair.

Problem 21.58. What are the three structural differences between DNA and RNA?

Ans. DNA contains deoxyribose and thymine, and is double-stranded while RNA contains ribose and uracil, and is single-stranded.

Problem 21.59. What enzyme must be encoded in the RNA of an RNA virus in order to replicate RNA upon invasion of host cells?

Ans. RNA replicase

Problem 21.60. What enzyme must be encoded in the RNA of an RNA virus in order to produce DNA upon invasion of host cells?

Ans. Reverse transcriptase

Problem 21.61. How can the same codon AUG code for both initiation and Met?

Ans. All polypeptide chains are initiated by a Met residue. Once a particular chain has been initiated, any subsequent AUG codon in the mRNA no longer codes for initiation, i.e., one mRNA codes for one polypeptide chain. Each mRNA has AUG at its 5′ end and one of the Stop codons at its 3′ end. (Postribosomal synthesis often removes the Met residue at the beginning of a polypeptide.)

Problem 21.62. Why are viral infections more difficult to treat than bacterial infections?

Ans. Antibiotics which successfully prevent bacterial growth interfere directly with their metabolism (including replication and protein synthesis) with little effect on the host's metabolism. Viruses, unlike bacteria, must replicate inside host cells and therefore spend little time outside host cells. It is much more difficult to design a drug that prevents the reproduction of the virus without interfering with host cell activities since the host cell's machinery is used for both tasks. The most successful approach to fighting viral infections is the use of *vaccines*. A vaccine, containing the virus in weakened form, stimulates the body's immune system to generate antibodies against the virus.

Problem 21.63. Consider the base triplet GAT ($5' \longrightarrow 3'$ direction) on the information strand of DNA. (*a*) What is the corresponding base triplet on the template strand of DNA? (*b*) What is the corresponding codon on mRNA? (*c*) What is the anticodon of the tRNA which base pairs with that codon? (*d*) What amino acid is carried by that tRNA?

Ans. (*a*) CTA ($3' \longrightarrow 5'$); (*b*) GAU ($5' \longrightarrow 3'$); (*c*) CUA ($3' \longrightarrow 5'$); (*d*) Asp

Problem 21.64. What is the relationship between a DNA double-stranded molecule, a gene, and a chromosome?

Ans. A chromosome is a DNA double-stranded molecule together with histone proteins. Many genes are contained in the DNA of a chromosome.

Problem 21.65. What genetic information is contained in a gene?

Ans. The amino acid sequence of one polypeptide molecule

Problem 21.66. What experimental evidence supports the base pairing of A with T and G with C in DNA?

Ans. The amount of purines (A + G) equals the amount of pyrimidines (C + T).

Problem 21.67. *Nucleosides*, precursors to nucleotides, are not phosphorylated. Show the formation of the nucleoside from ribose and cytosine.

Ans.

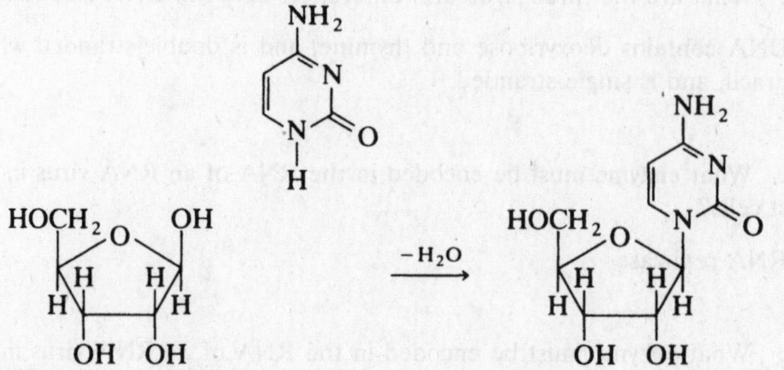

$\xrightarrow{-H_2O}$

Problem 21.68. What are the objectives of the Human Genome Project (HGP)?

Ans. HGP, funded by the National Institutes of Health and the Department of Energy, has identified the locations and base sequences of various human genes on specific chromosomes with the objective of building a database that would be used to cure various genetic diseases.

Problem 21.69. What are the main achievements of the Human Genome Project?

Ans. (*a*) About 98% of the 3.2 billion base pairs in the human genome appear not to be functional. Only 2% of the base pairs code for polypeptides. (*b*) The 2% of base pairs that code for polypeptides are contained in 21,000 genes. Since there are five times as many functional polypeptides as genes, the average gene gives rise to five different polypeptides through alternate splicing mechanisms. (*c*) 99.9% of the genome is the same for everyone. The difference among different people resides in only 1% of the genome. (*d*) The compositions and locations of specific genes on specific chromosomes have been determined. This offers the potential for improved treatments of various genetic diseases such as cystic fibrosis, sickle-cell anemia, Down syndrome, certain cancers, and Alzheimer's.

Problem 21.70. What is the basis for treatment of cancers by radiotherapy and chemotherapy?

Ans. Radiotherapy and chemotherapy use radiation and chemicals, respectively, to cause fatal mutations in the DNA of cancer cells. Cancer cells are damaged more by radiotherapy and chemotherapy than are normal cells because cancer cells grow more rapidly. Improved radiotherapy causes minimal damage to healthy tissues by delivering the minimum effective radiation dose. The radiation is confined as much as possible to the cancerous tissues by using collimated (narrowed and focused) radiation beams. Chemotherapy is similarly improved by choosing chemicals that attack cancer cells more selectively than healthy cells.

Metabolic Systems

22.1 INTRODUCTION

Biochemistry deals with the structure and function of cells and cellular components such as biopolymers, membranes, and ribosomes. The functions or physical and chemical processes involved in extracting energy from the environment and using that energy for sustaining life are known collectively as *metabolism*. Metabolism consists of the interaction of two major processes called *catabolism* and *anabolism*.

Problem 22.1. What does the process of catabolism consist of?

Ans. Catabolism comprises the chemistry involved in the degradation of energy-containing compounds leading to the capture of some of that energy, and decreasing the many types of energy-containing molecules to a limited number of simpler molecules used for the synthesis of components vital to cell structure and function. Catabolism consists of two linked processes, one of which is anaerobic, taking place in the absence of oxygen, and a second process which is aerobic, requiring oxygen.

Problem 22.2. What does the process of anabolism consist of?

Ans. Anabolism comprises the chemical processes which use key molecules and energy captured in catabolism to carry out cell functions and to synthesize cell structures.

Vertebrate cells consist of several separate compartments each surrounded by membranes of distinctive structure. These are principally the nucleus and the mitochondria. The rest of the inner portion of the cell outside both of these compartments but within the cell membrane, is called the *cytosol* (sometimes referred to as *cytoplasm*). Anaerobic processes take place in the cytosol, and aerobic processes occur in the mitochondia.

The efficient economy of cells is illustrated by the fact that all the different types of proteins and nucleic acids in the living world are synthesized from only 20 amino acids and eight nucleotide bases. These simple molecules are not only used for building structures, but have additional functions as well. For example, amino acids are the precursors for hormones, alkaloids, and porphyrins, and also serve as neurotransmitters. Nucleotides are precursors of energy carriers and of coenzymes. These are only a few of the functional compounds synthesized by cells.

Energy can neither be created nor destroyed. When glucose, for example, is oxidized by combustion to CO_2 and H_2O, its potential energy is lost as heat. However, cells transform the energy content of this reduced organic compound into a useful form by degrading it in small steps, many of which involve oxidation, to its final form, CO_2 and H_2O. Each small step is accomplished by enzymatic catalysis, and the steps are linked together as sequences of consecutive reactions. If two reactions take place in the same test tube as

$$A \longrightarrow B$$
$$C \longrightarrow D$$

they proceed independently and have no influence on each other. But if the two reactions in a common vessel are as follows:

$$A \longrightarrow B$$
$$B \longrightarrow C$$

they are linked by a common intermediate, B, so that the exothermic properties of the first may drive the succeeding endothermic reaction.

Problem 22.3. Suppose we have two separate reactions having a common intermediate, with a large equilibrium constant for the first reaction, and a small equilibrium constant for the second reaction. What will the overall equilibrium constant be for the total process?

Ans. We write the two processes along with their equilibrium constants as follows:

(1) $$A \longrightarrow B \qquad K_1 = \frac{[B]}{[A]}$$

(2) $$B \longrightarrow C \qquad K_2 = \frac{[C]}{[B]}$$

Now add the two reaction to obtain the overall reaction:

(3) $$A \longrightarrow C \qquad K_3 = \frac{[C]}{[A]}$$

The equilibrium constant for reaction (3), $[C]/[A]$, is the product

$$K_3 = \frac{[B]}{[A]} \times \frac{[C]}{[B]} = K_1 \times K_2$$

We have seen this formulation before in the derivation of

$$K_w = K_a \times K_b$$

in Chap. 9. The overall equilibrium constant for a set of reactions possessing a common intermediate can be expressed as the product of the equilibrium constants of all the individual steps. Let us now assume values for $K_1 = 1 \times 10^6$, and $K_2 = 1 \times 10^{-2}$. The overall equilibrium constant,

$$K_3 = K_1 \times K_2 = 1 \times 10^4$$

Remember, large K_{eq} means the reaction results in large quantities of product, and small K_{eq} means the opposite.

The energy derived from linked enzymatic pathways of catabolism and from photosynthesis in chlorophyll-containing organisms is used to synthesize ATP from ADP and P_i. ATP, adenosine triphosphate, is the storage form of energy captured by catabolism, and is also the principal fuel for anabolism. (Inorganic phosphates consisting of all species of phosphate, i.e., HPO_4^{2-}, $H_2PO_4^-$, PO_4^{3-}, are included under the symbol P_i for convenience in describing phosphorylation reactions in metabolism.) You will also see the expression sugar-PO_4. This is a common abbreviation for the word, sugar-phosphate, therefore the absence of a charge on the phosphate group. Any structural formula containing the phosphate group will show the appropriate charge. The linking of reactions into specific sequences is the key to channeling biomolecules to specific products in precisely the required amounts. The amounts and activities of the enzymes catalyzing each step can also be modified so as to keep all the metabolic pathways in harmonious operation.

22.2 ENZYMES, COFACTORS, AND COENZYMES

Although recent work has demonstrated modest catalytic activity by some nucleic acids, the overwhelming evidence is that enzymes are chiefly protein molecules possessing unique catalytic properties.

Problem 22.4. What are the unique aspects of enzymatic catalysis?

Ans. Catalysis occurs at temperatures consistent with the requirements for structural stability of functional proteins. Enzymatic catalysis takes place at *active sites* on the protein. The active site possesses two independent properties. One consists of a *binding site*; the other provides the catalytic activity. Cellular enzymes possess great selectivity, i.e., only molecules of specific structure will be accommodated at the binding site. This permits great selectivity among the myriad molecules present in the cell, and allows the efficient simultaneous operation of linked metabolic pathways.

Problem 22.5. What is the basis of the great selectivity of enzymes?

Ans. The selectivity of enzymatic binding sites is based on *complementarity*. This means that the structure of the binding site behaves as a lock does with its key. The origins of complementarity lie in structural and hydrophobic-hydrophilic considerations. The shape and charge of a molecule must be complementary to the shape and charge of the binding site. Furthermore, the affinity of a molecule for a binding site can be enhanced if the incoming molecule has a hydrophobic character which is matched by the hydrophobic character of the binding site.

Many enzymes are active in hydrolyses, and operate by acid-base catalysis. This is because the polypeptide chain can provide high concentrations of protons from readily available amino and carboxyl functional groups. However, many of the reactions which occur in metabolism involve chemical processes other than acid-base catalysis, which cannot be mediated by the side chains of amino acids. These include oxidation-reduction, group transfer, isomerization, and bond-breaking and bond-making reactions. Proteins consisting only of amino acid residues cannot catalyze these processes.

Problem 22.6. If simple polypeptide molecules cannot catalyze many of the reactions of metabolism, how is the necessary catalysis accomplished?

Ans. In order to accomplish these operations, the protein portion of an enzyme lacking catalytic properties, called an apoenyme, combines transiently with specialized small molecules called cofactors or coenzymes to form the catalytic holoenzyme. *Cofactors* are ions such as K^+, Mg^{2+}, or Zn^{2+}. Vitamins such as nicotinic acid (niacin), riboflavin, thiamin, and pyridoxine (B_6), serve as *coenzymes* in enzymatic redox, decarboxylation, and group transfer reactions.

Two coenzymes which are involved in most of the redox reactions of metabolism are nicotine adenine dinucleotide, NAD^+, and FAD, flavine adenine dinucleotide. Most metabolic oxidations are, in fact, dehydrogenations, and not reactions with oxygen. Nicotinamide is derived from nicotinic acid, and the isoalloxazine ring of FAD is derived from riboflavin. Thiamin is the principal cofactor in enzymatic decarboxylations. Many of the vitamins serve as coenzymes in a wide variety of cellular reactions.

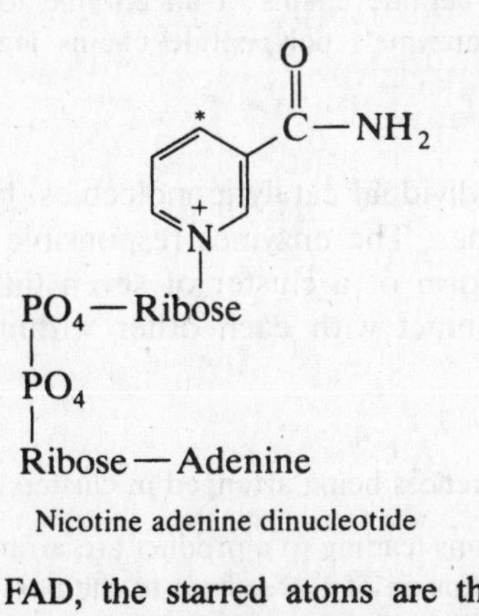

O ‖ *C—NH_2 ; $\overset{+}{N}$

PO_4—Ribose
|
PO_4
|
Ribose—Adenine

Nicotine adenine dinucleotide

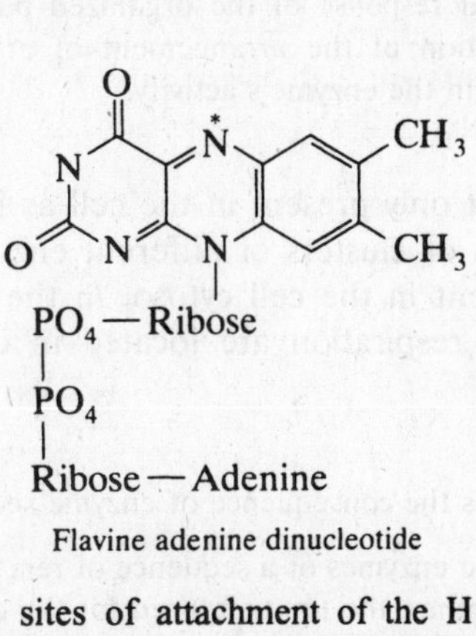

O ; *N ; N ; CH_3 ; O ; N ; N ; CH_3

PO_4—Ribose
|
PO_4
|
Ribose—Adenine

Flavine adenine dinucleotide

In both NAD^+ and FAD, the starred atoms are the sites of attachment of the H atoms removed from substrate by dehydrogenation. When carbon atom 2 of the ribose bound to the adenine portion of NAD^+ is phosphorylated, it becomes a new coenzyme, and is called $NADP^+$, nicotine adenine dinucleotide phosphate. The reduced forms of both NAD^+ and $NADP^+$ will be referred to as NADH and NADPH respectively. $NADP^+$ takes part only in biosynthetic reduction reactions. Biological dehydrogenations involve the transfer of a pair of H atoms with their electrons to NAD^+ or FAD. NAD^+ forms a loose, transitory complex with its apoenzyme, whereas FAD is bound very strongly to its apoenzyme, and remains bound to it as it undergoes reduction and oxidation.

Problem 22.7. Are there differences in the way NAD^+ and FAD behave as redox coenzymes?

Ans. The reaction of NAD^+ with a reduced compound can be represented as

$$RH_2 + NAD^+ \rightleftharpoons R + NADH + H^+$$

Only one of the pair of H atoms with both electrons reacts with NAD^+; the other enters the surrounding medium as an H^+ ion. The reaction of FAD with a reduced compound may be represented as

$$RH_2 + FAD \rightleftharpoons R + FADH_2$$

In this case, both H atoms with their electrons are transferred to the cofactor.

The activity of enzymes can be regulated by a number of means. The enzymes involved in the synthesis and hydrolysis of glycogen can be activated by phosphorylation, and deactivated by dephosphorylation. This is an example of *covalent modification*. Amino acids are synthesized by sequences of up to 15 separate enzymatically catalyzed reactions. If the end product is present in high concentrations, it combines with the first enzyme in the synthetic sequence and shuts it down. This is an example of *feedback inhibition*. Other kinds of feedback inhibition will prevent the synthesis of the enzyme itself by interfering with the transcription step producing messenger RNA.

Many enzymes are classified as *regulatory enzymes* whose activities can be controlled continuously (from 0 to 100 percent) through combination with specific activators or inhibitors called *modifiers*. For example, if citrate, a major component of aerobic metabolism, is present in high concentration, it will activate the enzyme system responsible for the synthesis of fatty acids. Such regulatory enzymes are called *allosteric* enzymes.

Problem 22.8. What is the mechanism of modification of the activities of regulatory enzymes?

Ans. The regulatory modifiers, such as ATP, citrate, as well as feedback inhibitors, do not combine with the active sites, but at other secondary binding sites having no catalytic activity. The secondary sites are complementary to the structures of the modifiers, and strongly bind them. There is an overall structural response of the organized polypeptide chains of an enzyme to this interaction. This modification of the arrangement of an enzyme's polypeptide chains leads to the consequent changes in the enzyme's activity.

Enzymes are not only present in the cell as individual catalytic molecules, but also combined in structural complexes or clusters of different enzymes. The enzymes responsible for the synthesis of fatty acids are present in the cell cytosol in the form of a cluster of seven different proteins. The enzymes of aerobic respiration are located in contact with each other within the membranes of mitochondria.

Problem 22.9. What is the consequence of enzyme sequences being arranged in clusters?

Ans. When the enzymes of a sequence of reactions leading to a product are arranged in close proximity to each other, the time required for the diffusion of a precursor to the next reaction site is greatly reduced. Therefore, the rate at which the end product of the reaction sequence appears is markedly enhanced by this arrangement.

Most of the steps in catabolic reaction sequences are characterized by small equilibrium constants, ca. 1×10^2, so that at such steps the reaction may be considered reversible. In terms of the Le Chatelier principle, a modest buildup of product will result in the synthesis of reactant. If this were the case for all the reactions in metabolic sequences, opposing cellular processes of degradation and synthesis would be chaotic, depending only on concentrations of metabolic components, and not on cellular requirements. However, in all such pathways, one or more of the reactions have very large equilibrium constants and are therefore irreversible, forcing the sequence to proceed to the end product.

Problem 22.10. If catabolic processes are irreversible, does that mean that opposing anabolic pathways are comprised of series of reactions totally different from those of catabolism?

Ans. No. Anabolic pathways utilize many of the same enzymatic reactions that occur in catabolism, but bypass the irreversible steps by employing different enzymatic reactions. This illustrates another form of regulation which effectively separates catabolic from anabolic processes, but at the same time is economical in that a minimum number of new proteins are required.

In addition to the regulatory mechanisms described above, metabolism is further regulated by *compartmentalization*. For example, the anaerobic phase of metabolism takes place in the cytosol, and the aerobic or respiratory phase occurs in the mitochondria.

Many of the reactions of metabolism involve substances called *high-energy compounds*. They are responsible for driving forward essentially unfavorable reactions. High-energy compounds like ATP have favorable equilibrium constants for group transfer and are common intermediates in reactions of metabolism which have small equilibrium constants. Let us examine the first reaction of anaerobic metabolism, the phosphorylation of glucose by ATP. The word reaction equation is

$$\text{D-Glucose} + \text{ATP} \xrightleftharpoons{\text{hexokinase}} \text{D-glucose-6-PO}_4 + \text{ADP}$$

There are two points to note carefully: (1) Phosphorylation occurs specifically at C-6, the carbon furthest from the carbonyl group (Fig. 22-2), and (2) only the D carbohydrate isomers are involved in metabolism. In the future, we will omit the conformational identity from these equations for convenience only. It is critical to note that this reaction, like all others of metabolism, require a specific enzyme in order to proceed at a rate appropriate to the temperature and time requirements of metabolism. The enzyme is specific with respect to two issues: (1) the transfer of a phosphate group (all enzymes specific for the transfer of a phosphate group are called *kinases*), and (2) the enzyme is selectively adapted to hexose sugars.

Problem 22.11. The equilibrium constant for the phosphorylation of glucose,

$$\text{Glucose} + \text{P}_i \rightleftharpoons \text{glucose-6-PO}_4 + \text{H}_2\text{O}$$

is 3.8×10^{-3}, quite unfavorable for the synthesis of the product which is central to metabolism. What mechanism does the cell use to synthesize this compound in useful concentrations?

Ans. In the cell, glucose is phosphorylated in an enzymatic reaction with ATP. To see how the cell accomplishes this apparently unfavorable reaction, we must calculate the equilibrium constant for the cellular reaction. The equilibrium constant for the hydrolysis of ATP is well known, so that our best approach is to consider the reaction of ATP with glucose as consisting of two reactions with a common intermediate, listing their equilibrium constants along with the reactions:

(1) $$\text{ATP} + \text{H}_2\text{O} \rightleftharpoons \text{ADP} + \text{P}_i \qquad K_1 = 2.26 \times 10^6$$

(2) $$\text{Glucose} + \text{P}_i \rightleftharpoons \text{glucose-6-PO}_4 + \text{H}_2\text{O} \qquad K_2 = 3.8 \times 10^{-3}$$

Sum the two equations:

(3) $$\text{Glucose} + \text{ATP} \rightleftharpoons \text{glucose-6-PO}_4 + \text{ADP}$$

Finally, multiply the two equilibrium constants to obtain the equilibrium constant for the sum of the two processes:

$$K_1 \times K_2 = K_3 = 8.6 \times 10^3$$

The overall equilibrium constant now favors the reaction in the direction of product.

Note that reaction (1), a highly favorable reaction, is a hydrolysis, or a transfer of the phosphate group to water. However, in the enzymatic process, there is a transfer of the terminal phosphate group of ATP to glucose instead of to water. Other high-energy compounds, in which group transfer is involved, encountered in metabolism include the following: thioesters like acetyl CoA:

$$CH_3-\underset{\underset{O}{\|}}{C}-S-R$$

acid anhydrides like acetyl phosphate:

$$CH_3-\underset{\underset{O}{\|}}{C}-O-PO_3^{2-}$$

and enol phosphates like phosphenolpyruvic acid:

$$CH_2=\underset{\underset{O-PO_3^{2-}}{|}}{C}-COO^-$$

The result of group transfer reactions of high-energy compounds, in which the transferred group is the common intermediate, is that chemically unfavorable reactions of metabolism can proceed and result in required quantities of products.

22.3 METABOLISM OF CARBOHYDRATES

Glucose is the principal reduced organic compound which fuels metabolism. The first pathway it encounters is anaerobic, located in the cell cytosol, and is called *glycolysis*. The steps in the glycolytic pathway are described in Fig. 22-1. The chemical structures of all the compounds named in the description of glycolysis may be found in Fig. 22-2.

Glucose entering the cell is phosphorylated, isomerized to fructose-6-PO_4, phosphorylated again to form fructose-1,6-bisphosphate, and cleaved into an equilibrium mixture of two three-carbon compounds which are in readily reversible equilibrium with each other. One of these, 3-phosphoglyceraldehyde, is oxidized, with the simultaneous reduction of the cofactor, NAD^+, to finally emerge as pyruvate. (Since the pH of cells is approximately 7.0, all organic acids are present in anionic form.) Because of shifts in the equilibrium due to the consumption of 3-phosphoglyceraldehyde, the other three-carbon compound, dihydroxyacetone phosphate, is continuously converted to 3-phosphoglyceraldehyde. An important result of this process is the generation of 2 mol of ATP from 1 mol of glucose. The ATP is generated by transfer of P_i to ADP from high-energy phosphate compounds emerging from the reaction sequence. This method of synthesis of ATP is called ***substrate level phosphorylation***.

Glucose-6-PO_4 is also used to synthesize pentose phosphates for the synthesis of nucleic acids and redox cofactors, and also the generation of NADPH, the reducing cofactor used in biosynthesis. This special enzyme pathway is called the *phosphogluconate shunt*.

The overall equation for the reactions of glycolysis is

$$\text{glucose} + 2\ \text{ADP} + 2\ P_i + 2\ \text{NAD}^+ \rightleftharpoons 2\ \text{pyruvate} + 2\ \text{ATP} + 2\ \text{NADH} + 2\text{H}^+$$

There are three important conclusions to be drawn from this equation: (1) No oxygen is required for the reaction, (2) 2 mol of ATP are produced by substrate level phosphorylation, and (3) 2 mol of NAD^+ are reduced to NADH. The fate of pyruvate depends on all three of these considerations, i.e., what mechanisms are available for the oxidation of NADH, what levels of ATP are present in the cell, and what is the concentration of oxygen in the cell.

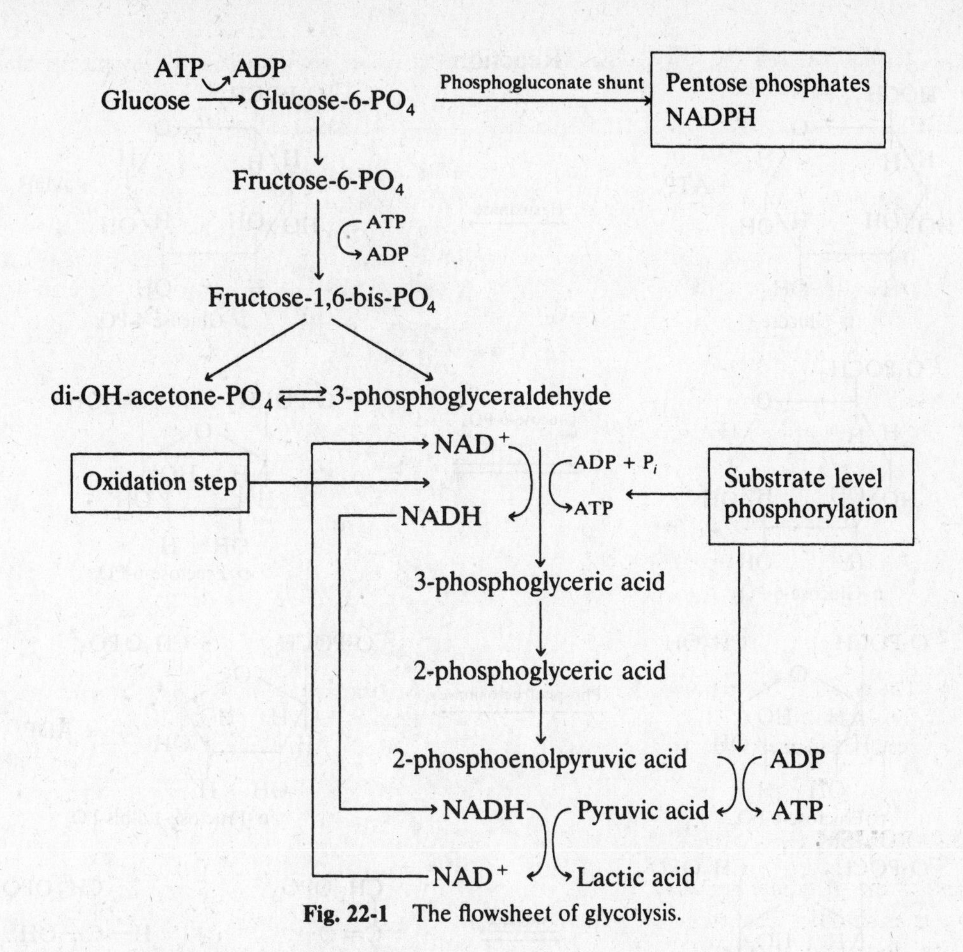

Fig. 22-1 The flowsheet of glycolysis.

Problem 22.12. What is important about the fate of the NADH produced by glycolysis?

Ans. Glycolysis depends upon the presence of NAD^+ in order for the oxidation step to occur. If the NADH is not recycled to NAD^+, glycolysis will stop at the oxidation step.

Problem 22.13. Would pyruvate be produced if all the glycolytic enzymes, ATP, ADP, and NAD^+, were incubated together?

Ans. No. Inorganic phosphate must be present if pyruvate is to be generated. If all the NAD^+ were to be used up, the process would also stop. To regenerate NAD^+, lactic dehydrogenase must also be present.

Pyruvate may now undergo either one of two processes. It can be reduced by the NADH generated in the oxidation step in glycolysis to produce lactate, or it can enter the mitochondria and proceed to the respiratory or aerobic phase of catabolism. The route taken depends on the type of cell in which glycolysis takes place.

Problem 22.14. What controls the fate of pyruvate produced by glycolysis?

Ans. In cells such as those making up brain tissue, where ATP is continuously required (no resting phase), respiration is the major route, so that in such tissue, pyruvate directly enters the mitochondria where the aerobic phase of catabolism takes place.

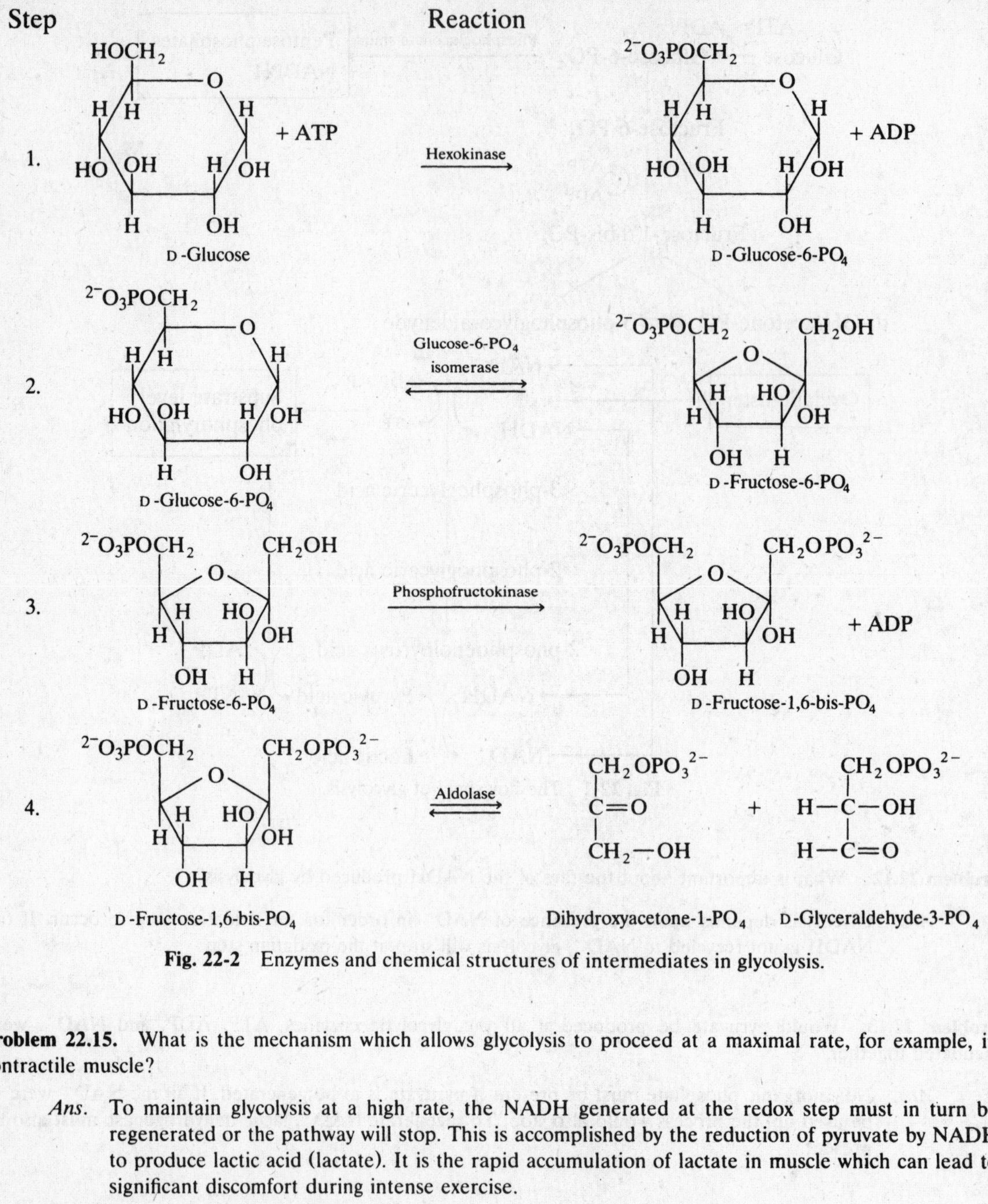

Step	Reaction
1.	D-Glucose + ATP $\xrightarrow{\text{Hexokinase}}$ D-Glucose-6-PO$_4$ + ADP
2.	D-Glucose-6-PO$_4$ $\xrightleftharpoons{\text{Glucose-6-PO}_4\text{ isomerase}}$ D-Fructose-6-PO$_4$
3.	D-Fructose-6-PO$_4$ $\xrightarrow{\text{Phosphofructokinase}}$ D-Fructose-1,6-bis-PO$_4$ + ADP
4.	D-Fructose-1,6-bis-PO$_4$ $\xrightleftharpoons{\text{Aldolase}}$ Dihydroxyacetone-1-PO$_4$ + D-Glyceraldehyde-3-PO$_4$

Fig. 22-2 Enzymes and chemical structures of intermediates in glycolysis.

Problem 22.15. What is the mechanism which allows glycolysis to proceed at a maximal rate, for example, in contractile muscle?

Ans. To maintain glycolysis at a high rate, the NADH generated at the redox step must in turn be regenerated or the pathway will stop. This is accomplished by the reduction of pyruvate by NADH to produce lactic acid (lactate). It is the rapid accumulation of lactate in muscle which can lead to significant discomfort during intense exercise.

Problem 22.16. What is the fate of the lactate produced in active muscle?

Ans. The accumulated lactate is rapidly transported out of muscle cells into the circulation. When it reaches the liver, it is converted back into glucose by a reversal of glycolysis. This latter process is called *gluconeogenesis*. (See Problem 22.19.)

After pyruvate enters the mitochondria, it encounters the second pathway, which is aerobic and cyclic, and is called the *Krebs* or *tricarboxylic acid cycle* (*TCA cycle*). The process is illustrated in Fig. 22-3, and the structures of intermediates and enzymes are illustrated in Fig. 22-4.

The first step is a decarboxylation of pyruvate and its conversion to acetyl-CoA, with the concommitant production of NADH. The next step is a condensation reaction between acetyl-CoA and oxaloacetate to produce the six-carbon compound, citrate. Citrate is isomerized to isocitrate,

$$\begin{array}{c} CH_2OPO_3^{2-} \\ | \\ H-C-OH \\ | \\ H-C=O \end{array} \rightleftharpoons \begin{array}{c} CH_2OPO_3^{2-} \\ | \\ C=O \\ | \\ CH_2-OH \end{array}$$

D-Glyceraldehyde-3-PO_4 Dihydroxyacetone-1-PO_4

5. $$\begin{array}{c} CH_2OPO_3^{2-} \\ | \\ H-C-OH \\ | \\ H-C=O \end{array} + NAD^+ + P_i \xrightleftharpoons[\text{dehydrogenase}]{\text{Glyceraldehyde-3-PO}_4} \begin{array}{c} CH_2OPO_3^{2-} \\ | \\ H-C-OH \\ | \\ O=C-OPO_3^{2-} \end{array} + NADH + H^+$$

1,3-Bisphospho-D-glyceric acid

6. $$\begin{array}{c} CH_2OPO_3^{2-} \\ | \\ H-C-OH \\ | \\ O=C-OPO_3^{2-} \end{array} + ADP \xrightleftharpoons[\text{kinase}]{\text{Phosphoglycerate}} \begin{array}{c} CH_2OPO_3^{2-} \\ | \\ H-C-OH \\ | \\ O=C-OH \end{array} + ATP$$

3-Phospho-D-glyceric acid

7. $$\begin{array}{c} CH_2OPO_3^{2-} \\ | \\ H-C-OH \\ | \\ O=C-OH \end{array} \xrightleftharpoons{\text{Phosphoglyceromutase}} \begin{array}{c} CH_3 \\ | \\ H-C-OPO_3^{2-} \\ | \\ O=C-OH \end{array}$$

2-Phospho-D-glyceric acid

8. $$\begin{array}{c} CH_3 \\ | \\ HO-C-OPO_3^{2-} \\ | \\ O=C-OH \end{array} \xrightleftharpoons{\text{Enolase}} \begin{array}{c} CH_2 \\ \| \\ C-OPO_3^{2-} \\ | \\ O=C-OH \end{array} + H_2O$$

Phosphoenolpyruvic acid

9. $$\begin{array}{c} CH_2 \\ \| \\ C-OPO_3^{2-} \\ | \\ O=C-OH \end{array} + ADP \xrightarrow{\text{Pyruvate kinase}} \begin{array}{c} CH_3 \\ | \\ C=O \\ | \\ O=C-OH \end{array} + ATP$$

Pyruvic acid

10. $$\begin{array}{c} CH_3 \\ | \\ C=O \\ | \\ O=C-OH \end{array} + NADH + H^+ \xrightleftharpoons[\text{dehydrogenase}]{\text{Lactic}} \begin{array}{c} CH_3 \\ | \\ H-C-OH \\ | \\ O=C-OH \end{array} + NAD^+$$

Lactic acid

Fig. 22-2 (*Continued*)

decarboxylated and oxidized in two steps to produce a four-carbon acid, and further oxidized to oxaloacetate to repeat the cycle. Because the cycle repeats in this cyclic manner it is sometimes referred to a catalytic cycle. During this process, 4 mol of NADH and 1 mol of $FADH_2$ are generated. The CO_2 from the metabolism of glucose is generated in the three decarboxylation steps involving pyruvate's conversion to acetyl-CoA and the two decarboxylations which reduce the six-carbon citrate to the four-carbon compound succinyl-CoA. An enzymatically complex reaction occurs here in which

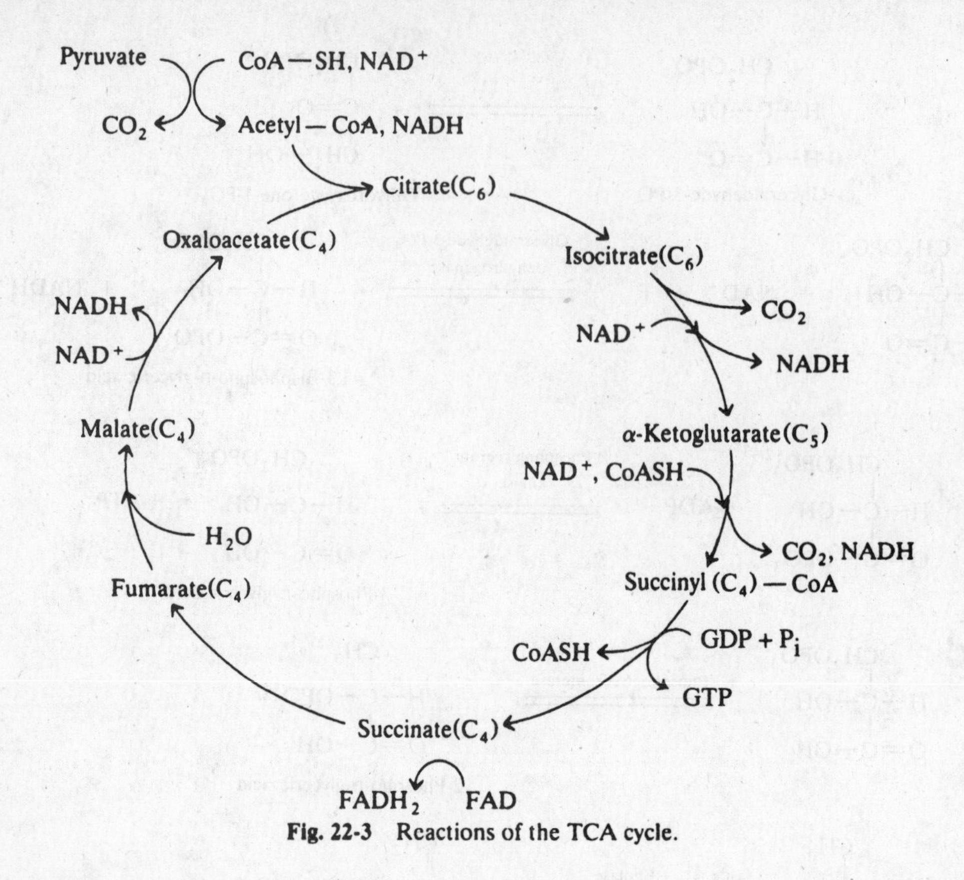

Fig. 22-3 Reactions of the TCA cycle.

GDP, guanosine diphosphate, is converted to GTP, guanosine triphosphate, by a substrate level phosphorylation, and CoA-SH is released. GTP is a high-energy compound equivalent to and interchangeable with ATP.

Problem 22.17. Are there any irreversible steps in the TCA cycle?

Ans. There are three irreversible steps. Each involves a decarboxylation in which CO_2 is lost. All the other steps are reversible, and the directions of the reactions are dependent on the energy state of the cell. This means that the TCA cycle can be used to synthesize metabolites.

Problem 22.18. In what way are the TCA cycle intermediates used to synthesize metabolites?

Ans. Each of the α-keto-acids can be transformed into a corresponding amino acid by transamination. (See Sec. 22.4.) Succinyl-CoA is used for porphyrin synthesis. Acetyl-CoA is a precursor of cholesterol and ketone bodies. Oxaloacetate can be transformed into pyruvate in the cytosol, and pyruvate into oxaloacetate in the mitochondria.

The hydrogen atoms and their accompanying electrons generated in the TCA oxidation steps captured by the cofactors NAD^+ and FAD now enter a third pathway called the *electron transport pathway*. Electrons are passed from protein to protein in this pathway in oxidation-reduction steps, and finally are combined with oxygen to form water. Compare these two separate processes for the production of CO_2 and H_2O with the simultaneous production of the same two substances from the combustion of glucose. The fate of H atoms and their electrons will be discussed in Sec. 22.6.

At this point we can return to the question of how and why glucose is regenerated by the anaerobic glycolytic pathway, or gluconeogenesis, which occurs in the liver.

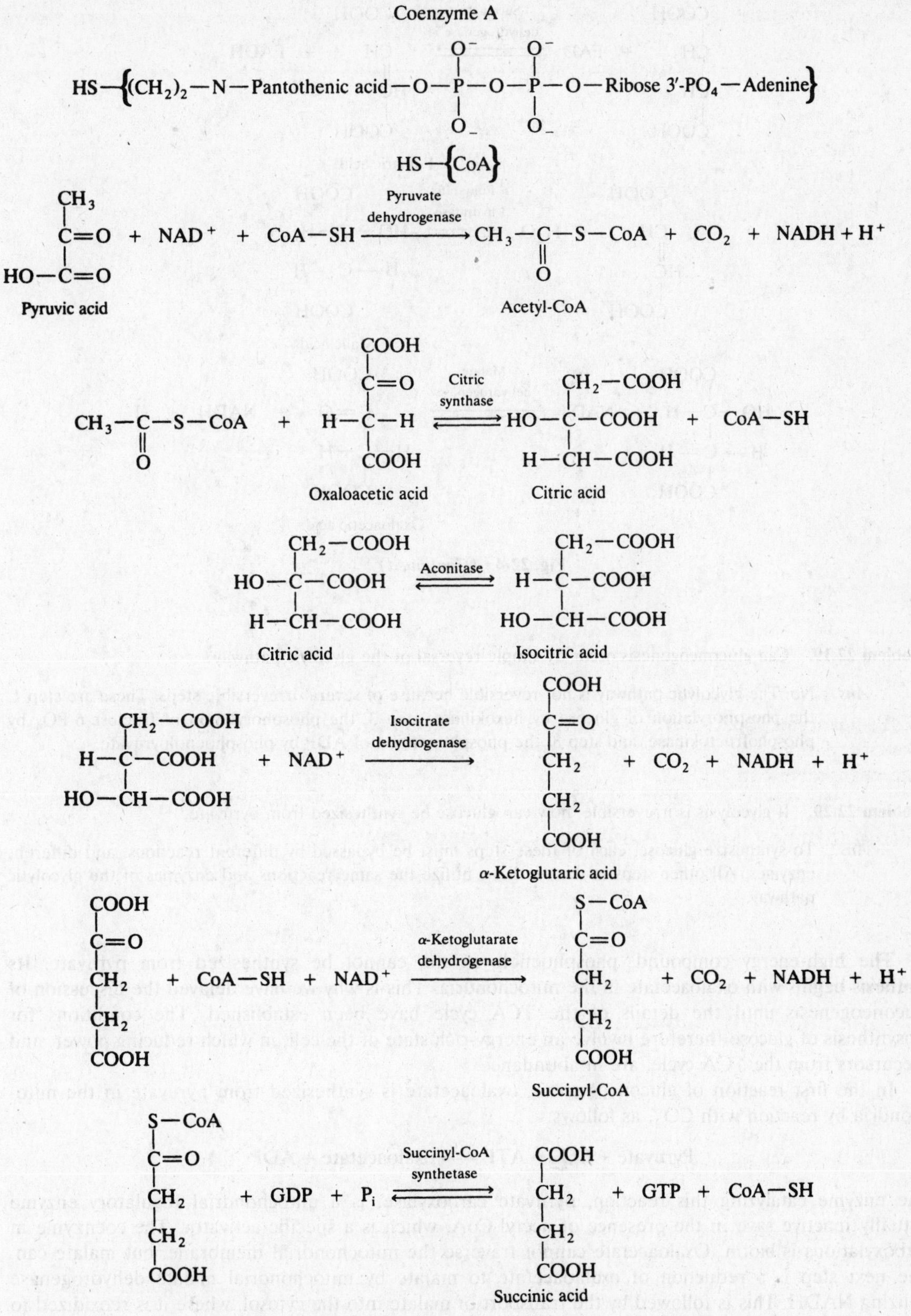

Coenzyme A

$$HS-\{(CH_2)_2-N-\text{Pantothenic acid}-O-P(O^-)(O)-O-P(O^-)(O)-O-\text{Ribose 3'-}PO_4-\text{Adenine}\}$$

$$HS-\{CoA\}$$

$$\underset{\text{Pyruvic acid}}{CH_3-C(=O)-C(=O)-OH} + NAD^+ + CoA-SH \xrightarrow{\text{Pyruvate dehydrogenase}} \underset{\text{Acetyl-CoA}}{CH_3-C(=O)-S-CoA} + CO_2 + NADH + H^+$$

$$CH_3-C(=O)-S-CoA + \underset{\text{Oxaloacetic acid}}{HOOC-C(=O)-CH_2-COOH} \xrightleftharpoons{\text{Citric synthase}} \underset{\text{Citric acid}}{CH_2(COOH)-C(OH)(COOH)-CH_2-COOH} + CoA-SH$$

$$\underset{\text{Citric acid}}{CH_2(COOH)-C(OH)(COOH)-CH_2-COOH} \xrightleftharpoons{\text{Aconitase}} \underset{\text{Isocitric acid}}{CH_2(COOH)-CH(COOH)-CH(OH)-COOH}$$

$$CH_2(COOH)-CH(COOH)-CH(OH)-COOH + NAD^+ \xrightarrow{\text{Isocitrate dehydrogenase}} \underset{\alpha\text{-Ketoglutaric acid}}{HOOC-C(=O)-CH_2-CH_2-COOH} + CO_2 + NADH + H^+$$

$$HOOC-C(=O)-CH_2-CH_2-COOH + CoA-SH + NAD^+ \xrightarrow{\alpha\text{-Ketoglutarate dehydrogenase}} \underset{\text{Succinyl-CoA}}{CoA-S-C(=O)-CH_2-CH_2-COOH} + CO_2 + NADH + H^+$$

$$CoA-S-C(=O)-CH_2-CH_2-COOH + GDP + P_i \xrightleftharpoons{\text{Succinyl-CoA synthetase}} \underset{\text{Succinic acid}}{HOOC-CH_2-CH_2-COOH} + GTP + CoA-SH$$

Fig. 22-4 Structures and reactions of TCA cycle intermediates.

$$\begin{array}{c} COOH \\ | \\ CH_2 \\ | \\ CH_2 \\ | \\ COOH \end{array} + FAD \xrightleftharpoons[\text{dehydrogenase}]{\text{Succinate}} \begin{array}{c} COOH \\ | \\ CH \\ \| \\ HC \\ | \\ COOH \end{array} + FADH_2$$

Fumaric acid

$$\begin{array}{c} COOH \\ | \\ CH \\ \| \\ HC \\ | \\ COOH \end{array} + H_2O \xrightleftharpoons[\text{hydratase}]{\text{Fumarate}} \begin{array}{c} COOH \\ | \\ HO-C-H \\ | \\ H-C-H \\ | \\ COOH \end{array}$$

L-Malic acid

$$\begin{array}{c} COOH \\ | \\ HO-C-H \\ | \\ H-C-H \\ | \\ COOH \end{array} + NAD^+ \xrightleftharpoons[\text{dehydrogenase}]{\text{Malate}} \begin{array}{c} COOH \\ | \\ C=O \\ | \\ H-C-H \\ | \\ COOH \end{array} + NADH + H^+$$

Oxaloacetic acid

Fig. 22-4 (*Continued*)

Problem 22.19. Can gluconeogenesis occur by simple reversal of the glycolyic pathway?

Ans. No. The glycolytic pathway is not reversible because of several irreversible steps. These are step 1, the phosphorylation of glucose by hexokinase, step 3, the phosphorylation of fructose-6-PO_4 by phosphofructokinase, and step 8, the phosphorylation of ADP by phosphoenolpyruvate.

Problem 22.20. If glycolysis is irreversible, how can glucose be synthesized from pyruvate?

Ans. To synthesize glucose, each of these steps must be bypassed by different reactions, and different enzymes. All other steps in the synthesis utilize the same reactions and enzymes of the glycolytic pathway.

The high-energy compound, phosphoenolpyruvate, cannot be synthesized from pyruvate. Its synthesis begins with oxaloacetate in the mitochondria. This is why we have delayed the discussion of gluconeogenesis until the details of the TCA cycle have been established. The conditions for biosynthesis of glucose therefore involve an energy-rich state of the cell, in which reducing power, and precursors from the TCA cycle, are in abundance.

In the first reaction of gluconeogenesis, oxaloacetate is synthesized from pyruvate in the mitochondria by reaction with CO_2, as follows:

$$\text{Pyruvate} + CO_2 + \text{ATP} \longrightarrow \text{oxaloacetate} + \text{ADP}$$

The enzyme catalyzing this reaction, pyruvate carboxylase, is a mitochondrial regulatory enzyme virtually inactive save in the presence of acetyl-CoA, which is a specific activator. The coenzyme in carboxylations is biotin. Oxaloacetate cannot traverse the mitochondrial membrane, but malate can. The next step is a reduction of oxalaoacetate to malate by mitochondrial malate dehydrogenase utilizing NADH. This is followed by the transport of malate into the cytosol, where it is reoxidized to oxaloacetate by cytosolic malate dehydrogenase.

Problem 22.21. What are the three steps of gluconeogenesis which bypass the irreversible reactions of glycolysis?

Ans. In the first of the bypass reactions, phosphoenolpyruvate is synthesized from the extramitochondrial oxaloacetate by the enzyme phosphoenolpyruvate carboxykinase:

(1) $$\text{Oxaloacetate} + \text{GTP} \longrightarrow \text{phosphoenolpyruvate} + CO_2 + \text{GDP}$$

Again, this occurs only under conditions of high ATP and NADH concentrations, which also drive the next several reversible steps to produce fructose-1,6-bisphosphate. The formation of this compound from fructose-6-PO_4 through glycolysis is irreversible; therefore, in the second bypass reaction, the gluconeogenesis pathway utilizes a new enzyme, fructose diphosphatase, to irreversibly hydrolyze the bisphosphate to fructose-6-PO_4, as follows:

(2) $$\text{Fructose-1,6-bis-}PO_4 + H_2O \longrightarrow \text{fructose-6-}PO_4 + P_i$$

The third bypass reaction requires the hydrolysis of glucose-6-PO_4 to glucose, which then can leave the liver and enter the blood for distribution to other tissues. This reaction requires a third new enzyme, glucose-6-phosphatase:

(3) $$\text{Glucose-6-}PO_4 + H_2O \longrightarrow \text{glucose} + P_i$$

Glucose-6-phosphatase is located in the liver, and is not found in any other tissue in vertebrates. This means that neither muscle nor brain can furnish free glucose to the blood.

Glucose-6-PO_4 can also be stored as a future energy source in the form of a carbohydrate polymer called *glycogen*. The process for the synthesis of glycogen, *glycogenesis*, requires a new set of enzymes which operate only under cellular conditions of high concentrations of ATP. Glycogen is synthesized only in the liver and muscle cells, and stored as glycogen granules. The first step in glycogenesis is an enzymatic isomerization in which the phosphate group attached to carbon 6 of glucose is transferred to carbon 1, i.e., glucose-6-PO_4 to glucose-1-PO_4. This is followed by activation of glucose-1-PO_4 in order for it to be incorporated into glycogen. It reacts with uridine triphosphate, UTP, to form uridine diphosphoglucose as

$$\text{Glucose-1-}PO_4 + \text{UTP} \longrightarrow \text{UDP-glucose} + PP_i$$

PP_i stands for pyrophosphate, which undergoes an enzymatic hydrolysis characterized by a large equilibrium constant. That secondary process drives the overall activation reaction:

$$^-O-\overset{\overset{O}{\|}}{\underset{\underset{O^-}{|}}{P}}-O-\overset{\overset{O}{\|}}{\underset{\underset{O^-}{|}}{P}}-O^- + H_2O \xrightarrow{\text{pyrophosphatase}} 2\,HO-\overset{\overset{O}{\|}}{\underset{\underset{O^-}{|}}{P}}-O^-$$

Pyrophosphate hydrolysis is a key step driving many biosynthetic reactions. Uridine triphosphate is similar to ATP and GTP, with the substitution of the nitrogen base uridine for either adenine or guanidine. Uridine diphosphate is abbreviated UDP, as were the diphosphates of adenosine diphosphate (ADP), and guanosine diphosphate (GDP).

Problem 22.22. In the reaction to form UDP-glucose, what is the origin of the PP_i?

Ans. The pyrophosphate derives from the two terminal phosphate groups which are cleaved from UTP, and replaced by glucose-1-PO_4 to form UDP-glucose.

The next step is the addition of the activated glucose to the end of a growing chain of glycogen. Glycogen consists of a chain of glucose molecules linked through α-1,4 bonds. The activated glucose is the substrate for glycogen synthase, a regulatory enzyme which is activated by glucose-6-PO_4. The reaction is as follows:

$$\text{UDP-glucose} + (\text{glucose})_n \longrightarrow (\text{glucose})_{n+1} + \text{UDP}$$

The degradation of glycogen to form glucose, *glycogenolysis*, proceeds by a route different from its synthesis. The chain terminal glucose is removed by phosphorolysis by the enzyme phosphorylase in its active form:

$$(\text{Glucose})_n + \text{P}_i \longrightarrow (\text{glucose})_{n-1} + \text{glucose-1-PO}_4$$

Glucose-1-PO_4 is converted to glucose-6-PO_4 by the enzyme phosphoglucomutase, and can be dephosphorylated in the liver to enter the blood stream for distribution. In muscle cells, it goes directly into glycolysis.

The activity of phosphorylase is also influenced by the hormones insulin, glucagon, and epinephrine. Insulin is involved with the transport of glucose into all cells except liver and red blood cells. Glucagon and epinephrine both influence the degradation of glycogen in the liver, and epinephrine alone in muscle cells.

Problem 22.23. What is the origin of the hormones insulin and glucagon, and what regulates their concentrations in the blood?

Ans. These hormones are polypeptides, and secreted by the pancreas. Insulin is synthesized and secreted by the β-cells, and glucagon by the α-cells. Their secretion into the blood is directly controlled by the blood glucose concentration. Above 80 to 100 mg percent, the pancreatic α-cells secrete insulin. Below that level, the α-cells secrete glucagon. That blood glucose concentration is optimum for brain function.

Problem 22.24. What are the specific effects of insulin and glucagon in carbohydrate metabolism?

Ans.

Insulin	Glucagon
1. Enhancement of glycogenolysis, and glucose uptake into muscle and fat cells	1. Stimulates glycogenolysis in liver only
2. Activates glycogen synthase	2. Deactivates glycogen synthase

Problem 22.25. What is the origin of epinephrine, and what are its specific effects in carbohydrate metabolism?

Ans. Epinephrine is a polypeptide synthesized in the adrenal cortex. It is released very quickly by any sudden environmental change sensed by the central nervous system, the fight or flight reflex. It stimulates glycogenolysis in both liver and muscle cells. It is also under control over long periods of time by cells in the hypothalamus of the brain which are sensitive to the blood concentration of glucose. Lowered glucose concentration will raise the epinephrine level.

Problem 22.26. How do the hormones epinephrine and glucagon affect carbohydrate metabolism?

Ans. The hormones epinephrine and glucagon cannot penetrate cell membranes. They affect metabolic processes by binding to specific receptors on the membrane, which receptors in turn activate a specific enzyme bound to the inner membrane surface, adenylate cyclase. This enzyme converts ATP to cyclic AMP (cyclic adenosine monophosphate), or c-AMP. The presence of c-AMP activates another enzyme, protein kinase, which phosphorylates and activates phosphorylase kinase. Phosphorylase kinase phosphorylates phosphorylase b (inactive) to form phosphorylase a (active) which in turn cleaves glucose from glycogen by phosphorolysis to yield glucose-1-PO_4.

This series of enzymatic steps, or cascade, causes an amplification of the initiating step of epinephrine binding to the membrane receptor of over a millionfold. In fractions of a second, the blood glucose concentration can be raised to 2 to 3 times its normal level. In outline form, these steps are as follows:

$$\text{Epinephrine} \longrightarrow \text{membrane receptor} \longrightarrow \text{membrane enzyme} \longrightarrow \text{c-AMP}$$

$$\text{Inactive protein kinase} \xrightarrow{\text{c-AMP}} \text{active protein kinase (K)}$$

$$\text{Inactive dephosphophosphorylase kinase (dPK)} \xrightarrow{\text{K}} \text{active phosphorylase kinase (PK)}$$

$$\text{Phosphorylase } b \text{ (inactive } K_b) \xrightarrow{\text{PK}} \text{phosphorylase } a \text{ (active } K_a)$$

$$(\text{Glucose})_n \xrightarrow{K_a} \text{glucose-1-}PO_4 + (\text{glucose})_{n-1}$$

The mechanism of glucagon's action is similar, but its role is to raise the blood glucose level to normal concentrations, and therefore it is active during fasting periods and between meals.

22.4 METABOLISM OF LIPIDS

The degradation of lipids involves the oxidation of fatty acids, which occurs in the mitochondria. Fatty acids in the cytosol must first undergo a complex enzymatic activation in which 2 mol of ATP are used in the conversion of the fatty acid to an acyl-CoA thioester.

Problem 22.27. Why does the activation process

$$\text{RCOOH} + \text{ATP} + \text{CoA-SH} \longrightarrow \text{CoA-S-COR} + \text{AMP} + PP_i$$

$$PP_i \longrightarrow 2\ P_i$$

which appears to require only 1 mol of ATP, actually consume 2 mol of ATP?

Ans. AMP, adenosinemonophosphate, cannot be phosphorylated by either substrate level or oxidative phosphorylation, but can react with another mole of ATP catalyzed by the enzyme adenylate kinase as:

$$\text{AMP} + \text{ATP} \longrightarrow 2\ \text{ADP}$$

The ADP thus generated now can be phosphorylated by substrate level or oxidative phosphorylation to form ATP. If we add the two equations, the result is

$$\text{RCOOH} + 2\ \text{ATP} + \text{CoA-SH} \longrightarrow \text{CoA-S-COR} + 2\ \text{ADP} + 2\ P_i$$

This CoA derivative then exchanges its CoA for another partner called *carnitine*. The carnitine-fatty acid complex, which is soluble in the mitochondrial membranes, is then transported from the cytosol past the inner mitochondrial membrane into the inner matrix, where it exchanges the carnitine for acetyl-CoA, to again become the fatty acyl-CoA derivative.

The first step in the oxidation (see Fig. 22-5) is a dehydrogenation between the α- and β- (carbons 2 and 3) carbons of the acyl-CoA derivative to produce the Δ^2-unsaturated acyl-CoA. The cofactor reduced in this reaction is FAD.

The next step is an enzymatic hydration of the double bond to form a β-hydroxyacyl-CoA. The enzyme which hydrates carbon atom 3 of this fatty acid derivative induces chirality at that carbon atom leading to the L-hydroxy configuration. This is important, because in the synthesis of fatty acids, the catabolic sequence is largely reversed. However, the two pathways are kept separated in part by using a different reductase which yields the D-hydroxy configuration.

The β-hydroxyacyl-CoA is next dehydrogenated, with NAD^+ as cofactor, to produce the β-keto-acyl-CoA.

At this point an enzymatic cleavage requiring a molecule of acetyl-CoA takes place. The result is the production of a molecule of acetyl-CoA derived from carbons 1 and 2 of the original fatty acid,

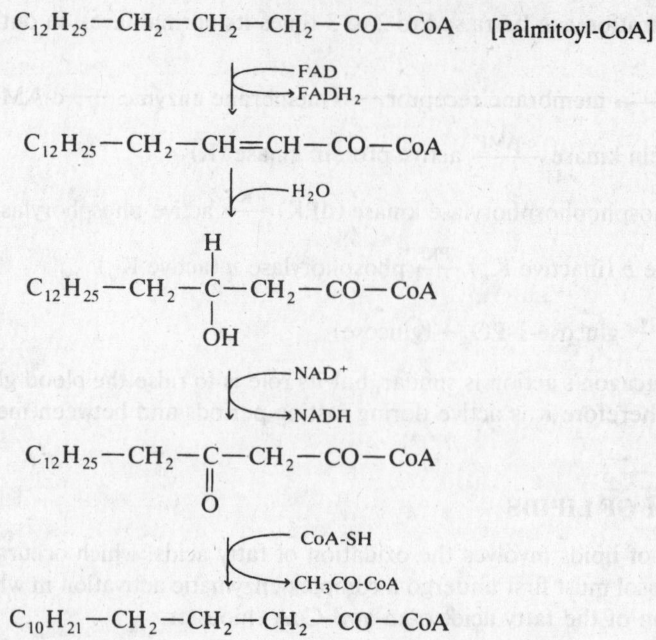

$$C_{12}H_{25}-CH_2-CH_2-CH_2-CO-CoA \quad \text{[Palmitoyl-CoA]}$$

$$\xrightarrow{FAD \rightarrow FADH_2}$$

$$C_{12}H_{25}-CH_2-CH=CH-CO-CoA$$

$$\xrightarrow{H_2O}$$

$$C_{12}H_{25}-CH_2-\underset{\displaystyle OH}{\overset{\displaystyle H}{C}}-CH_2-CO-CoA$$

$$\xrightarrow{NAD^+ \rightarrow NADH}$$

$$C_{12}H_{25}-CH_2-\underset{\displaystyle O}{\overset{}{C}}-CH_2-CO-CoA$$

$$\xrightarrow{CoA\text{-}SH \rightarrow CH_3CO\text{-}CoA}$$

$$C_{10}H_{21}-CH_2-CH_2-CH_2-CO-CoA$$

Fig. 22-5 The first oxidative cycle of fatty acid.

along with another fatty acyl-CoA molecule having two fewer carbon atoms than the original acyl-CoA derivative.

This new acyl-CoA undergoes the same sequence of reactions, ending again with the production of a second molecule of acetyl-CoA and another fatty-acyl-CoA derivative with two fewer carbon atoms in the chain. At each passage through this sequence of reactions, the fatty acid chain loses two carbons as acetyl-CoA, so that if we start with palmitoyl-CoA, a 16-carbon chain acid, seven of these sequences will have occurred, with the production of eight acetyl-CoAs, and 14 pairs of hydrogen atoms, seven of which emerge as $FADH_2$ and seven as $NADH + H^+$. Because the shortened fatty acid emerges as its acyl-thioester at the end of each round, it is necessary to prime it with ATP only once at the beginning of the sequence. The acetyl-CoA now enters the TCA cycle, and the hydrogens enter the electron transport chain where oxidative phosphorylation takes place. See Figs. 22-3 and 22-8. The synthesis of fatty acids essentially utilizes the chemical reactions of fat catabolism. It occurs, however, in the cell cytosol rather than in the mitochondria, and requires a different activation mechanism along with different enzymes and coenzymes.

Problem 22.28. Does acetyl-CoA undergo processes other than conversion to CO_2 and H_2O?

Ans. In liver mitochondria, acetyl-CoA may also be converted to *ketone bodies*. These are acetoacetate, $CH_3-CO-CH_2-COO^-$, acetone, $CH_3-CO-CH_3$, and 3-hydroxybutyrate, $CH_3-CHOH-CH_2-COO^-$. These arise through a process called *ketogenesis*. Ketogenesis occurs at a high rate when muscle activity is high, i.e., under conditions of high energy demand. Acetoacetate is rapidly converted to acetone and 3-hydroxybutyrate, which serve as energy sources for cardiac and skeletal muscle, supplying as much as 10 percent of their daily requirement.

Problem 22.29. Can acetyl-CoA be utilized to synthesize glucose in mammals?

Ans. No. However, in plants, acetyl-CoA reacts by way of unique enzymes to produce succinate. In this reaction, acetyl-CoA is not oxidized to CO_2, and there is a net synthesis of succinate which can be transformed into glucose by gluconeogenesis. In this way, plants can convert stores of seed fat to glucose and cellulose.

22.5 METABOLISM OF AMINO ACIDS

Amino acids are degraded in two ways. After a meal, the amino acid concentration in the gut is quite high. Blood rich in amino acids flows directly from the gut into the liver where a process called *oxidative deamination* takes place. The α-keto acids produced in these reactions enter the TCA cycle to be further degraded. The oxidative deamination reaction removes ammonia from all metabolites. However, ammonia is assimilated by one process, which produces glutamate. Essentially all amino acids are stripped of ammonia, which is collected by α-keto-glutarate to form glutamate by reversal of deamination. The following reaction illustrates the collection process, but recognize that the reaction is freely reversible, and written in reverse represents the more general reaction of deamination, which involves all incoming amino acids:

$$NH_4^+ + \alpha\text{-ketoglutarate} + NADPH \xrightleftharpoons[]{\text{glutamate dehydrogenase}} \text{glutamate} + NADP^+ + H_2O$$

The reaction represents the principal mode of assimilation of ammonia in the liver and kidney. The ammonia generated in this step may also be detoxified by conversion to urea. The glutamate produced by this reaction is used to synthesize amino acids from appropriate precursors, primarily α-keto acids from the TCA cycle. The transfer of ammonia to α-keto acids is accomplished by a second more general process called *transamination*. In this process, amino groups from one amino acid are transferred to an α-keto acid which is then converted to the corresponding amino acid, leaving behind an α-keto acid corresponding to the original amino acid. This process occurs in all cells, and is a key process in amino acid synthesis and transport.

Problem 22.30. What is the fate of ammonia which is not used for amino acid synthesis?

Ans. The ammonia derived from oxidative transamination is detoxified by a process in which urea is synthesized. It is a nontoxic water-soluble compound which is eliminated by secretion into the urine.

The processes of transamination, deamination, and urea formation are represented in outline form as follows:

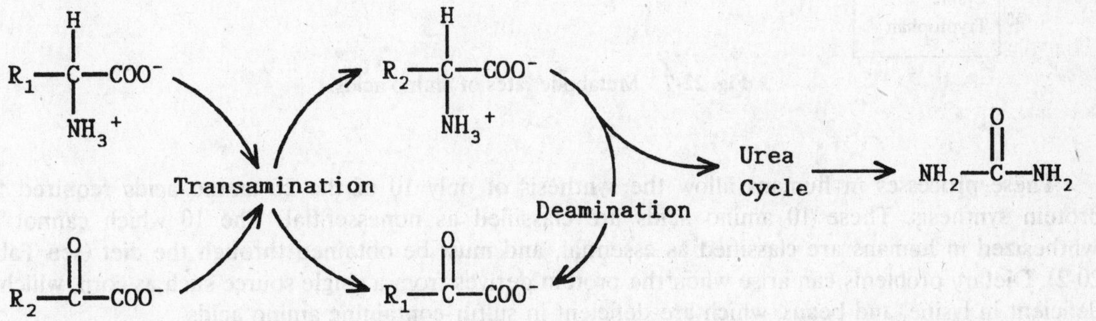

$R_1—CH(NH_3^+)—COO^-$ + $R_2—CO—COO^-$ → (Transamination) → $R_2—CH(NH_3^+)—COO^-$ + $R_1—CO—COO^-$

$R_2—CH(NH_3^+)—COO^-$ → (Deamination) → $R_1—CO—COO^-$ + Urea Cycle → $NH_2—CO—NH_2$

Transamination is effected by enzymes specific to particular amino acids. Transamination between aspartate and α-keto-glutarate illustrated in Fig. 22-6 results in the production of oxaloacetate and glutamate.

Two of the clinically important transaminases are alanine transaminase, also called glutamate-pyruvate transaminase, or GPT, and glutamate-oxaloacetate transaminase. (Fig. 22-6), called GOT. A heart attack, i.e., a myocardial infarction, results in tissue damage, and allows these enzymes, among others, to leak into the blood stream. The severity and stage of the heart attack can be monitored by measuring the concentrations of these enzymes in the blood. These measurements are known as the SGPT, and SGOT tests (S for *serum*).

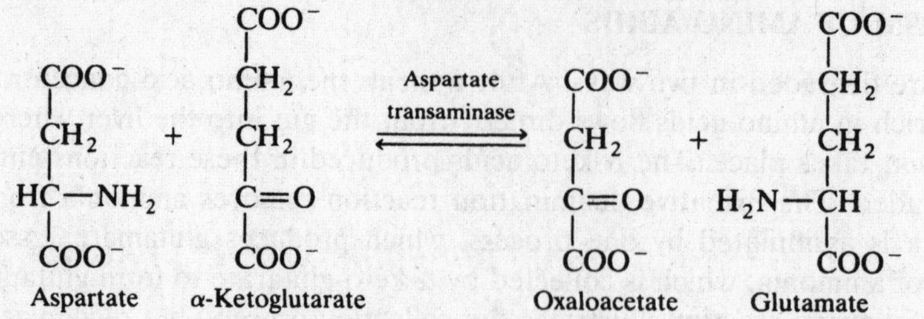

$$\underset{\text{Aspartate}}{\begin{matrix}COO^-\\ |\\ CH_2\\ |\\ HC{-}NH_2\\ |\\ COO^-\end{matrix}} + \underset{\alpha\text{-Ketoglutarate}}{\begin{matrix}COO^-\\ |\\ CH_2\\ |\\ CH_2\\ |\\ C{=}O\\ |\\ COO^-\end{matrix}} \xrightleftharpoons{\text{Aspartate transaminase}} \underset{\text{Oxaloacetate}}{\begin{matrix}COO^-\\ |\\ CH_2\\ |\\ C{=}O\\ |\\ COO^-\end{matrix}} + \underset{\text{Glutamate}}{\begin{matrix}COO^-\\ |\\ CH_2\\ |\\ CH_2\\ |\\ H_2N{-}CH\\ |\\ COO^-\end{matrix}}$$

Fig. 22-6 Transamination reaction between aspartate and α-ketoglutarate.

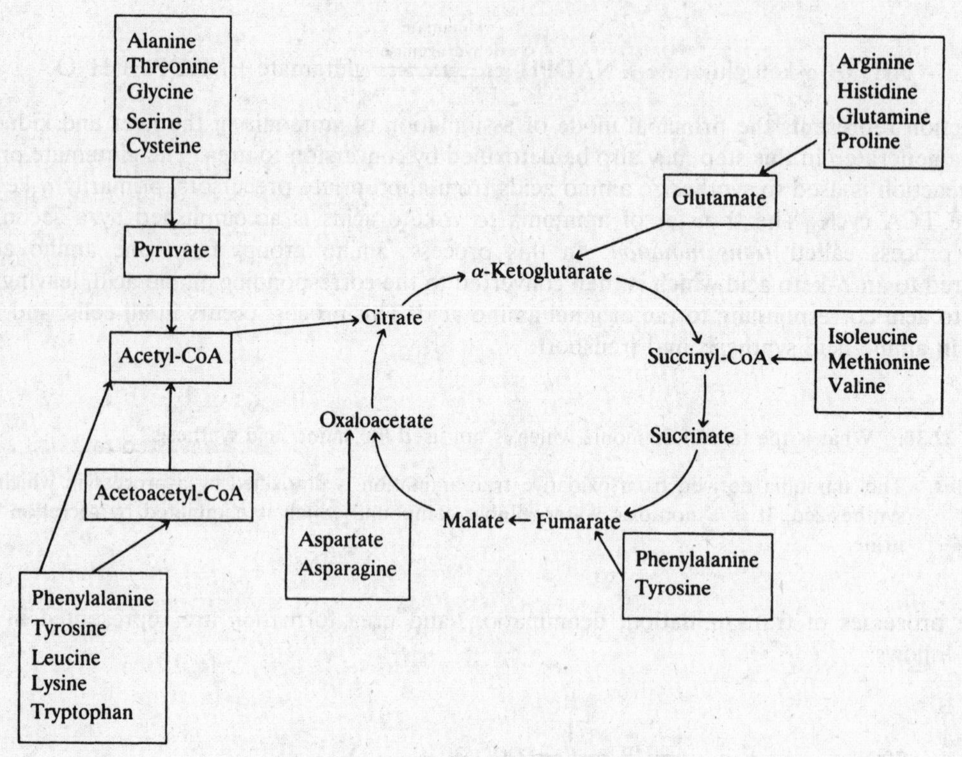

Fig. 22-7 Metabolic fates of amino acids.

These processes in humans allow the synthesis of only 10 of the 20 amino acids required for protein synthesis. These 10 amino acids are classified as nonessential. The 10 which cannot be synthesized in humans are classified as essential, and must be obtained through the diet (see Table 20-2). Dietary problems can arise when the protein derives from a single source such as corn, which is deficient in lysine, and beans, which are deficient in sulfur-containing amino acids.

The degradation of amino acids is complex, but eventually they all find their way into the TCA cycle for complete oxidation to CO_2 and H_2O. Ten amino acids are reduced to acetyl-CoA, five to α-keto-glutarate, three to succinyl-CoA, two to oxaloacetate, and two to fumarate. These are illustrated in Fig. 22-7. Five of the amino acids are converted to acetoacetyl-CoA. These yield ketone bodies in the liver, and are therefore called *ketogenic*. Large amounts of ketone bodies are produced from amino acids and also from lipids by the liver in cases of untreated diabetes mellitus. The 15 amino acids which can be converted to pyruvate, α-keto-glutarate, succinate, and oxaloacetate can in turn be converted to glucose, and are called *glucogenic*. Phenylalanine and tyrosine can be seen to be both ketogenic and glucogenic.

22.6 ENERGY YIELD FROM CATABOLISM

As hydrogen atoms with electrons are generated in the TCA oxidation steps, the hydrogen atoms are stripped of electrons and are secreted as H^+ ions to the outside of the mitochondria. The electrons find their way to oxygen within the mitochondria. However, the H^+ ions take a different route. The concentration of the H^+ ions on the outside of the mitochondria becomes so high that they must diffuse back into the mitochondria. This is allowed only at certain sites where an enzyme complex for the synthesis of ATP is located. The flow of H^+ ions into the mitochondria through this ATPase complex is the driving force for the synthesis of ATP from ADP and P_i. The mechanism of ATP synthesis in the mitochondria depends entirely on the integrity of the mitochondrial membrane. If it is disrupted to any extent, ATP synthesis ceases. The synthesis of ATP is inextricably linked to electron flow to oxygen to produce water, and is called *oxidative phosphorylation*, as opposed to *substrate level phosphorylation*. Stop the electron flow, and both ATP synthesis and oxygen consumption cease. The rate of ATP synthesis is controlled by the availability of ADP and P_i. When ADP concentration is high, the rate of oxygen consumption is increased, and when it is low, oxygen consumption decreases. The electron transport chain is illustrated in Fig. 22-8, the flow sheet or summary of respiration. This figure emphasizes the flow of hydrogen atoms and their electrons from pyruvate, amino acids, and fatty acids, and their ultimate reaction with oxygen to produce water. Note carefully that each of these nutrients is converted to the same substance, acetyl-CoA. ATP is synthesized at three sites along the electron transport chain. Experiments have shown that, for every pair of hydrogen atoms transported, 3 mol of ATP is synthesized, 1 mol at every site where a synthetic ATPase appears to be located.

The overall result of glycolysis may be written as follows:

$$\text{Glucose} + 2\ \text{ADP} + 2\ P_i \longrightarrow \text{lactate} + 2\ \text{ATP} + 2\ H_2O$$

Glucose is converted to 2 mol of lactate, 2 mol of ADP and phosphate are converted to 2 mol of ATP, and four electrons have been transferred from glyceraldehyde 3-PO_4 to pyruvate by way of NADH. Glucose must be primed by 2 mol of ATP, and 4 mol of ATP are generated by glycolysis resulting in a net synthesis of 2 mol of ATP per mole of glucose.

In the TCA cycle, from each entering pyruvate, four pairs of H atoms are transported via NADH, and one pair by $FADH_2$ (Figs. 22-3 and 22-8). Each NADH oxidized results in the generation of three ATP. $FADH_2$ enters the electron transport chain at site 2, and therefore results in the generation of only two ATP. There is one substrate level high-energy phosphate compound, GTP, which is equivalent to ATP, synthesized per entering pyruvate. In addition, when pyruvate enters the Krebs cycle, the NADH generated in the glycolytic oxidation step also enters the electron transport chain.

	Pair of H's / pyruvate	moles of ATP
Glycolysis	1 NADH	$3 \times 1 = 3$
Krebs cycle	4 NADH	$3 \times 4 = 12$
	1 $FADH_2$	$2 \times 1 = 2$
Total ATP from oxidative phosphorylation per pyruvate = 17		
	Substrate level ATP per pyruvate =	1
	Sum of ATP per pyruvate =	18
Since there are two pyruvate per mol of glucose:		
	Total ATP per mole of glucose =	36

This accounting tells us that the TCA cycle is 18 times as efficient as glycolysis in extracting energy from glucose. In total, the amount of energy extracted from glucose represents an efficiency of about 50 percent. The best that ordinary heat engines can manage is about 10 percent. Ignoring the energy-capturing processes of catabolism, the overall metabolic process of degradation of glucose can

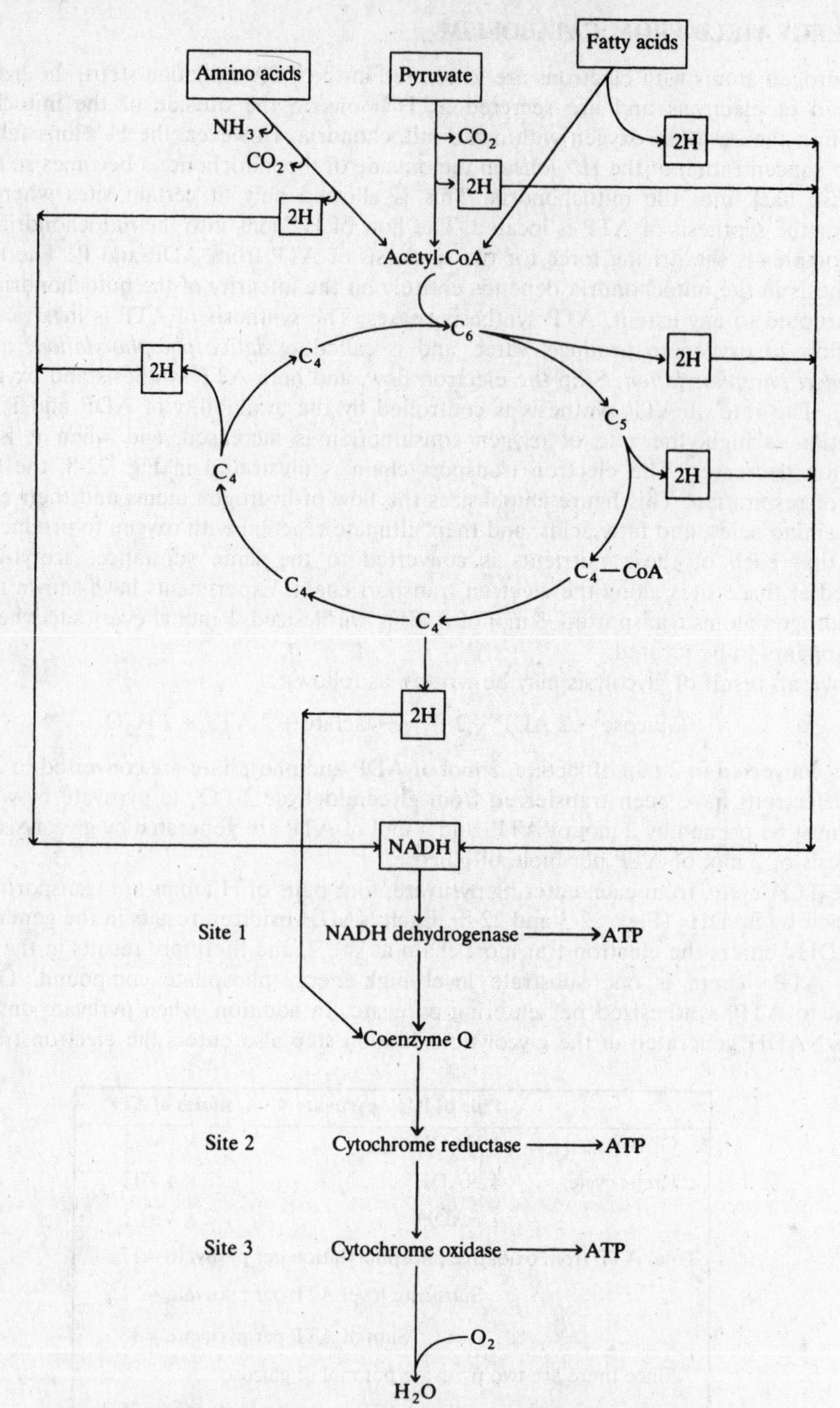

Fig. 22-8 The flow of H atoms resulting from metabolic dehydrogenations.

be described by the familiar reaction

$$C_6H_{12}O_6(aq) + 6\ O_2(g) \longrightarrow 6\ CO_2(g) + 6\ H_2O(l)$$

Note that CO_2 is a gas, and is eventually lost to the atmosphere, which means that the degradative process is irreversible.

Problem 22.31. Why is it often stated that fatty acids are a significant source of energy?

Ans. This can be substantiated by a rigorous analysis of ATP yield per palmitic (C_{16}) acid. Recall that there are 8 acetyl-CoA, 7 NADH, and 7 $FADH_2$ molecules generated per mole of palmitic acid. In addition, 2 mol of ATP must be used to activate the fatty acid, and 8 mol GTP (equivalent of 8 mol of ATP) are generated by the action of succinyl-CoA synthetase.

		ATP Yield
Fatty Acid Activation		-2
From Succinyl-CoA synthetase		8
From oxidation steps	7 NADH	21
	7 $FADH_2$	14
From 8 Acetyl-CoA	24 NADH	72
	8 $FADH_2$	16
	Total ATP per mol of palmitate	129

Furthermore, consider the fact that fatty acids are stored as triglycerides (TGA), which consist of 3 mol of fatty acid plus 1 mol of glycerol. The glycerol also enters the glycolytic pathway after being primed by ATP, and using the glycerol phosphate shuttle yields an additional 20 mol of ATP. The total yield of ATP per mol of tripalmitoyl glycerol is therefore

Fatty acid	$3 \times 129 = 387$
Glycerol	$= 20$
	Sum of ATP per mol = 407

We must now subtract the 2 mol of ATP required for activation of each fatty acid to the CoA-derivative

$$3 \times 2 = 6$$

Total ATP yield per mol of triglyceride = 401

In order to compare this yield with the 36 mol of ATP per mol of glucose*, we will calculate the ATP yield per gram of compound.

Formula weight Tripalmitoylglycerol: 807.3 g/mol

$$401 \text{ mol ATP}/807.3 \text{ g} = 0.497 \text{ mol ATP/g}$$

Formula weight glucose: 180.2 g/mol

$$36 \text{ mol ATP}/180.2 \text{ g} = 0.200 \text{ mol ATP/g}$$

The nutritional caloric estimates of energy/g for fats (palmitate) and carbohydrates (glucose) are 9.13 and 3.81 kcal/g respectively. Comparing the ATP per gram ratio from biochemical estimates with the nutritional ratios of energy per gram shows that it is valid to represent biochemical energy in terms of moles of ATP.

mol ATP/g TGA / mol ATP/g Glucose	kcal per gram palmitate/kcal per gram glucose
$\dfrac{0.497}{0.200} = 2.49$	$\dfrac{9.13}{3.81} = 2.40$

Problem 22.32. How many days can a 70-kg (ca. 150 lb) male live on stored fat?

Ans. **First calculate the mass of fat, assuming that 20 percent of body weight is fat: 70 kg × 0.2 = 14 kg fat. Next, assume all the fat is tripalmitoylglycerol, formula weight 855.**

***Under aerobic conditions, the 2 moles of NADH produced in glycolysis yield 6 additional moles of ATP upon oxidation.**

We know the following:

$$1 \text{ mol TGA} = 381 \text{ mol ATP}$$

$$1 \text{ mol ATP} = 7.3 \text{ kcal}$$

Set the use of energy = 1500 kcal/day (basal rate)

$$14000 \text{ g TGA}\left(\frac{1 \text{ mol TGA}}{855 \text{ g TGA}}\right)\left(381\frac{\text{mol ATP}}{\text{mol TGA}}\right)\left(7.3\frac{\text{kcal}}{\text{mol ATP}}\right)\left(\frac{1 \text{ day}}{1500 \text{ kcal}}\right) \approx 30 \text{ days}$$

If energy is consumed at about 1500 kcal/day, our slim fellow can live for about 30 days on his stored fat. You can see why stored fat is considered an excellent energy source.

ADDITIONAL SOLVED PROBLEMS

INTRODUCTION

Problem 22.33. Is there anything wrong with the description of a vertebrate cell as a collection of enzymes enclosed in a functionless envelope?

Ans. **Yes. The vertebrate cell consists of a number of compartments and subcellular structures in which enzymes and enzyme clusters or complexes function. The compartments are enclosed in membranes, as is the cell itself. All membranes have complex functions including specific transport by specialized proteins and enzymatic processes central to metabolism. For example, if the inner mitochondrial membrane is disrupted, synthesis of ATP will cease.**

Problem 22.34. Why is the cell conceived of as economical?

Ans. **From only 20 amino acids and eight nitrogen bases, cells have been able to create extraordinarily large numbers of different proteins, nucleic acids, and other compounds necessary to carry out metabolism and the specific functions of specialized tissues and organs such as the brain, kidney, and liver.**

Problem 22.35. Why is the energy of reduced organic compounds, ordinarily released to the surroundings when oxidized, not lost when oxidation occurs in living cells?

Ans. **Oxidation in cells occurs in sequences of reactions as opposed to a single combustion reaction. At particular steps of these sequences, evolved energy is partially captured through the synthesis of so-called high-energy compounds.**

Problem 22.36. What is the principal form of the energy captured by the cell as a result of oxidation?

Ans. **Captured energy is stored in the form of a unique compound called adenosine triphosphate, ATP.**

ENZYMES, COFACTORS, AND COENZYMES

Problem 22.37. What is the difference between an enzymatic active site, and a binding site?

Ans. **An enzymatic active site has two functions: It must not only catalyze a reaction, but it must also selectively bind the incoming substrate, i.e., it must consist of both a binding site and a catalytic site. Therefore, a binding site is part of the active site.**

Problem 22.38. What is meant by complementarity?

Ans. **Complementarity refers to the unique structure of an enzymes binding site. The selectivity of an enzyme is based on the structure of the binding site being complementary to that of the incoming substrate. It is the lock for the key of the substrate.**

Problem 22.39. Can polypeptides catalyze redox reactions?

Ans. No. The amino and/or carboxyl side chains of polypeptides cannot function as reversible electron sinks. Therefore redox enzymes must combine with other types of molecules called cofactors or coenzymes, which possess that ability.

Problem 22.40. Define apoenzyme and holoenzyme.

Ans. These words apply to enzymes which can only function when combined with cofactors or coenzymes. The protein portion of the enzyme lacking its cofactor or coenzyme is the apoenzyme, and the combination of enzyme plus cofactor is the holoenzyme.

Problem 22.41. Give an example of covalent modification of an enzyme.

Ans. In order for phosphorylase to be able to cleave glucose-1-PO_4 from glycogen, it must be phosphorylated itself by ATP. The reaction occurs at serine residues of the enzyme, and may be represented by the reaction

$$\text{HO}-\bigcirc-\text{OH} + 2\,\text{ATP} \longrightarrow {}^{2-}\text{O}_3\text{PO}-\bigcirc-\text{OPO}_3^{\,2-} + 2\,\text{ADP} + 2\,\text{H}_2\text{O}$$

Problem 22.42. What are allosteric enzymes?

Ans. Allosteric enzymes are a class of regulatory enzymes whose activities can be varied by binding specific small molecules at sites different from the enzymes' active sites. The binding of these modifiers changes the conformation of the protein so that it may become either more or less active.

Problem 22.43. What is a high-energy compound?

Ans. A high-energy compound is one characterized by a large equilibrium constant for transfer of a portion of the compound to water (hydrolysis).

METABOLISM OF CARBOHYDRATES

Problem 22.44. Describe the process of glycolysis.

Ans. Glycolysis is the first stage of degradation of glucose. It is an anaerobic process, i.e., it takes place without the use of oxygen. It can be described by the overall reaction

$$\text{Glucose} + 2\ \text{ADP} + 2\ \text{P}_i \longrightarrow 2\ \text{lactate} + 2\ \text{ATP}$$

Problem 22.45. What is meant by substrate level phosphorylation?

Ans. In substrate level phosphorylation, ADP is converted to ATP by direct chemical reaction with a high-energy compound containing a phosphate group.

Problem 22.46. What are the required conditions for glycolysis to be continuous?

Ans. There are two requirements: (1) There must be a ready supply of NAD^+, derived through recycling of the NADH produced in the oxidation step, and (2) inorganic phosphorus must be present, since it is continuously consumed.

Problem 22.47. Describe the Krebs or TCA cycle.

Ans. In the Krebs or TCA cycle, acetyl-CoA, derived from the oxidation of carbohydrates, fats, and some amino acids, is degraded to yield carbon dioxide and energy-rich hydrogen atoms.

Problem 22.48. What is the fate of the energy-rich hydrogen atoms produced by the TCA cycle?

Ans. The hydrogen atoms are fed into the electron transport system to eventually combine with oxygen to form water. This means that, in the oxidation of glucose to form carbon dioxide and water, the CO_2 is derived from decarboxylations in the TCA cycle, and the water is produced in the following stage at the end of the electron transport chain.

Problem 22.49. What is gluconeogenesis, and what is the initiating step of the process?

Ans. Gluconeogenesis is the process in which glucose is synthesized from pyruvate. It begins with the synthesis of oxaloacetate from pyruvate by carboxylation.

Problem 22.50. Can pyruvate be carboxylated in the cytosol to form oxaloacetate?

Ans. Pyruvate is transformed into oxaloacetate in the mitochondria by a regulatory enzyme which is active only in the presence of acetyl-CoA. The enzyme is not present in the cytosol.

Problem 22.51. What is glycogenesis and where does it occur?

Ans. Glycogenesis is the synthesis of glycogen or animal starch, which is a storage form of glucose. It takes place in the cytosol.

METABOLISM OF LIPIDS

Problem 22.52. True or false: The first step in the oxidation of a fatty acyl-CoA derivative requires NAD^+ as the coenzyme.

Ans. False. The required coenzyme is FAD.

Problem 22.53. True or false: The second oxidation step in fatty acid catabolism requires FAD as the coenzyme.

Ans. False. The required coenzyme is NAD^+.

Problem 22.54. True or false: The oxidation of a 12 carbon fatty acid (lauric acid) yields 5 acetyl-CoA, 6 $FADH_2$, and 6 $NADH + 6\ H^+$.

Ans. False. The complete oxidation of lauric acid yields 6 acetyl-CoA, 5 $FADH_2$, and 5 $NADH + 5\ H^+$.

Problem 22.55. What is the fate of the acetyl-CoA produced in lipid oxidation?

Ans. The acetyl-CoA from lipid oxidation adds to the pool of acetyl-CoA in the mitochondria, and enters the TCA cycle.

METABOLISM OF AMINO ACIDS

Problem 22.56. What are the processes undergone by dietary amino acids?

Ans. The two principal processes are oxidative deamination, for example,

$$\text{Glutamate} + \text{NADP}^+ \xrightleftharpoons[]{\text{glutamate dehydrogenase}} \text{NH}_4^+ + \alpha\text{-keto-glutarate} + \text{NADPH}$$

and transamination:

$$\text{Glutamate} + \text{oxaloacetate} \xrightleftharpoons[]{\text{aspartate transaminase}} \alpha\text{-keto-glutarate} + \text{aspartate}$$

Problem 22.57. Ammonia is toxic to cells. How is its toxicity ameliorated?

Ans. It is handled in two ways: (1) reversal of the glutamate dehydrogenase reaction, (2) synthesis of urea.

Problem 22.58. What is a glucogenic amino acid?

Ans. A glucogenic amino acid yields pyruvate as an end product of catabolism.

Problem 22.59. What is a ketogenic amino acid?

Ans. A ketogenic amino acid yields acety-CoA as an end product of its catabolism.

ENERGY YIELD FROM CATABOLISM

Problem 22.60. What is the meaning of oxidative phosphorylation?

Ans. During the passage of H atoms through the electron transport system, there is a simultaneous direct phosphorylation of ADP by P_i to form ATP. It is different from substrate level phosphorylation because phosphorylation is not effected by the transfer of a phosphate group from a high-energy intermediate.

Problem 22.61. What is the driving force behind oxidative phosphorylation?

Ans. During the passage of NADH through the electron transport system, protons are released to the outside of the mitochondrial membrane. This creates a proton gradient under which the external protons tend to diffuse back into the mitochondria. The structure of the membrane allows this to occur only at special sites where an ATPase is located. The passage of the proton through this structure is accompanied by the phosphorylation of ADP.

Problem 22.62. What is the fate of the electrons accompanying the H atoms through the electron transport system?

Ans. They finally react with oxygen to form water.

Problem 22.63. Is the rate of oxygen consumption under any controls?

Ans. The rate of oxygen consumption depends on the rate at which electrons reach the final enzyme in the electron transport system, cytochrome oxidase. There is an absolute coupling of electron transport with oxidative phosphorylation. Therefore the rate of oxygen consumption depends on the availability of ADP. The more ADP available, the greater the rate of oxygen consumption.

SUPPLEMENTARY PROBLEMS

Problem 22.64. Why is glycolysis irreversible?

Ans. It is irreversible because there are three irreversible reactions in its pathway.

Problem 22.65. What are the conditions necessary for gluconeogenesis to take place?

Ans. The cell must be in a high-energy state. The concentrations of mitochondrial pyruvate, ATP, and NADH, and cytosolic ATP must be high.

Problem 22.66. Can muscle or brain supply glucose to the blood?

Ans. No. To supply glucose to the blood, glucose-6-PO_4 must be dephosphorylated. Glucose-6-phosphatase is present only in the endoplasmic reticulum of liver cells.

Problem 22.67. Do the TCA intermediates play any role other than in the production of ATP through the oxidation of acetyl-CoA?

Ans. Many of the intermediates serve as precursors in the synthesis of other metabolites. For example, the α-keto acids can be transaminated to form amino acids, and succinyl-CoA is a precursor of porphyrin.

Problem 22.68. True or false: After priming a fatty acid to its CoA-derivative in the cytosol, it is transported into the mitochondria.

Ans. False. The fatty-CoA derivative cannot traverse the mitochondrial membrane. It must be converted to a fatty acyl-carnitine derivative which can penetrate the mitochondrial membrane. Once in the mitochondria, it is transformed back into the CoA derivative.

Problem 22.69. True or false: In the mitochondrial oxidation of fatty acids, the hydration of the enoyl-CoA derivative results in the chiral D-hydroxy derivative.

Ans. False. The L-hydroxy derivative is formed. It is in the synthesis of fatty acids in the cytosol that the D-hydroxy derivative is formed.

Problem 22.70. True or false: All acetyl-CoA is completely oxidized to CO_2 and H_2O.

Ans. False. Under high-energy states of the cell, some of the acetyl-CoA is converted to ketone bodies.

Problem 22.71. Can humans synthesize all the fatty acids necessary for cellular requirements?

Ans. No Fatty acids having more than one unsaturated bond are necessary for the synthesis of lipids such as prostaglandins. Humans have an enzyme system in the cytosol which can introduce only one unsaturation to form oleic acid from stearic acid. However, fatty acids like linoleic and linolenic acids, with two and three unsaturated bonds per molecule, must be obtained from the diet, and are called *essential fatty acids.*

Problem 22.72. What is phosphorolysis?

Ans. In a hydrolysis, the elements of water are introduced into an ester bond to cleave the ester into an acid, and an alcohol. In phosphorolysis, HPO_4^{2-} is used to cleave the ester bond as follows:

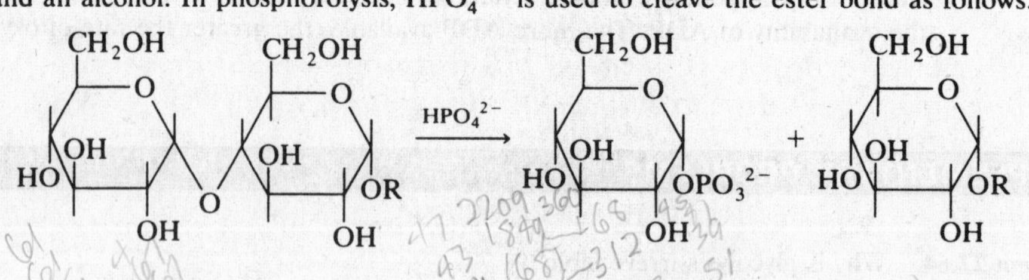

[Disaccharide (CH_2OH, OH, HO, OH, O-linked, OR) $\xrightarrow{HPO_4^{2-}}$ glucose-1-phosphate (CH_2OH, OH, HO, OPO_3^{2-}, OH) + glucose derivative (CH_2OH, OH, HO, OR, OH)]

Digestion, Nutrition, and Gas Transport

In this chapter we will take up some topics which have particular relevance to mammalian metabolism. These are digestion, nutrition, and the transport of oxygen to and carbon dioxide from metabolizing tissues.

23.1 DIGESTION

In digestion, foods are enzymatically degraded to low-molecular-mass components to prepare them for absorption in the gut.

Problem 23.1. Why is it necessary to hydrolyze foodstuffs to low-molecular-mass components?

Ans. This is necessary because the cells lining the intestine can only absorb relatively small molecules into the bloodstream.

The reduction of the major components of food—proteins, carbohydrates, and fats—to low-molecular-weight components begins in the mouth with mechanical action of chewing, and the secretion of a carbohydrase, amylase, in the saliva. The next stage occurs in the stomach, where secretion of the hormone gastrin is stimulated by the entry of protein into the stomach. Gastrin, in turn stimulates the secretion of pepsinogen by the chief cells of the gastric glands and HCl by the parietal cells of the stomach.

Problem 23.2. What is the function of the HCl secreted in the stomach?

Ans. The HCl brings the pH of the stomach to between 1.5 and 2.0, which serves two functions. The low pH kills most bacteria, and also denatures or unfolds globular proteins which makes internal peptide bonds accessible to enzymatic hydrolysis.

Pepsinogen is called a *zymogen* or inactive enzyme precursor, which is activated by the enzymatic action of trace amounts of pepsin already present in the stomach. Activation is achieved by the removal of a small terminal peptide. Pepsin attacks the peptide bonds of amino acids possessing hydrophobic side groups, which reduces proteins to mixtures of smaller peptides. Other enzymatic hydrolases which are found in the small intestine are also secreted as zymogens, and are activated by similar processing.

The low pH of the stomach contents stimulates the secretion of the hormone secretin, as it passes into the small intestine. This hormone is secreted into the bloodstream, and when it reaches the pancreas, it stimulates that organ to secrete bicarbonate into the gut which neutralizes the low pH of the entering stomach contents. Entering amino acids stimulate the secretion by intestinal cells of the specialized enzyme enterokinase. It specificity is directed toward the conversion of the zymogen, trypsinogen, secreted by the pancreas, to the active proteolytic enzyme, trypsin. Trypsin then converts chymotrypsinogen to the active chymotrypsin. These two enzymes reduce polypeptides to small peptides. The further hydrolysis of the small peptides to their constituent amino acids is accomplished by two other enzymes secreted by intestinal cells, carboxypeptidase and aminopeptidase. The mixture of amino acids is then transported across the intestinal cells, enters the blood, and is transported to the liver.

The principal carbohydrates of food are starch and cellulose of plant origin, and glycogen of animal origin. The hydrolysis of these carbohydrates, begun in the mouth, is completed in the small intestine chiefly by the action of pancreatic amylase. The β-1,4-linkages of cellulose cannot be hydrolyzed by mammalian enzymes. It passes through and out of the gut in the form of roughage. The hydrolysis of disaccharides to hexoses is accomplished by enzymes located within cells lining the small intestine. The mixture of hexoses is absorbed into the blood, and is brought to the liver.

Small amount of nucleic acids, DNA, and RNA, are present in food. These are hydrolyzed to nucleotides by pancreatic enzymes. Their further breakdown to free bases and monosaccharides occurs by enzymes from the epithelial cells lining the intestine.

The digestion of fats begins in the small intestine. The pancreas secretes a zymogen called *prolipase* which is activated by the proteases, and in the presence of a special protein called *colipase*, and bile salts synthesized in the liver and stored in the gall bladder, adsorbs to droplets of fat and begins the hydrolysis of fatty acids from triacylglycerols.

Problem 23.3. What is the function of the bile salts?

Ans. The bile salts act as emulsifying agents so that the churning action of the intestine produces a very fine suspension of fat droplets whose large surface area permits efficient enzymatic hydrolysis.

Generally, only one or two fatty acid chains are released by hydrolysis, and a mixture of the sodium and potassium salts of fatty acids along with monoacylglycerol are formed. This mixture is absorbed by intestinal cells, and there reassembled into triacylglycerols.

Problem 23.4. Are the triacylglycerols synthesized in the cells lining the lumen of the gut absorbed directly into the blood and brought to the liver as is the case with amino acids and hexoses?

Ans. No. The triacylglycerols do not penetrate the capillaries, but pass into the lymphatic system by way of small lymph vessels in the intestinal walls called *lacteals*. After a meal, the lymph in the lacteals, called *chyle*, has a milky appearance caused by the presence of suspended fat droplets called *chylomicrons*.

The chylomicrons are stabilized by adsorbed lipoprotein and phospholipids. The lymph enters the bloodstream at the juncture of the thoracic duct and the subclavian vein. This dietary fat is largely removed from the blood by the cells of adipose tissue.

23.2 NUTRITION

The result of digestion is a complex mixture of biomolecules, i.e., carbohydrates, fats, amino acids, vitamins, and minerals, These must be adequate to fulfill the requirements of biosynthesis, motion (muscle contraction), ion transport, and secretion.

The energy content of foodstuffs varies. The approximate caloric content of generic carbohydrates, fats, and proteins are listed in Table 23-1. These values are based on a varied diet, i.e., all components are present. Recall that humans cannot synthesize glucose from fat. In the absence of carbohydrates, therefore, fat metabolism becomes inefficient, and the caloric value of fat decreases.

Table 23-1. Energy Equivalents of Nutrients

Nutrient	Energy Equivalent, kcal / g
Carbohydrates	4.2
Fats	9.5
Proteins	4.3

Table 23-2. Recommended Daily Energy Allowances

	Age, years	Weight, kg	Energy, kcal
Infants	0.0–0.5	6	650
	0.5–1.0	9	970
Children	1–3	13	1300
	4–6	20	1700
	7–10	28	2400
Females	11–14	46	2200
	15–18	55	2100
	19–22	55	2100
	23–50	55	2000
	50+	55	1800
Males	11–14	45	2700
	15–18	66	2800
	19–22	70	2900
	23–50	70	2700
	50+	70	2400

23.2.1 Carbohydrates

Although carbohydrates are used as structural components of biomolecules, their principle role is that of suppliers of metabolic energy. The body's energy requirement at complete rest, and 12 h after eating, is called the *basal rate*. This is considered the energy required to maintain the basic "housekeeping" functions of the body. The basal rates for college-age males and females are 1800 and 1300 kcal/day, respectively. However, daily activity controls the total caloric requirement. Variations depend on the extent of muscular activity, body weight, age, and sex. Table 23-2 lists the recommended caloric intake as functions of sex and age. An adequate human diet is composed of five basic classes of nutrients. The components of each class are listed in Table 23-3.

Table 23-3. Nutrients Required by Humans

Energy sources	Vitamins	Minerals
Carbohydrates	Thiamin	Arsenic
Fats	Riboflavin	Calcium
Proteins	Nicotinamide	Chlorine
Essential amino acids	Pyridoxine	Chromium
Arginine	Pantothenic acid	Copper
Histidine	Folic acid	Fluorine
Isoleucine	Biotin	Iodine
Leucine	Vitamin B_{12}	Iron
Lysine	Ascorbic acid	Magnesium
Methionine	Vitamins A, D, E, K	Nickel
Phenylalanine		Molybdenum
Threonine		Phosphorus
Tryptophan		Potassium
Valine		Selenium
Essential fatty acids		Silicon
Linoleic acid		Sodium
Linolenic acid		Tin
		Vanadium
		Zinc

23.2.2 Proteins

A diet of only glucose would be adequate to fulfill all the energy and mass requirements for carbohydrates. However, the requirements for amino acids and fats are more stringent, because there are amino acids and fats which are considered essential from a dietary point of view.

Problem 23.5. Why are the amino acids listed in Table 23-3 considered essential?

Ans. Humans cannot synthesize either enough of or any of 10 of the 20 amino acids required for protein synthesis, and those 10 essential amino acids must be obtained from the diet.

Proteins are not required for their caloric value, but for their content of amino acids. This presents two nutritional problems: (1) Does a food protein contain the correct numbers of types of amino acids? (2) How accessible are the amino acids, i.e., how digestible is the food? These qualities of a dietary protein are expressed as its *biological value.*

Problem 23.6. What is the meaning of biological value of a protein?

Ans. If a given protein provides all the required amino acids in the proper proportions, and all are released upon digestion and absorbed, it is said to have a biological value of 100. The biological value of a protein will be high when only small daily amounts are required to maintain a person in *nitrogen balance.*

Problem 23.7. What is the meaning of nitrogen balance?

Ans. Nitrogen balance is achieved when the intake of protein nitrogen is equal to the loss of nitrogen in the urine and feces.

A protein's biological value will be less than 100 if it is incompletely digestible, or if it has only a small percentage of even one essential amino acid. In the latter case, one must ingest large quantities to obtain the essential amino acid, while those in abundance will be used calorically. A somewhat different, but experimentally useful classification is called the *chemical score.*

Problem 23.8. How is the chemical score of a protein determined?

Ans. The protein is completely hydrolyzed, and its amino acid composition is compared with that of human milk.

The biological value and chemical score of some food proteins are listed in Table 23-4.

Table 23-4. Chemical Scores and Biological Values of Some Food Proteins

Protein Source	Chemical Score	Biological Value
Human milk	100	95
Beefsteak	98	93
Whole egg	100	87
Cow's milk	95	81
Corn	49	36
Polished rice	67	63
Whole wheat bread	47	30

23.2.3 Fats

Problem 23.9. Why are the polyunsaturated fatty acids linoleic acid and linolenic acid considered essential?

Ans. Fatty acids containing more than one unsaturated bond past carbon 9 of a saturated chain counting from the carboxyl end cannot be synthesized by humans. Prostaglandins, a family of lipid-soluble organic acids which are regulators of hormones, are synthesized by mammals from arachidonic acid. Arachidonic acid is a polyunsaturated fatty acid which mammals can synthesize using the dietary polyunsaturated fatty acids of plant origin as precursors.

Deficiencies in these fatty acids are rare, since they are present in abundance in plant food, fowl, and fish. The caloric values of saturated and unsaturated fats are comparable; however, the proportion of saturated to unsaturated fats in the diet has significant physiological consequences.

Problem 23.10. What are the consequences of a high proportion of saturated to unsaturated fatty acids in the diet?

Ans. A great deal of statistical evidence has accumulated over the past 30 years which relates the decrease in concentration of high-density lipoproteins, along with the increase in concentration of both low-density lipoproteins and total blood cholesterol, in many individuals on diets rich in saturated fatty acids. The statistical studies also relate diets rich in saturated fats to a predisposition toward the development of coronary artery disease. Therefore, it is recommended that one increase the proportion of unsaturated fat in the diet.

The compositions of typical animal and plant fats is listed in Table 23-5.

23.2.4 Vitamins

Many vitamins, e.g., vitamin C or ascorbic acid, have been identified by the results of their dietary deficiency on humans and the subsequent cure of those conditions by their replacement in the diets of those affected.

Problem 23.11. What is an example of dietary deficiency and cure by replacement?

Ans. One of the earliest documentations of vitamin deficiency appears in the journals of Jacques Cartier, who explored North America in 1535. He described a disease which came to be known as scurvy. It was manifested in his sailors as terrible skin disorders accompanied by tooth loss. It took another 200 years before a British physician found he could cure scurvy by addition of lemons to the diet. The antiscurvy vitamin was isolated in 1932, and given the name of vitamin C.

Table 23-5. Fatty Acid Composition of Some Plant and Animal Fats

Plant and Animal Fats	Percentage of Total Fatty Acids		
	Saturated	**Monounsaturated**	**Polyunsaturated**
Butter fat	60	36	4
Pork fat	59	39	2
Beef fat	53	44	2
Chicken fat	39	44	21
Corn oil	15	31	53
Soybean oil	14	24	53
Soft Margarine	23	22	52

Other vitamins are known as growth factors because it can be shown that test animals will not grow if certain substances aside from carbohydrates, fats, and proteins are omitted from their diets. At present, vitamins have been chemically identified, and in many cases their metabolic roles as enzyme cofactors identified. In other cases, their precise biochemical function has not yet been made clear.

Vitamins can be divided into two classes, depending upon the significance of their deficiencies. In affluent countries, deficiencies of thiamin, riboflavin, niacin, ascorbic acid, and folic acid are relatively common but of marginal health significance, while in many parts of the world, deficiencies of these substances are significant and life-threatening. Deficiencies in pyridoxine, pantothenic acid, biotin, vitamin B_{12}, and the fat-soluble vitamins A, D, E, and K are rare. Because pyridoxine is required for transaminations, required amounts depend upon the quantity of protein in the diet. Biotin, pantothenic acid, and vitamin B_{12} are ordinarily not required in the diet. However, a deficiency in vitamin B_{12} occasionally occurs, and must be treated by injections of the purified vitamin.

Problem 23.12. Why are biotin, pantothenic acid, and vitamin B_{12} not ordinarily required in the diet?

Ans. These vitamins are not ordinarily essential because they are synthesized in adequate amounts by intestinal bacterial flora.

Table 23-6. Essential Vitamins, Coenzyme Form where Known, Biological Functions and Associated Deficiency Diseases where Known, and Recommended Daily Allowances (RDA) for Males, 23 to 50 years of Age

Vitamin	Coenzyme Form	Metabolic Role and / or Associated Deficiency Disease	RDA
Thiamine	Thiamin pyrophosphate	Dehydrogenase coenzyme; deficiency causes beriberi.	1.5 mg
Niacin	NAD^+	Dehydrogenase coenzyme; deficiency causes pellagra.	19 mg
Ascorbic acid	Unknown	Unknown: deficiency causes scurvy.	60 mg
Riboflavin	FAD	Dehydrogenase coenzyme.	1.7 mg
Pyridoxine	Pyridoxal phosphate	Transamination coenzyme.	2.2 mg
Folic acid	Tetrahydrofolate	1-carbon-group transfer; deficiency causes anemia	400 μg
Pantothenic acid	Coenzyme A	Fatty acid oxidation.	5–10 mg
Biotin	Biocytin	CO_2-transferring enzymes.	150 μg
Vitamin B_{12}	Deoxyadenosylcobalamine	Odd-numbered fatty acid oxidation; deficiency causes pernicious anemia	3 μg
Vitamin A_1	Unknown	Visual cycle intermediate; deficiency causes night blindness and xerophthalmia.	1 mg
Vitamin D_3	1,25-dihydroxy-cholecalciferol	Hormone, controlling calcium and phosphate metabolism; deficiency causes rickets.	10 μg
Vitamin E	Unknown	Protects against damage to membranes by oxygen; deficiency causes liver degeneration.	10 mg
Vitamin K_1	Unknown	Activation of prothrombin; deficiency causes disorders in blood clotting.	1 mg

Note: 1 mg = 1/1000 of a gram; 1 μg = 1/1,000,000 of a gram.

The fat-soluble vitamins are stored in body fat, and therefore need not be ingested daily. However, the water-soluble vitamins are excreted or destroyed during metabolic turnover, and must be replaced by regular ingestion. A listing of the vitamins, their coenzyme forms and their metabolic roles where known, and their recommended daily allowances for males 23 to 50 years old is found in Table 23-6.

23.2.5 Minerals

Carbohydrates, proteins, fats, and nucleic acids are composed of six elements: carbon, hydrogen, nitrogen, oxygen, phosphorus, and sulfur. In addition to these, many other minerals are required for experimental mammals, and presumed to be required for humans. These are divided into two groups, bulk and trace elements, and are found in Table 23-7. Table 23-8 lists some elements whose functions are known, or whose deficiencies result in well-recognized symptoms.

Table 23-7. Elements Required by Humans

Bulk Elements*	Trace Elements†
Calcium	Copper
Chlorine	Fluorine
Magnesium	Iodine
Phosphorus	Iron
Potassium	Manganese
Sodium	Molybdenum
	Selenium
	Zinc
	Arsenic‡
	Chromium‡
	Nickel‡
	Silicon‡
	Tin‡
	Vanadium‡

*These are required in doses in excess of 100 mg/day.

†These are required in doses of 1 to 3 mg/day.

‡These are known to be required in test animals and are likely to be required in humans.

Table 23-8. Minerals and Their Nutritional Functions

Elements	Nutritional Functions
Calcium	Bones, teeth
Phosphorus	Bones, teeth
Magnesium	Cofactor for many enzymes
Potassium	Water, electrolyte, acid-base balance; intracellular
Sodium	Water, electrolyte, acid-base balance; extracellular
Iron	Iron porphyrin proteins
Copper	Iron porphyrin synthesis and cytochrome oxidase
Iodine	Synthesis of thyroxin; lack leads to goiter
Fluorine	Forms fluoroapatite, strengthens bones and teeth
Zinc	Cofactor for many enzymes
Tin	Growth factor for mammals grown in ultraclean conditions
Nickel	Growth factor for mammals grown in ultraclean conditions
Vanadium	Growth factor for mammals grown in ultraclean conditions
Chromium	Growth factor for mammals grown in ultraclean conditions
Silicon	Growth factor for mammals grown in ultraclean conditions
Selenium	Component of the enzyme glutathione peroxidase
Molybdenum	Component of the enzymes xanthine, and aldehyde oxidases

23.3 METABOLIC GAS TRANSPORT

Oxygen is required by respiring cells, and carbon dioxide is generated by those cells. Tissues buried deeply within a multicellular organism cannot obtain atmospheric oxygen, nor can they dissipate carbon dioxide at a rate sufficient to sustain metabolism by simple diffusion. The gross diffusion problem has been solved by the evolution of circulatory systems and respiratory pigments.

23.3.1 Oxygen Transport

In terrestrial animals, oxygen enters the circulatory system at the respiratory membranes of the lungs.

Problem 23.13. What is the mechanism involved in oxygen entering the bloodstream from the lungs?

Ans. It diffuses across the membranes passively under a gradient in concentration, i.e., the concentration of oxygen in the alveolar spaces of the lung is greater than that in the circulatory fluid.

Problem 23.14. The solubility of oxygen in plasma is too low to support aerobic metabolism. How is sufficient oxygen brought to the tissue level by the circulatory fluid?

Ans. Red corpuscles contain large quantities of hemoglobin. This protein combines with oxygen so that it is not present as oxygen in solution but bound to the protein. This in effect serves to increase the blood concentration so that 100 mL of whole blood carries about 21 mL of oxygen. This represents an increase in oxygen concentration over that in plasma alone of about 50-fold.

Problem 23.15. How is oxygen transported by hemoglobin?

Ans. The oxygen is bound reversibly to Fe(II) in the heme molecule in hemoglobin.

Problem 23.16. What controls the amount of oxygen bound to hemoglobin?

Ans. The amount of oxygen combined with hemoglobin depends upon the partial pressure of oxygen present. The greater the oxygen partial pressure, the greater the amount of oxygen bound.

The oxygen binding curves for hemoglobin, and, for comparison, myoglobin, another oxygen binding protein, are presented in Fig. 23-1.

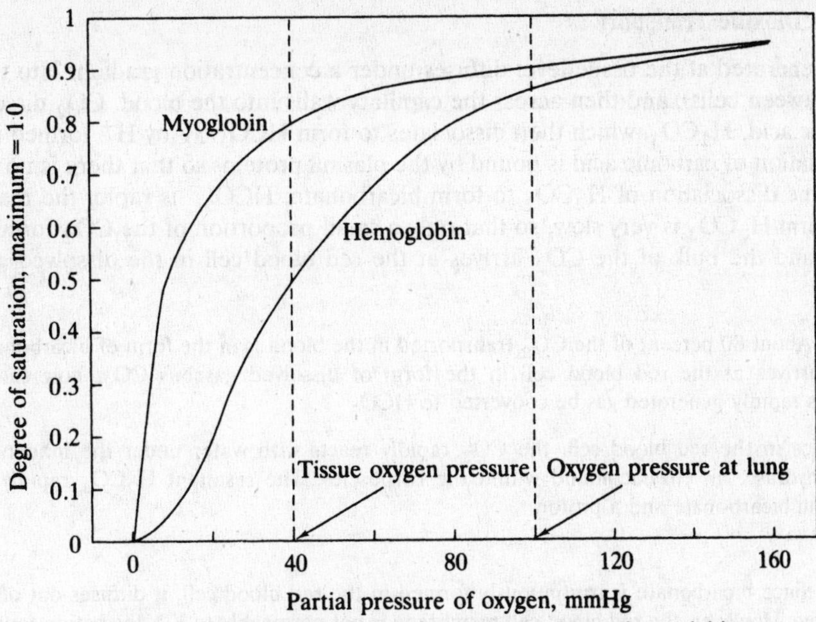

Fig. 23-1 Oxygen binding curves for hemoglobin and myoglobin.

Problem 23.17. Does the odd s-shaped curve characteristic of hemoglobin have physiological significance?

Ans. The oxygen binding curve of hemoglobin must be discussed in terms of the series of events occurring from the time of the loading of the oxygen in the lungs to its unloading at the tissue level. Oxygen is bound or loaded in the lungs at about 100 mmHg, at which pressure the hemoglobin becomes almost completely saturated with oxygen. The oxygen is then carried to the tissue level where the oxygen concentration is much lower than in the lungs. According to Fig. 23-1, if the partial pressure of oxygen at the cell level were about 60 mmHg, the hemoglobin could only give up about 20 percent of its load. However, at the actual partial pressure of oxygen at the cell level, 40 mmHg, the binding curve becomes quite steep. This means that, over only a narrow range of oxygen concentrations, hemoglobin can unload from, say, 80 percent degree of saturation to 50 percent saturation, half of its capacity. This allows a very efficient delivery of oxygen at the tissue level.

The acid-base properties of deoxygenated, HHb, and oxygenated, $HHbO_2$, hemoglobin are significantly different. $HHbO_2$ is a stronger acid than HHb. So upon oxygenation, the following reaction occurs:

$$HHb + O_2 \rightleftharpoons HbO_2^- + H^+$$

Problem 23.18. What is the physiological significance of the fact that oxygenated hemoglobin, $HHb0_2$, is a stronger acid than deoxygenated hemoglobin, HHb?

Ans. In physiological terms, this reaction is what occurs at the lungs where deoxygenated hemoglobin loads oxygen at 100 mmHg. Its importance lies in the fact that, when CO_2 arrives at the lungs, it is in the form of HCO_3^-. When O_2 loads, and H^+ ions become available, the reaction

$$HCO_3^- + H^+ \rightleftharpoons H_2CO_3 \rightleftharpoons CO_2 + H_2O$$

can proceed rapidly. So the binding of oxygen at the lungs increases the efficiency of releasing CO_2 to the atmosphere.

23.3.2 Carbon Dioxide Transport

The CO_2 generated at the tissue level diffuses under a concentration gradient into the interstitial space (space between cells), and then across the capillary walls into the blood. CO_2 dissolves in water to form carbonic acid, H_2CO_3, which then dissociates to form HCO_3^-. Any H^+ formed in the plasma from the dissociation of carbonic acid is bound by the plasma proteins so that there is no shift in blood pH. Although the dissociation of H_2CO_3 to form bicarbonate, HCO_3^- is rapid, the reaction of CO_2 with water to form H_2CO_3 is very slow, so that only a small proportion of the CO_2 forms bicarbonate in the plasma, and the bulk of the CO_2 arrives at the red blood cell in the dissolved gaseous state.

Problem 23.19. About 80 percent of the CO_2 transported in the blood is in the form of bicarbonate, HCO_3^-. If carbon dioxide arrives at the red blood cell in the form of dissolved gaseous CO_2, how can such a large proportion of this rapidly generated gas be converted to HCO_3^-?

Ans. Once in the red blood cell, the CO_2 rapidly reacts with water under the influence of carbonic anhydrase, an enzyme found within the corpuscles. The resultant H_2CO_3 rapidly dissociates to form bicarbonate and a proton.

Problem 23.20. Since bicarbonate is continuously formed in the red blood cell, it diffuses out of the red blood cell into the plasma. However, the red blood cell membrane is not permeable to K^+, the cation found within cells. How can an anion pass through a membrane impermeable to its accompanying cation?

Ans. Since the cation cannot leave with the anion, the anion must exchange locations with an anion external to, and permeable to, the membrane. When the HCO_3^- anion leaves, a Cl^- anion from the plasma enters the red blood cell. This is known as the *chloride shift*.

Problem 23.21. What is the fate of the proton generated by the dissociation of H_2CO_3 in the red blood cell?

Ans. Recall that the oxygenated hemoglobin arrives at the tissue level in the form of HbO_2^-. Its negative charge is balanced by the K^+ within the red blood cell. The proton generated by dissociation of H_2CO_3 reacts with HbO_2^- as

$$H^+ + HbO_2^- \rightleftharpoons HHb + O_2$$

which buffers any decrease in pH, and causes O_2 to leave its bound form and diffuse into the tissue under its concentration gradient.

The processes which occur in the red blood cell at the tissue level, are reversed at the lungs. In the following sequence of reactions, H^+ ions generated in the red blood cell at the tissue level, and eliminated in the lungs, are in bold type

Tissue level:
$$CO_2 + H_2O \rightleftharpoons H_2CO_3 \rightleftharpoons \mathbf{H^+} + HCO_3^-$$
$$\mathbf{H^+} + HbO_2^- \rightleftharpoons \mathbf{HHb} + O_2$$

Lungs:
$$\mathbf{HHb} + O_2 \rightleftharpoons HbO_2^- + \mathbf{H^+}$$
$$\mathbf{H^+} + HCO_3^- \rightleftharpoons H_2CO_3 \rightleftharpoons CO_2 + H_2O$$

This sequence of reactions illustrates the buffering of the blood by the carbonic acid-bicarbonate conjugate acid-base system. The protons generated at the tissue level by the ionization of carbonic acid are removed at the lungs when bicarbonate ion reenters the red blood cell and reacts with them to produce carbonic acid. H_2CO_3 then rapidly dissociates, under the influence of carbonic anhydrase, and the resultant CO_2 diffuses into the alveolar space because its concentration in the plasma is higher than in the alveoli.

Problem 23.22. What are the results of not removing CO_2 from the lungs at a rate equal to its arrival at the lung?

Ans. If for some reason, CO_2 is not removed from the lungs rapidly enough, the mass action system will "back up," the hydrogen ion will not be removed by reaction with HCO_3^-, and the blood pH will fall. This condition is called *respiratory acidosis*. On the contrary, if CO_2 is removed from the lungs faster than it arrives, the blood pH will rise. Such a condition is called *respiratory alkalosis*.

About 20 percent of the CO_2 in blood is carried as *carbamino* compounds formed by the reaction with amino groups of hemoglobin. The reaction may be represented as

$$R-NH_2 + CO_2 \rightleftharpoons R-NH-COOH$$

When CO_2 reacts in this way with oxygenated hemoglobin, it causes the dissociation of the oxygen as

$$CO_2 + HbO_2^- \rightleftharpoons HHbCO_2^- + O_2$$

In this way, the loading of CO_2 as a carbamino compound results in the additional unloading of O_2 at the tissue level.

The reactions involved in oxygen and carbon dioxide transport and their anatomical locations are summarized in Fig. 23-2

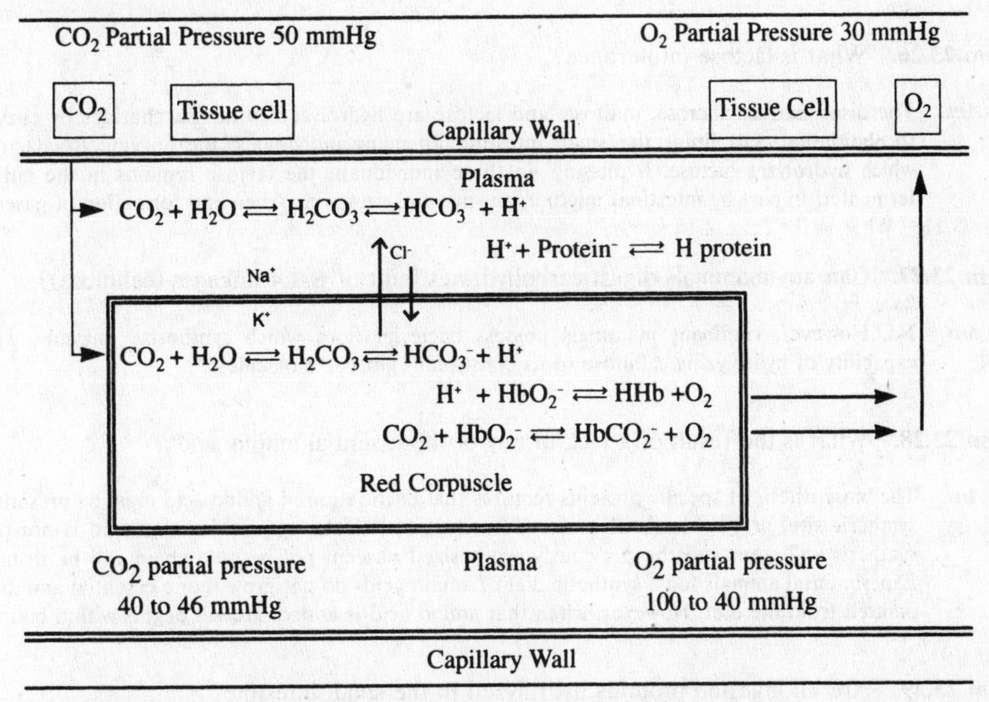

Figure 23-2 Summary of principal chemical reactions involved in making oxygen available to tissues, and transporting carbon dioxide from tissues. All gas exchange is controlled by gradients in gas partial pressure.

ADDITIONAL SOLVED PROBLEMS

DIGESTION

Problem 23.23. Why are digestive enzymes secreted as zymogens?

Ans. Secretion of hydrolytic enzymes as zymogens is necessary if the cells secreting these enzymes are not to undergo digestion themselves. Pancreatitis is a very serious malady in which inactive digestive enzymes are prevented from leaving the pancreas, and are converted to their active form prematurely within the pancreas itself leading to serious destruction of that organ.

Problem 23.24. What is the fate of the bile salts?

Ans. The bile salts, sodium salts of glycocholate and taurocholate, remain in the intestinal tract, are reabsorbed into the blood, and find their way to the liver to be recycled for repeated use. Bile salts are not only important for fat digestion, but are necessary for absorption of the fat-soluble vitamins A, D, E, and K as well.

Problem 23.25. The pH in the stomach is ca. 2.0, and in the intestine, ca. 7.0. How is the pH changed in the transition from stomach to intestine?

Ans. As the stomach contents pass into the intestine, the low pH causes the secretion of the hormone secretin into the blood. When the circulation reaches the pancreas, large quantities of HCO_3^- are secreted by that organ into the small intestine, which neutralizes the stomach contents.

NUTRITION

Problem 23.26. What is lactose intolerance?

Ans. The disaccharides sucrose, maltose, and lactose are hydrolyzed to monsaccharides by enzymes in the epithelial cells lining the small intestine. In many individuals, the enzyme β-galactosidase, which hydrolyzes lactose, is missing. In these individuals, the lactose remains in the gut and is fermented in part by intestinal microorganisms, and causes diarrhea and formation of gases.

Problem 23.27. Can any mammals digest carbohydrates built of β-1,4 linkages (cellulose)?

Ans. No. However, ruminant mammals possess bacterial flora which synthesize enzymes with the capability of hydrolyzing cellulose to its constituent glucose molecules.

Problem 23.28. What is the result of a lack of any of the essential amino acids?

Ans. The biosynthesis of specific proteins requires that each required amino acid must be present at the synthesis site, or synthesis will cease. If even one particular required amino acid is not present, synthesis will stop, and the previously synthesized nascent polypeptide chain will be dismantled. Experimental animals fed a synthetic diet of amino acids do not grow if one essential amino acid is omitted from the diet. However, when that amino acid is added, growth begins within hours.

Problem 23.29. Are all ingested proteins hydrolyzed in the small intestine?

Ans. No. Some fibrous animal proteins such as keratin are only partially digested. Proteins of plant origin which are surrounded by cellulosic husks, e.g., cereal grains, are incompletely digested.

Problem 23.30. How does the pancreas protect itself from autodigestion?

Ans. The pancreas secretes an array of proteolytic zymogens which must be activated in the small intestine by trypsin, itself secreted as a pancreatic zymogen, trypsinogen. It is possible for this zymogen to be prematurely activated to trypsin within the pancreas. However, the pancreas also secretes a specific trypsin inhibitor which binds strongly to and inactivates any trypsin prematurely activated.

Problem 23.31. If intestinal bacterial flora produce enough vitamin B_{12}, why does a deficiency in this vitamin occasionally occur?

Ans. Vitamin B_{12} deficiency, resulting in pernicious anemia, is ordinarily caused by the absence of *intrinsic factor*, a glycoprotein synthesized in the stomach. Vitamin B_{12} is transported across the intestinal cell membrane as a complex with intrinsic factor. Therefore in individuals who cannot synthesize this protein, the vitamin must be administered by injection directly into the bloodstream.

Problem 23.32. Describe the disease known as beriberi.

Ans. Beriberi was once thought to be an infectious disease. It is characterized by neurological disorders. The disease was unknown until the early 19th century when rice polishing machines were invented. These machines removed the brown outer hull of the rice seed. The cure was discovered when the addition of the outer hull to the diet completely reversed the symptoms. The critical component of rice hulls is thiamin, which is the cofactor in decarboxylations. The blood of those having the deficiency contains elevated levels of pyruvate, which must be decarboxylated in order to enter the TCA cycle.

Problem 23.33. Can any vitamin deficiency diseases occur in adults or children who ingest the correct vitamin RDAs?

Ans. Yes. Any ingested nutrient must be absorbed into the circulation if it is to reach metabolizing cells. In particular, those people suffering fat absorption disorders will be deficient in the fat-soluble vitamins A, D, E, and K.

Problem 23.34. Can any of the water-soluble vitamins be prevented from being absorbed?

Ans. Eggs are a rich source of biotin. However, egg white protein contains a protein called *avidin* which binds very strongly to biotin and this complex cannot be absorbed. A diet rich in egg white protein will cause a serious biotin deficiency.

Problem 23.35. What is the disease caused by vitamin A deficiency?

Ans. The disease is called *xerophthalmia*, or *night blindness*. It is caused by the insufficient synthesis of the visual pigment rhodopsin. Rhodopsin contains the retinal group, derived from vitamin A.

Problem 23.36. What differentiates between bulk and trace elements?

Ans. Bulk elements are required in amounts exceeding 100 mg/day, and trace elements are rarely required above 1 to 3 mg/day.

Problem 23.37. Why is zinc considered a required trace element?

Ans. Zinc deficiency in experimental animals is associated with profound disorders in DNA, RNA, protein, and mucopolysaccharide (connective tissue) metabolism.

METABOLIC GAS TRANSPORT

Problem 23.38. If both hemoglobin and myoglobin bind oxygen reversibly, why are their binding curves qualitatively different?

Ans. Myoglobin consists of a single polypeptide chain containing only one heme molecule. Hemoglobin consists of four polypeptide chains, each containing a heme molecule. The difference between the binding curves for the two proteins is that the oxygen molecule bound to a myoglobin molecule has no influence upon the binding of oxygen to another myoglobin molecule. However, when oxygen is bound to hemoglobin, the binding process changes the orientations of the four polypeptide chains with respect to each other. The binding of the first oxygen molecule changes the protein's structure in such a way as to make it easier to bind the next molecule. This is called *cooperative binding*, because the first bound molecule has an influence on succeeding binding reactions. It is a good model for the allosteric effects described in Chap. 22 in the discussion of enzyme regulation.

Problem 23.39. What is metabolic acidosis?

Ans. Metabolic acidosis is a lowering of the blood pH as a result of a metabolic disorder as opposed to the failure of the $H_2CO_3 - HCO_3^-$ buffer system. For example, there is a large and serious decrease in pH as a result of uncontrolled diabetes. The blood pH may fall from the normal 7.4 to as low as 6.8. The increased H^+ concentration is due to the large amounts of ketone bodies produced in the liver. The products are acetoacetic acid and β-hydroxybutyric acid. The bicarbonate buffer system attempts to compensate for the excess H^+, and the excess CO_2 must be eliminated at the lungs. However, so much CO_2 is lost by ventilation that the absolute concentration of the buffer system decreases, so the capacity of the buffer system is severely compromised and cannot reduce the metabolically produced excess H^+. In such cases, clinical treatment involves the intravenous administration of sodium bicarbonate to restore buffer capacity.

SUPPLEMENTARY PROBLEMS

Problem 23.40. What is the metabolic precursor of the bile salts?

Ans. Cholesterol

Problem 23.41. What are some of the specific functions of the trace element zinc?

Ans. It is a cofactor of almost 100 enzymes, e.g., NAD- and NADP-linked dehydrogenases, DNA- and RNA-polymerases, carbonic anhydrase, and intestinal carboxypeptidase.

Problem 23.42. What is an important biochemical function of manganese?

Ans. Urea, an end product of amino acid metabolism, is formed by hydrolysis of the amino acid arginine. The enzyme responsible for the hydrolysis, arginase, contains a tightly bound Mn^{2+} ion, critical for its activity.

Problem 23.43. What is the enzyme chiefly responsible for the hydrolysis of starch in the small intestine?

Ans. Although starch digestion begins in the mouth with the action of salivary amylase, only a small portion of dietary starch digestion is completed there. The bulk of the starch digestion occurs in the small intestine under the action of pancreatic amylase.

Problem 23.44. Plasma lipoproteins transport triacylglycerols in the blood. Do fatty acids ever appear in the blood?

Ans. Serum albumin strongly binds 2 to 8 mol of fatty acids per molecule of protein. This is transported in the blood to the heart and skeletal muscles, which absorb and use it as a major source of oxidative fuel.

Problem 23.45. What is the fate of nonphysiologic or foreign organic compounds which are ingested along with foodstuffs.?

Ans. Drugs, preservatives, and food additives are absorbed along with nutrients and carried to the liver. There the relatively insoluble compounds are detoxified by cytoplasmic oxidases. The reaction produces hydroxylated derivatives which are further combined by esterification to water-soluble compounds and along with originally water-soluble foreign compounds are eliminated in the urine.

Problem 23.46. What is metabolic alkalosis?

Ans. Metabolic alkalosis describes an increase in blood pH as a result of a metabolic disorder. This can occur when excessive amounts of H^+ ions are lost, for example, in continuous vomiting. H^+ ions are then borrowed from the blood, and the blood pH rises. Loss of K^+ because of kidney malfunction, or intake of excessive antacids, will also cause the blood pH to rise. Treatment involves the careful intravenous administration of ammonium chloride.

Basic and Derived SI Units and Conversion Factors

Quantity	SI Unit	Conversion Factors or Derived Units
Length	Meter (m)	1 m = 100 centimeters (cm)
		= 1.0936 yards (yd)
		1 cm = 0.3937 inch (in)
		1 in = 2.54 cm exactly
		= 0.0254 m
		1 angstrom (Å) = 1×10^{-8} cm
		1 mile = 1.6093 km
Mass	Kilogram (kg)	1 kg = 1000 grams (g)
		= 2.205 pounds (lb)
		1 lb = 453.6 grams (g)
Time	Second (s)	1 day (d) = 86400 s
		1 hour (h) = 3600 s
		1 minute (min) = 60 s
Volume (derived)	Cubic meter (m^3)	1 m^3 = 1×10^6 cm^3
		1 cm^3 = 1 mL
		1 liter (L) = 1000 mL
		1 liter (L) = 1×10^{-3} m^3
		= 1.057 quarts (qt)
Force (derived)	Newton (N)	N = m · kg/s^2
		1 dyne = 1×10^{-5} N
Pressure (derived)	Pascal (Pa)	1 atmosphere (atm) = 101325 Pa
		= 760 mmHg
		= 14.70 lb/in^2
Energy (derived)	Joule (J)	1 Joule (J) = N · m
		1 calorie (cal) = 4.184 J
		1 electron-volt (eV) = 96.485 kJ/mol
		1 L · atm = 101.325 J
		1 J = 1×10^7 ergs

Table of Atomic Masses

Element	Symbol	Atomic Number	Atomic Mass
Actinium	Ac	89	(227)
Aluminum	Al	13	26.98
Americium	Am	95	(243)
Antimony	Sb	51	121.75
Argon	Ar	18	39.95
Arsenic	As	33	74.92
Astatine	At	85	(210)
Barium	Ba	56	137.33
Berkelium	Bk	97	(247)
Beryllium	Be	4	9.012
Bismuth	Bi	83	208.98
Boron	B	5	10.81
Bromine	Br	35	79.91
Cadmium	Cd	48	112.41
Calcium	Ca	20	40.08
Californium	Cf	98	(251)
Carbon	C	6	12.011
Cerium	Ce	58	140.12
Cesium	Cs	55	132.91
Chlorine	Cl	17	35.45
Chromium	Cr	24	52.00
Cobalt	Co	27	58.93
Copper	Cu	29	63.55
Curium	Cm	96	(247)
Dysprosium	Dy	66	162.5
Einsteinium	Es	99	(252)
Erbium	Er	68	167.26
Europium	Eu	63	151.96
Fermium	Fm	100	(253)
Fluorine	F	9	19.00
Francium	Fr	87	(223)
Gadolinium	Gd	64	157.25
Gallium	Ga	31	69.72
Germanium	Ge	32	72.64
Gold	Au	79	196.97
Hafnium	Hf	72	178.49
Helium	He	2	4.0026
Holmium	Ho	67	164.93
Hydrogen	H	1	1.0079
Indium	In	49	114.82
Iodine	I	53	126.91
Iridium	Ir	77	192.22
Iron	Fe	26	55.85
Krypton	Kr	36	83.80
Lanthanum	La	57	138.91
Lawrencium	Lr	103	(262)
Lead	Pb	82	207.21
Lithium	Li	3	6.941
Lutetium	Lu	71	174.97
Magnesium	Mg	12	24.31
Manganese	Mn	25	54.94
Mendelevium	Md	101	(258)
Mercury	Hg	80	200.59
Molybdenum	Mo	42	95.94
Neodymium	Nd	60	144.24
Neon	Ne	10	20.18
Neptunium	Np	93	(237)
Nickel	Ni	28	58.69
Niobium	Nb	41	92.91
Nitrogen	N	7	14.01
Nobelium	No	102	(259)
Osmium	Os	76	190.2
Oxygen	O	8	16.00
Palladium	Pd	46	106.42
Phosphorus	P	15	30.97
Platinum	Pt	78	195.08
Plutonium	Pu	94	(244)
Polonium	Po	84	(209)
Potassium	K	19	39.10
Praseodymium	Pr	59	140.91
Promethium	Pm	61	(145)
Protactinium	Pa	91	231.04
Radium	Ra	88	(226)
Radon	Rn	86	(222)
Rhenium	Re	75	186.2
Rhodium	Rh	45	102.91
Rubidium	Rb	37	85.47
Ruthenium	Ru	44	101.07
Samarium	Sm	62	150.36
Scandium	Sc	21	44.96
Selenium	Se	34	78.96
Silicon	Si	14	28.09
Silver	Ag	47	107.87
Sodium	Na	11	22.99
Strontium	Sr	38	87.62
Sulfur	S	16	32.07
Tantalum	Ta	73	180.95
Technetium	Tc	43	(98)
Tellurium	Te	52	127.60
Terbium	Tb	65	158.93
Thallium	Tl	81	204.38
Thorium	Th	90	232.04
Thulium	Tm	69	168.93
Tin	Sn	50	118.71
Titanium	Ti	22	47.87
Tungsten	W	74	183.84
Uranium	U	92	238.03
Vanadium	V	23	50.94
Xenon	Xe	54	131.29
Ytterbium	Yb	70	173.04
Yttrium	Y	39	88.91
Zinc	Zn	30	65.41
Zirconium	Zr	40	91.22

For atomic mass values in parentheses, the mass is that of the most abundant isotope. For all other values, the atomic mass is the average of all the isotopes present, adjusted for the relative amounts of each isotope present.

Periodic Table

I	II											III	IV	V	VI	VII	VIII
1 **H** 1.0079																	2 **He** 4.0026
3 **Li** 6.941	4 **Be** 9.012											5 **B** 10.81	6 **C** 12.011	7 **N** 14.01	8 **O** 16.00	9 **F** 19.00	10 **Ne** 20.18
11 **Na** 22.99	12 **Mg** 24.31											13 **Al** 26.98	14 **Si** 28.09	45 **P** 30.97	16 **S** 32.07	17 **Cl** 35.45	18 **Ar** 39.95
19 **K** 39.10	20 **Ca** 40.08	21 **Sc** 44.96	22 **Ti** 47.87	23 **V** 50.94	24 **Cr** 52.00	25 **Mn** 54.94	26 **Fe** 55.85	27 **Co** 58.93	28 **Ni** 58.69	29 **Cu** 63.55	30 **Zn** 65.41	31 **Ga** 69.72	32 **Ge** 72.64	33 **As** 74.92	34 **Se** 78.96	35 **Br** 79.91	36 **Kr** 83.80
37 **Rb** 85.47	38 **Sr** 87.62	39 **Y** 88.91	40 **Zr** 91.22	41 **Nb** 92.91	42 **Mo** 95.94	43 **Tc** (98)	44 **Ru** 101.07	45 **Rh** 102.91	46 **Pd** 106.42	47 **Ag** 107.87	48 **Cd** 112.41	49 **In** 114.82	50 **Sn** 118.71	51 **Sb** 121.75	52 **Te** 127.60	53 **I** 126.91	54 **Xe** 131.29
55 **Cs** 132.91	56 **Ba** 137.33	57 **La** 138.91	* 72 **Hf** 178.49	73 **Ta** 180.95	74 **W** 183.84	75 **Re** 186.2	76 **Os** 190.2	77 **Ir** 192.22	78 **Pt** 195.08	79 **Au** 196.97	80 **Hg** 200.59	81 **Tl** 204.38	82 **Pb** 207.2	83 **Bi** 208.98	84 **Po** (209)	85 **At** (210)	86 **Rn** (222)
87 **Fr** (223)	88 **Ra** (226)	89 **Ac** (227)	† 104 **Unq** (261)	105 **Unp** (262)	106 **Unh** (263)												

*	58 **Ce** 140.12	59 **Pr** 140.91	60 **Nd** 144.24	61 **Pm** (145)	62 **Sm** 150.36	63 **Eu** 151.96	64 **Gd** 157.25	65 **Tb** 158.93	66 **Dy** 162.5	67 **Ho** 164.93	68 **Er** 167.26	69 **Tm** 168.93	70 **Yb** 173.04	71 **Lu** 174.97
†	90 **Th** 232.04	91 **Pa** 231.04	92 **U** 238.03	93 **Np** (237)	94 **Pu** (244)	95 **Am** (243)	96 **Cm** (247)	97 **Bk** (247)	98 **Cf** (251)	99 **Es** (252)	100 **Fm** (253)	101 **Md** (258)	102 **No** (259)	103 **Lr** (262)

Index